Resurgence and Revalorization of Indigenous Knowledge Systems in the Contemporary Society

Tlou Maggie Masenya
Durban University of Technology, South Africa

Monicca Thulisile Bhuda
University of Mpumalanga, South Africa

IGI Global
Scientific Publishing
Publishing Tomorrow's Research Today

Vice President of Editorial Melissa Wagner
Director of Acquisitions Mikaela Felty
Director of Book Development Jocelynn Hessler
Production Manager Mike Brehm
Cover Design Phillip Shickler

Published in the United States of America by
IGI Global Scientific Publishing
701 East Chocolate Avenue
Hershey, PA, 17033, USA
Tel: 717-533-8845
Fax: 717-533-7115
Website: https://www.igi-global.com E-mail: cust@igi-global.com

Library of Congress Cataloging-in-Publication Data

Names: Masenya, Tlou, 1972- editor http://id.loc.gov/vocabulary/relators/edt
http://id.loc.gov/authorities/names/n2022011786
http://id.loc.gov/rwo/agents/n2022011786 | Bhuda, Monicca, 1992- editor
http://id.loc.gov/vocabulary/relators/edt
http://id.loc.gov/authorities/names/n2025002626
http://id.loc.gov/rwo/agents/n2025002626
Title: Resurgence and revalorization of indigenous knowledge systems in the contemporary society / edited by: Tlou Maggie Masenya, Monicca Thulisile Bhuda.
Description: Hershey, PA : IGI Global Scientific Publishing, [2025] | Includes bibliographical references.
Identifiers: LCCN 2025019253 (print) | LCCN 2025019254 (ebook) | ISBN 9798337341262 ebook | ISBN 9798337341248 h/c | ISBN 9798337341255 s/c
Subjects: LCSH: Ethnoscience--Africa | Ethnoscience--India | Africa--Social life and customs http://id.loc.gov/authorities/subjects/sh2007100923 | India--Social life and customs
http://id.loc.gov/authorities/subjects/sh86007592
Classification: LCC GN476 (ebook) | LCC GN476 .R47 2026 (print) | DDC 305.800968 23/eng/20250--dc26
LC record available at https://lccn.loc.gov/2025019253

British Cataloguing in Publication Data
A Cataloguing in Publication record for this book is available from the British Library.

Editorial Advisory Board

Table of Contents

Detailed Table of Contents

Limpopo province covers 123 600 km2 and has a population of 5.3 million people, 89% of whom live in non-urban regions and 40–50% of whom are unemployed. It is undoubtedly South Africa's poorest province. The community of Magomani village for a long time has used commercial farming as the main basis for survival and sustaining livelihoods. However, because of the obstacles that local farmers have experienced over the years, there has been a reduction in commercial farming development which has led to the rise of poverty. These obstacles include: insufficient farming land which has been caused by population increase and land occupation; because of a variety of restrictions and constraints, smallholder farmers in rural areas face a lack of market; lack of investment from government and other potential sponsors, lack of working capital which result in poor veld condition and restriction to access of market. This study was qualitative in nature and aimed to investigate how commercial farming can be revived to develop the community of Magomani village.

In South Africa, documenting Indigenous Knowledge Systems (IKS) is a means of safeguarding and preserving these knowledges. Since IKS are tacit, largely passed down over generations and deeply rooted in cultural norms, it can be difficult to accurately document them. Moreover, IKS is in danger of extinction if no actions are taken to ensure their documentation. Due to these challenges, a call for documentation of IKS in universities is necessary and need urgent response. Therefore, the study aims to investigate ways South African universities document IKS. This study used secondary data and drew pre-existing data from academic works such as books, articles and government documents. The results show that there is progress made by universities to document IKS through collaborating with government and other stakeholders. Despite these successes, there are still challenges with IKS documentation, thus

their brothers' wives (varoora), diminishing authority, and competition from female church elders emerge as significant barriers. Additionally, geographical distance, limited resources, and educational gaps further complicate their ability to provide the mentorship and support that younger family members need. This study underscores the importance of addressing these challenges to empower aunts and enhance their contributions to familial and cultural continuity in the family.

A qualitative case study in orientation, the research used the Mulabalaba indigenous game to explore the forms of knowledge and pedagogies that underpin indigenous African curricula and the implications they hold for present-day higher education curriculum development and transformation. We observed ten purposively selected participants who played the Mulabalaba game during the School of Education's Staff Wellness Day in 2024. A key finding of the study was that the indigenous African curriculum is underpinned by diverse epistemic forms, values and skills like the 21st century skills that are highly sought in present-day higher education curricula. Other findings revealed that the underpinning epistemic forms, values and skills are interwoven seamlessly and comprehensively and delivered in captivating and engaging pedagogies. The study recommends future research into how lessons from indigenous African curricula can be incorporated to inform and strengthen present-day higher education curriculum development and transformation.

Gender-based violence has proven to be problematic worldwide, however, in South Africa, it has proven to be uncontrollable. In as much as national and provincial laws are enforced against gender-based violence, it continues to exacerbate. This has awakened South Africa to go back to African methods of addressing gender-based violence. This study aimed to investigate indigenous knowledge strategies for managing and addressing gender-based violence amongst the local citizens in KwaZulu-Natal. Important to note is that the rate of gender-based violence cases is witnessed to be increasing in KwaZulu-Natal. The reviewed literature has proven that Indigenous knowledge is capable of managing and addressing gender-based violence. Again, the findings of the study have demonstrated that indigenous knowledge strategies could become a roadmap for addressing gender-based violence not only in KwaZulu-Natal but in Africa as a whole.

The purpose of this chapter to explores the importance of acknowledging indigenous knowledge among landscape actors in managing changing environments. Global South is currently experiencing dramatic changes, due to climate change. Over the years, Western knowledge has been prioritised by landscape actors in their efforts to manage changing environments, while ignoring indigenous knowledge. Currently, indigenous knowledge is gaining recognition it the efforts to address changes in the environment. This study is grounded on the knowledge coevolution framework. The results of this study influence the need for landscape actors to adopt indigenous knowledge and manage changing environments. The study concludes that there is a need to ensure a working relationship between landscape actors. It is important to determine the management of power and its dynamics at different levels particularly between the government, local communities, and NGOs, to ensure interests are advocated for.

This study seeks to explore indigenous ways in managing informal diapers disposal in the Vhembe Biosphere Reserve. Disposable diapers are increasingly popular within the Vhembe biosphere reserve. However, the disposal of these diapers in areas without access to solid waste management infrastructure has become a scourge upon, low-income communities, presenting a multitude of health and environmental risks. As such, indigenous knowledge seems to have potential to address this challenge. This is because indigenous methods for waste management are environmentally friendly and sustainable. Indigenous knowledge systems play a crucial role in communities worldwide. However, the indigenous knowledge in communities is rarely known as people are more exposed to modern culture. The obviatives of this study was to determine causes of informal diapers disposal and to determine ways of managing informal diaper disposal in indigenous communities in the Vhembe Biosphere Reserve. The findings of this study provided awareness on the importance of indigenous ways in managing informal diapers disposal.

This chapter seeks to highlight the untouched tourism opportunities in Pushkar, Rajasthan and also to maintain sustainable practices. Many studies have been investigated on religious festivals in the Pushkar region, but there is a need for research that focuses on sustainability with event tourism. Therefore, this study aims to discover the potential and challenges of the Pushkar Fair in Rajasthan. The main objectives of this study are to identify diverse tourism opportunities in the Pushkar region of Rajasthan and to maintain sustainability at the time of such an event. This study is based on a detailed literature review, and data will be collected from reputed journals, Government documents, reports, and edited books published at national and international levels. The study concludes that despite the massive potential of religious and event tourism in Pushkar, Rajasthan, the role of sustainability and preservation of destinations have yet to be wholly harnessed.

This chapter aims to highlight the importance of nature and describe how it sustains Siswati indigenous knowledge systems (IKS) in Eswatini. Grounded in the Biocultural Diversity Framework and descriptive qualitative design, data for analysis was collected from secondary sources. Thematic analysis was employed to explore how nature supports traditional Siswati practices, beliefs, and knowledge systems. The results revealed that traditional food, medicine, attire and religious practices are embedded in natural features. The study also found that challenges such as land deforestation, wildfires and overharvesting contribute to the scarcity of plants and trees as well as animal products used for traditional medicines and food. Strategies that could alleviate some of these challenges such as environmental acts, establishing botanical gardens and involving conservatory stakeholders were proposed. Therefore, this study contributes to a deeper understanding on the relationship between biocultural diversity and Siswati IKS, offering insights into their maintenance.

This study is qualitative in nature and aims to analyze the Zulu perspectives on traditional healing system as part of the oral health practice, within the context of African traditional religion. This type of healing is very scientific within the context of African understanding, as it is very systematic, logically, and very relevant even today. It is also capable in healing different types of diseases including mystical illnesses which are African orientated. The healing practice entails the discipline of botany, zoology and anthropology. This kind of practice has been part of the African people for a very long time to restore life and to bring healthy balance between the human beings and the environment. It has been developed outside the sphere of influence of biomedicine of which it refers to the traditional scientific form of knowledge. Therefore, the healing concept in the Zulu perspective is orally done through consultation with the ancestral spirits which involve spirit mediums/traditional healers who are working interchangeable.

Naming is an evolving and continuous process that could also be influenced by changes and practices that have emerged due to the contact South African people had with colonization, Christianity, and Western education influence. The study investigated how the Batswana people in Gauteng province give name(s) to their newly born child/children through the different generational changes. The problem is that the customary and traditional practices of name giving are gradually disappearing as generations evolve. Data was purposively sampled through an interview schedule that was administered face-to-face and online. The findings amongst others revealed that the practices in name-giving vary across the generations hence Baby Boomers and Generation X this responsibility is given to the grandparents of the newly born baby. The names given by the Baby Boomers and Generation X are mostly inherited from the family pedigree. Whereas Generation Z prefers to give their children their names, they are trying to break the tradition of inheriting names.

This chapter examines the integration of Indigenous Technical Knowledge (ITK) with modern agricultural practices to promote sustainability, productivity, and resilience in farming systems. Rooted in centuries of adaptation to local environments, ITK encompasses diverse methods such as water harvesting, soil health management, animal health practices, and biodiversity conservation. Highlighting practices like the khadin system, johads, panchagavya, and sacred groves, this chapter showcases ITK's relevance in addressing contemporary agricultural challenges. It explores the potential of combining ITK with advanced technologies like the Internet of Things (IoT), Geographic Information Systems (GIS), and Machine Learning (ML) to optimize resource use, improve soil fertility, and conserve biodiversity. The chapter addresses the challenges of merging traditional wisdom with modern innovations, emphasizing this integration as a vital strategy for agricultural sustainability in the face of climate variability and resource constraints.

This chapter talks in depth about the Lepcha people's festivals, events, and traditional background in the Eastern Himalayas. The Lepchas, who were among the first people to settle in the area, are known for their ecological wisdom, spirituality, and cultural identity, which they exhibit via vivid rituals that honor the elements of Mother Earth. Chyu Rum Faat, Pang Lhabsol, and Tendong Lho Rum Faat are examples of festivals that have changed from animistic rituals to important parts of society. Travelers introduced native practices to the region, while foreign practices such as Tibetan Buddhism, British colonialism, and modern state governance created a fascinating combination. Modern festivals support the assertion of Lepcha heritage and provide a one-of-a-kind opportunity for cultural tourists to learn about Lepcha traditions and mythology. Lepcha festivals keep their traditional identity while changing with the times in terms of history and politics. The chapter also investigates the manner in which visitors perceive these events, as well as their engagement and enjoyment.

The Holistic Integration and Appreciation Model (HIAM) was developed to enhance educational outcomes for students from the Gond tribes by integrating their cultural heritage into the curriculum. This study aims to evaluate the effectiveness of the HIAM model in improving language proficiency, student engagement, cultural understanding, and community impact.An experimental research method was employed using a group post-test design. A purposive sample of one section of Class 9 students from Brain Child Academy in Mandla, Jabalpur, Madhya Pradesh, was selected as the intervention group. The conceptual framework of the tool was tested using Smart-PLS 4 software.The post-intervention data indicated significant improvements in language proficiency and student engagement. Students demonstrated a greater appreciation and understanding of their cultural heritage, reflected in increased self-esteem and sense of identity. The community also reported positive impacts, including heightened educational aspirations and greater involvement in school activities.

The call to have Indigenous Knowledge System (IKS) accredited programmes by universities is necessary to challenge existing Western Systems which have posed a threat to IKS in education. The aim of the study was to investigate the implementation of the IKS programmes by universities which were part of the 1996 indigenous technologies audit. The study used qualitative content analysis to extract information from a variety of sources, such as government documents, books, theses, and scholarly journals that are available on websites like Google Scholar, ScienceDirect, ResearchGate and university repositories. The findings of the study revealed that with assistance from government, there are universities who have successfully implemented IKS accredited programmes while others are still lacking behind since the 1996 audit. Therefore, the study recommends that universities which are yet to start this process need to benchmark and collaborate with those that are offering the IKS programmes. IKS postgraduates can also be sourced to be part of this process because they are equipped with expertise to implement these programmes.

Foreword

This book, *Resurgence and Revalorization of Indigenous Knowledge Systems in the Contemporary Society,* by Dr. Monicca Thulisile Bhuda and Prof. Tlou Maggie Masenya, brings together esteemed scholars to examine the meaning and role of Indigenous Knowledge Systems (IKS) in today's evolving world. This book is aligned with other academic contributions within the discussion of decoloniality that have made a point that IKS have long been the foundation of sustainable livelihoods, cultural identity, and scientific ingenuity. Rooted in centuries of lived experiences, these knowledge systems encompass indigenous medicine, agriculture, environmental conservation, governance, artistic expressions by offering invaluable insights into addressing contemporary global challenges. Yet, despite their richness and relevance, many Indigenous knowledge traditions have been marginalized, overlooked, or displaced by dominant Western paradigms of knowledge production. This book emerges at a critical time when societies are seeking more inclusive, holistic, and sustainable ways of understanding the world. It calls for a necessary re-examination of IKS, advocating for their rightful place alongside contemporary scientific and technological advancements.

The chapters within this book explore the intersection of Indigenous wisdom and modernity, shedding light on how traditional knowledge can inform and enrich fields such as climate resilience, health sciences, education, gender identities and innovation. In summary, this book underscores the importance of African scholarship in interrogating and engaging with these pressing issues. By drawing on lived experiences and real-life narratives, the book offers a form of "*human scholarship*" that challenges the academy to confront the core needs, challenges, and aspirations of humanity today. It critically examines traditional perspectives and their place in the context of modern technological advancements. Through this exploration, the book reaffirms the relevance of IKS, demonstrating their continued significance in shaping solutions for contemporary challenges. This contribution also fosters a dialogue that moves beyond appreciation toward meaningful action. It underscores the role of research, policy frameworks, and community engagement in revitalizing IKS within contemporary settings. By embracing both tradition and transformation, we can create a future where IKS informs global development in equitable and respectful ways. It is my hope that this book serves as a catalyst for further exploration, collaboration, and advocacy. May it inspire readers-whether academics, policymakers, or community leaders-to champion the integration of Indigenous Knowledge (IK), wisdom and practices in shaping a more just, sustainable, and knowledgeable world.

Luvuyo Ntombana
Sol Plaatje University, South Africa & Nelson Mandela University, South Africa

Preface

INTRODUCTION

We are pleased to present *Resurgence and Revalorization of Indigenous Knowledge Systems in the Contemporary Society*, a collective effort to recognize, reclaim, and reaffirm the profound value of Indigenous Knowledge (IK) in today's fast-paced, globalized world. This edited volume emerges from a deep sense of urgency—a need to bring to the forefront the cultural wisdom, lived experiences, and holistic worldviews that indigenous communities have cultivated and transmitted for generations.

Indigenous knowledge is far more than a body of facts—it is a way of life, an intimate understanding of the natural world, and a framework that binds communities together. It reflects the symbiosis between people and their environment, offering context-specific, time-tested solutions to everyday challenges, from sustainable agriculture to community health, environmental stewardship, and conflict resolution. Despite its relevance and richness, IK faces increasing threats from deterritorialization, modernization, colonization, globalization, and the pervasive influence of western-centric knowledge systems. This has led to an unfortunate loss of valuable cultural heritage and the marginalization of knowledge systems that are critical to local identity, survival, and sustainability.

This book is a response to that loss—a reaffirmation that indigenous knowledge must not only be preserved but also revalorized and integrated into contemporary education, governance, innovation, and development discourse. It explores a wide range of scholarly contributions that focus on the revaluation, reclamation, promotion, preservation, and dissemination of IK, drawing on case studies, theoretical analyses, and practical strategies from across the globe.

Our aim is to foster an inclusive and fair knowledge environment—one that honors the contributions of indigenous communities and promotes intercultural dialogue. The chapters in this book offer valuable insights into how indigenous knowledge can inform sustainable development, guide digital archiving practices, influence curriculum development, and enrich research and policy-making.

We hope that this volume will serve as a foundational resource for scholars, universities, students, policymakers, government departments, indigenous communities, librarians, archivists, IT professionals, data curators, and anyone invested in preserving the cultural and intellectual wealth of our world's first knowledge keepers. It is also our hope that academic institutions will embrace this book as a core reference in Indigenous Knowledge-related studies, enabling the next generation of thinkers to engage respectfully and constructively with indigenous epistemologies.

As editors, we express our sincere gratitude to the contributors whose research and reflections form the heart of this book. Together, we are working toward a more inclusive, respectful, and culturally grounded knowledge society.

ORGANIZATION OF THE BOOK

Chapter 1 explores agriculture as a central pillar for community development in South Africa's rural landscapes, with a focused case study on Magomani Village in Limpopo Province. The chapter investigates the decline of commercial farming due to land scarcity, market inaccessibility, lack of capital, and minimal government support. Through qualitative research, it advocates for the revitalization of farming practices as a pathway to reduce poverty and foster sustainable rural development.

Chapter 2 brings to the forefront the urgency of documenting Indigenous Knowledge Systems (IKS) within South African universities. Adopting a decolonial lens and secondary data analysis, the study uncovers current efforts and partnerships between academia and government, while calling for broader institutional commitment to preserving IKS amidst the threat of cultural erosion and knowledge extinction.

In Chapter 3, the authors address the marginalization of Indigenous Knowledge Systems in South Africa's basic education curriculum, particularly under the Curriculum and Assessment Policy Statement (CAPS). Drawing from Indigenous standpoint and constructivist theories, the chapter presents a compelling case for integrating IKS into subjects to provide culturally responsive education that aligns with learners' lived realities.

Chapter 4 assesses traditional food systems as viable tools to combat household-level food insecurity and promote food sovereignty in South Africa. Despite the country's nominal food security status, the chapter reveals significant nutritional deficits, particularly among rural and indigenous populations. It highlights traditional food systems as culturally grounded and nutritionally rich alternatives, advocating for their inclusion in policy frameworks.

Chapter 5 turns to Zimbabwe's Chipinge District, analyzing the traditional and evolving role of *vatete* (aunts) among the Ndau people. Despite their cultural importance as caregivers and moral guides, these women face diminishing authority due to social shifts, resource constraints, and contested roles within modern and religious family structures. The chapter calls for renewed recognition and support for these vital family actors.

Chapter 6 delves into the epistemic richness of indigenous African curricula through the case of the Mulabalaba game. Using qualitative methods, the study illustrates how this traditional game imparts values, logic, and 21st-century skills relevant to higher education. The chapter invites curriculum developers to consider indigenous pedagogies as transformative tools for decolonizing education.

Chapter 7 offers an insightful literature review on the application of indigenous strategies to address gender-based violence (GBV) in KwaZulu-Natal, South Africa. Noting the limitations of conventional legal approaches, the authors argue for the integration of African traditional mechanisms, which have shown potential in curbing GBV at the grassroots level, and which may serve as models across the continent.

Chapter 8 investigates the interplay between indigenous knowledge and landscape management amid environmental change in the Global South. Employing the knowledge coevolution framework, this chapter highlights the necessity of acknowledging local ecological knowledge among state and non-state actors, advocating for power-balanced collaborations to effectively address climate-related challenges.

Chapter 9 sheds light on indigenous methods for managing informal diaper disposal within the Vhembe Biosphere Reserve. As modern waste systems struggle to address environmental and health impacts in low-income areas, the chapter underscores the value of indigenous waste management techniques as sustainable, context-specific alternatives deserving broader attention and application.

Chapter 10 takes us to India's Pushkar Fair to examine how sustainability intersects with religious festival tourism. While this iconic event presents rich tourism opportunities, the chapter critiques the insufficient focus on environmental and cultural sustainability. It proposes a framework for sustainable tourism development grounded in policy alignment and heritage preservation.

Chapter 11 reflects on the integral relationship between nature and the preservation of Siswati Indigenous Knowledge Systems in Eswatini. Anchored in the Biocultural Diversity Framework, the chapter reveals how environmental degradation threatens cultural practices dependent on natural resources. It proposes strategic interventions, including conservation laws and educational programs, to ensure the resilience of Siswati traditions.

Chapter 13 examines generational shifts in name-giving practices among the Batswana people of Gauteng province, highlighting how colonialism, religion, and education have influenced the transition from inherited familial names to contemporary, self-selected ones. This evolving tradition reflects broader cultural transformations across generations.

In Chapter 14, the integration of Indigenous Technical Knowledge (ITK) with modern agricultural innovations is presented as a compelling model for sustainable farming. This chapter articulates how ancestral practices such as the khadin system and johads can be synergized with cutting-edge technologies to address pressing agricultural challenges, promote resilience, and conserve biodiversity.

Chapter 15 transports readers to the Eastern Himalayas, where the vibrant cultural heritage of the Lepcha people is brought to life through their unique fairs and festivals. It illustrates how cultural tourism can be a powerful tool for preserving indigenous identity while adapting to historical and modern influences.

Turning to education and literacy, Chapter 16 examines the revitalization of indigenous storytelling traditions in South Africa through the Vuselela Ilangabi initiative. Using participatory research methods, this chapter emphasizes the value of relationality and traditional narratives in promoting literacy and community engagement among tertiary students.

Chapter 17 critically assesses the lingering impact of linguistic imperialism and advocates for the decolonization of South African Sign Language (SASL). Grounded in an Afrocentric paradigm, the chapter outlines transformative strategies for empowering Deaf communities through language reclamation and culturally responsive policy.

The cultural symbolism and societal role of *umtfwalo* among the emaSwati are explored in Chapter 18. This chapter delves into the metaphoric and literal dimensions of this traditional "load," reflecting on its enduring significance and its inseparability from Swazi identity, as examined through an Afrocentric framework.

Chapter 19 underscores the centrality of oral literature in preserving Swazi cultural heritage. Through an analysis of proverbs, riddles, and other oral genres, the chapter offers insights into how traditional narratives continue to play a crucial role in transmitting values, history, and identity across generations.

In Chapter 20, the spiritual and ecological wisdom of the Shona people of Zimbabwe is explored, highlighting how religious beliefs serve as effective mechanisms for environmental conservation. Sacred sites and species are protected through deeply rooted cultural taboos, illustrating the power of belief systems in fostering ecological stewardship.

Chapter 21 turns to indigenous economic systems, focusing on how traditional farming stokvels can be adapted for the modern agripreneur. By examining the opportunities and limitations of integrating digital tools with communal practices, the chapter advocates for an inclusive model that embraces both heritage and innovation.

Chapter 22 presents the Holistic Integration and Appreciation Model (HIAM), which seeks to incorporate the vernacular literature of the Gond tribe into the English language classroom. Through a rigorous experimental design, the study demonstrates improved language proficiency and cultural pride among students, reinforcing the importance of culturally responsive pedagogy.

Finally, Chapter 23 investigates the implementation of the IKS programs by universities which were part of the 1996 indigenous technologies audit. Using qualitative content analysis, the study shows that with government assistance, universities can successfully implement IKS accredited programs.

Together, these chapters offer an insightful and timely contribution to the discourse on indigenous knowledge systems, cultural sustainability, and the enduring relevance of traditional practices in contemporary society.

CONCLUSION

In reflecting upon the wide-ranging contributions within this volume, it becomes abundantly clear that Indigenous knowledge systems remain vital, evolving, and deeply intertwined with the identity, sustainability, and cultural continuity of diverse communities. Each chapter offers a distinct lens through which readers may understand the resilience, wisdom, and innovation embedded in traditional practices—from healing and agriculture to language preservation, education, and environmental stewardship. What emerges is a powerful narrative of reclamation and revitalization, where oral traditions, spiritual beliefs, and cultural rituals are not relics of the past but dynamic forces shaping the present and future.

As editors, we are inspired by the rigorous scholarship and lived experiences shared throughout these pages. This collection reaffirms the necessity of interdisciplinary, Afrocentric, and community-rooted research that uplifts Indigenous voices and validates alternative epistemologies. It is our hope that this work not only contributes to academic discourse but also empowers policy-makers, educators, and practitioners to reimagine systems that honor and incorporate Indigenous ways of knowing. Let this volume be both a celebration and a call to action—towards preservation, innovation, and justice through cultural heritage.

Acknowledgement

I want to express my gratitude to all the authors who trusted Prof. Tlou Meggie Masenya and myself with their chapters and believed we could successfully finish this book. The authors of this book, who are experts in various fields, have done a remarkable job of contributing their expertise to make *Resurgence and Revalorization of Indigenous Knowledge Systems in the Contemporary Society* book a success. This has been a difficult but fascinating journey for me as an emerging editor and scholar. Without Prof. Tlou Meggie Masenya's mentoring and the ongoing support of the IGI Global development team, none of this would have been feasible. In addition, I want to express my gratitude to the reviewers for their insightful feedback and for taking time out of their hectic schedules to make sure that contributors' papers are strengthened by thorough evaluations. Furthermore, I want to express appreciation to my family for their unwavering support of my career and to the all-powerful God for providing me with the fortitude to complete this project effectively despite the obstacles I have encountered in both my personal and professional lives. Two academic women who are passionate in preserving, protecting, and disseminating Indigenous Knowledge Systems wrote this book. Throughout this process, they have remained steadfast and collaborated to make this book endeavor a success. There is a notable gender disparity in academia, with men frequently dominating and exerting greater influence than women. Thus, this book challenges the existing norms by demonstrating that women can do what Zozibini Tunzi, the 2019 Miss Universe, said: "Women, take up the space in society and cement yourselves! As the book's editors, Dr. Monicca Thulisile Bhuda and Prof. Tlou Meggie Masenya have made sure that the book accurately captures the essence of Indigenous Knowledge Systems and its role in modern society. Indigenous people and their knowledge systems, which are still evolving and adjusting to modern culture, are well represented in this book.

Monicca Thulisile Bhuda
University of Mpumalanga, South Africa

Introduction

Indigenous knowledge (IK) consists of local knowledge that has been created and passed down through socialization processes and word-of-mouth from one generation to the next. Indigenous knowledge also refers to the collective response of local populations to different environmental and social factors that enhance their capacity to adjust to their changing environment. According to the United Nations Educational, Scientific and Cultural Organization is a specialized agency (UNESCO) initiative on Local and Indigenous Knowledge Systems (LINKS), indigenous knowledge is defined as the understandings, abilities, and philosophies that have been formed by local groups with extensive histories and experiences of contact with their natural surroundings. Promoting and incorporating indigenous and local knowledge into international policy and practices, is the goal of UNESCO's Local and LINKS programme. This entails assisting researchers and Indigenous communities, encouraging interdisciplinary cooperation, and creating creative ways to use this knowledge. Additionally, the initiative strives to guarantee that Indigenous knowledge is acknowledged and appreciated in international processes, such as the Sustainable Development Goals of the United Nations.

Indigenous Knowledge Systems (IKS) are therefore the collective wisdom, abilities, customs, and beliefs that indigenous communities have amassed over many generations of habitation in a particular setting. As the community and its surroundings change, IKS also change because they are dynamic and adjust to new situations. Examples of IKS that exist which exist in indigenous communities are as follows: traditional medicine, agriculture and resource management, astronomy and navigation, oral traditions and storytelling, craft skills, language and linguistics, ecological knowledge, social structures, governance, and more. IKS are essential to modern society because they provide solutions to problems like healthcare, sustainable agriculture, and climate change. Moreover, IKS contributes to sustainable development and cultural heritage preservation. Therefore, IKS must be safeguarded in order to protect communities and traditional knowledge holders' livelihoods, boost economies, preserve the environment, and stop biopiracy. Without protection, IKS may be stolen, which would put nations and traditional knowledge holders at a social and economic disadvantage. Protecting KS supports local livelihoods, guarantees the continuation of customs, and enhances the welfare of both indigenous people and the public. It is therefore possible to safeguard IKS by combining documentation, legal frameworks, community involvement, and cultural preservation initiatives. This includes using intellectual property systems, creating sui generis laws and registries, recording and codifying IKS, and encouraging its dissemination within communities. IKS can be successfully safeguarded and advanced by combining these tactics, which will benefit knowledge holders and society at large.

Chapter 1
Agriculture as a Key Driver for Community Development: A Case of Magomani Village, Limpopo Province

Xholiso Ennocent Khazamula
https://orcid.org/0000-0002-1796-2156
Department of Sport, Arts, and Culture, Pretoria, South Africa

ABSTRACT

Limpopo province covers 123 600 km2 and has a population of 5.3 million people, 89% of whom live in non-urban regions and 40–50% of whom are unemployed. It is undoubtedly South Africa's poorest province. The community of Magomani village for a long time has used commercial farming as the main basis for survival and sustaining livelihoods. However, because of the obstacles that local farmers have experienced over the years, there has been a reduction in commercial farming development which has led to the rise of poverty. These obstacles include: insufficient farming land which has been caused by population increase and land occupation; because of a variety of restrictions and constraints, smallholder farmers in rural areas face a lack of market; lack of investment from government and other potential sponsors, lack of working capital which result in poor veld condition and restriction to access of market. This study was qualitative in nature and aimed to investigate how commercial farming can be revived to develop the community of Magomani village.

INTRODUCTION

South African agriculture is best described as a dual system in which commercial agriculture, which is well developed and capital demanding, contributes significantly to national food security. Smallholder farmers and subsistence farmers work in the second form of agriculture, which is less developed and less resourced. Despite the fall in agricultural activity among most rural South Africans because of urbanization, a close to 78% of rural households in South Africa continue to engage in agricultural activities (Statistics South Africa, 2012). Numerous data from Statistics South Africa (2012) show the 75 percent

DOI: 10.4018/979-8-3373-4124-8.ch001

of south African households are involved in smallholder farming, as well as how these people are among the poorest in South Africa, with little livelihood assets and vulnerability to household food insecurity.

The study commissioned by Machethe et al. (2004) concluded that farming is the utmost contributor to household income in the Limpopo Province. Agriculture has always been the backbone of many Indigenous communities' livelihood, not only in Limpopo but also in many other provinces across the country (Maponya, 2021). It is one of the most essential foundations of the provincial economy since it plays such a vital part in the province's economic growth (Mahopo 2023). Agricultural production is said to go through three stages in most cases: hobby farming, subsistence farming, and finally commercial farming (Maponya, 2021). Agricultural production is frequently divided into two systems: large-scale commercial farming and small-holder farming. These methods thrived owing to the advent of apartheid practices, in which white people performed large-scale commercial farming for decades and it was solely reserved for white people (Thamaga-Chitja & Morojele, 2014).

Even though agriculture is one of the province's economic contributors, many rural communities within the Limpopo province were and still live in poverty-stricken conditions. The situation in some village seemed to have worsen due to certain challenges that hinders them to break the poverty cycle and some of them include but not limited to insufficient farming land, which has been caused by population increase and land occupation; lack of secure title to the land, lack of investment from government and other potential sponsors, lack of working capital which result in poor veld condition and maintenance thus today, people have inadequate or damaged infrastructure. Ndlovu, Mwale and Zuwarinwe, (2021) submit that the above challenges do not only affect Limpopo, but these are challenges which are also common in other provinces such as Mpumalanga, Free State, and Eastern Cape (Ndlovu et al., 2021).

Due to lack of jobs in the country, farming becomes one of the key aspects which provide jobs for people at community level while ensuring that some households earn income. Araia (2019) also submits that one of the key roles of commercial farming at a community level is poverty alleviation through creating jobs for the community. It also plays an imperative role in terms of food provision which ensures food security as both community and province in totality.

According to Garidzarai (2018), it is evident that farming plays a key role in economic development at community level and reducing poverty. Although farming plays an important role, it is often overlooked thus facing a number of challenges which over the years have crippled commercial farming in Magomani village. The study focused on crop farming which is highly practiced by most Magomani village farmers. The study argued that farmers in Magomani village face a number of challenges in their commercial farming and require interventions in reviving their farming sector. The study argues that with government interventions, traditional leadership and other stakeholders, there is potential to revive local commercial farming with sustainable strategies that are sustainable and are adequate to be implemented.

PROBLEM STATEMENT

Agriculture has long been a fundamental pillar of rural economies, yet in Magomani Village, Limpopo Province, its potential as a driver for community development remains underutilized. Despite the village's rich agricultural resources and favorable climatic conditions, many local farmers struggle with outdated farming techniques, limited access to markets, and inadequate financial support. These challenges have led to low productivity, food insecurity, and a lack of economic opportunities, exacerbating poverty in the community. Without sustainable agricultural development, Magomani Village continues to experience high unemployment rates, especially among the youth, who are often forced to migrate to urban areas in search of better livelihoods.

This problem is particularly concerning because agriculture has the potential to create employment, enhance food security, and promote local economic growth if properly harnessed. However, the absence of modern farming methods, poor infrastructure, and limited governmental or private sector investment hinder its full contribution to community development. Without strategic interventions such as capacity-building programs, access to funding, and market linkages, Magomani Village risks missing agriculture's transformative power. Addressing these challenges is crucial to unlocking agriculture's role in sustainable development, empowering local farmers, and improving overall living standards in the village.

The research objectives framed for this study are to:

- To assess the current state of agriculture in Magomani Village and its contribution to community development.
- To identify the challenges faced by local farmers in improving agricultural productivity and accessing markets.
- To explore the socio-economic impact of agriculture on employment, food security, and poverty alleviation in the village.
- To recommend sustainable strategies that can enhance agricultural productivity and foster community development in Magomani Village.

COMMERCIAL FARMING IN A SOUTH AFRICAN CONTEXT

South Africa has a land reform program, which is one method the 'new' South Africa set out to correct apartheid's injustices and, by transferring land to black South Africans, to transform the structural basis of racial inequality. Materechera and Scholes (2021; 2022) argued that land reform has fallen well short of both public expectations and official aims throughout the first decade of democracy. Land policy has recently shifted from concentration on rural poverty to 'emerging' Black commercial farmers, which is congruent with shifts in macroeconomic policy and reflects shifting class alliances (Ndove et al., 2004). The initiative now is pursuing a limited de-racialization of commercial farming areas rather than an agrarian restructuring approach.

South Africa's market-led land reform is consistent with the government's macroeconomic orientation and conceptions of the state and market duties. As for Cousins (2013), land reform brought emphasis on entrepreneurship as a way of developing a Black middle class, with less direct involvement in the economy by the state and reliance on partnerships with the private sector. However as stated by Hall (2004), ownership of agricultural land is not a major objective for Black Economic Empowerment (BEE), because joint ventures and value-added companies in secondary agriculture offer more important potential for capital accumulation (Rusenga, 2022).

However, land reform has restrictions that relate not only to its extent but also to how resources are to be allocated, for what purpose, and to whom. The ongoing 'farm size issue' over the relative efficiency of large and small units was a prominent topic of contention in land reform negotiations in the early 1990s. World Bank experts lobbied for the economic efficiency of smallholder agriculture, emboldened by 'success' in Kenya (Hall & Williams, 2003). However, the South African National Department of Agriculture (NDA) and its provincial departments have been heavily criticized for failing to restructure their services to fit the needs of a new customer and to deliver this on a large scale to poorer land reform beneficiaries (Hall, 2004).

For example, the provincial department of agriculture in KwaZulu-Natal has developed income criteria to evaluate LRAD applications, requiring R20,000 profit per person within the first year of operation. This would imply achieving a 100 percent return on investment within the first year for impoverished applicants who are unable to get above the minimum grant amount, which is unrealistic in low-input or subsistence agriculture (Jacobs et al., 2003). Agriculture departments have also attempted to manage land usage through business planning, including limiting settlement growth on agricultural property and prescribing inheritance terms. Rather, officials have encouraged joint ventures and inverted rental markets as opportunities for Black people to own land but not produce it themselves, thereby satisfying both the demand for racial transformation and the imperative to maintain existing capital-intensive production patterns (Jacobs et al., 2003).

According to a poll conducted in the mid-1990s, 68 percent of poor Black people living in rural regions desired land, but the majority of these desired extremely small pieces of land, with half preferring one hectare or less (Marcus et al., 1996). While some may want to produce for local and distant markets or expand their size of operation, most are unable to sustain the risks associated with full-time commercial farming without substantial state backing (Andrew et al., 2003). Land occupations in urban and rural areas have increased the demand for land. The illegal occupation of peri-urban land at Bredell outside Johannesburg and Khayelitsha outside Cape Town in the winter of 2001 by shack dwellers whose dwellings had been flooded was a watershed moment in recent history. As a result, dominant discourses of commercial farming's economic efficiency and the inefficiencies of low-input agriculture by the poor continue to endure within the state bureaucracy. This ideological predilection for commercial farming is visible even among those charged with implementing land reform, where commercial farming criteria have been used to evaluate applications for land allocations.

Meanwhile, there have been reports of increased encroachment by the landless on white commercial farms where they have constructed their houses and, in some cases, begun to engage in some agricultural production of their own in portions of KwaZulu-Natal. In other cases, groups of restitution claimants have attempted to compel the state's hand by occupying both private and state-owned land (Hall, 2004; Hendriks, 2003). Land struggles are frequently intertwined with larger rural fights for survival and control of the countryside, as evidenced by the levels of violence on farms, which include killings of both farm workers and farm owners, the creation of farm commandos, and patterns of fence cutting and animal theft.

Socioeconomically, many smallholder farmers in South Africa are poor, uneducated, and live in rural settlements with little infrastructure, placing them in the so-called second economy Fenwick and Lyne (1999). Many of these villages are typically led by male traditional chiefs, even though most active producers are women (Ortmann & King, 2007). The collective ownership of land in these places tends to reduce its commercial worth, particularly in the context of South Africa's commercial farming, which promotes rivalry and uniqueness.

The poorest households in South Africa occupy about 60% and 52% of rural provinces in Limpopo and the Eastern Cape, respectively (Statistics South Africa, 2012), compared to the Western Cape and Gauteng provinces, which have the wealthiest families. Most South African households, particularly those in rural regions, use a combination of livelihood methods, with earnings and wages contributing to household income, followed by social subsidies, company income, and pension remittances (Statistics South Africa 2012). Ntshangase et al. (2018) argued that regardless of these livelihood options, agricultural operations continue to play a vital role in supplying much-needed sustenance, particularly food.

As per Chikazunga and Paradza (2012), the former South African government's agricultural economy flourished swiftly due to extensive state subsidies and assistance programs aimed at commercial farmers. Similar assistance programs and governmental subsidies were viewed as beneficial to the agricultural economies of the United States of America and Europe (Chikozho et al., 2020). Because South African agriculture is still strongly reliant on global markets for agricultural product marketing, most farmers in South Africa's emerging rural areas are still mostly poverty oriented. Efforts over the last two decades to turn them into commercial operators have failed (Makhura & Wasike, 2003), as evidenced by their poor salaries and unpredictable sources of livelihood. The new regime encourages formerly excluded farmers to join the mainstream agricultural industry. As a result, information clarifying the target group's actions will encourage the development and implementation of relevant policies (Khapayi & Celliers, 2015).

COMMON CHALLENGES FACED BY FARMERS IN SOUTH AFRICA.

The agricultural industry of South Africa has little place for new farmers and people living in rural areas. There is no significant support system in place to assist previously disadvantaged farmers (Chikazunga & Paradza, 2012), preventing such farmers from taking advantage of the various opportunities instituted by the South African government (Moloi, 2010). However, the following are some of the most typical issues that rural farmers encounter.

Lack of ownership has been a challenge since the fall of democracy, as few Black Africans own land. Most of the land in rural settlements is controlled by tribal authorities, who primarily focus on providing land for residential reasons. "No particular member of the community has individual rights or access to property, especially as no registered title documents have been provided at any point in time," (Van Averbeke et al., 1998). Without clear title to the land, it is difficult for local farmers to sustain their crops, which has resulted in several farmers leaving Magomani village throughout the years. Makhura (2001) vehemently argues that secure land tenure is an important pre-condition for a long-standing sustainability of farming practices. This means that without secure land tenure it becomes a challenge to maintain these farming practices.

Land is an important agricultural resource, however over the years the population has grown, which means that more land had to be distributed for residential purposes. This became a major challenge to farming as it meant that even the land which was used for farming purposes had to be utilized for residential purposes. This also forces farmers to limit their farming practices, which means they would not be as productive as they would want to be. It led to some farmers abandoning their farming land. Farmers in rural areas are allocated a small portion of land.

A study conducted by Khapayi and Celliers (2016) reported that farmers' land availability is a concern in market participation. Farmers must have enough acreage to produce if they are to engage in commercial agricultural markets. According to the report, 72 percent of farmers farm on less than 10 hectares of land. Only 28 percent of farmers farm on land larger than 10 hectares. This illustrates that insufficient land availability remains an issue for many new farmers in South Africa. This has significant implications for farm revenue and sustainability, particularly for livestock developing farmers who rely on land availability for grazing and livestock production expansion.

Currently, South African agriculture relies significantly on global markets to promote agricultural products (Chikazunga et al. 2012:3-4). Following the democratic transition in 1994, the removal of state subsidies to marketing boards, as well as deregulation of the agricultural sector, produced major issues for commercial farmers, particularly disadvantaged farmers. Interest rate and export subsidies had terminated fully by 1997, and by late 1998, all marketing control boards had been privatized, with only the sugar industry still receiving government price support (Chikazunga & Paradza, 2012). Gandure et al. (2013) argued that most rural farmers who are new struggle to gain access to formal agricultural markets. As a result, developing farmers are uninterested in conventional marketplaces. Lack of market engagement is a prevalent trait of emerging farmers around the world and identifies it as a limitation to emergent farmer development. Due to a variety of obstacles, developing farmers in South Africa's undeveloped rural areas find it difficult to compete in commercial markets (De Cock et al., 2013).

According to Baloyi (2010), farmers located in rural areas often lack infrastructure such as storeroom to put their products after harvest which makes them to mostly operate at a loss and make it difficult to sustain their farming practices. Lack of infrastructure also becomes a barrier for local farmers to enter the markets since buyers focus more on quality than quantity. Bienabe et al., 2004 clearly postulate that "Access to storage facilities increases farmers' flexibility in selling their products, as well as their bargaining power." Farmers in Magomani village often use their houses as storage which also compromises the quality of their products, and it hinders them from accessing the large market. Baloyi (2010) also contends that lack of infrastructure for most farmers in the province limits their chances to access the market and even when they access the markets, they sell their products as raw products.

Based on the listed challenges that have been highlighted above, it is the study's argument that similar challenges have been encountered by the community of Magomani village who seek majors to address them. The study claims that if the government's reform efforts and goals to promote agriculture for economic recovery are to be successful, agriculture can be used as a significant driver for community development in Magomani village. Making sure that farmers can scale up their operations over time to become more sustainable ought not to be the only goal.

THEORETICAL FRAMEWORK

This study employs the Sustainable Livelihoods Framework (SLF) to analyze the impact of commercial farming on community development in Magomani Village. SLF offers a comprehensive approach to understanding how households utilize various assets human, social, natural, physical, and financial capitals to achieve sustainable livelihoods. Through focusing on these assets, the framework facilitates an examination of the strategies households adopt to enhance their well-being and resilience. This approach is particularly relevant for assessing rural development initiatives, as it emphasizes the multifaceted nature of livelihoods and the importance of asset interconnectivity (Evwierhurhoma, 2024).

In the context of Magomani Village, the SLF allows for an in-depth exploration of how commercial farming influences the availability and utilization of livelihood assets. For instance, the introduction of commercial farming can lead to improved physical capital through better infrastructure, such as roads and irrigation systems, thereby enhancing market access for farmers. Additionally, it can bolster financial capital by increasing household incomes and providing opportunities for reinvestment in farming activities. The framework also considers the role of social capital, recognizing that community networks and cooperative structures can facilitate knowledge sharing and collective action, which are vital for sustainable agricultural practices (Cheng & Zhang, 2024).

Moreover, the SLF underscores the significance of external factors, such as policies, institutions, and vulnerability contexts, in shaping livelihood outcomes. Through incorporating these elements, the framework enables a holistic assessment of how commercial farming interacts with existing structures and the broader environment. This comprehensive perspective is essential for identifying potential challenges and opportunities associated with commercial farming initiatives. Furthermore, integrating the SLF with resilience assessment approaches can provide deeper insights into the adaptive capacities of farming systems, highlighting the importance of diversity, redundancy, and learning in building resilience against shocks and stresses (Darnhofer et al., 2022).

METHODOLOGY

A total of six Indigenous farming experts aged 45 to 63 years of age in Magomani village were sampled through expert purposive sampling based on their years of experience in commercial farming. Expert purposive sampling was best suited for the study as purpose of the study was to glean knowledge from farmers in the selected study area. The farmers who were sampled were all females who the study has identified that they participate in farming and commercialization compared to males. The researcher also employed the following instruments: Face-to-face interviews and thematic data analysis. These instruments are discussed below:

Face-to-face in-depth interviews often take the shape of a one-on-one informal conversation between the researcher and the participants. These interviews were semi-structured in nature which afforded the researchers an opportunity to probe any aspect further either for clarity or confirmation purposes. This tool was best suited for the study as it afforded the researchers an opportunity to not only record verbal responses from participants but also take note of their non-verbal gestures, and their body language to better understand their responses. As stated by Rutakumwa et al. (2020), face-to-face in-depth interviewing is a type of qualitative research or data gathering technique where information is gathered by direct communication between the participant and the researcher. Face-to-face in-depth interviews provide the chance to establish a connection with participants and learn more about their viewpoints and experiences.

The emerged data from the face to face in-depth interviews was manually analysed through a thematic analysis using the following protocols, as outlined by Braun & Clarke (2006): (1) This researcher read all the transcribed data from the in-depth interview and focus group interview thoroughly; (2) Once familiar with the data, she identified recurrent phrases and concepts that were considered important on which were highlighted and colour-coded in all the interviews; (3) concepts and phrases from the different data sets that were matched were then categorized into themes; (4) the researcher refined and defined the themes and potential subthemes within the data and similar themes emerged; and (5) the themes were reported as the results of the study. Castleberry and Nole (2018) agree that thematic analysis is a technique for analyzing qualitative data in which a set of data is read, and themes are discovered by looking for patterns in the meaning of the data. Alhojailan and Ibrahim (2012) added that a method of qualitative data analysis called thematic analysis includes reading over a data set (such as the transcripts from in-depth interviews or focus groups) and looking for patterns in the meaning of the data to generate themes.

The study area selected for this study is Magomani village. The study area is situated in the Vhembe district under Collins Chabane Local Municipality and it's about thirty kilometers from Malamulele town, five kilometers from Saselamani, and it is also situated approximately 214 kilometers north of the province capital city (Polokwane). The selected study area is a deep rural area which has a proximity of commercial farming land. The study area compromises of Indigenous people who have been depending on farming for survival and used indigenous methods of farming for generations. The community of Magomani village was chosen because it has invested in commercial farming as a means of creating jobs and capital, as well as because of its potential and ability to grow in farming and become an example to neighboring villages. In Magomani village, commercial farming could indeed flourish if the right sustainable practices are implemented which will benefit the community.

The study maintains that farming in Magomani village forms an integral part of their Indigenous knowledge and Indigenous people should commercialize their products. The commercialization of Indigenous knowledge for job and wealth creation has been supported by the IKS policy of 2004 and the IK Act of 2019. As the study has above mentioned, women participate in farming hence they were the main participants of the study. Indigenous women are recognized by the IKS policy of 2004 as the custodians of Indigenous knowledge systems, playing crucial roles in their preservation and intergenerational transmission.

Findings of the study

This section presents the findings of the study under themes which were drawn during data collection phase: commercial farmers' information needs; Farmer's access to commercial farming land; Government support for local farmers; Challenges to access to markets for products Strategies to revive commercial farming within the village; Education programmes aimed to target young people.

Challenges to access markets for products

Farmers who were the participants in the study have raised concerns that there is a poor market when it comes to their products. They have mentioned that they have their own markets to sell their products but, however, people prefer to buy from big established supermarkets. Participant one highlighted that she is actually working at a loss because so many products she sells get spoilt and that forces her to sell at half price in order to avoid further loss. Participant five stated that they are trying a new method of entering the market by being suppliers of big retail stores, which is a slow ongoing process since there are so many terms and conditions that need to be adhered to in order to get approval.

The participants stated that so far within the village and surrounding areas, Saselamani spar, Boxer supermarket as well as Shoprite Usave try to assist them access the market by means of buy their produce particularly fruits (Banana, Mango and oranges) and vegetables (Spinach, Cabbage, Tomatoes, Chill, Butternut, Rape and Onions). Participants also agreed to find strategies to attract local customers so that when they get disappointed by large retail stores, they can have an income backup from their successful markets. Participant three indicated that another way to access the market is to sell their produce to street vendors in nearby towns. Participant two agreed that there is indeed a good relationship with street vendors who are stationed in Saselamani, and they supply them with their produce from time to time.

Farmer's access to commercial farming land

The results of the study revealed that although traditional leaders are the custodians of the land, they often do not use it for the economic benefit of the economy. Instead, it is sold to external people who in return do not even employ locals. Another challenge that is faced in Magomani village is that there are land disputes and debates which do not get settled. In return, farmers get limited land to use and the chances of growing their land size are almost impossible without the intervention of traditional leadership and local government. Participant three acknowledged that some land is used for residential areas while some to develop a central business district (CBD). However, the participant further indicates that there is abundance of land in the area that can be used for farming which is a method for economic development. She has expressed that the land is in a healthy environment to explore farming which can even eliminate poverty in the community.

Investing in land and farming, according to participant five as well, will encourage students to enroll in studies that can help maintain the environment and boost the local economy through farming. Participant two argued that because there are farming opportunities in their area, schools should encourage pupils to pursue further education in related subjects because they will return as advisers and help the same community by working to extend the farming grounds and markets.

It is therefore participant two's advice that every member of the community has a role to play in order to ensure that commercial farming in Magomani remains sustainable and serves as an example to other surrounding villages. The participant stated that Limpopo is known for its reach environment and farming products. It is there the chance diverse communities capitalize and expand throughout the country. However, the strategies and initiative need to begin at a local level.

Government support for local farmers

Participants were asked about the government support they receive in their business, the kind of support and the time duration. Only two out of the six interviewed farmers have received government support in the form of workshops, seminars, funding, and mentorship through the Department of Small Business and Enterprise in partnership with Collins Chabane Local Municipality. The remaining four seemed to not have a clue that such support exists or how one can access them. What was gathered from this matter was that as much as the respondents are not receiving any support it could be that they do not have the right access to information. This became evident that lack of access to information for rural-based farmers is also one of the key challenges that hinder their business success, and access to large markets.

Participant six stated that most farmers are old women who have been using long standing traditions to market themselves and sustain their businesses. It is only now they are learning about other means to sell their products from young people in the community who are willing to assist. Therefore, due to lack of formal education, the farmers struggle to even follow correct channels to get funding. Community engagements are limited and when they are available, they do not speak to the local people. Instead, government officials from outside the village would be invited to speak and the content presented is not linked directly to the needs of the community, especially farmers who are elderly and have never attended any formal schooling.

Strategies to revive commercial farming within the village.

Farmers have devised strategies for sharing access to agricultural land, including irrigation plots. Work and produce are shared informally through arrangements based on responsibility for relatives and poor community members. The farmers decided to apply for grants that will help them sustain their businesses and have enough funds to buy land to farm. When land redistribution applicants receive funds to purchase ownership in a going firm, a farm worker share equity plan is formed. In theory, as co-owners, they will be able to participate in management decisions, have a stronger stake in the farming enterprise's success, and eventually be able to take over the farm entirely. Equity plans have been considered as particularly useful when land costs are exceptionally expensive (i.e., because grants would not be adequate to enable outright acquisition) and/or when the enterprise's complexity necessitates continuity of management.

Participants of the study believe that if they receive finance to sustain their enterprises, they will be able to produce food/crops in huge quantities and sell them at a low price. They will then make bids with local supermarkets and begin supplying them with affordable vegetables and fruits. They did, however, point out that money is required to provide an acceptable environment for agricultural and food production. Participants further expressed that to revive commercial farming in their village, the government should continue to support a land restitution and redistribution process that promotes agricultural production and land investment. In regarding the land redistribution, locals should be prioritized because they have witnessed those who are not from the community and are not South African by birth acquiring land for commercial interests while they continue to struggle. By putting more land to productive use and providing more South Africans with assets and opportunity for sustainable lives, the country is laying the groundwork for more inclusive, meaningful growth.

Participant 6 argued that proper land distribution is important in villages because vacant land is said to belong to the government or is for private use while residents have lived in that area before apartheid and, according to customary laws, the land should be of the traditional leaders, who usually distribute it to the worthy residents for various purposes. As a result, the unoccupied land must be reevaluated and investigated, and citizens must bring their perspectives on how the site should be used to improve the local economy. Furthermore, farmers will be able to expand their enterprises in a variety of ways because of this.

Education programmes aimed at targeting young people.

According to participant two, there is a need to encourage young people to pursue farming and that they should be aware of its economic importance. The greatest strategy for doing this is to have agricultural education programmes. These educational initiatives are crucial since they will instruct young people on how to start their own enterprises and on farming. In agreement with participant two, participant four said that elderly farmers without formal education also benefit from educational programmes since they will teach them about fundamental business principles and the procedures to follow in order to manage successful markets.

Participant six mentioned that her granddaughter attends the University of Limpopo and is enrolled in a study linked to agriculture; she claims that she served as her inspiration. She added that for them to give back to the same community, these young people need to be developed and guided. For young people to participate in agriculture and business courses and receive adequate financial support so that they may contribute back to their communities, it is the responsibility of the community, government, and traditional leadership. In response, participant five asserted that there are wealthy people who control the local economy of their community who can be willing to contribute. She further stated that they can aid in funding students if traditional authorities present them with well-structured plans since there comes a time when the government is unable to help people.

Participant one expressed that it would be extremely sad if the Magomani land is misused and left unutilized for economic gain. As a result, agriculture is a step forward and ought to be used as an illustration of how land should be managed to boost the local economy. She continued by saying that instead of waiting for jobs to be produced, individuals can start creating their own by making use of what they already have. With the help of the community, the government, and the traditional authority, commercial farming can be accomplished and put into practice on Magomani land.

Discussion

Magomani village has the potential to explore commercial farming since it has an abundance of land. It is this studies argument that even if that is the case, farmers who want to commercialize their products face numerous challenges that hinder their businesses from growing and being sustainable. The challenges that have been stated in the study have also been well discussed by participants who have solutions and suggestions. It was discussed that with well-structured and implemented solutions, farmers can have access to the market and there is a potential for commercial farming to expand in Magomani village. Similar studies conducted by Mpandeli and Maponya (2015) stated that access to market for local farmers particularly in Rabali and Tshakuma areas in Limpopo province is one of the key challenges that farmers in rural areas face because they lack knowledge and proper channels to guide them (Mpandeli & Maponya, 2015).

Mathinya et al. (2022), on the other hand, claimed that farmers' attempts to market their produce are primarily hampered by poor infrastructure, inadequate property rights, low education levels among farmers, limited access to credit, a lack of innovative production implements necessary to increase the yield of the produce produced, and a lack of the entrepreneurial skills necessary to make the farmers' efforts successful. For farming to strive in Magomani village, farmers need access to sufficient land which Van Averbeke et al. (1998) submited that "No particular member of the community has individual rights or access to land, especially as no registered title deeds have been given at any point in time". This shows that access to land for farming purposes remains a challenge. Without secure land tenure many farming practices will not stand the test of time as secure land tenure is very important for the sustainability of any farming practices. (Van Averbeke et al., 1998; Makhura, 2001).

Because the aim of commercial farming is to maximize profit margins, it must be extremely efficient and carried out on a huge scale in order to achieve the economy of scale. Materechera and Scholes (2022) agreed that another significant factor influencing agricultural production is financial support for farming systems. If the farming system does not have access to enough funding, it will not be able to continue. Khapayi and Celliers (2016) conducted a study on factors limiting and preventing emerging farmers to progress to commercial agricultural farming in the King William's Town area of the Eastern Cape Province, South Africa. They have stated in their discussion that in the previous 18 years (as defined by their 2016 study), the South African government has put into place a number of policies and programs and increased the money allocated to the agricultural sector in support of beginning farmers (Department of Agriculture, Forestry, and Fisheries, 2010). However, there is currently insufficient proof that these efforts were successful (Frequin et al., 2012).

For commercial farmers to grow their market and contribute to the agriculture industry and economy, there need to be sustainable strategies in place that can be implemented. With the help from different stakeholders and relevant government departments, underprivileged farmers can build an industry that can flourish for a long time. The current government can use the same system of the former South African government to guide them during the implementation process because the previous government had significant state subsidies and support programs for commercial farmers allowed the country's agricultural economy to expand quickly (Chikazunga & Paradza, 2012).

Currently, the marketing of agricultural products in South African agriculture is primarily dependent on global markets which needs to urgently change (Chikazunga & Oaradza, 2012). Prioritization and a plan are necessary for creating an agricultural transformation plan. Mmbengwa et al (2015) stated that this ought to concentrate on the adjustments most likely to spark rural economic expansion. Successful plans pinpoint objectives in a small number of value chains for crops and livestock, as well as in certain geographies and cross-cutting agricultural sector enablers.

The findings of this study reveal that agriculture has immense potential to drive community development in Magomani Village, yet several barriers hinder its full realization. One of the key challenges identified is inadequate infrastructure, particularly poor road networks, unreliable electricity supply, and limited irrigation systems. These deficiencies make it difficult for farmers to transport their produce to markets, access essential farming inputs, and practice modern irrigation techniques that could improve crop yields. As a result, many farmers struggle with low productivity and post-harvest losses, which ultimately affect their income and economic stability. Addressing these infrastructural challenges is crucial to unlocking the full potential of commercial farming in the village.

FUTURE RESEARCH DIRECTIONS

Future research should explore the long-term impact of commercial farming on community development in Magomani Village and similar rural areas. A longitudinal study could provide deeper insights into how sustained agricultural investments influence employment rates, household incomes, and food security over time. Additionally, research could examine the role of climate change adaptation strategies in enhancing agricultural resilience. Given the increasing unpredictability of weather patterns, studies focusing on sustainable farming practices and their effectiveness in mitigating climate risks would be valuable.

Furthermore, future studies could investigate the potential of agro-processing industries in adding value to agricultural products and creating additional employment opportunities. Research on the effectiveness of government policies and private sector involvement in promoting agricultural entrepreneurship could also be beneficial. Comparative studies between Magomani Village and other communities that have successfully leveraged agriculture for economic development may offer lessons on best practices. Finally, incorporating technological advancements such as artificial intelligence, precision farming, and digital marketplaces in future research could provide innovative solutions for improving agricultural productivity and market access.

CONCLUSION

This paper has highlighted the significant role that commercial farming can play in driving community development in Magomani Village, Limpopo Province. Despite the village's rich agricultural potential, various challenges, including inadequate infrastructure, limited financial support, and a lack of modern farming techniques, hinder the full realization of agriculture's benefits. Addressing these barriers is crucial in ensuring that agriculture becomes a key driver of economic growth, employment creation, and food security in the community. The study's findings emphasize the need for strategic interventions to maximize the potential of commercial farming for sustainable development.

One of the key takeaways from this study is the necessity of government and private sector collaboration in supporting local farmers. Investments in infrastructure, such as roads, irrigation systems, and storage facilities, can significantly enhance agricultural productivity and market access. Additionally, financial support mechanisms, such as grants, subsidies, and access to credit, can empower emerging farmers to adopt modern farming technologies and expand their operations. Without such interventions, the agricultural sector in Magomani Village will continue to struggle, limiting its contribution to community development. Furthermore, the paper underscores the importance of capacity-building initiatives for farmers. Training programs on sustainable and climate-smart agriculture can improve productivity while ensuring environmental conservation. Encouraging cooperative farming models can also enhance farmers' bargaining power, reduce input costs, and promote knowledge-sharing. Strengthening land tenure security is another critical aspect, as it provides farmers with the confidence to invest in long-term agricultural ventures.

In conclusion, the successful integration of commercial farming into community development strategies requires a multi-faceted approach that includes infrastructural development, financial empowerment, skill enhancement, and policy support. By addressing the identified challenges and implementing the recommended strategies, agriculture can significantly contribute to poverty alleviation, employment creation, and overall economic growth in Magomani Village. Future research should build upon these findings to explore innovative solutions that further enhance agricultural development and its impact on rural communities. it is evident that agriculture in Limpopo is very important in many ways and contributes immensely to both the local and provincial economy thus it has been classified as one of the key contributors to the provincial economy. Regardless of the status of agriculture in the province there are still a lot of challenges which farmers in Magomani village experience which need serious interventions and strategies for reviving. Participants' ideas show that techniques can be put in place to revitalize farming in Magomani village, but full engagement from residents, different stakeholders, traditional leaders, and the government is required. Regardless of their age group, participants are willing to take part in proposed initiatives to access the market and build a sustainable farm that has financial benefit. According to the results, reviving farming in Magomani village will strengthen the local economy and allow farmers to obtain the necessary assistance to expand their businesses.

RECOMMENDATIONS

Based on the findings of this study, several recommendations can be made to enhance the role of commercial farming in driving community development.

Firstly, the government and related stakeholders should prioritize infrastructure investment to facilitate commercial farming. Improved road networks, irrigation systems, and electrical connections will allow farmers to operate more efficiently, reduce post-harvest losses, and boost market access. These infrastructure developments will make a substantial contribution to the community's economic growth.

Secondly, financial support mechanisms should be established to assist emerging farmers. Access to credit and subsidies will enable farmers to invest in modern farming techniques, purchase high-quality seeds, and acquire essential machinery. Financial literacy programs should also be implemented to equip farmers with skills in managing farm finances effectively.

Thirdly, training and capacity-building initiatives should be introduced to equip farmers with knowledge on sustainable farming practices. Agricultural extension services should be expanded to provide technical support on soil management, pest control, and climate-smart agriculture. These interventions will ensure long-term productivity and environmental conservation.

Moreover, fostering cooperative farming models will enhance economies of scale and market competitiveness. Farmer cooperatives can negotiate better prices, reduce input costs, and collectively invest in processing facilities. Encouraging collaboration among farmers will lead to increased profitability and improved livelihoods. Additionally, policies that promote land tenure security should be strengthened to encourage investment in commercial farming. Farmers need assurance that they have long-term rights to the land they cultivate. Clear land ownership structures will enhance stability and encourage sustainable agricultural practices.

Finally, the integration of technology in farming should be encouraged to improve efficiency and productivity. Digital platforms can facilitate market linkages, provide weather forecasts, and enhance knowledge-sharing among farmers. Governments and private sector players should invest in digital literacy programs to ensure farmers can leverage these technological advancements. By implementing these recommendations, commercial farming can significantly contribute to community development, create employment opportunities, and enhance food security within the region.

REFERENCES

Alhojailan, M. I., & Ibrahim, M. (2012). Thematic analysis: A critical review of its process and evaluation. *West east journal of social sciences, 1*(1), 39-47.

Andrew, M., & Ainslie, A. C. & Shackleton. (2003). Land use and livelihoods. Evaluating Land and Agrarian Reform in South Africa, No. 8. Bellville: University of the Western Cape.

Araia, M. G. (2019). Ecological and Socio-Cultural Potential of Human Modified Forest Landscapes in Conservation of Tree Species Diversity: The Case of Vhembe Biosphere Reserve, South Africa (Doctoral dissertation, University of Pretoria (South Africa)).

Baloyi, J. K. (2010). An analysis of constraints facing smallholder farmers in the Agribusiness value chain: A case study of farmers in the Limpopo Province (Doctoral dissertation, University of Pretoria).

Bienabe, E., Coronel, C., Le Coq, J., & Liagre, L. (2004). *Linking smallholder farmers to markets: Lessons learned from literature review and analytical review of selected projects. Study Report, Final draft, March 2004. CIRAD & IRAM.* World Bank.

Braun, V., & Clarke, V. (2014). What can "thematic analysis" offer health and wellbeing researchers? *International Journal of Qualitative Studies on Health and Well-being*, *9*(1), 26152. DOI: 10.3402/qhw.v9.26152 PMID: 25326092

Castleberry, A., & Nolen, A. (2018). Thematic analysis of qualitative research data: Is it as easy as it sounds? *Currents in Pharmacy Teaching & Learning*, *10*(6), 807–815. DOI: 10.1016/j.cptl.2018.03.019 PMID: 30025784

Chikazunga, D., & Paradza, G. (2012). Can smallholder farmers find a home in South Africa's food-system? Lessons from Limpopo Province. The Institute for Poverty, Land and Agriculture Studies, (PLAAS) Blog. https://www.plaas.org.za/blog/can-smallholderfarmers-find-home-south-africa%E2%80%99s-food-system-lessons-limpopo-province. Accessed: 25 March 2023

Chikazunga, D., & Paradza, G. (2012). Smallholder farming: A panacea for employment creation and enterprise development in South Africa? Lessons from the Pro-Poor Value Chain Governance Project in Limpopo Province. *Strategies to overcome poverty and inequality: Towards Carnegie III.*

Chikozho, C., Managa, R., & Dabata, T. (2020). Ensuring access to water for food production by emerging farmers in South Africa: What are the missing ingredients? *Water S.A.*, *46*(2), 225–233.

Cousins, B. (2013). Land redistribution, populism, and elite capture: New land reform policy proposals under the microscope. *The Journal of the Helen Suzman Foundation*, *70*, 11–19.

De Cock, N., D'Haese, M., Vink, N., Van Rooyen, C. J., Staelens, L., Schönfeldt, H. C., & D'Haese, L. (2013). Food security in rural areas of Limpopo province, South Africa. *Food Security*, *5*(2), 269–282. DOI: 10.1007/s12571-013-0247-y

Department of agriculture forestry and fisheries. (2010). The Strategic Plan for South African Agriculture. Department of Agriculture, Directorate of Agricultural Information Services, Pretoria

Fenwick, L. J., & Lyne, M. C. (1999). The relative importance of liquidity and other constraints inhibiting the growth of small-scale farming in KwaZulu-Natal. *Development Southern Africa, 16*(1), 141–155. DOI: 10.1080/03768359908440066

Frequin, S., Anseeuw, W., & D'haese, M. (2012). *Demystifying contract farming: evidence from rural South Africa. Department of Agricultural Economics*. Ghent University.

Gandure, S., Walker, S., & Botha, J. J. (2013). Farmers' perceptions of adaptation to climate change and water stress in a South African rural community. *Environmental Development, 5*, 39–53. DOI: 10.1016/j.envdev.2012.11.004

Hall, R. (2004). A political economy of land reform in South Africa. *Review of African Political Economy, 31*(100), 213–227. DOI: 10.1080/0305624042000262257

Hall, R., Jacobs, P., & Lahiff, E. (2003). *'Final report,' Evaluating Land and Agrarian Reform in South Africa 10*. Programme for Land and Agrarian Studies, University of the WesternCape.

Hall, R., & Williams, G. (2003). Land reform in South Africa: problems and prospects. *From Cape to Congo: Southern Africa's evolving security architecture*, 97-129.

Hendriks, S. L. (2003). The potential for nutritional benefits from increased agricultural production in rural KwaZulu-Natal. *South African Journal of Agricultural Extension, 32*, 28–44.

Jacobs, T., Lahiff, E., & Hall, R. (2003). *Evaluating land and agrarian reform in South Africa: Land redistribution*). PLAAS, University of the Western Cape.

Khapayi, M., & Celliers, P. R. (2015). Issues and constraints for emerging farmers in the Eastern Cape Province, South Africa. *African Journal of Agricultural Research, 10*(41), 3860–3869. DOI: 10.5897/AJAR2015.9956

Khapayi, M., & Celliers, P. R. (2016). Factors limiting and preventing emerging farmers to progress to commercial agricultural farming in the King William's Town area of the Eastern Cape Province, South Africa. *South African Journal of Agricultural Extension, 44*(1), 25–41. DOI: 10.17159/2413-3221/2016/v44n1a374

Limpopo Department of Agriculture (LDA). 2008. Basic principles: Policy document. Available online at: www.lda.gov.za

Machethe, C. L., Mollel, N. M., Ayisi, K., Mashatola, M. B., Anim, F. D. K., & Vanasche, F. (2004). *Smallholder irrigation and agricultural development in the Olifants River Basin of Limpopo Province: Management transfer, productivity, profitability, and food security issues. Report to the Water Research Commission on the Project "Sustainable Local Management of Smaller Irrigation in the Olifants River Basin of Limpopo Province."*. Water Research Commission.

Makhura, M. N., & Wasike, W. S. (2003). Patterns of access to rural service infrastructure: The case of farming households in Limpopo Province. *Agrekon, 42*(2), 129–143. DOI: 10.1080/03031853.2003.9523616

Makhura, T. M. (2001). Overcoming transaction costs barriers to market participation of smallholder farmers in the Northern Province of South Africa. Unpublished PhD thesis, Pretoria, University of Pretoria.

Maponya, P. (2021). Opportunities and constraints faced by smallholder farmers in the Vhembe District, Limpopo Province in South Africa. *Circular Economy and Sustainability*, *1*(4), 1387–1400. DOI: 10.1007/s43615-021-00028-x

Marcus, T., Eales, K., & Wildschut, A. (1996). *Down to earth: land demand in the new South Africa. Land and Agriculture Policy Centre*. Indicator Press.

Materechera, F., & Scholes, M. C. (2021). Characterization of farming systems using land as a driver of production and sustainability in the Vhembe District, Limpopo, South Africa. *Agricultural Sciences*, *12*(11), 1352–1373. DOI: 10.4236/as.2021.1211087

Materechera, F., & Scholes, M. C. (2022). Understanding the Drivers of Production in South African Farming Systems: A Case Study of the Vhembe District, Limpopo South Africa. *Frontiers in Sustainable Food Systems*, *6*, 722344. DOI: 10.3389/fsufs.2022.722344

Mathinya, V. N., Franke, A. C., Van De Ven, G. W. J., & Giller, K. E. (2022). Productivity and constraints of small-scale crop farming in the summer rainfall region of South Africa. *Outlook on Agriculture*, *51*(2), 139–154. DOI: 10.1177/00307270221091839

Mmbengwa, V., Nyhodo, B., Myeki, L., Ngethu, X., & van Schalkwyk, H. (2015). Communal livestock farming in South Africa: Does this farming system create jobs for poverty stricken rural areas. *Sylwan*, *159*(10), 176–192.

Moloi, M. J. (2010). *A comparison of socioeconomic characteristics that determine the farm income of emerging lifestock and horticultural farmers in South Africa* (Doctoral dissertation).

Motsoaledi, P. A. (2003). Supplement to Nufarmer & African Entrepreneur (March 8, 2003, Vol. 2). Press release.

Mpandeli, S., Nesamvuni, E., & Maponya, P. (2015). Adapting to the impacts of drought by smallholder farmers in Sekhukhune District in Limpopo Province, South Africa. *Journal of Agricultural Science (Toronto)*, *7*(2), 115. DOI: 10.5539/jas.v7n2p115

Mpandeli, S., Simalenga, T., Siambi, M., Ramugondo, R., Mailula, N., & Liphadzi, K. (2008). Constraints and challenges to agricultural development in Limpopo Province, South Africa. *CGIAR Challenge Program on Water and Food*, 138.

Musyoki, A. (2012). The emerging policy for green economy and social development in Limpopo, South Africa (No. 8). UNRISD Occasional Paper: Social Dimensions of Green Economy and Sustainable Development.

Ndlovu, W., Mwale, M., & Zuwarimwe, J. (2021). Challenges and critical success factors for rural agrarian reforms in limpopo province, south africa. AGRARIS. *Journal of Agribusiness and Rural Development Research*, *7*(2), 191–206. DOI: 10.18196/agraris.v7i2.11272

Ndove, T. S., Whitbread, A. M., Clark, R. A., & Pengelly, B. C. (2004). Identifying the factors that contribute to the successful adoption of improved farming practices in the smallholder sector of Limpopo Province, South Africa. *Tropical legumes for sustainable farming systems in Southern Africa and Australia*, (115), 146.

Nesamvuni, A. E., Oni, S. A., Odhiambo, J. J. O., & Nthakheni, N. D. (2003). *Agriculture is the cornerstone of the economy of the Limpopo Province. A study commissioned by the Economic cluster of the Limpopo Provincial Government under the Leadership of the Department of Agriculture. University of Venda for Science and Technology*. Thohoyandou.

Ntshangase, N. L., Muroyiwa, B., & Sibanda, M. (2018). Farmers' perceptions and factors influencing the adoption of no-till conservation agriculture by small-scale farmers in Zashuke, KwaZulu-Natal Province. *Sustainability (Basel), 10*(2), 555. DOI: 10.3390/su10020555

Nyikai, R. A. (2003). Commercial and subsistence Farming: What is the future for smallholder Kenyan agriculture? *African Crop Science Conference Proceedings* 6:591-596.

Oni, S. A., Nesamvuni, A. E., Odhiambo, J. J. O., & Dagada, M. C. (2012). *The study of agricultural industry in the Limpopo Province*. Executive Summary.

Opdenakker, R. (2006). Advantages and disadvantages of four interview techniques in qualitative research. In Forum qualitative sozialforschung/forum: Qualitative social research (Vol. 7, No. 4).

Ortmann, G. F., & King, R. P. (2007). Agricultural cooperatives II: Can they facilitate access of small-scale farmers in South Africa to input and product markets? *Agrekon, 46*(2), 219–244. DOI: 10.1080/03031853.2007.9523769

Rusenga, C. (2022). Rethinking Land Reform and Its Contribution to Livelihoods in South Africa. *African Review (Dar Es Salaam, Tanzania), 14*(2), 125–150. DOI: 10.1163/09744061-bja10033

Rutakumwa, R., Mugisha, J. O., Bernays, S., Kabunga, E., Tumwekwase, G., Mbonye, M., & Seeley, J. (2020). Conducting in-depth interviews with and without voice recorders: A comparative analysis. *Qualitative Research, 20*(5), 565–581. DOI: 10.1177/1468794119884806 PMID: 32903872

Statistics South Africa. (2012). *General Household Survey Series Volume IV.Food Security and Agriculture (2002-2011): In Depth analysis of General Household Survey.* Statistics South Africa: Pretoria.

Thamaga-Chitja, J. M., & Morojele, P. (2014). The context of smallholder farming in South Africa: Towards a livelihood asset building framework. *Journal of Human Ecology (Delhi, India), 45*(2), 147–155. DOI: 10.1080/09709274.2014.11906688

Van Averbeke, W. M., Marete, C. K., Igodan, C. O., & Belete, A. (1998). An investigation into Food Plot Production at Irrigation into Food Plot Production at Irrigation Schemes in Central Eastern Cape. Water Research Commision Report, No. 719/1/98.

ADDITIONAL READING

Damonse, K. (2022). The unlocked power of the agricultural sector: An economy-wide approach in South Africa.

Hlatshwayo, S. I., Ojo, T. O., Modi, A. T., Mabhaudhi, T., Slotow, R., & Ngidi, M. S. C. (2022). The determinants of market participation and its effect on food security of the rural smallholder farmers in Limpopo and Mpumalanga provinces, South Africa. *Agriculture, 12*(7), 1072. DOI: 10.3390/agriculture12071072 PMID: 37701244

Kabini, D. M. (2022). The impact of the agricultural sector on economic growth and development in South Africa (Doctoral dissertation, North-West University (South Africa)).

Kubayi, V. R. (2022). *The Effects of Climate Variability and Change on Rural Livelihoods in the Tiyani Village, Limpopo Province*. University of Johannesburg.

Lavhelani, P. D. (2023). The influence of supply chain management systems on agricultural businesses in Vhembe District of Limpopo Province (Doctoral dissertation).

Ntuli, N. C. (2023). Adaptation strategies against drought: The case of rain-fed subsistence crop farmers in Mphego village in the Vhembe District of Limpopo province, South Africa.

Rankoana, S. A. (2022). Small-scale farmers' vulnerability to the impacts of climate change in Limpopo Province, South Africa: A review. *Int J Develop Sustain, 11*, 236–248.

Shikwambana, S. (2024). Smallholder farmer livelihood strategies for coping and adapting to drought in the Limpopo Province, South Africa (Doctoral dissertation, Cape Peninsula University of Technology).

KEY TERMS AND DEFINITIONS

Agriculture: The practice of cultivating soil, growing crops, and rearing animals to provide food, fiber, and other items for human survival and enhancement.

Commercial Farming: A type of agriculture focused on producing crops and livestock for sale in markets, often involving large-scale operations, modern technology, and profit-driven motives.

Community: A group of people living in the same area or sharing common values, interests, or identity, often connected by social, economic, or cultural ties.

Development: The process of improving the economic, social, and environmental conditions of a place or group, often measured through infrastructure growth, education, healthcare, and standard of living.

Education: The process of acquiring knowledge, skills, values, and attitudes through formal or informal methods, aimed at personal growth and societal contribution.

Government: The system or group of people with the authority to govern a country, state, or community, responsible for creating and enforcing laws, policies, and public services.

Chapter 2
Towards Documenting Indigenous Knowledge Systems Through South African Universities:
A Decolonial Approach

Monicca Thulisile Bhuda
https://orcid.org/0000-0002-8506-9562
University of Mpumalanga, South Africa

ABSTRACT

In South Africa, documenting Indigenous Knowledge Systems (IKS) is a means of safeguarding and preserving these knowledges. Since IKS are tacit, largely passed down over generations and deeply rooted in cultural norms, it can be difficult to accurately document them. Moreover, IKS is in danger of extinction if no actions are taken to ensure their documentation. Due to these challenges, a call for documentation of IKS in universities is necessary and need urgent response. Therefore, the study aims to investigate ways South African universities document IKS. This study used secondary data and drew pre-existing data from academic works such as books, articles and government documents. The results show that there is progress made by universities to document IKS through collaborating with government and other stakeholders. Despite these successes, there are still challenges with IKS documentation, thus the study recommends that other universities join this effort and seek out sustainable strategies to ensure IKS protection and accessibility.

INTRODUCTION

Indigenous knowledge Systems (IKS) are the customs and wisdom that indigenous people have accumulated over many generations of habitation in a specific area. Indigenous Knowledge Systems are used to explain occurrences in the social, cultural, physical, biological, and spiritual domains. As per Marker (2019), IKS comprises every human contact that may be recorded, communicated, and utilized for developmental goals. These knowledge systems differ and are distinguished from the knowledge

DOI: 10.4018/979-8-3373-4124-8.ch002

acquired in official educational settings. Indigenous knowledge Systems cover all facet of life, including environmental management.

For Indigenous communities who possess knowledge, it serves as the foundation for their survival (Azubuike & Aji, 2021). One way to preserve IKS for future generations, national development, and sustainable growth is through documentation. This is because the Indigenous Knowledge Systems are not static; they are dynamic and can be easily impacted and influenced hence the urgent need to document them. (Dei, 2000). Therefore, offering IKS in universities such as in research projects, collaboration with government and indigenous communities to capture this knowledge are some ways to document IKS (Bhuda & Gumbo, 2024). Makokotlela and Gumbo (2025) explained that because many South African universities have not consistently and widely employed IKS as a basis for teaching, learning, research and have community engagement initiatives.

it is crucial that universities in South Africa document these systems, considering the systemic bias and prejudice of a fundamental Eurocentric curriculum that is common, accessible, researchable, and normalized unlike IKS. Battiste (2018) clarified that academics should endeavor to develop IKS in universities which this will serve as a strategy for documentation. As pointed out by Battiste (2018), those persistent scholars will face a variety of discriminatory practices and systemic barriers as they pursue their goal of integrating IKS into universities. However, outcomes to document IKS can be attained via enhanced collaboration with the government, indigenous communities, stakeholders, and other universities that have initiated this process.

PROBLEM STATEMENT

Unlike explicit knowledge, IKS is tacit, which is knowledge that is hard to express, codify, or impart to others. Frequently, it is founded on practical competence, intuition, and experience. As observed by Kaya and Seleti (2013), oral traditions, rituals, and experiential learning are common ways for indigenous communities to transmit their knowledge, customs, and beliefs. It is therefore intrinsically tacit because it is not always simple to document. There is a significant chance that IKS will erode without steps taken to document them, depriving the next coming generation of accessibility. Furthermore, the studies that were undertaken by Makokotlela and Gumbo (2025) indicated the biggest challenge that exist in South Africa universities is the continuous marginalization of IKS while there are efforts to acknowledge Western Knowledge as best in education compared to IKS. Through teaching, research, technologies, and university documentation departments like libraries, South African universities have effectively recorded Western Knowledge Systems (WKS) over the years. Furthermore, a lot of institutional support has been provided to build a knowledge base for WKS, while IKS and its importance in South African education have been overlooked. This strategy used by South African universities demonstrates how IKS is still colonized and remains undermined in higher education (Mkhize & Ndimande-Hlongwa, 2014). Therefore, to decolonize higher education, these institutions must acknowledge the importance of IKS and endeavor to document them using a variety of techniques. Documenting IKS in universities is bringing the voices of indigenous Knowledge holders in the academy and enabling them to assert ownership of their knowledge systems. Acknowledging IKS also gives scholars and students the chance to carry out research that will support IKS's protection, promotion, and preservation (Makokotlela & Gumbo, 2025). Considering the mentioned above, this study examines how universities can and have documented IKS. Collaboration initiatives between South African universities and other institutions to document Indigenous Knowledge Systems will also be discussed in this study.

The research objectives formulated for this study are to:

- Investigate the purpose of documentation of IKS in South African universities.
- Investigate methods used by South African universities to document IKS.
- Explore other institutions and stakeholders' South African universities have collaborated with to document IKS.

  To shed light on how IKS is documented by South African universities, the researcher created the following study questions for this paper:
- Why should South African universities document IKS?
- How have South African universities documented IKS?
- Which other institutions and stakeholders are South African universities working with to document IKS?

DOCUMENTATION OF INDIGENOUS KNOWLEDGE SYSTEMS

As a means of ensuring the social, cultural, and economic interests of indigenous peoples and local communities, documenting IKS is currently a topic of much discussion. It has become a useful tool for preventing additional IKS loss, preserving IKS over time, facilitating benefit sharing between IKS holders and users, and keeping IKS safe from unauthorized usage. IKS has a high value, which makes preservation crucial for the benefit of indigenous communities who are knowledge holders and custodians of these knowledge systems. The study takes note that if there are current initiatives that are in place to document IKS, they can further be sustained through collaboration of universities, government, indigenous communities and other stakeholders who have passion for IKS.

Documenting has become a useful instrument for preventing further loss of IKS, preserving IKS over time, facilitating and enhancing knowledge exchange between IK holders and users, and ultimately safeguarding IKS from unauthorised uses (WIPO, 2017). Notably, IKS is gathered and arranged logically throughout the documentation period, adhering to prearranged plans and activities (Balogun & Kalusopa, 2022. (Balogun & Kalusopa 2021). The necessity of documentation is emphasized by the World Intellectual Property Organisation (WIPO, 2017) as part of a comprehensive plan for IKS protection. Furthermore, WIPO (2017) points out that IKS documentation might be in indigenous or other languages, use contemporary or more traditional technology (digital vs. written filing), and take many different forms, such as textual registers and files, video, photographs, and audio recordings (Bhuthelezi, 2024).

Due to the nature of IKS, information specialists are required to identify, categorize, and handle this data. Due to the fact that we now live in a global village that is always changing, it is necessary to pay attention to what the present generation logically requires and to frame IKS (Masenya, 2023). There is a greater interest in technology in the twenty-first century. Technology should therefore be used to plan IKS for sustainability. IKS is a tool that could help manage local affairs, prevent violence, grow more and better food, maintain healthy lifestyles, share money, and eventually help find global answers (Vilakazi et al., 2023). IKS has shielded communities affected by globalization from its negative impacts and promoted friendship. To achieve all the Millennium Development Goals, indigenous knowledge can help.

The documentation of IKS may yield the following outcomes:

- Monetary benefits of IK holders.
- IKS organized and systematized (preserved) for future generations.
- Collaboration and partnership among a wide range of actors.
- Identification and wider social recognition of local communities and indigenous peoples in relation to particular IKS.
- Capacity building and educational uses of databases or registers.
- Defensive intellectual property protection, i.e., the ability to stop the illegitimate acquisition of IKS intellectual property rights.
- Positive intellectual property protection for IKS or related products.

Kaya and Seleti (2013) stated that universities can play a role in ensuring that IKS is documented and is given the same recognition as Western Knowledge Systems (WKS). South African universities has provided enough resources and support in WKS preservation in return ignoring the significance of doing the same with IKS (Olaopa & Ayodele, 2022). A colonial legacy is still being navigated by universities in light of societal and global changes (Kaya, 2013; Nhalevilo, 2013). Therefore, there is a call by academics in South African universities for IKS to be accorded its rightful place in higher education which this is long overdue and requires urgent attention (Chabaya & Chabaya, 2023).

In essence, Maditsi and Bhuda (2023) discussed that universities must find ways to integrate and protect IKS which were suppressed, marginalized, and exploited by the legacy of colonialism. In the words of Bhuda and Koitsiwe (2022), documenting IKS in universities will guarantee their accessibility and protection while also contributing to the decolonization of education and documentation. Using a variety of approaches to document IKS will show that academic institutions are abandoning the idea that only WKS should be recognized and recorded in contrast to other knowledge systems such as IKS (Mubangizi & Kaya, 2015).

THEORETICAL FRAMEWORK

A decolonial theory that provides a framework of instruments and procedures to achieve decolonial futures informs this study. The first instrument for analyzing how colonialism still influences contemporary political, economic, cultural, social, educational, and knowledge systems is the analytical lens of colonialism. The second is the demand to break ties with colonialism (Mendoza, 2020). The institutions, norms, and values that resulted from western colonialism and still influence modern economies, communities, education, and cultures must be named, and they must be continuously unlearned, reimagined, and reshaped (Ajani & Gamede, 2021). Decolonial theory demands that Indigenous people and their knowledge be acknowledged in education, as they have historically been marginalized in an effort to elevate Western systems.

According to this theory, IKS should be acknowledged in education and safeguarded through established mechanisms that have protected Western Systems (Bhuda, 2024). Decoloniality, which centers Indigenous Knowledge Systems of oppressed people in education, is a key idea in postcolonial thinking. To document these knowledge systems and further the decolonial agenda, educational institutions like universities have an obligation to acknowledge IKS in their teaching, learning, and research (Chapman & Schott, 2020). They also must acknowledge and work with Indigenous Knowledge (IK) holders and ensure that they are not exploited. In order to relearn, rethink, and reshape the educational institutions, norms, and values that resulted from western colonialism, universities of South Africa should be at the forefront to document IKS and find strategies that will be sustainable.

METHODOLOGY

The secondary data which is existing literature was utilized in this article and was founded on a review of scientific literature pertaining to the specific topic discussed. How do universities document Indigenous Knowledge Systems? is the central question of this study. The method of existing literature provided a substantial response to this question. This method was used by the researcher to highlight how indigenous knowledge is documented and the role of universities in this process (Clyne et al., 2007). Methods of documenting Indigenous knowledge Systems were also discussed in this study, and they were linked to diverse Universities of South Africa.

Collecting existing literature data from a variety of digital and textual sources, including databases, books, journals, reports, and publicly available materials, then carefully organizing and integrating everything (Karppinen & Moe, 2012). To ensure a comprehensive analysis of the available resources, a variety of platforms were searched, including Google Scholar, JSTOR, ResearchGate, ScienceDirect and institution repositories. The ability of this method to provide a comprehensive analysis of the corpus of recent literature supports its use and makes it easier to fully understand the study issue.

The methods below by Van Dinter et al. (2021) were applied in order to provide a structured approach to reviewing and analyzing existing literature:

- Determine the research topic, objectives, and questions: these will serve as the study's guiding principles.
- Look for studies: Performing a thorough search for pertinent academic publications.
- Establish inclusion and exclusion criteria: To determine which studies are appropriate for the study and which are not. The purpose and goals of the research must be in line with the chosen studies.
- Extract data from included studies: This involves taking information that directly contributes to the objectives of the study.
- Data analysis: Combining results to provide meaningful insights.
- Present findings and evaluate the strength of the supporting data: Synthesize results to produce insightful conclusions.

Study Design

A systematic, descriptive review was used to analyze the documentation of IKS in South African universities (Kim et al., 2017). This literature research was conducted using a variety of information sources, including books, book chapters, dissertations, theses, and peer-reviewed journal articles that were relevant to this investigation. ResearchGate, ScienceDirect, Google Scholar, and university repositories were the sources of the information.

Literature selection process

Using certain keywords, a systematic search was carried out to locate content for this investigation. The search terms that were utilized were "IKS in universities," "documentation," "decolonization of universities," and "Indigenous communities." A sophisticated search function was used to find 97 distinct papers for the examination of this study. At least 64 papers were determined to be unrelated to the topic after the abstracts and titles were reviewed.

Triangulation Approach

A triangulation approach was used to ensure the reliability and accuracy of the results. Cross-referencing data from many sources and points of view is necessary to identify recurring themes and patterns (Noble & Heale, 2019). Using a triangulation approach, the study presents methods that universities use to document IKS. This method was required to determine the degree to which IKS has been sustained and documented by universities.

Ethical implications

To reduce biases in reporting, data interpretation, and source selection, the author took proactive steps. This required identifying any limitations or gaps in the research, thoroughly assessing the sources for potential biases or conflicts of interest and clearly presenting the findings. Furthermore, steps were taken to guarantee the authenticity and integrity of the study by adhering to academic integrity norms and ethical principles throughout the research process.

FINDINGS AND DISCUSSIONS

This section begins with providing historical evidence of the first initiative of integrating IKS into universities. Providing historical evidence and discussions is important to understand the journey that was undertaken to protect, promote and disseminate Indigenous Knowledge Systems through universities.

DOCUMENTING INDIGENOUS KNOWLEDGE SYSTEMS IN SOUTH AFRICAN UNIVERSITIES

Indigenous Knowledge Systems curriculum in South African Higher education

Mosimege (2004) explained in his study that in 1996, the Portfolio Committee of Arts, Culture, Science, and Technology assisted the Council for Scientific and Industrial Research (CSIR) in assigning nine historically underrepresented universities to conduct audits of indigenous technologies, aiming to document and innovate on technologies used for survival by marginalized communities. Higgs et al. (2002) explained that it was proposed by the government that the historically disadvantaged universities be targeted to be part of the project which will also benefit knowledge custodians of the identified technologies. The outcome of the project was to further explore the integration of IKS into the historically underrepresented universities.

Regarding what has been discussed above, Mosimege (2004,78) further elaborated following in his study:

The following nine universities were involved in the audit: University of Venda, University of the North, University of the North West, Vista University (Mamelodi, East Rand and Soweto campuses), University of South Africa, University of the North – Qwaqwa campus, University of Zululand, University of Transkei and University of Fort Hare. (The names of some of these universities have changed as a result of mergers with other universities and technikons. For instance, the Qwaqwa campus of the University of the North is now part of the University of the Free State.) The audits culminated in provincial workshops on indigenous knowledge systems (IKS) conducted by each university. These workshops were used as the basis for the first national workshop on IKS at the University of the North-West in September 1998.

After the successful workshops, a task team was developed by the Department of Arts, Culture, Science and Technology. The following initiatives were set to be implemented by the task team:

a) Drafting IKS legislation for the protection of indigenous knowledge, as well as the IKS Policy.
b) Drawing examples from countries which have worked on Indigenous Knowledge Systems for a very long period.
c) Collaborate with different stakeholders including knowledge holders in further developing the legislation and Policy.
d) To assist the Department of Science and Technology establish an IKS unit as a commitment to the recognition, affirmation, development, promotion and protection of IKS.
e) Support a on the call for the establishment of a programme for IKS in South Africa, especially in relation to current educational discourse in higher education.

The next section provides evidence of IKS programmes as means of preserving and protecting them. The study argues that the 1996 initiatives were the beginning of documenting IKS in education which for a long time were disparaged and deemed not fit to be part of education (Kaya, 2014). Therefore, recognizing IKS through programs is a response to decolonizing education and recognizing that education does not have to be only based on Western knowledge. IKS plays a role in education, and academic institutions need to develop ways for documenting it.

The North-West University.

In order to document IKS in South Africa, North-West University decided to incorporate these knowledges into their research, teaching, and learning since 2001. A Bachelor of Arts in Indigenous Knowledge Systems (BA.IKS) was introduced in 2001 by the North-West University (Bhuda & Gumbo, 2024). In 2013 a revised curriculum and a new degree was reintroduced by the North- West University (NWU) as a Bachelor of Indigenous Knowledge Systems (B.IKS) with 20 students as its first intake. According to Kaya and Seleti (2013), the only university in South Africa and possibly all of Africa that offers both an undergraduate and postgraduate-level certified IKS programs is NWU. Therefore, the NWUs IKS program suggests that in order to support local development, knowledge and technology pertaining to indigenous populations in a given geographic area should be investigated and recorded.

In support of Kaya and Seleti's (2013) findings, Higgs and Van Niekerk (2002) indicated that the IKS programs since its inception in 2001 also have a research component funded by the NRF and comprise six broad research focus areas as follows:

a) The nature of IKS and Indigenous Technology
b) Traditional medicine and health
c) Indigenous food systems
d) Socio cultural systems
e) Arts, crafts and materials

Thus, the IKS program offers a viewpoint on the potential for regaining an African identity as it shows itself in the effort to validate indigenous knowledge systems within an African setting. According to this viewpoint, the IKS program is an effort to situate African knowledge systems within the discourse of a holistic epistemology (Mulovhedzi & Luhalima, 2023).

The University of Venda

The work of Nevhudoli and Netshandama (2023) argued that the University of Venda (UNIVEN) also took initiative and necessary steps to introduce the Bachelor of Indigenous knowledge Systems (B.IKS) in 2014, a year after the North-West University. However, unlike the NWU, UNIVEN only has the degree at undergraduate level and not yet at a post graduate level. The UNIVEN's B.IKS program is a method of documenting AIKS and also to develop student with the necessary knowledge, skills that they will in their careers and South African development. The four-year B.IKS program is unique and pertinent to the UNIVEN's strategic plan in the present discourse of African Renaissance, decolonization, change, and cultural diversity because it is inter-, intra-, and multi-disciplinary in character (Nevhudoli & Netshandama, 2023).

Moreover, Natshandama and Nevhudol (2021) stated that the UNIVEN works with Indigenous Knowledge (IK) holders and practitioners in knowledge production, management, and protection to further document AIKS through the B.IKS program. This cooperation makes "knowledge production, innovation, and human capital development more pertinent to the development and sustainable livelihood of communities. This dismantles the boundaries that separate local communities, government, business, industry, and academia" (UNIVEN, 2025:1).

University of KwaZulu-Natal

A study by Balogun and Kalusopa (2021) discussed that the University of KwaZulu-Natal (UKZN) has a DST-NRF centre in IKS (CIKS) which serves as an IKS research hub in Africa and collaborates with some academic institutions like the North-West University (NWU), University of South Africa (UNISA), University of Limpopo (UL), and the University of Venda (UNIVEN) in documenting IKS through research which is from diverse disciplines. The directorate of Research Centers and Centers for Excellence (RCCE) oversees DSI-NRF CIKS. Through research, postgraduate training, community participation, networking, IKS curriculum studies, knowledge sharing, and service provision, the DSI-NRF CIKS is expected to advance, safeguard, and maintain IKS (Tshishonga, 2019).

Traditional medicine, biodiversity, food security, and environmental management are some of its main areas of attention. Additionally, the DSI-NRF CIKS has strategic partnerships with both domestic and foreign organisations. The DSI-NRF CIKS's initiatives align with the 2004 Indigenous Knowledge Systems policy. National IKS Policy of 2004 was enacted by the South African government (DSI) in 2004 as a means of identifying, promoting, and safeguarding IKS in South Africa (Kaya & Chinsamy, 2021). The DSI passed an additional IKS Bill in 2014, which became an IK Act in 2019, to further improve the management of IKS in South Africa.

University of South Africa, University of Limpopo and University of Mpumalanga

In addition to the two previously stated universities, the University of South Africa (UNISA), the University of Limpopo (UL), and the University of Mpumalanga (UMP) have also been found to have contributed slightly to the integration of IKS in teaching, learning and research. Only IKS modules that are integrated into other programs are offered by the universities listed. For example, as stated by the University of Mpumalanga (2025), the Bachelor of Arts degree at this university offers a fourth year (honors) curriculum on Indigenous Knowledge Systems in the Contemporary Society (CST 405).

For undergraduate students, UNISA (2025) offers a program called Indigenous Knowledge Systems and African Languages (offered also in indigenous languages). The African Religions Indigenous Knowledge Systems (HANA031) module is part of the UL's Cultural and Political Studies and the Bachelor of Arts program. UL also offers the Theories and Indigenous Knowledge Systems (HANB081) which is a fourth year (Honors) Anthropology module (University of Limpopo, 2025a; University of Limpopo, 2025b). Despite the lack of fully formed IKS programs at UNISA, UMP, and UL, their contributions to the documentation of these knowledges are acknowledged. They have complied with a request from the government to incorporate IKS into universities as a way to decolonize education.

The efforts made by universities to document IKS are acknowledged in this study, along with the gaps that still exist and present obstacles to the documentation's success. Therefore, the study contends that in order to effectively document IKS, universities must provide training to their staff and provide various solutions on how IKS can be documented through education. Options may vary from teaching, research, community engagement and even digitalizing IKS. Moreover, policy frameworks such as the IKS Policy of 2004 and IK Act of 2019 may also act as guidelines for universities. The IKS Policy of 2004 is a policy adopted by the South African Cabinet to recognize, protect, promote, develop, and affirm Indigenous Knowledge Systems (IKS) in the nation. It is one of two major policies that address IKS and can be utilized by universities when concentrating on IKS documentation.

The IKS Policy of 2004 indicates the following:

- Establishing mechanisms to regulate standards for information and material transfer agreements related to IKS
- Providing public resources according to identified priorities
- Developing and implementing infrastructure for IKS programs
- Leveraging funding for IKS laboratories and centers, practitioners, and holders
- Maintaining and disseminating information on IKS

Moreover, there is an IK Act of 2019 was established to focus on the Protection, Promotion, Development, and Management of Indigenous Knowledge. The objects of this Act are to do the following:

- Protect the indigenous knowledge of indigenous communities from unauthorised use, misappropriation and misuse.
- Promote public awareness and understanding of indigenous knowledge for the wider application and development thereof.
- Develop and enhance the potential of indigenous communities to protect their indigenous knowledge.
- Regulate the equitable distribution of benefits.
- Promote the commercial use of indigenous knowledge in the development of new products, services and processes.
- Provide for registration, cataloguing, documentation and recording of indigenous knowledge held by indigenous communities.
- Establish mechanisms for the accreditation of assessors and the certification of indigenous knowledge practitioners; and
- Recognise indigenous knowledge as prior art under intellectual property laws.

As mentioned before, South African universities can use these mentioned policies to guide term to establish their own IKS programmes that can be used as knowledge documenters in collaboration with the DSI, NRF, CSIR, DSI-NRF CIKS, indigenous communities and other stakeholders.

Indigenous Knowledge Systems Documentation Centers

In response to efforts to preserve Indigenous Knowledge Systems in South Africa, the Department of Science and Technology (DST), now known as the Department of Science and Innovation (DSI), initiated the National Recordal System (NRS) now known as National IK Recordal System (NIKRS) project to standardise and coordinate the gathering, preservation, maintenance, and dissemination of science and technology-related IKS data in the nation (Amechi, 2015). Indigenous Knowledge Systems documentation has opened up new avenues for the sharing, preserving, and using of this knowledge, but it has also brought forth new challenges for ensuring its long-term preservation and accessibility (Bagley, 2019). The Indigenous Knowledge Systems Documentation Centers (IKSDCs) were established as a result of the NIKRS project to collect, document, store, and disseminate IK and related activities. The IKSDCs are located in nine provinces e.g. Eastern Cape, KwaZulu-Natal, Free State, Gauteng, Mpumalanga, Limpopo, Northern Cape, North-West, and Western Cape (Kaya, 2013). Moreover, some IKSDCs are hosted by universities as follows:

- The North-West University is hosting a North-West Province IKSDC
- The University of Venda is hosting a Limpopo IKSDC
- The Walter Sisulu University is Hosting the Eastern Cape IKSDC
- The University of KwaZulu-Natal and the University of Zululand are hosting the KwaZulu-Natal IKSDC
- Cape Peninsula University of Technology is hosting the Western Cape IKSDC
- The University of Free State is hosting the Free State IKSDC

In addition to being in charge of gathering and arranging information resources, Balogun and Kalusopa (2021) argued that the IKSDCs actively participate in the creation of information that is shared, circulated, and distributed in collaboration with the communities involved. Regarding the IKS that was captured in the NIKRS, the IKSDC is also in charge of providing services to the public, local and national governments, and diverse communities.

Digitalization of Indigenous Knowledge Systems by NIKSO

The IKSDCs mentioned earlier in the study which are in nine provinces and are linked with universities are in collaboration with DSI's office which is National Indigenous Knowledge Systems Office (NIKSO) to ensure that IKS is documented through digitalization. The digitalization of Indigenous Knowledge Systems is achieved through the NIKSO which established by the Protection, Promotion, Development and Management of Indigenous Knowledge Act of 2019. To assist NIKSO achieve the goal of the NIKRS program which is to standardize and coordinate the gathering, preservation, maintenance, and dissemination of science and technology-related IKS data in the nation, the National Indigenous Knowledge Management System (NIKMAS) was created.

A web-based repository built on digital library technology, a standalone cataloguing application used offline in communities, and many paper-based questionnaires used as instruments during IK gathering are some of the components that make up the NIKMAS. Balogun and Kalusopa (2022) stated that National initiative of the DSI is to coordinate and standardize the capturing, storing, maintenance, and dissemination of science and technology-related data on IKS in South Africa is the web-based portal that South Africa uses to digitize and host its national heritage resources (Van Rooyen, 2011).

NIKMAS was developed as a semantic digital archive to assist the NIKRS effort. As stated by Amechi (2018), among other things, the NIKMAS is the technical framework for handling prior art declarations. NIKMAS was developed in tandem with the IK collection efforts in the communities to accommodate new procedures and needs that were recognised and established by and within the communities. Although there were some significant technological challenges during its first development, Khalala (2021) claimed that NIKMAS has the potential to become the best IK management system in the nation and the world.

Moreover, Khalala (2021) asserts that NIKMAS is much more than just a catalog of intellectual property resources; it also boasts an advanced digital repository with specially created meta-data schemata, a sophisticated cataloguing system, an advanced security model to safeguard and preserve the IK, an advanced semantic search engine, and an overall integration architecture that unifies the subsystems into a cohesive, practical whole. Balogun and Kalusopa (2021) claim that NIKMAS is the outcome of an ontological investigation into IK-related metadata, which embodies intelligent thinking. According to the national priorities of the DSI, the initial iteration of NIKMAS was created with an emphasis on African Traditional Medicine and Indigenous Foods. It is anticipated that NIKMAS will significantly benefit communities and scholars in both social and scientific ways.

University libraries and digitalization of IKS

Digital technologies must be used by universities to ensure the integrity and long-term accessibility of this invaluable knowledge while safeguarding cultural sensitivity and community rights. In order to digitize IKS, a technique for documentation and knowledge preservation, Dlamini and Ocholla (2018) emphasize the use of information and communication technology (ICT) tools. Masenya and Ngulube

(2019) stated that research results, theses, newspapers, images, historical documents, and more can all be digitalized and kept in university repositories. For the benefit of communities, digitizing IKS through university repositories should therefore be a feasible undertaking.

Based on Masenya's (2024) research, the preservation of IKS through university libraries by using digitalization technologies enhances the pride, dignity, and sense of belonging of the indigenous community in addition to creating jobs and generating income. In order to digitally preserve IKS, South African universities must make sure that the required materials are accessible and available. Dealing with the fast obsolescence of technology, making large amounts of digital data accessible, and maintaining reliable repositories that contain documentary proof of scholarship are just a few of the difficulties South African academic libraries face in their efforts to preserve their digital resources (Masenya, 2021). According to the same idea, Ntloko and Masenya (2024) noted that while Africans' IK is still a treasure trove, the problem is that they are currently lagging behind the rest of the world in terms of IK digitization for worldwide access.

According to a study by Buthelezi et al. (2024), the University of Zululand library has evidence of IKS documentation. The University of Zululand library is crucial to IKs' use and access, according to the study. Buthelezi et al. (2024) further elaborated that the University of Zululand Library's aim is to serve teaching, learning, and research activities as well as the needs of the local population, which can use the library without compromising the privileges of its main patrons. IK-related materials such as theses and dissertations can be found across the library, but especially in the Institutional Repository and the UZULU Collection area (Dlamini & Ocholla, 2018).

Notwithstanding the success the University of Zululand has had, IK-related documents can be found in the university library, Buthelezi et al. (2024) pointed out that it is unclear how the documentation process of IK-related materials is done, whose departments or disciplines are responsible, and what difficulties they encounter. Furthermore, the University of Zululand does not appear to have sufficient policy in place addressing the documentation of IK. Therefore, the It is uncertain how much and to what degree IK is used and accessed at the university. Furthermore, it is unknown how IK is developed, recorded, and disseminated.

University librarians should collaborate with institutional departments that have academic works in IKS in order to successfully and efficiently create, collect, categorize, organize, preserve, and disseminate IK in order to overcome obstacles like those faced by the University of Zululand. Libraries should train and engage communities in addition to collecting new information (Masenya, 2022). Chigwada and Ngulube (2024) concurred that libraries should administer, distribute, maintain, and safeguard indigenous knowledge. In agreement with Chigwada and Ngulube (2024), Masenya (2022) indicated that universities must take into consideration the following fundamental steps when documenting IKS through digitalization using institutional libraries.

a) Implement relevant policies and strategies:

Academic libraries cannot be said to be engaging in digital preservation without adhering to policies, hence universities must create policies to guarantee the durability of digital materials. Roya and Ngcobo (2023) stated that Policies pertaining to the selection and preservation of digital content are put into effect by libraries. Therefore, in order to preserve indigenous knowledge, appropriate knowledge management policies and processes must be put in place. According to Chigwada et al. (2024) preservation is a legitimate means of ensuring that indigenous knowledge is protected against biopiracy and other types of misuse and that communities are not disadvantaged due to the distinctive beliefs and folkways that shape their existence.

b) Proper allocation of resources:

For digital preservation to be sustainable, resources must be allocated appropriately. According to Okenjom and Asuquo's (2023) research, digital preservation is complicated and expensive. As a result, initiatives and activities involving digital preservation need a significant investment of resources in order to be sustained over time. For digital preservation to be successful, institutional commitment is required, which includes making sure that sufficient funding is available to support the project (Corrado & Moulaison, 2014).

c) Legal issues:

It is important to look into copyright concerns to make sure that the legal intellectual property rights needed to carry out the tasks required for long-term preservation have been given (Masango, 2010). Regarding the preservation of digital content and the dissemination of that information, South Africa's Copyright Act 98 of 1978 establishes restrictions and broad exceptions for libraries, archives, and museums on the protection of copyrighted works.

d) Collaboration and partnerships opportunities:

According to Dollar and Ashley (2014), an organization tasked with preserving electronic records benefits by upholding and encouraging cooperation among its numerous stakeholders. According to Okenjom and Asuquo (2023), libraries, archives, and museums employ cooperation tactics to bring together teams with the requisite knowledge to further digital preservation efforts and utilize resources within the broader library and digital preservation community. According to Asuquo eta al. (2023) establishing collaborations can maximize pooled resources and foster the development of collective knowledge in order to accomplish digital preservation objectives.

e) Technical expertise:

For an organization to have a successful digital preservation capability, it must have enough knowledge about digital preservation and electronic records management to support all of the necessary infrastructure and critical operations (Matlala, 2019). In order to ensure long-term accessibility of digital material, Segaetsho and Moloi (2019) claimed that technical expertise is essential for digital preservation, requiring knowledge of storage technologies and preservation techniques, audit frameworks, metadata standards, data migration, and electronic records management.

f) Institutional supporting factors

Initiatives for digital preservation must be in line with the institutional objectives and mission in order to be sustained. According to (Choudhury & Kalita, 2025), digital preservationists must be able to persuade management and other decision-makers of the value of digital preservation to the institutional overarching goal. According to Chigwada and Ngulube (2025), the implementation of sustainable digital preservation programs is hampered by a lack of finance, institutional support, experience, and knowledge of where to begin. Segaetsho and Moloi (2019) argued that sustainable digital preservation aims to maintain digital resources for the necessary amount of time while operating under resource constraints.

FUTURE RESEARCH DIRECTIONS

Scholars who wish to further explore the documentation of IKS in South Africa may find the findings offered in this study to be instructive and insightful. Universities can also use this study as a guide if they wish to contribute to IKS documentation. In order to guarantee that IKS is safeguarded, promoted, and shared through documentation, universities are currently working with other organizations and stakeholders (such as communities) to implement the 1996 requirement to acknowledge IKS and its value to higher education.

As a pioneer in IKS documentation, the DSI, working with the NRF, can carry out more research with South Africa's remaining public universities that have not yet begun the process of documenting IKS. By holding workshops with various institutions and letting DSI- NRF CIKS guide the process, DSI and NRF may delve deeper into the documentation of IKS. The DSI-NRF CIKS can also offer strategies and guidelines that universities can use to guarantee the effectiveness of IKS documentation.

CONCLUSION AND RECOMMENDATIONS

This study aimed to investigate methods used by South African universities in documenting IKS. This was achieved by using the secondary data where existing literature on similar studies was explored and discussed. The study has managed to provide evidence that indeed IKS has been documented through universities using various strategies which can be sustained. The results of the study show that the documentation of IKS is through the collaboration the universities have with other institutions such as the DSI, NRF, IKSDCs and indigenous communities. Despite the advancements in IKS documentation, the study indicated difficulties which may be encountered through these processes. However, there are several advantages to universities and other partners documenting IKS, such as protecting cultural assets, encouraging sustainable practices, and empowering communities by securing their knowledge and rights for future generations. Based on the findings, the study recommends the following:

- Department of Higher Education and Training (DHET) should work with the NRF, DSI, and DSI-NRF CIKS to develop guidelines and norms that universities should adhere to when documentation IKS.
- Institutions that have done extensive work in IKS documentation can also offer. workshops, seminars, and the required training across all 26 public universities.
- Indigenous communities should be included in the documentation process through cooperative methods to guarantee that their customs and values are upheld.
- Take note of the necessity of all-encompassing IKS legislation that addresses recognition, development, protection, effective implementation, and promotion.

- Universities must further explore collaboration opportunities in IKS documentation with provincial and local government.

REFERENCES

Ajani, O. A., & Gamede, B. T. (2021). Decolonising Teacher Education Curriculum in South African Higher Education. *International Journal of Higher Education, 10*(5), 121–131. DOI: 10.5430/ijhe.v10n5p121

Amechi, E. P. (2018). Traditional knowledge on the medicinal uses of plants, biopiracy and national patent measures in Africa: Exploratory reflections and comparative experiences. *Journal of Comparative Law in Africa*, *5*(1), 73–109.

Asuquo, M. E., Okenjom, G. P., Okpa, O. E., & Eyiene, A. E. (2023). Opportunities and challenges in digitization of indigenous knowledge and implication for educational management in the Nigerian Context. *Digital preservation and documentation of global indigenous knowledge systems*, 87-104.

Azubuike, F. C., & Aji, T. C. (2021). Role of information and communication technologies (ICTs) in repackaging indigenous knowledge: A 21st century perspective. *Library Philosophy and Practice*, 1-14.

Bagley, M. A. (2019). The fallacy of defensive protection for traditional knowledge. *Washburn Law Journal*, *58*, 323–363. DOI: 10.13140/RG.2.2.15115.34088

Balogun, T., & Kalusopa, T. (2021). A framework for digital preservation of Indigenous knowledge system (IKS) in repositories in South Africa. *Records Management Journal*, *31*(2), 176–196. DOI: 10.1108/RMJ-12-2020-0042

Balogun, T., & Kalusopa, T. (2022). Web archiving of indigenous knowledge systems in South Africa. *Information Development*, *38*(4), 658–671. DOI: 10.1177/02666669211005522

Battiste, M. (2018). Reconciling Indigenous knowledge in education: Promises, possibilities, and imperatives. *Dissident knowledge in higher education*, 123-148.

Bhuda, M., & Koitsiwe, M. (2022). The Importance of Underpinning Indigenous Research Using African Indigenous Philosophies: Perspectives From Indigenous Scholars. In *Handbook of Research on Protecting and Managing Global Indigenous Knowledge Systems* (pp. 223–248). IGI Global. DOI: 10.4018/978-1-7998-7492-8.ch013

Bhuda, M. T. (2024). Decolonising and indigenising African Indigenous research: Why Indigenous ways of knowing, being and living matter in research. In *Higher Education Transformation in Africa* (pp. 50–69). Routledge. DOI: 10.4324/9781003528319-4

Bhuda, M. T., & Gumbo, M. T. (2024). Developments Towards Integrating African Indigenous Knowledge in South African Universities. *UnisaRxiv*. DOI: 10.25159/UnisaRxiv/000082.v1

Buthelezi, S. C., Ocholla, D., & Dlamini, P. (2024). Strategies for documenting and disseminating indigenous knowledge at a South African university. *South African Journal of Information Management*, *26*(1), 1648. DOI: 10.4102/sajim.v26i1.1648

Chabaya, O., & Chabaya, R. A. (2023). Integration of African Indigenous Knowledge Systems into Higher Education Curriculum in Zimbabwe Universities: Exploring Experiences of Lecturers. *African Perspectives of Research in Teaching and Learning*, *7*(3), 625–638.

Chapman, J. M., & Schott, S. (2020). Knowledge coevolution: Generating new understanding through bridging and strengthening distinct knowledge systems and empowering local knowledge holders. *Sustainability Science*, *15*(3), 931–943. DOI: 10.1007/s11625-020-00781-2

Chigwada, J., & Ngulube, P. (2024). Librarians' role in the preservation and dissemination of indigenous knowledge. *IFLA Journal*, *50*(2), 242–256. DOI: 10.1177/03400352231217270

Chigwada, J., & Ngulube, P. (2025). Integrating CARE and FAIR Principles in the Digital Preservation of Indigenous Knowledge to Achieve the Sustainable Development Goals. In *Harnessing Indigenous Epistemologies for Sustainable Progress* (pp. 197–230). IGI Global Scientific Publishing. DOI: 10.4018/979-8-3693-8447-3.ch010

Chigwada, J., Ngulube, P., & Dewah, P. (2024). Preserving Indigenous Knowledge in Culture Houses in the Digital Era in Masvingo Province, Zimbabwe. *African Journal of Library Archives and Information Science*, *34*(2). Advance online publication. DOI: 10.4314/ajlais.v34i2.4

Choudhury, G., & Kalita, D. (2025). Collaborative stewardship: Sustainable approaches to indigenous knowledge in academic libraries. *Journal of Academic Librarianship*, *51*(2), 103007. DOI: 10.1016/j.acalib.2025.103007

Clyne, J., Mininni, P., Norton, A., & Rast, M. (2007). Interactive desktop analysis of high resolution simulations: Application to turbulent plume dynamics and current sheet formation. *New Journal of Physics*, *9*(8), 301. DOI: 10.1088/1367-2630/9/8/301

Dei, G. J. S. (2000). Rethinking the role of indigenous knowledges in the academy. *International Journal of Inclusive Education*, *4*(2), 111–132. DOI: 10.1080/136031100284849

Dlamini, P., & Ocholla, D. N. (2018). Information and communication technology tools for managing indigenous knowledge in KwaZulu-Natal Province, South Africa. *African Journal of Library Archives and Information Science*, *28*(2), 137–153.

Higgs, P., & Van Niekerk, M. P. (2002). The programme for indigenous knowledge systems (IKS) and higher educational discourse in South Africa: A critical reflection: Perspectives on higher education. *South African Journal of Higher Education*, *16*(3), 38–49. DOI: 10.4314/sajhe.v16i3.25215

Karppinen, K., & Moe, H. (2012). What we talk about when we talk about document analysis. *Trends in communication policy research: New theories, methods and subjects*, 177-193.

Kaya, H. O. (2013). Integration of African indigenous knowledge systems into higher education in South Africa: Prospects and challenges. *Alternation (Durban)*, *20*(1), 135–153.

Kaya, H. O. (2014). Revitalizing African indigenous ways of knowing and knowledge production. *Restoring indigenous self-determination, 105*.

Kaya, H. O., & Chinsamy, M. (2021). Combating world pandemics through complementarity of indigenous knowledge with other knowledge systems: A paradigm shift in healthcare research and education. *Alternation (Durban)*, *38*, 254–272. DOI: 10.29086/2519-5476/2021/sp38a11

Kaya, H. O., & Seleti, Y. N. (2013). African indigenous knowledge systems and relevance of higher education in South Africa. *International Education Journal: Comparative Perspectives, 12*(1).

Khalala, G. (2021). Protecting and benefiting from indigenous knowledge: The National Indigenous Knowledge Management System. *CSIR Science Scope*, *17*(1), 56–57.

Kim, H., Sefcik, J. S., & Bradway, C. (2017). Characteristics of qualitative descriptive studies: A systematic review. *Research in Nursing & Health*, *40*(1), 23–42. DOI: 10.1002/nur.21768 PMID: 27686751

Maasz, D., Winschiers-Theophilus, H., Stanley, C., Rodil, K., & Mbinge, U. (2018). A digital indigenous knowledge preservation framework: the 7C model—repositioning IK holders in the digitization of IK. *Digitisation of culture: Namibian and international perspectives*, 29-47.

Maditsi, M. E., & Bhuda, M. T. (2023). Decolonising and Humanising Pedagogies in South African Postgraduate Education: Lessons from Indigenous Knowledge Systems. In Masenya, T. M. (Ed.), *Digital Preservation and Documentation of Global Indigenous Knowledge Systems* (pp. 282–302). IGI Global. DOI: 10.4018/978-1-6684-7024-4.ch015

Makokotlela, M. V., & Gumbo, M. T. (2025). Decolonisation of the curriculum through the integration of indigenous knowledge in higher education. *South African Journal of Higher Education*, *39*(1), 237–254. DOI: 10.20853/39-1-6004

Marker, M. (2019). Indigenous knowledges, universities, and alluvial zones of paradigm change. *Discourse (Abingdon)*, *40*(4), 500–513. DOI: 10.1080/01596306.2017.1393398

Masango, C. A. (2010). Indigenous traditional knowledge protection: Prospects in South Africa's intellectual property framework? *South African Journal of Library and Information Science*, *76*(1), 74–80. DOI: 10.7553/76-1-88

Masenya, T. M. (2021). Trustworthiness of institutional repositories in academic libraries in South Africa. *Library Philosophy and Practice*, 1-36.

Masenya, T. M. (2022). Decolonization of indigenous knowledge systems in South Africa: Impact of policy and protocols. [IJKM]. *International Journal of Knowledge Management*, *18*(1), 1–22. DOI: 10.4018/IJKM.310005

Masenya, T. M. (2023). Revitalization and Digital Preservation of Indigenous Knowledge Systems for Sustainable Development of Indigenous Communities in South Africa. *The Serials Librarian*, *84*(5-8), 86–102. DOI: 10.1080/0361526X.2023.2277962

Masenya, T. M. (2024). Reterritorialization and Digital Preservation of Indigenous Knowledge Systems: Returning South African Indigenous Communities to Their Cultural Roots. In *Inclusivity and Indigeneity in Education for Sustainable Development* (pp. 64-77). IGI Global.

Masenya, T. M., & Ngulube, P. (2019). Digital preservation practices in academic libraries in South Africa in the wake of the digital revolution. *South African Journal of Information Management*, *21*(1), 1–9. DOI: 10.4102/sajim.v21i1.1011

Matlala, E. (2019). Long-term preservation of digital records at the University of KwaZulu-Natal archives. *Journal of the South African Society of Archivists*, *52*, 95–109.

Mendoza, B. (2020). Decolonial theories in comparison. *Journal of World Philosophies*, *5*(1), 43–60.

Mkhize, N., & Ndimande-Hlongwa, N. (2014). African languages, indigenous knowledge systems (IKS), and the transformation of the humanities and social sciences in higher education. *Alternation (Durban), 21*(2), 10–37.

Mosimege, M. (2004). Indigenous knowledge systems in South Africa: perspectives from the Department of Science and Technology: news and views. *Indilinga, 3*(1), 78–84.

Mubangizi, J., & Kaya, H. (2015). African indigenous knowledge systems and human rights: Implications for higher education, based on the South African experience. *International journal of African renaissance studies-multi-, inter-and transdisciplinarity, 10*(2), 125-142.

Mulovhedzi, S. A., & Luhalima, T. R. (2023). The role and place of african indigenous knowledge systems in the analyses of inclusive education. In *Using African epistemologies in shaping inclusive education knowledge* (pp. 29–48). Springer. DOI: 10.1007/978-3-031-31115-4_3

Natshandama, O. V., & Nevhudoli, N. D. (2021). Creating intergenerational learning spaces: A collaboration between UNIVEN Community Engagement Programme and Dzomo la Mupo. [CriSTaL]. *Critical Studies in Teaching and Learning, 9*(2), 39–63. DOI: 10.14426/cristal.v9i2.462

Nevhudoli, N. D., & Netshandama, V. O. (2023). What do Bachelor of Indigenous Knowledge Systems graduates say about their curriculum? A qualitative tracer study at the University of Venda. *Journal of Curriculum Studies Research, 5*(1), 141–158. DOI: 10.46303/jcsr.2023.11

Nhalevilo, E. Z. F. A. (2013). Rethinking the history of inclusion of IKS in school curricula: Endeavoring to legitimate the subject. *International Journal of Science and Mathematics Education, 11*(1), 23–42. DOI: 10.1007/s10763-012-9382-8

Noble, H., & Heale, R. (2019). Triangulation in research, with examples. *Evidence-Based Nursing, 22*(3), 67–68. DOI: 10.1136/ebnurs-2019-103145 PMID: 31201209

Nonaka, I. (1994). A dynamic theory of organizational knowledge creation. *Organization Science, 5*(1), 14–37. DOI: 10.1287/orsc.5.1.14

Ntloko, N. P., & Masenya, T. M. (2024). Enhancing access to electronic information through digital transformation in KwaZulu-Natal Department of health libraries in South Africa. In *Multidisciplinary Approach to Information Technology in Library and Information Science* (pp. 114–131). IGI Global Scientific Publishing.

Okenjom, G. P., & Asuquo, M. E. (2023). The Application of OAIS Model as a Framework for Digital Preservation of Indigenous Knowledge Systems: The Roles of Educational Managers. In *Digital Preservation and Documentation of Global Indigenous Knowledge Systems* (pp. 322–341). IGI Global. DOI: 10.4018/978-1-6684-7024-4.ch017

Olaopa, O. R., & Ayodele, O. A. (2022). Building on the strengths of African indigenous knowledge and innovation (AIK&I) for sustainable development in Africa. *African Journal of Science, Technology, Innovation and Development, 14*(5), 1313–1326. DOI: 10.1080/20421338.2021.1950111

Roya, W., & Ngcobo, S. (2023). The role of information communication technologies on African indigenous knowledge systems: Folktales. *International Journal of Research in Business and Social Science*, *12*(6), 46–55. DOI: 10.20525/ijrbs.v12i6.2662

Segaetsho, T., & Moloi, J. (2019). Integration of digital preservation knowledge, skills and competencies in the teaching curricula of library and information studies at the university of Botswana. *Mousaion: South African Journal of Information Studies*, *37*(1), 19–35. DOI: 10.25159/2663-659X/6436

Tshishonga, N. S. (2019). Incorporating Indigenous Knowledge in the Higher Education Sector for the Advancement of African Scholarship. In *The Formation of Intellectual Capital and Its Ability to Transform Higher Education Institutions and the Knowledge Society* (pp. 212–231). IGI Global. DOI: 10.4018/978-1-5225-8461-2.ch011

UNISA. (2025). Register to study through UNISA. https://www.unisa.ac.za/sites/corporate/default/Register-to-study-through-Unisa/Subjects-&-modules/All-modules/Indigenous-Knowledge-Systems-and-African-Languages-(Sesotho)-%E2%80%93-SSO3702

Univen. (2025). Department of Indigenous Knowledge Systems and Heritage Studies. https://www.univen.ac.za/faculties/humanities-social-science-and-education/department-of-indigenous-knowledge-systems-and-heritage-studies/

University of Limpopo. (2025a). School of Social Sciences: Sociology and Anthropology. https://www.ul.ac.za/faculty-of-humanities/school-of-social-sciences/sociology-and-anthropology/

University of Limpopo. (2025b). School of Social Sciences: Cultural and Political studies. https://www.ul.ac.za/faculty-of-humanities/school-of-social-sciences/cultural-and-political-studies/

University of Mpumalanga. (2025). https://www.ump.ac.za/Study-with-us/Faculties-and-Schools/Faculty-of-Economics,-Development-and-Business-Sci/School-of-Social-Sciences/Bachelor-of-Arts-Honours-in-Culture-and-Heritage-S.aspx

Van Dinter, R., Tekinerdogan, B., & Catal, C. (2021). Automation of systematic literature reviews: A systematic literature review. *Information and Software Technology*, *136*, 106589. DOI: 10.1016/j.infsof.2021.106589

Van Rooyen, B. (2011). Safeguarding the future of indigenous knowledge through ICT: ICT and society. *CSIR Science Scope*, *5*(2), 24–26.

World Intellectual Property Organization. (2017). *Documentation of traditional knowledge and traditional cultural expressions.* http:// www.wipo.int/edocs/pubdocs/en/wipo_pub_tk_9.pdf

ADDITIONAL READING

Ajani, Y. A., Oladokun, B. D., Olarongbe, S. A., Amaechi, M. N., Rabiu, N., & Bashorun, M. T. (2024). Revitalizing indigenous knowledge systems via digital media technologies for sustainability of indigenous languages. *Preservation. Digital Technology & Culture*, *53*(1), 35–44.

Bihari, S. (2023). Cultural heritage and indigenous knowledge: Reviving traditions for future generations. *Sustainable Development Goals in SAARC Countries: Key Issues. Opportunities and Challenges*, *1*, 24–32.

Eneh, A. C., & David, U. A. (2023). Indigenous Knowledge Management and ICT: The Intercepting Point. *Information Technology and Librarianship*, *3*(2), 96–107.

Liew, C. L., Yeates, J., & Lilley, S. C. (2021). Digitized indigenous knowledge collections: Impact on cultural knowledge transmission, social connections, and cultural identity. *Journal of the Association for Information Science and Technology*, *72*(12), 1575–1592. DOI: 10.1002/asi.24536

Mdhluli, T. D., Mokgoatšana, S., Kugara, S. L., & Vuma, L. (2021). Knowledge management: Preserving, managing and sharing indigenous knowledge through digital library. *Hervormde Teologiese Studies*, *77*(2). Advance online publication. DOI: 10.4102/hts.v77i2.6795

KEY TERMS AND DEFINITIONS

Decolonization: Decolonization, which includes political, cultural, and economic facets as well as the recovery of IKS, is the process by which a colonized area becomes independent of the governing colonial authority.

Documentation: Is the methodical process of locating, recording, and classifying knowledge inside an organization or setting to guarantee its availability and conservation.

Indigenous Knowledge Systems: Are the customs, beliefs, technologies, and methods of life that local communities have cultivated over many generations and that allow them to flourish in their natural surroundings.

Indigenous communities: Different social and cultural groups with ancestors' connections to the land are known as indigenous communities.

Higher education: Education at universities or other comparable educational institutions is referred to as higher education.

Universities: Universities are recognized for their emphasis on research and scholarship and provide a broad range of academic programmes, including undergraduate and graduate degrees.

Chapter 3

The Importance of Integrating Indigenous Knowledge Systems in South Africa's Basic Education:

Perspectives From Indigenous Scholars

Monicca Thulisile Bhuda
https://orcid.org/0000-0002-8506-9562
University of Mpumalanga, South Africa

Mothusiotsile E. Maditsi
https://orcid.org/0000-0001-5796-0174
North-West University, South Africa

ABSTRACT

Indigenous knowledge Systems (IKS) are practices, and beliefs of indigenous communities which include Indigenous people's abilities, experiences, and insights that are used to sustain their livelihoods. Such knowledge systems are important to be integrated in education because students need a culturally responsive education that relates to their cultural background However, the South African Curriculum and Assessment Policy Statement (CAPS), which is taught in basic education, does not give much emphasis on IKS and implementation mechanisms that it needs. Instead, IKS are marginalized and ignored in education, while Western Knowledge Systems are given priority in the classroom. The study applies a scoping literature review to investigate how IKS can be integrated into the CAPS with a focus on specific subjects. The indigenous standpoint theory and constructivist theory are applied in the study to demonstrate that there is no single objective world that all learners aim to know and understand. Instead, the learners make meaning of the world through their own cultural experiences.

DOI: 10.4018/979-8-3373-4124-8.ch003

INTRODUCTION

The realization of IKS as a vital tool to decolonize and provide alternative solid solutions to development agendas of South Africa should be realised. Since the existence of humanity, education has been a focal point of human development in various avenues of life. However, through colonialism and apartheid in the case of South Africa, the education system especially the one afforded to indigenous communities has been vastly unequal. This crisis of inequality was a byproduct of a systematic injustice of apartheid (Sonkqayi, 2021) and this resulted in IKS facing its demise within the education system.

The 1994 democratic dispensation saw a shift within the education paradigm which brought about new policies and education legislation being developed. This is evidenced by the DoE (1996) establishment of dual goals (Ndlovu-Gatsheni, 2018). The Department of Education (DoE) reflected in their mission statement that their vision as a department is of a South Africa in which all people have equal access to lifelong education and training opportunities which will contribute towards improving the quality of life and build a peaceful, prosperous and democratic society. For many scholars, this was the beginning of a crisis within the South African education system. Esteemed scholars in the area of IKS such as Odora-Hoppers, 2005; Green, 2008; Le Grange, 2018; Ndlovu-Gatsheni, 2018, have brought about the question of not only distributive but epistemic injustice.

This question raised by these IKS scholars has shown that there is a need for a paradigm shift within the mainstream education system of South Africa. There is a much-needed decolonization of the kind of education system that exists in South Africa. Dei (2008) argues that education is about equipping learners with knowledge, skill, and resources that allow them to improve their own conditions and to contribute to building healthy, sustainable communities. It is therefore vital to note that effective education should be able to solve existing problems within various local communities.

Dei (2008 further explained that education is a vehicle that should be used to drive sustainable development within local South African communities. This however cannot be achieved if the knowledge of these communities remains in the margins of the education system within the country. The paradigmatic lens that is applied by the education system in South Africa remains western and this has shown that this western lens fails to address persistent problems in South African communities. Scholars such as Khupe (2014) and Bhuda (2021) have voiced that the South African educational methodologies must be corrected as soon as possible since they are still dominated by the influence of the west.

The reality of providing for inclusion of IKS in South Africa's basic education is very complex (Prinsloo, 2001). This is because there have been massive challenges that have hindered the process of bringing fruitful principles of including and mainstreaming IKS in the formal education system of South Africa. Majority of these challenges are bound to the policy and legislative frameworks which have not been amended to reflect the role and value of IKS in the education system of the country. Therefore, there is a need to transform the South African education system and make sure that IKS is mainstreamed into this education system. If mainstreamed, IKS will be able to bring about a paradigm shift that is in line with the experiences of African people and thus portray a correct and consistent image of the identity of Africans generally and south Africa in particular (Dei, 2008; Sonkqayi, 2021).

The call for a paradigm shifts and epistemic justice in South Africa saw the first legislative step being the development of the 2004 IKS Policy. This legal instrument is viewed as the cornerstone framework that is driving IKS into the core business of education and other vital areas within South Africa. The IKS Policy of 2004 was developed to do away with the marginalization of IKS and the propagators of this knowledge during the apartheid South Africa (Bhuda, 2024). This policy further juxtaposed against the backdrop of centuries of oppression, exists an indisputable wealth of IK that has survived and, in some cases, has even grown within the protective confines of African societies and communities.

Within this policy, chapter two was focused on the relationship between IKS, the National System of Innovation (NSI) and the National Education System as all these three are an intimate and interwoven one (Bhuda & Pudi, 2020). This chapter was developed with the lens that education is part of culture and acknowledges that culture itself is transmitted through education. it is against this backdrop that IKS scholars view and place IKS at the center of knowledge decolonization and therefore also applies to driving epistemic justice within the South African education system (Bhuda, 2024).

Drawing from the previously mentioned discussions, the study focuses on how IKS can be included in the CAPS curriculum, which is taught in basic education, and goes on to address the subjects in which IKS can be implemented. While integration is of essence, the study will further depict how IKS can transform the curriculum in South Africa by incorporating indigenous methodologies into teaching and learning. It is therefore vital to illustrate that the integration of IK in the curriculum will behoove learners as they will be able to relate to the content being taught, establish connections to their own cultural backgrounds, and embrace education that is relevant to them.

REASEARCH SCOPE

This article offers a thorough examination of how IKS can be included in the CAPS curriculum, with a particular emphasis on a few subjects in basic education. The focus of the research is on articles articles, books and book chapters, scholarly articles, policy documents, and dissertations and theses that address the significance of incorporating IKS into basic education to deliver a culture-responsive education. It seeks to explore the integration of IKS into the CAPS curriculum while also highlighting the initiative's problems and support for changing basic education.

RESEARCH QUESTIONS

A study's comprehensive directions are provided by its research questions. They provide a clear direction for the data collection and analysis by helping to define the study's objectives and criteria. This study is crucial to demonstrate that IKS may be used in particular courses since, even if the CAPS curriculum promotes IKS's awareness in basic education, there are still difficulties integrating IKS into teaching and learning (Bhuda, 2024). In this article, four research questions were developed as follows:

a) Which subjects are relevant for IKS to be integrated into?
b) How can IKS be integrated into CAPS curriculum?
c) What are the challenges and prospects of integrating IKS into basic education?
d) What support is provided towards integrating IKS into the CAPS curriculum?

UNDERSTANDING INDIGENOUS KNOWLEDGE SYSTEMS: GENERAL PERSPECTIVE AS RELATES TO EDUCATION

IKS is defined by Grey (2014) as the expressions, practices, beliefs, understandings, insights, and experiences of Indigenous groups, generated over centuries of profound interaction with a particular territory. These knowledge systems are best viewed as a practical engagement which has been tested, augmented, and refined through generations of land-based practices. IKS is holistic in nature and cannot be considered apart from its applications, products, and articulations. This is knowledge that is peculiar to any given culture and therefore cannot be compartmentalized. This is so much the case that IKS as a body is often referred to as a knowledge practice or "praxis" (Battiste & Henderson, 2000). Although the terms are frequently used interchangeably, Indigenous Knowledge (IK) refers to the knowledge itself, whilst Indigenous Knowledge Systems (IKS) refers to the more general idea that includes the knowledge systems, customs, and beliefs of indigenous cultures.

Bhuda and Gumbo (2024) argued that IKS include a local community's traditional technology; social, economic and philosophical learning grounded in spirituality skills, practices and ways of being in nature (Keane, 2015). Its iterations and mechanisms are unique to each community, even where it shares certain features across groups by virtue of being embedded in a wider, common culture. In all locations IKS is the foundation of Indigenous governance, ecological stewardship, social, ethical, linguistic, spiritual, medical, food, and economic systems, so that the continual production and reproduction of local, land-based knowledge is the basis of Indigenous identity and sense of place in the world, as well as of Indigenous groups' very survival as distinct peoples (Grey, 2014).

Emeagwali (2020) points out that IKS is a synthesis of different disciplines and interconnecting epistemologies and value systems established by communities in ancestral lands that have developed paradigms and forms of life. This means that indigenous and local communities apply their own knowledge, epistemologies, and philosophies in the ways in which they learn, share, and produce their knowledge. One valuable example would be their philosophy and worldview of Africans which is enshrined in *Ubuntu.*

Ubuntu is viewed as the basis for community development in indigenous settings and this philosophy is applied across the board to inculcate the knowledge, skills and attitude to the broader communities based on humanity (Bhuda & Marumo, 2022). Scholars such as Lefa (2015) articulate that in an indigenous setting, the school outcomes are influenced by Ubuntu personality values that are embraced within the schools. Furthermore, it is evident that when the pupils/learners have Ubuntu there is greater desire to learn. This IKS is based on trial-and-error experimentation, as well as scientific methods and paradigms relevant to ecological, geographical, technological, social, and other traditions of life, and these applications have all contributed to the cumulative knowledge (Mashego et al, 2021).

IKS is currently taking a centre stage and there is a significant paradigm shift underway in which IKS and its ways of knowing are recognized as constituting complex knowledge systems with an adoptive integrity of their own (Ronoh, 2017). More and more teachers and indigenous scholars have now opted to applying indigenous methodological teachings, including indigenous games to foster teaching, and learning in schools. This has reflected massive paradigmatic shift to how learning is viewed contemporarily (Bhuda & Pudi, 2020; Emeagwali, 2014).

INDIGENOUS KNOWLEDGE SYSTEMS AND THE CURRICULUM: PROSPECTS FOR STEM EDUCATION IN SOUTH AFRICA

As stated in the United Nations Declaration of the Rights of Indigenous Peoples (United Nations General Assembly, 2007), indigenous peoples have the right to have the dignity and diversity of their cultures appropriately reflected in education (Article 15/1). The representations of IK /culture/peoples in education in formerly colonized countries often requires a largely non-Indigenous teacher workforce to effectively and respectfully engage with knowledge systems they may be unfamiliar with (Desmarchelier, 2020). This has therefore resulted in the marginalization of indigenous ways of knowing which can be deployed in classrooms. This marginalization has been vastly seen in Science, Technology, Engineering and Mathematics (STEM) education (Blose & Gumbo, 2024).

The IKS Policy of 2004 is viewed as the first and important milestone for South Africa in paving a way for IKS to be mainstreamed in the education system. The policy articulates that IK is dynamic in nature and changes its character as the needs of people change. It also gains vitality from being deeply entrenched in people's lives. As stated by Bhuda and Pudi(2020), the transformation of education syllabi from a primarily content driven approach to one of problem solving creates further impetus for the central recognition of IKS. What this means is that there is a need to indigenize the South African school curriculum. Bhuda (2021) notes that indigenization of this curriculum will result in doing away with the prevalence of culturally insensitive curricular models which will further require that appropriate methods and methodologies for mobilising IK in various learning contexts be identified and used.

Furthermore, the policy notes that the National Qualifications Framework (NQF) embraces the principle of lifelong learning, and recognises that it is not only in schools, colleges, and institutions where you can study and get a qualification (Bhuda, 2024). Accreditation by the NQF and related structures may be extended to many diverse forms of learning and should include the validation of knowledge learned and applied practically in indigenous communities; knowledge which is most frequently transmitted orally (Bhuda, 2024). The IKS Policy of 2004 pointed out that the Department of Education should take steps to begin the phased integration of IKS into curricula and relevant accreditation frameworks. This however has not come to full fruition as the marginalization and neglect of this knowledge has persisted. The knowledge has been subjected to assimilation by various other knowledge systems, and this further pushed its marginalization within the STEM education (Blose & Gumbo, 2024).

The failure to maintain IKS in the formal school curriculum is caused by various factors. Keane (2015:4) points out the following:

Efforts to integrate this knowledge into formal schooling via the science curriculum have focused on astronomy, living in nature, agriculture, technology, food, and plant uses; Absence of materials: The Science and Indigenous Knowledge Systems project at the University of the Western Cape has done pioneering work to produce teaching materials, however. Most school textbooks do not deal adequately or at all with indigenous knowledge; Teacher reluctance: several studies show science teachers of all cultures in South Africa have reservations about including IK into the classroom. Some teachers argue that science is a more powerful system of knowledge than indigenous knowledge. Besides, there are students from a range of cultural backgrounds in their classrooms, so whose indigenous knowledge should they teach?

Desmarchelier (2020) points out that IKS and science do not have to sit in opposition and can be seen as complementary rather than separate realities. This can be achieved by also involving the IK holders, developers, and propagators of this knowledge in the teaching STEM. This will help in demystifying the

STEM education to various indigenous learners. Kaya and Seleti (2013) posit that knowledge production and sharing at both community (IK) and global level (Western) in its cultural context by involving the community knowledge holders as producers and users in the core business of education should be echoed and advanced (Awour, 2007; Ronoh, 2017). That, teaching and learning of STEM education should be participatory by involving IK holders in the core business of education. That an African indigenous theory of knowledge based on the rich IKS history of ideas and intellectual development of Africa should be produced. There is strong governmental (DoE) support, human capital, finance and material support for the development of various activities associated with IKS as a core business and leg of education.

INDIGENOUS PEDAGOGIES AS A CATALYST OF CULTURALLY RESPONSIVE EDUCATION

Inclusive education refers to a broad philosophical position related to the educational rights of all children. In South Africa, inclusive education protects all children from discrimination and presents a commitment to creating access to and provision of education that accommodates the needs of children (Donald et al., 1997). Accordingly, inclusive education can be described as an educational system that adapts to the various requirements of students (Naicker, 1999). This kind of education system and methodology enables or fosters the teacher, learner relationship which in turn can advance the teaching and learning activities.

After decades of debate on the relevance of IKS and its suitability for integration in school curricula, there appears to be a shift in paradigm towards recognising indigenous ways of knowing and transforming curricula towards using inclusive, contextual and practical content and pedagogies that reflect the changing needs of African society (Ronoh, 2017). Nonetheless, post-colonial school curricula in Africa, which are mostly dominated by western values, knowledge, and pedagogies at the expense of IKS and epistemologies, remain a major area of concern in education.

However, Desmarchelier (2020) notes that teachers are often concerned with their abilities to successfully deliver classroom teaching inclusive of IKS and perspectives. Handayani et al (2018) support this statement and provides that it is indeed a challenge for educators and researchers to promote lessons and curricula to stay in synergy with the demands of the times without having to abandon local values. Teachers need to recognise that the IKS and classroom scientific knowledge can synchronise and be support for each other (Regmi & Fleming, 2012).

Desmarchelier (2016) argues that a more multifaceted understanding of epistemology drawn from diverse theoretical fields such as scientific considerations, personal epistemology and critical epistemology contributes to a more nuanced understanding of teacher positioning. According to Mkhwebane (2024), the inclusion of IKS in the classroom in terms of pedagogy serves multiple purposes. This is because indigenous peoples and communities have a robust and deep knowledge and practices that may be fruitful to apply in the classroom thereby cantering the teaching and learning in a culturally responsive manner. Based on the discussions and contributions made by scholars thus far, study outlines the following implications for a culturally responsive classroom and school:

- The readiness of educators to view themselves as students and work on gaining their own knowledge first. This paper sees this as a grassroots approach that allows teachers to see themselves as students and determine whether the teaching strategies they are using are effective. Acknowledging that knowledge of IKS and viewpoints is beneficial to all students. Since indigenous knowledge is simpler for students to get familiar with, it is worth and significance as a community-based knowledge cannot be overstated (Mkhwebane, 2024).

- Recognizing that educational systems are not valued neutral and critically analyzing whose opinions and knowledge have been prized in the system. Rather, cultural values are reflected in the way and content of instruction (Makokotlela & Gumbo, 2025). Thus, assisting students in comprehending this could aid them in navigating cultural differences. In order to recover and apply the knowledge that was left on the periphery of the knowledge economy to thrive in the same economy, the adage learning from the old to influence the new is crucial.
- Acknowledgement that IKS is linked to situations and making sure that IKS and/or viewpoints are meaningfully included in all subject areas (Naidoo, 2021). Teachers will be able to use and apply knowledge that is rooted in the learners' heritage as well as in their culture and traditions thanks to this. This will allow the teachers to ensure that the deliverables in class are met by encouraging meaningful engagements in the classroom (Mkosi et al., 2023).
- Consult any indigenous communities or organizations in the area before choosing any content to include in the classroom or school. Some people might be able to offer suitable resources. Additionally, acknowledging the potential of local Indigenous people as useful resources (Mehta et al., 2022). Building connections with the local community or organization(s) can help with this.

It is therefore of paramount importance that the education department/s be sensitized and encouraged to apply indigenous pedagogy in their classrooms. This is because indigenous pedagogy incorporates indigenous worldviews into engagements with information and learning. As pointed out by Burton and Point (2006), education is point specific, which means the tools used and pedagogy applied should inculcate stories and ceremonies which are essential in acquiring knowledge. Deloria and Wildcat (2001) provide that thoughtfully interwoven indigenous content and approaches must be informed by an understanding of indigenous epistemologies (how knowledge can be known) and pedagogies, (how knowledge can be taught).

Additionally, Kamasaki (2008) points out that foundations of Indigenous pedagogy are respect, mutual learning between student and teacher, and positionality or recognizing that everyone has different experiences that brought them to this learning. Bartlett et al (2012), Ragoonandan and Mueller (2017), Biermann and Townsend-Cross (2008) and Morcom and Freeman (2018) are some of the scholars who have presented that indigenizing the curriculum, applying indigenous pedagogies in the classroom and co-learning are vital to uprooting IK and implementing its practical, story based and culturally relevant pedagogy in schools.

UNPACKING CONSTRUCTIVISM AND INDIGENOUS STANDPOINT AS EDUCATION THEORIES

This study's arguments are sourced and motivated by the indigenous standpoint theory and constructivism theory which advocate for the inclusion of a cultural background and knowledge as part of the school curriculum for the benefit of leaners and their communities.

Indigenous standpoint theory

The Indigenous Standpoint Theory is based on incorporating IK into educational programs. IST, according to Winslett and Phillips (2005) is a complex process that fundamentally respects and embeds Indigenous community participation in the production and teaching of Indigenous standpoints and perspectives. This theory elevates Indigenous pedagogy by contextualizing the content into meaningful cultural and social perspective that Indigenous learners and their communities can relate to. This idea is significant because it advocates for the inclusion of IKS in the curriculum, claiming that including Indigenous information in educational curricula increases the visibility of IKS while also increasing self-esteem and enthusiasm in schooling (Nakata, 2004). Indigenous Standpoint Theory (IST) suggests that instructors and trainers incorporate IKS into the curriculum to make learning and application more relevant to Indigenous people (Foley, 2003).

Indigenous Standpoint Theory enhances Indigenous instruction by putting the curriculum into a cultural and social context that Indigenous students and communities can understand. "Instructional design cannot and does not exist without a consideration of culture," Henderson (1996, 86) claimed. According to McLoughlin and Oliver (2000), by acknowledging that learning is culturally and socially contextualized, the design process becomes anchored and positioned inside the communities and individuals for whom the learning materials are designed.

Constructivism theory

Constructivist theory acknowledges that the teacher is not a transmitter of knowledge, but rather a facilitator of learning. The teaching practices in the constructivist classroom also require that the learners must actively participate in the classroom activities, become more involved in the learning process, and take responsibility for their own learning (Jacobs, 2015). Therefore, the learners are not absorbers of knowledge, but active participants in constructing their own meaning based on strongly held preconceptions.

Constructivism is concerned with the implementation of learner centred curriculum in South Africa. According to Jacobs (2015) learner centeredness is more than just an approach to teaching, it is a philosophy which guides the teachers' teaching and learning practice – how they teach, how they interact with the learners, how they design their courses, learning activities and strategies.

METHODOLOGY

A scoping literature review is a research method used in this article to advance the aim stated in the introduction. A scoping literature review is a thorough method for examining the body of research on a particular topic or research issue (Munn et al., 2018). In order to give a comprehensive picture of the depth and breadth of information on the topic of interest, this research method attempts to map out important ideas, sources, and knowledge gaps (Pham et al., 2014). In order to find the current research, theories, and findings pertaining to the integration of IKS in CAPS curriculum, a scoping literature review of the topic of IKS in South African education entails methodically searching and examining relevant scholarly articles, books and book chapters, reports, and other sources.

Choosing a scoping literature review as a research method is important because it allows the researcher to map out the breadth and depth of current literature on the phenomena of interest, highlighting areas that need more investigation (Munn et al., 2018). This research method provides a comprehensive overview that directs the research initiative, enabling the development of more accurate and informed research questions and approaches. In this study, scoping literature reviews involved the following:

- Defining the scope of the review
- Identifying phenomenon of interest
- Conducting a thorough search to find pertinent literature utilizing relevant databases, including libraries, digital scholarly engines, keywords, and inclusion and exclusion criteria.
- Identified articles, books and book chapters, scholarly articles, policy documents, dissertations and theses were then screened.
- • To extract important information such relevant perspectives, primary findings, and developing themes, data extraction was done.
- Keywords were input into the databases and search engines of the library. Key phrases related to this study were "education," "curriculum," "Indigenous Knowledge Systems," "curriculum," "CAPS," and "culturally responsive."

ASPECTS OF INDIGENOUS KNOWLEDGE WHICH CAN BE INTEGRATED IN BASIC EDUCATION

Through the application of the scoping literature review, the study finds that IKS can be integrated in the following subjects: Natural Sciences and technology; Physical sciences; Mathematics; Life sciences and Agricultural science. These subjects are reported herein below as themes for discussion to unpack how IKS links and influences these. Furthermore, the study presents various indigenous pedagogical/ methodological areas in which can be applied in the teaching of these subjects.

Indigenous Knowledge Systems in Natural Sciences

Natural science is taught in grade 7- 9 and leaners can explain how science applies in their daily live and of the value that science adds to their lives. Examples can range from indigenous foods, healthy diet methods, nutrition and plants, medicinal plants etc According to van Wyk (2002), in natural science and technology education the Mode 1 and Mode 2 models of knowledge production (explained herein below) proposed by Gibbons (2000) are useful in terms of knowledge generation using local, indigenous expertise and resources (Posch, 2000). According to Gibbons (2000), major changes in knowledge generation are occurring outside of existing academic disciplines. The core or foundation of disciplinary and specialized knowledge is referred to as the Mode 1 strategy. With the increasing globalisation and shifting world order, there has been a significant shift towards Mode 2 knowledge production, in which educators can begin to recognize and validate the context in which such knowledge is produced.

For natural science and technology teachers, this may mean that as a result of recognizing the value of community involvement and the impact that the local context can have on knowledge production in science and technology, new and relevant knowledge could emerge. Although elements of existing knowledge have entered Mode 2, it may be an extremely useful platform for contextualizing the concept of IKS, because the latter is generated in the context of application rather than being developed first and then applied to the context later (van Wyk, 2002). In this way, the use of IKS in science and technology is sparked by the perception of shared interests rather than following pre-determined paths (Posch, 2000).

Scholars such as Mlotshwa and Tsakeni (2024) argue that the call for IKS to be mainstreamed into natural sciences studies within the grades that the subject is offered has always been echoed in South Africa. They point out that these calls for IKS integration led to the recommendations contained in CAPS for natural sciences Grades 7–9, which was first implemented in 2012. Since natural sciences encompasses biology, physics and chemistry, the theoretical and practical work of the subject would accommodate IKS integration as this form of knowledge provides practical solutions for problem-solving and sustainable development. IKS merges traditional wisdom from IK Holders and practitioners with the modern scientific inquiry and both complement one another (El Yazidi & Rijal, 2024).

Indigenous Knowledge Systems in Agricultural Science

Agricultural science in taught from grade 10-12 in South African basic education According to Odhiambo (1990), Africa's educational system has continuously overlooked the knowledge, skills, and survival methods of local farmers who have successfully managed their farmlands and stayed productive for millennia with little or no external input. This is because farmer education efforts in African rural areas have taken the shape of top-to-bottom, hand-me-down attempts to superimpose Western agricultural technologies. Materechera (2021) agrees with the above notion by illustrating that IK provides problem solving strategies for agriculture. He further points out that local people, through a systematic process of observing local conditions have provided the best mechanisms for addressing many agricultural challenges related to sustainability.

In contrast, as more people become aware of the dangers of so-called modern farming methods like synthetic fertilizers and toxic pesticides, European scientists are scouring Africa's communities for IKS, which they see as a "major untapped source for developing sustainable agriculture" (Odhiambo et al., 1990: 3). Odhiambo states that "Indigenous knowledge can reveal missing ecological keys, which may help scientists develop alternative agricultural technologies less dependent on non-renewable resources (e.g. fossil energy) and environmentally damaging inputs (e.g. chemical pesticides), than conventional technologies" (Odhiambo et al.,1990: 3).

It is evident that different practices can be used in incorporating IKS into agricultural science teaching. Scholars such as Agegbo and Ramnarain (2024) point out that building meaningful relationships with elders and knowledge keepers of Indigenous communities and having practical hands-on experience is an important strategy for teachers in incorporating IKS into their teaching practice. Within their systematic review, the authors indicate that South Africa has more evidence that IKS that was normally overlooked within various sciences is now being integrated within the CAPS curriculum. There is still much to be done within the IKS and agricultural science space but the practice that it is being slowly integrated shows that IKS is vital within South Africa's basic education sphere.

Indigenous Knowledge Systems in Life Sciences

IKS is described as a different way of knowing and an alternate manner of conveying concepts that are generally addressed using scientific knowledge in the National Curriculum Statement (NCS) policy document for Life Sciences (taught in 10-12). As a result, it promotes the interaction of many methods of knowing in formal education. Schmidt et al. (2019) found that students' perceptions of the usefulness of science in their lives were positively connected with relevance claims in the Life Science classroom. There has been a substantial quantity of literature published on ethnobotany, the study of human-plant relationships (Van Wyk & Gericke, 2000). This is likely owing in part to plants' natural medicinal properties and their significance in the pharmaceutical business. As a result, there has been some discussion about how to incorporate ethnobotany, specifically medicinal plants, as part of the IKS integration in the Life Sciences classroom (De Beer & Van Wyk, 2011).

Indigenous Knowledge Systems in Physical Sciences

IKS in a Physical Sciences can be included in both Physics and chemistry sections. In the Physics sections, different renewable energy sources can be a focus area and how they benefit living beings. These include for instance thee benefits of solar energy, wind energy, biomass energy, geothermal energy and hydropower energy. Traditional names of chemicals, food preservation, and fertilisers are all instances of IKS in the Chemistry portion of the Grade 10–12 CAPS (DBE, 2011b). Making fire from friction, utilizing clay pots to cool food and water, and animal behaviour related to earthquakes, tsunamis, or floods (in the context of electromagnetic radiation) are all examples of IKS in the Physics section. Brewing, iron and copper extraction, and the energy-efficient design of traditional homes are current IKS examples connected to the Physical Sciences for Grades 8 and 9 (DBE, 2011a).

Indigenous Knowledge Systems in Mathematics

Scholars such as Bhuda and Marumo (2021); Nxumalo and Mncube (2018); Owusu-Mensah and Baffour (2015); Kiran and Knights (2010); Nkopodi and Mosimege (2009) conducted studies on aspects of Indigenous knowledge that could be part of mathematics teaching and learning. They have further identified indigenous games as appropriate to learn mathematical concepts and ideas in the foundation phase.

A game is a key instrument, according to Yusuf et al. (2010), when it may be linked to the teaching of subjects in schools to help achieve this link. Instead of utilizing textbooks, employing indigenous games as part of a lecture allows a student to learn about mathematical ideas while learning. Furthermore, one of the Department of Education's recommendations for successful curriculum implementation was to avoid using textbooks and instead encourage teachers to build their own tools and teaching strategies (DoE, 2009). Games are crucial to add into mathematics teaching and learning for the following reasons, according to Bhuda and Marumo (2021, 119):

Give learners opportunities to utilize mathematical principles for situational problem-solving and strategic thinking, develop effective students ' interest in mathematics and passion by engaging them in pleasurable activities and challenges and Foster the notion that learning is as enjoyable as possible.

Games which have been identified to be relevant for mathematics teaching and learning in South African foundation phase are as follows: Jacks, Board game (*morabaraba* in Sotho and *umrabaraba* in IsiNdebele), Moruba (in sePedi) board game and string figure gates (Bhuda & Marumo, 2021; Nkopodi & Mosimege, 2009).

In addition to the ethnomathematical (indigenous) games provided by Bhuda and Marumo (2021) and Nkopodi and Mosimege (2009), Mosimege (2020) further adds that there are other games which compliments those already provided. He points out the following as games that can enhance mathematical learning within the education system of South Africa:

(i) Dibeke
(ii) (Kho-kho
(iii) Kgati
(iv) Diketo
(v) Jukskei
(vi) Moruba

In addition to studies on indigenous games, studies have been conducted on the exploration of the role and use of indigenous materials and activities in the teaching and learning of school mathematics (Mosimege, 2020). Moreover, Mosimege echoes that these indigenous games such as *Morabaraba* and *Moruba* when used within mathematics class they enhance the unpacking of various mathematical concepts such as "Identification of various quadrilaterals (squares), Ratio and proportion between the lines and the squares, Symmetry, Logical deductions, Counting and Addition and subtraction" (Mosimege, 2020:8).

CHALLENGES OF INTEGRATING INDIGENOUS KNOWLEDGE SYSTEMS IN BASIC EDUCATION

Western science has been taught in every country since colonial times as the most likely valid discipline for studying natural occurrences. As a result, some practices in science and related professions are considered standard, while others are considered alternative (Ogunniyi, 2009b). As a culture-free subject, school science was imposed on indigenous students. Western science's hegemony also ignores the socio-cultural contexts in which students and communities live, as well as the fact that scientific knowledge is socially created (Brayboy & Castagno, 2008). According to Brayboy and Castagno (2008), western science's required objectivity and universalism rationalizes our refusal to embrace alternative ways of knowing.

South Africa's education system underwent major adjustments following the 1994 political transition. This transition can be evidenced by the changes in the educational landscape which have been revisited to try and unpack and focus on the experiences of the indigenous community. These changes could be considered as an attempt to correct historical inequities in terms of educational access, equity, and quality. The Revised National Curriculum Statement (RNCS) for Grades R–9 was released in 2002; the National Curriculum Statement (NCS) for Grades 10–12 was released in 2003; and the CAPS for Grades R–12 was released in 2012. Moichela (2017) noted that the Department of Basic Education (DBE) endeavoured to create the national Curriculum in such a way that it was flexible and adaptable to local conditions and needs at the school level in all of these revisions. These curricular reviews were conducted with the important anticipation that the curricula would be interpreted and applied in various circumstances (Abah et al., 2015). The Life Sciences (LS), for example, which are mentioned in the 2003 National Curriculum Statements (NCS), promote and inspire learners' knowledge of and open-mindedness toward the existence and recognition of various points of view in a multicultural society.

Based on the foregoing, it is possible to conclude that space has been established in the NCS for the recognition of other Knowledge Systems, however there is still a sense of muteness in the Department of Basic Education (DBE)'s acknowledgement and recognition of IKS. In other words, there appears to be no notion of taking responsibility for the DBE's meaningful integration of IKS into the basic education curriculum. The department of basic education has somehow managed to dismiss the integration of IKS even though as Ogunniyi (2007) points out, the National Curriculum Statements demand IKS to be included in the curriculum, but the IKS that should be included have not been clarified. There has not been a clear or concise stipulation as to how IKS can be integrated as per the IKS Policy of 2004. Many departments including the DoE are signatory to the policy but have failed to provide relevant avenues to integrate this IKS into the curriculum.

The question today is how South African teachers utilized the space afforded by the CAPS for the functional application of IKS in the basic education curriculum in various circumstances? To answer that question, according to available research, teachers in Southern Africa have been trained in Western pedagogies, which call into doubt the legitimacy of IKS in the school curriculum. This context contributed to a lack of enthusiasm among teachers in most parts of the continent, including South Africa, for learning and understanding IKS. For example, Dziva et al (2011) found that a science perspective appears to be required as a bridge between leaners and a scientific worldview (Fakudze, 2004).

It is for these reasons that the argument for the integration of IKS into the curriculum for basic education in South Africa according to Moichela (2017, 112-113) is proposed because (a) incorporating IKS into the curriculum may help to reconcile two opposing perspectives on the world; (b) including IKS into thc curriculum may help to increase the visibility of IKS. (c) Its integration into the curriculum may help to boost students' self-esteem and interest in school; (d) its integration into the curriculum may act as a buffer against the collapse of local knowledge when confronted with the dominant paradigm of the West; and (e) its integration into the curriculum should humanize a curriculum that has been dehumanized by modernity. This work can be considered a contribution to the ongoing advocacy for the acknowledgment and acknowledgement of IKS in school curricula on this premise.

Teachers in rural areas of South Africa face additional challenges in their efforts to change the curriculum. This is because they have been left in the margins of decision making and their voices hardly listened to. The complexity of the terms 'rural' and 'urban' in relation to the historical context of South Africa's political past, as well as the overarching issue of teachers being trained in the Western continue to cast doubt on the validity of IKS in school curricula are among the challenges faced by rural teachers in integrating IKS into the curriculum. Based on the foregoing, it could be argued that incorporating IKS into the basic education curriculum will only be transformational if there is evidence of a paradigm shift in terms of accommodating and validating the diverse cultures that students, indigenous knowledge holders and teachers bring to the classroom (Fakudze, 2021). This would have an impact on the case for integrating IKS into the basic education curriculum in a postmodern era that embraces diversity that can only be realized in a newly created cognitive space that is “open,” as highlighted by Odora Hoppers (2002:4) and specifically as classified by Lather (1993:33):

- For educational process: constructionist, open text and reader/centred, focus on interpretations, ideally dialogical and democratic
- For educational structure: situated, contingent/partial/in flux, use of new information and communication technologies
- For educational content: reality always negotiated, interdisciplinary, focus on micro narratives and what is missing from the mainstream discourse, multiliteracies, multiple intelligences, "multiple sites from which the world is spoken" The terms suggested by Lather (1993) in relation to basic education are in line with the suggestion made earlier in this study about educating African learners in both 'rural' and 'urban' settings for the communal life and Ubuntu which are crucial for African thought and practice and the possibility of their realisation in a new cognitive space only.
- Ubuntu and communal life are African indigenous cosmologies that include the virtuous nature (moral roundedness or character leading to good behaviour) of co-operative and interpersonal skills that may render basic education both sustainable and competitive, as highlighted in Letseka (2000).

A challenge that has been identified by Bhuda (2021) in integrating IKS within the curriculum is the unwillingness and ignorance from policy makers, teachers, and trainers to act. Little efforts towards integrating IKS in the curriculum show the gap that still exists in acknowledging IKS as relevant to be part of the formal education. Taylor and Cameron (2016), in a similar spirit, remarked that conflict and inconsistencies are always observed between what is planned by the curriculum and what happens in the classrooms. O'Donoghue (2004) and Hlalele (2019) on the other hand, believe that it is difficult for African teachers to integrate IKS into a curriculum that is firmly based on Western scientific philosophy. Even though most educational policy papers recognize the value of IKS in classroom learning, Nel (2006) claimed that the application and monitoring of IKS in education are still unclear. IK cannot be applied at the postsecondary level unless it is used in secondary and primary schools.

Mosimege (2005) emphasizes the importance of good implementation in his analysis of the development of IKS Policy in South Africa from 1999/2000 to 2004. He claims that development programs for IKS are limited to higher education, with little activity in general education (Mosimege, 2005). According to a study of the literature in this field, thorough monitoring and evaluation are required for IKS to be properly implemented (Mosimege, 2005; Nel, 2006). In a similar line, Grange (2007) stated that for IKS to be successfully integrated into science education, educators must first comprehend the integration process and then be able to appropriately incorporate indigenous information into their science teaching.

THE SUPPORT FOR INTEGRATION OF IKS INTO THE SOUTH AFRICAN BASIC EDUCATION CURRICULUM

While the research on the incorporation of IKS into the South African school curriculum is mixed, it is crucial to note that the policy framework is in place. The mechanisms, rather than the virtues of integration, need to be examined and investigated, according to Onwu and Mosimege (2004). This is because the Life Science, Physical Science Natural Science Policy Documenst for Curriculum 2005 explicitly states that IKS must be included in the school curriculum. The notion of valuing and inclusivity supports the CAPS throughout the grades, and for all topics or learning areas (Onwu & Mosimege, 2004; Khupe, 2014). Although IKS are under attack in the face of globalization, van Wyk and Higgs (2011) argues that the production and value of Eurocentric knowledge and education is being standardized. However, a concerted integration effort is required to ensure that learners do not feel alienated from the IKS that surrounds them (van Wyk and Higgs, 2011).

While the updated National Curriculum Statement implicitly promotes IKS integration, Hewson et al. (2009) claim that one of the barriers has been instructors' lack of knowledge about the various IKS in South Africa. They can only teach what they know, and without a consistent IKS curriculum, the school system risks becoming fragmented. As a result, Mkhize and Ndimande-Hlongwa (2014) argued that integration tactics used in other nations could be altered to match the South African educational system. As a result, rules and models for aiding teachers and districts with IKS integration might be produced, as well as teacher orientation integration programs and an online database of teaching curricular materials for convenient access (Mkhize & Ndimande-Hlongwa, 2014)

Taylor and Cameron (2016) investigated the value of IKS in South African Physical Science curricula across time. Rather than a top-down approach, designing an IKS-valuing curriculum should involve collaboration and negotiation with IKS custodians (Khupe, 2014). Despite its shortcomings, the present South African secondary Physical Sciences curriculum (CAPS) includes eight appropriate examples of IKS as part of the obligatory content. Furthermore, the Grade 7–9 CAPS mandates that students comprehend "the various cultural settings in which IKS have formed" (DBE, 2011a: 9). This allows IKS to be taught as a whole system of thought in junior high. Learners, parents, and education departments, on the other hand, rate science teachers' efficacy based on their students' scientific exam results rather than their values, and, as previously stated, IKS is not included in the Grade 12 Physical Sciences examinations.

IKS have moved from the margins to the center in policy, research, higher education since the introduction of democracy in South Africa. IKS was first introduced to South African schools in 2001 and 2002 as part of national curriculum changes (Department of Education, 2002; 2003), and it has remained one of the guiding principles in multiple subsequent revisions of national school science curriculum policy (DBE, 2011a). Many people were wondering how IKS could be included into the curriculum before adequate study was undertaken. Researchers have discovered which Indigenous knowledges are appropriate for inclusion in the curriculum and how they might be integrated into teaching and learning. However, more research on how IKS provides a sustainable education not only in South Africa, but also across the African continent, is required (Bhuda, 2021).

The most recent reforms in Curriculum and Assessment Policy introduced in 2012 continue to provide for IKS, but they appear to acknowledge the difficulties for teachers by recommending examples linked to respective science topics and proposing that these topics be more closely related to learners' societal or cultural environment in order to minimize conflicts that may arise from learners' views of the world and those on which these science subjects are based. The curriculum policy (CAPS) also gives instructors specific instructions on how to value IKS by "acknowledging this country's rich history and tradition as significant contributors to cultivating the principles enshrined in the Constitution" (DBE, 2011a:5). Furthermore, the specific goals include training on "the history of scientific achievements and the link between indigenous knowledge and science" (DBE, 2011a:13). As Ogunniyi (2007, 964) points out, attempts to include IKS into science curricula are "not unique to South Africa. Many other African countries, Canada, the United States, Australia, the Middle East and Far East, as well as Central and South American countries, have attempted similar initiatives." However, the use of IKS in national curricula may be unique to South Africa.

In a series of research concentrating on science educators and IKS, Ogunniyi (2007) discovered that completing university courses geared to provide them with knowledge and abilities to combine science and IKS improved teachers' comprehension and enjoyment of IKS and science–IKS interaction. Teachers were 'less hopeful about its performance' in schools, even though they embraced IKS as part of the science curriculum, could distinguish between IKS and science, and were better aware of how to apply scientific or IKS worldviews (Ogunnyi & Hewson, 2008). Other studies, such as Matike's (2012), show that teachers, students, and school principals 'experience difficulties in understanding how IKS can be mainstreamed into all learning areas, particularly science-oriented subjects,' and that there was no consensus on how such mainstreaming could be accomplished. According to Khupe (2014), there is however hope that teachers can be able to integrate IKS in their lessons not only in science subjects only if proper training and sufficient leaning material is provided to them.

There is much desire to capacitate various educators on how to integrate IKS in their teaching. This is because there is still a negative teacher attitude and limited professional development for educators who do not understand the value of IKS (Akuma & Gaigher, 2021; Mlotshwa & Tsakeni, 2024). Despite the perceived benefits of teaching IKS and practical work in classrooms, teachers may still experience challenges that hinder implementation. Contemporarily, Teachers hold varied perceptions of IKS and not all of them are aware of the curriculum innovation (Bhuda, 2021; Mudaly & Ismail, 2013).

To help teachers transition from negative to positive perceptions of IKS, an intervention based on dialogic argumentation may be applied (Molise, 2025; Mlotshwa & Tsakeni, 2024). The DBE may implement these interventions by offering workshops and training to educators to offer directions on integrating IKS. Higher education institutions can also be encouraged by the Department of Higher Education and Training (DHET) to incorporate IKS into their curricula, which will provide students with the knowledge and skills they need to implement IKS and IKS in the classrooms once they get placed for work. In the past, DHET implemented IKS programs at the University of Venda (UNIVEN) and North-West University (NWU) with success (Nevhudoli & Netshandama, 2023).

FUTURE RESEARCH DIRECTIONS

This work adds to the body of knowledge of the importance of IKS in education. It shows that STEM subjects are suitable for IKS and that IKS should be incorporated into basic education. In addition to encouraging respect and understanding for various knowledge systems, IKS can help learners develop a sense of cultural identity and pride. Therefore, incorporating IK into CAPS curriculum such in STEM subjects supports inclusive and equitable education for all students and aids in the decolonization of education culture-responsive education. The Department of Basic Education and scholars interested in IKS, and curriculum studies are intended to find direction in this study.

CONCLUSION AND RECOMMENDATIONS

The purpose of the study was to look into the integration of IKS into basic education, with a particular focus on the significance of IKS in basic education. The research also identified areas in which IKS can be integrated in education from existing academic literature. According to the study's findings, IKS has the potential to be successfully integrated in a few subjects if teachers are given sufficient training on how to do so. More action and research on how IKS can be integrated is needed so that teachers can follow people's suggestions when they attempt it. Various indigenous scholars' work on the value of IK in education has been explored and offered in this study. The study also established theoretical frameworks that argue for the inclusion of a learner's cultural background in formal school teaching and learning.

The argument of this study is that the government can take additional steps forward and analyse how IKS components can be included in the curriculum based on proposals from varied indigenous scholars. According to the findings of this study, the Department of Basic Education has begun to take steps to incorporate various IKS into the curriculum. However, more work from the Department of Basic Education is needed in order to reach the final aim, and experts have identified problems that are impeding the successful integration of IKS in basic education. Other scholars have addressed some of the problems, while the Department of Basic Education must address others to enhance their policies on IKS integration.

The following recommendations are therefore made by the study:

- Educational legislation and policies should be amended to provide detailed information and on IKS in the CAPS curriculum.
- The value of IKS should be realized and incorporated in the teaching of STEM.
- Indigenous knowledge holders and educators should be valorised through the recognition of prior learning (RPL) process so that they can be incorporated in the teaching and learning of learners in communities.
- Teachers should receive training on how to integrate IKS as part of their teaching and learning. These trainings can be offered by DoE, schools and universities.

REFERENCES

Abah, J., Mashebe, P., & Denuga, D. D. (2015). Prospect of integrating African indigenous knowledge systems into the teaching of sciences in Africa. *American Journal of Educational Research, 3*(6), 668–673.

Ardill, A. (2013). Australian sovereignty, indigenous standpoint theory and feminist standpoint theory: First peoples' sovereignties matter. *Griffith Law Review, 22*(2), 315–343. DOI: 10.1080/10383441.2013.10854778

Assembly, U. G. (2007). United Nations declaration on the rights of indigenous peoples. *UN Wash, 12*, 1–18.

Awour, P. (2007). *Indigenous knowledge and sustainable crop production: A study of the millennium village in Nyanza province Kenya.* LAP Lambert Academic publishing.

Bartlett, C., Marshall, M., & Marshall, A. (2012). Two-eyed seeing and other lessons learned within a co-learning journey of bringing together indigenous and mainstream knowledges and ways of knowing. *Journal of Environmental Studies and Sciences, 2*(4), 331–340. DOI: 10.1007/s13412-012-0086-8

Bhuda, M. T. (2021). Making a Case for Indigenous Education Systems in South Africa. *African Journal of Development Studies,* 67-82.

Bhuda, M. T. (2024). Integrating Ethnomathematical Activities in the Foundation Phase as Means of Decolonisation: Insights from Indigenous Knowledge Systems. *African Journal of Development Studies (formerly AFFRIKA Journal of Politics, Economics and Society), 14*(4), 401-424.

Bhuda, M. T., & Gumbo, M. T. (2024). Developments Towards Integrating African Indigenous Knowledge in South African Universities. *UnisaRxiv*. DOI: 10.25159/UnisaRxiv/000082.v1

Bhuda, M. T., & Marumo, P. (2021). The Ndebele indigenous games pertinent to primary school mathematics learning: why indigenous games are a vital tool for mathematics teaching and learning. *Proceeding on the conference on the Conference on the Implications of Covid-19*, South Africa, 114-125

Bhuda, M. T., & Marumo, P. (2022). Ubuntu Philosophy and African Indigenous Knowledge Systems: Insights from decolonization and indigenization of research. *Gender & Behaviour, 20*(1), 19133–19151.

Bhuda, M. T., & Pudi, T. I. (2020). Indigenising mathematics in schools: Why ethnomathematics matters. *Developing teaching and learning in Africa: Decolonising perspectives*, 21-42.

Biermann, S., & Townsend-Cross, M. (2008). Indigenous pedagogy as a force for change. *Australian Journal of Indigenous Education, 37*(S1), 146–154. DOI: 10.1375/S132601110000048X

Blose, P., & Gumbo, M. T. (2024). Developing a framework for integrating IKS in technology education with sustainable development principles. *African Journal of Research in Mathematics. Science and Technology Education, 28*(3), 330–342.

Brayboy, B. M. J., & Castagno, A. E. (2008). Indigenous knowledges and native science as partners: A rejoinder. *Cultural Studies of Science Education, 3*(3), 787–791. DOI: 10.1007/s11422-008-9142-9

De Beer, J. J., & Van Wyk, B. E. (2011). An ethnobotanical survey of the Agter–Hantam, Northern Cape Province, South Africa. *South African Journal of Botany*, *77*(3), 741–754. DOI: 10.1016/j.sajb.2011.03.013

Dei, G. J. S. (2008). Indigenous knowledge studies and the next generation: Pedagogical possibilites for anti-colonial education. *Australian Journal of Indigenous Education*, *37*(S1), 5–13. DOI: 10.1375/S1326011100000326

Department of Education (2002). *Revised national curriculum statement for Grades R–9 (schools) natural sciences*. Pretoria: DoE.

Department of Education (2003). *National curriculum statement: Grades 10 –12 (general) physical sciences*. Pretoria: DoE.

Desmarchelier, R. (2016). *Whose knowledge?: Science education, Indigenous knowledges and teacher praxis* (Doctoral dissertation, University of Southern Queensland).

Desmarchelier, R. (2020). *Indigenous Knowledges and Science Education: Complexities*. Considerations and Praxis.

Donald, D. R., Lazarus, S., & Lolwana, P. (1997). *Educational Psychology in Social Context: Challenges of Development, Social Issues & Special Need in Southern Africa: a Teacher's Resource*. Oxford University Press.

Dziva, D., Mpofu, V., & Kusure, L. (2011). Teachers' conception of indigenous knowledge in science curriculum in the context of Mberengwa District, Zimbabwe.

El Yazidi, R., & Rijal, K. (2024). Science learning in the context of' indigenous knowledge' for sustainable development. *International Journal of Ethnoscience and Technology in Education*, *1*(1), 28–41. DOI: 10.33394/ijete.v1i1.10880

Emeagwali, G. (2014). Intersections between Africa's indigenous knowledge systems and history. In *African indigenous knowledge and the disciplines* (pp. 1–17). Brill. DOI: 10.1007/978-94-6209-770-4_1

Emeagwali, G. (2020). African Indigenous Knowledges and the Decolonization of Education in Africa. In *The Palgrave Handbook of African Education and Indigenous Knowledge* (pp. 335–349). Palgrave Macmillan. DOI: 10.1007/978-3-030-38277-3_17

Fakudze, C. (2021). The influence of local worldview presuppositions on learners' conceptions of Selected Mechanics topics. *South African Journal of Education*, *41*(2), 1885. DOI: 10.15700/saje.v41n2a1885

Fakudze, C. G. (2004). Learning of science concepts within a traditional socio-cultural environment. *South African Journal of Education*, *24*(4), 270–277.

Foley, D. (2003). Indigenous epistemology and Indigenous standpoint theory. *Social Alternatives*, *22*(1), 44–52.

Gibbons, M. (2000). Universities and the New Production of Knowledge: Some Policy Implications for Government. In Kraak, A. (Ed.), *Changing Modes*. Human Sciences Research Council Publishing.

Grange, L. L. (2007). Integrating western and indigenous knowledge systems: The basis for effective science education in South Africa. *International Review of Education, 53*(5-6), 577–591. DOI: 10.1007/s11159-007-9056-x

Green, L. J. (2008). 'Indigenous knowledge'and 'science': Reframing the debate on knowledge diversity. *Archaeologies, 4*(1), 144–163. DOI: 10.1007/s11759-008-9057-9

Grey, S. (2014). Indigenous knowledge. *Acta Paediatrica (Oslo, Norway), 91*(1), 1–8.

Handayani, R. D., Wilujeng, I., & Prasetyo, Z. K. (2018). Elaborating Indigenous Knowledge in the Science Curriculum for the Cultural Sustainability. *Journal of Teacher Education for Sustainability, 20*(2), 74–88. DOI: 10.2478/jtes-2018-0016

Henderson, L. (1993). Interactive multimedia computer courseware and culturally appropriate ways of learning. In Latchem, C., Williamson, J., & Henderson-Lancett, L. (Eds.), *Interactive multimedia: Practice and promise* (pp. 189–203). Kogan Page.

Hewson, M. G., Javu, T. T., & Holtman, L. B. (2009). The indigenous knowledge of African traditional health practitioners and the South African Science curriculum. *African Journal of Research in Mathematics. Science and Technology Education, 13*(1), 5–18.

Hlalele, D. J. (2019). Indigenous knowledge systems and sustainable learning in rural South Africa. *Australian and International Journal of Rural Education, 29*(1), 88–100. DOI: 10.47381/aijre.v29i1.187

Hoppers, C. A. O. (2005). *Culture, indigenous knowledge and development: The role of the university.* Centre for Education Policy Development.

Jacobs, K. R. (2015). *The classroom implementation of indigenous knowledge in the science curriculum by science teachers in the Western Cape province.*

Kaya, H. O., & Seleti, Y. N. (2013). African indigenous knowledge systems and relevance of higher education in South Africa. *International Education Journal: Comparative Perspectives, 12*(1).

Keane, M. (2015). Why indigenous knowledge has a place in the school science curriculum. Retrieved from https://www.researchgate.net/publication/280216361 Accessed: 10 February 2025

Khupe, C. (2014). *Indigenous knowledge and school science: Possibilities for integration* (Unpublished doctoral dissertation). Johannesburg, South Africa: University of the Witwatersrand

Kiran, A., & Knights, J. (2010). Traditional Indigenous games promoting physical activity and cultural connectedness in primary schools? Cluster Randomised Control Trial. *Health Promotion Journal of Australia, 21*(2), 149–151. DOI: 10.1071/HE10149 PMID: 20701567

Lather, P. (1993). Fertile obsession: Validity after poststructuralism. *The Sociological Quarterly, 34*(4), 33–34. DOI: 10.1111/j.1533-8525.1993.tb00112.x

Le Grange, L. (2018). Decolonising, Africanising, indigenising, and internationalising curriculum studies: Opportunities to (re) imagine the field. *Journal of Education (University of KwaZulu-Natal)*, (74), 4-18.

Letseka, M. (2000). African philosophy and educational discourse. In Higgs, P., Vakalisa, N. C. G., Mda, T. V., & Assie-Lumumba, N. T. (Eds.), *African voices in education.* Juta.

Macedo, D., Semali, L., & Kincheloe, J. (1999). Decolonizing indigenous knowledge. *What is indigenous knowledge: Voices from the academy*, 134-176.

Makokotlela, M. V., & Gumbo, M. T. (2025). Decolonisation of the curriculum through the integration of indigenous knowledge in higher education. *South African Journal of Higher Education, 39*(1), 237–254. DOI: 10.20853/39-1-6004

Mashego, S. L., Maditsi, M. E., & Bhuda, M. T. (2021). A Response to Covid-19: Perspectives From African Indigenous Knowledge Systems. *Gender & Behaviour, 19*(1), 17242–17254.

Matike, E. (2012). Knowledge and perceptions of educators and learners on the incorporation of indigenous knowledge into the school curriculum. *African indigenous knowledge systems and sustainable development*, 169-184.

Mehta, K., Alter, T. R., Semali, L. M., & Maretzki, A. (2022). AcademIK connections: Bringing indigenous knowledge and perspectives into the classroom. *Journal of Community Engagement and Scholarship, 6*(2). Advance online publication. DOI: 10.54656/XLUX1060

Mkhize, N., & Ndimande-Hlongwa, N. (2014). African languages, indigenous knowledge systems (IKS), and the transformation of the humanities and social sciences in higher education. *Alternation (Durban), 21*(2), 10–37.

Mkhwebane, L. N. (2024). Life sciences teachers' integration of indigenous knowledge: A vision for making science classrooms culturally responsive. *EURASIA Journal of Mathematics, Science and Technology Education, 20*(8), em2483.\ https://doi.org/DOI: 10.29333/ejmste/14859

Mkosi, N. N., Mavuso, M. P., & Olawumi, K. B. (2023). Using Ubuntu values in integrating African indigenous knowledge into teaching and learning: A review of literature. *International Journal of Learning. Teaching and Educational Research, 22*(5), 140–159. DOI: 10.26803/ijlter.22.5.7

Mlotshwa, N., & Tsakeni, M. (2024). Challenges Experienced by Teachers when Integrating Indigenous Knowledge Systems in Natural Sciences Practical Work. *African Journal of Research in Mathematics. Science and Technology Education, 28*(2), 300–313.

Mlotshwa, N. P., & Tsakeni, M. (2024). Exploring Indigenous Knowledge Systems Integrated in Natural Sciences Practical Work in Selected South African Schools. *International Journal of Learning. Teaching and Educational Research, 23*(9), 383–400.

Moichela, K. Z. (2017). Integration of indigenous knowledge systems in the curriculum for basic education: possible experiences of Canada. Unpublished doctoral dissertation, University of South Africa, Pretoria

Molise, H. V. (2025). Teachers' pedagogical knowledge of integrating indigenous knowledge systems in economics education curriculum. *Interdisciplinary Journal of Education Research, 7*(1), a04–a04. DOI: 10.38140/ijer-2025.vol7.1.04

Morcom, L., & Freeman, K. (2018). Niinwi-Kiinwa-Kiinwi: Building non-Indigenous allies in education through Indigenous pedagogy. *Canadian Journal of Education/Revue canadienne de l'éducation, 41*(3), 808-833.

Mosimege, M. (2005). National priorities in indigenous knowledge systems: Implications for research and curriculum development. *Indilinga*, *4*(1), 31–37.

Mosimege, M. (2020). The use of indigenous games in the teaching and learning of Mathematics. *Revemop*, *2*, e202009–e202009. DOI: 10.33532/revemop.e202009

Mudaly, R., & Ismail, R. (2013). Teacher learning through tapping into indigenous knowledge systems in the science classroom. *Alternation (Durban)*, *20*(1), 178–202.

Munn, Z., Peters, M. D., Stern, C., Tufanaru, C., McArthur, A., & Aromataris, E. (2018). Systematic review or scoping review? Guidance for authors when choosing between a systematic or scoping review approach. *BMC Medical Research Methodology*, *18*(1), 1–7. DOI: 10.1186/s12874-018-0611-x PMID: 30453902

Naicker, S. M. (1999). *Curriculum 2005: A space for all: An introduction to inclusive education*. Renaissance.

Naidoo, J. (2021). Integrating indigenous knowledge and culturally based activities in South African mathematics classrooms. *African Journal of Teacher Education*, *10*(2), 17–36. DOI: 10.21083/ajote.v10i2.6686

Nakata, M. (2007). *Disciplining the Savages: Savaging the Disciplines*. Aboriginal Studies Press.

Ndlovu-Gatsheni, S. J. (2018). *Epistemic freedom in Africa: Deprovincialization and decolonization*. Routledge. DOI: 10.4324/9780429492204

Nel, P. (2006). Indigenous knowledge systems, local community and community in the making. *Indilinga*, *5*(2), 99–107.

Nevhudoli, N. D., & Netshandama, V. O. (2023). What do Bachelor of Indigenous Knowledge Systems graduates say about their curriculum? A qualitative tracer study at the University of Venda. *Journal of Curriculum Studies Research*, *5*(1), 141–158. DOI: 10.46303/jcsr.2023.11

Nkopodi, N., & Mosimege, M. (2009). Incorporating the indigenous game of morabaraba in the learning of mathematics. *South African Journal of Education*, *29*(3), 377–392. DOI: 10.15700/saje.v29n3a273

Nxumalo, S. A., & Mncube, D. W. (2018). Using indigenous games and knowledge to decolonise the school curriculum: Ubuntu perspectives. *Perspectives in Education*, *36*(2), 103–118. DOI: 10.18820/2519593X/pie.v36i2.9

O'Donoghue, R. (2004). Marrying indigenous and western science through environmental education. In School science in Africa: Learning to teach, teaching to learn (pp. 180-190). Lansdowne: Juta Gariep (Pty) Ltd

Odhiambo, T., Kamp, J. R., & Kamp, R. (1990). You cannot fix indigenous knowledge. *Informationcentre for Low External Input Agriculture Newsletter*, •••, 11.

Odora Hoppers, C. A. (2002). Indigenous knowledge and the integration of knowledge systems. In Odora Hoppers, C. A. (Ed.), *Indigenous knowledge and the integration of knowledge systems: Towards a philosophy of articulation* (pp. 2–23). New Africa Books.

Odora Hoppers, C. A. (2015). Think Piece: Cognitive Justice and Integration without Duress. The Future of Development Education—Perspectives from the South. *International Journal of Development Education and Global Learning*, *7*(2), 89–106. DOI: 10.18546/IJDEGL.07.2.08

Ogegbo, A. A., & Ramnarain, U. (2024). A systematic review of pedagogical practices for integrating indigenous knowledge systems in science teaching. *African Journal of Research in Mathematics. Science and Technology Education*, *28*(3), 343–361.

Ogunniyi, M. B. (2007). Teachers' stances and practical arguments regarding a science– Indigenous Knowledge curriculum: Part 1. *International Journal of Science Education*, *29*(10), 1–19. DOI: 10.1080/09500690600931038

Ogunniyi, M. B. (2009). Implementing a science indigenous knowledge curriculum. The Western Cape experience. *Second national workshop on Science and Indigenous Knowledge System*, 21-27. University of the Western Cape. South Africa.

Ogunniyi, M. B., & Hewson, M. G. (2008). Effect of an argumentation-based course on teachers' disposition towards a science-indigenous knowledge curriculum. *International Journal of Environmental and Science Education*, *3*(4), 159–177.

Onwu, G., & Mosimege, M. (2004). Indigenous knowledge systems and science and technology education: A dialogue. *African Journal of Research in MST Education*, *8*(1), 1–12. DOI: 10.1080/10288457.2004.10740556

Owusu-Mensah, J., & Baffour, K. P. Q. (2015). Demystifying the myth of mathematics learning at the foundation phase: The role of Akan indigenous games (challenges and opportunities for indigenous knowledge systems (IKS) in the education system). *International Journal of Educational Sciences*, *8*(2), 313–318. DOI: 10.1080/09751122.2015.11890253

Pham, M. T., Rajić, A., Greig, J. D., Sargeant, J. M., Papadopoulos, A., & McEwen, S. A. (2014). A scoping review of scoping reviews: Advancing the approach and enhancing the consistency. *Research Synthesis Methods*, *5*(4), 371–385. DOI: 10.1002/jrsm.1123 PMID: 26052958

Posch, P. (2000). Community, school change and strategic networking. In Altrichter, H., & Elliott, J. (Eds.), *Images of educational change* (pp. 55–65). Open University Press.

Ragoonaden, K., & Mueller, L. (2017). Culturally Responsive Pedagogy: Indigenizing Curriculum. *Canadian Journal of Higher Education*, *47*(2), 22–46. DOI: 10.47678/cjhe.v47i2.187963

Regmi, J., & Fleming, M. (2012). Indigenous knowledge and science in a globalized age. *Cultural Studies of Science Education*, *7*(2), 479–484. DOI: 10.1007/s11422-012-9389-z

Ronoh, J. C. (2018). *Indigenous Knowledge in the School Curriculum: Teacher Educator Perceptions of Place and Position* (Doctoral dissertation, Nelson Mandela University).

Sonkqayi, G. (2021). The State of South African Education: A Critical Exploration of Distributive and Epistemic Injustices. *Think (London, England)*, *88*(3), 30–40.

Taylor, D. L., & Cameron, A. (2016). Valuing IKS in successive South African physical sciences curricula. *African Journal of Research in Mathematics. Science and Technology Education*, *20*(2), 35–44.

Van Wyk, B., & Higgs, P. (2011). The curriculum in an African context. *Indilinga, 10*(2), 171–181.

Van Wyk, B.-E., & Gericke, N. (2000). *People's plants.* Briza Publications.

Yusuf, M. W., Saidi, I., & Halliru, A. (2010). Ethnomathematics (A Mathematical game in the Hausa culture). *International Journal of Mathematical Science Education, 3*(1), 36–42.

ADDITIONAL READING

Aikman, S., & King, L. (2012). Indigenous knowledges and education. *Compare: A Journal of Comparative Education, 42*(5), 673–681. DOI: 10.1080/03057925.2012.706450

Barnhardt, N. R. (2014). Creating a place for indigenous knowledge in education: The Alaska native knowledge network. In *Place-based education in the global age* (pp. 137–158). Routledge.

Battiste, M. (2009). Naturalizing Indigenous knowledge in Eurocentric education. *Canadian Journal of Native Education, 32*(1).

Jacob, M. M., Sabzalian, L., Jansen, J., Tobin, T. J., Vincent, C. G., & LaChance, K. M. (2018). The gift of education: How Indigenous knowledges can transform the future of public education. *International Journal of Multicultural Education, 20*(1), 157–185. DOI: 10.18251/ijme.v20i1.1534

Owuor, J. (2007). Integrating African indigenous knowledge in Kenya's formal education system: The potential for sustainable development. *Journal of Contemporary Issues in Education, 2*(2). Advance online publication. DOI: 10.20355/C5Z594

Zidny, R., Sjöström, J., & Eilks, I. (2020). A multi-perspective reflection on how indigenous knowledge and related ideas can improve science education for sustainability. *Science & Education, 29*(1), 145–185. DOI: 10.1007/s11191-019-00100-x

KEY TERMS AND DEFINITIONS

Culturally responsive education: Creates inclusive learning settings that encourage participation, motivation, and academic success by recognizing and valuing students' many cultures, languages, and life experiences.

Curriculum: A curriculum is an interactive system of teaching and learning that has certain objectives, materials, tactics, metrics, and resources.

Department of Basic Education: In South Africa, the Department of Basic Education (DBE) oversees managing and supervising education from Grade R through Grade 12, as well as creating, preserving, and assisting the educational system in schools and guaranteeing adherence to national education regulations.

Indigenous Knowledge Systems: Indigenous knowledge Systems (IKS) are the knowledge, skills, and customs that indigenous groups have created and preserved over many generations.

Integration: Integration in education is the process of fusing various fields of study or disciplines into a single lesson or curriculum, highlighting linkages and bringing ideas together to produce a more comprehensive and significant educational experience.

Chapter 4
Traditional Food Systems as a Tool for Achieving Food Sovereignty in South Africa

Diane Thelma Molokwa
https://orcid.org/0000-0002-7352-6316
North-West University, South Africa

Mamokete Venolia Setshego
https://orcid.org/0000-0002-1925-5949
North-West University, South Africa

ABSTRACT

With a national food system characterized as being variety deficient, and a heavy reliance on commercially available food options, South Africa (SA) faces issues of food insecurity at a household level, despite being considered food secure at a national level. This chapter investigates the challenges attributed in national food systems, in spite of their promising status in literature, these food systems are marked with lack of dietary diversity, inadequate access to food, and nutrition insecurities, specifically, at a household level. The study aims to explore the potential of traditional food systems in ensuring food and nutrition security, and to examine their role in achieving food sovereignty for rural and indigenous communities in South Africa. A scoping literature review was applied to investigate how the traditional food systems in indigenous communities have so far managed to offer household food security, and how said food systems, with adequate policy and social efforts, could be applied on a broader scale, with the intention of achieving national food sovereignty in South Africa.

INTRODUCTION

Traditional food systems are collaborative that emphasize integrating sustainable food production, processing, distribution and consumption of food systems, containing variety of items from local and the natural environment that are culturally friendly (Duncan et al., 2021; Gutierrez et al., 2022; Feenstra & Campbell, 2014; Swiderska et al., 2022). In addition, Indigenous Climate Hub (2024) perceives traditional food systems as a framework that links people, land, culture and spirit. These food systems are not only a

DOI: 10.4018/979-8-3373-4124-8.ch004

complex network of relationships, but advocate for the use of native species, such as plants, insects and animals, including traditional farming techniques, indigenous food processing and preservation methods, localized food production and distribution (Indigenous Climate Hub, 2024).

Batal et al. (2021) on the other hand, identified that traditional food systems provide foods rich in nutritional value, which significantly improve minerals intake, such as iron, proteins and magnesium, when included in an individual's diet. Despite nutritional, cultural, spiritual and health value, traditional food systems are perceived by Swiderska et al. (2022) as an important component in maintaining environmental sustainability and mitigating climate change.

However, the benefits of traditional food systems are not acknowledged and included, particularly within national food system planning and agricultural policies, due to prevailing negative perceptions. As noted by Swiderska et al. (2022), these perceptions often depict traditional food systems as backward and unproductive. Furthermore, this negativity and related challenges led to a surge in household hunger across millennia (Stats SA, 2011). Stats SA (2021) reported further that about 2,1 million household, representing 11,6% of households in South Africa are perceived as food and nutrition insecure.

REASEARCH SCOPE

This chapter grants an essential structure for utilizing traditional food systems as a means to attain food sovereignty in South Africa, through a comprehensive review of scholarly articles, books, journals, food policies, and programs, dissertations, and theses. This chapter underscores the significance and necessity of integrating traditional food systems into South Africa's food system and policy frameworks. Furthermore, it proposes the development of policies and regulatory frameworks to facilitate the realization of food sovereignty and further highlights the importance disseminating indigenous knowledge of traditional food systems to ensure their effective incorporation into the national food system.

RESEARCH OBJECTIVES AND QUESTIONS

The chapter's objectives are as follows:

- To investigate the role of traditional food systems in achieving household food security.
- To explore how traditional food systems operate in local communities.
- To investigate the challenges associated with commercially available food options.
- To discuss how traditional food systems contribute to food sovereignty in local and indigenous communities.

This chapter's research questions are as follows:

- What role does traditional food systems play in local communities?
- How do local communities balance the use of indigenous foods in traditional food systems and commercially available (exotic) food options?
- What challenges are encountered with the use of traditional food systems?
- How can traditional food systems ensure food sovereignty?

TRADITIONAL FOOD SYSTEM S FOR FOOD SOVEREIGNTY IN SOUTH AFRICA

Despite their nutritional value, traditional food systems carry a socio-cultural value and have been found to contribute to the cultural, physical and spiritual wellbeing of locals (Batal et al., 2021; Kennedy et al., 2021; Gutierrez et al., 2022; Yung & Neathway, 2019). Kesa et al. (2023) supplement that traditional food systems might reduce diet-related diseases, since these food systems are easily adaptive to local climatic conditions, and are crop and pest resistant, therefore, supporting human health and maintaining a balance with nature, thus contributing to sustainable food production and consumption.

The indigenous knowledge that informs traditional food systems entails customs, historical and cultural elements, related to their utilization (Duncan et al., 2021; Gutierrez et al., 2022). Through these knowledge systems, communities can strengthen their food security and build resilience towards unforeseen stressors, by creating adaptive systems and techniques, using creative harvesting, processing, preservation technologies and resource management strategies (Singh et al., 2022). Food sovereignty as a concept is concerned with who the producers of food are, how and where it was produced, and who holds the power to make these decisions.

Moreover, instead of catering to the needs of corporations and markets, food sovereignty places the people who grow, distribute and consume food at the centre of food systems and policies (Jernigan et al, 2021). Herein lies the link between traditional food systems and food sovereignty, as both concepts are community centric, and they place emphasis on socio-economic growth, environmentally friendly methods, and all-around sustainability. The vision and intervention of food sovereignty is guided by its positioning as an alternative principle of food security, rooted in the democratic reconstruction of domestic agricultures where feasible, to counteract processes of increasing food dependency and systematically shifting indigenous communities from their traditional food systems to more commercial food systems, caused by corporate marketing (McMichael, 2014).

Food insecurity is a world crisis, and in Africa it is estimated that 868 million people experienced moderate to severe food insecurity in 2022, and indigenous peoples were said to be the most affected. Indigenous communities experience high rates of food insecurity as a by-product of settler-colonial activities, which includes forced relocation to rural reservation lands and degradation of traditional subsistence patterns (Jernigan et al., 2021). Africa is not making progress in achieving the SDGs goal, aimed to be achieved by 2030 (Food and Agricultural Organization of the United Nations, 2023). McMichael and Schneider (2011) argue further that the ability to feed the world through the illusion of food security via free trade, has lost legitimacy in the recent and continuing global food crisis, hence the necessity of revisiting the implementation of food sovereignty.

Food sovereignty recognizes the contribution of indigenous peoples, pastoralists, forest dwellers, workers and fishers to the food system, and ensures that food is produced in a culturally acceptable manner and in harmony with the ecosystem in which it is produced. This is how traditional food production systems have regenerated their soils, water, biodiversity and climatic conditions for generations (Food and Agriculture Organization of the United Nation, 2014; Siwior, 2021). Although the concept of food sovereignty is good, especially for indigenous people, the lack of resources, such as land, however, may constrain its implementation.

For instance, hunting is one of the practices in traditional food systems that indigenous people have used, and unfortunately in South Africa, hunter-gatherers now require a permit and licence to hunt, which restrict them from acquiring ''culturally accepted'' food. This study proposes that the inclusion of traditional food systems into the national food system, food programs and food policies, has the potential to achieve food sovereignty in South Africa, allowing vulnerable communities in the country an opportunity to achieve socio-economic growth and food security, simultaneously utilizing environmentally friendly practices that protect the environment and promote sustainable development at both household and national level in South Africa (McMichael & Schneider, 2011).

THEORETICAL FRAMEWORK

The Theory of Change (ToC) framework assists individuals in need of change to identify challenges (United Nations Development Group, 2022). In actual sense, this framework provides goals, preconditions, requirements, assumptions and interventions needed in a process of change (Belcher &Claus, 2020). This framework gives main actors, who are involved in a process of change, an opportunity to identify their actions in a process, and to further specify the reason for such change. This approach advocates for a structural change in any system. This approach, in fact, allows all stakeholders to partake in resolving root structural problems arising amongst communities, relationships, governments and organizations (United Nations Development Group, 2022; Denish et al., 2021).

According to Denish et al. (2021), this framework can be employed to transform food systems, with its focus on improving prioritizing and improving the pace and learning the potential of new interventions. In the context of this chapter, the ToC framework has been employed to facilitate a transformative shift on food sovereignty amongst indigenous and local communities in South Africa. Authors in this chapter are clear in their support for the implementation of this approach, which allows communities to regain their rights over their food system by enabling them to produce, distribute, define, and govern their own food systems in a manner that is culturally appropriate, sustainable, and equitable.

RESEARCH METHODOLOGY

This chapter employs scoping literature review methodology to identify gaps and mergers evidence on traditional food systems and their role in achieving food sovereignty in South Africa. Through a systematic and explicit studies search process, relevant studies meeting pre-specified criteria were identified to determine the necessity of literature review (Tricco et al., 2016; Munn et al., 2018). Moreover, the scoping review allowed for flexibility in incorporating a diverse ideal literature and studies using various methodologies, enabling the chapter to focus on relevant findings.

As alluded by Peterson et al. (2016) and Kazi et al. (2021), this research approach provides a comprehensive overview that informs the research initiative, facilitating the development of more precise and informed research questions and methodologies. Furthermore, this chapter highlights research gaps and purposes for future research directions for integrating traditional food systems into national food systems. In this chapter, the scoping literature review involved the following:

- Defining the scope of the review
- Identifying phenomenon of interest
- Conducting a thorough search to find relevant literature utilizing applicable databases, including libraries, digital scholarly engines, keywords, and inclusion and exclusion criteria.

- Identified articles, book chapters, journals, scholarly articles, food and agricultural policy documents, dissertations and theses were then screened.
- • To extract important information such relevant perspectives, primary findings, and developing themes, data extraction was done.
- Keywords were put into the databases and search engines of the library. Key phrases related to this study were "Food Sovereignty," "Traditional Food Systems," "Indigenous Knowledge Systems," "food policy," "Food Security," and "Environmentally friendly".

DISCUSSIONS

This section of the study provides discussions on traditional food systems as a tool for achieving food sovereignty in South Africa

TRADITIONAL FOOD SYSTEMS STRUCTURE

Nyuyen (2023) and FAO (2021) view food systems structure as diverse activities, which normally involve production, marketing, processing, consumption and disposal of goods, originating from forestry or fisheries, agriculture, including the inputs and outputs generated from all the liable activities. Rahayu et al. (2023) posits that traditional food systems are deeply entrenched in the identity, values and heritage of indigenous and local communities, comprising unique culinary practices, agricultural methods and dietary habits that been developed, shared and preserved over generations. In addition, Morris (2008) and Gutierrez et al. (2022) believe that the structure of traditional food systems is made-up of natural, environmental and societal elements, which deeply on the existence of indigenous knowledge systems.

The main elements of these food systems include water, air, land and soil, which are interconnected and predominantly maintained through local environment and cultural adaptations. Throughout centuries communities built up an understanding of local resources, traditional methods of crop growing, wild harvesting, hunting, preservation, and preparing the foods, and using these skills and knowledge to build health-promoting activities and to strengthen the local food system (McCune, Nuvayestewa & Kuhnlein, 2019; Gutierrez et al., 2022; Jonathan et al., 2022; Neufeld & Richmond, 2017; Marrero & Mattei, 2022).

It is further supported in various literature that the relationship that local communities build with their local biodiversity, encourages and strengthens management of the local environment (Neufeld & Richmond, 2017; Pengue et al., 2018; Nyuyen, 2018; Morrison, 2008; Johns et al., 2013). In fact, traditional food systems are based on sharing of indigenous knowledge, associated with local resources, as well as indigenous practices that are not exploitative and do not degrade and harm the ecosystem (McCune et al., 2019; Rua, 2024).

Traditional food systems, as alluded by Pengue et al. (2018), emphasize relationship between indigenous communities and their food system. These communities further engage in cultural events and activities that are tied to their traditions and were developed through long relationships with their local environment. Although indigenous communities still possess enormous traditional knowledge associated with traditional food system, the authors argue that these systems are still marginalized in food and agricultural policies, which further led to the threat on traditional food systems. Gutierrez et al. (2022)

endorse further that the use of traditional food systems is losing relevance, due to continuous deterioration of knowledge and local resources associated with these food systems.

INDIGENOUS FOOD SOVEREIGNTY PRINCIPLES

Traditional food systems allow rural communities to define their own food system and allow them to produce and distribute while consuming food in ways that are culturally and ecologically sound (Serrato, 2024). In recent years, organizations such as FAO, Native American Food Sovereignty, Intertribal Agricultural Council, as well as United State Food Sovereignty Alliance have been advocating for the importance of traditional food systems, and the ecosystems that provide them. Consequently, the use of this system can improve both health and wellbeing of local communities (FAO, 2024).

Based on the discussions provided in this review chapter, we affirm that traditional food systems share a sequence of principles with food sovereignty among which the importance and respect for local biodiversity stand out. This principle is essential, as it acknowledges the relationship that local communities have with their local environment. Marrero and Mattei (2022) stated that this unique relationship is emphasized on how communities learned to harvest and hunt without exploiting the ecosystem; hence these food systems are deemed inseparable with the land, air, water, animals and the environment. Another essential principle is the emphasis on culturally significance food sources. These comprise a diverse trace of edible plants, animals and insects that possess profound cultural and nutritional value (McCune, Nuvayestewa & Kuhnlein, 2019; Gutierrez et al., 2022; Jonathan et al., 2022; Neufeld & Richmond, 2017; Marrero & Mattei, 2022).

The third principle underlying traditional food systems is the emphasis on localized food systems, in which food is produced, harvested within local communities, promoting and emphasizing direct and positive relationship between local environment, and the land. The fourth principle emphasizes the extensive ecological knowledge possessed by local and indigenous communities. These communities have developed a comprehensive range of skills and practices related to food production, processing and preparation. Specifically, these skills encompass cultivation, planting, harvesting, preservation and culinary preparation techniques, which have been refined over generations through traditional knowledge and cultural transmission. Aligning these two frameworks will leverage the cultural significance and sustainability of traditional food systems, thereby promoting equitable and just food systems. These systems prioritize the rights and interests of indigenous and local communities, enabling them to exercise self-determination, control their food production, distribution, processing and consumption, and further promote the community's wellbeing (Science Forum South Africa, 2023).

POTENTIAL OF TRADITIONAL FOOD SYSTEMS TO FULFILL SDG2

Sustainable Development Goal 2 aims to end hunger, achieve food security, improve nutrition and promote sustainable agriculture (The Sustainable Development Goal Report, 2024). As alluded to by the United Nations Educational, Scientific and Cultural Organization (2016), the initiative planned by SGD 2 put emphasis on ensuring equal access to safe, nutritious and sufficient food for all. This encompasses promoting sustainable agricultural practices, increase in food production and productivity, reducing post-harvest losses and food waste, supporting small-scale farmers, and promoting equitable access to natural resources. Moreover, the initiative aims to address and resolve root causes of hunger, including poverty, inequality and related conflicts. The following discussion explores the potential of traditional food systems to contribute to the attainment of Sustainable Development Goal 2 (SDG 2):

Ending hunger

According to the Food and Agriculture Organization (2024), hunger is an uncomfortable or painful physical sensation caused by insufficient consumption of dietary energy. Unemployment, poverty and the energy crisis in South Africa are the main causes of hunger in South Africa (Stats SA, 2023). Stats SA (2024) suggests that 23.1% of households in South Africa experienced inadequate access to food in 2023, predominantly due to financial constraints within the national food systems. In essence, the monetary-driven nature of the national food system poses a substantial barrier to sufficient food access, as it prioritizes financial gains over equitable food distribution (Stats SA, 2024).

This chapter highlights that traditional food systems exhibit distinct operational characteristics that contrast national food systems paradigms. Notably, traditional food systems regularly utilize alternative strategies for food procurement, including local hunting, subsistence farming, fishing, and wild food harvesting. These practices are not only culturally acceptable, but also largely independent from monetary transactions, thereby providing a unique framework for achieving food security (Jonathan, 2022; Kom & Nethengwe, 2024).

A key aspect of these food systems is the incorporation of sustainable land management strategies, which not only regulate their right to the land use, territories and resources, but also ensure long-term viability of food system. By employing traditional and indigenous knowledge and community-based land management practices, these systems can contribute significantly to improving food security, improving nutrition, and promoting sustainable livelihoods (International Fund for Agricultural Development, 2022; Kom & Nethengwe, 2024).

It is pointed out by the International Fund for Agricultural Development (2022) that the resilience of traditional food systems can be endorsed to their fundamental nature, food diversity and sociocultural values and norms. These food systems often possess values of caring, reciprocity, social interactions and compassion, as shown by seeds and food sharing practices (International Fund for Agricultural Development, 2022).

Nutritional benefits

Recently, the cost of food has changed in the way that a healthy diet is five times more expensive than the diet that meets dietary energy requirement (FAO et al., 2022; Kenter et al., 2023). Most of South Africans consider price above all other aspects when purchasing food items in store (Ronquest-Ross et al., 2015). Nutritional value is among other unconsidered aspects when purchasing food, because healthier food is costly, and majority of South Africans cannot afford those (Ronquest-Ross et al., 2015). For instance, a study by Simelane (2024) indicates that over 50% of South Africans are classified as overweight or obese, due to malnutrition.

Additionally, the study suggests that urgent action is necessary to improve the nutritional intake of the communities, as approximately 23% of citizens are borderline malnourished. Furthermore, 18% of the population consumes an unacceptable diversity of food, leading to poor diets. To improve healthy and nutritious diets, a comprehensive approach is necessary, which encompass different aspects of food systems, including production, distribution, procurement, marketing, selection, preparation and consumption (Tartanac, 2021). This chapter stresses the significance of the traditional food systems in ensuring a comprehensive approach to nutritional intake amongst households in South Africa.

This is due to the unique knowledge and skills developed by indigenous and local communities in food production, selection, distribution, harvesting and processing, which enable efficient and sustainable food systems (Siwallace & Elder, 2009). However, Siwallance and Elder (2009) assert that the erosion of these traditional practices has compromised dietary health, leading to a continuous decline in nutritional security. Research by Kenter et al. (2023) emphasizes further that there isa potential of traditional food systems to provide optimal human nutrition, while maintaining an environmental balance.

Promoting sustainable agriculture

Agriculture is significant in the process of economic development and in achieving food security (Statistics SA, 2024). As viewed by Stats SA (2024) 11,7% of households in South Africa use agriculture as their main source of food and majority (75.5%) use it as an additional source of food; this is an indication that the main source of food for majority of South African citizens is the market. According to Von Braun et al. (2023), agricultural activities contribute to climate change, owing to extensive farming practices that cause soil degradation, pollute and deplete water supplies, and lower biodiversity. Traditional food systems, on the other hand, utilize methods that are sustainable, because they tend to use low external inputs intending to obtain food stability rather than increase in production (Pengue et al., 2018). The focus of traditional food systems is locally based, and consequently, the scale of production, processing, and trade within this system is influenced by knowledge of the environment (Pengue et al., 2018).

Simelane (2024) recommend the promotion of domestic food production, which does not only promote food security on a household level, but can promote sustainable agriculture, while encouraging the management of resources sinceindigenous people believe that they are the stewards (Kuhnlein& Receveur, 1996; Pengue et al., 2018). Kanter et al. (2023), which further explains that production and utilization of underutilized species can lead to sustainable agriculture and improve food security. Moreover, traditional food thrives with little care and without the use of external inputs and is invariably more resistant to pests and diseases, drought and other unfavourable growing conditions (Aworh, 2023). The preceding report indicates that the holistic nature of traditional food systems not only supports agricultural sustainability, but also fosters food sovereignty, because people get a chance to choose the food to cultivate.

THE DISPARITIES BETWEEN THE NATIONAL FOOD SYSTEMS AND TRADITIONAL FOOD SYSTEMS IN SOUTH AFRICA

Continuous transformation of food systems in many African countries, including South Africa has been profound since the commencement of democracy (FAO et al., 2022). However, the disparities between national and traditional food systems go beyond the democratic system. Ronquest-Ross et al. (2015) suggest that there are other complex sets of factors influencing disparities between these two food systems, such as food consumption patterns, which include education, demographic characteristics, geographical location, seasonal variations, disposal income, mainstream science policymakers, government policies, and other support services.

Furthermore, globalization, marketing, religion, culture, ethnicity, social networks, urbanization and consumer preferences may also affect traditional food consumption behaviors (FAO, 2014). Notably, any food system relies on interconnected activities, comprising production, aggregation, processing, distribution and consumption. These interlinked components equally control the structural outcome of food systems, emphasizing the need for an inclusive understanding of the multifaceted dynamics shaping food systems in South Africa (FAO, 2014).

Food production

The national food system in South Africa relies heavily on the intensive agricultural practices to produce food and is highly supported and included in agricultural and food policies. These practices are characterized by high external-inputs, such as genetically modified crop seeds, synthetic fertilizers, pesticides and fuel-based chemical inputs (Pengue et al., 2018). Nonetheless, even when these practices actively result in high productivity, they might also have harmful effects on individuals' health (Pengue et al., 2018; Kumar &Keshar, 2017).

In contrast, traditional food production systems emphasize local, sustainable, and cultivation of naturally acquiring foods (Sreedharan et al., 2023). These systems, according to Malli et al. (2023), encompass agricultural activities, such as hunting, fishing and foraging, which require regular access to the natural environment and biodiversity. In many African countries, including South Africa, there are laws and policies restricting communities' access to natural resources, creating barriers to consuming healthy and culturally acceptable foods. This chapter highlights the absence of meaningful collaboration and representation between national and traditional food systems in agricultural policies, precisely those related to food production, promotion and consumption.

Food access

National food systems offer high yield, which ensures food security and consumption (Pengue et al., 2018), but this system is predominantly influenced by financial considerations to ensure operational competence. As a result, these systems are often geared towards international markets where the access to food lies upon purchasing power (Pengue et al., 2018). Subsequently, households require a stable financial income to access food through market channels. This practice creates significant barriers for low-income and middle-class households, who often struggle with financial instability, thereby compromising their ability to have adequate access to foods.

Despite these challenges, the existing food system remains the primary source of food access, due to its stable marketing and supportive policies (Pengue et al., 2018). Access to traditional foods, on the other hand, faces distinct challenges. Climate change, relocation to urban areas, restricted access to hunting and food gathering areas, and declining natural resources hinder communities' ability to access their traditional food sources (Swiderska et al., 2022). Despite these challenges, traditional food systems remain essential for the food security and cultural heritage for many South African communities, who are still excluded in marketing and lack of supportive policies (Elliott et al., 2012). To address challenges faced by traditional food systems, Rathasinghe et al. (2019) cited in Kennedy et al. (2022) recommend that there should be more research on these systems, public awareness, as well as financial and political support to ensure that communities have access to this food system.

Food processing and preservation

National food systems utilize modern processing and preservation techniques, such as pasteurization, refrigerator, drying, vacuum packaging, irradiation, pascalization, bio-preservation, hurdle technology and modified atmosphere packaging (Kumar, 2019). However, these methods require significant financial security, as they often rely on store-bought equipment and require electricity, which can be costly and inaccessible to resource-constrained communities (Pengue, (2018). Furthermore, Pengue (2018) posits that numerous processed and packaged foods are often widely available, affordable and convenient to consume, but are ultimately harmful to health.

On the other hand, traditional food systems employ uncomplicated processing and preservation techniques, including sun-drying, crushing grains, salting, sugaring, underground pit storage, as well as sack storage and smoking (Pengue, 2018; Kumar, 2019). These methods rely immensely on indigenous knowledge, which has been passed down orally through generations. However, the erosion and potential extinction of indigenous knowledge pose a significant threat to the continued use of these traditional methods, due to lack of recognition and support from agriculture and food policies.

Food and agricultural policies

Agricultural and food policies are designed by historical inequities, further embedded in colonialism and exacerbated by apartheid policies and the country's isolation. In essence, these policies integrate political strategies into the planning of the country's food systems, thereby marginalising alternative food systems (FAO, 2014). This study emphasises that agricultural policies overlook and exclude traditional food systems from their agricultural planning and programs, thereby threatening their existence. The lack of guidelines and regulations governing the production, marketing, distribution and consumption of traditional foods, creates significant challenges for their sustainability. Additionally, Feenstra and Campbell (2014) stress the importance of a bottom-up approach, local initiatives and actions that can ensure a safer space for both traditional and national food systems to co-exist within the same policy framework.

Food consumption and diet

According to Ronquest-Ross and Sigge (2024) and Abdul et al. (2024), the national food systems' diet has undergone a drastic shift, transitioning from a healthy diet rich in nutrients to a diet predominantly identified by sugar-sweetened beverages, processed and packaged foods, and products with added calories and sweeteners. Moreover, this drastic shift has led to a decline in vegetable consumption, resulting in unhealthy eating behaviors. Despite the ill health associated with the national food systems, popularity and demand for these foods are still at its peak. Notwithstanding their potential to improve nutritional quality and promote healthy eating behaviors, traditional food systems often lack recognition and support from agricultural policies and political frameworks that shape healthy food consumption and diets (Abdul et al., 2024).

INCORPORATE TRADITIONAL FOOD SYSTEMS INTO SOUTH AFRICAN NATIONAL FOOD SYSTEM

There is an urgent necessity for an inclusive and transformative attempt to South Africa's national food system; one that prioritizes every individual (Science Forum South Africa, 2023; Ekwamu, 2021; Edwards et al., 2024). The development and adoption of collaborative and driven food systems are imperative for driving economic growth, reducing poverty, malnutrition, inequalities, unemployment and ensuring just and equitable food and nutrition security for all (Ekwamu, 2024).

This initiative is strengthened by the United Food System Summit 2021, which launched new actions aimed at delivering progress on all 17 Sustainable Development Goals (SGDs). The main key themes from the summit are to ensure healthier, more sustainable and equitable food systems. Furthermore, the said summit emphasised the need for collaborative action among all stakeholders to transform the way the world produces, distributes and consume their food (Ekwanu, 2024). In this chapter, authors provide recommendations on the elements of traditional food systems that, when supported and incorporated with national food systems, can support and improve food sovereignty among communities.

Promotion of traditional food systems into National Food Policy

This chapter asserts that for traditional food systems to be fully and effectively integrated into national food systems there isa need for targeted promotion and support of local indigenous food crops. Tarcza and Nemteanu (2015) remark that incorporation of traditional food into the national food system can actively address challenges related to dietary diversity, hunger and nutrition. Moreover, this integration will also promote the cultural heritage of the nation, as traditional foods are deeply embedded in the customs, rituals and histories of local communities (Ghosh et al., 2024; FAO, 2024).

To further promote traditional food systems, collaboration with national food policymakers and other relevant stakeholders, is essential. This can be achieved by developing community-supported agriculture programs where communities purchase shares at local farms, promoting direct sales and safe community engagements. Establishing local famers' market and traditional food festivals can be beneficial, allowing communities to celebrate their TFS, promote local cuisine and foster cultural exchange. This initiative can facilitate the successful incorporation of TFS into national food systems (Kuhnlein & Chotiriboon, 2022). In essence, this initiative is maintained by several pillars underpinning the national food policy, including the establishment of enhanced nutritional safety nets, alignment of agricultural investments with local economic development objectives, particularly in rural areas, and improved market access for the emerging agricultural sector, such as a smallholder farmer, through a government food purchase programme (Department of Agriculture, Forestry and Fisheries, 2014).

For instance, a study by Mbhebyane and Tambe (2024) examined the influence of community food environments on food insecurity in Limpopo Province, revealing that 91.8% of households engaged in agriculture and cultivated food for household consumption. This finding is further supported by Masekoameng and Malotja (2019), who noted that food availability in rural communities depends on land allocation, access to the wild, and the ability to utilize backyard gardens for food production, among other factors. In this instance, Mbhebyane and Tambe (2024), suggest that to integrate traditional food systems into national food systems, local household food producers could be targeted for implementation through subsistence farming programs or sustainable livelihood approaches.

In another study by Masekoameng and Molotja (2019) the role of indigenous foods and indigenous knowledge systems for rural households' food security in Sekhukhune District, revealed that traditional food plays a significant role in maintaining food security in the region. The research found that the natural environment surrounding villages provides a source of edible indigenous vegetables and fruits, which are either gathered in the wild or sold by local collectors. The study recommends that stakeholders, including food producers and policymakers, integrate traditional food and related practices into the national food industry and markets. Government support for the cultivation and promotion of neglected indigenous foods would not only enhance nutritional consumption among communities but also celebrate the diversity of traditional foods that reflect the country's rich cultural heritage, including Nguni, Basotho, VhaVenda and other cultures.

Indigenous food processing and preservation methods

Effective use of processing and preservation methods are critical in any food system to safeguard safety, an extended shelf-life, food availability and accessibility, and retention of nutritional value and food quality (Heaney et al., 2024). However, indigenous food processing and preservation methods are limited by lack of scientific evidence supporting their safety and efficacy. Incorporation of these methods into modern food processing techniques could potentially enhance their effectiveness (Heaney et al., 2024). Promoting and scientifically testing these methods could facilitate their adoption at a commercial scale, ultimately contributing to their sustainability and wider acceptance. Heaney et al. (2024) suggested further that adopting and exploring modern food processing methods could benefit indigenous food preservation and processing methods.

FUTURE RESEARCH DIRECTIONS

To further discover the potential of traditional food systems as a tool for achieving food sovereignty in South Africa, future research could attempt to investigate the following areas:

Impact of climate change on Traditional Food Systems: Examining the impact of climate change on traditional food systems and resilience of indigenous and local communities could further assist in informing strategies for adapting to climate change while maintaining and enhancing local food sovereignty.

Policy and institutional support for Traditional Food Systems: Thoroughly examining the role of policy and institutional support in promoting traditional food systems and food sovereignty could identify opportunities for advocacy and policy change.

Economic viability of Traditional Food Systems: Evaluating the economic viability of traditional food systems and identifying opportunities for sustainable livelihoods could further inform strategies of supporting local food economies.

Youth engagement and intergenerational knowledge transfer: Exploring the role of youth engagement and intergenerational knowledge transfer in preserving traditional food systems and associated knowledge could inform strategies for promoting intergenerational learning and cultural continuity.

CONCLUSION

In conclusion, this chapter contributes to the increasing debates on the fundamental role of traditional food systems in safeguarding food and nutrition security within South African communities. Furthermore, the chapter argues that community-led decision-making, regarding food choices, consumption patterns and production and distribution systems can foster food sovereignty within local communities, thereby empowering communities to exercise control over their food systems. This approach can ultimately foster food sovereignty, alleviate poverty, nutrition and food insecurity, and enhance self-determination. These efforts, as suggested in this chapter, require the establishment of policies and legal framework that is accommodating of local language, local traditional legal frameworks, such as customary law, norms and values of communities. Consequently, community involvement and participation can ensure good governance in food systems, the laws and regulations that are fair and effective, ensuring inclusiveness, even in decision-making.

RECOMMENDATIONS

Establish policy and regulatory frameworks

Food systems governance encompasses the set of rules, norms and processes that shape policies and decisions affecting food systems (International Development Law Organization, 2023). Communities developed a series of combined cultural, prohibitions, laws, norms and values, as well as social practices into the management of natural resources, ensuring sustainability and continuity (Pengue et al., 2018). Therefore, integrating both national agricultural and food policies with the relevant traditional legal framework will benefit the country by achieving sustainable food system, as well as conserving biodiversity and the protection of the national heritage and culture. International Development Law Organization (2023) advocates that to transform the food system, there is a need to strengthen food systems governance through sound legal and regulatory frameworks and effective institutions.

Dissemination of indigenous knowledge of traditional food systems

This knowledge serves as the basis for decision-making within communities, playing a chief role in agriculture, resource management, health, food and nutrition security, food processing, preserving, preparation, storage, as well as related activities (Bhuthelezi et al., 2024). Mdhluli et al. (2021) assert that the digitalization of traditional knowledge systems should not be limited to academic purposes, but should also utilize media platforms, such as television, radio and social media such as Facebook, YouTube, Tik Tok, and others as they appear to be suitable and effective. A system for disseminating traditional knowledge of these food systems through media is necessary, where people, especially the younger generation, can be taught and exposed to these knowledge systems and related food systems (Mdhuli et al., 2021; Kuhnlein & Chotiboriboom, 2022).

REFERENCES

Abdul, M., Ingabire, A., Lam, C. Y. N., Bennett, B., Menzel, K., MacKenzie-Shalders, K., & Van Herwerden, L. (2024). Indigenous food sovereignty assessment—A systematic literature review. *Nutrition & Dietetics: the Journal of the Dietitians Association of Australia, 81*(1), 12–27. DOI: 10.1111/1747-0080.12813 PMID: 37154011

Al-Khusaibi, M., Al-Habsi, N., & Rahman, M. (2019). Traditional Foods History, Preparation, Processing and Safety: History, Preparation, Processing and Safety. https://doiDOI: 10.1007/978-3-030-24620-4

Aworh, O.C. (2023). African traditional foods and sustainable food security. *Journal Food Control*, 145.

Belcher, B., & Claus, R. (2020). Theory of Change. td-net toolbox profiles. Swiss Academies of Arts and Sciences: Td-Net Toolbox for Co-Producing Knowledge. https://doi.org/doi.org/10.5281/zenodo.3717451

Buthelezi, S. C., Ocholla, D., & Dlamini, P. (2024). Strategies for documenting and disseminating indigenous knowledge at a South African university. *South African Journal of Information Management, 26*(1). Advance online publication. DOI: 10.4102/sajim.v26i1.1648

Carlile, R., Kessler, M., & Garnett, T. (2021). *What is food sovereignty?* TABLE Explainer Ser. DOI: 10.56661/f07b52cc

Department Of Agriculture, Forestry and Fisheries (2014). The national policy on food and nutrition security for the republic of South Africa. Government gazette: No. 37915.

Dinesh, D., Hegger, D. L. T., Klerkx, L., Vervoort, J., Campbell, B. M., & Driessen, P. P. J. (2021). Enacting Theories of Change for Food Systems Transformation under Climate Change. *Global Food Security, 32*, 100583. DOI: 10.1016/j.gfs.2021.100583

Duncan, J., Carolan, M., & Wiskerke, J. S. C. (2021). *Routledge Handbook of Sustainable and regenerative food systems*. Routledge.

Edwards, F., Sonnino, R., & Cifuentes, M. L. (2024). Connecting the dots: Integrating Food Policies Towards Food Systems Transformation. *Environmental Science & Policy, 158*, 103810. DOI: 10.1016/j.envsci.2024.103810

Ekwamu, A. (2021). Building Inclusive National Food Systems: Considerations for the United Nations Food Systems Summit. https://www.ruforum.org. Accessed: 15 Feb 2025

Elliott, B., Jayatilaka, D., Brown, C., Varley, L., & Corbett, K. K. (2012). "Are We not Being Heard": Aboriginal Perspectives on Traditional Foods Access and Security. *Journal of Environmental and Public Health, 6*, 130945. https://onlinelibrary.wiley.com/doi.10.1155/2012130945. DOI: 10.1155/2012/130945 PMID: 23346118

Feenstra, G., & Campbell, D. C. (2014). Local and Regional Food Systems. In Thompson, B. P., & Kaplan, D. M. (Eds.), *Encyclopedia of Food and Agricultural Ethic* (pp. 1345–1352). Springer.

Food and Agricultural Organization of the United Nations (FAO). (2014). Sustainable Food Systems Concept and Framework. https://openknowledge.fao.org/server/api/core/bitstreams/b620989c-407b-4caf-a152-f790f55fec71/content. Accessed: 15 Feb 2025

Food and Agricultural Organization of the United Nations (FAO). (2023). Alarm bells ignored as Africa continues to face deepening food crisis. https://www.fao.org/newsroom/detail/alarm-bells-ignored-as-africa-continues-to-face-deepening-food-crisis/en. Accessed: 15 Feb 2025

Food and Agriculture Organization (FAO), European Union (EU), CIRAD & DSI-NRF Centre of Excellence in Food Security (CoE-FS). (2022). Food Systems Profile – South Africa. Catalysing the sustainable and inclusive transformation of food systems. Rome, Brussels, Montpellier, France and Bellville, South Africa. DOI: 10.4060/cc0071en

Food and Agriculture Organization of the United Nation (FAO). (2014). Food Sovereignty Systems. https://openknowledge.fao.org/server/api/core/bitstreams/b620989c-407b-4caf-a152-f790f55fec71/content#:~:text=Food%20systems%20(FS)%20encompass%20the,the%20broader%20economic%2C%20societal%2. Accessed: 15 Feb 2025

Food and Agriculture Organization of the United Nations. (2021). Traditional Food Systems to support healthy and culture-base diet. http://www.fao.orggiahs/news/detail-events

Food and Agriculture Organization of the United Nations (FAO). (2024). Indigenous Peoples' Systems and Well-Being Interventions and policies for healthy communities. FAO Cine. https://www.fao.org/docrep/018/i3144e/i31144e. Accessed: 15 Feb 2025

Food and Agriculture Organization of the United Nations (FAO). (2024). Rediscovering African Foods: Promoting Native Foods for Improved Nutrition, Health and Food Security. Thirty-third Session Morrocco FAO Regional Conference for Africa. https://openknowledge.fao.org/server/api/core/bitstreams/2b445818-7a24-450f-9b4b-c9e392bf909a/content. Accessed: 15 Feb 2025

Ghosh, S., Meyer-Rochow, V. B., & Jung, C. (2023). Embracing Tradition: The Vital Role of Traditional Foods in Achieving Nutrition Security. *Foods*, *12*(23), 4220. DOI: 10.3390/foods12234220 PMID: 38231593

Gutierrez, C., von Glascoe, C. S., Bertran, M., Calderon, N., Villada-Canela, M., & Leyva, J. C. (2022). The Kumiai food system: Reconnecting nature, food, and health through ancestral knowledge. *Frontiers in Sustainable Food Systems*, 6, 824264. DOI: 10.3389/fsufs.2022.824264

Hawkes, C. (2022). Centre for Food Policy at City, University of London, and Results for Development (R4D), Taking Food System Approach to Policymaking: Resource for Policymaker. https://r4d.org/wp-content/uploads/R4D. Accessed: 15 Feb 2025

Heaney, D., Padilla-Zakour, O. I., & Chen, C. (2024). Processing and Preservation Technologies to Enhance Indigenous Food Sovereignty, Nutrition Security and Health Equity in North America. *Frontiers In Nutrition*, 19,11, 1395962. https://www.frontiersin.org/journals/nutrition/articles/10.3389/fnut.2024.1395962/full. Accessed: 15 Feb 2025

Indigenous Climate Hub. (2024). Traditional Food Systems and Climate Change: Adapting Indigenous Agriculture. https://indigenousclimatehub.ca. Accessed: 15 Feb 2025

International Development Law Organization. (2023). Rule of Law for Food Systems Transformation: Policy Briefing. https://www.idlo.int/sites/default/files/2023/other/documents/for_web_-_rule_of_law_for_food_systems_transformation. Accessed: 15 Feb 2025

International Fund for Agricultural Development (IFAD). (2022) Sustainable and resilient Indigenous Peoples' Food Systems for Improved Nutrition. https://www.ifad.org/en/w/publications/sustainable-and-resilient-indigenous-peoples-food-systems-for-improved-nutrition. Accessed: 15 Feb 2025

Jernigan, B. B. V., Maudrie, T. L., Nikolaus, C. J., Benally, T., Johnson, S., Teague, T., Mayes, M., Jacob, T., & Taniguchi, T. (2021). Food Sovereignty Indicators for Indigenous Community Capacity Building and Health. *Frontiers in Sustainable Food Systems*, 5, 704750. https://doiDOI: <ALIGNMENT.qj></ ALIGNMENT>10.3389/fsufs

Johns, T., Powell, B., Maundu, P., & Eyzaguirre, P. B. (2013). Agricultural biodiversity as a link between traditional food systems and contemporary development, social integrity and ecological health. *Journal of the Science of Food and Agriculture*, *93*(14), 3433–3442. DOI: 10.1002/jsfa.6351 PMID: 23963831

Jonathan, S. D., Maracle, K., & Wild, S. (2022). Exploring indigenous foods & food sovereignty. https://ecampusontario.pressbooks.pub/ifs1/

Kanter, R., Kennedy, G., & Boza, S. (2023). Local, traditional and indigenous food systems in the 21st century to combat obesity, under-nutrition and climate change. *Frontiers in Sustainable Food Systems*, *7*, 119574. DOI: 10.3389/fsufs.2023.1195741

Kazi, M.R., Chowdhury, N., Chowdhury, M., Turin, T.C. (2021). Conducting comprehensive scoping review to systemically capture the landscape of a subject matter. *Popul. Med*, 3(36). htpps://DOI: 10.18332/popmed/143831

Kennedy, G., Kanter, S. C., Covic, N., Delormier, T., Longvah, T., Maundu, P., Omidvar, N., Vish, P., & Kuhnlein, H. (2021). Traditional and indigenous fruits and vegetables for food system transformation. *Current Developments in Nutrition*, *5*(8), nzab092. DOI: 10.1093/cdn/nzab092 PMID: 34423230

Kesa, H., Kamgain, A. D. T., & Zuma, M. K. (2023). Knowledge, perceptions and consumption of indigenous foods in Gauteng Region, South Africa. *International Journal of Environmental Research and Public Health*, *20*(20), 6961. DOI: 10.3390/ijerph20206961 PMID: 37887699

Kom, Z., & Nethengwe, N. S. (2024). Indigenous and Local Knowledge: Instruments Towards Achieving SGD2: A Review in an African Context. *Sustainability (Basel)*, *16*(20), 9137. DOI: 10.3390/su16209137

Kom, Z., Nicolau, M. D., & Nenwiini, S. C. (2024). The Use of Indigenous Knowledge Systems Practices to Enhance Food Security in Vhembe District, South Africa. *Agricultural Research*, *13*(3), 599–612. DOI: 10.1007/s40003-024-00716-8

Kuhnlein, H. V., & Chotiboriboon, S. (2022). Why and how to strengthen indigenous peoples' food systems with examples from two unique indigenous communities. *Frontiers in Sustainable Food Systems*, *6*, 808670. DOI: 10.3389/fsufs.2022.808670

Kuhnlein, H. V., & Receveur, O. (1996). Dietary change and traditional food systems of indigenous peoples. *Annual Review of Nutrition*, *16*(1), 417–442. DOI: 10.1146/annurev.nu.16.070196.002221 PMID: 8839933

Kumar, A. (2019). Food Preservation: Traditional and Modern Techniques. *Acta Scientific Nutritional Health*, *3*(12), 45–49. DOI: 10.31080/ASNH.2019.03.0529

Kumar, R., & Keshar, D. (2017). Effects of Chemical Fertilizers on Human Health and Environment: A review. International Advanced Research Journal Science, Engineering and Technology, 4(6), 204. https://iarjset.com/upload/2017/june-17/IARJSET%2036.pdf

Mabhaudhi, T., Chibarabada, T. P., Chimonyo, V. G. P., Murugani, V. G., Pereira, L. M., Sobratee, N., Govender, L., Slotow, R., & Modi, A. T. (2019). Mainstreaming Underutilized Indigenous and Traditional Crops into Food Systems: A South African Perspective. *Sustainability*, *11*(1), 172. DOI: 10.3390/su11010172 PMID: 37681213

Malli, A., Monteith, H., Hiscock, E. C., Smith, E. V., Fairman, K., Galloway, T., & Mashford-Pringle, A. (2023). Impacts of Colonization on Indigenous Food Systems in Canada and the United States: A scoping review. *BMC Public Health*, *23*(1), 2105. DOI: 10.1186/s12889-023-16997-7 PMID: 37885000

Marerro, A., & Mattei, J. (2022). Reclaiming Traditional, Plant-based, Climate-resilient Food System in Small Islands. *The Lancet. Planetary Health*, *6*(2), 2. Retrieved February 15, 2025, from https://www.thelancet.com/journals/lanplh/article/PIIS2542-5196(21)00322-3/fulltext. DOI: 10.1016/S2542-5196(21)00322-3

Masekoameng, M. R., & Makwena, M. C. (2019). The Role of Indigenous Foods and Indigenous Knowledge Systems for Rural Households' Food Security in Sekhukhune District, Limpopo Province, South Africa. *Journal of Consumer Sciences, Special Edition Food and nutrition challenges in Southern Africa*, 4. https://www.ajol.info/index.php/jfecs/article/view/191563. Accessed: 15 Feb 2025

Mbhenyane, X. G., & Tambe, A. B. (2024). The Influence of Household and Community Food Environments on Food Insecurity in Limpopo Province, South Africa. *International journal of environmental research and public health*, 21(2), 125. https://pubmed.ncbi.nlm.nih.gov/38397616/ Accessed: 15 Feb 2025

McCune, L. M., Nuvayestewa, V., & Kuhnlein, H. V. (2019). Why and how to document Traditional Food Systems in your community: Report from breakout discussions at the 2017 Native American Nutrition Conference. *Current Developments in Nutrition*, *3*(2), 19–24. Retrieved February 15, 2025, from https://www.sciencedirect.com/science/article/pii/S2475299122136954. DOI: 10.1093/cdn/nzy089

McMichael, P. M. S., & Schneider, M. (2011). Food security politics and the Millennium development goals. *Third World Quarterly*, *32*(1), 119–139. DOI: 10.1080/01436597.2011.543818 PMID: 21591303

Mdhluli, T. D., Mokgoatšana, S., Kugara, S. L., & Vuma, L. (2021). 'Knowledge management: Preserving, managing and sharing indigenous knowledge through digital library'. *Hervormde Teologiese Studies*, *77*(2), a6795. Retrieved February 15, 2025, from https://hts.org.za/index.php/hts/article/view/6795/21040. DOI: 10.4102/hts.v77i2.6795

Morrison, D. (2008). BC Food System Network Working Group on Indigenous Food Sovereignty: Final Activity Report (Prepared for: Provincial Health Services Authority-Community Food Action Initiative, Interior Health-Community Food Action Initiative &the B.C. Food Systems Network-Working Group on Indigenous Food Sovereignty.

Munn, Z., Peters, M. D., Stern, C., Tufanaru, C., McArthur, A., & Aromataris, E. (2018). Systematic review or scoping review? Guidance for authors when choosing between a systematic or scoping review approach. *BMC Medical Research Methodology*, *18*(1), 1–7. DOI: 10.1186/s12874-018-0611-x

Neufeld, H. T., Richmond, C. A. M., & Health Access Centre, S. O. A. (2017). Impacts of place and social spaces on traditional food systems in Southwestern Ontario. *International Journal of Indigenous Health*, *12*(1), 93–103. DOI: 10.18357/ijih112201716903

Nyuyen, H. (2018). Sustainable food systems: Concept and framework. Food and Agriculture Organization of the United Nations. https://openknolwedge.fao.org/handle/20.500.14283/ca2079en. Accessed: 15 Feb 2025

Omer, S. (2024). Africa hunger crisis: Facts, FAQs, how to help. https://www.worldvision.org/hunger-news-stories/africa-hunger-famine- facts#:~:text=Fast%20facts%3A%20Africa%20hunger%20crisis,-More%20than%201&text=An%20estimate0868%20million%20people,the%20decades%20ha s%20abruptly%20reversed. Accessed: 15 Feb 2025

Pengue, W. A., Gemmill Herren, B., Balázs, B., Ortega, E., Acevedo, F., Diaz, D. N., ... & Westhoek, H. (2018). Eco-agri-food systems: today's realities and tomorrow's challenges.

Peterson, J., Ferguson, L. A., & Langford, C. A. (2016). Understanding scoping reviews: Definition, purpose, and process. *Journal of the American Association of Nurse Practitioners*, *29*(1), 12–16. DOI: 10.1002/2327-6924.12380 PMID: 27245885

Rahayu, Y. Y. S., Sujarwo, W., Irsyam, A. S. D., Dwiartama, A., & Rosleine, D. (2024). Exploring unconventional food plants used by local communities in a rural area of West Java, Indonesia: Ethnobotanical assessment, use trends, and potential for improved nutrition. *Journal of Ethnobiology and Ethnomedicine*, *20*(1), 68. DOI: 10.1186/s13002-024-00710-y PMID: 39030547

Ronquest-Ross, L. C., & Sigge, G. O. (2024). South Africa's food system: An industry perspective on past, present and future applications of science and technology. *South African Journal of Science*, *120*(7/8), 7–8. DOI: 10.17159/sajs.2024/16536

Ronquest-Ross, L. C., Vink, N. & Sigge, G. O. (2015). Food consumption changes in South Africa since 1994. *South African Journal of Science*, 111(9-10), 01-12.

Rua, I. (2023). Indigenous Peoples' Food Systems and Agroecology: Synergies and Convergences. https://www.culturalsurvival.org/net

Science Forum South Africa. (2023). Report: Africa Food Systems Transformation: Leave no one behind: A Case Study for South Africa. https://hsrc.ac.za/wp-content/uploads/2024/04/Africa-ood-Systems-_SFSA-side-event_Report_March-2024. Accessed: 15 Feb 2025

Serrato, C. (2024). The importance and challenges of indigenous food systems. https://www.forwardpathway.us/the-importance-and-challenges-of-indigenous-food-system

Simelane, T. (2024). Food survey shows need for area-specific interventions in SA. https://hsrc.ac.za/news/latest-news/food-survey-shows-need-for-area-specific-interventions-in-sa. Accessed: 15 Feb 2025

Simelane, T. (2024). National Food and Nutrition: South Africa. https://www.researchgate.net/publication/378304251_National_Food_and_Nutrition_Security_South_Africa_National_Report. Accessed: 15 Feb 2025

Simelane, T. (2024). Almost 50% of South Africans have been found to be over-weight or obese. https://www.researchgate.net/publication/381289579. Accessed: 15 Feb 2025

Singh, R. K., Turner, N. J., Reyes-Garcia, V., & Pretty, J. (2022). Socio-Ecological Diversity and Traditional Food Systems. Milton Park, Ox: CRC Press.

Siwior, P. (2021). Food Sovereignty: Definition, Origin of the Concept and Sustainable Food System. In: Encyclopedia of the UN Sustainable Development Goals. https://www.researchgate.net/publication/352127152_Food_Sovereignty_Definition_Origin_of_the_Concept_and_Sustainable_Food_System. Accessed: 15 Feb 2025

Sreedharan, S., Babu, V. S., Saritha, G. N. G., Sherly, S. V., & Kumar, A. (2023). A Cross-continental Survey of Traditional Food Systems That Are Based on Wild Food Plants. In Kumar, A., Singh, P., Singh, S., & Singh, B. (Eds.), *Wild Food Plants for Zero Hunger and Resilient Agriculture. Plant Life and Environment Dynamics*. Springer., DOI: 10.1007/978-981-19-6502-9_4

Statistics South Africa. (2011). Assessing Food inadequacy and hunger in South Africa in 2021 using the general household survey. https://www.statssa.gov.za/publications/03-00-20/03-00-202021. Accessed: 15 Feb 2025

Statistics South Africa. (2023). Focus on food inadequacy and hunger in South Africa in 2021. https://www.statssa.gov.za/?p=16235#:~:text=In%202021%2C%20about%202%2C1,and%20rising%20costs%20of%20living. Accessed: 15 Feb 2025

Statistics South Africa. (2024). General Household survey. https://www.statssa.gov.za/publications/P0318/P03182023. Accessed: 15 Feb 2025

Swiderska, K., Argumedu, A., Wekesa, C., Ndalilo, L., Song, Y., Rastogi, A., & Ryan, P. (2022). Indigenous Peoples' Food Systems and Biocultural Heritage: Addressing Indigenous Priorities Using Decolonial and Interdisciplinary Research Approaches. *Sustainability (Basel)*, *14*(18), 11311. DOI: 10.3390/su141811311

Tarcza, T. M., & Nemteanu, S. M. (2015). Promoting Traditional Food Products as Healthy Diet Product. Practical Application of Science, III, 3, 9. https://seaopenresearch.eu/Journals/articles

Tartanac, F. (2021). Traditional food systems to support healthy and culture-based diets. https://openknowledge.fao.org/server/api/core/bitstreams/ae59ca4d-4267-423a-821a- 154a4a2237a5/content. Accessed: 15 Feb 2025

Tefft, J., Jonasova, M., Adjao, R., & Morgan, A. (2017). Food Systems for an Urbanizing World: Knowledge product. https://www.semanticscholar.org/paper/Food-systems-for-an-urbanizing-world-%253A-knowledge-Tefft-Jonasova/ad6aa03d304c7f9e7fb0c24cf. Accessed: 15 Feb 2025

Tricco, A. C., Lillie, E., Zarin, W., O'brien, K., Colquhoun, H., Kastner, M., Levac, D., Ng, C., Sharpe, J. P., Wilson, K., Kenny, M., Warren, R., Wilson, C., Stelfox, H. T., & Straus, S. E. (2016). A scoping review on the conduct and reporting of scoping reviews. *BMC Medical Research Methodology*, 16 (1). https://doi DOI: 10.1186/s12874-0160116-4

United Nations. (2024). Sustainable Development Goal Report 2024. https://unstats.un.org/sdgs/report/2024/. Accessed: 15 Feb 2025

United Nations Development Group. (2022). Theory of Change: UNDAF Companion Guideline. https://unsdg.un.org/resources/theory-change-undaf-companion-guidance

United Nations Educational, Scientific and Cultural Organization. (2016). Goal 2 End Hunger, Achieve Food Security, and Improved Nutrition and Promote Sustainable Agriculture. https://unstats.un.org/sdgs/report/2016/goal-02/ Accessed: 15 Feb 2025

Von Braun, J., Afsana, K., Fresco, L. O., & Hassan, M. H. A. (2023). Food systems: seven priorities to end hunger and protect the planet. In *Science and innovations for food systems transformation* (pp. 3–9). Springer International Publishing. DOI: 10.1007/978-3-031-15703-5_1

Yung, K., & Neathway, C. (2019). Community champions for safe, sustainable, and traditional food system. *Current Developments in Nutrition*, *4*, 119. PMID: 32258999

ADDITIONAL READING

Byaruhanga, R., & Isgren, E. (2023). Rethinking the alternative: Food Sovereignty as a prerequisite for Sustainable Food Security. *Food Ethics*, *8*(16), 16. https://link.springer.com/articles10.1007. DOI: 10.1007/s41055-023-00126-6

McMichael, P. (2014). Historicizing food sovereignty. *The Journal of Peasant Studies*, *41*(6), 933–957. Advance online publication. DOI: 10.1080/03066150.2013.876999

Pimbert, M. P., & Barry, B. (2021). Let the people decide: Citizen liberation on the role GMOs in Mali's agriculture. *Agriculture and Human Values*, *38*(4), 1097–1122. DOI: 10.1007/s10460-021-10221-1

Sundram, P. (2023). Food security in ASEAN: Progress, challenges and future. *Frontiers in Sustainable Food Systems*, *7*, 1260619. DOI: 10.3389/fsufs.2023.1260619

Wittman, H. (2023). Food Sovereignty: An exclusive model for feeding the world and cooling the planet. *One Earth*, *6*(5), 474–478. DOI: 10.1016/j.oneear.2023.04.011

KEY TERMS AND DEFINITIONS

Traditional Food Systems: System that links people, land, culture and spirit. These food systems are not only a complex network of relationships, but advocate for the use of native species, such as plants, insects and animals, including traditional farming techniques, indigenous food processing and preservation methods, localized food production and distribution.

Food Sovereignty: Concept is concerned with who the producers of food are, how and where it was produced, and who holds the power to make these decisions.

Food Systems: Structure as diverse activities, which normally involve production, marketing, processing, consumption and disposal of goods, originating from forestry or fisheries, agriculture, including the inputs and outputs generated from all the liable activities.

Food Security Policy: Ensures global food security by providing all people have reliable access to sufficient, safe, and nutritious food, focusing on accessibility, affordability, utilization, and stability.

Chapter 5
Roles and Responsibilities of Vatete (Aunt) in the Family Among the Ndau People of Chipinge District, Zimbabwe: Challenges and Opportunities

Pindai Mangwanindichero Sithole
https://orcid.org/0009-0008-6881-9959
CeDRE International Africa, Zimbabwe

ABSTRACT

The findings of this study shed light on the essential yet often under-appreciated role of aunts within family structures, particularly in the context of Ndau people of Chipinge District, Zimbabwe. Aunts, traditionally regarded as key figures in nurturing and guiding younger generations, face a myriad of challenges that hinder their effectiveness in fulfilling these responsibilities. Issues such as mistrust from their brothers' wives (varoora), diminishing authority, and competition from female church elders emerge as significant barriers. Additionally, geographical distance, limited resources, and educational gaps further complicate their ability to provide the mentorship and support that younger family members need. This study underscores the importance of addressing these challenges to empower aunts and enhance their contributions to familial and cultural continuity in the family.

INTRODUCTION

This chapter focuses on the *vatete* (aunt) as a cultural heritage institution within families among the Ndau people of Chipinge District, Zimbabwe. The chapter is an outcome of a mixed methods study conducted by the author with 108 women across Chipinge District. *Vatete* is a Shona language term which means the same as aunt in English language. In this chapter, the two terms are used interchangeably with *vanatete* being the plural form of *vatete*. Vanatete are often referred to as 'vatete cultural heritage institu-

DOI: 10.4018/979-8-3373-4124-8.ch005

tion' because of the significant roles they play in preserving and transmitting family socio-cultural values, community knowledge and normative practices (Afolabi, 2005; Bwalya, 2022; Mkhize & Reddy, 2020).

While women, generally, are regarded as influential figures in their families from the perspective of the role they play and espouse as vanatete, this role is highly likely to be eroded soon if little or nothing is done to preserve it. In fact, the author was motivated to write this chapter to gain understanding of the role of vanatete and to explore any challenges they are facing in carrying out their roles and responsibilities in the family. Essentially, the chapter brings out the roles and responsibilities of vatete, challenges they are facing in the modern society and opportunities in preserving this crucial indigenous cultural institution among the Ndau people of Zimbabwe.

BACKGROUND

In the cultural context of the Ndau and Shona people of Zimbabwe, aunts play a significant social role within the family structure (Zimunya, 2021). Their responsibilities extend beyond mere familial ties; they are integral to the social fabric and cultural continuity of both their families and communities. Vanatete are often seen as advisors, caregivers, and mediators, embodying a blend of authority and nurturing that is respected and valued in their respective families and communities (Chikanda, 2017).

One of the primary roles of vanatete is to provide guidance and support to their siblings and their children. They are expected to advise their brothers on matters related to marriage and family life, often acting as intermediaries between their brothers and sisters-in-law (varoora) (Chikanda, 2017). This advisory role is crucial, as it helps to maintain harmony within the family and ensures that traditional values normative practices are upheld. It is clear that aunts' advisory role safeguards that family members' behaviours or actions are within the socio-cultural standards that are widely accepted and expected in the community.

Vanatete are also responsible for nurturing their nieces and nephews, playing a vital role in their upbringing, including, sometimes, even their education (Moyo, 2019; Nyoni 2020). They often take an active part in teaching children about cultural practices, social norms, and life skills, which are essential for their development and integration into society (Mavhunga, 2018; Moyo, 2019). Moreover, the role of vatete is characterized by a certain level of authority, particularly in family disputes. They are often called upon to mediate conflicts and provide resolutions, leveraging their experience and wisdom to restore peace within the family (Chikanda, 2017; Chivasa, 2022). This aspect of their role underscores the respect afforded to them within the family hierarchy. In many cases, the eldest tete may take precedence in discussions, reflecting the importance of age and experience in decision-making processes within the Shona culture (Mavhunga, 2018).

Vanatete also play a significant role in the socialization of children. They are responsible for instilling values and morals, often emphasizing the importance of community and collective responsibility (Nyoni 2020; Zimunya 2021). A common saying among the Shona is, "You do not educate your child for yourself alone; education is for society, by society," which highlights the communal approach to child-rearing (Mavhunga, 2018). Vanatete are seen as vital contributors to this process, ensuring that children grow up with a strong sense of identity and belonging.

In addition to their nurturing and advisory roles, vanatete are also expected to embody certain qualities that reflect their status within the family. They are often viewed as role models, and their behaviour can significantly influence how children perceive their roles in society. A good vatete is expected to balance playfulness with seriousness, demonstrating fairness and the ability to articulate thoughts clearly (Chikanda, 2017). This performance of their role is crucial, as it impacts how they are received by both family members and the broader community.

However, the role of a vanatete is not without challenges. They may face difficulties in their relationships with sisters-in-law or their brothers' partners, as family dynamics can sometimes complicate interactions. This implies that vanatete must carefully navigate these relationships, as they can be perceived as overstepping boundaries or being overly critical (Mavhunga, 2018). Despite these challenges, the position of vatete remains one of respect and influence, with the potential to effect positive change within their families and communities for common good of society. The challenges aunts face in fulfilling their roles and responsibilities within families can lead to several negative effects on family dynamics and relationships.

When aunts experience conflicts with sisters-in-law or other family members, it can create tension and division, undermining the unity that is crucial for harmonious family life (Chikanda, 2017). This discord may result in a lack of support for children, as the nurturing role of aunts becomes compromised; children may miss out on valuable guidance and cultural teachings that aunts typically provide, leading them to unacceptable behaviour in society (Mavhunga, 2018). Furthermore, if aunts feel undervalued or face criticism for their involvement in family matters, their motivation to engage positively may diminish, leading to emotional distance and disengagement from family matters (Nyoni, 2020). Such challenges can also strain communication patterns, making it difficult for family members to resolve conflicts effectively (Zimunya, 2021). Ultimately, these negative effects not only impact the relationships between aunts and their extended family but can also hinder the overall emotional and social development of the younger generation (Chivasa, 2022).

In summary, it is clear that the roles and responsibilities of aunts are multifaceted and deeply rooted in cultural traditions. Vanatete serve as advisors, caregivers, and mediators, playing a crucial role in the upbringing of children and the maintenance of family harmony and emotional wellbeing. Their influence extends beyond immediate family ties, contributing to the broader social fabric of their communities. However, the challenges they face in carrying out these roles and responsibilities have compromised integrity of family structures to an extent of dysfunctional families in some instances.

STATEMENT OF THE PROBLEM

The roles and responsibilities of vanatete among the Ndau people of Chipinge District, Zimbabwe, have not been sufficiently explored, despite their significant impact on family dynamics and cultural continuity. This study aims to address the gap in understanding how vanatete function as caregivers, advisors, and mediators within families, and how their contributions influence the upbringing of children and community cohesion. Additionally, it seeks to identify the challenges vanatete face in fulfilling their roles, including conflicts with other family members and societal expectations, which may affect their ability to support family harmony and cultural transmission. Understanding these dynamics is crucial for appreciating the broader social fabric of Ndau society and the vital role of aunts in fostering community resilience. In light of this, the study sought to seek how aunts are perceived, roles they play in their families, challenges they face and opportunities that exist in the aunt institution within families.

Chapter's Objectives

The objectives that guided the study were to:

- Explore roles and responsibilities of aunts in the family among the Ndau people;
- Identify challenges aunts face in carrying out their roles among the Ndau people; and
- Explore opportunities that can be considered to enhance the role and responsibilities of aunts in the families among the Ndau people.

LITERATURE REVIEW

The review of the literature focused on five aspects which are: role of vanatete, culture and heritage that underpin the authority of vanatete, perceptions of vatete in the family, methods vanatete use in mediation of conflicts in the family, and challenges vanatete face in carrying out their roles and responsibilities.

Brief description of the Ndau people

The Ndau people are an ethnic group primarily located in the eastern regions of Zimbabwe, particularly in the Chipinge District. The Ndau community is known for its rich cultural heritage, which includes traditional music, dance, and art forms that reflect their historical experiences and social values (Chikanda, 2017). The Ndau people primarily engage in agriculture, with subsistence farming being a significant aspect of their livelihood, alongside other economic activities such as trade and crafts (Zimunya, 2021). It is generally accurate to describe the Ndau people as sociable, tolerant, empathetic, and peace-loving.

These traits are often reflected in their cultural practices and community interactions, emphasizing strong familial bonds and communal support (Chikanda, 2017; Zimunya, 2021). The Ndau culture values harmony, cooperation, reciprocal relationship, and respect for one another, which contributes to their reputation for sociability and empathy (Mavhunga, 2018). Additionally, traditional conflict resolution methods often prioritize peaceful dialogue and mediation, further reinforcing their peaceful disposition (Nyoni, 2020). Religion plays a significant role in the lives of the Ndau people, influencing their daily practices and social norms. Traditional beliefs often coexist with Christianity, creating a unique spiritual landscape where ancestral worship and reverence for the spirits of the deceased are common (Mavhunga, 2018).

This blend of beliefs fosters a sense of community and continuity, as religious practices are often communal events that strengthen familial and social bonds (Zimunya, 2021). Respect is a cornerstone of Ndau culture, particularly within familial relationships. The Ndau people emphasize the importance of honouring elders and maintaining harmonious relationships among family members. This respect is manifested in various ways, including the use of polite language and adherence to traditional customs during family gatherings (Chikanda, 2017). Such practices not only reinforce familial ties but also promote a sense of belonging and identity within the community (Nyoni, 2020).

Role of Vanatete in the Family

The role of aunts (vanatete) in the family is multifaceted, encompassing responsibilities as caregivers, advisors, and mediators. According to Zimunya (2021), vanatete are integral to the family structure, providing emotional support and guidance to their siblings and their children. They often take on the role of educators, teaching cultural practices and social norms essential for the upbringing of children (Moyo, 2019). Chikanda (2017) emphasizes that aunts act as intermediaries in family matters, helping to maintain harmony and resolve conflicts, which underscores their importance in fostering family cohesion.

Culture and Heritage that Underpin the Authority of Vanatete in the Family

The authority of vanatete is deeply rooted in the cultural and heritage practices of the Ndau and Shona people. Mavhunga (2018) notes that aunts are respected figures within the family hierarchy, often regarded as role models due to their age and experience. This cultural reverence is reflected in the traditional belief that aunts play a crucial role in preserving cultural values and norms, which are vital for the continuity of the community (Nyoni, 2020). The saying, "You do not educate your child for yourself alone; education is for society, by society," illustrates the communal approach to child-rearing and the significant role aunts play in this process (Mavhunga, 2018).

Perceptions of Vatete in the Family

Perceptions of vanatete within the family context can vary significantly. While they are often viewed as authoritative figures, their relationships with other family members, particularly sisters-in-law, can be complex. Mavhunga (2018) highlights that aunts may sometimes be perceived as overstepping boundaries, leading to tensions within the family. However, when respected, vanatete are seen as vital contributors to family dynamics, providing guidance and support that enhances the well-being of children and the family unit as a whole (Chikanda, 2017).

Methods Vanatete Use to Mediate Conflicts in the Family

Vanatete employ various methods to mediate conflicts within the family. Chivasa (2022) suggests that aunts often leverage their experience and wisdom to facilitate discussions among family members, promoting open communication and understanding. They may use storytelling and cultural teachings as tools to convey important lessons and values, helping to resolve disputes amicably (Moyo, 2019). Additionally, vanatete often act as neutral parties, allowing them to navigate conflicts without bias, which is crucial for restoring harmony within the family (Zimunya, 2021).

Challenges Vanatete Face in Carrying Out Their Roles and Responsibilities

Despite their significant contributions, vanatete face numerous challenges that can hinder their effectiveness. Conflicts with sisters-in-law or other family members can create tension, undermining their authority and ability to fulfil their roles (Mavhunga, 2018). Nyoni (2020) notes that feelings of undervaluation or criticism can diminish aunts' motivation to engage positively in family matters, leading to emotional distance and disengagement. Furthermore, the evolving socio-economic landscape poses additional challenges, as traditional roles may be undermined by modern influences, complicating the dynamics within families (Chikanda, 2017).

METHODOLOGY

A mixed methods research methods approach (pragmatism) was employed utilizing both quantitative and qualitative data and this was helpful for validation of findings. The quantitative results provided trends, while qualitative responses explained the reasons behind the trends, thus enriching the overall analysis (Creswell & Plano Clark, 2017). The target study population were women from Chipinge District in Zimbabwe who are members of the United Church of Christ in Zimbabwe. Key selection criteria were that one is married, has at least one biological brother who is married too and has at least one child. The reason these selection criteria were considered was to ensure that the right people participated in the study to provide insights and views based on their lived experiences as aunts in their respective families.

In terms of data collection methods, a 5-point Likert scale questionnaire with two open-ended questions was employed in the study to assess understanding and challenges aunts face in Chipinge District. The questionnaire was voluntarily completed online following a Kobo Collect link sent to them through a church WhatsApp group whose membership is largely comprised of people from Chipinge District, of Ndau ethnic group. The link to the questionnaire was left open for two weeks to allow adequate time for the women to complete the questionnaire.

Finally, 108 women completed the questionnaire and this sample size was considered both adequate and representative for the study on the role of aunts in the family and the challenges they face. This is supported by Smith et al. (2020) guidelines which stress that a sample size of this magnitude is sufficient for achieving reliable statistical power in social science research like this one. Furthermore, the respondents were reflective of diverse demographics within the community, encompassing different ages, socio-economic backgrounds, which are crucial for capturing a comprehensive understanding of the aunts' roles and the challenges they face (Creswell & Creswell, 2017). Additionally, the qualitative insights gained from open-ended questions in the questionnaire allowed for a richer exploration of the respondents' experiences, aligning with the call for mixed-method approaches in family studies to enhance the depth of understanding (Bryman, 2016).

For qualitative dataset from the two open-ended questions, a common approach proposed by Patton (2015) to select between 10% to 30% of the total sample was applied. Given the study objectives and the complexity of the subject matter, a 20% (22) of the 108respondents was randomly selected for analysis of their qualitative data. This percentage is justified as it strikes a balance between obtaining a rich, diverse range of insights while remaining manageable for qualitative analysis (Creswell &Poth, 2018).

Furthermore, the approach aligns with qualitative research principles, which emphasizes depth over breadth, and it enables engagement with the data meaningfully (Braun & Clarke, 2019). Essentially, the sample of 22 qualitative scripts from 22 respondents was both practical and methodologically sound, providing valuable qualitative insights while maintaining rigor in the research process undertaken. In addition to the 22 women, three other women from Chipinge District, outside the 108, were purposively selected for in-depth interviews to gain more insights surrounding the role and challenges aunts face in the family.

FINDINGS AND DISCUSSION

This section focuses on the three key aspects, namely: (i) respondents' understanding of the role and responsibilities of aunts in the family; (ii) challenges aunts face in carrying out their responsibilities in the family; and (iii) strategies that can be considered to improve aunts' role in the family. For the 5-point Likert scale questions in the questionnaire, respondents were asked to indicate their level of agreement as follows: 1 = *Strongly disagree*; 2 = *Disagree*; 3. *Neutral*; 4 = *Agree*; and 5 = *Strongly agree*.

Roles and responsibilities of aunts in the family

The role of aunts in the family, particularly in the context of Chipinge District, is multifaceted and significant. Aunts serve as vital emotional and social support systems within families, contributing to the overall well-being and harmony of family units. The findings from the study indicate that aunts include active participation in family affairs, provision of emotional support, and play a crucial role in maintaining family harmony. These findings are displayed in Table 1 below.

Table 1. Roles and responsibilities of aunts in the family

The roles aunts play in the family			
Aunts' roles in the family	**N**	**Mean**	**Std. Deviation**
a) The active participation of aunts in the family affairs strengthens family bonds	108	4.67	.958
b) Aunts provide emotional support during challenging times in our family	106	3.86	1.108
c) Aunts play a key role in maintaining family harmony and emotional well-being	108	4.57	1.235
d) Aunts contribute significantly to the social well-being of the family	108	3.69	1.110
Overall score		**4.198**	**1.142**

Active Participation and strengthening family bonds

The data reveals that aunts' active participation in family affairs is perceived as a key factor in strengthening family bonds, with a mean score of 4.67, indicating that the respondents strongly agreed to the notion. This aligns with the findings of Mkhize (2019), who emphasizes that extended family members, including aunts, play a critical role in fostering familial relationships and cohesion. Aunts often serve as intermediaries between parents and children, facilitating communication and understanding within the family. This was expressed by some of the respondents who put it this way:

> Aunts provide mediation between family members in conflict (R2).
> Aunts teach brother's children ubuntu, reinforcing our culture in the girl child (R8).
> They provide information about family history, values, norms, and practices (R11)

It is clear from the quotations above that aunts' involvement in family gatherings, decision-making processes, and cultural rituals reinforces the interconnectedness of family members, promoting a sense of belonging, unity, and the common good. Moreover, the emotional support provided by aunts during challenging times is highlighted by a mean score of 4.57, revealing that the respondents agreed that aunts are play this role in the family. This finding resonates with the work of Kuo and Tsai (2020), who argue that extended family members, particularly aunts, are essential sources of emotional resilience. Aunts often provide a listening ear, offer guidance, and help navigate difficult situations, thereby enhancing the emotional well-being of family members. Their presence can alleviate stress and foster a supportive environment, which is particularly important in the face of societal challenges.

Maintaining family harmony and social well-being

The role of aunts in maintaining family harmony and emotional well-being is further underscored by a mean score of 4.57, denoting that the respondents agreed to this notion. Aunts often act as mediators in conflicts, helping to resolve disputes and restore peace within the family. This was emphasized by one of the respondents who stated that aunts "Provide emotional support to the family and give guidance to their brothers' children, especially girls" (R14). Another respondent buttresses, saying that aunts are a "key pillar of the social support system in the family" (R18). The finding that aunts maintain family harmony and social wellbeing resonates with Mkhize (2019), who notes that aunts are often seen as neutral parties who can provide objective perspectives during family disagreements. Their ability to navigate complex family dynamics contributes to a harmonious living environment, which is essential for the overall health of the family unit.

In addition to emotional support, aunts contribute significantly to the social well-being of the family, with a mean score of 3.69, signifying that aunts contribute to the social well-being of the family. This contribution can manifest in various ways, such as organizing family events, facilitating social interactions, and promoting cultural traditions. According to Mkhize (2019), aunts often play a pivotal role in the transmission of cultural values and practices, ensuring that younger generations remain connected to their heritage. This cultural continuity is vital for fostering a sense of identity and belonging among family members.

The findings discussed in the preceding sections undoubtedly underscore the vital roles and responsibilities that aunts play in family dynamics, providing emotional support, maintaining harmony, and contributing to social well-being. As families navigate the complexities of modern life, recognizing and empowering aunts can lead to stronger family bonds and improved emotional resilience.

CHALLENGES AUNTS FACE IN EFFECTIVELY PLAYING THEIR ROLES IN THE FAMILY

Despite the fact that aunts hold significant roles within family structures, serving as caregivers, advisors, and cultural custodians, the findings indicated that various challenges hinder their ability to fulfil these roles effectively. Table 2 on the next page outlines the challenges followed by the discussion with substantiation from relevant literature to contextualize the experiences of aunts in Chipinge District.

Table 2. Challenges aunts face in carrying out their roles

Challenges being faced by aunts to play their roles effectively in the family in Chipinge District, Zimbabwe			
Challenges that prevent aunts to play their roles in the family	**N**	**Mean**	**Std. Deviation**
a) Mistrust between the aunt and her *muroora* (brother's wife) can significantly hinder the aunt's ability to accomplish her role within the family, particularly in providing guidance and support to her brother's children and even *muroora* herself	108	4.85	.754
b) Increasingly, the authority and influence of aunts have significantly diminished in the families	108	4.41	.737
c) Female church elders are increasingly taking up the roles which were traditionally held by aunts, especially in courtship and marriage matters of their brothers' children	108	4.21	.774
d) The encroachment of foreign cultures is rapidly eroding the role of aunts in families	108	4.20	.818
e) Geographical distance significantly hinders aunt's ability to effectively fulfil her role in our family	108	4.05	.869
f) If an aunt has limited resources (i.e., material or financial) to provide support in the family, this can lead to feelings of inadequacy and frustration, compromising ability to effectively fulfil their roles	108	3.94	.926
g) Limited or lack of formal education often makes aunts ill-equipped to address modern issues that affect younger members of the family	108	3.74	1.179
h) I feel comfortable seeking advice from my aunt regarding personal issues	108	3.06	1.299
Overall score		**4.105**	**0.919**

Mistrust between aunts and varoora

One of the most significant barriers identified is the mistrust between aunts and their brothers' wives (*varoora*, being plural of *muroora*), which scored the highest mean score in the study (4.85). This mistrust can severely limit an aunt's ability to provide guidance and support to her brother's children and the muroora alike. The dynamics of family relationships in Zimbabwe are often complex, influenced by cultural expectations and gender roles. Muroora may perceive aunts as rivals, particularly in matters concerning the upbringing of children and family decisions, leading to tension and conflict (Mudzengi et al., 2021).

This situation is exacerbated by the patriarchal structure prevalent in many Zimbabwean families, where aunts may feel marginalized in their traditional roles. It is clear that mistrust between a daughter-in-law and vatete can significantly hinder her ability to accomplish her role within the family, particularly in providing guidance and support to her brother's daughters and the daughter-in-law herself. When mistrust exists, the aunt may feel sidelined or unwelcome, which can lead to a breakdown in communication and collaboration on family matters. The statement below highlights the fact that aunts should be involved in important family matters.

Families should involve their aunts in crucial matters of the family. Leaving them out in decision-making suggests they are of little value in the family life (R2).

This situation often creates an environment where the aunt hesitates to offer advice or assistance, fearing that her intentions will be misinterpreted or rejected. Consequently, the daughters may miss out on valuable mentorship and life lessons that the aunt could provide, limiting their understanding of cultural practices and familial responsibilities. The daughter-in-law, on the other hand, may perceive the aunt's involvement as intrusive, further exacerbating the divide.

This lack of trust can create a cycle of isolation, where the aunt feels disconnected from family dynamics, reducing her capacity to contribute positively. As a result, the collective wisdom and support that vanatete traditionally offer is diminished, leaving both the daughters and the daughter-in-law without essential guidance that could enhance their familial relationships and personal development. Ultimately, addressing this mistrust is crucial for fostering a collaborative family environment that allows vanatete to play their vital roles effectively.

Recent literature highlights the importance of trust and collaboration in family dynamics. For instance, a study by Naidu and Muchono (2019) emphasizes that effective communication and mutual respect between family members are crucial for maintaining harmony and fulfilling familial roles. When mistrust prevails, it not only affects the relationship between aunts and muroora but also impacts the overall family cohesion, leading to a breakdown in support systems that are vital for child-rearing and emotional well-being.

Barriers to effective communication and support

Despite the positive roles that aunts play, the findings also indicate challenges that hinder their effectiveness. The mean score of 3.06 for seeking advice from aunts suggests that there may be barriers to open communication. This finding is echoed by Kuo and Tsai (2020), who highlight that while aunts are often seen as supportive figures, there may be hesitance among younger family members to approach them for personal issues. The limited opportunities for aunts to get involved in the family issues is highlighted in the following quotation from one of the respondents:

Families should involve their aunts in crucial matters of the family. Leaving them out in decision-making suggests they are of little value" (R2).

Factors such as generational differences, perceived authority, and cultural shifts may contribute to this reluctance. Thus, generational differences in values further muddle the situation. Younger family members may adopt more progressive attitudes regarding gender roles and responsibilities, which can conflict with the traditional expectations held by vanatete. For example, while vanatete may emphasize the importance of familial duty and sacrifice, younger generations may prioritize individual aspirations and personal fulfilment. This divergence of worldview can create misunderstandings and resentment, as vanatete struggle to maintain their relevance and authority in a rapidly changing family social landscape.

Such generational gaps can hamper effective communication, making it challenging for vanatete to convey their wisdom rooted in indigenous knowledge, or to mentor younger relatives in ways that resonate with them. Furthermore, the diminishing authority and influence of aunts within families, as indicated by a mean score of 4.41 raises concerns about their ability to fulfil their roles effectively. The encroachment of foreign cultures and changing societal norms can undermine traditional family structures, leading to a decline in the recognition of aunts' contributions. This aligns with the observations of Mkhize (2019), who notes that globalization and cultural exchange can dilute the roles of extended family members, including aunts.

Diminished authority and influence

The findings also reveal that the authority and influence of aunts have significantly diminished, with a mean score of 4.41. Traditionally, aunts played a pivotal role in family decision-making and cultural practices. However, contemporary shifts in family structures and societal norms have reduced their status. This decline can be attributed to various factors, including urbanization and changing gender roles, which have altered the dynamics of authority within families (Mugadza et al., 2017). As a result, aunts may find it challenging to assert their influence, limiting their ability to guide and support younger family members.

The erosion of authority among aunts is echoed in the work of Dube and Guveya (2014), who argue that the changing socio-economic landscape in Zimbabwe has led to a reconfiguration of family roles. Aunts, once seen as key figures in the upbringing of children, are increasingly sidelined by other authority figures, such as female church elders and community leaders. This shift not only diminishes the role of aunts but also reinforces a hierarchical structure that may not align with traditional practices, further complicating their relationships within the family.

Infiltration by female church elders

The emergence of female church elders taking on roles traditionally held by aunts is another significant challenge, reflected in a mean score of 4.21. In matters of courtship and marriage, church elders often assume authoritative positions, overshadowing the contributions of aunts. This shift can lead to feelings of inadequacy among aunts, who may perceive their roles as diminished in the face of these new authority figures (Benjamin, 2017). The involvement of church elders in family matters can create a divide between traditional practices and contemporary religious influences, complicating the role of aunts in guiding younger generations.

This rise of female church elders who often take on the roles traditionally held by vanatete, especially in courtship and marriage matters really complicates and suffocates the space of vanatete institution. While these church female leaders can provide valuable support and guidance, their involvement frequently leads to competition and tension within families and, by extension, in the community. Some of vanatete have expressed that they feel their traditional authority is undermined by the church elders, which potentially have an opportunity to create friction and conflict in the community. This unconventional trajectory where some women in the church have taken it upon themselves to play the role of vanatete is clearly expressed in the quotation below from one of the respondents:

I feel so frustrated and angry that female elders are stepping into the roles that should belong to us aunts. They think they know better about courtship and marriage, but they don't understand our family dynamics like we do. It feels like we're being pushed aside, and our voices are ignored when it comes to important decisions. We have traditions and wisdom to share, but it's like they don't trust us anymore. It really hurts to see our roles diminished like this (R17).

This infiltration of female church elders and the and the socio-cultural dynamic it brings, not only affects the relationships between vanatete and church leaders but can also disrupt the entire family structure. This is because there are high likelihood that the female church elders bring different values and expectations contrary to those of the families. The elders may promote modern views that clash with traditional expectations and practices, leading to confusion and disunity among family members.

Research by Dhliwayo et al. (2020) supports this observation, noting that the increasing influence of religious institutions in family matters often leads to a redefinition of roles within the family. Aunts, who traditionally provided guidance in matters of courtship and marriage, may find themselves excluded from these discussions, leading to a loss of cultural continuity and support for younger family members. From a different perspective, the infiltration by female church elders is seen as a solution to a vacuum left by some aunts who may not be playing out their role and responsibilities in the family and this is emphasized in the two statements below.

Aunts should be role model; they need to lead an exemplary life and avoid bias in executing their roles (R14).

The aunt should be sincere and act as a role model within the family(R9).

It also emerged that one of the explanations why some aunts leave a vacuum in their traditional role in the family is limited or lack of knowledge and skills of what they should do as illuminated in the quotations below.

Teach the role of Vatete at an early stage to maintain her relevance in the family (R15).

Empower vatete to understand her role and responsibilities in the family (R2).

The quotations above clearly indicate that while the role of an aunt is often viewed as a natural extension of familial relationships, it is essential to recognize that fulfilling this role effectively requires deliberate training and mentorship. Aunts must navigate complex family dynamics, cultural expectations, and modern challenges, which necessitate a deeper understanding of their responsibilities. Through investing in education and support, we empower aunts to embrace their roles with confidence and competence, ensuring they can provide meaningful guidance and care within the family.

Encroachment of Foreign Cultures

The encroachment of foreign cultures has resulted in a rapid erosion of traditional roles, with a mean score of 4.20. Globalization and exposure to different cultural norms can undermine the significance of local customs and practices that the aunts traditionally upheld. This cultural shift can lead to a loss of identity for both aunts and younger family members, who may feel disconnected from their heritage (Mudzengi et al., 2021). As traditional practices wane, the unique contributions of aunts are often overlooked in favour of more contemporary approaches to family life. The impact of globalization on family structures is well-documented in the literature. For instance, Macheka (2016) argues that the influence of Western values and practices has led to a decline in the importance of extended family roles, including that of aunts. This shift not only affects the roles of aunts but also has broader implications for community cohesion and cultural preservation.

Geographical Distance

Geographical distance poses another barrier to effective aunt roles, reflected in a mean score of 4.05. With increased migration and urbanization, many aunts may live far from their families, limiting their ability to participate in family affairs. The centrality of dense social interaction between the family and aunt is explicitly captured in the statement below.

Encourage communication and frequent visits between family members so that there is improved understanding and closeness(R1).

In fact, it has been noted that physical separation can hinder emotional connections and diminish the practical support aunts can provide, especially during critical life events such as weddings, funerals, or family gatherings (Dube &Guveya, 2014). The challenges posed by geographical distance are compounded by the lack of communication infrastructure in rural areas, which can further isolate aunts from their families. As noted by Ngarivhume et al. (2015), the inability to maintain regular contact can lead to feelings of disconnection and frustration among aunts, who may feel that their roles are diminished due to their physical absence.

Limited Resources

Aunts with limited material or financial resources may experience feelings of inadequacy, with a mean score of 3.94. The ability to provide support, whether emotional, financial, or logistical, is integral to the role of an aunt. When resources are scarce, aunts may feel frustrated and less effective in fulfilling their responsibilities, which can lead to diminished self-esteem and a sense of isolation within the family context (Mugadza et al., 2017). Moreover, the stressed socio-economic context in Chipinge District exacerbates this challenge. High levels of unemployment and economic instability limit the resources available to families, placing additional stress on aunts who may already be shouldering significant burdens in the families where they are married.

The expectations from aunts to provide support to the brother's family during times of financial hardship can lead to feelings of inadequacy and frustration, further complicating their ability to fulfil their roles effectively. As a result, many aunts may find themselves caught in a cycle of stress and anxiety, which detracts from their ability to nurture and guide the younger generation. The literature highlights the importance of resource availability in fulfilling familial roles. For example, Benjamin (2017) emphasizes that economic constraints can significantly impact the ability of aunts to provide support, leading to feelings of inadequacy and frustration. This situation can create a cycle of dependency and disempowerment, further complicating the role of aunts in family dynamics.

Lack of Education

Lastly, the limited or lack of formal education among aunts, scoring a mean of 3.74, can leave them ill-equipped to address modern issues affecting younger family members. As societal challenges evolve, the need for knowledge and skills to navigate these complexities becomes crucial. Aunts may struggle to provide relevant guidance on contemporary issues such as health, education, and career choices, which can impact their perceived value within the family (Mudzengi et al., 2021).

Limited access to education for some aunts is quite a significant challenge, which has historically restricted many of them from obtaining the skills and knowledge necessary to navigate complex family dynamics and provide adequate support to younger generations. While some aunts possess valuable life experience and indigenous knowledge about family matters, the lack of formal education can leave them ill-equipped to address modern issues, such as financial management or health education, which are increasingly relevant in today's society. This educational gap often perpetuates cycles of poverty and dependence, making it difficult for vanatete to assert their authority and influence effectively within the family structure.

The relationship between education and familial roles is well-established in the literature. Dube and Guveya (2014) argue that education is a critical factor in empowering aunts to fulfil their roles effectively. Without access to education and training, aunts may find themselves ill-prepared to address the needs of younger family members, leading to a further erosion of their authority and influence. It is clear that the findings illustrate significant challenges faced by aunts in fulfilling their roles and responsibilities within families. Mistrust, diminishing authority, the rise of alternative women outside the family structure, cultural erosion, geographical distance, limited resources, and educational gaps all contribute to the difficulties which aunts encounter. Addressing these challenges is essential for empowering aunts and enhancing their contributions to family life.

The findings presented speak alongside scholars of cultural heritage and family structures who are convinced that vatete should be regarded as a cultural heritage institution because of five specific and complementary reasons. One, vanatete are pivotal in passing down cultural or indigenous knowledge, stories, and rituals to younger generations in the family, thereby ensuring that traditions are maintained and respected (Afolabi, 2005). In this regard, they play a role of cultural transmission in the family. Two, vanatete serve as role models, providing guidance and support in line with the desired norms and practices, shaping the values and behaviours of children and adolescents (Houghton,2020).

This means that they are regarded as role models and mentors in the family. Three, vanatete help in bridging the generational gap, assisting younger family members in navigating and applying indigenous knowledge and cultural practices within the modern context (Bwalya, 2022; Mkhize& Reddy, 2020; Tansley,2021). This role underscores that vanatete are a mediation between generations to ensure indigenous knowledge and cultural practices are carried forward through generations, thereby ensuring cultural heritage resilience in the modern society. Four, beyond the family, vanatete actively engage in community activities that promote cultural heritage and participating in communal traditional rituals, thereby fostering a sense of community identity and solidarity (Olupona, 2006).

This enhances their status as cultural custodians through community involvement. Finally, aunts provide emotional and social support, contributing to the well-being of family members and reinforcing familial bonds which are central to cultural identity (Moyo,2012). Put together, it can be seen that vanatete play a significant role in preserving and transmitting cultural-family values, indigenous knowledge, traditions, and normative practices within families and communities.

FUTURE RESEARCH DIRECTIONS

Four areas of research could be considered in the future towards promotion of the roles and responsibilities of aunts among the Ndau people or any other ethnic groups in Zimbabwe:

- **Impact of Educational Programmes**: Future research could investigate the effectiveness of educational and mentorship programs aimed at empowering vanatete. Studies could assess how these initiatives influence their ability to fulfil traditional roles and improve family dynamics.
- **Intergenerational Perspectives**: Exploring intergenerational attitudes towards the roles of aunts within the Ndau community could provide insights into how cultural perceptions are evolving. Research could focus on the perspectives of both younger and older family members regarding the value of aunts.
- **Role of Men in Supporting Aunts**: Investigating how men and boys perceive and support the roles of vanatete could highlight strategies for enhancing collaboration within families. This research could explore the effectiveness of targeted programs that educate males about the importance of aunts in family life.
- **Cultural Revitalization Initiatives**: Examining the impact of cultural revitalization efforts that celebrate the roles of aunts may reveal effective strategies for reinforcing their significance within the community. Research could analyze the outcomes of such initiatives on social cohesion and family support systems.

CONCLUSIONS AND RECOMMENDATIONS

The discussions have clearly shown that aunts are upheld as key players in the family life. It has also become clear that the challenges vanatete in Chipinge District face regarding the effective carrying out of their traditional roles are multifaceted and deeply rooted in socio-cultural, educational, socio-economic, and trust. Limited education, competition with female church elders, generational differences, and mistrust all contribute to a complex landscape that can hinder their ability to fulfil their responsibilities. These challenges highlight the need for targeted support and resources to empower vanatete in their traditional roles, enabling them to contribute effectively to their families and communities.

In view of the challenges revealed, efforts to improve the situation could include community programs that foster trust and collaboration among family members, sustainable educational or training initiatives including mentorship to equip aunts with the necessary competences. In addition, cultural revitalization initiatives that celebrate the roles of extended family members like ants are important for consideration towards improvement of the aunt's role and responsibilities. In addition, church programs could incorporate teaching girls and women on the role and responsibilities of vatete in the family. At the same time, men and boys could also be taught on the importance of involving vatete in crucial family matters including the role men should play to support their wives to carry out their roles as aunts. The training initiatives for women, men, girls and boys in the church or other platforms on the aunt trajectory should also foster intergenerational dialogues, acknowledging and supporting the vital role that vanatete play in the family. These recommendations indicate that a holistic approach is required to enhance the role of vatete in the family leading to stronger familial bonds and social resilience within the greater community.

REFERENCES

Afolabi, E. A. A. (2005). Cultural Practices of African Women. In African Women: APolitical Economy (pp. 97-115). Zed Books.

Braun, V., & Clarke, V. (2019). Reflecting on reflexive thematic analysis. *Qualitative Research in Psychology, 16*(2). Advance online publication. DOI: 10.1080/14780887.2019.1628806

Bryman, A. (2016). *Social Research Methods* (5th ed.). Oxford University Press.

Bwalya, F. (2022). The Role of Aunts in Family Dynamics: Bridging Generational Gaps in African Contexts. *Journal of Intergenerational Relationships, 20*(3), 245–260.

Chikanda, A. (2017). The Role of Aunts in Zimbabwean Families: A Cultural Perspective. *Journal of Southern African Studies, 43*(2), 345–360.

Chivasa, S. (2022). Aunts as Cultural Custodians: The Role of Tete in Shona Society. *Journal of African Cultural Studies, 34*(1), 56–70.

Creswell, J. W., & Creswell, J. D. (2017). *Research Design: Qualitative, Quantitative, and Mixed Methods Approaches* (5th ed.). SAGE Publications.

Creswell, J. W., & Poth, C. N. (2018). *Qualitative Inquiry and Research Design: Choosing Among Five Approaches* (4th ed.). SAGE Publications.

Dube, L., & Guveya, E. (2014). Technical Efficiency of Smallholder Out-grower Tea. (Camellia Sinensis) Farming in Chipinge District of Zimbabwe. *Greener Journal of Agricultural Sciences, 4*(8), 368–377. DOI: 10.15580/GJAS.2014.8.091114354

Houghton, A. (2020). Aunts in the Family: The Role of Extended Family in Child Development and Socialization. In Family Relations and Social Capital: New Perspectives (pp. 110-125). Routledge.

Hsiao, C. H., Lee, C. Y., & Yang, C. C. (2020). Exploring the effectiveness of WhatsApp for data collection: A systematic review. *International Journal of Social Research Methodology, 23*(5), 475–486.

Kuo, C. Y., & Tsai, Y. F. (2020). The role of extended family in emotional support: A study of aunts and uncles. *Journal of Family Studies, 26*(2), 123–138.

Mavhunga, C. (2018). Family Dynamics and the Role of Women in Zimbabwean Society. *African Journal of Gender and Development, 6*(1), 45–60.

Mkhize, N. (2019). The importance of extended family in African societies: A focus on aunts and uncles. *African Journal of Social Work, 9*(1), 45–58.

Mkhize, N., & Reddy, P. (2020). Cultural Transmission and Mediation in African Families: The Role of Aunts. In African. *Journal of Family Studies, 5*(1), 45–58.

Moyo, T. (2012). Emotional Support and Family Dynamics in African Contexts. In African Family Dynamics: Perspectives and Challenges (pp. 45-60). African Books Collective.

Moyo, T. (2019). Cultural Practices and Family Structures in Zimbabwe: The Case of the Shona People. *Zimbabwean Journal of Sociology*, *12*(3), 78–92.

Mudzengi, B. K., Gandiwa, E., Muboko, N., Mutanga, C. N., & Chiutsi, S. (2021). Ecotourism resilience: The case of Mahenye community project, Chipinge. District, Zimbabwe. *African Journal of Hospitality, Tourism and Leisure*, *10*(2), 459–471. DOI: 10.46222/ajhtl.19770720.111

Naidu, M., &Muchono, W. (2019). Male Perceptions Toward the Cultural Practice of Chinamwali/Khomba Among the Mahenye. Oriental Anthropologist: A Bi-annual International Journal of the Science of Man, 19(1), 25–40. DOI: 10.1177/0972558X19835381

Nyoni, T. (2020). The Influence of Extended Family on Child Development in Zimbabwe. *Journal of Child and Family Studies*, *29*(4), 1023–1035.

Olupona, J. K. (2006). African Traditional Religion in the Modern World. In The Study of Religion: A Critical Survey (pp. 123-136). Routledge

Patton, M. Q. (2015). *Qualitative Research & Evaluation Methods* (4th ed.). SAGE Publications.

Smith, P. L., Smith, R. J., & Jones, L. M. (2020). Sample size determination for social science research: A practical guide. *Journal of Social Research*, *34*(4), 112–130. DOI: 10.1080/12345678.2020.1234567

Tansley, C. (2021). Intergenerational Relationships and Family Dynamics: The Role of Extended Family. *Journal of Family Studies*, *27*(2), 123–140.

Zimunya, T. (2021). Women's Roles in Zimbabwean Society: A Historical Perspective. *African Studies Review*, *64*(2), 123–140.

ADDITIONAL READING

Hays, S. (2015). *Aunts and uncles: The forgotten family members*. Family Press.

Johnson, L. (2018). Family dynamics and the role of non-parental figures. *Journal of Family Studies*, *24*(3), 45–60.

Patel, R. (2019). Cultural perspectives on the role of aunts in family life. *International Journal of Cultural Studies*, *15*(2), 112–128.

Smith, J. A. (2016). The role of extended family in child development. *Child Development Perspectives*, *10*(1), 23–29.

Thompson, K. (2020). The impact of extended family on child well-being. *Journal of Child Psychology and Psychiatry, and Allied Disciplines*, *61*(4), 456–467. PMID: 31912483

KEY TERMS AND DEFINITIONS

Cultural heritage: It refers to the tangible and intangible legacy of people in a geo-social location which include physical artifacts, traditions, normative practices, languages, and values that are passed down through generations.

Family structure: This is the composition and organization of a family unit, including the relationships between its members. It can vary widely, encompassing nuclear families (parents and children), and extended families (including other relatives).

Familial Ties: These are the connections and relationships that exist between family members. These ties are often characterized by emotional bonds, shared responsibilities, and mutual support, playing a crucial role in shaping individual identities and influencing social interactions within families.

Mediate: This is to intervene in a dispute or negotiation to facilitate communication and resolution between conflicting parties. A mediator helps to clarify issues, promote understanding, and guide the conversation toward a mutually acceptable agreement.

Socio-cultural: It refers to the interplay between social and cultural factors that influence human behaviour and interactions within a society. It encompasses the norms, values, beliefs, and practices that shape individuals' experiences and identities in a given cultural context.

Sense of identity and belonging: A sense of identity and belonging refers to an individual's perception of themselves in relation to their social or cultural group. It encompasses feelings of connection, acceptance, and shared values with others, contributing to emotional resilience and well-being.

Chapter 6
Indigenous African Curricula and the Question of Epistemic Diversity:
The Case of the Mulabalaba Game

John Nyambe
University of Namibia, Namibia

Taimi Nghikembua
University of Namibia, Namibia

Mishack Thiza Gumbo
https://orcid.org/0000-0001-6760-4341
University of South Africa, South Africa

ABSTRACT

A qualitative case study in orientation, the research used the Mulabalaba indigenous game to explore the forms of knowledge and pedagogies that underpin indigenous African curricula and the implications they hold for present-day higher education curriculum development and transformation. We observed ten purposively selected participants who played the Mulabalaba game during the School of Education's Staff Wellness Day in 2024. A key finding of the study was that the indigenous African curriculum is underpinned by diverse epistemic forms, values and skills like the 21st century skills that are highly sought in present-day higher education curricula. Other findings revealed that the underpinning epistemic forms, values and skills are interwoven seamlessly and comprehensively and delivered in captivating and engaging pedagogies. The study recommends future research into how lessons from indigenous African curricula can be incorporated to inform and strengthen present-day higher education curriculum development and transformation.

DOI: 10.4018/979-8-3373-4124-8.ch006

INTRODUCTION AND BACKGROUND

Using the Mulabalaba game, the study reported in this chapter explored the forms of knowledge that underpin indigenous African curricula, the pedagogies used to teach these forms of knowledge, and the implications indigenous African curricula hold for higher education curriculum development and transformation. Mulabalaba game, is an indigenous African game widely played among different African communities.

The game carries different nomenclatures according to different languages and cultural groups. In Namibia, the game is called Mulabalaba among the Lozi-speaking language group in the far north-eastern Zambezi Region of the country. In contrast, among the Awambo-speaking groups in north-central Namibia, it is called Owela. According to Meeran et al. (2024, p. 378), among some Sotho-speaking communities in South Africa, the game is called 'Mmela' while its South Sotho name is 'Morabaraba', and among the Nguni speakers, it is called 'Umulabalaba'.

As free leisure is discouraged in most traditional African settings, the Mulabalaba game is considered one of the most acceptable occupations for the young and elderly in the afternoons while resting from their daily chores of ploughing or herding cattle or goats. Sometimes, when herding cattle or goats, young men digress from their main task of herding and engage in the Mulabalaba game, something that elders do not approve of as the cattle or goats end up getting lost or may end up in the fields. The game may not interfere with work.

In the traditional community setting, the game is designed by anyone familiar with it before they start to play. The game is drawn up or designed on the ground under a tree, preferably on hard-surfaced ground, to ensure it can be reused again the following day(s). If the game is played only for that day, it could also be designed or developed on soft sand. In modern days, the game can be designed on hard boards (see Figures 1 and 2 below) or on a computer.

Traditionally, the design involves digging up 24 or more shallow holes in equal rows of 4, each with six holes or more. The rows are designed in even numbers. The rows can be used to represent the multiplication table. In the traditional setting, the depth and width of the holes are measured in terms of appropriateness as opposed to numerical specifications. However, despite the specifications or measurement stated in qualitative terms of appropriateness, the holes always maintain a standard size irrespective of the community where the game is played.

Figure 1. Mulabalaba game board (Goggle search, 2024)

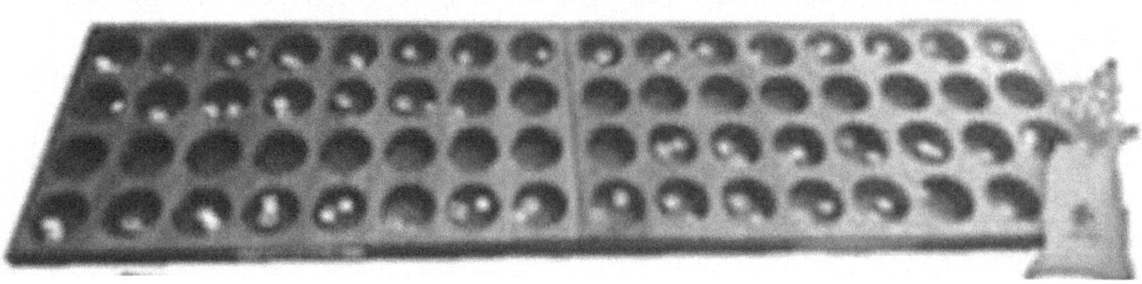

Figure 2. Mulabalaba game board (Goggle search, 2024)

The number of pebbles, seeds, or stones used to play the game varies depending on the number of rows. However, the pebbles, seeds or stones used may not occupy all the holes to allow players space to move. Traditionally, the pebbles represent cows, bulls and calves (Mosimege & Nkopodi, 2017). They also represent seeds. The holes represent kraals for cattle or holes for sowing seeds. The game can either be played as in Figure 1 or Figure 2 above. The game is played by two individuals, or two teams seated on opposite sides of the game. Each player or team owns the two rows on their side.

The holes on both sides are populated with two pebbles or cows per hole, as depicted in Figure 1(a), except for the ones in the middle, which are left to allow the two parties to move. Each side proceeds by playing the game counterclockwise by picking two cows from a kraal of their choice, dropping each cow into the subsequent holes, picking some cows along the way and dropping them off until the player lands one cow into an open kraal. Suppose the player lands one cow into an open kraal in the front row, and the open kraal is adjacent to two other kraals of the opponent, which contain cows. In that case, those cows are captured and placed into the opponent's kraals by dropping one cow per kraal in a forward-moving direction. The cows are again captured if the last cow drops into a front-row kraal adjacent to the opponent's two kraals that contain cows. An opponent may only capture if the adjacent front row contains a cow and not if it is only the adjacent rear kraal that contains a cow or cows. The game is won when one side loses its cows to its opponent and no longer has cows remaining to play with meaningfully. The game is also lost when their opponents completely block one side so that any manoeuvring made results in their cows being captured.

STATEMENT OF THE PROBLEM

Over the years, Euro-American-centric epistemologies and pedagogies have dominated curriculum development across the African continent. Rarely have the official circles of curriculum development given any consideration for including African epistemologies and pedagogies in curriculum development. Consequently, as Knaus et al. (2022, p. 4) rightly observes, African higher education has largely remained committed in shape, scope and form to Western approaches. According to Kaya and Seleti

(2013), the higher education system is still too academic and distant from the developmental challenges of African local communities. Most African higher education institutions' organizational structure, operations, climate and milieu have remained typical mimicries of the Western higher education model.

Higher education curriculum development has thus remained a space for epistemic colonial domination, relegating local indigenous knowledge to subjugated forms of knowledge that are expelled and dismissed from the curriculum space. Le Grange (2016, p. 4) describes this relegation and subjugation of indigenous knowledge as "the murder" or "epistemicide" of indigenous knowledge. Consequently, the epistemicide means that most of the irreplaceable indigenous knowledge may be lost forever unless some form of reclamation, promotion, preservation and dissemination is devised. This chapter is an attempt at the reclamation, promotion, preservation and dissemination of indigenous knowledge.

Shahjahan et al. (2021, p. 76) further highlight the epistemicide of local indigenous knowledge by locating the phenomenon within the global geopolitics of knowledge, which is underpinned by "a set of knowledge/power relations that privileges a certain gaze or representation of the world deemed universal, delocalized and applied unquestioningly". Shahjahan et al. (2021) further observe that those who work and live in the metropolitan centers of knowledge and learning (e.g., Anglo-Euro-American contexts) have the epistemic privilege to articulate and shape global discourses that culminate in higher education curricula. Conversely, those who live in the periphery of the global knowledge system (e.g., indigenous communities) are denied the privilege to participate in and shape the global discourse that forms higher education curricula.

Relevantly, Ndlovu-Gatsheni (2021) argues that the underlying cause of cognitive conditioning is epistemicide. He further argues that epitemicide kills existing endigenous knowledge, linguicides kills existing indigenous languages by imposing colonial languages, culturecides kills indigenous cultures by weighing cultural imperialism, and exiles the natives from their languages, histories, cultures, and even from themselves. It is for this reason that scholars call for epistemic disobedience (Mignolo, 2011), defiant scholarship (Daley & Murrey, 2022) and radical indigenism (Hart, 2010).

Given the discontent towards the Euro-American-centric curricula, over the past few decades, we have witnessed an exponential rise in the calls for decolonization of higher education curricula across African higher education institutions. Most notable are the 2015 #FeesMustFall and the #RhodesMustFall student protest movements at South African higher education institutions. The calls made by student protests to decolonize higher education have been followed by a proliferation of studies by several scholars (e.g., Martinez-Vargas, 2020; Ntshoe, 2020; Knaus et al., 2022; Gumbo et al., 2022; Isabirye, 2022) echoing similar calls to decolonize African higher education.

A qqualitative case study in orientation, the study reported in the chapter used Mulabalaba, an African indigenous game, to illustrate how African epistemologies and pedagogies can inform curriculum development in African higher education. In particular, the study explored whether the Mulabalaba game is underpinned by any knowledge and pedagogy that can inform official curriculum development in higher education. The study's key question was: What forms of knowledge and pedagogies underpin the Mulabalaba indigenous game, and what implications do these forms of knowledge and pedagogies hold for curriculum development in African higher education contexts?

RESEARCH OBJECTIVES

The objectives of the research were to investigate:

i) the forms of knowledge that underpin indigenous African curriculum as seen in the case of Mulabalaba game;
ii) the pedagogies used to teach these forms of knowledge;
iii) the implications of these pedagogies for enhancing learning; and
iv) the implications the forms of knowledge and pedagogies in indigenous curricula hold for higher education curriculum development and transformation.

LITERATURE REVIEW

To gain a deeper understanding of the phenomenon under investigation, we focus, in this section, on clarifying the concept of curriculum in the higher education context, its epistemic or knowledge foundations, knowledge structures and facilitation at the level of teaching and learning. Literature (Tyler, 1981; Fraser & Bosanquet, 2006; Bovill, 2021; Liang, et al 2024) highlights the diverse curriculum definitions and conceptualizations underpinned by what Habermas (1968), in his seminal work on *Knowledge and Human Interests,* termed the knowledge-constitutive human interests.

Underpinned by the knowledge-constitutive human interests revolving around technical control, the product orientation conceptualizes curriculum as a product, a finished product on paper, a plan, a guide, or a formal academic plan that is defined and recorded on an official template of some sort and is handed over to students at the start of the semester (Fraser & Bosanquet, 2006; Bovill, 2021). The student simply receives the finished product in a take-it-or-leave-it approach, with little or no contribution from the student side. This conceptualization views the teacher as a transmitter of knowledge, while knowledge is viewed as an object that can be packaged and handed over to the student.

The second conceptualization views curriculum as a process, a step-by-step interactive and flexible process that is informed by, and centers on student learning. While the teacher is viewed in this conceptualization as a reflective practitioner who constantly reflects to improve practice, knowledge is seen as socially co-constructed with the student. This conceptualization of curriculum is underpinned by emancipatory knowledge-constitutive human interests. Lastly is the conceptualization of curriculum as praxis or transformative action that emancipates students to engage in transformative actions intended to transform and shape human society. Curriculum as praxis is underpinned by empowerment and emancipatory knowledge-constitutive human interests (Fraser & Bosanquet, 2006; Bovill, 2021).

While product orientation has gained hegemonic dominance in most higher education institutions, deeply saturating our consciousness as the only viable option that can (Apple, 2004), it is possible that conceptualizations may co-exist in a mixed mode in a given higher education context. These conceptualizations inform and shape the manner of curriculum development. Rooted in the product orientation tradition, Tyler (1981, p. 18) views curriculum development as "developing the plans for an educational program, including the identification and selection of learning experiences, the organization of the learning experiences, and the evaluation of the educational program".

Significant to the present study are the epistemic foundations or knowledge foundations that underpin curriculum in higher education contexts. Adopting a socio-epistemic approach to the Sociology of Knowledge, Bernstein (1990; 1996; 2000) and Muller (2009) provided insights into the epistemic foundations and knowledge structures underpinning curriculum. In particular, Muller (2009) highlighted the issue of disciplinary knowledge and discipline by using the notion of paradigms to illustrate the social connectedness or the degree of consensus within a particular field regarding theory, methodology and techniques for solving problems.

Muller (2009) linked the epistemic foundations of curriculum to disciplinary areas within the hard pure sciences that are abstract and reflective (e.g., natural sciences), the hard applied sciences that are abstract and pragmatic (e.g., science-based professions like engineering), the soft pure sciences that are concrete reflective (e.g., social sciences), and lastly, the soft applied sciences such as social professions (e.g., social work) (Muller, 2009). Bernstein (2000) classified the epistemic foundations into two basic forms of knowledge, namely: mundane knowledge (which is the everyday knowledge accessed through experience) and esoteric knowledge (principled knowledge accessed through systematic investigation). These forms of knowledge and the knowledge paradigms are further distinguishable on the principle of coherence (Bernstein, 2000; Muller, 2009; Muller & Gamble, 2010).

The hard pure sciences/esoteric sciences are characterized by internal conceptual coherence appropriate to the epistemological core of their discipline. The soft pure sciences/mundane sciences, on the other hand, are characterized by contextual coherence, with adequacy to a given context. Bernstein (2000) further characterized these knowledge structures as vertical knowledge structures (hierarchical knowledge structures where lower-level forms of knowledge are a prerequisite to higher-level forms of knowledge) and horizontal knowledge structures, which are context-bound and segmented with segments positioned next to one another in a specific context. In the segmented horizontal knowledge structure, Knowledge of one segment is not a prerequisite to the knowledge of the next segment. Also to be noted is the observation that diverse knowledge forms may converge to form regions (Muller, 2008) or new fields of knowledge comprised of a cluster of disciplines which may combine both vertical and horizontal knowledge forms (e.g. business studies).

Significant to the present study is what Bernstein Bernstein (1990; 1996; 2000) termed the epistemic device. The epistemic devices are the rules and principles by which knowledge is converted or pedagogised for practice in the classroom context (Singh, 2002). Key to the pedagogic device is the principle of contextualization, which is the principle by which knowledge is appropriated, dislocated and relocated to a new context as it moves through the pedagogic fields of knowledge production, recontextualization and reproduction for use in the classroom level. Recontextualization of knowledge to facilitate teaching and learning in higher education curricula has tended to follow an epistemically diverse curriculum model, which has embodied four ways of knowing, namely: propositional knowledge, practical knowledge, experimental knowledge and epistemic knowledge (Luckett, 2001; Vorster & Quinn, 2012).

Luckett (2001) viewed the recontextualization at the classroom level in terms of quadrants, which included: i) *propositional knowledge,* which represents the traditional cognitive learning where a learner gains foundational competence by learning basic concepts or disciplinary knowledge through demonstration, listening, observation and induction to gain competence, ii) *practical knowledge,* where the learner gains practical competence by applying disciplinary knowledge through learning by doing or apprenticeship, iii) *experiential knowledge,* where the learner gains personal competence through learning by engaging personally and thinking reflexively, and iv) *epistemic knowledge,* where the learner develops reflexive competence and gains meta-cognitive competence through thinking epistemically, contextually and systemically. Practical and experiential knowledge represent the practical part of learning, while propositional and epistemic knowledge represents the theoretical part. While the dominant conceptualization of epistemic diversity, and recontextualization, as seen in the foregoing presentation, still take Western-centric orientations, we argue that epistemic diversity, and knowledge recontextualization at the classroom level, should be reconceptualized to mean epistemes that center indigenous African knowledge.

From the perspective of epistemic diversity that centers indigenous knowledge, Bhuda (2020; 2024), for instance, argued for a culturally responsive recontextualization of mathematics teaching through indigenous games that promote ethnomathematics in the foundation phase schooling in South Africa.

We take heed of Apple's (2004) fundamental question about knowledge, namely: whose knowledge is included in the curriculum and why, which and whose knowledge is left out and why? How does this knowledge contribute towards ethical and moral goals intended to transform society for betterment? These questions apply to the present study. Epistemic diversity would, therefore, refer to epistemes that recognize African indigenous knowledge. This entails that any recontextualization of knowledge from Bernstein's (1990; 1996)'s epistemic field of knowledge production to the field of classroom reproduction of knowledge should locate the knowledge within a particular cultural context, in this case, the African indigenous cultural context, as argued by Bhuda (2021).

Lastly, Bernstein (1990; 1996) was not only concerned about the *what* (knowledge) but also the *how* (pedagogy) of knowledge transmission as elaborated in his theory of pedagogy. He maintained that various forms of pedagogic practice (e.g., student-centered or teacher-centered pedagogy) can be generated using the internal rules constituting the inner logic of pedagogy. For instance, where the rules of hierarchy are explicit, and we have clear hierarchical relations of power between teacher and student, we are likely to have traditional pedagogy rooted in teacher authority. When the rules of hierarchy are blurred, and the student has control over the instructional aspects of pedagogy (selection, sequencing, pacing and evaluation), we are likely to have a student-centered pedagogy. We argue that the latter is more aligned with African indigenous pedagogy, namely *Ubuntugogy* (Bangura, 2005), an art and science of teaching and learning undergirded by humanity towards others.

METHODOLOGY

Given the nature of the research question, the qualitative case study research design was deemed appropriate. The case study enabled the researchers to explore the phenomenon in-depth in its real-life context (Yin, 2003; Johnson & Christensen, 2014). While the findings from case studies may not be generalisable, Patton (2002, p. 46) maintains that "one can learn from them – and learn a great deal, often opening up new territory for further research".

Sampling procedure

To gain an in-depth understanding of the Mulabalaba game, purposive sampling was adopted to select a sample that guaranteed an information-rich case, which Patton (2002) views as a case from which the researcher can learn a great deal about issues of central importance to the purpose of the research. Purposive sampling thus entailed sampling individuals who knew the rules of the game as well as how to play the game. Thus, ten participants were purposively selected to participate in the study from a very informed position about this type of game. The sample members were university academics, two females and eight males. All sample members had continued playing the game since they grew up in the villages.

Data generation

Techniques for generating data

Data for the study were generated through two techniques, namely naturalistic participant observation and conversational interviews (Patton, 2002). The naturalistic participant observation technique was deemed appropriate, allowing the researchers to observe the Mulalabala game in its natural setting. The naturalistic setting enabled the researchers to understand and capture the essence of the game. Our observation role fluctuated, at some point getting fully engaged in experiencing the game as players while at other points in time, observing and talking to other participants about the game.

Participants were interviewed to augment observation data. Initially, a semi-structured interview guide was prepared. However, this was abandoned, given the need for spontaneity and flexibility in our interview process as the game unfolded. An unstructured conversational interview (Patton, 2002; Johnson & Christensen, 2014) approach was finally adopted. This entailed talking with one or more individuals during the game. The interview technique allowed us to "enter into the other's perspective" and to "find out what is on someone else's mind, to gather their stories" about the Mulabalaba game (Patton, 2002, p. 341). The interview technique was important to us as we could not observe everything, including the feelings and thoughts of our research participants.

Procedure for generating data

Following an announcement of the end-of-year "Staff Wellness Day" in the School of Education, two of us requested the programme organisers to include the Mulabalaba game as one of the games to be played during the Staff Wellness Day in 2024. Our intentions to learn about the game and collect data were announced to the organisers. The organizers then tasked us to organise the game. We identified and invited ten colleagues who understood and knew how to play the game. We asked these colleagues to set aside time for the game during the Staff Wellness Day. We explained our role as researchers, and our colleagues granted us consent to learn about the game and collect data during the game through observation and interviews. One of the colleagues with a Mulabalaba board offered to bring it to the Staff Wellness Day. Data were collected during six sessions that lasted between 45 and 60 minutes each. Two of the six sessions involved single players that played against each other, while the rest involved teams that played against each other.

Data analysis

Thematic data analysis technique (Miles & Huberman, 1994), which entailed observing patterns in the data, assigning codes, and generating themes and sub-themes, was used to analyse the data. Following this technique, themes emerged that enabled the organised presentation of the findings. The analytical framework presented in the preceding section was used to enable us to make sense of the data.

Ethical considerations

As alluded to in the preceding section, participant consent was sought and granted by the ten purposively selected sample members after the purpose of the study was explained to them. To protect and ensure the safety of participants, they preferred that no personal names and identities be used throughout the report. Participants were thus assigned pseudonyms. Throughout the report, the following pseudonyms are used: Mwale, Shivute, Nganga, Chuma, Sikute, Simana, Nekongo, Iipumbu, Pingana and Mbambo. The study was also approved, and permission was granted by the organizers of the Staff Wellness Day. The Mulabalaba game was then included in the activities for the day.

FINDINGS AND DISCUSSION

Logical and careful planning

One of the key findings that emerged through observation and interview data is that the Mulabalaba game involves careful logical planning and thought. We observed that before a player or a team moved their cows and played them by dropping one cow in the subsequent kraals, there was a lot of careful thinking and planning to be done. The player or team could take about five minutes before they moved their cows, and after, the opposing side almost started complaining about the delay.

The planning also involved determining which kraal to move, as the cows were not moved haphazardly. One of the female participants, Mwale, indicated that *you need to look after your cattle, move them only when it is proper to do so, keep them safe from the drought.* Mwale continued, *"you should be careful before you move, otherwise your cows will be captured, or they will die from drought"*. We also learned from Shivute that *a move cannot be reversed once made.* Therefore, given that any move made cannot be reversed, it was important for the players to ensure that careful thinking was applied before the cows were moved.

We also observed that when playing in teams, the group reprimanded a team member who moved a cow or cows on his or her own before the group agreed to do so. A group needed to agree before their members could move a cow. However, since the board was too long, stretching for about 64 holes or kraals, the team members jointly played, with one team member dropping the cows only as far as his hands could reach and then the other team member picking it up from there, to continue dropping the cows. To this effect, Nganga indicated: *the team must agree before any move is made; otherwise, they risk being beaten by their opponents.* Given the carefulness and consideration demanded by the game, the sessions were quite longer, some lasting for 45 minutes while others could go beyond an hour.

Constant computation/counting

In addition to being careful, it was also observed that players would at times pick one kraal and count the number of cows in the kraal, count the subsequent kraals where the cows should be dropped off and establish whether the final landing could yield them any capturing of the opponent's cows. Some players went as far as counting their opponents' kraals or cows. This move was, however, not so welcomed by the opposing side as they felt that the opponent should not count their kraals. The counting/computation involved adding cows, subtracting kraals, and multiplying. Chuma advised: *before your cows depart from a kraal, make sure the last cow is dropped in a kraal that can enable you to capture more cows from your opponent, or it is dropped in a kraal where you can block your opponent.*

Sikute concurred with Chuma: *You need to make strategic moves, count the number of cows you intend to move, determine where you will end up, whether you will capture and are still safe, also assess the opportunities the opponent can use against you.* When asked whether the goal is to capture many cows from your opponent, Mwale was quick to advise: *one needs to learn not to be greedy; being motivated to capture more cows can lead you into trouble. Rather take moves that are winning.* She further maintained: *big numbers are difficult to calculate. It is difficult to capture with big numbers. Always consider moving kraals with few cows. Then, you can compute easily if you are going to capture your enemy's cows.* From our observations, the game involved advanced cognitive processes where players had to compute and think strategically about how to manoeuvre and outsmart their opponents. The computing process lengthened the game.

An addictive and disruptive game

One of our observations was that it was difficult to leave or stop once a player or team started playing the game. The ten players who we initially requested to play the game for us had opportunities to participate in the other games planned for the Wellness Day, such as soccer, athletics, dance and water games as per the programme. Even when they left for other games, we observed that the Mulabalaba players did not stay longer with other games and returned to the Mulabalaba game.

As researchers, we also tried to join the other games, but we were also caught up in the addiction to the Mulabalaba game. Occasionally, other players not initially selected for our sample kept joining, leaving and returning. One of the female participants, Chuma remarked: *when we were growing up in the village, pots used to burn on the fire.* Chuma alluded to the fact that the game was so absorbing that one would forget to attend to the pot that one was cooking on the fire. Simana concurred: *that is why elders sometimes did not like the game. At times the boys would play, and the cattle would get lost or go into the fields.*

We observed that the game organisers struggled to control staff members playing the Mulabalaba game, especially when they needed them to join the other games. One of the organisers reprimanded the players out of frustration and asked one of us (researchers) to stop the game for a while so that the Mulabalaba players could go to other games. We tried to pause the game but with no success. Eventually, the Staff Wellness Day organisers decided to diversify and have various games running parallel for different groups. This enabled the Mulabalaba group to stay with their game. We learned that the game was disruptive in this regard.

While some staff members referred to the game as addictive, we concluded from our observations that it was rather captivating, capturing the attention of its players. The players seemed to enjoy the game; hence it was not easy to make them leave for other games. This is confirmed by the level of noisy conversation and loud laughter from the Mulabalaba table compared to other games.

Amenability to certain knowledge/content areas

Another feature that emerged in our conversations with the players of the Mulabalaba game is that it was amenable to certain knowledge or content areas. For instance, some players kept referring to their pebbles or stones as cows throughout the game. Players also referred to the holes on the board as kraals.

They also spoke about ensuring that one's cows survive the drought. These references suggested that the Mulabalaba game can be used in the teaching of Agriculture.

While most players referred to the stones as cows, Shikongo referred to the stones as seeds. Shikongo was also frequently heard referring to the holes not as kraals for keeping cattle but also for sowing the seeds. When we asked Shikongo why he referred to the stones as seeds and not cows, he shared his background growing up in a crop farming community and how the game was used to plan for their crops. We concluded from this interaction with Shikongo and the other players that if it were to be used to teach an agricultural lesson, the lesson using the game would have cultural relevance.

Apart from teaching Agriculture, it was evident that the Mulabalaba game could be used for facilitating Mathematics lessons. The logical and careful reasoning and the constant calculation or computation throughout the game suggested that the game could be used in teaching mathematical concepts such as addition, subtraction, multiplication or division. Chuma, who is from a mathematics background, kept referring to the game as mathematical. We asked Chuma if she had been using the game in her classes. Unfortunately, Chuma had not been using the game although she planned to secure the Mulabalaba board and start using the game in her lessons.

During the initial interview phase, when we requested players to describe the game to us, some used descriptors such as, the game resembled the multiplication table and that the total number of holes should be an even number. The mathematical act of calculating or counting was evident throughout the game as various players calculated before they made their moves. The players also described the game in Figure 1(b) about the Mulabalaba board that was brought along was that in Figure 1(a). However, it was evident from the descriptions that Figure 1(b) could be used to teach other mathematical concepts such as shapes. Listening to the conversations during the game, as researchers, we felt that the game also had the potential to be used in a language teaching class as team members needed to communicate with each other to play.

A skills-based game

Some of the research participants narrated the skills involved in the game. For instance, Chuma referred to *problem-solving* as one of the game's core skills – problem-solving is one of the critical skills targeted in Technology Education; hence, Mulabalaba can also be used in teaching this subject. Chuma argued that to win the game, one needed to solve the problem, or challenges set by the opponent. The game entailed setting traps for each other, representing problems that needed resolution. Pingana narrated what skills he thought the game taught, i.e. *perseverance, endurance and persistence.*

The skills narrated by Pingana were evidenced by the fact that the game could last for hours, and the players never gave up. In some cases, a player who lost a game still wanted to continue with the game. It appeared that losing served as motivation to engage further. The players, therefore, persevered, endured and persisted with the game. For Mbambo, the game taught the skills of *teamwork, competition and task orientation*. Teamwork was evident through several teams that played against each other. Teamwork entailed team planning, calculating and making moves – teamwork and planning are skills well sought after in Technology Education, and calculation in Mathematics. It was further evident that teamwork made the game stronger and more challenging than when individual players played against each other. Competition and task orientation were evident because the teams or individual players remained focused on the task and were geared to win the game. Lipumbu referred to *planning* and *communication* as some of the skills that one learns through the game. We concluded that players picked up several skills as they played the game.

From the foregoing presentation of findings, we conclude that indigenous African curricula, as seen in the case of the Mulabalaba game, are underpinned by conceptual, disciplinary or propositional knowledge available in the domains of hard pure sciences, hard applied sciences, soft pure sciences as well as the soft applied sciences. The Mulabalaba game represents a process of recontextualization which, as Bernstein (1990; 1996) argued, is a process of appropriating a discourse from the epistemic field of knowledge production, relocating the discourse, and de-locating it in a new context for classroom use.

Through the recontextualization process, the Mulabalaba game is predisposed towards teaching concepts in Agriculture, Mathematics, Language and Technology Education. The game is predisposed towards the teaching of disciplines with vertical knowledge structures (e.g., Mathematics), horizontal knowledge structures (Languages) as well as regionalized knowledge structures such as Agriculture and Technology Education. Reference made by players to cows, kraals, drought, and seeds during the game suggests that the game is underpinned by conceptual knowledge existing in given disciplinary areas, in this case, the disciplinary area of agriculture.

Conceptual or disciplinary knowledge in Mathematics was evident throughout the games. As alluded to earlier, the players engaged in mathematical activities such as calculations, computation, logical and careful thinking, subtraction, addition, division, and multiplication – all that suggested that the Mulabalaba game renders indigenous African curricula underpinned by propositional or foundational knowledge drawn from diverse knowledge structures. We learned through a literature review that scholars in Mathematics (e.g., Igbal Nst & Batubara, 2024; Matsekoleng et al., 2022; Meeran et al., 2024; Mosimege & Nkopodi, 2017; Nkopodi & Mosimege, 2009; Bhuda, 2024; Putri et al., 2024; Wulandari, 2024) argue that the incorporation of indigenous cultural games into Mathematics teaching is called ethnomathematics. Nkopodi and Mosimege (2009) argued from an ethnomathematical perspective that the Mulabalaba game is ideal for students:

…to learn the language and vocabulary of Mathematics, develop mathematical skills, develop ability with mental Mathematics, and devise problem solving strategies and to be the generator of mathematical activity at a variety of different levels .. the use of the game also leads to discovery of patterns, decision making, and logical reasoning, deduction and skills (Nkopodi & Mosimege, 2009, p. 380).

Nkopodi and Mosimege further analyzed the game and found the following concepts embedded in it, particularly the Mulabalaba game presented in Figure 1(b) design. The following mathematical concepts are found in the analysis of *Morabaraba*:

- Identification of various quadrilaterals (squares) and similarities and differences between them;
- Ratio and proportion between the lines and the squares making the complete *Morabaraba* board;
- Symmetry: symmetry is observed in at least three different instances, namely the different sides of the board, within each side of the board, and the placement of tokens and repetitive movements of the tokens on the board;
- Logical deductions in the execution of the various steps of the game;
- Counting of the tokens;
- Addition and subtraction of the tokens until a game is won based on the remaining number of tokens (Nkopodi & Mosimege, 2009, p. 385).

In addition to the conceptual knowledge that emerged in the conversation with the participants and our own observation of the processes involved in the Mulabalaba game, we also concluded that learning the rules of the game represented foundational knowledge. The rules and other discipline-specific concepts represented traditional cognitive learning that is taught differently.

In addition to conceptual or propositional knowledge, we further conclude that observation and interview data gleaned through the Mulabalaba game illuminated how knowledge was being appropriated from the epistemic field of knowledge production, and recontextualized or pedagogised (Singh, 2002) into practical knowledge as curriculum knowledge foundation akin to what happens in modern day higher education curriculum. We observed that some players who dropped in from the other Wellness Day games to play the Mulabalaba game had to learn the rules (conceptual and procedural knowledge) and were also allowed to apply the rules in practice (practical knowledge) by playing together with the seasoned mentors.

The game entailed gaining practical knowledge to move one's cows around the kraals and capture the opponent's cows. As a participant observer, one of us who had played the game years back in primary school had to re-learn the game's rules and undergo some apprenticeship of a kind, learning by doing, however, learning as a teammate with an experienced and knowledgeable player. This enabled the researcher to gradually gain practical competence in playing the game. We conclude that in addition to foundational knowledge, indigenous African curricula are further underpinned by practical knowledge, as seen in the case of the Mulabalaba game.

In addition to the skills above concerning Technology Education, we observed that creativity and criticality can also be used. Viewed as a technological solution to problem-solving, as is the case in the hard applied sciences, the design of the Mulabalaba game can hone those creative and critical thinking skills and several other skills listed by Matsekoleng et al. (2022) in the subject. Conceptual (knowing that) and procedural knowledge (knowing how) also play a critical role in problem-solving through design in Technology Education. Conceptual knowledge is about the relationships among knowledge items, such as systems concepts (McCormick, 2004). Procedural knowledge refers to design, problem-solving, planning systems, analysis (or systems approach), optimization, modelling and strategic thinking (McCormick, 2004). Hence, Gumbo (2019) coins teknowledgé to describe indigenous knowledge because it is mainly practical.

The richness of communication involved in the Mulabalaba game presents an opportunity to teach language such that it provides alternative approaches to Western approaches. The teacher could organize the lesson around the game if brought into the learning situation. He or she could let the players play and, together with learners, identify parts of speech, grammatical principles, expressions of kinds, etc, to teach language.

Our experience as researchers and long-time players of the Mulabalaba game, who had to re-learn the game as participant observers, was that once we demonstrated practical competence gained through apprenticeship training, we were sometimes left to try the game independently. We also observed players who dropped by and were learning the game, underwent the same process. Players learning the game for the first time had to undergo apprenticeship learning. They later tried to play the game on their own, through what Luckett (2001, p. 55) refers to as "learning by engaging personally, thinking reflexively" to gain "personal competence" in the Mulabalaba game. We draw parallels here with the case of a teacher education student, who first learns conceptual or foundational knowledge in a teacher education program, and practices how to teach the concepts under apprenticeship training by a teacher educator or a schoolteacher to gain practical competence. Lastly, the teacher education student is placed in a practicum situation where he tries out teaching alone to gain personal competence.

Furthermore, from the sophisticated maneuvering, and the complicated computations that players engaged in to outsmart their opponents, we conclude that the Mulabalaba game illuminated "the development of meta-cognition", enabling the player to "think epistemically, contextually and systemically" (Luckett, 2001, p. 55; Muller, 2008). We argue that the advanced players whose games lasted an hour, or more were operating at this level of competence. For instance, we observed teams that had lost more than half of their cows and were on the verge of losing the game. Still, such teams could bounce back and win the game by applying advanced cognitive abilities, thinking epistemically and meta-cognitively exploring alternative ways to play the game and outsmart their opponents.

Figure 3. Forms of knowledge (Adapted from Luckett, 2001)

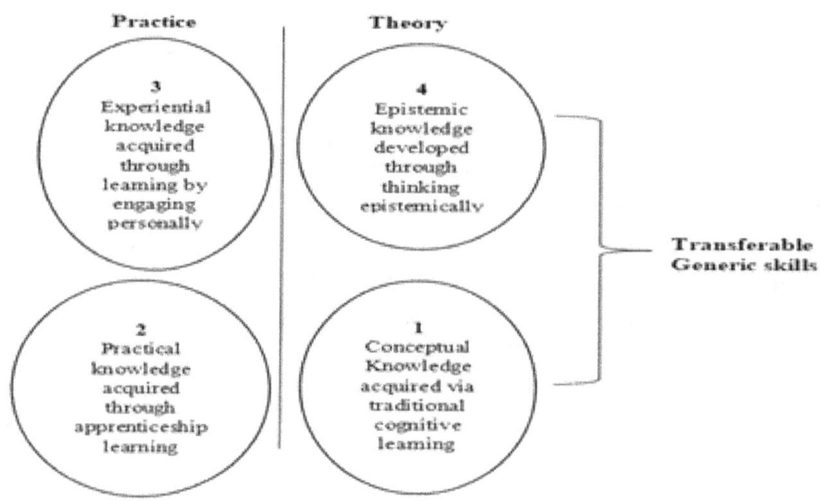

As presented in Figure 3, we conclude that the indigenous African curricula, as manifested through the Mulabalaba game, is an epistemically diverse curriculum. In the same vein as most official curricula, the diverse epistemic foundations in indigenous African curricula are a result of a recontextualization process that has pedagogised knowledge into four types of knowing for use at the classroom room.

We further conclude that the indigenous African curricula, as evidenced through the Mulabalaba game, are underpinned by several transferable skills, which, although not directly taught, are acquired by game players. Once learned or acquired through the Mulabalaba game, the skills can be applied in other contexts. The skills are meant to be transferable from one context to the other. The same principles apply to learning 21st century skills in academic programs today. For instance, problem-solving as a skill, once acquired, can be used in different contexts. We argue that these skills are like modern-day 21st century skills, even though the Mulabalaba game has come a long way over the past generations. The skills that the game promotes are the kind of skills that most employers and 21st century industry demand from higher education graduates. This makes indigenous knowledge a fit in the modern era, offsetting the notion that it is outdated and primitive.

As it came out in the interviews, the Mulabalaba game promotes, among others, the following attributes, which are 21st century skills: problem-solving, teamwork, communication, task orientation, perseverance and endurance, and competition. It has become mandatory, particularly in the Namibian context, that every higher education program should outline the graduate attributes or skills it intends to develop among the graduates. These skills should align with 21st century skills. The skills are also referred to as employability attributes, given that these are the kinds of skills that employers are looking for. We conclude, in this respect, that African indigenous curricula advance the development of 21st century skills, as demanded from higher education programs.

It was also evident in the data that in addition to skills, the Mulabalaba game developed values in the game participants. For instance, when asked whether the ideal is to capture kraals with many cattle, Mwale, one of the female participants, advised that: *One needs to learn not to be greedy, being motivated to capture more cows can lead you into trouble*. Therefore, in addition to skills, the indigenous African curricula are underpinned by values education. Our findings in this regard are consistent with Meeran et al. (2024, p. 2), who argued that "the games also develop values such as patience and respect, and skills such as strategic thinking, communication and decision making". These values invoke Botho (Ubuntu), emphasizing unity, respect, care, community ties, etc. According to Letseka (2013), Botho prescribes how human beings should relate to one another. Teamwork is critical for attaining wellness and performing activities, such as in Technology Education, where learners mostly engage in group activities. According to Maphalala (2017), Botho is expressed in the sayings, *cross a river in a crowd and the crocodiles will not eat you*. Moalosi et al. (2007) and Moalosi (2007), applies these values to Technology Education in the context of Botswana.

We further conclude from the data that indigenous African curricula are underpinned by captivating and engaging pedagogies that capture the participants' attention and interest. The engaging and captivating nature of the games was described in the data as *"addictive"*. While "addictive" may suggest something negative, in this regard, it was apportioned positive connotations. For instance, we heard in interviews from Chuma, who shared her experiences with pots that would burn on the fire while she and her mates were engaged in the game. Similarly, Simana shared his experience of cattle getting lost or going into the fields where they are not supposed to go.

The pedagogy also suggested an "unstoppable" learning process. The organizers of the Staff Wellness Day struggled to stop the game to allow players to go into the other sports activities for the day. However, this was achieved with difficulty. The players enjoyed the game, and it was not easy for them to leave. Noisy conversations and loud laughter, something uncommon in modern-day classrooms, were also heard from the players. If we espouse social constructivism in teaching and learning, then it is highly sensible that it cannot be divorced from play; games can enhance learning and make it enjoyable as we learned in Mulabalaba. We conclude that the pedagogical underpinnings of the African indigenous curricula, as seen from the case of Mulabalaba, could be instructive for higher education curricula. Putri et al. underscored the effectiveness of the indigenous pedagogies in Mathematics when they argued that:

...this approach consistently leads to substantial improvements in students' mathematics learning outcomes compared to traditional learning models. This aligns with the hypothesis that culturally integrated learning enhance students' comprehension and retention of mathematical concepts (Putri, et al., 2024, p. 118).

Similarly, Wulandari et al. (2024, p. 204) aver that pedagogies incorporating cultural components into teaching can inspire and improve students' comprehension of mathematical ideas. We conclude that indigenous African curricula align, through the process of recontextualization with propositional, practical, experiential, and epistemic knowledge, as expected in most higher education programs. In addition, the indigenous African curricula are underpinned by certain core values (which perfectly align with Botho) and develop skills required for survival in the 21st century. Indigenous African curricula are based on pedagogical approaches that are not only captivating of the student's interest but are also more culturally relevant, thereby enabling meaningful learning among students. An epistemically diverse modern curriculum ought to include indigenous knowledge.

Lastly, we conclude from the foregoing discussion that African indigenous curricula are underpinned by conceptualization of curriculum as a process that is student-centered, as in *Ubuntugogy* (Bangura, 2005), with the progression of learned determined by the pace at which the student acquires the competencies during the apprenticeship process.

Implications for higher education curriculum development

Based on the discussion, we argue that several lessons could be learned from the indigenous African curricula to inform today's curriculum development and transformation in higher education. One of the key lessons is how the indigenous African curricula, as seen through the Mulabalaba game, weave together seamlessly, through the recontextualization process, the underpinning knowledge foundations together with the teaching of skills and values in a captivating and more effective manner. In present-day higher education curricula, the relation between content knowledge, values and skills is not so easily discernible, particularly at the curriculum's classroom implementation level. These elements can be easily seen on paper in curriculum design. However, one may struggle to see or recognize the values and skills the curriculum promotes. This seamless weaving of knowledge forms, values, and skills in one comprehensive design and delivery is a key lesson for present-day higher education curriculum development and transformation that can be learned from the indigenous African curricula.

Furthermore, we find the culturally responsive pedagogies used by the indigenous African curricula as another lesson that could be learned to enable meaningful development and transformation of present-day higher education curricula. As indicated earlier, higher education curricula across the African higher education system are centered around Western pedagogies based on cultures foreign to the African student. This poses a serious challenge for students relating to foreign ways of learning. The integration of local cultural aspects, as seen in the indigenous African curricula, is strongly recommended for African higher education curriculum development and transformation.

Next is the captivating and engaging pedagogies where learning is almost unstoppable, providing significant lessons for higher education curricula, where there is a struggle to gain student interest in learning. The enthusiasm, perseverance, endurance, interest, noise and laughter associated with the Mulabalaba game could play a greater role if the higher education curriculum followed similar lines as the indigenous African curriculum.

Lastly, the open-environment-based game, such as Mulabalaba, situates teaching and learning in the real environment. This brings an important perspective to learning. The walled classrooms are not meant to imprison learners by confining them in a closed environment, which is also cognitively imprisoning. Education must connect with the real environment. Taking learners outside only for extramural activities rather than for learning is an anomaly.

In conclusion, we argue that several lessons can be drawn from the indigenous African curricula to inform curriculum development and transformation in higher education.

FUTURE RESEARCH DIRECTIONS

One key direction for future research emerging from the chapter is to investigate how cultural responsiveness can be promoted in higher education learning and teaching. The chapter suggests that a culturally responsive curriculum will likely generate student excitement and unstoppable interest in learning, as we saw in the Mulabalaba game. However, modern-day higher education curriculum falls short of the aspect of cultural responsiveness, especially in the African context, which pushes an agenda of Euro-American-centric culture. The failure to be culturally responsive could explain the low student completion rates in African higher education institutions.

CONCLUSION AND RECOMMENDATIONS

This study has contributed to the ongoing search for ways to decolonize African higher education by centering indigenous African knowledge. In particular, the study has provided a case of the Mulabalaba game as a concrete example to aid the frontiers of knowledge and practice in the decolonial discourse. Furthermore, the study has demystified the belief that indigenous curricula are haphazard and that any systematic body of knowledge and pedagogy does not underpin them. Yet, we have seen in the presentation and discussion of findings that indigenous African curricula command better coherence and seamlessness in their design without romanticizing it. They provide avenues for alternative forms of knowledge and pedagogies, as could be noticed in the Mulabalaba game.

We have provided implications for these forms of knowledge and pedagogies for curriculum development in African higher education contexts. We recommend that future studies focus on how lessons from indigenous African curricula can inform curriculum development and transformation in African higher education institutions. We also recommend a concerted effort by the educational authorities to ensure the integration of indigenous knowledge into the university curricula. As designers and implementers of the curriculum, academic staff have a big advantage in transforming the curricula by centering African Indigenous knowledge so that universities become truly African universities.

REFERENCES

(2022). African geographies. *Singapore Journal of Tropical Geography*, *43*, 159–176.

Apple, M. W. (2004). *Ideology and curriculum* (3rd ed.). Routeledge Taylor & Francis. DOI: 10.4324/9780203487563

Bangura, A. K. (2005). Ubuntugogy: An African educational paradigm that transcends pedagogy, andragogy, ergonagy and heutagogy. *Journal of Third World Studies*, *22*(2), 13–53. https://www.jstor.org/stable/45198556

Bernstein, B. (1990). *The structuring of pedagogic discourse: class, codes and control*. Routledge.

Bernstein, B. (1996). *Pedagogy, symbolic control and identity: theory, research, critique*. Taylor & Francis.

Bernstein, B. (2000). *Pedagogy, symbolic control and identity: Theory, research, critique*. Rowman & Littlefield. DOI: 10.5771/9781461636205

Bhuda, M. T. (2021). Making a case for indigenous education systems in South Africa. *African Journal of Development Studies (AJDS)*, 67-82, https://doi.org/DOI: 10.31920/2634-3649/2021/SIv1a4

Bhuda, M. T. (2024). Integrating ethnomathematical activities in the foundation phase as a means of decolonisation: Insights from indigenous knowledge systems. [AJDS]. *African Journal of Development Studies*, *14*(4), 401–424. DOI: 10.31920/2634-3649/2024/v14n4a18

Bhunda, M. T., & Pudi, T. I. (2020). Indigenising mathematics in schools: Why ethnomathematics matters. In Msila, V. (Ed.), *Developing Teaching and Learning in Africa: Decolonising Perspectives* (pp. 21–42). SUN Press., DOI: 10.18820/9781928480716/02

Bovill, C. (2021). *Curriculum conceptualisation and the University of Edinburg Institute for Academic Developmen*t, University of Edinburg, docs.hss.ed.ac.uk

Daley, P. O., & Murrey, A. (2022). Defiant scholarship: Dismantling coloniality in contemporary African geographies. *Singapore Journal of Tropical Geography*, *43*(2), 159–176.

Fraser, S. P., & Bosanquet, M. B. (2006). The curriculum? That is just a unit outline, isn't it? *Studies in Higher Education*, *31*(3), 269–284. DOI: 10.1080/03075070600680521

Gumbo, M. T., Gasa, V., & Knaus, C. (2022). Centring African knowledges to decolonise higher education. In Knaus, C. B., Mino, T., & Seroto, J. (Eds.), *Decolonising African higher education* (pp. 21–36). Routledge. DOI: 10.4324/9781003158271-2

Habermas, J. (1968). *Knowledge and human interests*. Beacon Press.

Hart, A. M. (2010). Indigenous worldviews, knowledge, and research: The development of an indigenous research paradigm. *Journal of Indigenous Voices in Social Work*, *1*(1), 1–16. http://hdl.handle.net/10125/12527

Igbal Nst, M., & Batubara, N. K. (2024). The importance of ethnomathematics in education: A systematic literature review. *Proceedings of the International Conference on Education Teacher Training & Education Faculty*. Universitas Serambi Mekkah.

Isabirye, J. (2022). Reclaiming indigenous epistemes: *Entenga* drums revival at Kyambongo University. In Knaus, C. B., Mino, T., & Seroto, J. (Eds.), *Decolonising African higher education: Practitioner perspectives from across the continent*. Taylor & Francis Group. DOI: 10.4324/9781003158271-6

Johnson, R. B., & Christensen, L. (2014). *Educational research: Quantitative, qualitative, and mixed approaches (5th Ed.)*. Los Angels: Sage

Kanus, C. B., Mino, T., & Seroto, J. (2022). Decolonising higher education: Definitions, conceptualisations, epistemologies. In Knaus, C. B., Mino, T., & Seroto, J. (Eds.), *Decolonising African Higher Education: Practitioner perspectives from across the continent*. Taylor and Francis Group.

Kaya, H. O., & Seleti, Y. N. (2013). African indigenous knowledge systems and relevance of higher education in South Africa. *The International Education Journal: Comparative Perspectives*, *12*(1), 30–44.

Le Grange, L. (2016). Decolonizing the university curriculum. *South African Journal of Higher Education*, *30*(2), 1–12. DOI: 10.20853/30-2-709

Liang, Y., Dai, K., & Mathews, K. E. (2024). Followers, customers, or partners in Australian, Mainland Chinese, and Hong Kong universities. *Higher Education*, *88*, 2251–2267. DOI: 10.1007/s10734-024-01215-z

Luckett, K. (2001). A proposal for an epistemically diverse curriculum for South African higher education in the 21st Century. *South African Journal of Education*, *15*(2), 49–61.

Martinez-Vargas, C. (2020). Decolonizing higher education research: From a university to a pluri-versity of approaches. *South African Journal of Higher Education*, *34*(2), 112–128. DOI: 10.20853/34-2-3530

Matsekoleng, T. K., Mapotse, T. A., & Gumbo, M. T. (2024). The role of indigenous games in education: A Technology and Environmental Education perspective. *Diaspora, Indigenous, and Minority Education*, *16*(4), 68–82. DOI: 10.1080/15595692.2022.2160436

McCormick, R. (2004). Issues of learning and knowledge in Technology education. *International Journal of Technology and Design Education*, *14*(1), 21–44. DOI: 10.1023/B:ITDE.0000007359.81781.7c

Meeran, S., Kodisang, S. M., Moila, M. M., Davids, M. N., & Makokotlela, M. V. (2024). *Ethnomathematics in intermediate phaser: Reflections on the Morabaraba game as indigenous mathematical knowledge*. African Journal of Research in Mathematics, Science and Technology Education., DOI: 10.1080/18117295.2024.2340095

Mignolo, W. D. (2011). I am where I think: Remapping the order of knowing. In Lionnet, F., & Shih, S. (Eds.), *The creolization of theory* (pp. 159–192). Duke University Press.

Miles, M. B., & Huberman, A. M. (1994). *Qualitative data analysis*. Sage.

Moalosi, R. (2007). The impact of socio-cultural factors upon human-centred design in Botswana. Unpublished doctoral thesis. Queensland University of Technology.

Moalosi, R., Popovic, V., & Hickling-Hudson, A. (2007). Product analysis based on Botswana's post-colonial socio-cultural perspective. *International Journal of Design*, *1*(2), 35–43.

Mosimege, M., & Nkopodi, N. (2017). The use of indigenous game of Morabaraba in mathematics education. In M. T. Gumbo & V. Msila (Eds.). African voices on indigenization of the curriculum: Insights from practice, pp. 155-180. Westville: Reach Publishers (Self-Published).

Muller, J. (2009). Forms of knowledge and curriculum coherence. *Journal of Education and Work*, *22*(3), 205–226. DOI: 10.1080/13639080902957905

Muller, J., & Gamble, J. (2010). Curriculum and structuralist sociology: The theory of codes and knowledge structures. In Petersen, P., Baker, E., & McGaw, B. (Eds.), *International Encyclopedia of Education* (Vol. 1, pp. 505–509). Elsevier. DOI: 10.1016/B978-0-08-044894-7.00090-7

Ndlovu-Gatsheni, S. J. (2021). The cognitive empire, politics of knowledge, and African intellectual productions: Reflections on struggles for epistemic freedom and resurgence of decolonisation in the twenty-first century. *Third World Quarterly*, *42*(5), 882–901. DOI: 10.1080/01436597.2020.1775487

Nkopodi, N., & Mosimege, M. (2009). Incorporating the indigenous game of morabaraba in the learning of mathematics. *South African Journal of Education*, *29*(3), 377–392. DOI: 10.15700/saje.v29n3a273

Ntshoe, I. M. (2020). Ontological social justice in decolonized and post-apartheid settings. *South African Journal of Higher Education*, *34*(3), 263–280. DOI: 10.20853/34-3-3283

Padayachee, K., Matimolane, M., & Ganas, R. (2018). Addressing curriculum decolonization and education for sustainable development through epistemically diverse curricular. *South African Journal of Higher Education*, *32*(6), 288–304. DOI: 10.20853/32-6-2986

Patton, M. Q. (2002). *Qualitative research and evaluation methods* (3rd ed.). Sage.

Putri, L. I., Begimbetova, G., Sa'idah, N., & Murfi, A. (2024). Evaluating the impact of ethnomathematics on mathematics achievement: A meta-analysis of studies from 2014- 2024. *Global Educational Research Review*, *1*(3), 110–122. DOI: 10.71380/GERR-10-2024-12

Shahjahan, R. A., Estera, A. L., Surla, K. L., & Edwards, K. T. (2021). "Decolonizing" curriculum and pedagogy: A comparative review across disciplines and global higher education contexts. *Review of Educational Research*, *92*(1), 73–113. DOI: 10.3102/00346543211042423

Sign, P. (2002). Pedagogising knowledge: Bernstein's theory of pedagogic device. *The British Journal of Sociology*, *23*(4), 571–582.

Tyler, R. W. (1981). Specific approaches to curriculum development. In Giroux, H. A., Penna, A. N., & Pinar, W. F. (Eds.), *Curriculum & Instruction*. McCutchan Publishing Corporation.

Vorster, J. A., & Quinn, L. (2012). Privileging knowledge, creating knowers: an analysis of a formal programme for university lecturers. In Quinn, L. (Ed.), *Re-imagining academic staff development: spaces for disruption*. Sun Press.

Wulandari, D. U., Marianna, N., Moh, W., & Amien, S. (2024). Integration of ethnomathematics teaching materials in mathematics learning in elementary school. *IJORER: International Journal of Recent Educational Research, 5*(1), 204-218. https://journal.ia-education.com/index.php/ijorer

Yin, R. K. (2003). *Case study research design and methods* (3rd ed.). Sage.

ADDITIONAL READING

Apple, M. W. (1993). *Official knowledge: Democratic education in a conservative era*. Routledge.

Bonoh, J. C. (2017). *Indigenous knowledge in the school curriculum: Teacher educator perceptions of place and position*. MEd dissertation. Qheberha: Nelson Mandela University.

Ezeanya-Esiobu, C. (Ed.). (2019). *Indigenous knowledge and education in Africa*. Springer. DOI: 10.1007/978-981-13-6635-2

Knaus, C. B., Mino, T., & Seroto, J. (2022). *Decolonising African Higher Education: Practitioner perspectives from across the continent*. Rotledge Tylor & Francis. DOI: 10.4324/9781003158271

Kugara, S. L., & Mdhluli, T. D. (2023). Integrating African indigenous education in the curriculum: A learning curve for South Africa. *Journal of Curriculum Studies Research*, *5*(3), 131–143. DOI: 10.46303/jcsr.2023.35

Tarisayi, K. S. (2024). Integrating indigenous knowledge in South African Geography education curricula for social justice and decolonization. *E-Journal of Humanities* [EHASS]. *Arts and Social Sciences*, *5*(7), 1195–8949.

KEY TERMS AND DEFINITIONS

Curriculum: formal learning experiences to which a student is exposed at an institution of learning. As seen in the chapter, these experiences could be conceptualised from a product orientation, a process orientation or a praxis orientation.

Indigenous knowledge: knowledge rooted in local cultures, values, beliefs and practices.

Indigenous African knowledge: a body of knowledge, beliefs, values and culture practiced daily in agriculture, health, environment, technology, etc; it is maintained by African communities over a long period.

Knowledge: a body of theories, practices and conclusions arrived at through a systematic process.

Epistemic diversity: an approach that recognises and includes several forms of knowledge, including indigenous African knowledge, as opposed to an epistemic approach that recognises and includes only one form or knowledge.

Pedagogy: is the art or science of teaching that deals with the facilitation of learning.

Decolonisation: the process of undoing colonial structures and epistemies.

Chapter 7
The Application of Indigenous Knowledge Strategies in Managing and Addressing Gender-Based Violence in KwaZulu-Natal, South Africa:
A Literature Review

Petros Nhlavu Dlamini
https://orcid.org/0000-0002-0014-419X
University of Zululand, South Africa

Mbongeni Shabalala
University of Zululand, South Africa

ABSTRACT

Gender-based violence has proven to be problematic worldwide, however, in South Africa, it has proven to be uncontrollable. In as much as national and provincial laws are enforced against gender-based violence, it continues to exacerbate. This has awakened South Africa to go back to African methods of addressing gender-based violence. This study aimed to investigate indigenous knowledge strategies for managing and addressing gender-based violence amongst the local citizens in KwaZulu-Natal. Important to note is that the rate of gender-based violence cases is witnessed to be increasing in KwaZulu-Natal. The reviewed literature has proven that Indigenous knowledge is capable of managing and addressing gender-based violence. Again, the findings of the study have demonstrated that indigenous knowledge strategies could become a roadmap for addressing gender-based violence not only in KwaZulu-Natal but in Africa as a whole.

DOI: 10.4018/979-8-3373-4124-8.ch007

INTRODUCTION

Gender-based violence (GBV) is a widespread global issue that transcends borders, cultures, and socioeconomic backgrounds. It is recognised as a severe violation of human rights, impacting individuals regardless of gender. However, women and young girls remain the most affected, with statistics showing that one in three women worldwide has experienced physical or sexual violence at some point in their lives (Mudau, Matshidze & Nthane, 2024). This persistent crisis highlights the urgent need for effective intervention strategies to combat GBV and protect vulnerable populations.Although GBV is often framed as an issue predominantly affecting women, studies have revealed that men also suffer from abuse, particularly at the hands of their intimate partners (Mpani & Nsibande, 2015; Oosthuizen, Martin & de Villiers, 2024). Despite its existence, GBV against men remains underreported and inadequately addressed due to deeply entrenched societal norms surrounding masculinity. These norms often discourage men from coming forward with their experiences, fearing stigma, ridicule, or disbelief (Mngomezulu, Makhonza & Nzima 2019). Graaff (2021) argues that violence against men is frequently masked by traditional gender roles, making it difficult to distinguish from general acts of aggression. Furthermore, instances where men do not conform to patriarchal norms often result in them becoming targets of gendered violence.

The United Nations Children Fund (UNICEF) (2019) defines GBV as any harmful act perpetrated against an individual's will, driven by socially constructed gender roles and power dynamics. This form of violence encompasses a broad spectrum of abuse, including physical, psychological, emotional, mental, and financial harm. GBV can occur in both private and public spaces, affecting victims in domestic settings, workplaces, educational institutions, and communities at large. The consequences of GBV are devastating, often leading to long-term psychological trauma, economic instability, and even loss of life (Oosthuizen, Martin & de Villiers, 2024). In South Africa, and particularly in the province of KwaZulu-Natal, GBV remains a pressing issue. The prevalence of physical and sexual violence in this region continues to rise despite existing legal frameworks aimed at curbing the crisis. Studies indicate that in Africa, approximately 45.6% of women have experienced GBV at least once in their lifetime, a figure that surpasses the global average of 35% (Mudau et al., 2024; Oosthuizen, Martin & de Villiers, 2024). The high incidence of GBV in the region underscores the need for comprehensive, culturally relevant solutions that address both the root causes and consequences of gender-based violence.

Understanding the scope and impact of GBV is crucial for developing effective policies and interventions. Efforts to combat GBV have put in place strategies such as educational programs, legal reforms, community-driven initiatives, and increased support for survivors. Currently, there is a growing recognition of the role that Indigenous Knowledge Systems (IKS) can play in addressing GBV by incorporating traditional values and community-based approaches. By acknowledging and integrating indigenous strategies alongside contemporary legal and policy frameworks, South Africa can work towards a more inclusive and effective response to gender-based violence (Oosthuizen, Martin & de Villiers, 2024). It is widely acknowledged that domestic violence remains a pervasive issue in Southern Africa, particularly in the Kingdom of Eswatini, where gender-based and sexual abuse are deeply entrenched within a patriarchal societal structure (Human Rights Watch, 2024). In this cultural context, local customs and traditional laws often hold greater authority than regional and international legal frameworks, making it difficult to implement and enforce policies aimed at addressing gender-based violence effectively (Oosthuizen, Martin & de Villiers, 2024). The deeply rooted belief systems that sustain gender inequalities contribute to the normalisation of violence against women and other vulnerable groups, further complicating efforts to protect victims and hold perpetrators accountable.

Despite these challenges, Eswatini has taken significant steps toward recognising and addressing the crisis of domestic violence. The government, alongside civil society organizations and international bodies, has worked to develop legislative measures and intervention strategies aimed at combating gender-based violence (Human Rights Watch, 2024). One of the key approaches being explored is the integration of Indigenous Knowledge Systems (IKS) into prevention and response efforts. Traditional knowledge and community-driven practices have long played a role in conflict resolution, social cohesion, and the preservation of cultural values, making them valuable tools in the fight against domestic violence (International Commission of Jurists (ICJ), 2015; Manthwa, 2024). By incorporating indigenous knowledge techniques, Eswatini seeks to provide culturally relevant solutions that resonate with local communities, ensuring greater acceptance and effectiveness. These approaches may include traditional mediation practices led by community elders, storytelling as a means of educating younger generations about respect and equality, and the reinforcement of communal responsibility in protecting vulnerable individuals. When combined with modern legal frameworks and advocacy initiatives, these indigenous strategies have the potential to foster sustainable change by addressing the root causes of gender-based violence and challenging harmful societal norms. Moving forward, there is a need for ongoing research and policy development to refine and strengthen the application of IKS in addressing gender-based violence. Greater collaboration between traditional leaders, government agencies, legal experts, and advocacy groups will be essential in ensuring that culturally informed interventions align with human rights principles and contribute to a more equitable society

Similarly, Zimbabwe has made notable legal strides in addressing domestic violence, recognising it as a critical issue requiring legislative intervention. After years of resistance to formal legal measures in domestic affairs, the country introduced the Domestic Violence Act in 2007. This landmark legislation marked a significant shift in Zimbabwe's approach to gender-based violence, acknowledging the need for legal protection and justice for survivors. However, the implementation of the Act was not without challenges. It followed 16 years of intense debate, during which it was often perceived as an unwarranted intrusion into private family matters. This resistance stemmed from deeply ingrained societal norms that traditionally viewed domestic issues as private affairs to be resolved within the home, rather than matters requiring state intervention. While the enactment of the Domestic Violence Act was a major step forward, its impact has been hindered by various factors, particularly difficulties in enforcement. Despite legal provisions aimed at protecting victims and prosecuting offenders, gender-based violence remains widespread. Many incidents continue to go unreported due to stigma, fear of retaliation, lack of trust in law enforcement, and financial dependency on abusers. Even when cases are reported, the rate of prosecution and conviction remains low, limiting the effectiveness of the legal framework in deterring perpetrators and delivering justice (International Commission of Jurists (ICJ), 2015).

These examples highlight the ongoing struggle in Southern Africa to balance legal reforms with deeply entrenched cultural norms. While legislative measures provide a crucial foundation for addressing domestic violence, their success ultimately depends on broader societal acceptance, effective enforcement mechanisms, and comprehensive support systems for survivors. Legal interventions must be complemented by community-driven strategies, including Indigenous Knowledge Systems (IKS), which can offer culturally relevant approaches to addressing gender-based violence. Traditional conflict resolution practices, community engagement initiatives, and education rooted in indigenous values can play a crucial role in shifting mindsets, fostering accountability, and promoting long-term social change. Moving forward, a holistic approach that integrates legal frameworks with indigenous knowledge and grassroots activism is essential to combating gender-based violence in Zimbabwe and the broader Southern African region. By leveraging both modern legal protections and culturally resonant strategies, societies can create more effective and sustainable solutions to this pervasive issue.

In KwaZulu-Natal (KZN) where this study is conducted, gender-based violence (GBV) continues to escalate. According to the 2024 Second Quarter Crime Statistics, Inanda and Umlazi police stations report the highest number of rape cases in the country. KZN also ranks second, after Gauteng, in rapes occurring on educational premises where children are the victims. The existence of such statistics is a stark indictment of society; even a single case of child sexual assault is an unacceptable violation of humanity. At the root of GBV is systemic inequality, deeply embedded in patriarchy. This system upholds the belief that men have power over women, reinforcing gender disparities in both explicit and subtle ways (DA KwaZulu-Natal, 2024; Dlamini, 2022). Its influence is so pervasive that women are still advised to change their behavior, dress differently, or avoid certain spaces for their "safety" as if they, rather than perpetrators, are the problem. Similar to the rest of the world, South Africa acknowledges gender-based violence as a serious violation of human rights. In an effort to heal and rebuild South Africa, the South African government has invested extensively in addressing gender-based violence as well as many other types of violence since 1994 (Mudau et al., 2024; CSVR, 2016). The Crime Prevention Strategy, which was enacted in 1996, recognized gender-based violence as the primary concern. The South African government has ratified a number of regional and international treaties to address gender-based violence, including the Maputo Protocol, the Beijing Declaration of 1995, and the Southern African Development Community Policy on Gender (CSVR, 2016). South Africa continues to be referred to as "the rape capital of the world" in spite of all the efforts made to address sexual violence through the establishment of laws and forums (Mudau et al., 2024; CSVR, 2016). Moreover, South Africa has introduced the GBV Index to monitor and track progress in implementing the National Strategic Plan (NSP) on Gender-Based Violence and Femicide (GBVF). This tool helps stakeholders identify GBV trends, allocate resources effectively, and assess the NSP's impact. The Commission for Gender Equality (CGE) will use the index to evaluate progress. It provides data on GBV at national, provincial, district, and municipal levels, with plans to include demographic details such as age and gender (Oosthuizen, Martin & de Villiers, 2024; Manthwa, 2024). We argue that the index also captures both criminal and non-criminal forms of GBV, supporting data-driven decision-making to enhance GBV response efforts.

In February 2020, during the State of the Nation Address, the president of South Africa promulgated the following stricter measures to mitigate gender-based violence in all municipalities including that of KwaZulu-Natal where the present study is conducted (Anene & Njoku, 2020; Mudau et al., 2024). These measures are as follows:

· Imposing stricter least punishments for offences against women and children, refusing parole and bail to those who commit crimes like rape and child murder, and reducing recidivism by adding additional rehabilitation facilities to the penitentiary system. To address violence against women and children as well as marginalized groups, emergency response teams will bring together SAPS, the Department of Health, Education, Social Development, and Justice.
· All offenders of gender-based violence will be listed in the national register for offenders of gender-based violence.
· increasing funding for campaigns against gender-based violence through community projects, workplace regulations, and educational programs (Mudau et al., 2024).

Despite a general increase in criminal activity, the annual police crime statistics consolidated report in Figure 1 below reveals a concerning trend in sexual offences: from 2021/2022 to 2022/2023, there was a rise of 1041 cases (2.5%) in rape and 349 additional cases (17.2%) in attempted sexual offences.

Table 1. South African police service annual crime statistics breakdown between 2013-2023

Crime category	2013-2014	2015-2016	2017-2018	2019-2020	2021-2022	2023
Murder	34 828	37 689	41 661	45 153	52 675	27494
Sexual offences	110297	101555	102528	99507	106192	53498
Attempted murder	34526	36332	37213	37615	47226	25131
Assault with an intention to inflict bodily harm	364889	353549	338331	310113	332216	169374
Common assault	327567	321408	318255	314936	355337	185374
Common robbery	107615	107528	102495	89473	88657	47057
Robbery with aggravating	248008	273483	278396	263831	278913	146125
Sexual Offences breakdown						
Rape	88544	81331	81618	78619	84519	42780
Sexual assault	12684	12483	15186	14774	15281	7483
Attempted sexual offences	5554	4646	4212	3876	4403	2376
Content sexual offences	3515	3095	2475	2238	1989	859

Source: (Masiphephe Network, 2024)

This chapter advocates the use of indigenous knowledge strategies to address and alleviate gender-based violence in KwaZulu-Natal. In particular, it focusses on how IKS could become a roadmap in addressing violence regardless of gender. It has been noted that many studies have placed emphasis exclusively on women and children only. However, this study also expands its focus to include men. To effectively address and contribute to the reduction of gender-based violence in KwaZulu-Natal, this study sought to explore a suitable alternative that incorporates local knowledge specifically, indigenous knowledge. The aim of this study aligns with the Afrocentric paradigm, a model developed by Molefi Kete Asante (1989), which emphasises the use of African-centered approaches to solving African problems. The Afrocentric paradigm is particularly relevant in addressing gender-based violence within the African context, as it challenges the dominance of Western knowledge systems that may not fully capture the complexities of African societies. Asante (1989) argues that Western perspectives are often insufficient in resolving African-specific issues, and therefore, Afrocentricity should be embraced as a guiding principle for formulating effective solutions. He further emphasises that this approach must be applied in a way that prioritises the unique interests, values, and worldviews of African communities more specifically, those of KwaZulu-Natal in this case.

PROBLEM STATEMENT

The rising prevalence of gender-based violence (GBV) in South Africa, particularly in KwaZulu-Natal, calls into question the effectiveness of Eurocentric methodologies in addressing this crisis. Numerous studies suggest that reliance on foreign knowledge has led to the marginalisation of indigenous African approaches to problem-solving. Dlamini and Ocholla (2018) argue that indigenous knowledge has been sidelined by African societies in favor of external solutions, often failing to effectively address African challenges. Similarly, Pophiwa and Saidi (2022) contend that foreign methodologies have proven inadequate in resolving African social issues. Olaopa and Ayodele (2021) emphasise the potential of indigenous knowledge in addressing Africa's interconnected developmental challenges, asserting that the survival of African communities depends on its application. Given these perspectives, this study explores the use of Indigenous Knowledge Systems (IKS) as a strategy for alleviating GBV in KwaZulu-Natal. Indigenous knowledge offers valuable insights into cultural traditions, values, beliefs, and worldviews (Shokane & Masoga, 2018). By examining the application of IKS strategies, this research aims to assess whether these traditional approaches can provide more effective solutions to the persistent issue of GBV in the region.

In fulfilling the aim of the study, this chapter is guided by the following three objectives:

- To analyse the effectiveness of Afrocentric approaches in addressing GBV. This objective aims to assess the relevance and impact of Afrocentric models in shaping community-driven interventions against GBV, free from Western influences.
- To critically assess the negative influence of Indigenous Knowledge Systems (IKS) on gender-based violence (GBV).
- To explore the application of indigenous knowledge strategies in combating GBV.

CONCEPTUALISATION/THEORETICAL MODEL

According to Asante (1989), Afrocentricity refers to the intellectual and philosophical perspective that places African culture, traditions, and experiences at the center of analysis and understanding. It is a framework that challenges Eurocentric narratives by asserting that Africans should be the primary subjects of their own histories, social realities, and intellectual discourse rather than being viewed through an external or colonial lens. Developed as a response to the marginalisation of African contributions in global knowledge systems, Afrocentricity seeks to reclaim African identity by promoting indigenous knowledge, traditions, and philosophies. This perspective emphasises self-determination, cultural pride, and the application of African-centered epistemologies in problem-solving and decision-making. In the context of research and scholarship, Afrocentricity encourages the use of African worldviews, languages, and methodologies to address social, political, and economic challenges facing African people. It recognises that Africa's historical, cultural, and intellectual heritage holds valuable solutions to contemporary issues, including governance, education, and social justice (Asante, 1989). By foregrounding African voices and perspectives, Afrocentricity contributes to the decolonisation of knowledge and the reassertion of Africa's place in the global intellectual landscape. It serves as a vital tool in fostering cultural awareness, identity formation, and the empowerment of African communities in their pursuit of sustainable development and social transformation.

It can be emphasised that Afrocentricity represents a fundamental shift in thinking, advocating for the recognition and application of indigenous knowledge as a primary resource in problem-solving, yet also not totally eliminating foreign knowledge in addressing problem faced by Africa when it comes to gender-based violence. This perspective suggests that African knowledge systems should not be viewed as secondary to or in competition with Western ideologies but should instead be the leading framework for addressing societal challenges, including gender-based violence. Asante (2006), supports Asante's

(1991) stance, reinforcing the idea that Afrocentric knowledge should not be forced to compete with foreign knowledge within the African context. In practical terms, this means that indigenous knowledge in KwaZulu-Natal should be given precedence when developing strategies to combat gender-based violence. Rather than relying on external frameworks that may not fully resonate with local realities, solutions should be rooted in culturally relevant methodologies that reflect the lived experiences of African communities. By centering indigenous expertise, this approach ensures that responses to gender-based violence are not only effective but also sustainable and deeply connected to the values and traditions of the people they aim to serve.

The Afrocentric model is highly relevant in addressing gender-based violence (GBV) as it emphasizes the importance of African people relying on their indigenous knowledge systems rather than Western frameworks. This approach aligns with the study's focus by encouraging a perspective rooted in African traditions, values, and problem-solving methods. The Afrocentric model challenges the dominance of Western ideologies and promotes a return to African-centered ways of thinking how Africans would have conceptualized and addressed societal issues before colonial influence. By reclaiming indigenous perspectives, this model provides culturally relevant solutions to GBV, fostering community-driven strategies that resonate with African societies and their historical context.

METHODOLOGY

This study is a desktop review of existing literature on indigenous knowledge (IK) and gender-based violence (GBV). In line with established research methodologies on IK, the study sourced literature from academically accredited journals, books, and reputable databases, as recommended by Neuman (2014:69). By relying on scholarly sources, the study ensured credibility and academic rigor. Additionally, since the authors themselves have published multiple works on indigenous knowledge systems (IKS) and gender-based violence, they incorporated their expertise to enhance the study's depth and contextual understanding.

A systematic literature review was conducted using the generic methods proposed by Templier and Paré (2015). These methods provided a structured approach to reviewing and analysing literature, which included:

1. Formulating the study topic and research questions – Defining clear objectives to guide the literature review.
2. Examining existing literature – Conducting a broad search for relevant scholarly work.
3. Screening for inclusion – Selecting studies that aligned with the research objectives.
4. Evaluating the caliber of primary studies – Assessing the methodological quality of the reviewed research.
5. Obtaining data relevant to the study's aims – Extracting information that directly supported the study's purpose.
6. Analysing the data – Synthesising findings to generate meaningful insights.

Study Objectives and Literature Selection Process

The study's objectives were explicitly stated in the review to establish a clear framework for the evaluation process. These objectives helped in:

- Identifying specific information required for analysis.
- Providing guidance and direction for further research.
- Informing the search and selection of relevant literature.

The authors ensured that both published and unpublished studies were included in the review to provide a comprehensive body of knowledge. Additionally, all literature was carefully screened for relevance to the study's objectives. This involved evaluating the applicability and rigor of primary studies, ensuring that only high-quality sources were included. Assessing the scientific quality of research designs and methodologies allowed the authors to filter out unreliable sources and select studies that would provide credible insights. This meticulous screening process ensured that variations in quality did not compromise the reliability of the findings but instead guided the analysis of data and literature.

Search Strategy and Data Extraction

To extract relevant information, the authors used specific search terms such as:

- Indigenous knowledge (IK)
- Local knowledge
- Indigenous knowledge strategies
- Impact of IK on GBV

These key search terms yielded an extensive body of literature. For instance, a search on Google Scholar retrieved approximately 3,500,000 records. To refine the results, the authors applied filters, selecting only publications from reputable academic journals, which narrowed the results to 300,000 records. Among these, 50,000 records contained Indigenous Knowledge Systems (IKS) content. However, only five articles specifically discussed IK strategies for reducing GBV.

To further refine the dataset, the authors:

- Excluded studies published before 2014, reducing the dataset to fewer than 35,000 articles.
- Eliminated articles without authors and publication years to ensure proper citations and verification of sources.

This rigorous selection process ensured that only high-quality, relevant, and academically verified sources were included in the final review, enhancing the reliability and credibility of the study's findings.

THE CONCEPT OF INDIGENOUS KNOWLEDGE

The phrase "indigenous knowledge" (IK) or "local knowledge" is frequently mistaken, often due to its narrow definitions that are only partially explored by academics and researchers. This has led to a variety of definitions, each emphasising different aspects of IK, reflecting its diverse and complex nature across various communities and disciplines (Dlamini, 2024). The term Indigenous Knowledge (IK) is not static; rather, it is a dynamic and evolving system that integrates traditional wisdom, cultural practices, and contemporary experiences to address real-world challenges. Different institutions and scholars define IK based on their worldviews, disciplines, and research focus. For instance, UNESCO (2017) defines IK as the knowledge, skills, and ideologies developed by societies with a long history of interaction with their natural environment. This definition highlights the environmental and ecological aspect of IK, emphasising how indigenous communities have historically adapted to and managed their surroundings using traditional wisdom and sustainable practices. On the other hand, the Masiphephe Network (2024) takes a broader approach, defining IK as a complex system of knowledge unique to Indigenous Peoples. This perspective acknowledges that IK is not limited to environmental interaction but also includes the cultural, linguistic, historical, political, and legal frameworks that shape indigenous societies. This definition aligns with the understanding that IK is deeply intertwined with a community's identity, governance structures, and way of life.

The authors argue that indigenous Knowledge Systems (IKS) are not fixed or outdated; instead, they evolve within the framework of contemporary society while drawing from the wisdom and experiences of past generations. Over time, indigenous communities have refined and adapted their knowledge to address emerging social, economic, and environmental challenges. For example, many indigenous practices in agriculture, medicine, and conflict resolution remain relevant today and are increasingly recognised for their contributions to sustainable development. According to Masiphephe Network (2024), Shokane and Masoga (2021), IK also plays a crucial role in promoting health and well-being. It encourages holistic approaches to wellness, emphasising the importance of natural remedies, traditional healing practices, and harmonious relationships with nature. Furthermore, IK is community-centered, ensuring that knowledge is preserved, passed down through generations, and adapted to meet the evolving needs of indigenous peoples.

In essence, Indigenous Knowledge is a multidimensional and evolving system that encompasses a wide range of practices, beliefs, and traditions. It is a valuable resource that provides alternative perspectives on various issues, including sustainability, governance, health, and cultural preservation. Recognising and respecting IK can enhance global knowledge systems by integrating local wisdom with modern scientific approaches, ultimately contributing to more inclusive and sustainable solutions for contemporary challenges. Some authors consider Indigenous Knowledge (IK) as spiritual knowledge, deeply intertwined with cultural traditions, beliefs, and practices. For example, Dlamini (2024) argues that IK is not just practical wisdom but also a sacred, spiritual resource that people access through a connection with its original custodians elders and knowledge keepers. Dlamini (2024) further described indigenous knowledge (IK) as a spiritual knowledge accessed through a connection with its original owners. In many cultures, elders serve as caretakers of this knowledge, preserving it through oral traditions and practices (Dlamini & Ocholla, 2018). The World Intellectual Property Organisation (WIPO) (2018) defines IK as a living body of knowledge passed down from generation to generation within communities. It is implicit in nature and plays a crucial role in shaping cultural and spiritual identities. According to Shokane and Masoga (2021), IK is essential for survival, particularly in specialised domains such as environmental conservation, food security, and resilience during crises. It provides communities with practical strategies for sustainability while preserving their heritage and traditions.

Rural communities actively utilise Indigenous Knowledge (IK) for a wide range of essential purposes, including traditional medicine, soil conservation, weather forecasting, education, conflict resolution, and various other applications (Khanyile, Dlamini & Masenya, 2023). This knowledge is deeply embedded in the daily lives of rural populations, guiding them in sustainable living, resource management, and community well-being. Africa, in particular, heavily relies on traditional medical knowledge and indigenous research technologies to provide healthcare solutions (James, Wardle & Steel, 2018). Traditional medicine plays a pivotal role in addressing the health needs of people who cannot afford conventional medical treatments. It has significantly contributed to the advancement of primary healthcare, especially for rural communities that lack access to modern medical facilities. As a result, Mokhesi and Modjadji (2022) recognise traditional medicine as the primary source of healthcare for African rural populations. Despite its vast contributions, the application of indigenous knowledge in solving local and national issues has not received sufficient recognition or attention. Indigenous knowledge is not just a collection of practical skills it is an integrated system that embodies the distinct cultures, languages, values, histories, governance structures, and legal traditions of Indigenous Peoples. This knowledge serves as a foundation for problem-solving in various societal contexts, yet it remains underutilised in formal decision-making processes at national levels. Given the urgency of addressing gender-based violence (GBV), Indigenous Knowledge Systems (IKS) have the potential to offer context-specific, community-driven solutions. The prevalence, prevention, and intervention strategies related to GBV require approaches that are place-based, cumulative, and dynamic. This makes Indigenous Knowledge a complex but valuable resource for tackling societal challenges, including GBV, in a holistic and culturally relevant manner. Recognizing and integrating IK into broader policy frameworks can lead to more inclusive, effective, and sustainable solutions.

The chapter asserts that Indigenous Knowledge (IK) preserves valuable insights and traditions that have been passed down through generations. According to Masiphephe Network (2024) and Manthwa (2024), South Africa can utilise its cultural heritage to develop sustainable and effective solutions by integrating Indigenous Knowledge Systems (IKS) into gender-based violence (GBV) prevention efforts. This suggests that the successful application of IK strategies in KwaZulu-Natal could benefit the entire country, offering a model for GBV management. The chapter aims to analyse the impact of IKS on GBV in KwaZulu-Natal and explore potential solutions that could enhance existing interventions.

THE NEGATIVE INFLUENCE OF INDIGENOUS KNOWLEDGE SYSTEMS (IKS) ON GENDER-BASED VIOLENCE (GBV)

Indigenous knowledge, despite being historically overlooked (Dlamini & Ocholla, 2018), is increasingly recognised as a valuable tool for addressing domestic and societal issues in various countries. While Indigenous Knowledge Systems (IKS) offer valuable cultural insights and community-based solutions, some traditional beliefs and practices within IKS can contribute to gender-based violence (GBV) by reinforcing patriarchal norms and harmful gender roles (Murovhi, Matshidze, Netshandama & Klu, 2018). Below are some ways in which IKS can negatively influence GBV:

Patriarchal Structures and Gender Inequality

- Many traditional African societies are structured around patriarchal norms, where men hold authority over women. This hierarchy often justifies male dominance and subordination of women, limiting their autonomy and decision-making power.
- Women may be expected to endure abuse in silence due to cultural norms that discourage speaking out against male authority (Murovhi, Matshidze, Netshandama & Klu, 2018).

Harmful Traditional Practices

- Certain cultural practices, such as forced marriages, bride price (lobola), and widow inheritance, reinforce the notion that women are property rather than individuals with rights.
- Female genital mutilation (FGM), practiced in some African communities, is another harmful tradition that violates women's bodily autonomy and perpetuates gender-based violence (Burayu & Degefa, 2024).

Silencing of GBV Survivors

- In some traditional settings, family or community elders handle cases of domestic violence instead of legal systems. This can result in victims being pressured to reconcile with abusers rather than seek justice.
- Victims may be discouraged from reporting GBV cases due to the stigma attached to challenging cultural norms (Rigual et al., 2021).

Beliefs in Witchcraft and Superstitions

- Some communities associate GBV with supernatural causes, blaming spirits or curses rather than holding perpetrators accountable.
- Women accused of witchcraft are often subjected to violence, exclusion, or even death in extreme cases (Gershman, 2022).

Cultural Expectations of Masculinity and Femininity

- Traditional masculinity ideals encourage men to be dominant and aggressive, while femininity is linked to submission and obedience. These stereotypes normalise violent behavior and discourage men from expressing emotions in non-violent ways.
- Men who do not conform to traditional masculinity norms may also face gender-based violence, further reinforcing rigid gender roles (Young et al., 2024).

Indigenous knowledge practices have suppressed men

- Additionally, gender-based violence against men remains largely overlooked. Unlike cases involving women and children, abuse against men in intimate relationships is often dismissed or ridiculed. Many men fear social stigma and humiliation, preventing them from reporting abuse.

This highlights the need for a balanced approach that acknowledges the strengths of IKS while addressing its limitations, ensuring that gender equality, justice, and human rights are upheld for all (Mngomezulu, Mkhonza & Nzima, 2019).

It can be said that indigenous knowledge systems (IKS) hold potential for positive contributions to social cohesion and problem-solving, it is essential to critically examine and challenge its aspects that perpetuate GBV. Modern approaches to GBV prevention should integrate beneficial indigenous knowledge while discarding harmful traditions that sustain gender inequality and violence. Balancing cultural preservation with human rights advocacy is key to addressing GBV effectively. Again, over time, certain indigenous practices have either evolved, lost their original meaning, or become misinterpreted, leading to unintended harmful consequences. Many historical customs within IKS have traditionally favored men over women, reinforcing gender inequality and excluding women from decision-making processes, which has resulted in their disempowerment.

Certain outdated cultural practices continue to perpetuate GBV by preventing women from seeking justice, education, and healthcare. In some communities, discussing or challenging harmful traditions is taboo, leading to underreporting of GBV cases. This silence has contributed to rising HIV/AIDS infections, emotional distress, and a lack of justice for survivors, directly conflicting with Sustainable Development Goal 16, which advocates for equal access to justice (Masiphephe Network, 2024).

INDIGENOUS KNOWLEDGE SYSTEMS (IKS) STRATEGIES FOR ADDRESSING AND ALLEVIATING GENDER-BASED VIOLENCE

Although Eurocentric knowledge systems continue to dominate and often suppress Indigenous Knowledge Systems (IKS), it is crucial to prioritise IKS due to its potential in addressing social issues such as gender-based violence (GBV). For example, indigenous knowledge offers culturally relevant, community-driven solutions that can help develop sustainable and effective interventions. South Africa, in particular, possesses a rich cultural heritage that can be leveraged to create long-term strategies for GBV prevention and response. By integrating traditional conflict resolution mechanisms, community engagement practices, and indigenous values, the country can establish solutions that are more deeply rooted in societal structures (Kaya, 2013). This approach not only enhances community participation but also ensures that interventions are culturally sensitive and widely accepted.

- Firstly, indigenous Knowledge Systems (IKS) emphasise the importance of cultural sensitivity in addressing gender-based violence (GBV) by integrating local traditions, beliefs, and social behaviors into intervention strategies. This culturally aware approach fosters trust, strengthens community involvement, and creates a supportive environment for survivors, making them more likely to seek help and report abuse (Sonke Gender Justice, 2023). By recognising and respecting cultural norms, IKS can be a powerful tool in challenging harmful gender stereotypes while promoting gender equality. Communities that embrace indigenous wisdom can develop holistic and sustainable solutions to combat GBV while ensuring that interventions align with their cultural values and lived experiences (Manthwa, 2024; Masiphephe Network, 2024; Friedman, 2020).
- Secondly, addressing gender-based violence (GBV) requires a comprehensive and culturally relevant approach that not only seeks justice for survivors but also promotes their long-term healing

and reintegration into society. Indigenous Knowledge Systems (IKS) emphasise holistic healing, recognizing that true recovery extends beyond just the physical harm and must also include mental, emotional, and spiritual well-being. Holistic healing is rooted in the belief that individuals are deeply interconnected with their families, communities, and natural surroundings (Mukuka, 2022). This perspective encourages healing not just for survivors, but for entire communities, fostering restoration, balance, and social harmony. Survivors of GBV often experience trauma, isolation, and emotional distress, which Western legal and medical systems may not fully address. However, indigenous healing practices provide alternative pathways that consider the survivor's personal, familial, and spiritual needs. IKS incorporates traditional counseling, ceremonies, and rituals as part of the healing process. Community elders, spiritual leaders, and traditional healers play a vital role in offering support and guidance. Traditional ceremonies such as cleansing rituals, storytelling, or reconciliation gatherings help survivors reclaim their sense of dignity, belonging, and self-worth. These practices provide a space where survivors can share their experiences in a culturally safe and supportive environment, reducing shame and stigma often associated with GBV.

Additionally, restorative justice practices within IKS focus on rehabilitation and reconciliation rather than solely punitive measures. Unlike conventional justice systems, which often fail to address the root causes of GBV, indigenous approaches emphasise restoring relationships, accountability, and community healing. Some cultures engage in peace circles, mediation, and community dialogues, allowing both survivors and perpetrators to seek resolution, transformation, and behavioral change under the guidance of elders or community leaders. By integrating traditional knowledge with modern legal and psychological support systems, South Africa can develop more inclusive, culturally sensitive, and long-lasting solutions for GBV survivors. Recognising the value of indigenous healing practices ensures that survivors receive well-rounded care, empowering them to move forward with strength, resilience, and community support (Masiphephe Network, 2024; Mukuka, 2022).

- Thirdly, Indigenous Knowledge Systems (IKS) play a vital role in education and empowerment, serving as key tools in the fight against gender-based violence (GBV). These systems emphasize community-led programs that promote gender equality, encouraging active participation in violence prevention initiatives. By integrating traditional wisdom with modern education, IKS fosters awareness and behavioral change, ensuring that knowledge is passed down intergenerationally. In communities, especially among young people, there is a need for education that fosters an appreciation of teachings that actively oppose gender-based violence (Siwila, 2016). Siwila (2016, pp.152-153) further argues that mutual respect between men and women is a fundamental rite that validates their social status within marriage. This rite, from its inception, has been regarded as a prerequisite for marriage, reinforcing the principle that both men and women should be equally valued and respected. Such teachings play a crucial role in shaping societal attitudes toward gender equality and in mitigating gender-based violence by promoting respect and recognition of the dignity of both genders. Empowering women and marginalized groups through skills development, leadership training, and advocacy programs helps challenge societal norms that perpetuate violence. Additionally, elders and cultural leaders can play a crucial role in educating communities about gender justice, reshaping harmful traditions, and reinforcing values that protect survivors.

Through education and community-driven empowerment, IKS contributes to long-term solutions for preventing GBV (Masiphephe Network, 2024; Musuku, 2019).

· Fourthly, Indigenous Knowledge Systems (IKS) emphasize community accountability in preventing and responding to gender-based violence (GBV). By fostering collective responsibility, communities create safe spaces, challenge harmful norms, and support survivors. IKS promotes local support networks, including elders and community organizations, to provide counseling, resources, and advocacy. Additionally, restorative justice mechanisms ensure accountability and healing. These community-driven solutions make GBV prevention sustainable, inclusive, and culturally relevant (Masiphephe Network, 2024; Mukuka, 2022).

· Fifthly, Collaborative partnerships play a crucial role in leveraging Indigenous Knowledge Systems (IKS) to combat gender-based violence (GBV). IKS emphasizes the significance of cooperative relationships between indigenous communities, non-governmental organizations (NGOs), and government agencies to create inclusive and culturally relevant prevention strategies. These collaborations help tailor interventions to the specific needs of indigenous groups, ensuring that GBV responses are not only effective but also respectful of cultural traditions. By pooling resources, sharing best practices, and integrating indigenous wisdom with modern approaches, these partnerships contribute to long-term, sustainable solutions that enhance both GBV prevention efforts and support systems for survivors (Masiphephe Network, 2024). In KwaZulu-Natal in particular, Indigenous Knowledge Systems (IKS) play a vital role in fostering peace and security through home-grown philosophies and indigenous languages. These knowledge systems provide culturally rooted approaches to conflict resolution and social harmony, making them valuable tools for combating gender-based violence (GBV) in KwaZulu-Natal. By integrating indigenous knowledge into GBV prevention strategies, communities can promote respect, equality, and the principles of *Ubuntu*—which emphasizes compassion and interconnectedness. For instance, the University of KwaZulu-Natal recently hosted a forum highlighting the relevance of IKS in promoting peace between men and women, reinforcing that respect and *Ubuntu* should be at the forefront of efforts to address GBV (University of KwaZulu-Natal, 2021).

· Sixthly, Indigenous Knowledge Systems (IKS) emphasise the importance of conventional healing methods in addressing the holistic needs of survivors of gender-based violence (GBV). This approach recognises that healing extends beyond physical recovery and must also address emotional, mental, and spiritual well-being. By incorporating traditional medicine, rituals, ceremonies, and culturally relevant counseling, IKS-based healing programs provide survivors with alternative and complementary therapies that align with their cultural beliefs and lived experiences (Mukuka, 2022). These healing methods often draw from ancestral wisdom and community support, helping survivors find inner strength, resilience, and a sense of belonging. For instance, traditional ceremonies and rituals are used to help survivors process trauma, regain self-confidence, and reconnect with their communities. Spiritual and cultural practices, such as herbal remedies, cleansing rituals, and storytelling circles, foster an environment where survivors feel understood, supported, and empowered. Moreover, traditional counseling often provided by elders, spiritual leaders, or community healers helps survivors navigate their experiences while reinforcing cultural values of healing and restoration (Mudau, Matshidze & Nthane, 2024). Integrating these conventional healing methods within modern GBV intervention strategies ensures that survivors receive culturally appropriate and effective support. It also fosters greater acceptance and participation in healing programs, as many individuals feel more comfortable engaging with practices rooted in their own traditions. By acknowledging and legitimising indigenous healing techniques, communities can create more sustainable and impactful solutions to address GBV (Masiphephe Network, 2024; Musuku, 2019).

· Seventhly, Indigenous Knowledge Systems (IKS) emphasize the critical role of family and community support in the healing process of gender-based violence (GBV) survivors. A strong, inclusive support system can significantly enhance a survivor's emotional, psychological, and social well-being, fostering a sense of belonging, validation, and security. By encouraging the active participation of family members, elders, and community leaders, IKS promotes a collective approach to healing. Elders and traditional leaders, who are often seen as custodians of wisdom and cultural values, play a crucial role in guiding survivors, offering advice, and facilitating reconciliation where necessary (Mukuka, 2022). Their involvement validates the survivor's experience, reduces feelings of isolation, and helps combat stigma and victim-blaming. Family support is equally vital in ensuring GBV survivors rebuild their confidence and regain stability. A nurturing home environment fosters emotional strength and provides the encouragement needed for survivors to heal. When family members stand in solidarity with survivors, they help to break the silence surrounding GBV and challenge harmful societal norms that often protect perpetrators (Mudau, Matshidze & Nthane, 2024). Additionally, community-led interventions, such as healing circles, peer support groups, and mentorship programs, create safe spaces where survivors can share their experiences, learn from one another, and receive reassurance. These communal practices, deeply rooted in African traditions, help restore dignity and self-worth to survivors. Ultimately, when families and communities unite in supporting GBV survivors, they strengthen the healing process and contribute to broader efforts to eliminate GBV within society (Masiphephe Network, 2024; Musuku, 2019).

· Eighthly, African Indigenous societal lifestyles prioritise social harmony over social control, with the concept of Ubuntu playing a central role in fostering unity and interdependence (Olowu, 2017). Ubuntu, though known by different names across various African communities, emphasises that individuals and society are interconnected, making both personal and collective well-being equally important. Olowu (2017) further supports this perspective, highlighting that throughout African history, communalism has been the foundation of the African worldview. This deep-rooted sense of community influences all aspects of life, ensuring that societal cohesion and mutual support remain key values within African cultures (Friedman, 2022).

Experts agree that social control must be regulated to achieve social harmony. Historically, the Ubuntu ideology played a crucial role in uniting societies. However, in modern times, the absence of legal frameworks governing social control has led to a decline in the practical application of Ubuntu. Individualism and power struggles have replaced the traditional emphasis on human equality and interdependence. This shift is evident in rising societal issues such as crime, femicide, gender-based violence, and serial killings, which demonstrate that Ubuntu is often discussed in theory but not practiced in daily life (Mukuka, 2022). The African communal mindset has weakened, with personal gains taking precedence over collective well-being. A stark example of this decline is rape, a prevalent form of gender-based violence. Offenders, rather than acknowledging the harm caused, often take pride in their actions, leading to repeated violations against victims (Friedman, 2022).

· Ninthly, African Indigenous Child-Rearing Practices: According to Murovhi et al. (2018), child-rearing methods in African communities are deeply rooted in cultural traditions. These techniques shape children's behaviors and expectations throughout their lives, influencing their development from infancy to adolescence and even guiding their own parenting approaches in adulthood. Initiation schools have historically played a significant role in the enculturation of children. In Vhavenda culture, for example, boys attend vhutuku and mula, while girls participate in domba, musevhetho, and vhusha. These traditional schools are integral to child development, teaching cultural values, life skills, and social responsibilities. However, some practices within initiation schools can have harmful consequences. For instance, certain songs taught to boys in Mula contain messages that disparage women, which may reinforce gender inequality and negative perceptions upon their return from initiation (Murovhi et al., 2018). Murovhi et al. (2018) further emphasised the role of initiation schools as key channels of enculturation in African societies. These schools significantly shape children's upbringing by passing down cultural knowledge and values. However, the nature of the teachings provided is crucial, as they can influence both positive and negative behaviors. While initiation schools play an essential role in guiding adolescents into adulthood, some teachings may instill harmful behaviors, which are often mistakenly attributed to natural adolescent development. According to Murovhi et al. (2018), beyond the specific examples of disparaging songs, negative influences from initiation schools can lead to general disrespect in homes, schools, and communities. In extreme cases, these behaviors may escalate, contributing to serious social issues such as assault, rape, mugging, and burglary. Given these concerns, there is a need to critically review the teachings offered at initiation schools in KwaZulu-Natal, particularly regarding their direct and indirect impact on gender-based violence (GBV) and negative behavioral patterns. Ensuring that initiation schools promote respect, equality, and positive societal values is vital for curbing violence and fostering responsible citizenship.

Frosch, Schoppe-Sullivan and O'Banion (2019) argue that the quality of parent-child interactions is fundamental in shaping a child's emotional, psychological, and social development from early childhood through adolescence. The way parents engage with their children during formative years has a lasting impact on how children express themselves, build relationships, and handle challenges in life. These interactions influence their self-perception, the way they relate to others, and how they navigate conflicts in different settings, including family, school, and peer relationships. Furthermore, Frosch, Schoppe-Sullivan and O'Banion (2019) emphasise that social learning theory suggests that children learn behaviors through observation and imitation. If a child is exposed to violence, aggression, or emotional neglect within the home environment, they may internalize these experiences and view violence as a normal response to conflict. Mudau, Matshidze and Nthane (2024) explain that children who witness or experience domestic abuse are at a higher risk of developing aggressive tendencies or becoming victims of abuse later in life. These early experiences shape their understanding of interpersonal relationships, often leading to difficulties in trust, emotional regulation, and conflict resolution as they grow.

The long-term impact of parental behavior on children's development and well-being has been widely studied, with many scholars reinforcing the importance of nurturing, supportive, and non-violent parenting practices (Jeong, 2021). Research highlights that positive reinforcement, emotional support, and effective communication foster healthier relationships and help children develop empathy, problem-solving skills, and emotional intelligence. Conversely, exposure to hostility, neglect, or abuse increases the likelihood of children developing low self-esteem, anxiety, depression, and behavioral issues. Given these findings, it is essential for parents, caregivers, and society as a whole to prioritise creating safe, nurturing, and supportive environments for children. Educational programs and awareness campaigns should be implemented to guide parents in using positive discipline techniques, constructive communication, and emotional support strategies. By addressing harmful parenting practices and promoting healthy family dynamics, society can break the cycle of violence, fostering emotionally resilient, well-adjusted individuals who contribute to safer and more harmonious communities.

CONCLUSION AND RECOMMENDATIONS

This study aimed to explore indigenous knowledge strategies for addressing gender-based violence (GBV) in KwaZulu-Natal. A review of global literature revealed that the application of indigenous knowledge in combating GBV in the region is still in its early stages. Additionally, there is limited academic research specifically linking indigenous knowledge systems (IKS) to GBV prevention. However, the available literature suggests that integrating IKS into GBV interventions offers a promising approach to addressing the issue in KwaZulu-Natal. The increasing prevalence of gender-based violence (GBV) in KwaZulu-Natal has sparked renewed interest in Indigenous Knowledge Systems (IKS) as a viable solution. Literature indicates that the normalization of GBV due to its persistent occurrence necessitates collective action for its eradication. Thus, the University of KwaZulu-Natal (2021) emphasizes that community forums promoting local knowledge are essential in addressing GBV. This chapter concludes that every community member plays a critical role in combating GBV through education, advocacy, and support for survivors. Asante's (1996) model emphasizes the importance of promoting African Indigenous Knowledge Systems (IKS) as a means of addressing challenges faced by African communities. This perspective suggests that KwaZulu-Natal communities should rely on their locally developed and culturally relevant knowledge to effectively combat gender-based violence (GBV). By integrating indigenous wisdom, traditions, and values into GBV prevention strategies, communities can create solutions that are both practical and deeply rooted in their social and cultural contexts.

The chapter recommends further research to explore the potential of Indigenous Knowledge Systems (IKS) in addressing gender-based violence (GBV) in KwaZulu-Natal. It highlights the importance of raising awareness about the value of local knowledge and underscores the role of indigenous wisdom in shaping effective GBV prevention strategies. Key approaches include educating communities on IKS-based solutions, guiding survivors toward legal and social support services, and leveraging the guidance of community elders. By integrating these elements, communities can develop culturally relevant and sustainable interventions to combat GBV in KwaZulu-Natal.

FUTURE RESEARCH DIRECTIONS

his study reviewed existing literature on indigenous knowledge strategies for addressing gender-based violence (GBV) in KwaZulu-Natal. However, as the study was based solely on a literature review, certain aspects could not be fully explored. To bridge these gaps, further research is recommended in the following areas:

- **Empirical Study with Indigenous Knowledge Holders:** A similar study should be conducted in KwaZulu-Natal, involving direct consultations with owners of indigenous knowledge. This would allow them to share their perspectives and experiences regarding the potential of indigenous knowledge systems (IKS) in addressing gender-based violence.
- **Engagement Between GBV Policymakers and Indigenous Knowledge Holders:** Another study should be undertaken to conduct a survey-based investigation that facilitates discussions between GBV policymakers and IK holders. This would help in exploring how local knowledge can be effectively integrated into strategies aimed at combating GBV.

These further studies would contribute to a more comprehensive understanding of the role of indigenous knowledge systems in addressing gender-based violence and provide valuable insights for policy development and intervention strategies.

REFERENCES

Anene, C. P., & Njoku, C. U. (2020). *African indigenous knowledge systems: Problems and perspectives. Communal oral historians as reservoirs and curators of indigenous historical scholarship.* Retrieved May 10, 2023, from https://acjol.org/index.php/jassd/article/view/1497/1479

Asante, M. K. (1991). The Afrocentric Idea in Education, Spring. *The Journal of Negro Education.* 60(2), 170-180. Retrieved June, 11, 2024, from https://www.jstor.org/stable/pdf/2295608.pdf

Asante, M. K. (2006). *The Afrocentric Paradigm.* Africa World Press.

Burayu, E. T., & Degefa, B. D. (2024). Exploring harmful traditional practices and its associated factors among puerperal mothers who delivered in public health facilities in southwestern Ethiopia. *Scientific Reports, 14*(1), 27065. https://pmc.ncbi.nlm.nih.gov/articles/PMC11544142/. DOI: 10.1038/s41598-024-75344-x PMID: 39511244

CSVR. (2016). *Mapping local gender-based violence prevention and response strategies in South Africa.* Retrieved May 10, 2023, from http://www.csvr.org.za

DA KwaZulu-Natal. (2024). *Debate on the KZN Premier's Report: To combat GBV, we need men to be part of the solution – not the problem.* Retrieved June 20, 2025, from https://kzn.da.org.za/2024/11/debate-on-the-kzn-premiers-report-to-combat-gbv-we-need-men-to-be-part-of-the-solution-not-the-problem

Dlamini, N. Z. (2022). Unmasking Christian women survivor voices against gender-based violence: a pursuit for a feminist liberative pastoral care praxis for married women in the Anglican Church of Southern Africa. University of KwaZulu-Natal (Doctoral Dissertation).

Dlamini, P., & Ocholla, D. N. (2018). Information and communication technology tools for managing indigenous knowledge in KwaZulu-Natal Province, South Africa. *African Journal of Library Archives and Information Science, 28*(2), 137–153.

Dlamini, P. N. (2024). Acquiring and transferring of indigenous knowledge among its owners: A literature review. In Ocholla, D., Onyancha, O. B., & Adesina, A. O. (Eds.), *Information, Knowledge, and Technology for Teaching and Research in Africa. Synthesis Lectures on Information Concepts, Retrieval, and Services.* Springer., DOI: 10.1007/978-3-031-60267-2_8

Friedman, C. (2022). *African indigenous knowledge systems in the CAO sector in SA.* Retrieved May 20, 2024, from https://idl-bnc-idrc.dspacedirect.org/server/api/core/bitstreams/5fee0a83-36f0-4237-be0d-4483a169cbeb/content

Frosch, C. A., Schoppe-Sullivan, S. J., & O'Banion, D. D. (2019). Parenting and Child Development: A relational health perspective. *American Journal of Lifestyle Medicine, 15*(1), 45–59. DOI: 10.1177/1559827619849028 PMID: 33447170

Gershman, B. (2022). Witchcraft beliefs around the world: An exploratory analysis. *PLoS One, 17*(11), e0276872. Retrieved October 12, 2024, from https://pmc.ncbi.nlm.nih.gov/articles/PMC9683553/. DOI: 10.1371/journal.pone.0276872 PMID: 36417350

Graaff, K. (2021). The implications of a narrow understanding of gender-based violence. *Feminist Encounters*, *5*(1), 1–12. DOI: 10.20897/femenc/9749

Human Rights Watch. (2024). Eswatini: Address violence against women. Retrieved March 20, 2025, from https://www.hrw.org/news/2024/01/29/eswatini-address-violence-against-women

International Commission of Jurists (ICJ). (2015). *Sexual and gender-based violence, fair trial rights and the rights of victims: Challenges in using law and justice systems faced by women human rights defenders*. Retrieved November 5, 2023, https://www.refworld.org/docid/57ee7f724.html

James, P. B., Wardle, J., Steel, A., & Adams, J. (2018). Traditional, complementary and alternative medicine use in Sub-Saharan Africa: A systematic review. *BMJ Global Health*, *3*(5), e000895. Advance online publication. DOI: 10.1136/bmjgh-2018-000895 PMID: 30483405

Jeong, J., Franchett, E. E., Ramos de Oliveira, C. V., Rehmani, K., & Yousafzai, A. K. (2021). Parenting interventions to promote early child development in the first three years of life: A global systematic review and meta-analysis. *PLoS Medicine*, *18*(5), e1003602.

Kaya, H. O. (2013). Integration of African indigenous knowledge systems into higher education in South Africa: Prospects and challenges. *Alternation (Durban)*, *1*, 135–153.

Khanyile, N. C., Dlamini, P., & Masenya, T. M. (2023). Techniques in preserving traditional medicinal knowledge in uMhlathuze in KwaZulu Natal, South Africa. *The Serials Librarian*, *84*(5–8), 148–164. DOI: 10.1080/0361526X.2023.2286234

Lut, M., Linkova, M., & Strid, S. (2023). Theorising gender-based violence policies: A 7P Framework. *Social Sciences (Basel, Switzerland)*, *12*(7), 385. DOI: 10.3390/socsci12070385

Manthwa, A. (2024). The transformative nature of indigenous knowledge systems to address gender inequality. *Journal of Law. Social Development*. Advance online publication. DOI: 10.25159/2520-9515/14759

Masiphephe Network. (2024). *Indigenous knowledge systems and gender-based violence in South Africa*. Retrieved November 10, 2024, from https://www.masiphephe.org.za/wp-content/uploads/2024/04/Perspective-2-Indeginous-Knowledge-systems-.pdf

Mngomezulu, T., Mkhonza, L., & Nzima, D. (2019). Causes of men abuse in the Province of Kwazulu-Natal, South Africa. *Gender & Behaviour*, *17*(4), 14327–14339.

Mokhesi, T., & Modjadji, P. (2022). *Usage of traditional, complementary and alternative medicine and related factors among patients receiving healthcare in Lesotho*. Retrieved June 14, 2024, from https://purerims.smu.ac.za/en/publications/usage-of-traditional-complementary-and-alternative-medicine-and-r

Mpani, P., & Nsibande, N. (2015). *Gender-based violence*. April 21, 2025. https://www.soulcity.org.za/projects/advocacy/gbv/resources/understanding-gender-policy-and-gender-based-violence-in-south-africa-a-literature-review

Mudau, T. J., Matshidze, P., & Nthane, T. (2024). Investigating African indigenous ways to mitigate gender-based violence in Bushbuckridge of Mpumalanga Province in South Africa. [ERATS]. *E-Journal of Religious and Theological Studies*, *10*(6), 130–142. DOI: 10.38159/erats.202410613

Mukuka, S. (2019). *A quest for embracing indigenous knowledge systems in the united church of Zambia: Pastoral care and imbusa, submitted in fulfilment of the degree requirement for the degree of PhD, in practical theology*. University of KwaZulu-Natal, Pietermaritzburg.

Mukuka, S. (2022). Affirming the traditional teaching of imbusa as a tool against gender-based violence. *Journal of International Women's Studies,* 24(4), (journal). Retrieved November 20, 2024, from https://vc.bridgew.edu/jiws/vol24/iss4/4

Murovhi, A., Matshidze, P., Netshandama, V., & Klu, E. (2018). Traditional child rearing practices in Vhavenda families South Africa. *Journal of Gender* [JGIDA]. *Information and Development in Africa, 7*(1), 1–14.

Murovhi, A., Pfarelo, M., Vhonani, N., & Ernest, K. (2018). Traditional child rearing practices in Vhavenda families South Africa. *Journal of Gender* [JGIDA]. *Information and Development in Africa, 7*(1), 21–37.

Neuman, W. L. (2014). *Social research methods: qualitative, quantitative approaches* (7th ed.). Pearson Education Limited.

Olaopa, O. R., & Ayodele, O. A. (2021). Building on the strengths of African indigenous knowledge and innovation (AIK & amp; I) for sustainable development in Africa'. *African Journal of Science, Technology, Innovation and Development,* •••, 1–14. DOI: 10.1080/20421338.2021.1950111

Olowu, O. J. (2017). *Indigenous approaches to conflict resolution in Africa: A study of the Barolong people of the North-West Province, South Africa*. Retrieved May, 16, 2024, from http://digitallibrary.aun.edu.ng:8080/xmlui/handle/123456789/416

Oosthuizen, M., Martin, L., & de Villiers, D. (2024). Towards a gender-based violence index for South Africa. Retrieved March 10, 2025, from https://www.parliament.gov.za/storage/app/media/OISD/Reports/Commission_for_Gender_Equality/2024/01-08-2024/GBV_Index_Report.pdf

Pophiwa, N., & Saidi, U. (2022). Approaches to embedding indigenous knowledge systems in made in Africa evaluations. *African Evaluation Journal, 10*(1), 1–9. DOI: 10.4102/aej.v10i1.623

Rigual, C., Myrttinen, H., Rahmawati, A., & Achakpa, M. (2021). Interlude. The silencing of gender-based violence. *International Development Policy | Revue Internationale de politique de développement.* Retrieved October 11, 2024, URL: https://journals.openedition.org/poldev/4625DOI: https://doi.org/.DOI: 10.4000/poldev.4625

Shokane, A. L., & Masoga, M. A. (2018). African indigenous knowledge and social work practice: Towards an Afro-sensed perspective. *Southern African Journal of Soc Work Soc Dev, 30*(1), 1–18. DOI: 10.25159/2415-5829/2320

Siwila, L. (2016). The role of men in women's sexual reproductive health rights, towards pragmatic model. In Chitando Ezra, C., & Nyambula, N. (Eds.), *Abundant Life, The Church and Sexuality* (pp. 151–156). World Council of Christian Publications.

Sonke Gender Justice. 2023. *Working with traditional leaders to combat Gender Based Violence*. Retrieved November 10, 2024, from https://genderjustice.org.za/news-item/working-with-traditional-leaders-to-combat-gender-based-violence/

United Nations Children Fund (UNICEF). (2019). *Gender-based violence in emergencies operational guide*. Retrieved May 20, 2024 from https://www.unicef.org/sites/default/files/2020-05/Gender-Based-Violence-in-Emergencies-Operational-Guide-May-2019.pdf

University of KwaZulu-Natal. (2021). *African indigenous knowledge systems, gender dimensions and youth in conflict resolution and peace building highlighted in high-level colloquium*. Retrieved July 10, 2024, from https://ndabaonline.ukzn.ac.za/UkzndabaStory/Vol9-Issue56/African%20Indigenous%20Knowledge%20Systems%20Gender%20Dimensions%20and%20Youth%20in%20Conflict%20Resolution%20and%20Peace%20Building%20Highlighted%20in%20High%20Level%20Colloquium/

Young, A. M., Moji, M., Duby, Z., Tenza, S., Atujuna, M., Palanee-Phillips, T., Minnis, A. M., & Montgomery, E. T. (2024). The ideal man: How gender norms and expectations shape South African men's masculinity, sexual identities, and well-being. *Journal of Men's Studies*, *0*(0). Advance online publication. Retrieved December 12, 2024, from. DOI: 10.1177/10608265241303734

KEY TERMS AND DEFINITIONS

Indigenous knowledge systems: It refers to the knowledge, skills, and ideologies developed by societies with a long history of interaction with their natural environment. This knowledge is typically passed down through generations and is deeply embedded in cultural traditions, practices, and belief systems.

Indigenous knowledge holders: Are individuals within indigenous and local communities who possess specialized expertise in various aspects of indigenous knowledge (IK). They are recognized for their deep understanding of traditions, cultural practices, beliefs, and skills, which are often passed down through generations.

Gender-based violence: It is as any harmful act inflicted upon an individual against their will, rooted in socially constructed gender roles and unequal power dynamics. It encompasses various forms of violence, including physical, sexual, psychological, and economic abuse, disproportionately affecting women and girls, though men and boys can also be victims.

Gender-based violence policymakers: Are individuals, organizations, or governmental bodies responsible for developing, implementing, and enforcing policies aimed at preventing and addressing gender-based violence within a country. Their role includes creating legal frameworks, advocating for survivor support services, promoting gender equality, and ensuring accountability for perpetrators.

Chapter 8
The Importance of Acknowledging Indigenous Knowledge Systems Among Landscape Actors in Managing Changing Environments

Andani E. Budeli
https://orcid.org/0000-0001-9725-4311
University of Venda, South Africa

ABSTRACT

The purpose of this chapter to explores the importance of acknowledging indigenous knowledge among landscape actors in managing changing environments. Global South is currently experiencing dramatic changes, due to climate change. Over the years, Western knowledge has been prioritised by landscape actors in their efforts to manage changing environments, while ignoring indigenous knowledge. Currently, indigenous knowledge is gaining recognition it the efforts to address changes in the environment. This study is grounded on the knowledge coevolution framework. The results of this study influence the need for landscape actors to adopt indigenous knowledge and manage changing environments. The study concludes that there is a need to ensure a working relationship between landscape actors. It is important to determine the management of power and its dynamics at different levels particularly between the government, local communities, and NGOs, to ensure interests are advocated for.

BACKGROUND OF THE STUDY

Knowledge is an important part of culture, carrying what is understood to be true. According to Gupta (2015), different cultures have their perspectives about what is known and how one acquires knowledge. This knowledge is maintained by Indigenous people, and such is referred to as Indigenous Knowledge (Gupta, 2015). Indigenous Knowledge (IK) is the actual knowledge of a given society demonstrating experiences based on traditions and without neglecting recent experiences with modern technologies (Anand et al., 2025). This knowledge is specific to a society or culture as well as the environment peo-

DOI: 10.4018/979-8-3373-4124-8.ch008

ple live in (Ncube, 2018). Indigenous knowledge is not developed through formal education but based on the day-to-day experiences of people and their cultural beliefs. This indigenous knowledge works systematically, thus known as Indigenous Knowledge Systems (IKS). Indigenous Knowledge Systems is the structured organisation of the Indigenous knowledge acquired by local people through intellectual reasoning in their daily lives (Gupta, 2015). IKS is also associated with experiments on people (trial and error) (Gupta, 2015).

Indigenous knowledge is consumed through observation and repetition of activities (Barnhardt & Kawagley 2005). This is because of its roots in the oral tradition's spiritual and holistic understanding of nature. However, this contrasts with the dominant reductionist approach of Western knowledge that is on cynicism and phenomena isolation (Chapman & Schott, 2020). This has led to arguments and perceptions that indigenous knowledge should be separated from other knowledge systems. As a result, indigenous knowledge has been ignored in efforts to conserve and manage landscapes while Western scientific knowledge is prioritized (Edgar et al., 2022). This creates an impression that indigenous knowledge is not an important aspect of landscape management. Therefore, it is important to ensure there are concerted efforts to promote landscape management knowledge that recognises indigenous knowledge.

Indigenous Knowledge reflects the relationship between humans, their environment, and all that lives or is found in that environment (Wilder et al., 2016). Indigenous knowledge (Adade et al., 2020). This knowledge is intertwined with social norms, sociocultural governance, spiritual beliefs, management of natural resources, and historical and contemporary practices (Cajete, 2020). The importance of indigenous knowledge lies in its creation of approaches to the contemporary and prospects of indigenous communities (Sheenah, 2004). This is because the development of indigenous knowledge is context-specific, with sustainability as an objective (Wolfgramm, 2007). A study conducted by Adade et al. (2020) highlights the interest of several scholars in the integration of indigenous knowledge and local perspectives into sustainable landscape management from the past 10 years. This is due to the realisation that indigenous knowledge is significant in landscape and biodiversity conservation. As a result, indigenous knowledge has been in many discussions around both rural and urban ecosystems conservation and sustainable agriculture.

The need to give attention to IKS for landscape management has been stressed in the 1987 World Commission on Environment and Development (WCED). According to Adade et al. (2020), the ignorance of IK would lead to the loss of skills, practices, and understanding required for the sustainable development of ecosystems. The 1993 United Nations Conference on Environment and Development (UNCED) was positioned toward achieving sustainable development through exploring the contribution of IKS (Adade et al., 2020). The World Summit on Sustainable Development that took place in South Africa in 2022 provided highlight on the intrinsic relationship between indigenous people and their land (McGregor, 2014).

The risk of climate change is severe throughout the world, including indigenous communities. This is happening despite the least greenhouse emissions indigenous people contribute (Cajete, 2020). Indigenous people are asymmetrically exposed to higher risks of dramatic consequences of climate change due to global warming (IPCC, 2018). The risk of indigenous peoples is exacerbated by their dependence on natural resources for various purposes (Salick & Byg, 2007). Scholars believe that addressing the changes in the environment is largely dependent on the collaboration of knowledge systems (Reid et al. 2006; Armitage 2008; Berkes 2012). Therefore, it is vital to identify ways for different systems of knowing to co-exist to address environmental changes. Scholars such as Smith (1999), Wilson (2008), and Bohensky & Maru (2011) submit that diverse sources of knowledge have the capability to influence the management of the environment.

Historically, indigenous people have endured extreme weather events (Cajete, 2020). In the face of the current challenges associated with climate change, they have been seen to be resilient, however, their knowledge has been ignored due to Western dominance (Bohensky & Maru, 2011). The knowledge of indigenous people has been successful in strengthening communities while mitigating challenges embedded in climate change. The significance of indigenous knowledge is reflected in its application in many rural communities. This is due to their dependence on natural resources for the security of their livelihood. According to Adade et al. (2020), the evidence of increasing interest in the incorporation of knowledge systems and livelihood is rooted in landscape management. Adade et al. (2020) recognise the nexus between people and ecosystem services offered by the landscape, thus the need for its management.

STATEMENT OF THE PROBLEM

Indigenous knowledge has been recognised by governments for its potential in addressing landscape changes. As concerns around landscape change grow, it is necessary to give precedence to sustainable practices (Adade et al., 2020). IKS have thus been included in policies; hence, the submission that landscape actors should include it in their activities. However, several conventional approaches to landscape management are disengaged from the land, thus causing ecological harm and destroying biodiversity (Justo, 2023). This is associated with a lack of representation of IK in the activities of landscape planning and management. The recognition and acknowledgment of IK has the potential to offer perceptions that could improve current landscape design and management practices. This is particularly because indigenous knowledge offers locally appropriate and applicable solutions to landscape and climate change challenges (Gómez et al., 2013). It is, therefore, important that the bottom-up approach be encouraged in landscape management that acknowledges context and local indigenous knowledge (Justo, 2023).

CHAPTER OBJECTIVES

- To appraise the role of indigenous knowledge systems in landscape management
- To determine the importance of integrating indigenous knowledge systems in landscape management.

LITERATURE REVIEW

Overview of the Landscape Changes

Landscape is commonly understood as a primary sphere in which human beings and the environment interact (Bu¨ rgi et al., 2005). Landscapes are a central part of the discourse about environmental changes, sustainable development, and climate change (Turner et al., 2007; Fragkias & Boone, 2016). Musakwa and Wang (2018), report it is fundamental to understand the drivers to landscape change. The interaction between human beings and the environment allows them to derive useful resources for their sustainability. In this regard, land uses that stress the environment have negative effects on how the ecosystem functions (Luvuno et al., 2016; Fox & Rowntree, 1996).

Landscape change has recently attracted attention, due to its closeness with climate change, and its impact on food security (Musakwa & Wang, 2018). Plieninger, et al. (2016) present climate change, urbanization, and intensification of agriculture as reflectors of the landscape changes. Musakwa and Wang (2018) observe that there is limited knowledge about land use and landscape evolution dynamics, as well as the socio-economic factors contributing to landscape evolution. As a result, in Southern Africa, small communities, particularly in rural areas, are usually at the mercy of the negative effects of landscape changes.

Socioeconomic drivers of landscape changes are associated with poverty and unemployment, forcing people to excessively use natural resources. The majority of people in Zimbabwe cut down trees for sale as firewood for commercial gain (Cavendish, 2001; Nyahunzvi, 2015). Correspondingly, youth in South Africa leave the elders to tend the fields as they migrate to urban areas. Musakwa and Niekerk (2013) opine that the migration to urban areas exposes the agricultural land in rural areas to urban use. This means that land designated for agricultural activities may be converted into a different use, due to lack of people working on it. It can be argued that poverty also influences agricultural intensification due to consumer demand. This means brings a question of landscape sustainability due to agricultural intensification (Musakwa & Wang 2018).

In South Africa, there is a growing need for food due to population growth (Simane et al., 2025). This is exacerbated by the decline in profits due to a lack of employment (StatsSA, 2023). This leads to the intensification of farming to feed the populations; however, if not properly managed can lead to land degradation. Modern farming, which includes overuse of chemical inputs for quick production due to the demand of food, may also negatively impact the environment (Musakwa & Wang 2018). There is an urgent need to explore factors associated with landscape change. The understanding of these factors has the potential to inform the roles different stakeholders can play to address landscape change.

Conception of Landscape and the Importance of Indigenous Knowledge

Landscape refers to everything one can see when looking across the land (Pearson & Gorman, 2023). The understanding of landscapes recognizes the culture found within or around these environments. Therefore, Brabyn and Mark (2011) argued that this means that landscape is not just something people look at for beauty and aesthetics that can just be used by people for economic and personal profit. Freeman et al, describes landscapes as the biophysical- presenting landscapes as just a natural phenomenon the anthropogenic where the landscape is seen as nature with some human constructions; the intangible which views the landscape as being based on individual or social perceptions or interpretations; and the coupled social-ecological or integrated interpretation where the landscape is seen as a whole incorporating the natural, human and spiritual dimensions (Pearson & Gorman, 2023). These perspectives denote the complex way people interact with and understand their environment. It presents a deep relationship between people and the land that they see and use. This means that one's cultural background determines how they relate to their landscape. This may be associated with indigenous people's intimate relationship with their landscapes, as they do not see a separation from them. (Heimer, 2005) opines that people's experiences in a particular environment is likely to present a strong relationship with the place.

Indigenous knowledge is understanding landscapes as complex systems (Crumley, 2017). Pearson, & Gorman (2023) present an example of Maori people from New Zealand, whose worldview sees no separation between physical landscapes and ancestral and cultural activities. The aboriginal culture also presents the significance of cultural and spiritual importance of landscapes as sacred to the people (Pearson & Gorman, 2023). Toussaint et al. (2005) emphasize the connectivity that indigenous people have with water sources. This relationship is stressed through teaching good environmental practices to children and passing on the importance of practicing water etiquette in the form of particular songs and chants to use when approaching a site (Pearson & Gorman, 2023).

The above demonstrates that throughout time, human beings have been known to have cultural networks with the environment to fulfill their survival and cultural needs (Lemus et al., 2024). These networks are commonly known for subsistence harvesting and medicinal use while also extending to cultural expressions and intangible connections, thus making the environment important for spiritual and religious purposes. This positions indigenous knowledge as a vital part of how human beings understand the world they live in (Congretel et al., 2020). The oral transmission, cultural expressions, and rituals echo the approaches and practices developed through the people's experiences of and understanding their environment (Diver et al., 2019).

Lemus, et al. (2024) present indigenous knowledge as a strategy of life within local communities, indicating its significance as an element of human cultural capital. According to Jessen et al. (2022), although this indigenous knowledge is important in the conservation and management of landscape, several studies present indigenous knowledge as a complement to scientific studies, which are Western-oriented. However, Trisos et al. (2021) argue that such studies have shortfalls in presenting indigenous knowledge in its essence. This exposes perceptions that may pose limitations in the activities of landscape management (Kadykalo et al., 2021). Lemus et al. (2024) are of the view that cultural landscapes represent the intimacy between biological diversity and human culture. They embody important relationship between people, their heritage, and the natural environment.

Indigenous Knowledge in Changing Environments

The world, particularly the Global South, is currently experiencing dramatic changes due to climate change, while different areas in the world are home to many biodiversity hotspots (Loch & Riechers, 2021). This is attributed to globalisation and market pressures, as well as poor integration and recognition of indigenous knowledge in contemporary academia and different spheres of life. Various studies have shown the need to overcome these challenges through indigenous knowledge in management and research. According to Lalonde (1991) the maintenance and recognition of indigenous knowledge is gaining momentum in the science of sustainability. Indigenous knowledge is considered to play an intrinsic role in understanding local social-ecological systems and providing ways to cope with current global changes. Loch and Riechers (2021) emphasize that there is a need to include indigenous knowledge in conservation approaches in areas where landscape challenges are advanced. Loch and Riechers (2021) further conclude that future research is important in positioning indigenous knowledge and management of landscapes

Indigenous knowledge is consumed through observation and repetition of activities (Barnhardt & Kawagley 2005). This is because of its roots in the oral tradition, spiritual and holistic understanding of nature. However, this contrasts with the dominant reductionist approach of Western knowledge, which is based on cynicism and phenomena isolation (Chapman & Schott, 2020). This has led to arguments and perceptions that indigenous knowledge should be separated from other knowledge systems. As a result, Indigenous knowledge has been ignored in efforts to conserve and manage the landscape while Western scientific knowledge is prioritised (Edgar et al., 2022). This creates an impression that indigenous knowledge is not an important aspect of landscape management. Therefore, it is important to ensure there are concerted efforts to create landscape management knowledge that recognises indigenous knowledge.

Recognition of Indigenous Knowledge in Conservation Efforts

There has recently been a growth in the recognition of the reputation of indigenous knowledge in the conservation of biodiversity. Scholars and practitioners have started acknowledging indigenous people for the knowledge gathered through their existence and interaction with natural environments (Gautam, 2019). Indigenous knowledge is important in addressing the modern conservation challenges. Nevertheless, the application of indigenous knowledge with contemporary science and policies has its challenges. These challenges are related to a lack of recognition, the question of legitimacy, and the risk of diluting indigenous practices (Gautam, 2019).

Indigenous knowledge systems have been sidelined and demoted while accommodating Western science and policies. According to Gautam (2019) colonialism and the increase of industrialisation have resulted in the displacement of Indigenous people, thus negatively impacting the environment. The practices of indigenous people have been viewed as backward, and as a result, they were not consulted when developing policies in conservation. As mentioned above, there is currently a shift in attitude regarding activities around and about conservation (Gautam, 2019). There seems to be increased recognition of indigenous knowledge and its contribution to sustainable resource management. The practices of Indigenous people have been seen to offer solutions to some environmental challenges (Moller et al., 2004; Ndagala, 1990).

Climate Change-Related Practices of Indigenous People

Chanza et al. (2024) present the attention given to Climate Smart Agriculture. This agricultural approach that ensures the production of food is linked to building resilience to adapt to climate change (FAO, 2013). This approach acknowledges traditional farming practices that are climate-resilient and generate low emissions. According to Sereenonchai and Arunrat (2020), the application of Climate Smart Agriculture is through avoiding burning rice residues using rice straw compost and alternative wetting and drying amongst farmers in Thailand. In Ethiopia, such practices are known to include soil conservation, water-efficient irrigation, and fuel wood conservation (Regasa & Akirso, 2019).

The combination of aspects of forestry and agriculture, known as agroforestry, is common in many rural communities (Chanza et al., 2024). This practice is known to ensure that ecosystem services support the livelihoods of rural communities while conserving the environment (Galang & Vaughter, 2020). According to Chanza et al. (2024), this practice involves ecological networks and cultural influences to ensure the environment and people benefit. This relationship can be fathomed and promoted with the use of indigenous knowledge. This is because of an acknowledgment of the indigenous knowledge systems of the environment that has allowed communities to conserve and benefit from it (Avila-Bello et al., 2023).

Scholars have widely reported indigenous forest conservation as important to lessening the effects of climate change. This is commonly through the conservation of areas designated and known as sacred sites, rich in tree species (Edgar et al., 2022; Chanza et al., 2024). Communities protect these areas for cultural and spiritual uses, and the protection is usually against the interference of the people. The existence of sacred sites makes them peculiar as opposed to other areas that have been legally conserved. Taboos, myths associated with cultural beliefs impose strict rules and sanctions, thus maintaining the condition of such environments (Maru et al., 2023). The protected areas are key to protecting rich biodiversity, providing access to medicines, and regulating climate.

Chanza and Musakwa (2021) report that in Zimbabwe, people are more interested in conserving ecosystems if there are benefits to be derived. These benefits can be in the form of food, availability of sites to observe seasonal and climate change, as well as spiritual benefits. This means that there is a need to ensure people are made aware of ecosystem conservation. One can submit that people are often responsible for something that they can interact with, thus, a need for education that emphasizes the relationship between humans and nature.

The Maasai and Grassland Management in East Africa

A study by Gautam (2019) presents the Maasai people of Kenya and Tanzania, who have practiced pastoralism in the Eastern part of Africa for years. Maasai herders have practiced rotational grazing to prevent overgrazing. This has allowed vegetation to regenerate while maintaining biodiversity for future grazing and other uses. It is indicated that the Maasai communities usually have a higher biodiversity level than adjacent lands used for commercial agriculture or livestock ranching (Reid, 2012). This practice of the Maasai's of controlled burning, and reduces the risk of wildfires, while promoting new growth. Modern conservationists have been seen to recognize this as an approach to ecosystem management.

Water Conservation and Management Amongst Vhavenda People of South Africa

The study by Edgar et al. (2022) found that African indigenous knowledge of water conservation and management was effective. This is because this knowledge was not only used to communicate lessons about water conservation but also to transfer knowledge of all attributes of life in an African indigenous society. The findings of the study revealed that taboos were influential in regulating and curbing behaviours that would initiate damage to water bodies, such as pollution, cutting trees, and killing aquatic animals.

The systematic way of accessing water sources at different times effectively allowed human beings, livestock, and wild animals to have their fair share of the water provided by their environment. Women and children used to fetch water from the rivers and springs in the morning for domestic use. Midday and the rest of the day were given to the livestock till the sun set when women and children would be fetching water for the last time. Night hours and the early morning before sunrise were allocated for wild animals. This allowed for the reduction of water pollution and competition for water that would possibly result in death and injuries among human beings and animals.

Oral tradition, particularly African Indigenous games, and storytelling played an intrinsic role in educating community members, especially children. Through oral tradition, the worldview of African indigenous people in all characteristics of life, including water conservation and management, was explained. This was a vital method of conserving and managing water as it permitted children and other members of the African indigenous communities to attain African Indigenous perspectives, principles, beliefs, and practices of water conservation and management that allowed them to share a common worldview with the rest of the communities they inhabited (Edgar et al., 2022).

COLLABORATED UNDERSTANDING OF THE ENVIRONMENT

There is a difference between the way in which scientific science and indigenous knowledge is perceived (Weiss et al., 2013). Over the years, the focus of the differences between these knowledge holders has neglected the similarities existing in these knowledge systems and how they contribute to the management of changing environments (Wilson, 2008). This has led to the disconnect in understanding what the systems can contribute to understanding the changes in their environments (Weiss et al., 2013).

Scholars such as Riedlinger & Berkes (2001); Laidler (2006) present these systems of knowing and their efforts to address changing environments as complementary. However, this matters not in cases where only one system of knowing is revered and accommodated in efforts to address challenges posed by environmental challenges. A study conducted by Rathwell, et al. (2015) presents those users of the local resources, who in this study would be referred to as indigenous people, use their resources in response to their experiences and changes encountered in their environments. According to Rathwell et al., (2015), these local ways of doing things have proved to work against some environmental challenges.

It is important to recognize that indigenous knowledge is communally shared based on cultures of the local communities (Rathwell et al., 2015). Thus, its recognition in the climate change discourse offers local people an opportunity to contribute to the discourse. A submission is that being exposed to different ways of understanding the world is an opportunity to capacitate communities, while one way of knowing limits an opportunity to find solutions (Miller et al., 2008).

THEORETICAL FRAMEWORK

A theoretical framework is a map for a research enquiry that guides and supports a study (Lederman & Lederman, 2015). A theoretical framework provides a structure that defines how one applies philosophies to approach the study (Grant & Osanloo, 2014). The importance of a theoretical framework provides a foundation from which knowledge is understood in a study (Arafat, 2024). It provides structure and support for the justification of the rationale of the study, problem statement, and objectives. The theoretical framework also provides ground for the literature review. According to Grant and Osanloo (2014), the vision of the study is unclear without the structure provided by the theoretical framework. This chapter adopted a knowledge coevolution framework.

Knowledge Coevolution

This study will be grounded on the knowledge coevolution framework, which recognises the need for indigenous knowledge to be part of the knowledge generation for the communities and researchers (Chapman & Schott, 2020). Knowledge coevolution is referred to as a process of generating information through integrating knowledge systems to facilitate and promote self-determination and cultural resilience of communities (Schott et al., 2020). Knowledge co-evolution is presented by Chapman & Schott, (2020) as a concept that develops knowledge on co-production and two-eyed seeing due to its ability to empower existing systems of knowing. This necessitates acknowledging diversity equilibrium towards accessing resources, respect, and power (Haverkort & Reijntjes, 2010).

Chapman and Schott (2020) present knowledge coevolution as a framework that strengthens indigenous knowledge, allowing it to advance in its rights to benefit indigenous stakeholders. At the same time, this framework makes efforts to provide room for different knowledge systems to integrate regardless of their differences. This ensures a holistic approach rather than having one knowledge absorbed into the other. Co-evolution deconstructs the dynamics of power associated with Western structures and norms while recognising indigenous knowledge and the context of its application (Baggio, 2021). This brings reflects efforts to decolonisation, that ensures there is acknowledgment of knowledge systems embedded in culture of the local people (Smith, 1999), while taking advantage of their worldviews as landscape actors. Knowledge coevolution will allow for the study of indigenous knowledge and the development of a context-specific approach to landscape management.

METHODOLOGY

This chapter employed a desktop study methodology depending on the review and analysis of existing secondary data. Desktop studies are usually the first step in various studies. Desktop studies are conducted through a review and analysis of information that has already been published (Kiely, 2025). This method is important for understanding context and providing answers to research questions through the utilization of available information (Kiely, 2025).

This approach involved a comprehensive search of reports and publications (Hewett, 2023) about and around landscape management and indigenous knowledge to gather relevant information. The desktop study was fashioned in line with the aim and objectives of the chapter. The author reviewed literature from academic papers, articles, books, online publications, and government agencies. The author selected reliable and relevant data from these sources, drew conclusions, and discussed them to achieve the chapter's aim and objectives (Kiely, 2025). The data collected underwent a thorough process of critical evaluation and interpretation to identify key themes. These themes are developed in relation to the objectives of this chapter. This methodology enabled the addressed acknowledgement of Indigenous knowledge in landscape management through the lens of related established knowledge. The following is a summary of how desktop was adopted;

- Definition of the chapter aim to understand what has to be achieved
- Identification of relevant keywords to determine the type of information searched
- Selection of reliable sources and information
- Collection or extraction of relevant data from the sources

- Analysis of the data while identifying and developing themes from patterns and key findings.

DISCUSSION OF FINDINGS

Integrating Indigenous Knowledge Systems to Inform Policies and Frameworks on Environmental Management

The current environmental needs require approaches that do not only prioritise scientific responses in mitigating the effects of climate change and promoting adaptation to it. According to Madzivhandila (2024), indigenous knowledge systems serve as a source of information and decision-making for many local communities. This means indigenous knowledge systems are important and should be considered in the efforts to address climate change, particularly in areas rich with it (Madzivhandila, 2024). In this regard, practitioners should consider indigenous knowledge systems should be considered for the provision of solutions to changing environments.

It is reported that the lives of indigenous people are greatly influenced by their knowledge acquired through experience and adapting to the changing environment (IPCC, 2014). These experiences of indigenous people are fashioned through culture, thus influencing their perceptions about life (Mafukata, 2015). This should allow for commitments to assist communities to integrate their local indigenous practices for sustainability. Madzivhandila (2024) emphasizes this with a South African saying, "Nothing for us without us," indicating the importance of involving local communities.

It is important to note that for years, indigenous people in the world have successfully developed and implemented strategies to adapt and address environmental issues (The & Ahmad 2021, in Tshifhumulo & Makhanikhe 2021). According to Madzivhandila (2024), the understanding of climate change effects requires a holistic approach due to their interdisciplinary nature. According to Thinh et al. (2007), there is evidence that development is usually limited in its success when local knowledge is not included in policies and planning. As a result, governments need to consider the perspectives of local people when developing policies. The discussion below considers three aspects as important in ensuring IK is acknowledged in landscape management.

Acknowledging IK for Contextual Sustainable Approaches

The government of South Africa has made its investment in the recognition of indigenous knowledge. However, its appreciation and familiarity are still missing in both urban and rural areas. This is because, in most if not all cases, people seem to only entertain indigenous knowledge when it favours them to do so. However, there are important aspects of it that landscape actors should consider.

Indigenous people possess ecological knowledge about local ecosystems and interactions. This knowledge is intertwined with their belief systems and practices (Walentowski, 2024). Thus, indigenous people have spiritual connections with their environment, and this helps inform their strategies of conservation. Therefore, it becomes important to consider what is known, how that is known by the local people when and before initiatives are introduced in local communities (Walentowski, 2024).

The practices of conservation of indigenous people emphasize the sustainable use of natural resources. This is because they depended on these resources for years. The relationship between people and their environment is two-way, meaning they protect their environment to benefit from it. As such, the sustainable use of natural resources is primary to indigenous people, and their methods must be appreciated. Indigenous people also have areas of priority or biodiversity hotspots, sacred areas (Edgar et al., 2022). Maintaining their agricultural land through various methods such as rotational farming, selective harvesting, and prescribed burning has for years proved to be sustainable.

Walentowski (2024) stated that Indigenous peoples have a strong bond with their ancestral lands and tend to prioritize their stewardship. Their land management practices, such as controlled burning or agroforestry, contribute to the maintenance of ecological balance, soil fertility, and the protection of critical habitats for numerous plant and animal species. These approaches often promote biodiversity conservation through sustainable land use practices. As mentioned above, indigenous knowledge often includes cultural and spiritual values that recognize the intrinsic value of nature. This deep connection with the environment fosters a holistic perspective and a sense of responsibility toward conserving biodiversity beyond utilitarian purposes (Walentowski, 2024). Thus, an important need to consider it in landscape management.

Shaping Power Dynamics in Landscape Management

The management of landscape is a demanding effort that requires commitment from all landscape actors. The changing environments have proved to be stressful, thus the need for coordinated efforts. This is a step towards an inclusive decision-making process, which has previously been an ignored aspect in landscape management (Favretto et al., 2021). This is because indigenous knowledge was marginalized, regardless of the success Indigenous people have demonstrated throughout their time of existence. African Indigenous Knowledge Systems cannot be ignored in the current state of changing environments. African indigenous knowledge values perspectives and experiences developed over generations. Its incorporation enhances the understanding of ecological processes, sustainable practices, and effective management. It is also important in the provision of a holistic and balanced approach to biodiversity/ landscape management (Favretto et al., 2021). This leads to improved outcomes and sustainable solutions that benefit both people and the environment.

Shaping power dynamics requires revisiting the sources of power in society. This is important in understanding why certain things happen or do not happen in landscape management. It is important to note that most areas that are focal points for conservation and management due to rich biodiversity are rural. Therefore, one would expect people residing in these areas to be more involved in managing their landscape. In South Africa, the government, through the Department of Forestry, Fisheries and Environment (DFFE), oversees the administration of activities related to landscape management. As per Siangulube et al. (2023), this department works through local municipalities, with contact with the local communities. In these communities, Traditional leaders are the custodians and adjudicators of matters on land and natural resources management. Traditional leaders work through their possession of indigenous knowledge, supported by members of their community. Civil society, particularly the Non-Governmental Organisation, exist to shape ideologies, ensure compliance, and support communities and government initiatives through the implementation of projects and programmes.

The mismatch between these landscape actors is usually a result of approach, where members of the community are not involved in decision-making regarding the management of their biodiversity (Adade et al., 2020). This, in a way, perpetuates the historical, political injustices that relegated the views of the people regarding what concerns them. Thus, when community members have to act, they don't as a result of lack of interest and sometimes thinking they have nothing to contribute due to exclusion.

"Some 'economically powerful and influential entrepreneurs and staff of government agencies enter these spaces and influence decisions (Siangulube et al., 2023)

In some cases, where the government and NGOs make efforts to ensure the management of changing environments, there is too much introduction of modern or foreign methods. These methods are, at times careless about context and indigenous knowledge. This is usually perpetuated by the idea that indigenous knowledge is of the past. A study conducted by Siangulube et al. (2023), indicates that the intersections between these landscape actors expose that power disproportions are the results of actors' social belongingness, situatedness, and settlement histories.

The following are some factors visible in relation to power dynamics and influence to decision making.

- Colonial Legacy: displacement and marginalization of indigenous communities leading to lack of access to resources and decision-making process. This continues to influence power dynamics today.
- Socioeconomic Factors: Communities with economic resource advantages have more influence and decision-making power compared to those with limited resources. This can create an uneven playing field.
- Historical Injustices: Past injustices, such as land dispossession or forced displacement, can lead to power imbalances among different groups. These historical events shape the relations and dynamics between actors in the landscape, influencing who gets to speak and participate in decision-making processes.
- Cultural and Ethnic Context: Dominant cultural or ethnic groups may have historically held more power and influence, while minority groups may have been marginalized. Issues of tribalism also come into play. It is important to appreciate diversity and not see it as division.

Collaborating Knowledge Systems Amongst Landscape Actors

It is important to note that these actors of landscape management come from different schools of thought. Much of what they know has been learnt through formal education and experience, which may not be influenced by indigenous perspective. As such, there is a need to encourage collaboration of knowledge systems possessed by landscape actors. This means that there is a need for a holistic approach to landscape management and control, which can be achieved through integrating African indigenous and modern knowledge (Tengö, 2014).

Collaboration between two of these knowledge systems in landscape management activities helps build and rebuild the relationship between people and their natural environment (Adade et al., 2020). This would serve as a way to close the gap between African indigenous and modern knowledge while ensuring that African indigenous knowledge does not disappear but takes a stand in comprehensive engagements on landscape management.

Scholars believe that integrating African indigenous knowledge and Western science is an excellent approach to resource management. Collaborating African indigenous and modern knowledge in landscape management may have great potential for establishing sustainable solutions to challenges brought about through changing environments (Adade et al., 2020). This can be achieved by acknowledging both knowledge systems for their intrinsic and diverse roles in biodiversity management.

It is essential not to forget the competition between indigenous knowledge and Western science. There must be a remedy to close the gap that keeps African indigenous knowledge and Western science at odds (Favretto et al., 2020). Integrating African indigenous knowledge and modern knowledge in landscape management should be based on solving problems related to changing environments and not a competition between knowledge systems (Favretto et al.,2021).

FUTURE RESEARCH DIRECTIONS

The evolution of landscape changes requires innovative research that embraces different knowledge systems to enhance environmental stability. In this regard, future research should further investigate the interaction between people and landscapes in the contemporary time to ascertain how that impacts landscape. This is important in ensuring that landscapes are responsive to emerging challenges and opportunities. There is also a need for future studies to explore community engagement between different stakeholders. Community engagement is vital in its accommodation of participatory approaches that have the potential to foster local stewardship and ownership.

CONCLUSION AND RECOMMENDATIONS

The discussion above highlights the existence of different landscape actors in South Africa. These actors act to achieve the same or similar objectives. However, there are areas in which they miss each other in their activities. This is usually visible in cases where local communities and indigenous knowledge are not considered in efforts to address changing environments. This chapter presented the exclusion of IKS in policy implementations in landscape management. Mainstream approaches have been reported to be out of context in many rural areas, resulting in ecological disturbance. The chapter recommends the bottom-up approach to landscape management, which acknowledges local knowledge and practices. There is a need to ensure there is a working relationship between these actors. This relationship should allow for the integration of knowledge systems possessed for sustainable management. It is, therefore, important to determine the management of power and its dynamics at different levels, particularly between the government, local communities, and NGOs, to ensure interests are advocated for.

This chapter recommends the acknowledgement of IK in landscape management. There is a need for landscape actors to identify and use IK relevant to their work. This study recommends collaboration between stakeholders to manage changing landscapes. This is important in ensuring the participation of local communities in stages of landscape management. Collaboration is also important for knowledge sharing. This chapter also recommends investment to capacitating both indigenous people and landscape actors or professionals. The government must facilitate capacity building to ensure that best practices are adopted.

REFERENCES

Adade Williams, P., Sikutshwa, L., & Shackleton, S. (2020). Acknowledging indigenous and local knowledge to facilitate collaboration in landscape approaches—Lessons from a systematic review. *Land (Basel)*, *9*(9), 331. DOI: 10.3390/land9090331

Anand, S., Kumar, H., Gagan, A., & Mayala, V. K. (2025). Indigenous Knowledge and Water Conservation Practices in South Africa: A Systematic Literature Review. *Journal of Environment and Earth Science*.

Arafat, A. A. D. (2024). Introduction: Theoretical Framework. In *Human Security in the Middle East and North Africa* (pp. 1–46). Springer Nature Switzerland. DOI: 10.1007/978-3-031-70196-2_1

Armitage, D. (2008). Governance and the commons in a multi-level world. *The International Journal of the Commons*, *2*(1), 7–32. DOI: 10.18352/ijc.28

Ávila-Bello, C. H., Hernández-Romero, Á. H., Vázquez-Luna, D., Lara-Rodríguez, D. A., Martínez-Jerónimo, A., Meneses-García, B. N., & Sánchez-Sandoval, X. M. (2024). Design of complex agroecosystems: Traditional and formal knowledge to conserve agrobiodiversity in the Santa Marta Mountains, Veracruz, México. *Environment, Development and Sustainability*, *26*(3), 7129–7161. DOI: 10.1007/s10668-023-03002-9

Baggio, J. A. (2021). Knowledge generation via social-knowledge network co-evolution: 30 years (1990–2019) of adaptation, mitigation and transformation related to climate change. *Climatic Change*, *167*(1), 13. DOI: 10.1007/s10584-021-03146-5

Barnhardt, R., & Oscar Kawagley, A. (2005). Indigenous knowledge systems and Alaska Native ways of knowing. *Anthropology & Education Quarterly*, *36*(1), 8–23. DOI: 10.1525/aeq.2005.36.1.008

Berkes, F. (2009). Community conserved areas: Policy issues in historic and contemporary context. *Conservation Letters*, *2*(1), 20–25. DOI: 10.1111/j.1755-263X.2008.00040.x

Berkes, F., & Davidson-Hunt, I. (2008). The cultural basis for an ecosystem approach. *The ecosystem approach: Complexity, uncertainty, and managing for sustainability*, 109.

Bohensky, E. L., & Maru, Y. (2011). Indigenous knowledge, science, and resilience: What have we learned from a decade of international literature on "integration"? *Ecology and Society*, *16*(4), art6. DOI: 10.5751/ES-04342-160406

Brabyn, L., & Mark, D. M. (2011). Classifying landscape character. In Stea, D. (Ed.), *Landscape in Language: Transdisciplinary Perspectives; Mark, D.M., Turk, A.G., Burenhult, N* (pp. 395–409). John Benjamins Publishing Co. DOI: 10.1075/clu.4.21bra

Bürgi, M., Hersperger, A. M., & Schneeberger, N. (2005). Driving forces of landscape change-current and new directions. *Landscape Ecology*, *19*, 857–868. DOI: 10.1007/s10980-005-0245-3

Cajete, G. A. (2020). Indigenous science, climate change, and indigenous community building: A framework of foundational perspectives for indigenous community resilience and revitalization. *Sustainability (Basel)*, *12*(22), 9569. DOI: 10.3390/su12229569

Cavendish, W. (2000). Empirical regularities in the poverty-environment relationship of rural households: Evidence from Zimbabwe. *World Development*, *28*(11), 1979–2003. DOI: 10.1016/S0305-750X(00)00066-8

Chanza, N., Musakwa, W., & Kelso, C. (2024). Overlaps of indigenous knowledge and climate change mitigation: Evidence from a systematic review. *Frontiers in Climate*, *6*, 1344931. DOI: 10.3389/fclim.2024.1344931

Chapman, J. M., & Schott, S. (2020). Knowledge coevolution: Generating new understanding through bridging and strengthening distinct knowledge systems and empowering local knowledge holders. *Sustainability Science*, *15*(3), 931–943. DOI: 10.1007/s11625-020-00781-2

Crumley, C. L. (2017). Historical ecology and the study of landscape. *Landscape Research, 42*(sup1), S65-S73.

Diver, S., Vaughan, M., Baker-Médard, M., & Lukacs, H. (2019). Recognizing "reciprocal relations" to restore community access to land and water. *The International Journal of the Commons*, *13*(1), 400. DOI: 10.18352/ijc.881

Edgar, B. A., Matshidze, P. E., & Lee, K. S. (2022). Exploring Vhavenda indigenous methods of water conservation and management in Tshidzivhe village, Limpopo province, South Africa. *South African Journal of Cultural History*, *36*(2), 1–22. DOI: 10.54272/sach.2022.v36n2a1

FAO. (2013). *Climate-Smart Agriculture Sourcebook. The Food and Agricultural Organisation of the United Nations*. FAO.

Favretto, N., Afionis, S., Stringer, L. C., Dougill, A. J., Quinn, C. H., & Ranarijaona, H. L. T. (2020). Delivering climate-development co-benefits through multi-stakeholder forestry projects in Madagascar: Opportunities and challenges. *Land (Basel)*, *9*(5), 157. DOI: 10.3390/land9050157

Favretto, N., Shackleton, S., Sallu, S. M., & Hoffman, T. (2021). Editorial for special issue:"collaboration and multi-stakeholder engagement in landscape governance and management in Africa: lessons from practice". *Land (Basel), 10*(3), 285. DOI: 10.3390/land10030285

Fragkias, M., & Boone, C. G. (2016). Modern political economy, global environmental change and urban sustainability transitions. *Current Opinion in Environmental Sustainability*, *22*, 63–68. DOI: 10.1016/j.cosust.2017.04.007

Galang, E. I. N. E., & Vaughter, P. (2020). Generational local ecological knowledge on the benefits of an agroforestry landscape in Mindanao, Philippines. *Asian Journal of Agriculture and Development*, *17*(1), 90–108. DOI: 10.37801/ajad2020.17.1.6

Gautam, S. K. (2019). The role of indigenous knowledge in biodiversity conservation: Integrating traditional practices with modern environmental approaches. *Environment Reporter*, *1*(2), 1–3. DOI: 10.51470/ER.2019.1.2.01

Gómez-Baggethun, E., Corbera, E., & Reyes-García, V. (2013). Traditional ecological knowledge and global environmental change: research findings and policy implications. *Ecology and society: a journal of integrative science for resilience and sustainability, 18*(4), 72.

Grant, C., & Osanloo, A. (2014). Understanding, selecting, and integrating a theoretical framework in dissertation research: Creating the blueprint for your "house". *Administrative issues journal, 4*(2), 4.

Gupta, A. D. (2012). Way to study indigenous knowledge and indigenous knowledge system. *Research Journal of Recent Sciences ISSN, 2277*, 2502.

Haverkort, B., & Reijntjes, C. (2010). Diversities of knowledge communities, their worldviews and sciences: On the challenges of their co-evolution. *Traditional knowledge in policy and practice: Approaches to development and human well-being*, 12-30.

Heimer, H. (2005). Topophilia and quality of life: defining the ultimate restorative environment.

Hewett, M., Butler-Kruger, L., Foflonker, K., Edge, J., Botha, M. H., & Seedat, S. (2023). Desktop analysis and qualitative investigation of programmes and initiatives for the prevention of and response to gender based violence at South African universities with medical campuses. *South African Journal of Higher Education, 37*(5), 142–157. DOI: 10.20853/37-5-5181

Jessen, T. D., Ban, N. C., Claxton, N. X., & Darimont, C. T. (2022). Contributions of indigenous knowledge to ecological and evolutionary understanding. *Frontiers in Ecology and the Environment, 20*(2), 93–101. DOI: 10.1002/fee.2435

Justo (2023). Indigenous Views on Landscape Design and Maintenance for Sustainability. *Native Tribe Info*.

Kadykalo, A. N., Cooke, S. J., & Young, N. (2021). The role of western-based scientific, indigenous and local knowledge in wildlife management and conservation. *People and Nature*, *3*(3), 610–626. DOI: 10.1002/pan3.10194

Kiely, T. J. (2025). *What Is Desk Research? Meaning, Methodology, Examples*. Melwater.

Laidler, G. (2006). Inuit and Scientific Perspectives on the Relationship Between Sea Ice and Climate Change: The Ideal Compliment? *Climatic Change*, *78*(2–4), 407–444. DOI: 10.1007/s10584-006-9064-z

Lalonde, A. (1991). African indigenous knowledge and its relevance to environment and development activities. *Canadian International Development Agency*, 1-12.

Lederman, N. G., & Lederman, J. S. (2015). What is a theoretical framework? A practical answer. *Journal of Science Teacher Education*, *26*(7), 593–597. DOI: 10.1007/s10972-015-9443-2

Lemus, K., Barrasa, S., & McCall, M. K. (2024). Wildlife and cultural landscape management from a local knowledge approach: A review. *Human Dimensions of Wildlife*, •••, 1–13. DOI: 10.1080/10871209.2024.2384747

Loch, T. K., & Riechers, M. (2021). Integrating indigenous and local knowledge in management and research on coastal ecosystems in the Global South: A literature review. *Ocean and Coastal Management*, *212*, 105821. DOI: 10.1016/j.ocecoaman.2021.105821

Luvuno, L. B., Kotze, D. C., & Kirkman, K. P. (2016). Long-term landscape changes in vegetation structure: Fire management in the wetlands of KwaMbonambi, South Africa. *African Journal of Aquatic Science*, *41*(3), 279–288. DOI: 10.2989/16085914.2016.1177482

Madzivhandila, T. S. (2024). Integrating Indigenous Knowledge Systems (IKS) to Climate Change Adaptation in South Africa: Lessons from the "Past". *South African Journal of Higher Education*, *38*(5), 159–168. DOI: 10.20853/38-5-6417

Mafukata, A. M. (2015). "Mythical leopards (Panthera Pardus) of the Vhavenda." International. *Journal of Information Research and Review*, *2*(7), 836–852.

Maru, Y., Gebrekirstos, A., & Haile, G. (2023). Indigenous sacred forests as a tool for climate change mitigation: Lessons from Gedeo community. *Southern Ethiopia. J. Sust.Journal of Sustainable Forestry*, *42*(3), 260–287. DOI: 10.1080/10549811.2021.2007490

McGregor, D. (2014). Lessons for Collaboration Involving Traditional Knowledge and Environmental Governance in Ontario, Canada. *Alternative*, *10*(4), 340–353. DOI: 10.1177/117718011401000403

Miller, T. R., Baird, T. D., Littlefield, C. M., Kofinas, G., Chapin, F. S.III, & Redman, C. L. (2008). Epistemological pluralism: Reorganizing interdisciplinary research. *Ecology and Society*, *13*(2), art46. DOI: 10.5751/ES-02671-130246

Moller, H., Berkes, F., Lyver, P. O. B., & Kislalioglu, M. (2004). Combining Science and Traditional Ecological Knowledge: Monitoring Populations for Co-management. *Ecology and Society*, *9*(3), 2. DOI: 10.5751/ES-00675-090302

Musakwa, W., & Niekerk, A. V. (2013). Implications of land use change for the sustainability of urban areas: A case study of Stellenbosch, South Africa. *Cities (London, England)*, *32*, 143–156. DOI: 10.1016/j.cities.2013.01.004

Musakwa, W., & Wang, S. (2018). Landscape change and its drivers: A Southern African perspective. *Current Opinion in Environmental Sustainability*, *33*, 80–86. DOI: 10.1016/j.cosust.2018.05.001

Ncube, B. (2018). Insights into indigenous knowledge strategies for coping and adapting to drought in agriculture: A Karoo scenario, South Africa. *Indilinga*, *17*(1), 92–108.

Ndagala, D. K. (1990). Pastoralists and the state in Tanzania. *Nomadic Peoples*, •••, 51–64.

Nepstad, D., Schwartzman, S., Bamberger, B., Santilli, M., Ray, D., Schlesinger, P., Lefebvre, P., Alencar, A., Prinz, E., Fiske, G., & Rolla, A. (2006). Inhibition of Amazon deforestation and fire by parks and indigenous lands. *Conservation Biology*, *20*(1), 65–73. DOI: 10.1111/j.1523-1739.2006.00351.x PMID: 16909660

Nyahunzvi, D. K. (2015). Negotiating livelihoods among Chivi curio traders in a depressed Zimbabwe tourism trading environment. *Anatolia*, *26*(3), 397–407. DOI: 10.1080/13032917.2014.974065

Pearson, D., & Gorman, J. (2023). Acknowledging Landscape Connection: Using Sense of Place and Cultural and Customary Landscape Management to Enhance Landscape Ecological Theoretical Frameworks. *Land (Basel)*, *12*(4), 729. DOI: 10.3390/land12040729

Plieninger, T., Draux, H., Fagerholm, N., Bieling, C., Bürgi, M., Kizos, T., Kuemmerle, T., Primdahl, J., & Verburg, P. H. (2016). The driving forces of landscape change in Europe: A systematic review of the evidence. *Land Use Policy*, *57*, 204–214. DOI: 10.1016/j.landusepol.2016.04.040

Rathwell, K. J., Armitage, D., & Berkes, F. (2015). Bridging knowledge systems to enhance governance of the environmental commons: A typology of settings. *The International Journal of the Commons*, *9*(2), 851–880. DOI: 10.18352/ijc.584

Regasa, D. T., & Akirso, N. A. (2019). Determinants of climate change mitigation and adaptation strategies: An application of protection motivation theory. *Rural Sust.Res*, *42*(337), 9–25. DOI: 10.2478/plua-2019-0007

Reid, W. V., Berkes, F., Wilbanks, T. J., & Capistrano, D. (2006). Bridging scales and knowledge systems: concepts and applications in ecosystem assessment.

Schott, S., Qitsualik, J., Van Coeverden de Groot, P., Okpakok, S., Chapman, J. M., Lougheed, S., & Walker, V. K. (2020). Operationalizing knowledge coevolution: Towards a sustainable fishery for Nunavummiut. *Arctic Science*, *6*(3), 208–228. DOI: 10.1139/as-2019-0011

Sereenonchai, S., & Arunrat, N. (2020). Practical agricultural communication: Incorporating scientific and indigenous knowledge for climate mitigation. *Kasetsart Journal of Social Sciences*, *41*(1), 60–67.

Sheehan, N. (2004). Indigenous knowledge and higher education: Instigating relational education in a neocolonial context.

Siangulube, F. S., Ros-Tonen, M. A., Reed, J., Djoudi, H., Gumbo, D., & Sunderland, T. (2023). Navigating power imbalances in landscape governance: A network and influence analysis in southern Zambia. *Regional Environmental Change*, *23*(1), 41. DOI: 10.1007/s10113-023-02031-4

Simane, B., Kapwata, T., Naidoo, N., Cissé, G., Wright, C. Y., & Berhane, K. (2025). Ensuring Africa's Food Security by 2050: The Role of Population Growth, Climate-Resilient Strategies, and Putative Pathways to Resilience. *Foods*, *14*(2), 262. DOI: 10.3390/foods14020262 PMID: 39856928

Smith, H. A., & Sharp, K. (2012). Indigenous Climate Knowledges. *Wiley Interdisciplinary Reviews: Climate Change*, *3*(5), 467–476. DOI: 10.1002/wcc.185

Smith, L. (1999). *Decolonising methodologies: research and indigenous peoples*. Zed Books.

Stats, S. A. (2023). Focus on food inadequacy and hunger in South Africa in 2021. https://www.statssa.gov.za/?p=16235

Tengö, M., Brondizio, E. S., Elmqvist, T., Malmer, P., & Spierenburg, M. (2014). Connecting diverse knowledge systems for enhanced ecosystem governance: The multiple evidence base approach. *Ambio*, *43*(5), 579–591. DOI: 10.1007/s13280-014-0501-3 PMID: 24659474

Thinh, D., Mu, K., & Mul, X. (2007). The role of indigenous knowledge In sustainable development: A case study of the Vietnam mountain regions. In *Finds and Results from the Swedish Cyprus Expedition: A Gender Perspective at the Medelhavsmuseet* (p. 262). Kamla-Raj Enterprises.

Tilman, D., Fargione, J., Wolff, B., D'antonio, C., Dobson, A., Howarth, R., ... & Swackhamer, D. (2001). Forecasting agriculturally driven global environmental change. *science, 292*(5515), 281-284.

Toussaint, S., Sullivan, P., & Yu, S. (2005, March). Water ways in Aboriginal Australia: An interconnected analysis. [). Taylor & Francis Group.]. *Anthropological Forum, 15*(1), 61–74. DOI: 10.1080/0066467042000336715

Trisos, C. H., Auerbach, J., & Madhusudan, K. (2021). Decoloniality and anti-oppressive practices for a more ethical ecology. *Nature Ecology & Evolution, 5*(9), 1205–1212. DOI: 10.1038/s41559-021-01460-w PMID: 34031567

Tshifhumulo, R., & Makhanikhe, T. J. (Eds.). (2022). *Handbook of Research on Protecting and Managing Global Indigenous Knowledge Systems*. IGI Global. DOI: 10.4018/978-1-7998-7492-8

Turner, B. L.II, Lambin, E. F., & Reenberg, A. (2007). The emergence of land change science for global environmental change and sustainability. *Proceedings of the National Academy of Sciences of the United States of America, 104*(52), 20666–20671. DOI: 10.1073/pnas.0704119104 PMID: 18093934

Walentowski, H. (2024). Integrating Indigenous Knowledge and Landscape Ecology for Sustainable Land Management. *Journal Biodiversity Bioprospecting Development, 10*, 114.

Weiss, K., Hamann, M., & Marsh, H. (2013). Bridging Knowledge: Understanding and Applying Indigenous and Western Scientific Knowledge for Marine Wildlife Management. *Society & Natural Resources, 26*(3), 285–302. DOI: 10.1080/08941920.2012.690065

Wilder, B. T., O'meara, C., Monti, L., & Nabhan, G. P. (2016). The importance of indigenous knowledge in curbing the loss of language and biodiversity. *Bioscience, 66*(6), 499–509. DOI: 10.1093/biosci/biw026

Wilson, S. (2008). *Research Is Ceremony: Indigenous Research Methods*. Fernwood Publishing.

Wolfgramm, R. M. (2007). *Continuity and vitality of worldview (s) in organisational culture: Towards a Maori perspective* (Doctoral dissertation, ResearchSpace@ Auckland).

ADDITIONAL READING

Milder, J. C., Hart, A. K., Dobie, P., Minai, J., & Zaleski, C. (2014). Integrated landscape initiatives for African agriculture, development, and conservation: A region-wide assessment. *World Development, 54*, 68–80. DOI: 10.1016/j.worlddev.2013.07.006

Nawrotzki, R., & Kadatska, P. (2010). Addressing climate change with indigenous knowledge. *The International Journal of Climate Change, 2*(1), 33–47. DOI: 10.18848/1835-7156/CGP/v02i01/37290

Saunders, D. (2023). Embedding Indigenous Knowledge in the Conservation and Restoration of Landscape: *Insight Report.*

Shiferaw, A., Hurni, H., & Zeleke, G. The role of Indigenous Knowledge in Land Management for Carbon Sequestration and Ecological Services in Southern Ethiopia. *Soil-Based Ecological Services and Potentials for Sequestering Soil Organic Carbon (SOC) in Ethiopia*, 79.

Van Oosten, C., Buck, L. E., Hart, A. K., & Diaw, M. C. (2015). Strengthening Capacities for Collaborative Landscape Management in Africa.

Yanou, M. P., Ros-Tonen, M. A., Reed, J., Moombe, K., & Sunderland, T. (2023). Integrating local and scientific knowledge: The need for decolonising knowledge for conservation and natural resource management. *Heliyon*, *9*(11), e21785. DOI: 10.1016/j.heliyon.2023.e21785 PMID: 38027861

KEY TERMS AND DEFINITIONS

Biodiversity: Biodiversity refers to the different kinds of life on earth. It comprises different plants, animals, fungi, bacteria, and microorganisms, as well as genetic variations within each species and areas in which they live and interact.

Climate change: Climate change refers to long-term alterations in temperatures and patterns of weather. These shifts can occur naturally or through man-made activities such as the burning of fossil fuels.

Indigenous Knowledge: Indigenous Knowledge refers to knowledge, beliefs, practices, and innovation developed by local communities through experience and interaction with their environment for many years.

Changing environments: Changing environments refer to shifts in the conditions of an ecosystem or planet over a long period. These changes have impacts on living organisms and ecological processes. They can naturally occur or be caused by human beings.

Natural Resource Management: Natural resource management is about the planning, allocation, and use of resources to meet the desired objectives in a sustainable manner that minimises waste and negative impacts. It encompasses forests, land, minerals, and water.

Landscape management: Landscape management is concerned with caring and the sustainable use and development of landscapes to meet anticipated environmental, social, and economic needs. It recognises natural features, human activities, and cultural values to fashion its development based on the desired direction

Chapter 9

Exploring Indigenous Ways of Managing Informal Diapers Disposal in the Vhembe Biosphere Reserve

Mbulaheni Oskiedo Nedondwe
https://orcid.org/0009-0007-2570-3609
University of Venda, South Africa

Andani A. Budeli
University of Venda, South Africa

ABSTRACT

This study seeks to explore indigenous ways in managing informal diapers disposal in the Vhembe Biosphere Reserve. Disposable diapers are increasingly popular within the Vhembe biosphere reserve. However, the disposal of these diapers in areas without access to solid waste management infrastructure has become a scourge upon, low-income communities, presenting a multitude of health and environmental risks. As such, indigenous knowledge seems to have potential to address this challenge. This is because indigenous methods for waste management are environmentally friendly and sustainable. Indigenous knowledge systems play a crucial role in communities worldwide. However, the indigenous knowledge in communities is rarely known as people are more exposed to modern culture. The obviatives of this study was to determine causes of informal diapers disposal and to determine ways of managing informal diaper disposal in indigenous communities in the Vhembe Biosphere Reserve. The findings of this study provided awareness on the importance of indigenous ways in managing informal diapers disposal.

PROBLEM STATEMENT

Waste disposal is a growing crisis in many developing countries (Ferronato & Torretta, 2019). In South Africa, the lack of access to reliable and efficient waste disposal systems continues to pose human and environmental health risks, as is the case in many rural communities in the developing world (Muranko, 2021). Diapers are one of the waste streams contributing to increasing landfill inputs in both

DOI: 10.4018/979-8-3373-4124-8.ch009

urban and rural areas. In most areas, there are poor or no proper waste management systems for diapers, thus contributing to the pollution of the environment.

Disposal of nappies (or children's faecal waste) continues to be a major environmental health issue in low-income communities and particularly informal settlements in SA (Haywood et al., 2021). The disposal of used disposable diapers has become an uncontrolled environmental health issue. These diapers are often discarded informally and illegally, littering open spaces across various areas. The informal disposal of diapers seems to be worsened by the lack of African ways of understanding waste management that have been useful in the indigenous communities for years. As a result, there is an urgent need to address illegal dumping of diapers as it possesses significant challenge to the well-being and livelihoods of the people.

OVERVIEW ON THE MANAGEMENT OF WASTE

Many local authorities in the developing world lack the adequate means to handle and dispose of waste in an environmentally friendly manner (Wambui & Makindi, 2015). According to Wambui and Makindi (2015), because of this, cities are grappling with the management of used disposable diapers, often referred to as an absorbent hygienic product. Whereas UNICEF (2019) says that "it is common to see these absorbent hygienic products thrown away at undesignated places. The practice has created unsightly images in town, especially in the developing world".

According to Manyele, (2018), the use and disposal of soiled diapers in rural areas represent a lack of Environmental Health Literacy. This is usually observed in disadvantaged communities with low socio-economic status and lack of environmental health risk awareness programs. In Sub-Saharan Africa (SSA), especially in the rural areas, waste management has become a serious challenge (Kumar & Smith, 2017) This is evident in most areas where communities lack appropriate services and find their own means of managing waste, which has left these communities with serious environmental health risks (Agyei-Mensah & Oteng-Ababio, 2012)

South Africa has established various policies and laws aimed at environmental waste management, as highlighted by (Seopa, 2023). According to Seopa (2023), the South African constitution guarantees everyone the right to an environment that is not harmful to their health. Furthermore, it places an obligation on the government to enact legislation and implement policies to protect this right (DEAT, 2007). Key legislation supporting waste management includes the Waste Act of 2008 (Act No. 59 of 2008), the Local Government Transition Act of 1993 (Act No. 209 of 1993), the Municipal Structures Act of 1998 (Act No. 117 of 1998), and the Municipal Systems Act of 2000 (Act No. 32 of 2000). Despite these legal frameworks, it is concerning that the management of used disposable diapers as solid waste remains inadequately addressed in South Africa.

Disposable diapers are increasingly popular in rural areas in Ghana and other African countries (Oteng-Ababio, 2018). Kordecki et al. (2022) emphasize the urgency of an enhanced understanding of purchasing and disposal practices, including economic, socio-behavioral, and cultural factors, to guide interventions that seek to curb the use of disposable diapers or promote more sustainable practices regarding these diapers. Indigenous knowledge is attached with valuable insights into sustainable waste management practices that can be adapted to address contemporary challenges, such as diaper disposal (Zazu, 2007; Kosoe, 2019).

THEORETICAL FRAMEWORK: TWO-EYED SEEING APPROACH

According to Creswell (2020), the term "theoretical framework" consists of two words: "theory" and "framework." The term "theoretical" comes from the term "theory," which refers to studying a certain process, event, phenomena, or population to better understand it (de Vos & Strydom, 2011). According to Hornby (2015), a framework is a structure that helps shape an action or notion. A theoretical framework provides a detailed explanation and comprehension of a theory that directs a study (Grant & Osanloo, 2014). The theoretical framework aligns the study's findings with the selected theory (Webster & Watson, 2002). According to Grant et al. (2014) Theoretical framework refers to a design that encompasses an exhaustive explanation and thorough understanding of a theory that guides a study. The purpose of the theoretical framework is to anchor the study's findings with the selected theory (Webster et al., 2002). This study was grounded by the Two-eyed seeing approach.

Two-Eyed Seeing Approach was developed and introduced by Mi'kmaq Elders, Albert, and Murdena Marshall, from Unama'ki (Cape Breton), Nova Scotia, Canada, in 2004, stresses the importance of viewing the world through other practice and Indigenous worldviews (Bartlett et al., 2012). Initially, the use of Two-Eyed seeing was developed and promoted as a strategy to encourage the participation of Mi'kmaq university students in pursuing the study of science, valuing and including both Indigenous and other ways of knowing in relevant science curricula (Bartlett et al., 2012).

While its roots are Mi'kmaq, Two-Eyed seeing seeks to bring together various Indigenous and other perspectives and ways of knowing and is not exclusive to Mi'kmaq worldviews (Bartlett et al., 2012). Since its development, the use of Two-Eyed Seeing as a guiding framework for delivering education to, as well as conducting research with, Indigenous people has been advocated by numerous organizations and institutions. The Two-Eyed Seeing approach acknowledges the significance of indigenous knowledge and cultural traditions in solving environmental challenges (Bartlett et al., 2012). In this study, indigenous knowledge and cultural practices connected to diaper disposal are critical in understanding how to manage informal diaper disposal in a sustainable and culturally appropriate way (Martin, 2017).

The Two-Eyed Seeing approach is important because it allows a deep exploration of both indigenous ways and the modern ways for the benefit of indigenous communities (Berkes, 2012). Furthermore, the approach highlights the importance of combining indigenous and contemporary knowledge to acquire a better understanding of informal diaper disposal and to develop better solutions based on both indigenous and modern knowledge. In this study, combining indigenous knowledge with modern knowledge about diaper disposal provides a comprehensive understanding of the issue and identifies effective solutions that can be used in indigenous communities to manage informal diaper disposal.

SOLID WASTE MANAGEMENT PRACTICES WITHIN INDIGENOUS COMMUNITIES

From ancient times, before the arrival of colonization, indigenous waste management practices have been instrumental in preventing, disposing, reducing, and reusing waste material within indigenous communities. Before the introduction of mainstream sciences facilitated through colonial means (Madonsela et al., 2024). Wahab et al. (2004) highlighted that indigenous waste management practices have been at the centre of sanitation, holding together sustainable environmental health issues within rural communities. In indigenous communities, the practices associated with waste management are not exclusively designated to certain agencies (Assuah, 2023). Considering the absence of funds for waste management in indigenous communities. Indigenous peoples are aware of the dangers and effects that come with improper waste management in their communities. Indigenous groups actively take it upon themselves to prevent waste concerns by applying their indigenous knowledge to lessen the effects of inadequate waste management. Therefore, within indigenous society, it is the responsibility of every individual to manage waste effectively (Madonsela et al., 2024).

To address waste issues, indigenous societies have occasionally combined communal labour. But as human civilization has progressed, indigenous waste management techniques have been ignored and marginalized by the present generations (Kosoe et al. 2019). Indigenous communities have previously been acknowledged as having effective waste management practices. Despite this observation, in recent times, experts of waste management approaches have not recognized the role of informal indigenous groups and community-based organizations in addressing the problem of waste management (Boateng 2016).

THE USE OF DISPOSABLE DIAPERS IN SOUTH AFRICAN COMMUNITIES

Disposable diapers are made from cloth or synthetic materials containing absorbent chemicals that enable discreet urination and defecation (Eja, 2014). They have become essential in daily baby care routines (Kamaat & Malkani, 2003), with 90% to 95% of diapers used in developed countries being disposable (Thaman, 2014). On average, a baby uses 4 to 5 disposable diapers per day (Thaman, 2014), highlighting their widespread adoption as an alternative to cloth (reusable) diapers in both developed and developing regions (Remigos, 2014). The rise of disposable diapers led to the decline of traditional cloth diaper usage.

Historically, the term "diaper" referred to a specific cloth pattern rather than its function. It originally described a diamond-shaped pattern on white cotton or linen fabric. A century ago, diapers were made from woven materials like cotton or wool, treated with wax to prevent leaks. These materials, often shaped into triangles, were folded and wrapped around the baby. Inner liners were later developed to improve absorbency. By the early 20th century, waterproof rubber pants were worn over cotton diapers fastened with pins. Early disposable diapers featured three components: a removable top sheet, a poly-back sheet film, and a fluff pulp core (Modak et al., 2015).

POPULAR DISPOSAL OPTIONS FOR SOILED DISPOSABLE DIAPERS

Aumonier & Collins (2020) highlight that "soiled disposable diapers are solid waste and are therefore disposed using popular disposal methods for solid waste. These include methods such as burning, composting, landfilling, and open dumping. However, the very primitive open dumping seems the most popular method of disposable diaper disposal". A study by Remigios (2019) reported that most women using disposable diapers resorted to either burning the diapers, burying them in the ground, flushing them in the toilet, or simply dumping them in open spaces in the evening when no one was watching.

Burning

Incinerators are designed for satisfactory burning of combustible refuse, provided air pollution standards are met. However, the use of incinerators for disposing solid waste is now obsolete even in Western countries because of the high capital and operating costs involved. Air pollution also makes it unattractive. Incineration is still not the best option even when heat recovery is possible in this method, and volume reduction and residue sterilization are real (Eja, 2014; Tembo & Charazeni, 2017). Burning of disposable diapers is not done under controlled conditions as provided by incinerators but open burning of diapers along other refuse is popular comprising 10.34% and 40% of total disposal methods in Kenya (Wambui & Muchiri, 2015) and Zimbabwe (Muia, 2018), respectively. The wet condition of disposable diapers due to it being soiled with faeces makes it very difficult to be burnt by fire (Remigos, 2014).

Burning of disposable diapers (single use diapers), it is a great way of managing diaper disposal in our communities, it has some good advantages like making it easier for the disposable diapers to be disposed after being burnt also it destroys pathogens, gems, bacteria's, viruses which can have some health implications on animals and human beings. However, with burning method it also comes with some bad outcomes to our environment, lifestyles and health. With the burning of disposable diapers, it releases some chemicals that contribute to air pollution. When burning, it occurs when anything that releases a smoke is being burnt, it increases greenhouse gases emissions which also cause climate change and currently climate change is a major problem that needs to be addressed as diapers disposal also forms part in the causes of climate change.

Burying

Some users bury soiled disposable diapers in the ground. According to Remigios (2014), some women revealed that they dug holes in their yards to bury these diapers. It accounts 4.60% of total disposal methods in Kenya (Wambui, Muchiri, 2015). Most of the disposable diapers that are disposed with household garbage, in waste (Tembo 2017; Wambui, 2015; Muia, 2018,) are eventually taken to the municipal landfill. Landfill (or controlled tipping) is a system of waste disposal that is based on an arrangement in which solid waste are shifted from one area to another and dumped into trenches which are covered up at the end of each day or as one day's collection from a particular locality is completed (Eja, 2014). In this method of refuse disposal, refuse is dumped in accordance with a preconceived plan, compacted and covered during and at the end of each day. It is not an open dump, and if it is poorly managed, it may result in spontaneous ignition accompanied by smoke and smell. It might also result in leachates that pollute underground water (Eja, 2014).

Although the burying method has been used since then, this diaper disposal management is good and has been used for centuries, mostly in indigenous communities. The burial method is cost-effective because the diapers are buried in the yard, as it can be associated with taboos in African indigenous communities. This method is also easy to implement as it is just a matter of burying them down. The burial method also has some arguments with regard to the environment.

OPEN DUMPING

Disposable diapers are indiscriminately thrown away or dumped openly at various points (Jesca, 2015; Excelsior, 2020). According to Tembo and Chazireni (2016), about 50% of caregivers in the Mberengwa district, Zimbabwe, dispose of soiled diapers in open places, especially in bushes and on roadsides. This practice is evident in Nigerian rural and urban communities (Mutura, 2020). Even when some refuse receptacles have been erected at collection points, the refuse dumped into such closets may not be removed over a long period. Some rural communities in Nigeria hold the noxious believe that infant excreta, if disposed in the bush, are blessing to their farms and this encourages the dumping of soiled disposable diapers into vacant plots (Excelsior,2020), being completely unaware of the near non-degradability of those materials (Ogundele et al., 2018).

It is so worrying and painful to see disposable diapers (single-use diapers) being disposed around the environment in an open space. It becomes a major problem and threatens human health and day-to-day lives, not to mention how disturbing it is when it comes to animals' lives. Domestic animals like goats, sheep's and pigs end up eating the diapers that are disposed in an open space and as such, animals end up dying as it cannot be digested, same applies when the diapers are being disposed in the rivers, there are aquatic animals that lives in water, there are people that depends on the water from rivers for survival and as such it is life threatening for them to drink the water that is from rivers as now we are having a problem of disposable diapers being thrown in rivers (EPA, 2012).

Kumar & Smith (2017) define an open dump as a site where solid waste is discarded without adequate protection for the surrounding land or ecosystem. In many developing countries, open-pit dumping remains the most common method of disposing of medical waste, mainly because it is cost-effective and there are few alternative solutions. However, this method poses significant environmental risks. Wambiu et al. (2015) note that in some rural communities, there is a belief that disposing of diapers in the bush brings benefits to farmland, leading to the practice of dumping used diapers on vacant plots. To minimise these dangers, careful selection and management of waste disposal are essential to prevent fires and other hazards. Moreover, open dumpsites should be in areas where they are unlikely to cause complaints from nearby residents (Koliyabandara et al., 2024).

THE IMPACT OF USED DISPOSABLE DIAPERS ON THE ENVIRONMENT

The disposal of disposable diapers has various environmental impacts, largely linked to the materials used in their production. These effects extend across physical, social, and economic aspects of the environment.

PHYSICAL IMPACT OF DISPOSAL OF DISPOSABLE DIAPERS

The disposal of disposable diapers significantly impacts the environment in several ways, especially in developing countries where they are often discarded in open areas or along roadsides (Mudau et al., 2023). The illegal dumping of soiled diapers in these open sites exposes nearby communities to health risks, including diarrheal diseases and unpleasant odours. Furthermore, the air pollution caused by the production of disposable diapers is more harmful than that from the manufacturing and use of cloth diapers. During production, chemicals like chlorine, dioxins, and sodium polyacrylate are released into the atmosphere (Šmajgl & Obhođaš, 2015). Each disposable diaper contributes around 630 kg of greenhouse gases, which is equivalent to the emissions from driving a car for 2,896 kilometres. Additionally, infant waste from discarded diapers can seep into the groundwater, leading to contamination and the spread of communicable diseases when placed in landfills. While disposable diapers may decompose in approximately five months, the chemicals from decomposing waste can leak from dumpsites and landfills, contaminating groundwater, soil, and nearby water sources, which can cause waterborne diseases like cholera (Šmajgl & Obhođaš, 2015).

Millions of tons of untreated waste from disposable diapers are added to landfills each year. The major environmental impacts occur during the transportation, treatment, and disposal of diapers in landfills and incinerators. However, the environmental effects during commercialization and use are generally less harmful, limited mainly to energy consumption and emissions related to transportation and infrastructure maintenance. It is worth noting that 35% of the cellulose pulp used in baby diapers is derived from coniferous trees, contributing to deforestation, soil degradation, and loss of biodiversity, especially when the cellulose comes from plantations (Šmajgl & Obhođaš, 2015).

SOCIAL IMPACT OF DISPOSAL OF DISPOSABLE DIAPERS

Human faeces release viruses that can lead to serious health concerns. In many cases, household waste, including used disposable diapers, is disposed of in compost pits or littered in public spaces, posing a significant infection risk to anyone who meets them. Used disposable diapers contain harmful substances, such as dioxins, which are persistent environmental pollutants. When these chemicals enter the environment, they can accumulate in the human body, potentially leading to developmental delays, immune system suppression, hormonal imbalances, skin diseases, and certain cancers (WHO, 2010).

The absorbent layer of diapers is often treated with chemicals that can cause allergic reactions in babies. Moreover, used disposable diapers, which carry human waste and urine, contain pathogens like viruses and bacteria that can cause infectious diseases in humans (WHO, 2024). The WHO (2014) also highlighted that human waste is linked to numerous infectious diseases, including cholera, hepatitis, polio, typhoid, cryptosporidiosis, schistosomiasis, and ascariasis. The heat trapped in disposable diapers can also lead to fertility problems in males. Research has shown that prolonged use of diapers can significantly increase scrotal temperatures in male infants, which may contribute to a decline in sperm production later in life (Maluni, 2020).

Diaper rash tends to worsen when the skin remains moist for extended periods. The condition is influenced more by diapering practices than by the type of diaper used. Caregivers employ various cleansing methods, such as bathing and applying lotions, creams, ointments, or powders. Cultural beliefs, social status, and emotional factors greatly influence diapering habits (Maluni, 2020). According to Modak et al. (2015), irritant diaper dermatitis is a term used to describe a range of symptoms in the diaper area caused by inflammatory skin reactions. Though the severity and frequency of diaper rash have decreased over time, it remains a common issue. Clinical and laboratory studies have identified several causes of diaper rash, including skin moisture, irritants, and increased pH from exposure to excreta.

Mothers often feel distressed and guilty when their babies develop severe diaper dermatitis (Heimall., 2012). In severe cases, prolonged diaper rash can lead to accusations of child neglect or abuse (Scheinfeld, 2005). In some cases, mothers may resort to inappropriate toilet training methods to reduce the cost of diapers. Moist skin is more susceptible to damage and irritation, and increased friction from irritants can cause further harm. These findings underscore the importance of managing skin wetness and pH to maintain healthy skin in the diaper area (Atherton, 2004).

ECONOMIC IMPACT OF DISPOSAL OF DISPOSABLE DIAPERS

Disposable diapers are essential but costly items that help parents maintain their baby's health and hygiene. Newborns require frequent diaper changes, sometimes as often as every hour, while older children typically need a change every three to four hours. Paediatric nurses generally advise parents to expect six to eight wet diapers per day, which indicates that the baby is receiving enough breast milk or formula (Shin, 2005).

For American infants, who use an average of 6.3 diapers daily, diaper manufacturers estimate an annual cost of about $1,500 (roughly R20,000) (Browne, 2011). This cost can be even higher for low-income families who have limited access to the internet for online purchasing or who face transport challenges to supermarkets, discount stores, or warehouses where diaper prices are lower. As a result, they often rely on local convenience stores to buy disposable diapers (Neff, 2011). While cloth diapers may seem like a more cost-effective option since they can be reused, low-income families may not have the necessary resources for purchasing enough cloth diapers or facilities for daily washing. Additionally, many childcare centres require infants to wear disposable diapers, and most mothers consider them a necessity (Raver et al., 2010).

THE ENVIRONMENTAL IMPACT OF DIAPER DISPOSAL.

Baby diapers are vital products that ensure access and convenience to hygiene in day-to-day life (Schenck, 2023). These disposal diapers are used only once before they are disposed (Thaman & Eichenfield, 2014). According to Thomas et al. (2024), the use of disposable diapers is common in many low- and middle-income countries, while waste management services are minimal. This leads to unmanaged disposal of diapers in open spaces and water bodies. The UN-Habitat. SDG 11 Synthesis Report (2018) presents this as a threat to the achievement of the Sustainable Goals, particularly SDG 11.6 on addressing the inclusivity, safety, resilience, and sustainability of human settlements and SDG 6.2 ON sanitation and hygiene.

Studies show that the single use of disposable diapers has a drastic impact on climate change. This is due to the emission of green greenhouse gasses from landfills, usually when such diapers are not properly managed and are dumped in water bodies and open spaces. A study conducted by Schenck et al. (2023) indicates that in most rural areas, there are no waste management measures. This reflects negatively in that improper disposal of diapers contributes to the toxicity of humans and ecology. According to Cabrera and Garcia (2019), there are limited studies conducted to determine the impact of mismanaged disposal of diapers, particularly when dumped in water bodies or open spaces.

Despite increased concern about environmental sustainability, the significance of Indigenous Knowledge in diaper disposal management is largely ignored (Datta & Das, 2022). While acknowledging the value of Indigenous Knowledge in environmental management, there is still a considerable gap in understanding its applications in diaper disposal techniques. The prevailing Western viewpoint on waste management has traditionally excluded Indigenous knowledge, sustaining a lack of representation and recognition of Indigenous epistemologies (McGregor, 2019).

METHODOLOGY

This study employed the use of qualitative research design. The primary objective of this qualitative study was to investigate the significance of indigenous knowledge in the context of diaper management within the Vhembe Biosphere Reserve. The qualitative research design was particularly well-suited for this exploration, as it enables an in-depth examination of the perspectives, experiences, and practices of the individuals with expertise in this area, providing rich insights into the role of indigenous knowledge in addressing environmental challenges.

This study also explored the Indigenous ways of managing informal diaper disposal in Vhembe Biosphere Reserve, and the exploratory research design allowed the researcher to ask different questions to gain more knowledge the questions such as how, why, what, and where. The researcher collected data in Tshiungani village, within Vhembe Biosphere Reserve, in the Limpopo province of South Africa. A non-probability sampling was used to purposively select nine (9) participants: a Traditional Leader, two Traditional Birth Attendants, two Young Mothers, and four Elderly members of the community who possess indigenous knowledge and experience about the management of diaper disposal.

Data was collected through semi-structured interviews and analysed using thematic analysis. Thematic analysis is a method of analysing data in qualitative research that focuses on identifying, assessing, and recording forms or themes within data (Hammersley, 2015). The researchers paid attention to the following ethical considerations: informed consent and voluntary participation, permission, and confidentiality and privacy. This was done to ensure that no one was forced to be part of the study, and participants were advised and informed about the rights that they had when the data was being collected.

DISCUSSIONS OF FINDINGS

This part of the write-up focuses on the discussions of the findings that the researcher found from participants when conducting the data collection. These discussions focus on the ways that indigenous people used to use while exploring the indigenous ways that can be used to manage informal diaper disposal in the contemporary world.

OVERVIEW ON DIAPER DISPOSAL IN AFRICAN COMMUNITIES

The study shows that in the earliest years, there were not much or a lot of resources that could be used particularly for baby diapering purposes. Participants indicated that as diapering resources were not available, caregivers had to use the natural resources that were available, such as cotton-like products of trees *(Madzudzu)* that were found in the forest on top of and under stones. These cottons were also found on indigenous trees such as *Museri* (Eugenia Natalitia), *Muvhavhanyana* (Acanthospermum Hispidum), *Muelela* (Albizia Adianthifolia), and *Mugupe* (Diospyros Mespiliformis). As these cotton-like products were commonly used in African communities, when visitors came to see the newborn child, they used to carry cotton-like products with them and give them out as a present.

Participants present this as a cultural practice that was followed also to prevent a number of people coming to visit the baby, the baby is not supposed to be seen by a lot of people when the baby is still a newborn. The use of cottons as a present when visiting the baby and as to entry was because it was not easy to find or to have access to the cottons as they were found in the forest. This was a practice made in indigenous communities that they used to limit people visiting the baby as other people would visit the baby when they are from having sex, others they have diseases, and it would bring bad lucks to the baby and transmission of different diseases. This practice was also used to allow caregivers to be at home taking care of the baby rather than going out to the forest looking for cottons that they will be used for diapering purposes.

Participants stated that African communities depend on agriculture for survival; thus, during the time of harvesting maize meals, they used to take out the silks that are found on top of the corn, which are known as *malelebvudzi*. Traditional Birth Attendants (TBAs) would collect the silks from corns and keep them in a safe place as they will be used for wiping the baby's faeces. When they want to use *malelebvudz*i, they will just put water on them as they will be dry, then they wipe the baby's buttocks. These practices were used because of a lack of resources for diapering purposes. However, these methods are believed by participants to be environmentally friendly and do not result in pollution as after wiping, caregivers would collect and put them inside a sack for safe disposal.

As things were changing because of time, modernisation, and developments, there was an introduction of cotton clothes in South Africa. Even African communities adopted the use of cotton clothes. This also brought about changes in traditional clothes, where traditional clothes were made of cottons that were introduced with the existence of foreign cultures in African communities. This also influenced diapering practices, where African communities started to use old clothes that people wore and cloth-based traditional attires as diapers. They would take them and tire it around the hips of the baby; this is called it was putting on the kids the traditional clothes that had impacted on covering the kids and they would be tied like ropes on their waists which is called u "*tswingisa nwana*", "*zwo vhofha sa ludede kha khundu*" this used to apply in African indigenous communities when using old cloth as diapering method, they would tie the cloth around the hips. When the baby has pooped, they would take the waste to the hole that was dug for waste disposal and the cloth they would take it and wash it inside a material created which was called "*Dongo*" then they wash the cloth used.

With the Western practices being advanced in South Africa and its indigenous communities, the study shows that African communities started having access to cloth diapers that were directly made for diapering purposes. African indigenous communities adopted the use of cloth diapers as they were commonly used at that time. The majority of the cloth diapers there were white. Participants mentioned that the good thing about the use of cloth diapers is that they can only buy the cloth diaper once in a while, which also reduces the cost.

Participant:

''Cloth diapers are convenient as caregivers will not always be buying the diapers all the times, this reduces financial burden of thinking about buying cloth diapers every now and then"

Participants also alluded that single-cloth diapers can be used for more than 5 years with different generations using the same cloth diapers. This was also a good in that it did not result to pollution most particularly land pollution, cloth diapers were environmentally friendly. Participants also mentioned that caregivers such as mothers in some cases would be lazy to wash the diapers and the diapers will get damaged as the majority are white, and they will start having stains that they cannot be removed. Other arguments were that cloth diapers are time-consuming. This is because they first have to be soaked before parents, particularly mothers, Traditional Birth Attendants and also the grandmothers, as they were there for guidance, can wash them. Then, after washing the cloth diapers, the caregivers still had to wait for them to dry, which is time-consuming. Study shows that is not easy to travel with a baby when using cloth diapers, which made people stop using cloth diapers, particularly with the introduction of disposable diapers (Mawere, 2015)

Participants alluded that the advancement of Western knowledge introduced the use of Disposable diapers, and most people started using them. The disposable diapers are currently commonly used in indigenous African communities. The following is what one participant had to say,

Participant:

Every household that has a newborn baby currently uses disposable diapers.

When these disposable diapers were introduced, African community members were only aware of one familiar brand, which is the ''pampers,'' and to date, community members still call all the diapers ''ma-pampers'' without checking the diaper brand name. This resulted because it was the first diaper that African communities were familiar with. Studies show that the use of disposable diapers over the past method that was used before causes a lot of pollution; it causes land pollution, air pollution, and water pollution in communities (Nkomazana, 2017; Mawere, 2015). This will be discussed below in other subjects of this study.

THE USE AND MANAGEMENT OF DISPOSABLE DIAPERS

The adoption of disposable diapers has changed the practice of childcare globally, more especially in African indigenous communities. In recent years, disposable diapers have become so common that which majority of communities in African communities are using disposable diapers. Even in indigenous communities where cloth diapering methods were used, they have moved from the use of cloth diapers, and they are now using disposable diapers.

Commonly, cloth diapers, after use they have to be washed as mentioned above in the overview of diaper disposal. However, there are a lot of dynamics that are related to the use and management of disposable diapers. Previously, people in communities would dig holes with the purpose of disposing diapers. However, there were measures followed when digging the holes. The holes are not just dug everywhere, but caregivers would identify a space inside the homestead that will be of use, and that space will specifically be used for digging holes for diaper disposal. Others would dig next to the outside toilet that was used in indigenous communities as the space near the toilet will not be used for anything, and when they are dug in the hole, they won't be exposed everywhere, and that would mean that a huge part land won't be polluted by these diapers.

Moreover, diapers will also not be exposed to the water, and if it is like that, the water will not be polluted, too. Participants also said that caregivers also used to burn the used disposable diapers. They would identify a space which they can go to and burn the used disposable diapers, which is far from where people live to avoid causing smoke to people and the smell that the diapers will release. According to participants, these methods were effective as each and every household will be responsible for disposing the diapers of their own children, and as such, there will be a limitation of pollution in their communities.

At the moment, there is a concern in Indigenous communities about the current ways of using and managing the used disposable diapers. This is because nowadays, people seem to just throw diapers around the open spaces under the sun, in the water sources, in the bush, and also alongside the road. This shows that people in communities no longer follow the environmentally friendly ways of managing solid waste used by their forefathers. The current ways of disposing of diapers show that African cultural ways of seeing life and beliefs are no longer followed (Le Roux, 2018).

Participants also mentioned that their cultural setup does not promote the haphazard disposal of diapers. It is believed in the community that diapers cannot just be disposed of everywhere, they can be used to bewitch the baby. It is believed that it can also bring bad luck to the baby. In the community, as the children grow, they are also told that they are not supposed to pee and release their waste anywhere. This emphasises the idea that baby faeces are not supposed to be everywhere. The following is what one participant mentioned:

Participant:

" *Malatwa a nwana ha tei u soko latiwa fhethu hunwe na hunwe nga maanda duvhani hune a dovha a khou swa nga duvha zwiita uri nwana a dzule a khou swa thumbuni musi duvha li tshi khou fhisa mafhambuwa a nwana zwa dovha zwa ita uri muvhili wa nwana u dzule u khou fhisa*"

Translation: A child's feces cannot be thrown out anywhere, especially in areas where they will be exposed to direct sunlight, as this can cause stomach problems and, if the feces are disposed of everywhere while burning, can cause fever.

As the indigenous people are worried and concerned about the current ways of diaper disposal, one can conclude that haphazard disposal in the open spaces like in the bushes, gardens, the playgrounds, alongside the roads and the water streams have a negative impact on the human beings' health and also the environmental implications. The next part of the write-up of the study looks at the causes of the unmanaged diaper disposal in the communities.

CAUSES OF INFORMAL DIAPER DISPOSAL PRACTICES IN COMMUNITIES

This part of the write-up explores various causes of improper diaper disposal practices. In this regard, participants detailed that improper diaper disposal is often attributed to a lack of knowledge, particularly because currently there is no longer awareness and knowledge shared and raised by the Traditional Birth Attendants. Traditional Birth Attendants are caregivers specifically for pregnant women, new mothers, and young and newborn babies that are found mostly in indigenous communities. Traditional Birth Attendants focus on raising the babies, taking care of the babies, as well as teaching and educating mothers on how to take care of the baby, how to protect the baby waste, and the materials they use when taking care of the baby. Nowadays, there is a lack of knowledge on how mothers and caregivers should dispose and handle the diapers that have baby waste, and this is happening because in most indigenous communities, Traditional Birth Attendants are no longer available.

Participants indicated that currently, the majority of people think that a diaper is just something that can be used once and then later thrown away. This is because they lack knowledge about the importance of managed diaper disposal and baby waste management. This is also caused because in African indigenous communities, there are no longer elderly people who assist with guidance with baby care as the mothers lately are young mothers; hence, they face these challenges due to lack of guidance on how to manage the diapers.

Participants mentioned that many individuals are unaware or, to a certain extent, ignorant of the environmental and health impacts of improper diaper disposal; they just dispose of used diapers without checking if the diapers are disposed of in a good way that will not harm the environment. Because community members do not have access to information on proper diaper disposal methods, caregivers resort to harmful practices such as burning, burying, or throwing diapers into waterways and water streams,

Furthermore, waste management services like creating dumping areas for disposable diapers, having municipality trucks that come to collect waste every week and also having project that are based on collecting diapers and mitigating the disposal of disposable diapers, are being costly and not available at Tshiungani, leading caregivers to seek alternative disposal methods that are the ones that caused the problem of unmanaged diaper disposal as they dispose diapers in open spaces. The options that caregivers use currently contribute to the disruption of the environment, increased climate changes, and is also causing problems on animals, the aquatic animals and also domestic animals.

Inadequate waste management infrastructure is a major contributor to improper diaper disposal in modern times. In many communities, waste collection services are irregular or unreliable, and not available in rural areas, thus forcing caregivers to seek alternative disposal methods that are not supposed to be used such as disposing diapers in the bush, water streams and alongside the road which affects the Traditional Health Practitioners when they want to harvest medicine to heal people, they will find diapers near the stem of the plant which leaves them not harvesting the medicine.

Lack of access to recycling facilities, composting programs, and proper waste treatment plants increases the issue of unmanaged diaper disposal as the disposable diapers that are being used take time to decompose because the materials used are not environmentally friendly. There is also an issue of not having access to facilities like bins and areas where it is only for disposing solid waste, and with lack of those facilities like bins they will throw the diapers everywhere.

Participants also bordered that some cultural and social norms significantly influence diaper disposal practices. In some communities, diaper disposal is viewed as a taboo, leading to inadequate discussion and education. Common practices of indigenous people, such as burning or burying diapers, may be passed down through generations, perpetuating harmful disposal methods (Mawere, 2015). Furthermore, a lack of community engagement and collective action to address the issue perpetuates the problem.

EFFECTS ON HUMAN AND ANIMAL HEALTH

Participants responded to the research with a desire to help the community, about the diaper disposal. The study established that it was believed that baby waste is not supposed to be thrown in the bush, forest and open space where they will be exposed to the sun, it causes "*tshilala*" to a baby. *Tshilala* is a sickness that affect with their stomach, it disturbs their digestive system, and they will always have problems with their stomach.

Tshilala is caused when the baby's waste is thrown and exposed to the sun. As the sun burns it also causes the baby's body temperatures to be high and also gives the baby endless running stomach. The baby's waste must be handled with care to avoid causing health issues, the diapers that are disposed in the sun not only that they cause problem with the digestive system but they also makes the baby to be restless as the temperatures will be high, the body will be hot all the times as the waste that contains his being are thrown in the sun. Participants mentioned that caregivers should try to dispose of them under a shadow, or when disposing, they must dig down and dispose them down there. After that then they cover, so that it can be cool there as it will help the baby's body to be cool, not hot. That's something we grew up knowing and abiding by, but this modern generation doesn't really follow that.

Some implications come with unmanaged diaper disposal. According to the participants, unmanaged diaper disposal can harm the baby's life because the baby's waste is scattered around, and people often have access to the baby's life because of the waste that is based on the participant's cultural belief. This group of people holds that if the baby's waste is passed by everyone, it would have problems or cause taboos. According to the participants, today's children have a problem when growing up as their diapers are just thrown around. "We no longer have or can control those children when they are growing up as their diapers are just thrown where the sun has access to them, and it causes health implications on their stomach". Today's children have endless sicknesses, and they are not strong enough to fight any diseases. They always have a runny stomach too because of the way they dispose of diapers; this is also attributed to informal waste as informed by the indigenous beliefs in the community.

Participants explained that unmanaged diaper disposal has negative effects on the environment, human health, and society. Used diaper disposal contaminates the air, water, and soil. It also has an impact on a person's health. Toxic substances released by diapers can cause illnesses in humans. Since plastic takes a long time to break down and might hinder successful cultivation, improperly managed diaper waste also adds to the problem of plastic pollution.

Since many villages use river water for cooking and laundry, poorly handled diapers can have a variety of negative effects on people's lives. Throwing dirty diapers into the river simply indicates that individuals won't do any of these things, which could result in them purchasing the most expensive water. Due to the disposal of dirty diapers, people will have to pay money to purchase water that is in the river but cannot be utilised.

Participant:

> "Stated that disposal of diapers practice sends out Health hazardous to the people and to the animals, the disposable diapers carry waste which must be disposed and when they are not, there are animals that eat those diapers because of the smell, and such animal will end up dying because the diaper that it ate cannot be digested. The is also a problem to people it causes cholera to people who depends on river water that drinks and depend on that water for use."

The study found out that disposable diapers contain a gel on them and that gel has chemicals. it means that when those chemicals are eaten by aquatic animals, they end up dying as it is not consumable. For species like fish, there will be a formation of algal bloom, which is formed after the waste is washed inside the river from the diapers. The algal bloom takes all the energy and oxygen for aquatic animals which majority of them end up dying because of lack of oxygen. This also mess up the food chain for the aquatic animals, for instance, if the fish dies, animals that depends on the fish for survival they will die, and also the will be a loss of species, and those that are available will be in large numbers which causes the imbalance of organisms in the water.

When diapers are dumped into rivers, communities that depend on the water for irrigation, drinking, or bathing suffer terrible effects. They are susceptible to waterborne illnesses such as *gastroenteritis, cholera, typhoid, and dysentery* because of the tainted water. Common dangers include bacterial, viral, and parasite infections such as *Giardiasis, Salmonella, E. Coli, and Rotavirus*. Extended exposure can raise the risk of cancer, harm the nervous system, interfere with reproduction, and affect a child's development. The financial cost is high because of medical costs, lost wages, and fewer options for employment. The elderly, children, and pregnant women are among the vulnerable groups that are disproportionately impacted. Furthermore, there are health crises in the community, a decline in quality of life, and higher death rates (EPA,2020). It is imperative to stop diaper contamination in waterways to protect public health and welfare (WHO, 2018).

WAYS TO MITIGATE INFORMAL DIAPER DISPOSAL IN AFRICAN INDIGENOUS COMMUNITIES

This part of the write-up presents ways to mitigate informal diaper disposal in African indigenous communities. This discussion brings insight into innovative ways on African indigenous communities can manage diaper disposal. The study found that there are currently different institutions and organisations that are working and dealing with the problem of informal diaper disposal, even in some of the indigenous communities. The discussions below bring up the ways that can be adopted and used to mitigate the informal disposal of diapers in African indigenous communities.

ESTABLISHMENTS OF DIAPER DISPOSAL CENTRES

There are different companies and individuals that come to indigenous communities and villages that collect non-used materials like metal and plastic in exchange for money. The collection of unused materials also serves as a way of reducing solid waste within these communities, which also reduces pollution. Such is practised also when there are events wherein different people will be picking up empty juice bottles, used takeaways, and plastics.

This study sees the need for such practices to be employed with disposable diapers. The participants mentioned that it raises questions if there is nothing that can be done, just like how it is being done with plastics and metals that are no longer used. Indigenous communities must develop initiatives within the diaper disposal sectors where they will take used disposed diapers that are supposed to be disposed in the landfills and use as something that will uplift and benefit the community.

With the introduction of disposable diapers, there is a problem with how communities dispose the used disposable diapers. The introduction and establishment of diaper disposal centres can be a great initiative in addressing the diaper disposal problem more, especially in rural areas and in indigenous communities. Diaper disposal centres offer safe disposal practices where each solid waste will have its own section where it will be disposed. Diaper disposal sectors also reduce environmental pollution such as land pollution, water pollution as community members will stop disposing used disposable diapers in the rivers and other water sources. Also, community members will limit or fully stop disposing of diapers in open spaces around the environment.

In indigenous communities, there are a lot of diapers that are disposed in open spaces, water streams, and in the bushes. The centres for diaper disposal help the indigenous communities and indigenous people as they can dispose of diapers in the designated places or containers that diapers are supposed to be disposed in. The use of trash bins for diaper disposal assists the indigenous communities with the disposal of diapers as when the trash bags are full will be taken to the dumpsite and diapers will be thrown there. A participant outlined that back then there was a dump area for all solid waste that was far away from where people are living, this can also be adopted in our current time and create a dumping site that will be far from the people and away from water sources, this helps in managing diaper disposal in the indigenous communities.

PRODUCTION OF BIODEGRADABLE DIAPERS

Diaper disposal in indigenous societies is a difficult issue that requires innovative ways and sustainable solutions. The production of biodegradable diapers can decrease the environmental implications of diaper disposal and improve living conditions. Biodegradable single-use diapers are made from materials that can break down naturally in the environment, such as plant-based plastics, cornstarch, or sugarcane (Chen, 2020). According to Aumonier & Collins (2020), Biodegradable diapers are made from natural materials that are easily decomposed, decreasing waste sent to landfills and lowering the danger of environmental damage. Indigenous communities can profit economically by producing biodegradable diapers, creating job opportunities, and promoting self-sufficiency. This article investigates the possibilities of biodegradable diaper manufacturing in indigenous cultures.

Biodegradable diapers provide a sustainable solution to pollution and environmental deterioration. Biodegradable diapers degrade naturally, reducing the amount of waste transported to landfills and lowering plastic pollution in seas and rivers. Furthermore, biodegradable diapers take less energy to create and disintegrate, resulting in lower greenhouse gas emissions. They also reduce the usage of harsh chemicals, lowering the likelihood of chemical contamination in the soil, air, and water. Overall, biodegradable diapers encourage sustainable waste management methods, which save natural resources and safeguard the environment, which is one of the goals of indigenous people to conserve and protect the natural environment in indigenous communities.

As biodegradable diapers are made from natural materials, mostly, the biodegradable diapers must have their own designated place where they will be disposed. With the use of biodegradable diapers that are made out of natural materials, there can be a great change also in the agricultural sector as indigenous people in indigenous communities depend on agriculture for survival. The biodegradable waste will be disposed in a space for Agric where they will be decomposed fast and they will make the soil to be more fertile due to the natural materials used when making the biodegradable diaper and also the baby waste that will be inside the diaper.

THE USE OF CLOTH DIAPERS

Although cloth diapers are not indigenous, they have been used by many African indigenous communities, and they have turned to be of great use. For generations, indigenous cultures throughout the world have used traditional and sustainable childcare techniques, including the usage of cloth diapers. These traditional cloth diapers not only decreased waste and environmental effect but also encouraged healthy skin growth and hygiene habits. Furthermore, the use of cloth diapers was frequently entrenched in cultural and spiritual traditions as they protected the baby faeces rather than throwing them, they would just wash the cloth diapers after being in use.

Participants have contributed greatly regarding implementation that ab be taken to mitigate the disposal of diapers practice. It was learnt from the participants that one of the best ways to mitigate they mentioned the use of cloth diapers, cloth diapers are reusable they can be used more than once. Also, cloth diapers they do not cause any pollution, they are environmentally friendly as they are washed after being used rather than being thrown. Between disposable diapers and cloth diapers participants alluded that they would choose the cloth diapers, judging how the disposable diapers are impacting our environment and the world, cloth diapers are good because it is just a matter of washing it when they are dirty and the good thing about cloth diapers is that you would just buy cloth diapers once, and you can still use them in the five (5) coming years.

FUTURE RESEARCH DIRECTIONS

With the current problem of unmanaged diaper disposal, this calls for need on research based on how indigenous knowledge was used back then, the practices and beliefs associated with unmanaged diaper disposal. There must be an in-depth exploration about how and what indigenous people used to use back then as a way of managing diaper disposal, considering the reasons that are there. Indigenous people and the indigenous communities have a connection with the nature and the environment, research will assist in knowing why the indigenous people used to have the cultural and traditional practices of diaper disposal.

This study did not produce African indigenous ways of managing or mitigating informal diaper disposal. As such, future researchers should consider the African indigenous ways of managing or mitigating diaper disposal mostly in indigenous communities. This is important in understanding whether the ways used are friendly to the environment or not. There is also a need to study the relevance of common ways used to dispose such as burning and burying in the contemporary society.

CONCLUSION AND RECOMMENDATIONS

Revitalization of indigenous knowledge needs and requires documentation, preservation of traditional practices, beliefs and values. The effective management of diaper disposal in Indigenous communities requires a collaborative approach that honour traditional knowledge and practices. As such, there is a need to bridge the gap between modern solutions and indigenous ways, sustainable, culturally sensitive strategies can be developed. Collaboration is crucial among various stakeholders, including indigenous communities, governments, researchers, and organizations. This collaborative approach can empower indigenous communities, improve environmental health, preserve traditional knowledge, and foster innovative, sustainable solutions. It is essential to recognize the value of Indigenous knowledge and its potential to transform waste management practices.

The study recommends that there be educational programs to raise awareness about the effects of diaper disposal, and the programs must also provide indigenous ways that can be used to address informal diaper disposal. The study recommends that there should be educational programmes within indigenous communities that are led by elderly people and the traditional birth attendants educating the communities about the importance of protecting the baby waste, also raising awareness on the environmental, social impacts, and economic impacts of informal diaper disposal.

REFERENCES

Agyei-Mensah, S., & Oteng-Ababio, M. (2012). Perceptions of health and environmental impacts of e-waste management in Ghana. *International Journal of Environmental Health Research, 22*(6), 500–517. DOI: 10.1080/09603123.2012.667795 PMID: 22428915

Asomani-Boateng, R. (2016). Local networks: Commodity queens and the management of organic solid waste in indigenous open-air markets in Accra, Ghana. *Journal of Planning Education and Research, 36*(2), 182–194. DOI: 10.1177/0739456X15604445

Assuah, A. (2023). "A lot of people ignore our culture when it comes to waste management": Examining the impacts of culture on solid waste management in two Canadian First Nations. *Alternative, 19*(2), 466–474. DOI: 10.1177/11771801231163635

Bennett, M., & James, P. (2016). *Environmental Management: Principles and Practice*. Routledge.

Browne, D. (2011), Is the poor economy to blame for the rise in diaper rashes? Retrieved from http://www.cnn.com/2011/09/12/health/pooreconomy-diaper-rash-

Buxner, S., Davis, L. F., Anides Morales, A., Sandhaus, S. A., Abrell, L., Kilungo, A., . . . Morales, F. (2018, December). Project Harvest: Engaging Diverse Citizen Scientists in Rainwater Quality Monitoring. In *AGU Fall Meeting Abstracts* (Vol. 2018, pp. NH43B-1036).

Davis, M. (2019). *Indigenous Knowledge and Biodiversity*. Cultural Ecology Journal.

DEA. (2008). National Environmental Management: Waste Act, 2008 (Act No. 59 Of 2008), Draft Health Care Risk Waste Management Regulations. [Online]. Available:https://www.gov.za/sites/default/files/gcisdocument/201409/35405gen452.pdf Accessed 10/02 2022.

DEAT. (2007), Assessment of the Status of waste service delivery and capacity at local government level, Pretoria, Department of Environmental Affairs and Tourism. Accessed at https://www.deat.gov.za

Edana (2007). Disposable Diapers: A Technical Review. Edana, Brussels, Belgium

Evans, J., & Thomas, C. (2023). *Environmental governance*. Routledge. DOI: 10.4324/9781003334699

Excelsior News., (2017), The ugly truth behind disposable diapers. (Accessed on 06 June 2024)

Ferronato, N., & Torretta, V. (2019). Waste mismanagement in developing countries: A review of global issues. *International Journal of Environmental Research and Public Health, 16*(6), 1060. DOI: 10.3390/ijerph16061060 PMID: 30909625

Fruitema, M. L. (2015). *A Political Ecology of Solid Waste Management in Niadub, Panama* (Doctoral dissertation, University of Miami).

Hammersley, M. (2015). Sampling and thematic analysis: A response to Fugard and Potts. *International Journal of Social Research Methodology, 18*(6), 687–688. DOI: 10.1080/13645579.2015.1005456

Haywood, L. K., Kapwata, T., Oelofse, S., Breetzke, G., & Wright, C. Y. (2021). Waste disposal practices in low-income settlements of South Africa. *International Journal of Environmental Research and Public Health, 18*(15), 8176. DOI: 10.3390/ijerph18158176 PMID: 34360468

Heimall, L. M., Storey, B., Stellar, J. J., & Davis, K. F. (2012). Beginning at the bottom: Evidence-based care of diaper dermatitis. *MCN. The American Journal of Maternal Child Nursing*, *37*(1), 10–16. DOI: 10.1097/NMC.0b013e31823850ea PMID: 22072018

Huang, Y., Liu, C., & Chen, Y. (2018). Environmental impact of disposable diapers: A life cycle assessment. *Waste Management (New York, N.Y.)*, *75*, 123–132.

Indigenous and Northern Affairs Committee. (2017). Report on Indigenous Waste Management.

Jones, R. (2019). *The Environmental Impact of Disposable Diapers*. Waste Management Studies.

Keske, C., Mills, M., Tanguay, L., & Dicker, J. (2018). Waste management in Labrador and northern communities: Opportunities and challenges. *The Northern Review*, *47*(47), 79–112. DOI: 10.22584/nr47.2018.005

Kimani, E. W., Muchiri, J., & Makindi, S. (2015). Soiled diapers disposal practices among caregivers in poor and middle income urban settings.

Koliyabandara, P. A., Preethika, D. D. P., Cooray, A. T., Liyanage, S. S., Siriwardana, C., & Vithanage, M. (2024). The environmental pressure by open dumpsites and way forward. In *Technical Landfills and Waste Management: Volume 1: Landfill Impacts, Characterization and Valorisation* (pp. 171-204). Cham: Springer Nature Switzerland. DOI: 10.1007/978-3-031-52633-6_7

Kordecki, H., Antrobus-Wuth, R., Uys, M. T., van Wyk, I., Root, E. D., & Berrian, A. M. (2022). Disposable diaper waste accumulation at the human-livestock-wildlife interface: A one health approach. *Environmental Challenges*, *8*, 100589. DOI: 10.1016/j.envc.2022.100589

Kordecki, H., Antrobus-Wuth, R., Uys, M. T., Van Wyk, I., Root, E. D., & Berrian, A. M. (2022). Disposable diaper waste accumulation at thc human-livestock wildlife interface.

Kosoe, E. A., Diawuo, F., & Osumanu, I. K. (2019). Looking into the past: Rethinking traditional ways of solid waste management in the Jaman South Municipality, Ghana. *Ghana Journal of Geography*, *11*(1), 228–244.

Le Roux, C. (2018). African indigenous knowledge systems and sustainable development. *Journal of Sustainable Development*, *11*(2), 1–12.

Madonsela, B. S., Semenya, K., & Shale, K. (2024). A review of indigenous knowledge systems and their application in sustainable solid waste management. *WORLD (Oakland, Calif.)*, *5*(2), 219–239. DOI: 10.3390/world5020012

Maluni, J. K. (2020). Disposal of disposable child diapers by caregivers and their environmental health implications in Kenya: A review. *J Res Innovation Implications Educ*, *4*(1), 22–28.

Manyele, S. V., Kivela, M., & Maziba, W. (2018). Environmental health risks associated with waste disposal in rural Tanzania. *Journal of Environmental Science and Health.*

Mawere, M. (2015). African environmental ethics, sustainable development and the future of humanity. *Journal of Environmental Science and Health. Part B, Pesticides, Food Contaminants, and Agricultural Wastes*, *50*, 1–10.

Mawere, M. (2015). *Humans, other beings and the environment: Harurwa (edible stinkbugs) and environmental conservation in southeastern Zimbabwe*. Cambridge Scholars Publishing.

Mudau, L. S., Thumbathi, A. M. D., & Nkosi, D. V. (2023). Poor disposal of soiled diapers in Sub-Saharan Africa: A gap in environmental health literacy in rural communities. In *Environmental Health Literacy Update-New Evidence, Methodologies and Perspectives*. IntechOpen.

Muranko, Ż., Tassell, C., Zeeuw van der Laan, A., & Aurisicchio, M. (2021). Characterisation and environmental value proposition of reuse models for fast-moving consumer goods: Reusable packaging and products. *Sustainability (Basel)*, *13*(5), 2609. DOI: 10.3390/su13052609

Nkomazana, F. (2017). The impact of modernization on traditional African cultural practices. *Journal of Cultural Studies*, *19*(1), 1–12.

Ogundele, O. M., Rapheal, O. M., & Abiodun, A. M. (2018). Effects of municipal waste disposal methods on community health in Ibadan-Nigeria. *Polytechnica*, *1*(1), 61–72. DOI: 10.1007/s41050-018-0008-y

Onyancha, O. B. (2022). Indigenous knowledge, traditional knowledge and local knowledge: What is the difference? An informetrics perspective. *Global Knowledge, Memory and Communication*, *73*(3), 237–257. DOI: 10.1108/GKMC-01-2022-0011

Oteng-Ababio, M. (2018). Crossing conceptual boundaries: Re-envisioning coordination and collaboration among women for sustainable livelihoods in Ghana. *Local Environment*, *23*(3), 316–334. DOI: 10.1080/13549839.2017.1418847

Remigios, M. V. (2014). *The significance of culture as a household energy determinant: the case of Chiwundura Communal Area*.

Rivers, L. (2020). Assessing the Lifecycle of Diapers. *Environmental Research Letters*.

Scheinfeld, N. (2005). Diaper dermatitis: A review and brief survey of eruptions of the diaper area. *American Journal of Clinical Dermatology*, *6*(5), 273–281. DOI: 10.2165/00128071-200506050-00001 PMID: 16252927

Seopa, S. H. M., Letsoalo, J. M., & Mashaba, M. J. (2021). An investigation of practices and effects of disposable infant diapers on the environment: a case study of Mashashane Village.

Shin, H. T. (2005). Diaper dermatitis that does not quit. *Dermatologic Therapy*, *18*(5), 124–135. DOI: 10.1111/j.1529-8019.2005.05013.x PMID: 15953142

Simbine, S. L., & Le Roux, L. (2021). Indigenous mentoring and monitoring system amongst Vatsonga speaking people towards child protection. *Journal of Social Development in Africa*, *36*(2), 7–32.

Šmajgl, D., & Obhođaš, J. (2015). Occurrence of tin in disposable baby diapers. *X-Ray Spectrometry*, *44*(4), 230–232. DOI: 10.1002/xrs.2609

Smith, A. (2021). Principles of Environmental Management. *The Journal of Environmental Education*.

Smith, L. T. (2019). *Decolonizing Methodologies: Research and Indigenous Peoples*. Zed Books.

Smith, M. V., Kruse, A., Weir, A., & Goldblum, J. (2013). Diaper need and its impact on child health. *Pediatrics*, *132*(2), 253–259. DOI: 10.1542/peds.2013-0597 PMID: 23897910

Tembo, E., & Chazireni, E. (2016). The negative environmental impact of disposable diapers: The case of Mberengwa District, Zimbabwe. *International Journal of Health Sciences*, *4*(2), 2158–2161.

Theodorson, G. A. (1969). *Modern Dictionary of Sociology*. Methuen.

Thompson, G. (2022). Sustainable Environmental Practices. *Journal of Sustainability Studies*.

UNICEF. (2019). Diarrhea and wastewater.

United Nations. (2019). Indigenous Peoples and the 2030 Agenda.

Wahab, M. A., Yousuf, M., & Hossain, M. E. (2004). *Some indigenous medicinal knowledge for treating Jaundice in Chittagong Hill Tracts*.

Welford, R. (2016). *Corporate environmental management 1: Systems and strategies*. Routledge. DOI: 10.4324/9781315825120

Welman, C. F., & Michell, B. (2005). *Research methodology*. Oxford University Press.

White, J. (2021). Indigenous Practices and Biodiversity. *Global Ecology and Conservation*.

WHO. (2018). Water, sanitation and hygiene.

Wilson, D. C., Rodic, L., Modak, P., Soos, R., Carpintero, A., Velis, K., & Simonett, O. (2015). *Global waste management outlook*. Unep.

World Health Organisation. (2010). *A handbook of medical waste management guideline*.

World Health Organization., (2016), Population Growth and Diapers. Accessed on 02/06/2024.

World Health Organization. (2018). *Water*. Sanitation, Hygiene, and Waste Management.

Zazu, C. (2007). Exploring Opportunities and Challenges for achieving the integration of Indigenous Knowledge Systems into Environmental Education processes: A case study of the Sebakwe Environmental Education programme (SEEP) in Zimbabwe. Unpublished Master thesis. Rhodes University.

ADDITIONAL READING

Bennet, S. M., Machete, M., & Karabo, S. (2024). Indigenous Knowledge Systems of Solid Waste Management in Bushbuckridge Rural Communities, South Africa. *Waste*, *2*(3), 293–311. DOI: 10.3390/waste2030017

Cathetina, J. S., Takunda, Y. C., Hung, T., & Andrea, C. (2023). Disposable Diaper Usage and Disposal Practices in Samora Machel Township, South Africa. *Sustainability (Basel)*, *15*(12), 9478. DOI: 10.3390/su15129478

Diana, A. P. (2014). The role of indigenous solid waste management practices (iswmp) in the management of solid waste in urban centres in png. *University of Goroka. Dissertation.*

Heather, R., Breanna, A., & Clair, N. (2015). Disposing of children's diapers with solid waste: a global concern? *Waterlines*, Vol. 34, No. 3, *Community-Led Total Sanitation and Sanitation Marketing*, pp. 255-268 (14 pages).

Mpinane, F. S., Agnes, M., & Suzan, O. (2022). A Critical Analysis of Indigenous Systems and Practices of Solid Waste Management in Rural Communities: The Case of Maseru in Lesotho. *International Journal of Environmental Research and Public Health*, *19*(18), 11654. DOI: 10.3390/ijerph191811654 PMID: 36141920

Mudau, N. (2024). *Knowledge and Practices of Diaper Waste Management in a Rural Village of Limpopo Province of South Africa. University of Johannesburg*. ProQuest Dissertations & Theses.

KEY TERMS AND DEFINITIONS

Indigenous knowledge- Indigenous knowledge is the unique knowledge confined to a particular culture or society. It is also known as local knowledge: folk knowledge, people's knowledge, traditional wisdom, or traditional science. This knowledge is generated and transmitted by communities over time to cope with their own agro-ecological and socio-economic environments.

Informal diaper disposal- it refers to the practice of disposing used disposable diapers in an unofficial manner: often without proper waste management or proper sanitation. This includes disposing of diapers in open spaces, water sources.

Indigenous communities-: refer to a group of people who have a long history of living in a particular region, often with their own distinct culture, language, and traditions.

Environment- this refers to the natural world around us: including air, water, land, and all the living and non-living. It is the foundation of our ecosystem and provides us with essential resources and provides us with essential resources such as food, water, and air.

Vhembe Biosphere Reserve-: A UNESCO-recognized area in north east South Africa, encompassing diverse ecosystems, cultural sites, and biodiversity hotspots, aiming to balance conservation with sustainable development.

Chapter 10
Role of Sustainability in Religious Festival and Event Tourism in Rajasthan:
Insights From Pushkar Fair, India

Aruditya Jasrotia
https://orcid.org/0000-0002-8274-5609
Amity University, India

Kuldeep Singh
https://orcid.org/0000-0002-7999-1585
Amity University, India

ABSTRACT

This chapter seeks to highlight the untouched tourism opportunities in Pushkar, Rajasthan and also to maintain sustainable practices. Many studies have been investigated on religious festivals in the Pushkar region, but there is a need for research that focuses on sustainability with event tourism. Therefore, this study aims to discover the potential and challenges of the Pushkar Fair in Rajasthan. The main objectives of this study are to identify diverse tourism opportunities in the Pushkar region of Rajasthan and to maintain sustainability at the time of such an event. This study is based on a detailed literature review, and data will be collected from reputed journals, Government documents, reports, and edited books published at national and international levels. The study concludes that despite the massive potential of religious and event tourism in Pushkar, Rajasthan, the role of sustainability and preservation of destinations have yet to be wholly harnessed.

INTRODUCTION

The role of sustainability in religious festivals is increasingly recognized as communities around the world strive to balance cultural and religious traditions with environmental responsibility (Thomas, 2023). The number of events and festivals is increasing in tourist destinations, especially rural areas, but sustainability is needed. There are minimal studies on sustainability in religious tourism; still, some

DOI: 10.4018/979-8-3373-4124-8.ch010

studies explicate the influence of sustainable tourism development on events and festivals (Sharpley, 2020). Therefore, this study focuses on sustainable practices in religious festivals and events in Pushkar, Rajasthan, India.

Events are crucial in developing the tourism industry and economic benefits (Seraphin, 2020). Mohanty et al . (2020) discussed the global events during COVID-19. They focused on the systematic review while describing the implications of COVID-19 on event tourism. The New Zealand Tourist and Publicity Department (1987) reported: ''Event tourism is an important and rapidly growing segment of international tourism''. Lopez (2013) noted that resources should be preserved in religious and pilgrimage tourism, and growth in sustainability should create memorable experiences for visitors. According to Rowen(2020), festivals are an essential segment of the tourism industry and provide a rich heritage and culture for tourists. The event provides a platform for promoting and creating awareness regarding the host region.

Tourism generated by festivals and fairs in India significantly influences cultural practices, often leading to both positive and negative impacts. While these events attract large numbers of visitors, they can also result in the commercialization of traditions, altering their authenticity to cater to tourist expectations (Chhabra, 2010). Sacred rituals may be shortened or modified for convenience, and traditional attire, music, and performances may be adapted to appeal to a broader audience (Singh, 2014; Bhuda, & Koitsiwe, 2022). Additionally, overcrowding and increased demand for souvenirs can lead to the commodification of cultural artifacts, sometimes reducing them to mere attractions rather than meaningful traditions (Cohen, 1988). Cultural heritage events in India are diverse, encompassing traditional music, dance, art, craft, cuisine, and folklore, in addition to religious activities (Ministry of Tourism, 2020). These events serve as platforms for preserving and showcasing the country's rich heritage while also promoting tourism. Each cultural heritage event is organized with careful planning, involving various stakeholders such as local communities, government bodies, tourism departments, and private organizations (UNESCO, 2011).

CONCEPTS OF EVENTS AND FESTIVALS

Getz (2008) defined an *event* as a planned and organized occurrence, typically scheduled at a specific time and location, involving a gathering of people. They are often designed to fulfil specific goals like celebration, education, promotion, entertainment, or networking. It is to be noted that event management has a vast potential market for tourists all over the globe, and it is essential to focus on sustainability. The role of stakeholders becomes crucial in this regard.

Der Wagen (2001) described the events as given below:

- Offer participants of tourists and locals and create a memorable experience
- Need proper planning and organizing in costing
- Be prepared for a short period
- Need careful and extensive planning and strategy
- It should be planned once in a year
- Need coordination of many stakeholders and event manager.

According to Salem, Jones, and Morgan (2004, p. 19), events are *"a unique blend of activities, which are the tools for achieving the overall event aims and satisfying customer needs"*.Van der Wagen (2007, p. 5) states that the event can be best described as *"a complex social endeavour characterized by sophisticated planning with a fixed deadline, often involving numerous stakeholders."*

Falassi stated that the definitions and meanings of festivals in literature have been made in various ways. The definitions made by Falassi (1995:37) are:

- Celebrations based on holy and cultural traditions and customs
- Celebrations of a crucial people or an event and cultivation of a vital crop
- Public fairs and feasts

Figure 1. Typology of planned event (Getz, 2008)

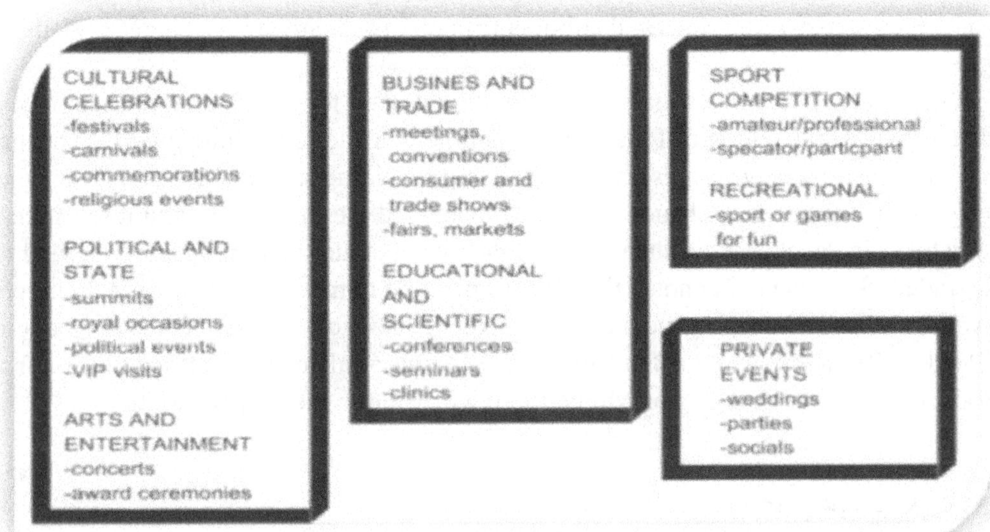

LITERATURE REVIEW

Thomas (2023) analyzed the involvement of the community in rural tourism events and discussed that due to rural fairs and festivals, economic growth and employment opportunities can be increased. Hall (2019) focused on the role of the tourism ecosystem and human environment in promoting sustainability in tourism. Boluk et al. (2019) revealed that multiple factors and the involvement of various stakeholders play a crucial role in the growth of sustainable tourism in religious events. Roxas et al. (2020) noted that due to pilgrimage tourism, there is an overall sustainable development and all-inclusive approach.

Gazolla et al. (2018) discussed the sustainable development and community involvement that supports the cultural resources and management of the tourism destination.

Okech (2011) analyzed the importance of sustainable events in Kenya and focus on local economic development and opportunities for sustainable development. There is a need for strategic management and proper planning for sustainable events and festivals. (Getz, 1997; Goldblatt, 1997; Hall & Macionis, 1998). Romanelli et al. (2021) analyze the role of sustainable tourism in religious events and the importance of the role of local communities in religious destinations. Ivanova (2017) discussed the impacts of event tourism in Bulgaria and emphasized the concept of sustainable tourism more. She focuses on both negative and positive impacts on event tourism. Tolkach and King (2015) said that community attachment positively influences support for community tourism development.

Getz (1997) stated that a festival is a celebration or event usually characterized by performances, activities, and rituals, often centered around a specific theme, cultural tradition, religious observance, or significant aspect of life. Moscardo (2007) confirmed that festival and event tourism revolved around four central themes, i.e. monetary effects, better promotion, accessibility and quality of administration and the primary focus is on the showcasing of culture and heritage. Fox Gotham (2005) revealed that due to festivals and events, many traditions and cultures from the past can be rejuvenated and also enhance economic growth.

Bairwa and Patwari (2021) investigated the growth and need for sustainable tourism through fairs and festivals in the Marwar region of Rajasthan, which helps foster economic growth. Sharma (2021) revealed that solid waste management is crucial while implementing sustainable strategies in the context of Ajmer, Rajasthan. Mathur and Chaudhary (1999) discussed the misuse of natural resources and the role of sustainability in Pushkar, Rajasthan (India). Due to mass tourism, tourism products and resources are being degraded. Sharma (2021) noted that due to the large number of tourists and their activities, a substantial volume of solid waste is generated, and there is a need for sustainable management in Ajmer, Rajasthan. Parekh (2010) revealed that fairs and festivals from rural areas can attract tourists from urban areas and boost the local economy by providing them with employment opportunities.

MATERIAL AND METHODS

This study has adopted a case study methodology to observe the opportunities and challenges of tourism in Pushkar, Rajasthan. The research covers a detailed literature review from various research papers, Articles, Government policies, reports, Books, Magazines and Newspapers. Yin (2003) recommended that the case study approach for specific situations made an obvious case for applying a case study methodology for the situations that are specific to answer 'why' and 'questions'. For better clarity and deeper understanding, case study approaches can be applied successfully in tourism and hospitality research (Hudson & Ritchie, 2009, p. 222). Also, in this study, the involvement of stakeholders from Pushkar Rajasthan and their discussion on sustainability issues were considered. The present study also analyzes the potential and challenges of Pushkar fairs on the social and cultural growth, and researchers also take views of pilgrims, tourists, and locals like shopkeepers, tour operators, transporters, etc and try to implement their suggestions.

Study Site

Rajasthan is famous for many fairs and festivals, and this state is very rich in heritage and culture. Many entertainment activities include music, folk dance, puppet dance, etc. This state focuses more on celebrating and welcoming guests from all over the globe and creating positive, memorable experiences in the minds of tourists. Pushkar Camel Fair is one such great fair which showcases the rich heritage of Rajasthan (Akhtar & Farooq, 2023).

Pushkar is a town located in the Ajmer district of Rajasthan, India. Renowned for its picturesque setting and cultural significance, Pushkar is often called Tirth Raj, meaning the "King of Pilgrimage Sites." The town is particularly famous for the Pushkar Lake, a sacred body of water surrounded by numerous ghats and temples. One of the critical attractions of Pushkar is the Brahma Temple, dedicated to Lord Brahma, the creator of Hindu mythology. This temple is one of the few dedicated to Lord Brahma, making it a significant pilgrimage site for Hindus. Aside from its religious importance, Pushkar is also known for hosting one of the world's largest and most vibrant camel fairs, the Pushkar Camel Fair. This annual event attracts travelers, traders, and photographers from across the globe, adding a colorful and lively atmosphere to the town.

The streets of Pushkar are lined with vibrant markets, offering a wide array of traditional Rajasthani handicrafts, textiles, and jewelry. The town's charm lies in its narrow lanes, adorned with vibrant colors and a blend of modern and traditional elements. Beyond its cultural and religious richness, Pushkar's tranquil surroundings and desert landscape add to its appeal, providing visitors with a serene and spiritually uplifting experience. Whether one seeks spiritual solace, cultural exploration, or the excitement of a lively fair, Pushkar offers a unique blend of experiences in the heart of Rajasthan.

Figure 2. Map of Pushkar

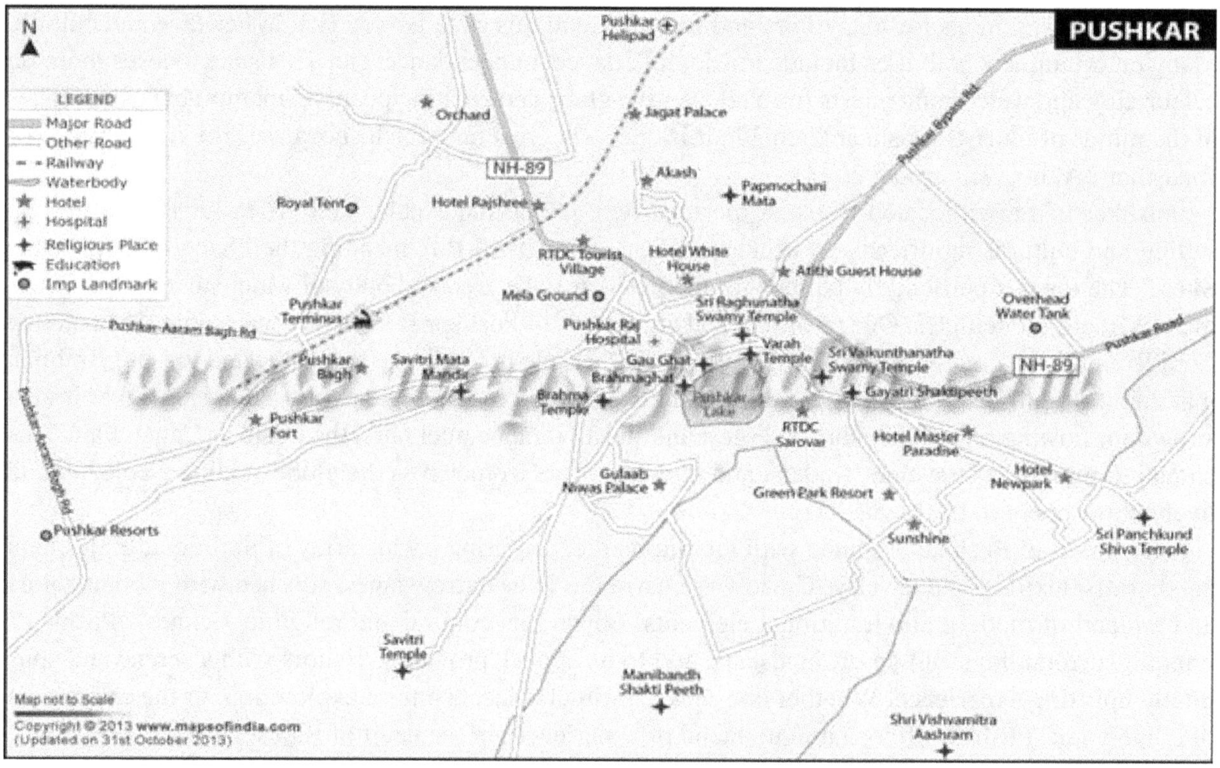

Case Study: Pushkar Fair

Pushkar Fair, held annually in the town of Pushkar in the state of Rajasthan, India, is one of the country's most renowned and vibrant cultural events. This case study explores the historical background, economic impact, cultural significance, and challenges associated with the Pushkar Fair. The Pushkar Fair has a history dating back several centuries. It is believed to have originated from a pact between Lord Brahma (the creator in Hindu mythology) and the local traders. The fair coincides with the Kartik Purnima, the moon night in the Hindu lunar month of Kartik, attracting pilgrims and traders from across the country. The primary objective of the Pushkar Fair is to facilitate the trading of livestock, mainly camels, while also serving as a cultural and religious gathering. Pilgrims dip in the holy Pushkar Lake during this time, believing it to cleanse them of sins.

Economic Impact:

- **Livestock Trading:** The fair is a crucial hub for trading camels, horses, and cattle. It serves as a significant source of income for the local communities and attracts buyers and sellers from various parts of India.

- **Business Opportunities:** Beyond livestock trading, the fair provides a platform for diverse businesses, including handicrafts, textiles, jewelry, and traditional Rajasthani items. Local artisans and traders find lucrative opportunities to showcase and sell their products.
- **Tourism Boost:** The fair draws many domestic and international tourists, contributing significantly to the local economy. The influx of visitors increases demand for accommodation, food, transportation, and various services.

Cultural Significance:

- **Religious Practices:** Pilgrims visit Pushkar during the fair to take a ceremonial dip in the Pushkar Lake, considered auspicious during Kartik Purnima. The town's numerous temples witnessed increased activity during this time.
- **Traditional Performances:** The fair showcases Rajasthan's rich cultural heritage, featuring traditional music, folk dances, and competitions. These performances add to the festive atmosphere and attract enthusiasts worldwide.

The Pushkar Fair is a unique blend of commerce, culture, and spirituality, attracting people from different walks of life. While it has significantly boosted the local economy and brought international recognition to Pushkar, ongoing efforts are essential to address challenges and ensure the sustainability of this iconic cultural event. The fair serves as a testament to the enduring traditions of Rajasthan and the harmonious convergence of economic and cultural interests.

Major Tourist Destinations in Pushkar:

a) Pushkar lake
b) Brahma temple
c) Gurudwara Singh Sabha
d) Varaha temple
e) Savitri temple
f) Rangji temple
g) Pap mochini temple
h) Sri panchkund shiva temple
i) Atmateshwar temple
j) Man mahal

DISCUSSIONS

Due to festivals and events, visitor demand is high, directly affecting sustainability. Therefore, it is challenging for the local people, government, and other stakeholders to focus on sustainability and preserve the resources (Quinn, 2006). To achieve sustainable growth, benefits are given to the local people, and more employment opportunities must be provided (Yu & Turco, 2000). Promoting local culture and socio-economic benefits is essential to maintain sustainability in tourism destinations (Richards &

Hall, 2000). The community's development and awareness are crucial for promoting sustainability, and the local community should focus on managing carrying capacity at the destination (Richards & Hall, 2000). Bruce et al. (2001) emphasized the role of local communities' role in developing an agenda for promoting sustainable tourism. Event tourism studies are newer in concept, and there is still scope for growth in event and festival tourism. Formica (1998) recognized marketing impacts, advertising, forecasting of the events to promote it.

Cultural tourism plays a significant role in shaping indigenous knowledge in India, influencing both its preservation and transformation. Indigenous knowledge, deeply rooted in traditional practices, beliefs, and customs, is often shared with tourists through storytelling, handicrafts, rituals, and performances (Chatterjee, 2017). While tourism provides a platform for indigenous communities to showcase their heritage, it also poses challenges that can lead to the dilution or modification of traditional knowledge (Salazar, 2012).

One major impact of cultural tourism is the commercialization of indigenous practices. As communities adapt their customs to meet tourist expectations, there is a risk of oversimplification or alteration of traditional knowledge to make it more marketable (Cohen, 1988). For instance, rituals that hold deep spiritual significance may be shortened or modified for the convenience of visitors. Similarly, traditional crafts, once created with cultural and spiritual meanings, may be mass-produced to cater to tourist demand, often compromising their authenticity (Chhabra, 2010). Moreover, cultural tourism can contribute to the erosion of indigenous knowledge if local communities prioritize economic benefits over the transmission of traditional skills and wisdom to younger generations (UNESCO, 2017). Younger members may be drawn to tourism-related jobs, leading to a decline in the intergenerational transfer of indigenous knowledge (Grydehøj, 2019).

However, tourism also offers opportunities for revival and sustainability. By creating demand for authentic cultural experiences, it encourages communities to document and safeguard their traditions (George & Reid, 2005). Initiatives such as community-based tourism and heritage tourism help in balancing economic benefits with cultural preservation. To ensure that cultural tourism supports indigenous knowledge rather than diminishing it, policies and community-driven approaches must emphasize ethical tourism practices (Scheyvens, 1999). Encouraging responsible tourism, promoting local involvement, and implementing cultural preservation programs can help maintain the integrity of indigenous traditions while allowing communities to benefit from tourism-driven economic opportunities.

RAJASTHAN FAIR AND FESTIVALS

Rajasthan, the largest state in India, is known for its rich cultural heritage, vibrant traditions, and a calendar filled with colorful fairs and festivals that reflect the state's diverse and lively spirit. Here is an overview of some of the major fairs and festivals celebrated in Rajasthan:

a) Pushkar Fair (Pushkar Camel Fair):
   - When: Annually in the Hindu month of Kartik (October-November)
   - Where: Pushkar
   - Highlights: Pushkar Fair is one of the largest camel fairs in the world, drawing visitors with its vibrant atmosphere. It uniquely blends religious fervor, cultural events, and a bustling

livestock market. Pilgrims also gather to take a dip in the sacred Pushkar Lake during this time.

b) Desert Festival:
- When: Usually in February
- Where: Jaisalmer
- Highlights: Celebrated amidst the golden sands of the Thar Desert, this festival showcases the rich cultural heritage of Rajasthan. Visitors can enjoy traditional music and dance performances, camel races, turban-tying competitions, and the illuminated Jaisalmer Fort.

c) Teej Festival:
- When: Typically in July or August
- Where: Celebrated throughout Rajasthan with particular enthusiasm in Jaipur
- Highlights: Teej is a monsoon festival primarily celebrated by women. It involves fasting, traditional dances, swings, and various cultural events. Women dress in colorful traditional attire, and processions featuring the goddess Parvati are taken out in many cities.

d) Gangaur Festival:
- When: Usually in March or April
- Where: Celebrated throughout Rajasthan, prominently in Udaipur
- Highlights: Dedicated to Goddess Gauri, this festival is particularly significant for married women who pray for the well-being and longevity of their husbands. Processions with idols of Gauri, traditional dances, and other cultural events are integral to the celebration.

e) Elephant Festival:
- When: Usually in March
- Where: Jaipur
- Highlights: An enchanting festival featuring a grand procession of decorated elephants, showcasing the majestic animals' beauty and grace. The event includes elephant polo, races, and cultural performances.

f) Marwar Festival:
- When: Held in September or October
- Where: Jodhpur
- Highlights: Celebrating the music and dance of the Marwar region, this festival is a showcase of traditional Rajasthani folk arts. Visitors can witness folk performances camel tattoo shows and participate in various competitions.

g) Kite Festival (Makar Sankranti):
- When: January
- Where: Celebrated throughout the state
- Highlights: The festival marks the sun's transition into the zodiac sign of Capricorn. People engage in kite-flying competitions, and the sky is filled with colorful kites of various shapes and sizes.

These festivals add vibrancy to Rajasthan and provide a platform for locals and tourists to experience the state's rich cultural tapestry, traditional art forms, and warm hospitality. Each celebration is a reflection of Rajasthan's historical legacy and the enduring spirit of its people.

CHALLENGES OF SUSTAINABILITY IN PUSHKAR

Festivals and fairs in India are deeply rooted in cultural heritage, religious traditions, and social customs. They serve as a means of preserving and passing down traditions from one generation to another (Singh, 2002). However, the influx of tourists during these events has led to significant transformations, some of which threaten the authenticity and sustainability of cultural practices. One major challenge is the commercialization of rituals and traditions. Festivals like Pushkar Camel Fair, Kumbh Mela, and Hemis Festival attract large numbers of domestic and international tourists, leading to the modification of rituals and performances to cater to visitor expectations (Chhabra, 2010). Sacred ceremonies that were once conducted for spiritual purposes are now often adjusted for convenience, sometimes losing their original significance (Cohen, 1988). For instance, traditional performances such as Kathakali in Kerala or Ramlila during Dussehra are sometimes shortened or altered to make them more appealing to tourists, undermining their traditional depth and value (Richards, 2007). Another pressing concern is the commodification of cultural artifacts and practices.

Many indigenous crafts and traditional goods, such as Madhubani paintings, Banarasi silk, and tribal jewelry, are increasingly mass-produced for the tourism market. This not only dilutes their authenticity but also disrupts the livelihood of artisans who follow age-old, labor-intensive production techniques (UNESCO, 2017). Furthermore, the demand for souvenirs has led to the exploitation of cultural symbols, where sacred objects are sold as decorative items, stripping them of their cultural and religious significance (Smith, 2009). The sheer volume of visitors during major festivals also puts immense pressure on local communities and heritage sites (UNWTO, 2018). Overcrowding at events like the Kumbh Mela or Durga Puja leads to environmental degradation, waste mismanagement, and damage to heritage sites. Pilgrimage sites such as Varanasi and Puri, which experience a surge in tourists during religious festivals, often struggle to maintain their cultural and ecological balance (Scheyvens, 1999). Local traditions and daily life are disrupted as spaces meant for community gatherings are overtaken by tourism infrastructure. Additionally, cultural homogenization poses a threat to diversity.

In an effort to cater to a global audience, festival organizers and local stakeholders may modify or standardize traditional practices, leading to a loss of regional uniqueness (Salazar, 2012). For example, some folk performances and rituals are adapted to fit modern stage settings, altering their traditional elements. This shift can make cultural expressions more performative rather than meaningful, affecting their transmission to future generations (Timothy & Boyd, 2003). To mitigate these threats, it is essential to promote responsible tourism practices that respect and preserve cultural heritage. This includes involving local communities in decision-making, implementing sustainable tourism policies, and educating tourists on the significance of traditional practices (Ministry of Tourism, 2020). By striking a balance between tourism development and cultural preservation, India can ensure that its festivals and fairs remain authentic and meaningful while still benefiting from the economic advantages of cultural tourism.

There are many negative impacts of event tourism on sustainability. Some of them are listed below:

- The traffic problem is widespread in Pushkar, Rajasthan, due to overcrowding of the tourists.
- Disposable waste and natural resources are also significant concerns due to their harmful effects on tourists' health.
- There is also road infrastructure damage due to high traffic jams and difficulty in finding proper parking places in Pushkar, Rajasthan.

- There is also a high risk of terrorism activity in Pushkar due to religious activities and the need for safety and security in such places.
- Vandalism is also very prominent in the Pushkar region, and many tourists disturb artificial resources like monuments, buildings, and hotels due to vandalism.
- Riots in religious festivals of Rajasthan also disturbs tourist all over, which damage resources.

Negative Impacts on the Environment

Due to the Pushkar fair and many religious festivals in Rajasthan, there are some negative impacts on the environment. Noise pollution due to many vehicles hurt the environment and sustainability. There is also a risk of damaging the environment to the host community. Also, due to the high amount of litter in Pushkar, the environment has a negative impact. There is also high air pollution due to many newly constructed buildings and industries. There is also a negative impact on Pushkar Lake due to deforestation and water shortage. Many fish and marine animals die every year due to harmful environmental activities. Verma and Summarwar (2012) analyzed water degradation in the Pushkar region and discussed the shortage and hardness of water after the Pushkar fair.

Strategies to overcome the negative environment

Sustainability in the context of the Pushkar Fair involves addressing environmental, social, and economic aspects to ensure the long-term viability of the event without compromising the well-being of the local community, the environment, or the cultural heritage. Here are several considerations for promoting sustainability in Pushkar:

1. **Waste Management:**
    a. Implement effective waste management strategies to handle the influx of visitors. This includes recycling initiatives, waste segregation, and proper disposal methods.
    b. Encourage the use of eco-friendly materials and discourage the use of single-use plastics during the fair.
2. **Water Conservation:**
    a. Promote water conservation practices, especially around Pushkar Lake. This may involve educating visitors about the importance of water conservation, installing water-saving devices, and monitoring water usage.
3. **Energy Efficiency:**
    a. Explore opportunities for energy efficiency in event infrastructure. This could include using renewable energy sources, energy-efficient lighting, and encouraging participants to use sustainable energy practices.
4. **Cultural Preservation:**
    a. Ensure that the cultural heritage and traditions associated with the Pushkar Fair are respected and preserved. This involves promoting responsible tourism practices and discouraging any activities that could harm the historical and cultural significance of the event.
5. **Community Engagement:**

a. Involve the local community in the planning and execution of the fair. This ensures that the benefits of the event are distributed equitably and that the local population actively participates in decision-making processes.

6. **Economic Impact:**

   a. Analyze and optimize the economic impact of the fair to ensure that it contributes positively to the local economy. This may involve supporting local businesses artisans, and ensuring fair trade practices.

7. **Infrastructure Development:**

   a. Invest in sustainable infrastructure development that can accommodate increasing visitors without degrading the natural environment.

SUGGESTIONS

For sustainable tourism development and to promote event tourism, the role of host communities is crucial (Ivanova, 2017). Development of infrastructure and proper management is needed to attract more tourists. There should be more social and cultural exchange to promote event tourism and festivals. In addition, the price of the tickets to the event for tourists should be reduced. There should be employment opportunities for local people, and there should be some skill development training programs.

The role of public and private organizations is very crucial for the maintenance of sustainability in events and festivals in Rajasthan and required to enhance tourism-based services. To protect the environment, there must be limitations in tourism activities, especially in delicate areas. There should be proper education and community awareness for environmental protection in destinations where events occur frequently. (Jain & Kaur, 2020). There should be proper employment opportunities for the local people and all age groups. Human resources should be used in various tourist destinations to provide them with training to make an event successful without any harmful effects. Cooperative societies should be encouraged to make natural agriculture products to promote the tourism business, especially in the food and beverage sector. The government should promote eco-friendly practices in Rajasthan through proper marketing and advertising. Eco-friendly vehicles should be used so that there can be less pollution during events and festivals in Rajasthan. Proper website and internet use should be used to showcase the importance of sustainability in Rajasthan.

FUTURE RESEARCH DIRECTIONS

Future research on sustainability in religious festival and event tourism in Rajasthan, particularly in the context of the Pushkar Fair, should explore interdisciplinary approaches that integrate environmental science, cultural studies, and tourism management. There is a need for longitudinal studies that assess the long-term ecological and socio-cultural impacts of recurring events like the Pushkar Fair. Future inquiries could focus on the effectiveness of current sustainability measures implemented by local authorities and tourism stakeholders, including waste management systems, crowd control mechanisms, and community involvement strategies. Additionally, research should investigate the perceptions and attitudes of different stakeholder groups—such as local residents, pilgrims, tourists, and vendors toward sustainability initiatives. Comparative studies between Pushkar and other religious festivals across India or globally can also offer valuable insights into best practices and culturally sensitive models of sustainable tourism. The role of technology and digital tools in promoting awareness and managing large-scale events sustainably is another promising area for exploration. By addressing these dimensions, future research can contribute to the development of more inclusive, resilient, and culturally respectful tourism frameworks for religious and cultural festivals.

CONCLUSION AND RECOMMENDATIONS

Pushkar is one of the most famous attractions in India, and tourists from all over the world like to visit this destination. Many tourists prefer this destination post-COVID-19 due to its natural attractions and sustainability (Mishra & Kumar, 2022). This destination is also famous because of the annual international cattle fair in November. This paper focuses on the sustainability issue in Pushkar, Rajasthan and how to manage it. To sustain the tourism business in Pushkar, there is a need to control pollution, overcrowding, waste management, sewage treatment, lake cleanliness, and the preservation of the natural environment (Chowdhary & Mishra, 2020). There are many challenges for the tourism stakeholders in Pushkar, which were discussed thoroughly in this paper. Carrying capacity should be adequately maintained during the Pushkar fair. There is also a need to focus on the negatives for better sustainable development, like Western influence on local culture, cultural degradation, loss of tradition and culture, etc. Due to many business activities and a large number of tourists, it is observed that there is a loss of natural resources and cultural identity in Pushkar, Rajasthan. Therefore, the Ministry of Tourism, State Government, and Local Government are vital to maintaining sustainability in Pushkar, Rajasthan. Also, local people play an essential role in managing the destination, and their awareness of managing events and festivals is crucial.

The government's role is crucial for promoting culture in Pushkar, Rajasthan. Pushkar Fair is one such event that focuses on culture and heritage and provides employment to the local people. There is a need to promote the Pushkar fair through proper marketing, training of locals and exports of tourism products. Cooperation with tour operators and local participation in trade fairs enhance the branding of the Pushkar fair globally. Proper planning by stakeholders and tourism managers can increase the economic growth in the Pushkar region (Verma & Summarwar,2012).

LIMITATIONS AND FUTURE SCOPE OF THE STUDY

There are also some limitations of the study; as Rajasthan is the largest state in the country, it is not easy to cover other cities, and this study is confined only to Pushkar. Researchers can also cover other popular cities like Jodhpur, Udaipur, and Jaipur, but they need more time. Researchers can also use quantitative and qualitative data in this study, and insights from travel experts, tourists, and social media can help future scholars and practitioners conduct periodic research in event and festival tourism in Rajasthan. This research will help scholars and government bodies implement sustainable strategies in the context of some religious events and festivals.

REFERENCES

Akhtar, S., & Farooq, N. (2023). Folk Dance and Music of Rajasthan: A Study of their Role in Fairs and Festivals. *Social Science Journal for Advanced Research, 3*(4), 59–64. DOI: 10.54741/ssjar.3.4.9

Bairwa, M., & Patawari, S. (2021). Festivals for Sustainable Tourism Development: A Case Study of Marwar Region of Rajasthan, India. *Local Food and Community Empowerment through. Tourism (Zagreb)*, •••, 82–96.

Bhuda, M. T., & Koitsiwe, M. (2022). The Covid-19 Pandemic and The Challenges Encountered by The Southern Ndebele Crafts Entrepreneurs Due to Closure of The Tourism Sector. *Gender & Behaviour, 20*(1), 19188–19202.

Boluk, K. A., Cavaliere, C. T., & Higgins-Desbiolles, F. (2019). A critical framework for interrogating the United Nations Sustainable Development Goals 2030 Agenda in Tourism. *Journal of Sustainable Tourism, 27*(7), 847–864. DOI: 10.1080/09669582.2019.1619748

Bruce, D. M., Jackson, M. J., & Cantallops, A. S. (2001). PREPARe: A model to aid the development of policies for less unsustainable tourism in historic towns. *Tourism and Hospitality Research, 3*(1), 21–36. DOI: 10.1177/146735840100300103

Chatterjee, S. (2017). Indigenous knowledge and cultural tourism in India: Challenges and prospects. *Journal of Heritage Management, 2*(1), 45–58. DOI: 10.1177/2455929617706439

Chhabra, D. (2010). Authenticity and tourism: Productive debates, constructive dialogues and a way forward. *Tourism Recreation Research, 35*(3), 255–258. DOI: 10.1080/02508281.2010.11081639

Chowdhary, N., & Mishra, S. (2020). Tourism development and environmental sustainability in Pushkar, Rajasthan. *Indian Journal of Tourism and Hospitality Management, 11*(2), 89–102.

Cohen, E. (1988). Authenticity and commoditization in tourism. *Annals of Tourism Research, 15*(3), 371–386. DOI: 10.1016/0160-7383(88)90028-X

Der Wagen, L. (2001) Event Management: For tourism, cultural, business and sporting events. Sydeny: Hospitality Press Melbourne

Falassi, A. (1987). *Time Out of Time: Essays on the Festival.* University of New Mexico Press.

Formica, S. (1998). The development of festivals and special events studies. *Festival Management & Event Tourism, 5*(3), 131–137.

Fox Gotham, K. (2005a). Theorizing urban spectacles: Festivals, tourism and the transformation of urban space. *City (London, England), 9*(2), 225–245. DOI: 10.1080/13604810500197020

Gazzola, P., Pavione, E., Grechi, D., & Ossola, P. (2018). Cycle tourism as a driver for the sustainable development of little-known or remote territories: The experience of the Apennine regions of Northern Italy. *Sustainability (Basel), 10*(6), 1863. DOI: 10.3390/su10061863

George, E. W., & Reid, D. G. (2005). The power of community tourism in destination development. *Journal of Travel Research, 43*(5), 523–532. DOI: 10.1177/0047287505278296

Getz, D. (1997). *Event Management and Event Tourism.* Cognizant Communications Corp.

Getz, D. (2008). Event Tourism: Definition, Evolution and Research. *Tourism Management, 29*(1), 403–428. DOI: 10.1016/j.tourman.2007.07.017

Goldblatt, J. (1997). *Special Events: Best Practice in Modern Event Management.* Van Nostrand.

Grydehøj, A. (2019). Cultural sustainability: Indigenous knowledge and island tourism development. *Island Studies Journal, 14*(1), 165–182. DOI: 10.24043/isj.74

Gupta, S. K., Tiwari, S., & Voda, M. (2021). Stakeholders'satisfaction Towards Sustainable Tourism Development In Pushkar Region Of Rajasthan. *Geographia Technica, 16*(1).

Hall, C. M. (2019). Constructing sustainable tourism development The 2030 agenda and the managerial ecology of sustainable tourism. *Journal of Sustainable Tourism, 27*(7), 1044–1060. DOI: 10.1080/09669582.2018.1560456

Hall, C. M., & Macionis, N. (1998). Wine tourism in Australia and New Zealand. In Hall, C. M., & Jenkins, J. (Eds.), *Tourism and Recreation in Rural Areas.* Wiley.

Ivanova, M. INNOVATION AS A PREREQUISITE FOR SUSTAINABLE BUSINESS DEVELOPMENT. *ИНОВАЦИИТЕ КАТО ПРЕДПОСТАВКА ЗА УСТОЙЧИВО РАЗВИТИЕ НА БИЗНЕСА,* 18.

Jain, T., & Kour, G. (2020) Destination events and campaign planning for Repositioning: Case of Rajasthan, India.

Lopez, L. (2013) La Imagen de Santiago de Compostela y del Camino en Italia. Una aproximación desde la geografía cultural. Universidad de Santiago de Compostela. Available at: http://dspace.usc.es/handle/10347/7111 (accessed 20 April 2013).

Mathur, P., & Chowdhary, N. R. (1999). ENVIRONMENTAL DEGRADATION AT PUSHKAR VALLEY (INDIA): RESTRAIN TOURISM OR PROMOTE TOURISM? *Biologica, 37,* 117–123.

Ministry of Tourism, Government of India. (2020). *Incredible India: Cultural festivals and heritage tourism.* https://tourism.gov.in

Ministry of Tourism, Government of India. (2020). Sustainable tourism guidelines for cultural destinations. https://tourism.gov.in

Mishra, S., & Kumar, R. (2022). Post-COVID tourism recovery and destination resilience: A case of Pushkar, Rajasthan. *Journal of Tourism and Sustainability, 4*(1), 44–56.

Mohanty, P., & Choudhury, R. (2022). Events tourism in the eye of the COVID-19 storm: Impacts and implications. In *Event Tourism in Asian Countries* (pp. 97–114). Apple Academic Press. DOI: 10.1201/9781003161134-6

Moscardo, G. (2007). Analysing the role of festivals and events in regional development. *Event Management, 11*(1-2), 23–32. DOI: 10.3727/152599508783943255

Okech, R. N. (2011). Promoting sustainable festival events tourism: A case study of Lamu Kenya. *Worldwide Hospitality and Tourism Themes, 3*(3), 193–202. DOI: 10.1108/17554211111142158

Parekh, M. A. (2010). Fair-Based rural tourism: A potential growth engine for rural economy in Gujarat. *IUP Journal of Infrastructure*, *8*(1/2), 83.

Quinn, B. (2006). Problematising festival tourism: Arts festivals and sustainable development inIreland. *Journal of Sustainable Tourism*, *14*(3), 288–306. DOI: 10.1080/09669580608669060

Richards, G. (2007). *Cultural tourism: Global and local perspectives*. Haworth Press.

Richards, G., & Hall, D. (Eds.). (2000). Tourism and Sustainable Community Development, Ritchie, J. B., & Hudson, S. (2009). Understanding and meeting the challenges of consumer/tourist experience research. *International Journal of Tourism Research*, *11*(2), 111–126.

Romanelli, M., Gazzola, P., Grechi, D., & Pollice, F. (2021). Towards a sustainability-oriented religious tourism. [Routledge, London.]. *Systems Research and Behavioral Science*, *38*(3), 386–396. DOI: 10.1002/sres.2791

Rowen, I. (2020). The transformational festival as a subversive toolbox for a transformed tourism: Lessons from Burning Man for a COVID-19 world. *Tourism Geographies*, *22*(3), 695–702. DOI: 10.1080/14616688.2020.1759132

Roxas, F. M. Y., Rivera, J. P. R., & Gutierrez, E. L. M. (2020). Framework for creating sustainable tourism using systems thinking. *Current Issues in Tourism*, *23*(3), 280–296. DOI: 10.1080/13683500.2018.1534805

Salazar, N. B. (2012). Community-based cultural tourism: Issues, threats and opportunities. *Journal of Sustainable Tourism*, *20*(1), 9–22. DOI: 10.1080/09669582.2011.596279

Salem, G., Jones, E., & Morgan, N. (2004). An overview of events management. In Yeoman, I., Ali-Knight, M. J., Drummond, S., & McMahon-Beattie, U. (Eds.), *Festival and EventsManagement* (p. 14). Elsevier ButterworthHeinemann. DOI: 10.1016/B978-0-7506-5872-0.50007-7

Scheyvens, R. (1999). Ecotourism and the empowerment of local communities. *Tourism Management*, *20*(2), 245–249. DOI: 10.1016/S0261-5177(98)00069-7

Seraphin, H. (2020). *COVID-19: an opportunity to review existing grounded theories in event studies.* Paper presented at the Journal of Convention & Event Tourism

Sharma, R. (2021). Solid Waste Management a Case Study of Ajmer. Adn Envi Was Mana Rec, 4 (3): 226-228. *Keywords: Municipal solid waste, Biomedical waste, SW (Solid waste) management, Ajmer city, Landfill.ISSN*, *2641*, 1784.

Sharpley, R. (2020). Tourism, sustainable development and the theoretical divide: 20 years on. *Journal of Sustainable Tourism*, *28*(11), 1932–1946. DOI: 10.1080/09669582.2020.1779732

Singh, S. (2002). Tourism and culture in India. *International Journal of Tourism Research*, *4*(5), 423–435.

Singh, S. (2014). Tourism in India: Policy pitfalls. *International Journal of Tourism Cities*, *1*(1), 26–42.

Smith, M. K. (2009). *Issues in cultural tourism studies* (2nd ed.). Routledge. DOI: 10.4324/9780203869857

Thomas, A. (2023). Rural Event Tourism: Developing Sustainable Rural Event Tourism for the Development of the Community. In *Event Tourism and Sustainable Community Development* (pp. 279-300). Apple Academic Press.

Timothy, D. J., & Boyd, S. W. (2003). *Heritage tourism*. Pearson Education.

Tolkach, D., & King, B. (2015). Strengthening Community-Based Tourism in a new resource-based island nation: Why and how? *Tourism Management*, *48*, 386–398. DOI: 10.1016/j.tourman.2014.12.013

Toneva, P. I. (2017). Studying the impacts of event tourism on the host communities in Bulgaria.

UNESCO. (2017). Intangible cultural heritage and sustainable tourism. United Nations Educational, Scientific and Cultural Organization. https://ich.unesco.org

UNWTO. (2018). Tourism and culture: Investing in future generations. United Nations World Tourism Organization. https://www.unwto.org

Van der Wagen, L. (2007). *Human resource management for events: Managing the event workforce*. Routledge.

Verma, S., & Summarwar, S. (2012). Analysis of water of Pushkar Lake before, during and after the Pushkar Fair. *International Journal of Life Sciences Biotechnology and Pharma Research*, *1*(4), 35–41.

Yin, R. K. (2003). Applied social research methods series: Vol. 5. *Case study research design and methods* (3rd ed.).

Yu, Y., & Turco, D. (2000). Issues in tourism event economic impact studies: The case of the Albuquerque International Balloon Fiesta. *Current Issues in Tourism*, *3*(2), 138–149. DOI: 10.1080/13683500008667870

ADDITIONAL READING

Bhuda, M. T., & Kgwadi, B. (2023). The economic measures established to rebuild the South African cultural heritage tourism after the COVID-19 impact. *Gender & Behaviour*, *21*(1), 21398–21412.

Sharma, A., & Bansal, S. (2021). Cultural erosion and natural resource depletion in Pushkar due to tourism activities. *The International Journal of Environmental Studies*, *78*(3), 345–359. DOI: 10.1080/00207233.2021.1873930

KEY TERMS AND DEFINITIONS

Heritage: Heritage refers to the legacy of physical artifacts and intangible attributes inherited from past generations, maintained in the present, and preserved for future generations.

Religious Festival: A Religious Festival is a celebration or observance that is rooted in religious beliefs, traditions, and practices.

Event Tourism: Event Tourism refers to the planning, promotion, and management of events as a key attraction to draw tourists to a destination.

Chapter 11
Nature as a Means of Preserving Siswati Indigenous Knowledge Systems

Muzi Matfunjwa
https://orcid.org/0000-0003-4553-3225
North-West University, South Africa

Respect Mlambo
https://orcid.org/0000-0002-0348-6726
North-West University, South Africa

ABSTRACT

This chapter aims to highlight the importance of nature and describe how it sustains Siswati indigenous knowledge systems (IKS) in Eswatini. Grounded in the Biocultural Diversity Framework and descriptive qualitative design, data for analysis was collected from secondary sources. Thematic analysis was employed to explore how nature supports traditional Siswati practices, beliefs, and knowledge systems. The results revealed that traditional food, medicine, attire and religious practices are embedded in natural features. The study also found that challenges such as land deforestation, wildfires and overharvesting contribute to the scarcity of plants and trees as well as animal products used for traditional medicines and food. Strategies that could alleviate some of these challenges such as environmental acts, establishing botanical gardens and involving conservatory stakeholders were proposed. Therefore, this study contributes to a deeper understanding on the relationship between biocultural diversity and Siswati IKS, offering insights into their maintenance.

INTRODUCTION AND BACKGROUND

In an era of rapid globalisation, cultural diversity faces increasing pressure as indigenous knowledge systems (IKS) become vulnerable to erosion (Blench, 2001; Magocha, et al., 2019). This has led to people adopting modernisation and technology while neglecting IKS practices. For many African communities, indigenous knowledge is a foundational element, encompassing generations of wisdom, beliefs, and practices developed through a close relationship with nature (Chikaire, 2012). IKS have played a

DOI: 10.4018/979-8-3373-4124-8.ch011

critical role in enhancing the resilience of African communities in the face of health crises, serving as a vital resource across different historical periods, including during and after the end of colonial rule (Mzimela & Moyo, 2024).

These knowledge systems are deeply embedded in cultural practices, stories, languages, oral traditions, songs, dances, ceremonies, and daily life activities, shaping both individual identities and the shared heritage of communities (Bruchac, 2014). Key characteristics of IKS include affordability, cultural relevance, experiential creativity, inclusiveness, adaptability, support for local decision-making, and oral transmission across generations (Mokhutso, 2021). Within this framework, nature stands as both a repository and a vehicle for preserving indigenous knowledge. Through interactions with the natural world, communities such as those speaking Siswati are able to protect and pass down their cultural knowledge to future generations.

Siswati, one of the official languages of Eswatini and South Africa, holds an extensive store of indigenous knowledge woven into its language, rituals, and environment-based cultural practices (Muswazi, 2001; Manyatsi, 2011; Matfunjwa, 2022). For Siswati-speaking communities, nature is more than a mere setting; it is a vibrant force, rich with spiritual, medicinal, and practical importance that underpins human existence (Sherman, 2020). Social, economic and cultural life of a society is influenced by its immediate environment such as flora, vegetation, animals, rivers, waterfalls, their climate, the sea, the moon and the sun. It is through nature that inhabitants find meaning and healing through norms, cultural and religious practices. These natural elements influence belief systems, seasonal practices, livelihoods, and communal rituals. Through engagement with nature, individuals and communities derive meaning, healing, and identity, which are expressed and preserved through Siswati norms, cultural and religious practices.

STATEMENT OF THE PROBLEM

The components of Siswati IKS, including traditional food, medicine, attire, cultural beliefs and practices, are deeply reliant on nature for their survival. Elements of natural environment, such as plants, trees, animals, waterfalls and rivers, serve as essential resources for sustaining these practices. In particular, plants, trees, and animals provide raw materials necessary for producing traditional medicine and facilitating healing practices. Despite the importance of IKS in Eswatini, there is limited documentation of indigenous knowledge, especially concerning the use of plants and trees for medicinal purposes (Dlamini, 2001). The dominance of oral tradition has played a major role in the under-documentation of IKS because it has been transmitted orally for several decades. Indigenous knowledge was generated through lived experiences such as rituals, songs, storytelling, and socio-economic practices, and it was preserved through word of mouth (Manyatsi, 2011). The heavy reliance on verbal transmission led to IKS not being written down (Malapane, Chanza & Musakwa, 2024). This factor was also accompanied by illiteracy as some people who practised IKS could not write, making it less likely to be written but preserved through lived experiences. Despite this, Africans have used traditional plants and wild animals as medicine for many years, long before modern medicine arrived. This interdependence between nature and Siswati culture has endured for centuries. However, irresponsible actions by members of communities, such as human settlement, deforestation, and overhunting, directly threaten the relationship between emaSwati and their ecosystems. Similarly, Vilakati (2001) observes that Eswatini's biodiversity faced degradation due to factors like commercial agriculture and deforestation. Stears et al. (2016) note that the demand for bushmeat and animal skins through overhunting, endangered the survival of certain species in Southern Africa. Unfortunately, these activities are detrimental to the preservation of Siswati indigenous knowledge. This chapter aims to highlight the importance of plants, trees, animals and waterfalls as aspects of nature and describe how they sustain Siswati indigenous knowledge, specifically focusing on traditional food and medicine, as well as cultural practices in Eswatini.

RESEARCH OBJECTIVES

The main objectives of this chapter are:

- to identify natural resources that hold significant cultural and historical value in Siswati indigenous knowledge.
- to determine how the integration of natural elements such as plants, trees, animals and waterfalls contribute to the preservation and transmission of Siswati indigenous knowledge.
- to explore challenges faced by these natural elements that affect Siswati IKS as well as to propose approaches to overcome them.

This chapter contributes significantly to the indigenous knowledge systems of emaSwati. It highlights how Siswati IKS such as traditional food, medicine, attire, cultural beliefs and practices are all maintained by nature. The importance of conserving nature as an integral part of Siswati indigenous knowledge is emphasised. These cultural expressions are not merely symbolic but are actively maintained through the responsible use and preservation of natural resources. The chapter emphasises the urgent need to conserve nature, not only for ecological purposes but also as an essential component of the survival and continuity of Siswati indigenous knowledge. The study also demonstrates that nature and indigenous knowledge are intertwined. This relationship reflects a holistic worldview in which environmental stewardship is both a cultural obligation and a means of safeguarding traditional knowledge for future generations.

THEORETICAL FRAMEWORK

The theoretical foundation of this study is based on the Biocultural Diversity Framework (BDF), developed and popularised by anthropologist and linguist Luisa Maffi in the early 2000s. This framework integrates biological, cultural, and linguistic diversity, emphasising their interconnectedness within ecological systems and cultural practices (Maffi & Dilts, 2014). Central to the BDF is the assertion that preserving nature within African cultures is essential not only for biodiversity conservation but also for maintaining cultural identities and practices (Elands et al., 2019). By preserving nature, communities sustain IKS intricately linked with local ecosystems (Nemogá et al., 2022). This theory's underpinnings enable the researchers to uncover complex relationships between nature, cultural practices, and environmental sustainability in Eswatini.

This approach not only promotes conservation efforts but also respects and preserves the rich cultural heritage associated with nature in Swati communities. This theory was adopted in this study because it provides a comprehensive lens for understanding the vital relationship between human beings, plants, trees, animals and waterfalls, which are central to the preservation of IKS in Eswatini. It also relevant as it underscores the interconnectedness of natural environments and cultural practices, offering a way to explore how traditional ecological knowledge is passed down through generations. By adopting the BDF, the chapter aims to highlight the importance of plants, trees, animals and waterfalls as aspects of nature and describe how they sustain Siswati indigenous knowledge.

LITERATURE REVIEW

Dlamini (2007) discussed the importance of IKS in Eswatini and its benefits to emaSwati. The author demonstrated that IKS is not confined to traditional and medicinal practices but extends to socio-economic activities like agriculture. The study assessed the strengths and limitations of traditional agriculture, the transmission of indigenous knowledge, and its relevance to sustainable development. It underscored a shift from traditional to modern agricultural practices driven by perceptions of traditional methods as primitive and insufficient for feeding large populations. This transition contributed to environmental degradation and the erosion of indigenous knowledge critical for sustainable development. Dlamini (2007) identified sustainable practices such as mixed cropping, seed selection, and post-harvest handling techniques that are environmentally friendly and enhance food security. Indigenous agricultural methods were found to align with local ecosystems, incorporating strategies for soil fertility maintenance and wildlife conservation. Social values and religious beliefs were integral to the transmission and application of indigenous knowledge within traditional agricultural systems.

Despite challenges posed by climate change and the necessity for irrigation in certain regions, traditional agriculture demonstrated sustainability in ensuring food security and environmental conservation. Ultimately, Dlamini (2007) concluded that IKS offer valuable insights for sustainable agriculture and development. Documenting and understanding these practices present opportunities to integrate them into modern agricultural systems, thereby enhancing sustainability and food security.

Sone (2014) explored the profound relationship between Siswati oral literature and the ecological wisdom of the emaSwati. The study argued that Siswati oral literature, encompassing proverbs, songs, stories, riddles, and family praise names, serves as a repository for expressing and preserving deep-rooted values and observations of the natural environment. This literature not only reflects the emaSwati's keen observations and respect for nature but also plays a pivotal role in environmental conservation. The study highlighted several Siswati proverbs that use nature as a metaphor to convey essential life lessons and values. For instance, the proverb *Inhlanti ishelwe ngemanti* (The fish's water has dried up) underscored the importance of water conservation and resilience in adversity. Regarding songs,

Sone (2014) noted that Siswati songs abound with environmental references, exemplified by *Ekhaya angiphumi ndzawo* (my home is nowhere), where a woman laments her plight and highlights the interconnectedness between humans and nature. This song emphasises the vulnerability of both humans and nature when not safeguarded. In stories, such as *Ingcavulele* (a huge water animal), cautionary tales against exploiting rare animals resonate with contemporary conservation challenges like rhino poaching, emphasising the imperative to protect wildlife. Riddles, like *Ngikuphica ngengubo yami lemhlophe yembese tintsaba* (I cover the mountains with my white blanket), showcase emaSwati's acute observation and reverence for the natural environment, fostering logical thinking and observation skills. Sone (2014) also noted that family praise names use natural imagery to describe individuals, reinforcing the connection between people and their environment. For example, the Bhembe clan is praised as *loluhlata njenge ncoshana* (green like a cypress), highlighting both their beauty and intelligence. Sone (2014) concluded that Siswati oral literature serves as a potent tool for environmental conservation, encapsulating the ecological wisdom of the emaSwati. The study revealed the importance of preserving and integrating traditional environmental knowledge into conservation efforts to ensure the sustainability of Eswatini's environment and rich cultural heritage.

Taringa (2014) examined the role and significance of animals in the religious and cultural practices of the Shona society in Zimbabwe. The study challenged anthropocentric perspectives by demonstrating that animals hold integral roles in Shona spirituality, health, and identity. Various animals were discussed and their symbolic meanings and roles in omens, medicine, and rituals were described. For example, baboons and monkeys guide travellers, lions and leopards are seen as royal and ancestral mediums, and elephants, wild pigs, pangolins, and snakes have significant interpretations related to safety, health, and communication with ancestors. Birds predict joy, luck, rain, or witchcraft, and insects are key to forecasting weather and seasons, impacting agricultural practices. Taringa (2014) argued for the inclusion of this indigenous knowledge in environmental education to preserve the symbiotic relationship between humans and nature. The study suggested that the Shona's biocentric attitude, valuing animals and birds for their intrinsic worth and as messengers, could enhance environmental conservation efforts. It concluded by advocating for a biocentric approach to nature and the integration of indigenous knowledge into modern education to address environmental degradation and climate change.

Mahlalela (2014) focused on the economic valuation and natural resource rent as tools for wetland conservation, with a case study conducted on the Lawuba wetland in Eswatini. The study aimed to assess the local community's knowledge of the wetland's benefits and threats, their attitudes towards conservation, and to estimate the economic value and resource rent of the wetland's fibre provisioning services. Using a random sample of respondents, data was collected on the benefits, threats, attitudes, and the annual economic value of harvested fibre. The study revealed that households in Lawuba possessed a high level of knowledge about the benefits and threats to the wetland and held positive attitudes towards its conservation.

Mahlalela (2014) also reported that the economic value of fibre provisioning services was estimated to range between US $20,310 and US $32,673 per annum, with a per capita value of US $70 per year. Resource rent, calculated via the net price method, increased across the value chain and was positive, indicating that the wetland provides economic benefits. Despite the positive resource rent and community attitudes towards conservation, Mahlalela (2014) noted the absence of institutions capable of capturing and reinvesting rent for sustainable wetland conservation. Therefore, it was recommended that establishing suitable resource management institutions to ensure the sustainable use and conservation of the wetland was necessary. It also proposed that households benefiting from the resources should reinvest a portion of the resource rent back into the wetland for conservation purposes.

Sone (2016) investigated the role of Siswati oral literature in promoting environmental sustainability, revealing the intrinsic connection between nature, culture, and indigenous knowledge. The study advocated for incorporating traditional knowledge systems into modern conservation practices, underscoring the significance of indigenous environmental expertise. Traditional Siswati society, as illustrated in the research, practised environmental consciousness through sustainable methods in hunting, fishing, and land cultivation, ensuring resource renewal and continuity of human life. Siswati oral literature, notably the collection *Tinganekwane Tesiswati: Siswati Folktales*, serves as a vital medium for conveying environmental knowledge and fostering eco-friendly attitudes. Stories like *The Origin of Hills and Woodlands* and *Ingcavulele* teach moral conduct and environmental ethics, emphasising preservation and the protection of rare species. Sone (2016) highlighted how integrating indigenous knowledge within

Siswati oral literature can enrich eco-literature and eco-criticism, particularly in the African context. The study concluded that Siswati oral literature is a crucial resource for understanding and advocating environmental sustainability, calling for the integration of traditional knowledge into contemporary conservation efforts to enhance biodiversity and sustainable development.

Pato (2018) explored the role of Siswati literature in preserving and promoting Siswati cultural heritage amidst contemporary challenges such as globalisation, religious influence, and educational shifts that threaten traditional practices and values. The study examined various literary forms such as prose, drama, poetry, and oral narratives to analyse their cultural significance, including themes of traditional ceremonies, mourning customs, and societal roles within Siswati communities. Highlighting literature as a medium for both educating Siswati and non-Siswati audiences about Siswati traditions, the study underscored its pivotal role in cultural preservation. Pato (2018) advocated for translating Siswati cultural experiences into English to broaden accessibility, especially among immigrants and non-Siswati speakers. Additionally, the study recommended integrating Siswati language and culture into school curricula to ensure their continuity. Emphasising the importance of cultural awareness, Pato (2018) called for further investigation into the intersection of tourism and cultural preservation, proposing collaborative efforts to promote and revive Siswati indigenous practices through literary and educational initiatives.

Mandillah and Ekosse (2018) explored the significance of African totems in environmental conservation, emphasising their role as cultural heritage symbols that promote sustainability. They highlighted the critical role of indigenous knowledge in addressing health, agriculture, education, and environmental issues on local, developmental, and global levels. The authors examined the socio-cultural belief systems of selected tribes in Kenya and South Africa, showing how totems and taboos have historically preserved biodiversity. Mandillah and Ekosse (2018) categorised totems into animals, plants, and insects, providing examples such as the peacock (inner wisdom), vulture (dignity and perseverance), and owl (wisdom and magic). The study acknowledged the challenges indigenous communities face in maintaining their traditions within a Western-dominated world and stressed the importance of integrating indigenous knowledge into modern conservation efforts. Mandillah and Ekosse (2018) concluded that indigenous African belief systems, particularly totems, have effectively reduced wildlife and biodiversity loss. They recommended supporting educational and research centres controlled by indigenous communities, integrating modern laws with traditional customs, and teaching totemic beliefs to children for the preservation of these traditions.

Phaka et al. (2023) studied the intersection of frog and reptile conservation within South Africa's cultural practices, highlighting the role of ethnoherpetology in assessing traditional influences on herptile populations. The study identified protective cultural practices, such as totemism and the association of certain species with rain or deities, alongside risky behaviours like the use of herptiles in traditional medicine and negative cultural perceptions leading to their killing. The authors advocated for integrating cultural insights into modern conservation frameworks, aligning with South African legislation and global biodiversity policies such as the Kunming-Montreal framework. Phaka et. al. (2023) proposed leveraging protective cultural practices to enhance conservation efforts and addressing cultural-driven threats through collaborative conservation initiatives. The study categorised cultural practices into gastronomy, medicine, rituals, and symbolism, discussing their implications for herptile conservation. It concluded that while cultural practices offer conservation opportunities, challenges remain, including potential increases in herptile use for traditional medicine and negative attitudes towards these species. The authors recommended collaborative efforts between conservationists and custodians of traditional culture to advance integrated conservation approaches supported by South African environmental laws.

METHODOLOGY

This chapter adopts a descriptive qualitative approach wherein data were collected from secondary sources using purposive sampling. Etikan et al. (2016) posited that in purposive sampling, researchers make a critical judgement on what must be known and select data that best answers their research questions. To answer the research questions of this study, books, articles, theses, and internet resources on culture and traditional practices in Eswatini were used to contextualise how nature supports Siswati IKS. The researchers began by conducting a systematic literature search on existing aspects of nature from the secondary sources that pertain to the use of these aspects by emaSwati to sustain their socio-cultural way of life. The importance of a systematic literature search is to conduct a methodical inquiry and transparent search from publications on a specific subject that provide evidence about an existing phenomenon (Cooper, 2018). In addition to secondary sources, the researchers drew on their intuitive knowledge as native Siswati speakers to enhance data collection and analysis, gaining deeper insights into how aspects of nature such as plants, trees, animals and waterfalls contribute to the sustainment of Siswati IKS. Anderson (2004) noted that intuitive knowledge entails researchers exploring an interesting phenomenon that relates to their lives to understand it better. By following these processes, the aspects of nature that support Siswati IKS were then identified and analysed to determine how they enhance the preservation of IKS.

The researchers chose Siswati IKS elements, namely traditional medicines, traditional food, cultural beliefs, and practices, which they perceived to rely predominantly on nature. This selection was done after a triangulation to validate the data from the various sources. According to Cohen et al. (2007:114), triangulation refers to adopting more than one method to gather data for research purposes on aspects of people's behaviour. This method was conducted by comparing the aspects of nature found in the different sources and those known by the researchers as practitioners of IKS to establish the validity of the data. Noble and Heale (2019) supported that the credibility and validity of research results are increased by triangulation. Relevant examples of these aspects of nature, such as images, were mainly provided by the researchers, while other materials were sourced from credible online sources to support arguments made. The results were then presented thematically and discussed with these specific illustrations.

FINDINGS AND DISCUSSION

Nature is the main pillar that sustains Siswati IKS including traditional medicines, traditional food, cultural beliefs and practices. Siswati IKS afford emaSwati considerable local therapeutic autonomy, which subsequently makes them social capital. They support Siswati cultural values and practices in the natural environment of emaSwati. The following sections focus on how aspects of nature such plants, trees, animals and waterfalls sustain Siswati IKS elements namely traditional medicines, socio-cultural practices and traditional food.

Traditional Medicine

Traditional medicine is the foundation of Siswati IKS and the lives of emaSwati depend on it (Twala, 2005). The medicine is derived from a variety of plant and tree species as well as animals distributed in different ecosystems in Eswatini. Although traditional medicine is typically prepared by mixing herbs with animal parts, most of this medicine is taken from plants (Green & Makhubu, 1984; Francia & Stobart, 2014). Traditional medicines are widely used as remedies for both chronic and acute conditions by the emaSwati in their environmental context. The knowledge and use of medicinal plants have been practised in Southern Africa to treat various ailments way before the introduction of Western medicines (Dlamini, 2001). This knowledge about useful medicinal plants, trees, roots, and animal parts is preserved and transmitted from generation to generation through the practices of Traditional Health Practitioners (*tinyanga; sangoma, lugedla, umfembi, umprofethi* and *gobela*) and elderly people in Siswati societies (Twala, 2005; Nann, 2021). Kasenene (1993) also affirms that Siswati communities still have women and men with vast knowledge on the use of traditional medicines.

From the biodiversity of Eswatini, people have been able to stock plants, tree roots and barks to address health and spiritual issues. For example, the two types of *inhlaba* (aloe vera), presented in Figures 1 and 2 are mainly used as an antiviral and to treat wounds as well as sores (Dlamini, 2001). The *inhlaba* is also used to wean infants from breast-feeding women. The bark of umkhanyakude (acacia xanthophloea tree) that is shown in Figure 3 is used as remedy to alleviate fever (Long, 2005).

Figure 1. Inhlaba lencane 'Aloe vera'

Figure 2. Inhlaba lenkhulu 'big aloe vera'

Figure 3. Umkhanyakude 'Acacia xanthophloea'

Nature has also proven to be a reliable provider of health solutions during pandemics in which people use their indigenous knowledge to sustain themselves. This is typically facilitated by the flora which is conserved in nature. During the Covid-19 pandemic, in which modern medicine seemed to be failing, people resorted to their IKS to explore what nature could provide them. For example, specific wild plants among the existing flora in Eswatini were carefully selected to treat Covid-19 symptoms like fever. These plants included *umhlonyane* (artemmisia afra) which is presented in Figure 4. Amusan et al. (2007) note that *umhlonyane* is used to treat flu symptoms and fever whereas the roots of *umntulu* (vangueria infausta) shown in Figure 5, are used to make a decoction for treating chest pain.

Figure 4. Umhlonyane 'Artemmisia afra'

Figure 5. Umntulu 'Vangueria infausta'

The use of indigenous knowledge to prepare a wide range of traditional medicines shows that the flora in their environment supports the emaSwati's knowledge about useful traditional medicines. This shows that the knowledge would not have been preserved and transmitted over time from the colonial era to the postcolonial and modernisation periods if the plants and trees were extinct. Mics (2024) support that plants and trees serve as symbols and repositories of cultural and spiritual knowledge in many indigenous communities, with their properties, uses, and healing methods being passed down from elders to younger generations. Therefore, if these plants and trees were extinct, the cultural practices associated with them could have been lost, causing a gap in the knowledge that has been transmitted over time.

The availability of wildlife in Eswatini has always facilitated the existence of traditional medicine and IKS on the use of specific wild animals to treat sickness. According to Adeola (1992: 125), "wild animals and their by-products (hooves, skins, bones, feathers, and tusks) are widely used for the preparation of curative and preventive medicines". Lev (2003: 107) also observes that "since ancient times animals and products derived from different organs of their bodies have constituted part of the inventory of medicinal substances used in various cultures; such uses still exist in ethnic folk medicine". Like other nations, emaSwati also use wild animals and their products to treat basic and complicated ailments. In Eswatini, Zootherapy is part of their indigenous knowledge. Friant et al. (2022: 2) define Zootherapy as "the use of animals for traditional medicine and related cultural purposes (e.g., healing, witchcraft, rituals, charms, ceremonies, and festivals), involves medicinal and symbolic use of animals and animal by-products (e.g., excreta, fur, feather, bones, glands, etc.)". Therefore, various wild animals and their products such as snakes and crocodiles among others are used to develop traditional medicines for ailments in Eswatini. *Imamba* snake that is presented in Figure 6 is used to treat spiritual attacks such as *emafufunyane, tive, sitfutfwane* and *kuphambanelo ngumcodvo* (mental illnesses) while the fats of *ingwenya* (nile crocodile) in Figure 7 are used to neutralise *sidliso* as well as its brain and liver to poison an individual (Twala, 2005). *Sidliso* refers to 'black poison' which is the evil deeds of witches (Ashforth, 2002: 121).

Figure 6. Imamba (South African National Biodiversity Institute)

Figure 7. Ingwenya 'nile crocodile' (Wikimedia commons)

Cultural Beliefs and Practices

EmaSwati distinguish themselves by their Siswati traditional religion, cultural beliefs and practices. The cultural beliefs and practices are often mixed with Christianity in Eswatini (Kumalo, 2013). Kasenene (1993) asserted that emaSwati's life is immersed in their religion and cultural beliefs as they interact in the societies, observing sacred entities such as rivers, mountains, plants, and animals that are prominent in their religious settings. This reveals that the natural environment is important in supporting Siswati traditions, religious and cultural activities, thus preserving the emaSwati's indigenous knowledge. In Eswatini, certain plants, trees and animal products are used for specific cultural practices such as the *Incwala* 'first fruits ceremony'.

According to Masango (2008: vii), *Incwala* refers to "the first fruit sacred ceremony and national prayer involving the king". It is a sacred ceremony that takes place over six days, during which every person is welcome to participate. This annual ritual revitalises Siswati society, politics, and the natural environment (Ebewo, 2011). For instance, the stem and branches of *lusekwane* (dichrostachys cinerea) in Figure 8 are used during the *Incwala* ceremony for covering *inhlambelo*, a sacred area within the cattle kraal where main rites are conducted during this ceremony (Loffler & Loffler, 2005; Vilakati & Msibi, 2006). Ndwandwe (2019) explains that all substances used in the *inhlambelo*, a sacred enclosure at the cow byre, are tangible expressions of emaSwati's sacred covenant with both nature and the creator, underscoring their profound commitment to ecological principles. This is one of the reasons everyone is expected to attend the *Incwala*, clad in appropriate attire (Ndlovu, 2019). Moreover, traditional attires made of animal skins and products such as *umdada* (leopard skin skirt), *sigeja* (cow tails) and *inyoni* (ostrich feathers) are worn during the *Incwala* ceremony as presented in Figure 9.

Figure 8. Lusekwane 'dichrostachys cinerea' at Incwala ceremony (Independent News Eswatini)

Figure 9. Traditional attire during Incwala ceremony (Independent News Eswatini)

Umhlanga, also known as the reed dance ceremony and illustrated in Figure 10, serves as a compelling example of how the natural environment plays a critical role in sustaining IKS in Eswatini. This ceremony is an eight-day event in which unmarried females with no children cut reeds and offer them to the Queen Mother of Eswatini after which they dance (Eswatini National Trust Commission, 2024). This cultural event involves the collection and use of reeds from the natural landscape, reinforcing traditional practices and communal identity while demonstrating the interconnectedness between ecological resources and cultural heritage.

Figure 10. Umhlanga 'reed dance ceremony' (Eswatini National Trust Commission, 2024)

After being harvested during the reed dance ceremony, the *umhlanga* are subsequently used in the construction of *emaguma* (traditional enclosures), which serve as protective structures for *boguca-sithandaze* (sacred huts) located at royal residences, as depicted in Figure 11. This practice not only reflects the utilitarian value of natural resources within Swati culture but also highlights the continued relevance of indigenous architectural techniques and spiritual traditions rooted in the sustainable use of the environment.

Figure 11. Umhlanga used to construct emaguma (Flickr 2006)

Therefore, nature serves as a vital resource for the preservation and continuation of cultural practices and IKS in Eswatini. The depletion or eradication of key elements of biodiversity such as specific plants, trees, and animals poses a serious threat to the sustainability of Siswati IKS. This is because many traditional ceremonies, including the and *Umhlanga*, are deeply rooted in the availability and symbolic significance of these natural resources. Without access to these ecological components, the cultural rituals and knowledge transmission processes that define and sustain the Siswati identity would be severely compromised.

Nature's physical features such as waterfalls are used for religious practices such as spiritual cleansing and healing while sacred mountains are used as burial sites for kings, queen mothers and chiefs. These features of nature enabled the emaSwati for centuries to conduct their traditional beliefs and customs effectively as they depended on their existence. Figure 12 shows the *Mvutjini* waterfall in Eswatini which is used for spiritual cleansing and healing purposes. Spiritual healing is a purposeful and cautious action conducted by anyone who wants to enhance their well-being (Rafii, et al., 2020). The Mvutjini waterfall is believed to possess supernatural powers enabling emaSwati to cleanse bad luck and ward off evil spirits.

Figure 12. Mvutjini waterfall in Eswatini

In Eswatini, the Mdzimba and Mbilaneni mountains are sacred burial sites where immediate royal families such as kings are buried in caves found in these mountains (Kuper, 1963; Hall, 2009; Sone, 2014; Russell, 2022). These sacred mountains present fundamental means for the emaSwati to preserve and conduct burial practices for the royal family. Kasenene (1993: 42) postulated that "these mountains are of great religious significance and during crises, such as drought, epidemic or any other catastrophe, the king visits the mountains to intercede with the ancestors on behalf of the nation". Dube (2021) reports that after the political unrest in Eswatini in which protestors lost their lives, the Royal Family offered 15 cows to the late King Sobhuza 11 at Mbilaneni mountain, seeking protection and guidance during these unrests. These examples of physical natural features found in Eswatini affirm that nature has been at the centre of the IKS.

Moreover, some plants and trees are essential in conducting the burials of people based on Siswati traditional religion. Certain tree species are utilised to construct burial materials for royal people such as *masweti* (manonthotaxis caffra) (Singwane & Beckedahl, 2019). A tree branch of *umlahlabantfu* (ziziphus mucronata) presented in Figure 13 is believed to command the spirit of a deceased person back to their home after they die to prevent their spirit from tormenting living people. Alternatively, *umhlanga* can be used to bring back home the spirit of the dead (Vilakati & Msibi, 2006). *Umlahlabantfu* is also used to make *ematinta* (wooden necklace) for breastfeeding women associated with the royal family (Singwane & Beckedahl, 2019).

Figure 13. Umlahlabantfu 'Ziziphus mucronata'

Additionally, some tree species are also used to build kraals and traditional houses such as huts. The use of these trees supports the implementation of socio-cultural lifestyles as the emaSwati used their indigenous knowledge and architecture in constructing their traditional home structures. Singwane and Beckedahl (2009) aver that there are several trees suitable for kraal construction such as *imbondvo lemnyama* (combretum molle), *lusekwane* (dichrostachys cinerea), *umphahla* (brachylaena rotundata) and *umhlume* (adina spp.) which are also used for building huts. The use of various tree species reveals that they are significant in the socio-cultural life of emaSwati, and they apply their indigenous knowledge to utilise these trees in supporting their specific needs.

Furthermore, some indigenous plants are used to heal physical illnesses and address religious problems. Africans discovered these plants' usefulness centuries ago and they continue to constitute fundamental indigenous knowledge. *Imphepho* (helichrysum petiolare and helichrysum odoratissimum) presented in Figure 14 is one of the plant species found in all parts of Eswatini that has multiple functions in Siswati IKS.

Figure 14. Imphepho 'Helichrysum petiolare and Helichrysum odoratissimum'

EmaSwati use *imphepho* as a catalyst to call up ancestors, to purify their sacred places and to chase away evil spirits. This is done through burning the dry *imphepho* so that the smoke can fill the intended area. Bhuda and Marumo (2020) and Belani et al. (2022) also observe that *imphepho* is traditionally used as a ritual incense during the ancestral offerings and evocation as well as to ward off harmful spirits. Medicinally, the *imphepho* is used "to relieve gastrointestinal and respiratory conditions, fever and urinary tract infections, implying that they produce antibacterial compounds" (Serabele et al., 2021: 117). The remedy is prepared by boiling the leaves of this plant and drinking it as a tea (Bhuda & Marumo, 2020). The multifaceted usage of this plant demonstrates the various ways in which indigenous plant species sustain elements of IKS.

Traditional Food

The natural flora found in Eswatini is the main source of traditional food. Before the widespread use of modern food brought by commercial farming and modernisation, the emaSwati used indigenous knowledge to explore which plants were edible, how they could cook or store these plants and for how long to ensure that they did not starve. Even today, emaSwati still gather indigenous plants from their natural environment and use their indigenous knowledge to prepare food as well as preserve it. Knorr and Augustin (2023: 9520) note that "food preparation and processing today has been built on historical knowledge acquired through the ages, including how to cook, transform, preserve and store food safely". In the context of Eswatini, many wild plants are used for food, for example, *emahala* (aloe maculata), *ligusha* (corchorus) 'natural and cooked' and *chuchuza* (blackjack) that are presented in Figures 15, 16, 17 and 18 respectively are some of the edible wild plants that are cooked as food.

Figure 15. Emahala 'Aloe maculata'

Figure 16. Ligusha 'corchorus'

Figure 17. Cooked ligusha 'corchorus'

Figure 18. Chuchuza 'blackjack'

These leafy traditional foods are typically associated with good health, promote blood circulation, and are thought to alleviate specific health issues (Fakudze, 2023). They are often preserved in summer so that they can be used during the dry season. The common indigenous preservation method used by the emaSwati is called *kufusa,* a process of drying the leaves of the plants, grinding and storing them in a sealable container. Figure 19 presents *inshubaba* 'momordica foetida' which is being prepared for preservation.

Figure 19. Inshubaba 'momordica foetida' preservation

Therefore, without the natural flora in Eswatini, the availability of traditional food can be under threat. Many indigenous edible plants, which are deeply embedded in local diets and cultural practices, rely on healthy ecosystems for their growth and regeneration. As such, the conservation of the natural environment is not only essential for ecological balance but also for preserving food security, cultural heritage, and the intergenerational transmission of IKS related to food cultivation, preparation, and consumption.

Challenges Faced by Nature and Indigenous Knowledge Systems

In Eswatini, wild plants, trees and animals that contribute to the sustainment of Siswati IKS face multifaceted challenges. Land deforestation due to commercial farming and urbanisation have contributed immensely to the eradication of useful wild plants, indigenous trees and wild animals. Manyatsi (1999) and Manyatsi and Hlophe (2010) concur that many indigenous trees were removed from Eswatini communal land when people cleared the land for agriculture and cut down the trees for building material as well as fetching firewood. The demand for firewood has resulted in many people cutting a variety of tree species for commercial purposes and this has led to a reduction of trees such as Faurea galpinii which is classified as a 'vulnerable flora' protected by the Flora Protection Act of 2001 (Manyatsi & Hlophe, 2010). The extensive axing of trees directly harms nature and the practising of IKS such as the construction of traditional homesteads and structures like kraals. Manyatsi and Hlophe (2010: 231) caution that "the firewood collection can have a detrimental impact on [Eswatini] Swaziland's wildlife and the tree species that are popular sources of firewood may decline if the industry is not controlled".

Land deforestation has also contributed to the scarcity of plants and tree products used as traditional medicines and food. The establishment of big commercial farms in the Manzini and Lubombo regions in places such as Nyonyane and Ndzevane resulted in the removal of many indigenous trees that bear wild fruits such as *emaganu* (marula), *tinenyi* (wild berries) and *emahlala* (kind of strychnos fruits). This also eradicated edible plants such as *inshubaba*. Therefore, the impact on the physical environment has a direct influence on the IKS of the emaSwati.

Wildfires which are often experienced during winter seasons have negative effects on some plant species. They also lead to displacement of wild animals which constitute traditional medicine products. The destruction contributes to a massive loss of the biodiversity (Food and Agriculture Organisation, 2007). The wildfires experienced in Eswatini play a major role in the reduction of indigenous wild plant use for medicinal purposes such as *imphepho*. This does not merely affect the natural flora but has undesirable consequences on the practice of traditional healing which is a threat to IKS (Bhuda & Mdhluli, 2024).

Overharvesting indigenous plants, trees and animals to make traditional medicines is a direct threat to the indigenous medicinal knowledge. The harvesting of traditional medicines in Eswatini for export to other neighbouring countries such as South Africa has contributed to the overutilisation of medical plants and animal products. Dold and Cocks (2002: 589) state that "the trade in traditional medicines forms part of a multi-million rand 'hidden economy' in southern Africa stimulated by high population growth, rapid urbanisation, unemployment, and the high cultural value of traditional medicines". Singwane and Shabangu (2012: 19) claim that approximately 219 tonnes of medicinal plants and tree products are used per year from Eswatini. Hence, *imphepho* is the most exported traditional medicine from Eswatini to South Africa. This shows the overt threat to our traditional medicines. If the use and harvesting of these useful medicinal plants, trees and animals are not regulated well, they will become extinct. This will have detrimental effects because indigenous knowledge about effective traditional medicines will not be preserved for future generations (Bhuda & Khazamula, 2022).

Approaches to Preserve Siswati Indigenous Knowledge Systems

Several approaches can be implemented to promote the role of nature in preserving Siswati IKS. Traditional leaders can play a vital role in the preservation of indigenous elements such as the medicinal plants, trees and animals. These natural elements are overharvested in communal land under the supervision of chiefs. If the traditional leaders do not enforce conservation laws made by the government, future generations will not have IKS useful references. Currently, acts such as the National Trust Commission Act of 1972, drafted by the Eswatini National Trust Commission strongly protect and promote the conservation of indigenous fauna and flora mainly on natural parks and reserves (Zwane et al., 2011). This makes the fauna and flora found in communal land more vulnerable to overexploitation. Dlamini and Lupupa (1995) also advocate for strict conservatory strategies in which the Eswatini traditional systems can restrict the axing of wild trees and extensive harvesting of plants such as those found in wetlands.

Another approach to enhance nature and the preservation of IKS is to promote societies to establish botanical gardens. These gardens could be placed in appropriate communities where indigenous plants and trees used for medicine, food, religious and cultural purposes would be planted. Currently, private commercial companies and research centres breed indigenous plants and trees. However, they do not benefit underprivileged communities which cannot afford to buy this bred flora. If communities own the botanical gardens, this will benefit nature and the practice of IKS. Eyong (2007) notes that indigenous species are sold illegally to European countries. It therefore comes at no surprise that African countries have fewer botanical gardens compared to the west, yet they are the source of a large variety of these species. This observation underscores the imperative need to establish botanical gardens in local communities to preserve nature and IKS practices.

The preservation of nature and IKS is not only a responsibility for an individual, but it needs multiple stakeholders and entities to collect and disseminate information on the importance of nature in sustaining valuable Siswati IKS. Muswazi (2001) emphasises that indigenous library and information services, information and communication technology and IKS experts need to work together in gathering, storing and sharing indigenous knowledge. This approach will facilitate a holistic effort in preserving nature and IKS.

FUTURE RESEARCH DIRECTIONS

For future work, the authors will investigate the role played by other natural aspects, such as mountains, caves, rivers, and dams, in preserving Siswati IKS. This research will provide an in-depth insight into the relationship between IKS and all natural aspects in Eswatini and enhance the contextualisation of these phenomena in the modern era.

CONCLUSION AND RECOMMENDATIONS

This study has discussed how nature sustains Siswati IKS. It was discovered that the natural environment in which emaSwati live enables them to partake in and preserve their religious, cultural and medicinal practices. The role played by nature's physical features such as mountains and waterfalls to sustain socio-cultural and religious indigenous knowledge systems has been described. The study also showed that wild plants, trees and animals contribute immensely to the preservation of Siswati indigenous knowledge. Challenges facing nature and the sustainment of indigenous knowledge systems with regard to land deforestation, wildfires and overharvesting, contribute to the scarcity of plant, tree as well as animal products used for traditional medicines and food in Eswatini. Several approaches that can be implemented to promote the preservation of IKS through nature were also discussed. In light of the findings, this study puts forward the following recommendations:

- Local communities should develop and manage botanical gardens that focus on cultivating indigenous plants and trees used in traditional medicine, food, and rituals. This will ensure sustainable access to critical natural resources and serve as a living archive for future generations.
- Traditional leaders (chiefs and elders) should take an active role in enforcing environmental protection laws and regulating the use of indigenous flora and fauna in their communities. Their authority can help curb overharvesting and deforestation while promoting respect for cultural beliefs and practices tied to nature.
- A structured programme should be initiated to record, transcribe, and digitise oral traditions, medicinal practices, cultural beliefs and practice. This will preserve valuable indigenous knowledge

currently transmitted orally, protecting it from being lost due to generational gaps or external influences.

- Siswati IKS should be incorporated into school curricula at primary, secondary, and tertiary levels. Teaching learners about their cultural heritage will foster appreciation, cultural identity, and responsibility toward conserving both nature and indigenous practices.
- Government agencies, non-governmental organisations, and cultural institutions should organise awareness campaigns, workshops, and media programmes to educate the public about the value of preserving natural resources such as plants, trees, animals and waterfalls as a means to sustain Siswati IKS. This would help build a conservation-minded society that values nature and its indigenous roots.

REFERENCES

Adeola, M. O. (1992). Importance of wild animals and their parts in the culture, religious festivals, and traditional medicine of Nigeria. *Environmental Conservation, 19*(2), 125–134. DOI: 10.1017/S0376892900030605

Amusan, O. O., Sukati, N. A., Dlamini, P. S., & Sibandze, F. G. (2007). Some Swazi phytomedicines and their constituents. *African Journal of Biotechnology*, *6*(3), 267–272.

Anderson, R. (2004). Intuitive inquiry: An epistemology of the heart for scientific inquiry. *The Humanistic Psychologist, 32*(4), 307–341. DOI: 10.1080/08873267.2004.9961758

Ashforth, A. (2002). An epidemic of witchcraft? The implications of AIDS for the post-apartheid state. *African Studies*, *61*(1), 121–143. DOI: 10.1080/00020180220140109

Belani, A., Oladimeji, K. E., Hyera, F. L., Matubatuba, C., & Juzang, I. J. (2022). Can the use of Impepho leaves extend beyond traditional to modern medicine? An opinion. *Africa Journal of Nursing and Midwifery*, *24*(2), 1–4. DOI: 10.25159/2520-5293/11872

Bhuda, M., & Khazamula, E. (2022). Towards finding a cure for COVID-19: The role of African Traditional Medicine as an indicator of development. *African Journal of Development Studies*, *2022*(2, si1), 137–153. DOI: 10.31920/2634-3649/2022/SIv2a8

Bhuda, M. T. & Marumo, P. (2020). African traditional medicine and healing in South Africa: Challenges and prospects before and during COVID 19. *Ife Centre for Psychological Studies/Services, 18*, 16710-16723.

Bhuda, T. M., & Mdhluli, T. D. (2024). *African traditional medicine as a COVID-19 treatment and the South African government response*. Pharos Journal of Theology. DOI: 10.46222/pharosjot.105.42

Blench, R. M. (2001). *Globalisation and policies towards cultural diversity. Natural Resource Paper, 70*. Overseas Development Institute., http://www.oneworld.org/odi/nrp/70.html

Bruchac, M. M. (2014). Indigenous Knowledge and Traditional Knowledge. In Smith, C. (Ed.), *Encyclopedia of Global Archaeology* (pp. 3814–3824). Springer. DOI: 10.1007/978-1-4419-0465-2_10

Chikaire, J., Osuagwu, C. O., Ihenacho, R. A., Oguegbuchulam, M. N., Ejiogu-Okereke, N., & Obi, K. U. (2012). Indigenous knowledge system: The need for reform and the way forward. *Global Advanced Research Journal of Agricultural Science*, *1*(8), 201–209.

Cohen, L., Manion, L., & Morrison, K. (2007). *Research methods in education* (6th ed.). Routledge. DOI: 10.4324/9780203029053

Cooper, C., Booth, A., Varley-Campbell, J., Britten, N., & Garside, R. (2018). Defining the process to literature searching in systematic reviews: A literature review of guidance and supporting studies. *BMC Medical Research Methodology*, *18*(1), 1–14. DOI: 10.1186/s12874-018-0545-3 PMID: 30107788

Dlamini, G. M., & Lupupa, T. (1995). Swaziland: Country report to the FAO international technical conference on plant genetic resources. Food and Agricultural Organization.

Dlamini, P. M. (2001). Gender, biodiversity and local knowledge systems: Highlights of the seminar discussions. In: Dlamini, P.M. & Vilane, B.R.T. (Eds), *Proceedings of the National Seminar on Institutions Working on Gender, Biodiversity and Local Knowledge Systems in Swaziland*. April, pp. 10-20.

Dlamini, R. M. (2007). *Investigation of sustainable indigenous agricultural practices: A systems approach* [Unpublished master's dissertation]. Stellenbosch University, Stellenbosch, South Africa.

Dold, A. P., & Cocks, M. L. (2002). The trade in medicinal plants in the Eastern Cape Province, South Africa. *South African Journal of Science*, *98*(11), 589–597.

Dube, E. (2021, July 23). King ask Sobhuza 11 to save him from Swazis, present over 15 cows at his cave. *Swati Newsweek*. https://sdnewsweek.wordpress.com/2021/07/23/king-ask-sobhuza-ii-to-save-him-from-swazis-present-over-15-cows-at-his-cave/

Ebewo, P. (2011). Swazi Incwala: The performative and radical poetics in a ritual practice. *South African Theatre Journal*, *25*(2), 89–100. DOI: 10.1080/10137548.2011.636976

Elands, B. H., Vierikko, K., Andersson, E., Fischer, L. K., Goncalves, P., Haase, D., Kowarik, I., Luz, A. C., Niemelä, J., Santos-Reis, M., & Wiersum, K. F. (2019). Biocultural diversity: A novel concept to assess human-nature interrelations, nature conservation and stewardship in cities. *Urban Forestry & Urban Greening*, *40*, 29–34. DOI: 10.1016/j.ufug.2018.04.006

Eswatini National Trust Commission. (2024). *Umhlanga (The Reed dance)*.https://entc.org.sz/umhlanga-the-reed-dance/

Etikan, I., Musa, S. A., & Alkassim, R. S. (2016). Comparison of convenience sampling and purposive sampling. *American Journal of Theoretical and Applied Statistics*, *5*(1), 1–4. DOI: 10.11648/j.ajtas.20160501.11

Eyong, C. T. (2007). Indigenous knowledge and sustainable development in Africa: Case study on Central Africa. *Tribes and Tribals*, *1*(1), 121–139.

Fakudze, M. (2023, May 6). Eating ligusha, chuchuza could kill you – health expert. *Times of Swaziland*. http://www.times.co.sz/sports/140028-eating-ligusha%2C-chuchuza-could-kill-you-%E2%80%93-health-expert.html

Flickr. (2006). Swazi huts. (https://www.flickr.com/photos/50194168@N00/115207528)

Food and Agricultural Organization. (2007). FAO Forestry Paper: Vol. 151. *Fire Management-Global Assessment 2006: A thematic study prepared in the framework of the Global Forest Resources Assessment 2005*. Food and Agricultural Organization.

Francia, S., & Stobart, A. (Eds.). (2014). *Critical approaches to the history of Western herbal medicine: From classical antiquity to the early modern period*. Bloomsbury Academic. DOI: 10.5040/9781474210577

Friant, S., Bonwitt, J., Ayambem, W. A., Ifebueme, N. M., Alobi, A. O., Otukpa, O. M., Bennett, A. J., Shea, C., Rothman, J. M., Goldberg, T. L., & Jacka, J. K. (2022). Zootherapy as a potential pathway for zoonotic spillover: A mixed-methods study of the use of animal products in medicinal and cultural practices in Nigeria. *One Health Outlook*, *4*(5), 1–21. DOI: 10.1186/s42522-022-00060-3 PMID: 35216623

Green, E. C., & Makhubu, L. (1984). Traditional healers in Swaziland: Toward improved cooperation between the traditional and modern health sectors. *Social Science & Medicine, 18*(12), 1071–1079. DOI: 10.1016/0277-9536(84)90167-9 PMID: 6463675

Hall, J. (2009). *Sangoma: My odyssey into the spirit world of Africa*. Sterling Publishing Company, Inc.

Independent News Eswatini. (2020, December 20). King commissions Ingatja, Incwala holiday on Saturday, January 2. https://independentnews.co.sz/local-news/5584/king-commissions-ingatja-incwala-holiday-on-saturday-january-2/

Kasenene, P. (1993). *Swazi traditional religion and society*. Websters.

Knorr, D., & Augustin, M. A. (2023). Preserving the food preservation legacy. *Critical Reviews in Food Science and Nutrition, 63*(28), 9519–9538. DOI: 10.1080/10408398.2022.2065459 PMID: 35442825

Kumalo, S. (2013). *Religion and Politics in Swaziland*. University of Johannesburg Press. DOI: 10.18820/9781920382230

Kuper, H. (1963). *The Swazi: A South African Kingdom*. Holt, Rinehart and Winston.

Lev, E. (2003). Traditional healing with animals (zootherapy): Medieval to present-day Levantine practice. *Journal of Ethnopharmacology, 85*(1), 107–118. DOI: 10.1016/S0378-8741(02)00377-X PMID: 12576209

Loffler, L., & Loffler, P. (2005). *Swaziland tree atlas - including selected shrubs and climbers*. (Southern African Botanical Diversity Network Report No. 38). SABONET.

Long, C. (2005). *Eswatini flora - Siswati names and uses*. http://eswatinibiodiversity.com/flora/clusagelist.asp?uid=2&pg=21

Maffi, L., & Dilts, O. (2014). *Biocultural diversity toolkit: An introduction to biocultural diversity*. Terralingua.

Magocha, M., Soundy, P., Muchie, M., & Magocha, B. (2019). Reviewing the applications of indigenous knowledge systems (IKS) in innovative crop production. *Indilinga, 18*(2), 229–244.

Mahlalela, L. S. (2014). *Economic valuation and natural resource rent as tools for wetland conservation in Swaziland: The case of Lawuba wetland* [Unpublished Master's dissertation]. University of Pretoria, Pretoria, South Africa.

Malapane, O. L., Chanza, N., & Musakwa, W. (2024). Transmission of indigenous knowledge systems under changing landscapes within the vhavenda community, South Africa. *Environmental Science & Policy, 161*, 103861. DOI: 10.1016/j.envsci.2024.103861

Mandillah, K. L., & Ekosse, G. I. (2018). African totems: Cultural heritage for sustainable environmental conservation. *Conservation Science in Cultural Heritage, 18*, 201–218.

Manyatsi, A. M. (1999). Development of a computerised rangelands resource information system for Swaziland. *University of Swaziland Journal of Agriculture, 8*(1), 12–22. DOI: 10.4314/uniswa.v8i1.4612

Manyatsi, A. M. (2011). Application of Indigenous Knowledge Systems in Hydrological Disaster Management in Swaziland. *Current Research Journal of Social Sciences*, *3*(4), 353–357.

Manyatsi, A. M., & Hlophe, E. T. (2010). Contribution of sale of firewood towards rural livelihood in Swaziland, and its environmental sustainability. *Current Research Journal of Social Sciences*, *2*(4), 226–232.

Masango, L. P. (2008). Reading the Swazi reed dance (Umhlanga) as a literary traditional performance art [Unpublished Doctoral dissertation]. University of the Witwatersrand, Johannesburg, South Africa.

Matfunjwa, M. N. (2022). *A pragmatic exploration of Siswati naming practices* [Unpublished Doctoral dissertation]. University of South Africa, Pretoria, South Africa.

Mics, F. (2024). The sacral significance of forests and trees in the religions, philosophies, and spirituality. *Opuscula Theologica et Scientifica*, *2*(2), 23–65. DOI: 10.59531/ots.2024.2.2.23-65

Mokhutso, R. J. (2021). The impact of African indigenous knowledge system on healthcare system in South Africa: The Covid-19 perspective. *Gender & Behaviour*, *19*(1), 17713–17728.

Muswazi, P. (2001). Indigenous knowledge management in Swaziland: Perspectives. *Information Development*, *17*(4), 250–256. DOI: 10.1177/0266666014241079

Mzimela, J. H., & Moyo, I. (2024). On the Efficacy of Indigenous Knowledge Systems in Responding to the COVID-19 Pandemic: Unsettling Coloniality. *International Journal of Environmental Research and Public Health*, *21*(6), 1–12. DOI: 10.3390/ijerph21060731 PMID: 38928977

Nann, S. (2021). How beliefs in traditional healers' impact on the use of allopathic medicine: In the case of indigenous snakebite in Eswatini. *PLoS Neglected Tropical Diseases*, *15*(9), e0009731. DOI: 10.1371/journal.pntd.0009731 PMID: 34499648

Ndlovu, H. L. (2019). Joint worship ceremonies of Africanists and Christians in the kingdom of Swaziland. In C. Lewis & D. Cohn-Sherbok, (Eds.), *Interfaith worship and prayer: We must pray together* (pp 58-67). Jessica Kingsley Publishers.

Ndwandwe, J. (2019, December 15). Incwala third phase indigenous sacredness. *Eswatini Observer*. http://new.observer.org.sz/details.php?id=11125

Nemogá, G. R., Appasamy, A., & Romanow, C. A. (2022). Protecting indigenous and local knowledge through a biocultural diversity framework. *Journal of Environment & Development*, *31*(3), 223–252. DOI: 10.1177/10704965221104781

Noble, H., & Heale, R. (2019). Triangulation in research, with examples. *Evidence-Based Nursing*, *22*(3), 67–68. DOI: 10.1136/ebnurs-2019-103145 PMID: 31201209

Pato, J. S. J. (2018). *Role played by Swazi literature in preserving and promoting the culture of Swazi people* [Unpublished Master's dissertation]. Stellenbosch University, Stellenbosch, South African.

Phaka, F. M., Hugé, J., Vanhove, M. P., & du Preez, L. H. (2023). Frog and reptile conservation through the lens of South Africa's nature-based cultural practices. *African Journal of Herpetology*, *72*(2), 190–206. DOI: 10.1080/21564574.2023.2261021

Rafii, F., Eisavi, M., & Safarabadi, M. (2020). Explaining the process of spiritual healing of critically-ill patients: A grounded theory study. *Ethiopian Journal of Health Sciences*, *30*(4), 579–588. PMID: 33897218

Russell, T. (2022). Symbolic kraals: Subterranean food stores, hidden wealth and ethnographic errors. *Journal of Social Archaeology*, *22*(3), 317–337. DOI: 10.1177/14696053221117467

Serabele, K., Chen, W., Tankeu, S., Combrinck, S., Veale, C. G., van Vuuren, S., Chaudhary, S. K., & Viljoen, A. (2021). Comparative chemical profiling and antimicrobial activity of two interchangeably used 'Imphepho' species (Helichrysum odoratissimum and Helichrysum petiolare). *South African Journal of Botany*, *137*, 117–132. DOI: 10.1016/j.sajb.2020.09.023

Sherman, J. H. (2020). Reading the Book of Nature after Nature. *Religions*, *11*(4), 1–12. DOI: 10.3390/rel11040205

Singwane, S. S., & Beckedahl, H. R. (2019). The distribution and utilization of benefits derived from plantation-style community forests in Eswatini: A case of Ngcayini and Ezikhotheni chiefdoms. *Journal of Sustainable Development in Africa*, *21*(3), 188–203.

Singwane, S. S., & Shabangu, N. (2012). An examination of the utilization and management of natural woodlands in Swaziland - A case of Ka Bhudla Community. *Journal of Sustainable Development in Africa*, *14*(1), 34–45.

Sone, E. M. (2014). Nature in crisis: An ecological construction and conservation of the environment in Swazi oral literature. *South African Journal of African Languages*, *34*(2), 195–205. DOI: 10.1080/02572117.2014.997056

Sone, E. M. (2016). Swazi oral literature, Eco-culture and environmental apocalypse. *Southern African Journal for Folklore Studies*, *26*(1), 44–58. DOI: 10.25159/1016-8427/1264

South African National Biodiversty Institute. (n.d.) *Black Mamba.*https://www.sanbi.org/animal-of-the-week/black-mamba/)

Stears, K., Shrader, A., & Castley, G. (2016). A conservation assessment of Equus quagga. In M. F. Child, L. Roxburgh, E. Do Linh San, D. Raimondo & H.T. Davies-Mostert, (Eds.), *The Red List of mammals of South Africa, Swaziland and Lesotho*. (pp. 1-8). South African National Biodiversity Institute and Endangered Wildlife Trust.

Taringa, N. T. (2014). *The sacred duty of animals in African traditional religion and culture.*https://www.bayreuth-academy.uni-bayreuth.de/en

Twala, J. M. (2005). *Tangoma nemisebenti yato ekutfutfukiseni lulwimi lweSiswati.* [Unpublished master's dissertation]. University of Zululand, Richards Bay, South African.

Vilakati, J. D. (2001). Biodiversity management in Swaziland. In P. M. Dlamini, & B.T. Vilane, (Eds.), *Proceedings of the National Seminar on Institutions Working on Gender, Biodiversity and Local Knowledge Systems in Swaziland* (pp. xxxv-xlii). Food and Agricultural Organization.

Vilakati, T., & Msibi, T. (2006). *Giya sigiye*. Macmillan Boleswa. Manzini.

Wikimedia commons. (n.d.). Nilecrocodile https://commons.wikimedia.org/w/index.php?curid=66978

Zwane, P. E., Masarirambi, M. T., Seyoum, T., & Nkosi, B. S. (2011). Natural fibre plant resources of economic value found in wetlands of Swaziland: A review. *African Journal of Agricultural Research*, *6*(4), 773–779.

ADDITIONAL READING

Nwauche, E. (2025). *The protection of indigenous knowledge in Africa*. Springer Nature., DOI: 10.1007/978-3-031-71264-7

KEY TERMS AND DEFINITIONS

EmaSwati: refers to people who speak Siswati mainly as a mother tongue and are connected by their culture, social, cultural and economic activities.

Siswati Indigenous Knowledge Systems: refers to the extensive repository of traditional knowledge, cultural practices, values, and beliefs cultivated and practiced by emaSwati across generations, which are rooted in their interaction with nature, culture, and spirituality.

Preservation: refers to active maintenance and protection of something in its original state, particularly ecosystems, species, cultural heritage, and natural resources, to prevent damage, decay, destruction, extinction, exploitation, or degradation for the benefit of future generations.

Nature: refers to the physical world, encompassing all living and non-living elements that occur naturally, including organisms, ecosystems, and natural phenomena, and holds spiritual and cultural significance in indigenous contexts.

Chapter 12
The Concept of Healing in the Zulu Worldview Within the Context of Traditional Religion

Mthokozisi C. Myeza
https://orcid.org/0009-0002-6475-9967
University of South Africa, South Africa

ABSTRACT

This study is qualitative in nature and aims to analyze the Zulu perspectives on traditional healing system as part of the oral health practice, within the context of African traditional religion. This type of healing is very scientific within the context of African understanding, as it is very systematic, logically, and very relevant even today. It is also capable in healing different types of diseases including mystical illnesses which are African orientated. The healing practice entails the discipline of botany, zoology and anthropology. This kind of practice has been part of the African people for a very long time to restore life and to bring healthy balance between the human beings and the environment. It has been developed outside the sphere of influence of biomedicine of which it refers to the traditional scientific form of knowledge. Therefore, the healing concept in the Zulu perspective is orally done through consultation with the ancestral spirits which involve spirit mediums/traditional healers who are working interchangeable.

INTRODUCTION

This chapter will analyze the Zulu perspective on traditional healing system within the context of African traditional religion. The chapter will explicate the traditional healing of mystical illness which refers to some kind of pollution. The healing will be done through compound medicines which focus not on physiological aspect of illness but on symbolic world as well as spiritual world. This will be explicated within the African context, of which traditional religion has been used to refer to the relationship between human and God through the utilization of natural environment (Okeke, 1998).

The main aim of this type of healing is to bring healthy balance between human beings and the environment. This type of healing has been part of the African people and has been done orally through the consultation with the ancestral spirits. This includes both divine healers (*izangoma)* through diagnostic and herbal doctors *(izinyanga)* through therapeutic. Mashele et al. (2021) discussed that the main link

DOI: 10.4018/979-8-3373-4124-8.ch012

between traditional medicinal practitioners and African religion in healing context are medicinal plants from the natural environment. Therefore, the power of nature in medicine underlines the interconnectedness and interdependence between humanity and the rest of God creation (Metz & Molefe, 2021).

Cordeiro-Rodrigues and Metz (2021) stated that it should also be noted that for African people, religion is a necessity and not an option. The reason for that is because it provides them with an opportunity to relate closely to their natural environment and seek to establish harmony with it. Therefore, it should be borne in mind that the Zulu traditional healing system is part of the oral health practice, which is very scientific within the context of African understanding, as it is very systematic, logical, and relevant even today. It is also capable in healing different types of diseases including mystical illness that are African orientated (Jwara, 1998).

LITERATURE REVIEW

Physical environment, the perspective and healing among the Zulu nation

Henriques (2013) pointed out that the perspective of the Zulu people displays a close connection between the living and the dead which forms environment in which they live. Therefore, Zulu perspective when it comes to illness, reflects the greater knowledge of herbal for healing purposes. As a result, the Zulus were able to harness their local flora for health benefit of their community. It should be noted that the knowledge and appreciation of local flora does not constitute pantheism, where nature is worshipped.

One should bear in mind that this simply constitutes a healthy respect for flora and fauna which is religious in nature for its potential health benefits to the local community (Donda, 1997). Therefore, in the sense, the Zulus served as earth keepers who respect the plant life around them as well as animals as there are parts that were respected for having special qualities to enhance human healing and health. This is the Zulu perspective which affirmed that Zulus are very much concerned about their environment as the signs in nature and in dreams which give them messages regarding decision to be taken in their daily life (Hutchings, 2013).

Okon (2006), describes a people's perspective as the complexity of their beliefs and attitude concerning the origin, nature, and structure of the universe. According to Okon, the perspective of African Traditional Religion is divided into two categories, namely the material and religious perspectives. It is also linked with many African societies who believe that the universe is made up of the visible (material) and the invisible (immaterial). According to Emefie Ikenga-Metuh in Okon (2006), the material perspective sees the cosmos as a product of chance which has no meaning and has no end. On the contrary, the religious worldview sees the cosmos as having a great meaning and purpose for which every happening is interpreted based on what the people perceive as its meaning or purpose. The latter worldview is subscribed by the Zulu traditional healing system.

According to Onyeidu (1999) the African worldview is purely religious, it therefore means that religion and religious practices which are orally practiced, are their waking thought. It is for this reason that the entire culture of African is permeated with religious practices. Ezeanya (1980) agrees that Africa life is so much wrapped in religion, and religion is life. This is why most Africans perceive the world around them as a religious world, understand and interpret the world mainly through the prism of religion. They also see the universe as created by the Supreme Being where every life event is traceable to a particular spirit which must be appeased through ceremonies because it binds and forms religious values that govern the family and society. Therefore, traditional religion guides all activities of the African life be it social, health or economic in nature.

METHODOLOGY

Research paradigm

Paradigms are practices and philosophies that regulate investigation within a discipline by providing structures and procedures through which exploration is accomplished. Therefore, the paradigm for this research paper will be based on constructivism and interpretivism as these theories are related and sometimes used interchangeably to define a paradigm for research. These theories share the goal of understanding the multifaceted world of lived experience from the point of view of those who live it. Therefore, the basic rules that regulate the constructivist paradigm is that knowledge is socially assembled by those people who are active in the research process (Schurink, 2001).

The research approach

To investigate the concept of healing from the Zulu perspective which exist within the terrain of oral health and African Traditional Religion, this study is qualitative in nature. The qualitative approach puts emphasis on collecting data in naturally occurring settings and can be classified as interactive or non-interactive (Leedy & Ormrod, 2005). The qualitative research approach is linked to constructivism and interpretivism because it seeks to understand a phenomenon by focusing on it in totality.

The qualitative approach was chosen because the study was essentially based on literary sources which include books, unpublished materials, and some interviews that were done with diviners and indigenous healers, as well as journals and other archival material that could be found. This was done to gain a deeper understanding of the concept of healing mystical illness in the Zulu perspective so that one could weigh objectively issues raised in the literature and the views of diviners and indigenous healers. A qualitative approach was relevant for this study because of its intention to explore the knowledge of indigenous healing system as part of oral health that exists within the indigenous knowledge systems (McMillan & Schumacher, 2010).

This method has been helpful because interviews were conducted. This method has been chosen to use the method of sampling called nonprobability. This method is associated with minor sampling which includes simple random, quota and purposive sampling (Kumar, 2014). This type of random sampling has been used because there were no people who were systematically selected (Bernard, 2002). The researcher used the people that he identified himself and there were no ground rules that required them to follow.

That is why they were advised that they were allowed to withdraw if they felt tired. The use of purposive sampling is where the subjects of the research were chosen because they have identified with the kind of knowledge they have (Schmidt & Brown, 2017). This kind of sampling has been chosen because the researcher realized that not all the herbal healers will take part in the study. It is only those who agree to answer the questions. The focus group discussions were held to allow the participants to voice out what they know about the methods involved in the traditional healing process within the Zulu perspective.

THE ROLE OF AFRICAN TRADITIONAL RELIGION IN TRADITIONAL HEALING

The African traditional religion

According to Mbiti (1961) Africans are notoriously religious as each people has its own religious system with a set of beliefs and practices. African Traditional Religions are not primarily for individual, but for the community of a person is part. This means that to be human is being part and belong to the whole community by orally participating in the beliefs, ceremonies, rituals, and festivals of that community. Religion permeates into all the departments of life so fully that it is not easy or possible to isolate it. It remains a vital force in Africa where it greatly influences and continues to influence the lives of the African people. This is the reason the Zulu traditional medicine is often connected with religion. The practice of medicine is considered by Zulus to be a gift from the creator and is dispensed through the agency of the ancestors. All Zulu medicinal healers seem to acknowledge God as the real healer.

Divine and ancestral sanctions are considered necessary before and during the preparation of the medicine and its application (Mashego et al., 2021). This is because life is understood in a holistic or integral way of which it has been the central theme and preoccupation of African people from the beginning of creation. God in His love gave every people knowledge about life, that is, how to promote, protect, transmit and heal it. This means Africa's attention to life has always been holistic (Jwara, 1998). This is why belief and action in African traditional society cannot be separated.

What people do is motivated by what they believe, and what they believe springs from what they do and experience (Nkonge, 2014). This means that they belong to a single whole. This is the reason Zulus believed that when something happens to one part, it affects the other parts. Like other Africans, the Zulus hold the view that the human being is a composite of material and immaterial substances (Jwara, 1998). In other words, every human being comprises body and soul. Therefore, the integrity of the body must be in good relationships in society, because there can be no individual development outside the group that supports the individual (Phiri, 2009; Marques et al., 2021).

In addition, Jwara (1998) pointed out that in African cosmology, human life is lived in three interrelated environments, i.e. embryonic, terrestrial and after-life. Therefore, the reality of life is that sickness in all its aspects can be experienced in any of these stages. When that happens, Africans turn to traditional medicine as they have done time immemorial. Traditional medicine restores the holistic wellbeing of everyone. There are two reasons for that, traditional medicine is more accessible to people because it is less expensive than western medicine, it can be easily found everywhere, and it is organic in nature with no side effects (Hill & Mathews, 2021; Adeleye et al., 2021).

That is why Zulus attribute religiously an equal importance to the horizontal relationship among the living and the vertical relationship between human beings and the living dead (ancestors) and God. The African traditional religion remains the vital force in Africa where it greatly influenced and continue to influence the lives of the African people and the Zulus in particular (Nkonge, 2014). It has been largely responsible for shaping the character and culture of the Africans throughout the centuries. It is for this reason according to the Zulu perspective generally that they believed that nothing could go well with human beings in the society unless good relations are maintained between them and their fellow human beings and between them and the powers (force of gravity, moon, sun, rivers, sea and nature) that control the universe which refers to the physical environment (Jwara, 1998). This means broken relationships may result in sickness or even death.

To restore good relations, the Zulus believe that they need God and the ancestor's spirits to heal them. On the spiritual part they need God, but mainly ancestor's spirits and on the physically part they need the power of the traditional healer to heal them, who depends on the healing powers of Creator (God) through the products of nature which includes flora and fauna with the healing ingredients (Van der Zeijst et al., 2021). Bryant (1966) pointed out that it can be affirmed that the AmaZulu are very much connected to their environment as a sign in nature and in dreams give them messages regarding decisions to be taken in their daily life. This is why according to Dube (2011), in Africa there are rich categories of holistic healers who deal with the entire spectrum of life issues: biological, social, mental, psychosomatic, spiritual and so on. People's belief in immorality has therapeutic power, an appropriate attitude toward death is necessary for spiritual heal (Shange & Ross, 2022).

To put matters in proper context, Africans, according to Kalu (2000), operate with a three-dimensional perception of space: the sky, the earth (land and water) and ancestral or spirit world, which is located under the earth. The testimony to this is displayed during the funerary rites and the pouring of liberations on the ground. It is for this reason, according to Magesa (1997), that no way is anything in African Traditional Religion understood apart from the context of God, the ancestors, and spirits. It is important to note that the power of nature in medicine underlines the interconnectedness and interdependence between humanity and the rest of creation. The roots, leaves or parts of animals boiled in water or pulverized in fire to form the basic ingredients of medicine. This is the reason that all elements of plants, animals, water, and fire represent the major forces of nature. When these elements are boiled, burned, and applied to the human body, they generate the power to protect and heal. Magesa (1997) further pointed out that in applying them to the human body, this is how the link between nature and humanity is religiously established. This is where the spirit mediums become vital and necessary.

THE ZULU TRADITIONAL MEDICAL PRACTIONERS

The Zulu traditional medical practitioners are divided into herbal practioner male (*inyanga*) and diviner female (*isangoma*). The nature of becoming one of the two types of medical practioner is not the same. The details of the difference between the two practitioners will be dealt with by looking at these practitioners separately. This is very important when it comes to medical practicing to understand the medical approach from the Zulu perspective (Jwara, 1998).

The diviner (*isangoma*)

The stem of the noun *isangoma* (ngoma) refers to dance and *ingoma* as a noun refers to the song sung by diviners (*izangoma*). There is also a sense related to the administering of medicine in the verb stem *-goma*. Therefore, there is thus an indication that the terms denoting to medicine and dance are diachronically related, as the diviner (*isangoma*) through dancing and singing leading to divining and directing of the patient to an herbalist (*inyanga*) to *goma* the patient (Donda, 1997). The practice of *isangoma* is divining (*ukubhula*) which is associated with the hit of a bag of sorghum (*ukubhula amabele*) using the stick. The main objective is to obtain corn from husks. When the diviner is divining the devotees before her/him used to come with sticks to the diviner. The expression that comes to play is to consult the diviner (*ukuthatha izinduku*). When the diviner is divining, the devotees hit with the sticks on the ground. This means the truth is to be separated from the lies. It should be noted that a person does not choose to become a diviner (*isang* she is chosen by her ancestors, who bestow upon her clairvoyant powers. The ancestral spirits do not take possession of the body, but they are close to the diviner as they sit on her shoulders and whisper into her ears. It should be noted again that diviners (*izangoma*) are differently classified into four main divination techniques namely:

- The head diviner (*isangoma sekhanda*).
- The bone thrower (*esichitha amathambo*).
- The whistling diviner (*umlozi*).
- The diviner who uses a mirror (*esisebenzisa isibuko*).

The focus of all these diviners is to do divination to assist the sick people by identifying the cause of illness and providing health solutions.

Herbal healer (*inyanga*)

This type of traditional health practitioner is optional because people have power to choose this line of medicinal job, although some medicinal healers are chosen by their ancestors to possess the gift of healing people. To African society (Ngubane, 1977) the medicinal healer or herbalist are the greatest gift, and the most useful source of help. They are called by different names as medicine man, herbalist, traditional doctors or *waganga* in Kiswahili. According to Donda (1997) medicinal healer, *inyanga* in IsiZulu, is associated with the moon that brings light at night. The power of darkness exercised by sorcerers is associated with death. Therefore, according to the Zulu worldview *inyanga* is the sign of life with light whereas the sorcerer is the symbol of death with dark. The traditional healer *(inyanga)* and the sorcerer are thus antipoles which both fight over the individual in different ways. It is from this background that *(inyanga)* medicinal healer lights the way through healing so that the patient does not get into the dark of the grave whereas the sorcerer contrasts with the medicinal healer by enticing the patient to the grave.

According to Mbiti (1969) these are the specialists who have suffered most from European and American writers and speakers who so often and wrongly called them witch doctors. The medicinal healer is the friend of the community and accessible to everybody. What should be noted is that there is no fixed rule governing the calling of someone to become a medicine man. It may come when he is young and unmarried. In other cases, the medicinal healer passes on the profession to his son or other younger relatives (Jwara, 1998). The profession mostly is dominated by men although there are few cases of women in this profession. Their personal qualities are expected to be trustworthy, upright morally, friendly, willing, and ready to serve. They should be able to discern people's needs and not to be exorbitant in their charges. This attitude was centered around the inner value and dignity of the human personality, which is nothing else, but humanness instilled in medicinal healers (Savory,1988).

Participants provided insights into the fact that the training in medicine is always oral, and it can be either formal or informal and it is long. Part of the training requires the candidates to acquire knowledge in matters pertaining to medicinal value, quality and mixing of different herbs, leaves, roots, fruits, barks and grass. They should also know and understand various objects like minerals, dead insects, bones, feathers, powders, and the flesh of different wild animals. Be able to understand and know the causes, cures and the prevention of diseases and other forms of suffering and how to combat these sufferings, handling of bad spirits and living dead as well as various secrets some of which may not be divulged to outsiders. The payment for training also varies and is made from time to time since the training in nature is long because different phases regarding different knowledge of medicinal plants need to be acquired by the candidate.

Medicinal healers are therapeutics in practice because they are more concerned with treating the physical effects of the affliction and the symbolic affliction although they also depend on the spirit of their ancestors to find the appropriate cure (Magesa, 1997). Some healers are both diagnosticians and therapeutics. Therefore, the spirit may convey information usually in dreams during the night to clearly demonstrate which herbs to look for, and where and how to apply them (Magesa, 1997). As is the case with the diviners and for the benefit of the readers, there are different classifications of medicinal healers. Three types of medicinal healers *(izinyanga)* can be distinguished as follows:

- Villagers *(amaxhwele)*
- Herbal vendors *(ochitha)*
- Traditional healers *(izingedla)*

The main role of the above herbal healers is to provide healing through therapeutic using different medicines.

THE ROLE OF SPIRIT MEDIUMS IN THE FIELD OF HEALTH

According to Chuvunduku (2001) Spirit mediums generally are defined as traditional healers who practice oral healing through natural (organic) medicinal plants, this is why they appear to play an important role in the field of health. Spirit mediums or traditional healers are mainly divided into two, that is diviner *(isangoma)* and herbal healer *(inyanga)*. It should be noted that not everyone can become a spirit medium because it is a very complex field of health within the context African worldview. Firstly, a person must be possessed by the spirit of his or her ancestors and is called *ithwasa* (Myeza, 2017).

Even if a person displays signs of possession, there are several oral tests and training prescribed by accredited Spirit medium which may take place over number of years determined by the ancestral spirits and ending with more searching tests conducted by a senior Spirit medium who is the leader of the branch of traditional healers of which is called *(iziko)* branch. These trainings are based on the divination which is done orally by the senior diviner *(ugobela oyisangoma)* and effective oral administration of medicine by senior traditional healer or herbalist *(ugobela oyinyanga)* which include sustainable harvesting and proper mixing. The main function of the diviner is diagnosis.

Participant 2 stated that the diviner is concerned with the spiritual causes of the affliction whereas the traditional healer or herbal doctor is very therapeutic and more concerned with the treatment of physical effects of the affliction. The traditional healer also depends on his ancestor's spirit to find the appropriate cure, which may be conveyed usually in dreams. This happens in terms of which herbs to look for, where and how to apply them (Myeza, 2017). This tends to agree with Western Anthropologist through ethnographic research that discovered that people with different cultural traditions often had ideas of their own about the ways of dealing with health challenges (Olupona, 2000). This is evident to the Zulu traditional healing system.

THE ZULU CONCEPT OF HEALTH AND SICKNESS/DISEASES

It is widely believed among AmaZulu as people that a person cannot just get sick out of the blue without any external force unless it has to do with the ordinary breakdown of the body or has to do with misfortune or old age. According to Ngubane (1977) the Zulu perspective of health and sickness is to be understood in the light of the following categories: sickness *(isifo)* and sicknesses *(izifo)*. Ngubane (1977) further pointed out that according to the Zulu people, sickness *(isifo)* is a generic term that refers to all forms of illness and to various forms of misfortune and being vulnerable to disease.

Disease is a serious derangement of health, a disordered state of an organism or an organ or any form of illness with special symptoms and names. According to Donda (1997) each of the term *isifo* (illness) and *ukufa* (death) can both be used for illness and death e.g.:

- *Ukufa kokuwa.*
- *Isifo sokuwa*

In the above examples 1 and 2 *ukufa* and *isifo* according to Donda (1997) have been used to refer to disease or illness.

In the following example below 3 and 4 *ukufa* and *isifo* have been used to refer to death.

- *Ukufa kumenele.*
- *Isifo simenele*

When the terms *ukufa* and *isifo* are isolated, they can be used as synonyms, but they can change their semantic value in certain contexts such as in the example (1) and (2) as well as in (3) and (4) above.

Donda (1997) further pointed out that a disease based on Zulu cosmology and its causality based on African culture is called *ukufa kwabantu* means man made. This includes *inter alia disease* like *umeqo* illness caught after walking over the path that has been manipulated with path trap medicine,disease of the kidneys caught after illicit sexual intercourse with a treated partner.

According to Jwara (1998) the Zulu worldview reveals that the disease and death among young people is caused by misfortunes and witchcraft *(amashwa nabathakathi)* while death for aged people is believed to be going home *(bayagoduka)*. It is for this reason that Berglund (1990) asserted that Zulu nation is a community of survivors and shades. It is because the ancestors play a major role in the clan, and they make their will know by signs which the survivors need to interpret. This is the world of the AmaZulu as people where there is a close connection between the living and the dead which forms the environment in which they live.

Like Jwara (1998) and Nwoye (2017) affirms that the basic tenets of Afrocentric worldview which include an important belief, namely: that there are two realms of existing within the Afrocentric paradigm which refers to the physical and spiritual realms. These two realms are interlinked, interdependent and interacting. The belief that AmaZulu as people live and die physically and then live again spiritually as ancestors and the belief in a higher power is imperative in the Zulu perspective (Mwikamba, 1991; Thabede, 2008). This belief reveals an understanding that people are guided by their elders even after those elders have physically died. Therefore, the spiritual world can communicate with the physical world through signs which could be virtues or punishments or even protection against something.

These signs communicate whether the ancestors or higher powers are at peace, acknowledging something or are enraged with an act by people in the physical realm (Olumbe, 2008). Therefore, both Nwoye (2017) and Olumbe (2008) agree that this understanding of living in a world with two realms means that the good and the bad are not only in physical form but also spiritual form. The spiritual realm, like the physical realm, holds not only good spirits that guide the living but also bad spirits that have the potential to harm the living. Therefore, AmaZulu strongly belief that the world consists of good and evil (Jwara, 1998). This created understanding to AmaZulu that people are vulnerable to dangers and evils which are physical and spiritual. Therefore, it emerges important to AmaZulu to understand the origin and causes of illness. This is what will be dealt with in the next section.

THE CAUSES OF ILLNESSES

Health is one of the primary concerns of Zulu people. Their traditional religious belief system identifies numerous and varied causes of illness and diseases. Many times, certain illnesses and diseases have a distinct reason for appearing. Therefore, it is the duty of the divine healer (*isangoma*) to diagnose the causes of illness through diagnosis which will assist the herbal doctor to heal the illness through therapeutic activity.

(a) Traditional causes of illnesses

According to the Zulu perspective three distinction are made between categories of causes of illness and diseases (Donda, 1997):

(i) Natural/biological
(ii) Mystical processes
(iii) Spirits

The focus of this chapter will be on the healing of mystical causes of illness that may lead to death. This illness is commonly called mystical illness, (*ukufa okuletwe umthakathi)* which needs to be mystical and traditionally healed using compound medicinal plants.

(b) Mystical (social) causes of illnesses

People are sometimes considered sick not because of any organic disorder, but because they undergo certain life crises and are regarded as stricken with pollution. Pollution according to participant 3 is a mystical concept that cannot be demonstrated according to the laws of cause and effect. According to Zulu perspective, pollution is a mystical force which shows no somatic symptoms. It cannot be demonstrated according to the laws of cause and effect. It is essentially a happening associated with birth, on the one hand and death, on the other (Peters, 2021). Both birth and death are mysteries associated with the other world from which people are believed to come and to which people return in spirit form. According to Jwara (1998) this world and the other world are viewed as two separate entities, the beginning of life, whose source is believed to be in the other world, happens in this world, the cessation of life in this world is believed to mean continuity of life in the other world. Notionally there is an area of overlap between the two worlds. Such an area is marginal and dangerous. This is the area which the sorcery exploits to cause the physical disorder condition which lead to the mystical disease which is created in an area of overlap between the two worlds, the area, which is marginal and dangerous, eventual permits to death (Jwara, 1998).

To be mystically ill or sick is culturally acquired and is caused by social agents such as sorcerers and spirits of various kinds. The power of nature (sun, moon, and night) is used during the creation of these illnesses (Dube, 2011). These types of diseases are characterized by the element of bewitching, this means the involvement of sorcerer *(umthakathi/abathakathi* as explained by participant 1. The few types of illnesses involved are the following: *ukukhahleliswa* (stroke), *ukuthwetshulwa* (converted), *Umeqo* (walk trap), *isibhobo* (muscle sharp pain) and *amashwa* (misfortunes). These types of illnesses are not somatic in nature and are healed not by focusing on the cause of illness. The mystical illness and disease do not occur by chance but have a definite cause which is diagnosed by diviner healer *(isangoma).*

TRADITIONAL DIAGNOSIS

Since the Zulu experience illness and diseases as a threat to their lives, so diagnosis, usually is administered by divine healer (*isangoma*) as a sprit medium. As it is essential to identify the cause and prescribe appropriate medication for the medicinal healer. According to the Zulu culture it seems that a divine healer (*isangoma*) is not supposed to dispense any medicine to patients. Since the task of a divine healer (*isangoma*) is to diagnose the disease and then refer the patient to a good medicinal healer (*inyanga*) suggested by his (*isangoma*) ancestors (Jwara, 1998). At the same time the medicinal healer (*inyanga*) is also not supposed to do divination because it is the task of the divine healer (*isangoma*).

Bhuda and Marumo's (2020) study shows the major difference between the two is that the divine healer (*isangoma*) does not necessarily need to learn all kinds of herbs because his main vocation is to be in touch with the ancestors to be able to do the divination. The same applies to the medicinal healer (*inyanga*) to the contrary who is supposed to acquire more knowledge about the medical herbs. It is for this reason that when the patient comes to the medicinal healer (*inyanga*) for medical assistance, the patient explains to him all about his health complaints, whereas the *isangoma* is the one who will explain to the patient the whole condition of his health. In scientific terms the medicinal healer works like a medical doctor whereas divine healer *(isangoma)* works like a psychiatrist (Myeza, 2013).

It should be noted that diagnosis is the monopoly of divine healer *(isangoma*) alone. Therefore, these practitioners employ a wide range of methods at their disposal. Participant 2 stated that the average person therefore knows who to approach. In cases of grave illness, especially if it is believed that spirits are responsible for the problem. According to Bhuda and Marumo (2020), the divine healer (*isangoma*) is expected to discover why the spirits afflict the victim with illness and how a remedy may be prescribed. So, to do this, different divine healers employ several different methods of diagnosis. Most diviners (*izangoma*) in the Zulu nation are usually women, it is the women who brought (*ubungoma*) vocation to the in laws. It appeared that the ancestors do not take possession of the body, but they are close to the diviner, they sit on her shoulders and whisper into her ears.

According to Donda (1997) possession is one significant means of diagnosis used by divine healers. The divine healer performs a ritual which involves snuff-taking. This is followed by putting on symbolic clothes, the color of symbolic clothes depends on the type of ancestral spirits that entered the diviner healer during the initiation period. Other items include animal skin and bird feathers. The hide of a lion or leopard serves as a mat. All such items are expressions of power. During this process he wields in his right hand a spear. During singing and dancing, ancestral or alien spirits, which are causing illness in a patient, are induced to possess him and explain the reason for the affliction (Metz & Molefe, 2021).

The whistling diviner (*umlozi*) as per participants of this study is in possession of divining by listening to the ancestral spirit. The ancestral spirits in this case communicate directly with the clients by whistling out words which are meaningful to the listener. The whistling sound comes directly from the rafters of the thatched roof upper part of the rondavel or *iqhugwana* hut opposite the doorway. During this process, participant 1 stated that the diviner sits almost at the center of the hut facing its upper part. Some stand up inside the hut facing its upper part of the hut. A diviner possessed by such ancestors is said to have great ancestors *(amakhosi amakhulu)*.

Whereas the bone thrower (*isangoma esichitha amathambo*) will throw the bones and have different meanings, which he interprets according to the way they fall. It is also believed that through the throwing of bones, the diviner healer establishes contact with her ancestral spirits (Bhuda et al., 2022). The configuration of the bones tells the diviner the unknown secrets of the diseases and problems of the patient. This is an indication that the diagnosis of illness is centered on the diviner healer who applies different methods to determine the cause of the problem. The Zulu system of diagnosis has a logic and respect for the abilities of diagnosticians who identify the source of affliction and refer the patient to the medicinal healer who would be able to prescribe medication to effect healing (Dube, 2011).

Participant 1 stated that dreams also constitute another means of diagnosis which is done by the medicinal healer who is possessed specifically by *isindawe* spirits. Such medicinal healer is very exceptional for male because it is not for everyone to be possessed in this manner. Dube (2011) stated that in doing the diagnosis the medicinal healer prepares himself with certain rituals before going to sleep to read and interpret dreams properly. He bathes himself with medicine and puts on his ritual attire with leopard colors and pours some snuff on the pillow, accompanying his actions with incantations. The ritual is intended to establish contact with ancestral spirits. Such diagnosis does not take place at any time during the day but invariably at dawn, which are special times for effective communication with spiritual world (Dube, 2011). AmaZulu strongly believe that dreams forewarn or indicate that illness and disease and even misfortune and death will occur. The striking factor in the reading of dreams is that they are believed to come from the ancestral spirits who when contacted reveal the cause of illness. The Zulu system of diagnosis has a logic and respect for the abilities of diagnosticians who identify the source of affliction and prescribe medication to effect healing by herbal healer (Jwara, 1998).

TRADITIONAL HEALING

It is widely believed among the Zulu people that a person cannot just get sick out of the blue without any external force unless it has to do with the ordinary break down of the body or has to do with old age (Jwara, 1998). It is from this background that Zulus can heal illness and diseases through a variety of methods that include rituals and treatments administered by the medicinal healer (*inyanga*). The healing of illness and disease in Zulu system is centered on the medicinal healer (*inyanga*) who specializes with the natural medicinal plants and parts of animals. Participant 2 explained that the treatment of illness and disease among the Zulus is handled according to various categories. The minor diseases that cause illness such as headaches, flu and stomach problems are usually healed through herbal medicine administered by the elders of the family as mentioned above.

A participant of this study expressed that serious and mystical illnesses, which are characterized by physical disorders, are caused by mystical diseases that are usually cured by various methods that are on the same level as the illness. Compound medicine is usually employed for the treatment of serious mystical illness. The physical disorder caused by sorcery through witchcraft are regarded as stricken with pollution. Pollution is a mystical concept that cannot be demonstrated according to the laws of cause and effect. The correction of the condition which involves mystical means is healed by using symbolic or compound medicine (Jwara, 1998).

THE ZULU PERSPECTIVE ON THE DIVISION OF MEDICINE

The healing in the Zulu perspective is orally done through divination and therapeutic activities in as far as the administration of medicines is concerned. It has been developed outside the sphere of biomedicine, of which it refers to the traditional scientific form of knowledge. According to the Zulu perspective the natural medicines (*imithi/amakhambi*) is divided into herbal medicines (non-ritualized medicines) and compound medicines which is comprised by more than one medicine (ritualized medicines). Herbal medicines only require common knowledge and are very potent. They are used particularly for minor ailments as household remedies (Bryant, 1970 cited in Ngubane, 1977). Compound medicines are very sophisticated and administered in a ritual context and have a symbolic meaning. Therefore, they are prepared only by professional practitioners and highly trained traditional healers. According to Ngubane (1977) compound medicines are mainly used to correct the cause of illness rather than to cure organic symptoms. This means that what compound medicines healed is not simply the physiological aspect of illness, but the symbolic world as well as the spiritual world.

Ritualised medicine (Compound medicines)

Such medicines according to Ngubane (1977:09) can be divided into three colors categories as follows:

- Black medicines (*imithi emnyama*) represent darkness which is associated with death.
- Red medicines (*imithi ebomvu*) represent transformation that is the process of progressing away from darkness of night to the goodness of daylight.
- White medicines (*imithi emhlophe*) emphasize light and life.

Participants of this study stated that these types of medicines are administered in a ritual context and have a symbolic meaning which is culturally interpreted. That is why even treatment with such medicines is very symbolic according to the Zulu perspective. According to the Zulu perspective, symbolism plays a central role in the healing using compound medicines responding to mystical illnesses (Henriques, 2013). The color symbolism of compound medicines is also related to the cosmic order of the day and night. On other hand it is also conceptualized as similarly related to the bodily functions of eating and defecating.

THE HEALING OF MYSTICAL ILLNESS

To be mystical ill means that a symbolic condition for pernicious illness and eventual death is created by sorcerer using black and red medicines to push the victim to an area of overlap between the two worlds (Jwara, 1998). According to participant 2, the treatment of mystical disease employing compound medicines follows the same pattern of bodily functions of eating and defecating to establish a balance between a person and the environment which has been caused by mystical illnesses, which are culturally acquired to reintroduce good health to the patient rather than to cure the organic symptoms.

The pollution cause mystical illness are mysteriously associated with other world from which the people are believed to come that is birth and to which people return in spirit form that is death. Jwara (1998) further indicated that there is an area of overlap between the two worlds. Such an area is marginal and dangerous. This is the area which the sorcery exploits with black and red medicine to causing the mystical disease which eventual lead to death (Bhuda & Khazamula, 2022).

Such illness is interpreted on cultural bases whereby one is likened to this symbolic condition by sorcerer where the movement of the targeted person will be reinforced away from the daylight, the world of birth into the dimness of the sunset, the world of spirit marked by death and on into the dark of the night means death by sorcery. During this process it is believed that the sorcerer (*umthakathi*) uses the energies of the earth to strengthen the black medicines which represent the darkness of the night to exploit the overlap area between the two worlds mentioned above (Peters, 2021). Therefore, if the sick person final enters this area through the forces of mystical medicines used by the sorcerer, together with the black sheep to turn around or manipulate water into black *(ukuphehla amanzi amnyama)* that will cause the ancestors to abandon the targeted person/patients, eventual death is imminent.

To correct the effects of such illness, caused by sorcery and drive a patient out of mystical illness means that the treatment that will be performed is known as *(ukuguqula amanzi amnyama)* changing of black water. Participant 3 explained that the treatment process requires the use of symbolic medicines to correct the cause of illness, and which is characterized by a notion of taking out the dark by discarding it out of the system. This type of treatment is very deep and critical, that is why it requires the medicine-doctor (*inyanga*), through therapeutic method of healing, to use black, red, and white medicines to re-introduce good health to the patients. The treatment is performed to reverse the black water, which means to change the effects of black watery medicine to allow the ancestors of the patient to look after and carry on protecting them.

According to Ngubane (1977) during the process the boiled black medicines is accompanied by the black sheep, which is killed by suffocation represent natural human death, where a person naturally dies without being wounded. It is buried like a corpse. The whole treatment is performed during the darkness of the night of which darkness represent death. During the treatment, the patient abstains from social activities because the type of medicine being used represents danger and the patient himself is considered as having an intensified form of pollution being excreted, of which it is dangerous to other people as this form of pollution is transferable. As stated by Myeza (2017), the extraction of what is bad from the body by vomiting over the dead sheep and discarding it secretly away from home for three days, is related to excretion. After treatment for three days with black medicines in the wilderness, the patient does not look back. This means leaving behind what is bad and proceeding to the future which is good (Bhuda & Mdhluli, 2024). The logic behind using the black sheep is to counterattack the symbolic condition created by the sorcerer with the black sheep is to take the illness, close it and lock it down with the black sheep so that it will stop and never come back again (Myeza, 2013).

The findings of this study show that is why this black medicine is followed by red or white medicine which serves to purify and restore good health. It is for this reason these medicines are prescribed and prepared only by a professional practitioner's medicinal doctor (*izinyanga*). The use of boiled red medicine represents transformation where day and night are divided by the twilight of sunrise and sunset. In addition, participant 4 elaborated that this treatment is done through steaming and vomiting outside the homestead before dawn. This treatment is repeated for three consecutive days. On the third and last day what remains of the medicine is thrown away at the frequented cross pathways where the dirt of the illness will be symbolic taken by different people. (Peters, 2021). This will inhibit the same illness to trace itself back to the original patient. Therefore, twilight is a reddish dim illumination, which at sunset becomes progressively dimmer and at sunrise becomes progressively brighter. Myeza (2013) stated that this means red medicine represents good and less danger, so the patient at this stage is pushed to the reddish twilight of the sunrise of which progressively brighter. At this stage the stringent abstinence that accompanies black medical treatment is relaxed, which suggests that less danger is associated with red medicines.

The treatment using red medicines is followed by white medicines which are not boiled. It is taken cold after the head of the homestead stirs (*uyawuphehla*) the white medicine until it froths. While he stirs the medicine he speaks to the ancestors, reporting all the problems and requesting them to look after all his dependents. Every adult member smears the white froth on the face or limbs and will drink the medicine to vomit anywhere else within the premises (Shange & Ross, 2022). Ngubane (1977) posited by illuminating that the daylight represents life and good health, therefore the use of cool white medicines in the third phase which accompanies the sacrificial white goat serves as the antithesis of the black sheep used during the first phase with black medicines. It represents life as people come together and share the meal. It represents daylight and social activity, and it brings together in a special way both the living and the dead. The white medicines in this context like the white goat emphasize light and life. Because they are meant to bring and reinforce good health into the system as well as all the good things of life. Participants indicated that this is why vomiting happens at home in the morning to bring the good things within reach. Eating together the meat of the sacrificial goat, and performing white treatment in the open, in daylight, represents eating itself, which is life giving, good and an antithesis of excretion represented by black treatment (Ngubane, 1977).

In addition, Myeza (2017) stated that on the set of symbolic meanings that operates within the color symbolism of medicines. There is another segment of the notions of hot and cold. According to participant 2, illness is associated with heat; it is for this reason that the black and red medicines are boiled or burnt. These boiled medicines are used in three different ways, through steaming, bathing, and vomiting. When they are burnt on an enamel plate, the focus will be on smoking it, dipping and then sucking the fingers (*ukucinda*). The logic behind is the restoration of the balance which is the removal from the darkness into the light. The cooling of heat of the illness which heat is represented by black and red medicines while cooling is represented by white medicines. They serve to re-establish the natural state of life (Myeza, 2017).

The sequence of using these compound medicines is culturally bound as mentioned above, is rigid and is never reversed as it starts with black followed by red and concluded with white. The reason for this, is because these medicines are responsible for healing something which is very deep and critical, which means the patient is always between life and death. According to Ngubane (1977) what needs to be healed regarding mystical illness in Zulu context, is not simply the physiological aspect of illness, but the two worlds. The two worlds consist of the symbolic world and the spiritual world. With the symbolic world, daylight represents life, and darkness represents the night that is danger. So, the spiritual world refers to the departed who are also invited to assist during the healing process as the water has been polluted with black medicine. It is for this reason that the treatment with such colored medicines is intended to establish and restore a balance between the person and the environment.

IMPLICATION FOR FUTURE RESEARCH

This study will contribute to understanding of healing practices that have been transmitted orally and Zulu healing practices are presently in the realm of oral tradition. The aim of this chapter has been to show that Zulu ideas and beliefs relating to illness and its treatment are part of a coherent body of knowledge that leads itself to being interpret analyzed in a framework of sociological theory. This is a cultural re-awakening where the Zulu healing ideas were supposed to put into proper perspective within the Nguni group with an aim to see to what extent what have been said for Zulu applies to the other Nguni societies. This has been hampered by the insufficient literature in the field of health and disease. This suggest that more Zulu ethnographic data that need to be collected so it will speak to unthinking some of the presumptions of knowledge that has been polluted by Eurocentrism so to escape from what Paulin Hountondji (1990) termed scientific dependence. This also to discourage a form of dependence where African scholars are engaged in scientific research that is of direct service to coloniality.

CONCLUSION AND RECOMMENDATIONS

The main aim in this chapter has been to show that the Zulu perspective on ideas and beliefs relating to the illness and its treatment are part of a coherent body of knowledge that has been interpreted and analyzed in a framework of healing within the context of traditional religion. It is important to note that the Zulu culture is rich in medicinal cures. Therefore, it has been displayed that within the Zulu perspective medicine is powerful and can be used for good or for bad. It is for this reason that the importance of medicine and morality all have the role to play in the practice of healing among the Zulus. Because according to the Zulu perspective, medicine only works if the patient is in good moral standing and is free of negative emotions. Different causes of illness have been established for the future healing of patients to take place. Witchcraft has been noted to appear to take center of stage in the paradigm of all evil and anti-social behaviour through mystical causes of illness. Furthermore, it has been revealed that a holistic approach is required to affect the healing among the Zulus. This entails the use of spirit mediums who are divided into diviners and herbal healers. They work jointly, starting with divination which is performed by the diviner healer who identifies the causes of illness and later the herbal doctor who does the therapeutic healing practice using compound medicines. It was also revealed that religion and environment are interconnected. This is why it was postulated that it is difficult to separate African traditional medicine from African religion, because Africans in general, and Zulus in particular, are connected to the God's creation, i.e. the sky and earth.

Based on the findings of this study, the following are recommended

- Zulu traditional healing should be reinforced because such practice has been undermined, and attempts have been made in encouraging the complete uprooting these aspects of African way of health.
- Traditional healing methods needs to be distinguished from the faith healing emphasized by the church because for the church it draws inspiration from the Bible. The traditional healing on the other side centers around the power of medicine which is reversing the health condition created where the same medicine has been used. This type of healing has nothing to do with sin and sickness from the context of the Bible.

This contribution is postulating that African should invest his trust in the cultural institution that have saved his society on health issues so well from time immemorial. It becomes very important that culture on medicinal healing practice must be propagated to ensure that the people develop a sense of themselves, dignity and personality.

REFERENCES

Adeleye, O. A., Femi-Oyewo, M. N., Bamiro, O. A., Bakre, L. G., Alabi, A., Ashidi, J. S., & Fakoya, G. (2021). Ethnomedicinal herbs in African traditional medicine with potential activity for the prevention, treatment, and management of coronavirus disease 2019. *Future Journal of Pharmaceutical Sciences, 7*, 1–14.

Berglund, A. I. (1990). *Zulu thought patterns and symbolism*. David Philiph.

Bernard, R. H. (2002). *Research methods in anthropology: Qualitative methods and quantitative methods*. Altamira Press.

Bhuda, M. T., & Marumo, P. (2020). African traditional medicine and healing in South Africa: Challenges and prospects before and during COVID 19. Ife Centre for Psychological Studies/Services. PO Box 1548. *Ile-Ife Osun State, Nigeria, 18*(4), 16710.

Bhuda, M. T., Marumo, P., & Motswaledi, T. (2022). Behavioral Analysis of The Impact of African Traditional Approaches on Curing and Restoring of Health During Covid-19. *Gender & Behaviour, 20*(1), 19152–19166.

Bhuda, M., & Khazamula, E. (2022). Towards finding a cure for COVID-19: The role of African Traditional Medicine as an indicator of development. *African Journal of Development Studies, 2022*(si1), 137.

Bhuda, T. M., & Mdhluli, T. D. (2024). *African traditional medicine as a COVID-19 treatment and the South African government response*. Pharos Journal of Theology.

Bryant, A. T. (1970). *The Zulu Medicine and Medicine-Men, C.* Struik.

Bryant, A. T. (1966). *The Zulu Medicine and Medicine Men*. Centaur Publishers.

Chuvunduku, G. (2001). The African Religion in Zimbabwe Today, Occasional Paper no.1, Crossover Communication dr.merten@web.de Harare/Munster.

Cordciro-Rodrigucs, L., & Metz, T. (2021). Afro-communitarianism and the role of traditional African healers in the COVID-19 pandemic. *Public Health Ethics, 14*(1), 59–71.

Donda, G. N. (1997). *A Study of Zulu Concepts, Terms and expression Associated with Umuthi, Master of Arts*. University of Vista.

Dube, L. (2011). *AFRICAN INITIATIVES in healing ministry*. Unisa Press.

Ezeanya, S. N. (1980). The Contribution of African Traditional Religion to Nation Building. In Amucheazi, E. C. (Ed.), *Reading in Social Sciences*. Fourth Dimension.

Henriques, A. C. (2013). *Environment, Worldview and healing among the Zulu Christians (19th and 20th centuries)*. UKZN.

Hill, C. E., & Mathews, H. (2021). Traditional health beliefs and practices among southern rural Blacks: A complement to biomedicine. In *Perspectives on the American South* (pp. 307–322). Routledge.

Hutchings, A. (1996). *Zulu medical plants*. University of Natal Press.

Jwara, S. M. (1998). *Healing Ministry in The Zulu Context: The Potential Role of Zulu Traditional Medicine in the Catholic Church's Healing Ministry, with Special Reference to the Marianhill Mission. Unpublished MTh in Theology*. University of Natal.

Kalu, O. U. (2000). *Ancestral Spirituality and Society in Africa*. Crossroad Publishing Company.

Kumar, R. (2014). *Research methodology: A step-by-step guide for beginners*. Sage.

Leedy, P. D., & Ormrod, J. E. (2005). *Practical research-planning and design*. McMillan Publishing.

Magesa, L. (1997). African Religion, USA: Orbis Books, Maryknoll.

Marques, B., Freeman, C., & Carter, L. (2021). Adapting traditional healing values and beliefs into therapeutic cultural environments for health and well-being. *International Journal of Environmental Research and Public Health, 19*(1), 426.

Mashego, S. L., Maditsi, M. E., & Bhuda, M. T. (2021). A response to Covid-19: Perspectives from African indigenous knowledge systems. *Gender & Behaviour, 19*(1), 17242–17254.

Mashele, N. M., Thompson, L. J., & Downs, C. T. (2021). Traditional health practitioners' and other community members' perceptions of vultures in the Kruger to Canyons biosphere region, South Africa. *The Journal of Raptor Research, 55*(3), 340–358.

Mbiti, J. S. (1969). *African Religion and Philosophy, London.Ibadan*. Heinemann.

McMillan, J. H., & Schumacher, S. (2010). *Research in education: Evidence-based enquiry* (8th ed.). Pearson Education.

Metz, T., & Molefe, M. (2021). Traditional African religion as a neglected form of monotheism. *The Monist, 104*(3), 393–409.

Mwikamba, C. M. (1991). Search for an African identity. *Wajibu, 6*(4), 6–10.

Myeza, M. C. (2013). *Ucwaningo lezincazelo zamagama emithi ngesiZulu, umqulu ongashicilelwe weqhuzu leMaster of Arts*. University of South Africa.

Myeza, M. C. (2017). *Imimoya yamathongo. Pinetown*. Pearl Publisher.

Ngubane, H. (1977). *Body and Mind in Zulu Medicine, An Ethnography of health and disease in Nyuswa-Zulu thought and practice*. University of Edinburg.

Nkonge, D. K. (2014). The Relationship between Religion and Science in Mission: Reflection from an African Christian Perspective. Research on Humanities and Social Sciences, International Institute for Science, Technology and Education Journal, Vol.4, No.2

Nwoye, A. (2017). An African Theory of Human Personhood: Psychology in Society, 42-66.

Okeke, V. I. (1998). Religion and its Place in Human Society. In Iffih, B. N. (Ed.), *Modern Textbook on Social Sciences*. Enungu Joen Publishers.

Okon, E. E. (2006, August). "Ekpu-Oro: Studies in Oron Cosmology and Cultural History" in African Journal of Religion. *Cultura e Scuola, 1*(1), 2–26.

Olumbe, D. (2008). African Worldview: an introduction. https://watumishiwaneno files wordpress.com/2014/08/African Worldview introduction pdf.

Olupona, J. K. (2000). African Spirituality: Forms and Meanings and expressions. New York: Herder & herder Books.

Onyeidu, S. O. (1999). African Traditional Religion: The problem of definition. Nigeria.

Parrinder, G. (1958). *The World Living Religions*. Pen Books.

Phiri, G. A. (2009). *Social-Cultural Anthropology: Communication with the African Society*. Stock Publishers.

Redfield, R. (1952). The primitive world view. *Proceedings of the American Philosophical Society, 96*(1), 30–36.

Ritter, E. R. (1955). Shaka Zulu: The rise of the Zulu Empire: South Africa.

Schurink, C. B. (2001). *Qualitative data analysis and interpretation. Research at grass roots for the social sciences and human services professions* (4th ed.). Van Schaik.

Shange, S., & Ross, E. (2022). "The Question Is Not How but Why Things Happen": South African Traditional Healers' Explanatory Model of Mental Illness, Its Diagnosis and Treatment. *Journal of Cross-Cultural Psychology, 53*(5), 503–521.

Shoko, T. (2011). *AFRICAN INITIATIVES in healing ministry*. Unisa Press.

Thabede, D. (2008). African worldview as the basis of practice in the helping professions. Social Work/Maatskaplike Werk Journal, 44 (3), 233-245.

Van der Zeijst, M., Veling, W., Makhathini, E. M., Susser, E., Burns, J. K., Hoek, H. W., & Susser, I. (2021). Ancestral calling, traditional health practitioner training and mental illness: An ethnographic study from rural KwaZulu-Natal, South Africa. *Transcultural Psychiatry, 58*(4), 471–485.

Peters, A. (2021). Cultural conceptualisations of witchcraft and traditional healing in Black South African English herbalist classifieds. Cultural linguistics and world Englishes, 333-359.

ADDITIONAL READING

Cock, I., Mavuso, N., & Van Vuuren, S. (2021). A review of plant-based therapies for the treatment of urinary tract infections in traditional Southern African Medicine. *Evidence-Based Complementary and Alternative Medicine : eCAM, 2021*(1), 7341124.

Mudau, T. E., Olowoyo, J. O., & Amoo, S. O. (2022). Ethnobotanical assessment of medicinal plants used traditionally for treating diabetes in Vhembe district, Limpopo Province, South Africa. *South African Journal of Botany, 146*, 304–324.

Mutola, S., Pemunta, N. V., & Ngo, N. V. (2021). Utilization of traditional medicine and its integration into the healthcare system in Qokolweni, South Africa; prospects for enhanced universal health coverage. *Complementary Therapies in Clinical Practice*, *43*, 101386.

Nann, S. (2021). How beliefs in traditional healers' impact on the use of allopathic medicine: In the case of indigenous snakebite in Eswatini. *PLoS Neglected Tropical Diseases*, *15*(9), e0009731.

Sundararajan, R., Langa, P. V., Morshed, T., & Manuel, S. (2021). Traditional healers as client advocates in the HIV-endemic region of Maputo, Mozambique: Results from a qualitative study. *SAHARA J : Journal of Social Aspects of HIV/AIDS Research Alliance*, *18*(1), 77–85.

KEY TERMS AND DEFINITIONS

African epistemology: is the knowledge which is more oriented with African knowledge.

Illness: is a health condition that refers to the unbalance health condition between a person and the environment.

Divine Healer: is a person who has powers to sniff out hidden sources and is more concerned with the spiritual causes of the afflictions with the assistance of ancestral spirits

Herbal healer: is a person with deep knowledge of herbs, roots, even fruits with power to prevent or cure disease or other afflictions.

Mystical illness: is culturally acquired and likened to pollution that moves a targeted person away from the daylight into the dimness of the sunset and on into the night.

Social illness: is caused by social agents who range from spiritual forces, witchcraft and sorcery.

Spirit medium: is a person who can communicate with the dead people while in a state of spirit possession.

Spiritual world: is the world of the dead is created by sorcery for death objective.

Symbolic world: does not exist, and it has been created by sorcery using symbolic color medicines.

Chapter 13
Name Giving in the Batswana Culture:
A Generational Change Analysis

Welcome Mswazi Kubeka
https://orcid.org/0009-0005-2514-3424
University of Johannesburg, South Africa

ABSTRACT

Naming is an evolving and continuous process that could also be influenced by changes and practices that have emerged due to the contact South African people had with colonization, Christianity, and Western education influence. The study investigated how the Batswana people in Gauteng province give name(s) to their newly born child/children through the different generational changes. The problem is that the customary and traditional practices of name giving are gradually disappearing as generations evolve. Data was purposively sampled through an interview schedule that was administered face-to-face and online. The findings amongst others revealed that the practices in name-giving vary across the generations hence Baby Boomers and Generation X this responsibility is given to the grandparents of the newly born baby. The names given by the Baby Boomers and Generation X are mostly inherited from the family pedigree. Whereas Generation Z prefers to give their children their names, they are trying to break the tradition of inheriting names.

INTRODUCTION

The Batswana people form part of the Sotho group in South Africa which is composed of Northern Sotho (Sepedi) spoken mostly in the Limpopo province; Southern Sotho (Sesotho) spoken predominantly in the Free State province; and (Setswana) spoken both in the Northwest and Northern Cape provinces. All these three groups of languages are also spoken in the Gauteng province because some of these people from the abovementioned provinces have settled in this province (Gauteng) to seek job opportunities. The essence of this chapter focuses on the Batswana people who are residing in the Gauteng province.

The Batswana people are one of the cultural groups are still observing their cultural upbringings. They (Batswana people) always use "proverbs" to understand their surroundings and the changes that are happening around them. Wikipedia (n.d.) defines a proverb (from Latin: *proverbium*) as, "*a simple*

DOI: 10.4018/979-8-3373-4124-8.ch013

and concrete saying, popularly known and repeated, that expresses a truth based on common sense or experience". They are often metaphorical since they are a figure of speech that directly refers to one thing by mentioning another for rhetorical effect (Wikipedia, n.d.). Giving a person(s) a name(s) in the Batswana culture is not only a cultural phenomenon but also a linguistic phenomenon. The meaning that is attached to a person's name also defines the person behind the name "personhood". It is also believed that the given name to a person doesn't only serve as an "identity" but also determines the type of person the individual will be in the future (Guma, 2001). The proverb in Setswana which says, "*Leina lebe seromo*" (literally translated as, "a person's name will shape his or her future") qualifies the notion that the name may influence the character of the person.

STATEMENT OF THE PROBLEM

Name giving in the Batswana culture have now taken a different direction across generations in so far as customs and traditions. The problem is that this phenomenon of name giving that used to be a collective responsibility of the family has become an individualistic task. The exclusion of the key immediate family members in this process of naming the newly born children has also resulted in them not adhering to the traditional cultural practices that are supposed to be performed.

The researcher has been observing and noticing that this cultural norm so far as name-giving is gradually disappearing as generations evolve. Hence, in this study, the researcher wanted to understand the reasons why certain practices with respect to the name given by the Batswana people in Gauteng Province are no longer applicable as per generation (zooming specifically to these four generational changes Baby Boomers, Generation X, Millennial, Generation Z).

AIM OF THE STUDY

The study aimed to investigate how the Batswana people in Gauteng province give name(s) to their newly born child/children through the different generational changes. To achieve this aim, the following questions were posed:

- Who named your child/children?
- Is your child/children's name(s) given or inherited name(s)?
- What meaning is associated with your child/children name(s)?
- Was any cultural procedure followed when you gave or inherited your child/children's name(s)?

THEORETICAL FRAMEWORK

The theory that underpins this study is the Generational theory of Mannheim (1952), Howe and Strauss (1997). Mannheim (1952) stems this Generation theory from Marxist thoughts such as the idea that someone's beliefs are shaped by the type of group that person belongs to and associates with. Hence this argument is that all truths and ideas are influenced by the social context upon which they emanate. In understanding this Generational theory, Mannheim (1952) defines generations as a cohort of a population who experienced the same events during their lived period. In this Generational theory, Mannheim (1952) further posits that the social context is developed by the older generation, and this is the first generation that the new generations make first contact with. It is this new generation that always affords opportunities that could enhance the culture and cultural continuity. The specific generation forms a cohort with specific social identity because of common experience, this leads to identical influences of individuals.

In understanding this theory of generation, a simple relationship between it and this study on name giving can be positioned across generational mixes. This relationship can be explained through the notion of social context and culture, especially to the older generations such as the Baby Boomers (1946-1964) and Generation X (1965-1980) who believe that the newly born babies are supposed to be named after the society's events and cultural norms in which family legacies are supposed to be observed.

In expanding this Generational theory of Mannheim (1952), Howe and Strauss (1997) based their assertion that the category of value is the key component when each generation's timeframe is conceptualized. Some of the core values such as certain social, economic, and political circumstances are essential not considering the dates of birth. They argued that generations are equated to the lifespan of human beings which is between 80 to 90 years. It is also notable that the human being during this lifespan goes through these different stages which are awakening, ascension (flowering), destruction, and crisis. Through all these stages a new generation appears.

LITERATURE REVIEW

As an introduction to this section of the literature review, it is imperative to understand what Indigenous Knowledge is since name giving is also part of practices that are aligned with it. In his writings, Ogunniyi (2021, p.37) points out that the knowledge contributions emanate from indigenous people. Indigenous Knowledge is generated from one generation to another as this has a closer tie with the cultural values of that society.

According to Diyal (2021) there are numerous definitions of Indigenous Knowledge but in all of these, the central theme is that it consists "*of cultural beliefs and traditions being passed on from their forefathers to the present generation for the purpose of survival while still living in harmony with ecosystems. Traditional knowledge is something that is learned during a lifetime*" (p. 48). Further in his assertation Diyal (2021, p. 57) notes that Indigenous African people over the years had their peculiar approaches, traditions, and norms that have been sustained over centuries in so far as the management of the environmental practices.

Indigenous Knowledge is central to the body of knowledge that communities have accumulated through life experience, unofficial research, and observation. As it is Indigenous Knowledge is gradually fading away and could be lost forever. It is against this background that this Indigenous Knowledge should be reevaluated and preserved for future generations.

Name as a concept

According to Chauke (as cited in Koma, 2012) names play a significant role since it remind the next generation that a long time ago there lived a particular person or group of people in a particular place. He also indicated that names serve as a symbol that will forever be in societies or family's lives. De Klerk's

(1996) opinion on naming is that it is a linguistic act, intimately linked with values, traditions, hopes, fears, and events in people's lives hence it is clear that naming is concerned with the culture of people. Naming is an evolving and continuous process that could also be influenced by changes and practices that have emerged due to the contact that South African people had with colonization, Christianity, and Western education influence.

Sequences in name giving

The naming of children in other countries outside the continent of Africa had a particular sequence in the past. In Argentina in the past, first-born babies were named after their parents but now the focus is on distinctiveness hence parents choose the names that they like or that are fashionable (Pocket Cultures, n.d.). In addition, children in the past were named after historical and political events or sports celebrities, which nowadays is not the case. Further, children had two names, the first and middle name, which is not happening now (Pocket Cultures, n.d.).

In Costa Rica in the past, parents were giving their children three or four names after their corresponding saints (Pocket Cultures, n.d.). Whereas, in India the common practice in naming children is that of using "family names" hence a first male child is named after the paternal grandfather (Pocket Cultures, n.d.). The Indian also use a common system of naming their children after a God or Goddess.

Children born by Spanish parents are traditionally given one or two names that are followed by two surnames. The first surname given to the newly born child is that of the father's first surname, and the second surname is that of the mother's first surname (Pocket Cultures, n.d.). The child will carry two surnames.

Naming (places) in Southern Africa provided an important framework for language planning. This (naming) has been characterized by a shift from European place names to indigenous names of African languages that reflects a shift in terms of policy towards African languages. Though most of these place names were officially in English, the Africans typically referred to the places using their African names (Makoni, Makoni & Mashisi, 2007). A typical example in this instance is Salisbury was always Harare amongst Africans. This is clear that policy and practice were in sync after the changes as far as place names are concerned.

Distinction between given and inherited names

This distinction between "given" and "inherited" name(s) be put into perspective. A "given" name(s) is a name that is given to a person during birth (Campbell & Campbell, 1996). Whereas the "inherited" name(s), is a name related to the family name and is carried from one generation to another, is regarded as a traditional name(s) (Campbell & Campbell, 1996). The "given" and "inherited" name(s) in Batswana culture can be categorized as follows, namely:

- Ancestors name(s) - child (children) are given name(s) of their grandparents to perpetuate the names(s) of the ancestor(s) with a notion that the "given" or "inherited" name(s) will link the child with the ancestor(s). It is also believed that the child or children named after their grandparent(s) will inherit the virtues of their grandparents.
- Prominent or famous person(s) name(s) – these name(s) may promote a positive self-image of the bearer.

- Historical events name(s) – naming people after historical events serve both the psychological and emotional needs of the family since it serves as a "recording system" in which the individual will aspire to live up to the expected behavior.
- Virtue name(s) – naming a child (children) after a long childless marriage relationship is a traditional belief that "*badimo*" (ancestors) have finally answered the prayers of the family. Name(s) that are given based on personal beliefs mostly expressing happiness (Manion, 2005).
- Christian name(s) – is given to the newly born child (children) by the parents who are "Christians" hence they adopted English name(s) as second name(s) given to their siblings.
- Socio-cultural, lived-in experience name(s) - these are the name(s) given or inherited due to society or family circumstances.
- Teknonyms – is the name(s) given to newlywed brides by their in-laws which they will use in addressing their "*ngwetsi*" (bride) since it is taboo to refer to her (bride) with maiden name(s).
- Initiation name(s) – are name(s) given to boys and girls who have been through the initiation process since they are regarded as accomplished "persons" accepted as human beings in society.
- Totem name(s) – are referred to as praise name(s) hence it is regarded as a guardian spirit that inhabits certain animals and protects the tribe or clan (Brochman, as cited in Koma, 2012).

Characteristics of the different generations

To understand the naming across generational changes, the characteristics of the Baby Boomers, Generation Z, Millennials, and Generation Z are briefly summarized in this Table 1.

Table 1. Characteristics of the generations

Summary of the characteristics			
Baby Boomers (1946-1964)	**Generation X (1965-1980)**	**Millennials (1981-1996)**	**Generation Z (1997-2012)**
They value individuality and are self-absorbed and self-centered generation.	They were regarded as the generation without identity and hence others refer to them as the "lost generation"	This individual of the generation was protected and sheltered by their parents.	This generation is regarded as the "Digital natives" since they were born when technological innovation was at its peak. Contrary to Gen. X and Millennials who are referred to as the "Digital migrants"
They grew up during a time of economic prosperity.	They were shaped by the social trends and cultural events that occurred during their first 20 years (Hedden, as cited in Govitvatana, 2001).	They showed high confidence, and they were not used to experiencing failure.	They are at the advantage of advancing their knowledge base since they are exposed to an abundance of information.
Their high expectations were met but disappointment came in around the 80's and 90's (Okrent, 2000; Smith, 2000).	They are the generation that grew up during the times of fast food and instant satisfaction (Alverson, 1999).	They are referred to as the "Web" generation, technologies such as computers, smartphones and social media are at the center of their lives.	Though technology has its advantages, there is also negativity regarding its availability since it can create disorders such as lack of socialization, low self-esteem, depression, and anxiety.

continued on following page

Table 1. Continued

Summary of the characteristics			
Baby Boomers (1946-1964)	**Generation X (1965-1980)**	**Millennials (1981-1996)**	**Generation Z (1997-2012)**
They always want to be involved and be part of decision-making. They value meaningful contributions (Broderick, 1992).	They experienced many social changes such as divorces, drugs, homelessness, a decline in income and troubled education system.	Doing things simultaneously "multi-tasking" is their way of life. (www.youtube.com).	They end up being the "loneliest generation" since most of the time they spend it alone online.
They are highly resistant to change and are afraid of trying something new. Hence, they are not technologically savvy (Zetlin, 1995)	These generations come from a group of two-income families, though some social aspects were noticed such as the apart drifting of the families.'	They embrace diversity and inclusive attitudes. Hence, they easily accommodate and accept other people's cultures, lifestyles, and beliefs.	They are more focused and competitive; this drives them to be "impatient" since they always want to get what they want "now".
	This generation grew up during the era of the computer revolution.	They are prone to regeneration. When they are praised and complimented, they feel motivated to do more.	

Aspects considered in name giving

In naming their children each generation considers many aspects such as in the case of the **Baby Boomers,** they take into consideration *stability and timelessness* hence they prefer names that have been there over time to those that are trendy. **Generation X** prefers aspects such as *cultural heritage* with the notion that the name should reflect the clan's cultural background. This will be a base to connect with family roots. This generation further believes in *family legacy* they prefer to name their children after their relatives or ancestors, and this makes them preserve their history and keep the lineage (family tree). The **Millennials** generation considers *cultural heritage* and *individuality* as the key to name-giving. Through these aspects the child's identity is positioned. They also believe that a *socially accepted* name is easily resonating with the communities at large. Another consideration is the *balance of tradition and modernity* in name-giving which contributes to creating meaningful choices when blending traditional (old) names with contemporary (new) names. **Generations Z** thinks that a name should consider *cultural significance*. They consider the *meaning* and *symbolism* of the name which could convey positive meaning and aspirations. They believe that names should convey values associated with strength, resilience, and hope amongst others. These generations also see *uniqueness* as a key for name-giving since they prefer to have distinctive or names that stand out though reflecting a specific desire. As a generation that is accessible to a vast amount of content and information, they consider trends and influences when giving names. The current trends and celebrities who are considered role models are commonly used as choice names for their children. Hence these chosen names will contribute to *personal connections* and will demonstrate great personal significance.

RESEARCH METHODOLOGY

This study used a Qualitative research methodology whereby interviews were conducted. In this study sampled participants were twenty (20). These sampled participants were purposively sampled, and the characteristics were associated with the four identified generations that are Baby Boomers (Born between 1946 and 1964), Generation X (Born between 1965 and 1980), Millennials (Born between 1981 and 1996), and Generation Z (Born between 1997 and 2012).

The research design is a master plan which outline how the study will be conducted so that the research questions are addressed. Yin (2003) states that "a research design is an action plan for getting from *here* to *there,* where 'here' may be defined as the initial set of questions to be answered and 'there' is some set of (conclusions) answers" (p. 19).

This study adopted an interpretivism stance in which the different generational participants may present their worldview in so far as name giving of the newly born babies. A deeper understanding of the phenomenon was conducted through a structured interview schedule in which participants provided their responses. This also aligned with the contention by Creswell (1998) that "a phenomenological study describes the meaning of the lived experiences for several individuals about a concept or phenomenon" (p. 51).

DATA COLLECTION AND ANALYSIS

Data was collected through an interview schedule. This was done online (using Microsoft Teams) and face-to-face interviews. It is important to note that through this interview process, no recording was done. This was to allow the Respondents to be themselves and feel free when responding to the interview questions. Permission was requested from the Respondents to allow the interviewer to take down the notes.

The data collected was manually analyzed. The first part of the data collection instrument was about the gender representation of the Respondents and the average number of children (per household) across the four generations. The descriptive statistics for the first part were presented in which the frequencies and the percentages were indicated (see Tables 2 and 3) respectively.

In analyzing the second part of the data collected, the researcher did consider these steps as presented by Saldanah (2009). Based on the Respondent's responses (captured from the generated notes) the researcher did develop a transcribed manuscript. This transcribed manuscript was read thoroughly, starting with one manuscript per Respondent and per interview question. The next process was to identify the text segments, then select the text segment and highlight it with color. The next step was to assign the code word which was followed by clustering the codes into categories. The last step was to develop the themes from the categories.

ETHICS

Ethical considerations were the key when conducting data collection. Respondents were asked to complete the consent form. In order not to put the Respondents under strain and pressure, they were asked only to provide their signatures (voluntary), no personal details such as their names and surnames were requested.

RESULTS AND DISCUSSION

The results presented are a sample of 20 participants. Part One of the results presents the respondent's representation as per gender and the number of children per generational mix.

In Part Two the respondents' responses (*verbatim*) as per interviewees are presented. The themes that emerged after the data coding and categorizing are presented in the discussion.

The researcher would like to put it up front that, the analysis of the meanings of names is based on what the researchers think the meaning in simple English, not taking into consideration their deep meaning as per Setswana languages. For instance, the name "Basetsana" (means girls), whereas the Setswana deep analysis means that "the family has born only girls, there are no boys".

Part One

Table 2. Gender representation of the respondents (n=20)

Generation	Gender	Frequency	Percentage
Baby Boomers	Female Male	2 3	40 60
Generation X	Female Male	3 2	60 40
Millennials	Female Male	1 4	20 80
Generation Z	Female Male	3 2	60 40

The gender representation of the respondents as per generation revealed that there was a consistent so far as the responses of both Generation X and Generation Z in which 60% of the responses were female and 40% males. In so far as the results that were received further revealed that more Males (60%) respondents and 40% of the responded were Females. Finally, the data revealed that from the Millennials generation, a majority (80%) of the males responded.

Table 3. The average number of children

Generation	The average number of children (per household)
Baby Boomers	5
Generation X	4
Millennials	4
Generation Z	2

The results in Table 3 point towards a decline in the average number of children per household across generations. The Baby Boomers (born between 1946 and 1964) had an average of five (5) children per household and this aligns with the point that the group of parents were born during the times when there was economic prosperity. Contrary Generation Z (born between 1996 and 2012) has an average of two (2) children per household. These results could also be prone to the fact that some of these parents (Generation Z) were born during the time of economic meltdown (2008) that resulted in the cost of living and rising unemployment. Further, the reason for having this average of two children per household could be confined by the characteristics that Generation Z is career focused.

Part Two

In Part Two, to better present the Respondent's responses these abbreviations will be used, namely: Baby Boomers (BB); Generation X (Gen X); Millennials (Mille); Generation Z (Gen Z) and Participant (P).

Theme 1: Practices in name-giving

The (United Nations [UN], 1989) Convention on the Rights of the Child (1989) Article 7 states that "*a child shall be registered immediately after birth and shall have the right from birth to a name, the right to acquire a nationality*". This is also aligned with the African Charter on the Rights and Welfare of the Child (1999) in Article 6 which proclaims that "*Every child shall have the right from birth to a name. Every child shall be registered immediately after birth* and that *Every child has the right to acquire a nationality*". The (Republic of South Africa [RSA],1996) Act no 108 on Section 28 of the Bill of Rights states that "*Every child has the right to a name and a nationality from birth*"

The study results revealed that the responsibility of name-giving varies across generations. The results revealed that the responsibility of name giving evolves as per generation. It is noted that this responsibility of name giving for the Baby Boomers is a parental responsibility in which both parents (both the father and mother of the child) are involved.

The process that sees more parental involvement of grandparents from both sides of the child's parents is gradually not noticed, especially in the Millennials and Generation Z parents. A collective approach in name giving is noticed in the Baby Boomers and Generation X parents whereby there is a high participation of grandparents from both sides of the families.

In asserting that position BB P1 stated that, "*... my parents are the ones who named all my children ...*". Though this response from Baby Boomer P1 suggests that there was a collective agreement from both the families, that is the grandparents of the children. Whereas the tone "*my parents*" from the later response could indicate that the other family especially the bride's family (since Baby Boomer P1 is a male) were not part of the discussion when giving names to children. This position is also justified by the responses received from Gen X P2 who alluded that, "*... my children were given names by my parents ... that is my mother and father ...* ".

It is clear that a certain pattern in so far as name giving has been noticed from these two generations, that is Baby Boomers and Generation X whereby the responsibility is split amongst the two families. In which the bride's family are the ones who name the firstborn child, and this becomes a norm if the groom (father of the child) has not performed any traditional or cultural requirements. In this case, BB P5 has this to say, "*since when I had my first child ... we were still young ... and my boyfriend ... who then became my husband later ... had not paid any family requirements ... he was supposed to pay ... [tshenyo] ... as a sign to indicate that ... he made me pregnant before we marry ... my parents named our child ...*".

This pattern of splitting the responsibility of naming children amongst both families takes another dimension if both the biological parents of the children are married (either in civil or customary). The groom's parents are the ones who are given the upper hand opportunity to name their firstborn (grand-child). Then the bride's parents are responsible for naming the other children and that also depends on the number of children who are born. Hence, Gen X P5 pointed out that "*... my parents ... my father and mother named our first-born child .... my in-laws were given an opportunity ... for naming the second born child ...*".

Noticeably from the respondents' responses that there is a bit of relation in this process of name giving from these generations, that is the Baby Boomers and Generation X. Whereby the biological parents of the children are also allowed to name their children. In these instances, the grandparents of the children always show their dominance by affording the biological parents to name their last-born children, whereas firstborn are named by them (grandparents). This stance is captured by Gen X P4 who said, "*... my late husband and I ... were given permission ... by our parents to give name to the 6th child ... our lastborn ...* ". The same sentiment is echoed by BB P2 who indicated that "*... the lastborn was named by us ... myself and my wife ...*".

Theme 2: Meaning in the name

The study by Tomanek (2023) points out that parental involvement plays a great influence in choosing and naming of children. This could be based on the influences that traditions have on these premises whereby some of the names given could be inherited from the family's pedigree or history. Sometimes these names may emerge from the biological parents who are merely given new names. This could further uphold the parent's beliefs and values that are influenced by personal preferences and societal trends.

In their studies, Baranov, Albitsky and Ivanova (2012) opinioned that in most cases when children's names have a reflection of the families' cultural heritage, they are prone to contribute to their identities and interactions within society. This is also evident from the responses that were received from these generations, that is the Baby Boomers and Generation X. BB P1 stated that "*... one of my children ... was given the inherited name of my uncle ... Relebogile ... means we are thankful ...*". Sharing the same sentiment BB P2 alluded to this "*... most of the names ... our parents gave to our children are inherited .... They represent the family members ...*". This notion of having inherited names was further captured by Gen X P2 who added that "*... the first born and the second born names were inherited ... firstborn was named after my aunt ... the second one after my uncle*". In the same breath Gen X P4 indicated that "*... the first born was an inherited name .... my in-laws named him ... after my husband's grandfather ...*".

In so far as the Millennials' generation the results revealed that some of the names are inherited, and others are given. Though these given names are not even new ones they have to be within the family pedigree. This is clear that there is no direct involvement of the grandparents in name-giving (it is the responsibility of the biological parents). This was noted by Mille P2 who mentioned that" *... first child name is inherited ... named after our clan's totem ...*"

It is clear that from these generations the understanding of inheriting names even has boundaries in which there is a certain protocol that is supposed to be followed within the families as per the family's pedigree.

According to Hoorens (2023), parents who give their children new and unknown or uncommon names contribute to those names having positive psychological impacts which could contribute positively to their self-esteem. The response that was received from Gen Z P1 is clear that their parent's preferred name for their daughter was an inherited one as per family pedigree hence this "… *my aunt's name is Moiponi … imagine if we have agreed to inherit that name …*". In probing the responses trying to find out if the suggested name by the parents was given to the child, what could have happened, the Gen Z P1 response was "… *my child will be growing with a name … is not fashionable … not trending … embarrass us as parents …*". In alignment with the other responses, Gen Z P2 added "*Both our children's names are given … we did not want to inherit any name from both our parents' sides … are not good … they encourage sadness …*".

The researcher argues that naming children based on names that are popular and trending with the broader evolution that is experienced in societal changes, in the long run, could have an impact on how those children perceive those given names. The Researchers' argument is further closely associated with the responses that were received from Gen Z P4 who said "… *our children's names are given … not taking from any family members …we were avoiding similar names …are common … a bit boring … shows lack of creativity …*". In probing the response further trying to find out, what is wrong with giving the same names to your children as those of your parents, the response from her, that is Gen Z P4 was "… *we wanted to be unique … having names which have more meaning … and project the future …*". The same understanding could be shared from the response that was given by Gen Z P5 who contended that "*Some of our elders' names are depressing … are making a person not taken seriously … you know if you feel that your name is not taken seriously … you end up having an inferiority complex …* ".

The argument has been posed by Kiema (2010) who said "The names imposed on us have resulted in self-condemnation, self-estrangement, submission, powerlessness, apathy, dependence, land loss and erosion of our identity. Our self-esteem, potential for self-actualization and dignity ironed out of us" (p. 71. This assertion could assist us in understanding the meaning of some of the names that have been given or inherited by the newly born children.

The results revealed that the meanings that are associated with the names of the inherited and given names of the Baby Boomers and Generation X reflect cultural background and honoring ancestral tradition whereby there is a connection with the family roots. The meaning of these names also reflects the family legacy whereby names come from relatives so that family history can be preserved. The Baby Boomer's names are inherited hence some of their meanings are associated with the family tree tradition, events, and circumstances. In this case BB P3 stated " … *first born … named after my wife's uncle … named Boikanyo* (means someone you can depend upon) … *second born a girl named after her Rrakgadi* (Aunt) … *Lesedi … to retain and keep our grandmother's name alive … one of our children was named Moiponi* (meaning that through her we can reflect ourselves) … *as we were not allowed to call her by that name … we nicknamed her Ouma …*". In preserving the family pedigree in so far as names are inherited BB P4 said "*Our daughter was named after Rrakgadi, Motlapepule* (means she came with rain) … *even though she was born in winter and there was no rain …*".

In so far as Generation X most of the meanings of names given to their children are aligned to the lived-in experience of the biological parents as per social-cultural family circumstances. This came clearly from Gen X P2 who said "… *we named him Relebogile … this name is associated with the happiness … going through in our marriage … also be given an opportunity not to inherit a name from family …* ". The response shows that there is an excitement of affirmation of breaking the tradition of inherent names. Another respondent, Gen X P1 had this to say, "*we named our first-born child Lerato* (means love) … *since we longed for this child … a longer period …*". Gen X P3 indicated that "… *named the second child Khumo* (means wealth) … *since we were anticipating … he would look after our wealth when we are gone …*". Another meaning that is associated with a name given by this Generation X is the response that has been received from Gen X P4 who mentioned that "… *we were already staying in our own home … we named her Reitumetse* (meaning we are happy) … *we were celebrating our own place …*". This respondent further said, "… *last born child … cementing our marriage … we named him Mosalagae … meaning that he is the one who will look after our house … when everyone is gone … make sure that this family house remains … e seke ya nna matlotlo* (literally means not to be an abandoned house or ghost house) …".

The meanings of names that are given by Generation Z parents are mostly associated with expectations and gratitude, affirmation of breaking the tradition to inherit names, projecting good things, prosperity, love, projecting brighter future and progress. This has been captured from the responses by Gen Z P1 who said "… *we named our only child Palesa* (meaning a flower) … *the meaning of her name suggests that she is a flower … everything which we have … we are going to protect and nurture her … so that beautiful things will come from her …* ". Gen Z P3 responded as "… *named the elder child … our boy Lebogang* (means be thankful) … *we were giving thanks to God … bless us with a baby boy … our second child daughter we named Keletso* (means a wish) … *we discussed and planned … to have only two children … when we receive the second child … we said our wishes has been met …* ".

The meanings of the names that are given by the Millennial generations are associated with the state of well-being, living conditions, appreciation of life in general, signifying the Gods' gifts and acknowledging of bringing a "seed" bearing a child. Mille P1 has this to say "… *my daughter was named Warona* (means our child) … *named our son Kamogelo* (means God is Gracious/Merciful) … *I appreciated and welcomed all the lord has given me …. Named our second daughter Bahumi … resembling the point that we are better off now … we are to live beyond the poverty state …*". The same position was alluded by Mille P5 who mentioned that "… *named our last born Dimpho* (means Gifts) … *as we were acknowledging all the gifts that we have received from God … as a family …*". Mille P2 also added that "… *the meaning that is associated with the name of the first child Kubu* (means hippopotamus) … *we are being referred as the "Bakubu" … because we are associated to an animal … Kubu … this is our totem … this name resembles our clan … we honor … we respect that animal …*".

Theme 3: Cultural practices with the name-giving

Traditionally, name-giving has been a collective responsibility of the families hence there were also specific cultural practices that had to be followed. This further dictate that the immediate family members such as the grandparents, parents (biological), aunts, and uncles are to be involved in these cultural practices. Though these cultural practices are gradually infused as the generations evolve, modernization and westernization are experienced.

The cultural practices that are associated with the name given by the Baby Boomers and Generation X are assigned this responsibility of name giving to the newly born and this is followed by family celebrations. Further, these generations, that is the Baby Boomers, and Generation X believe in the cultural practices of confirming the newly born child. This responsibility of confirming the child is assigned and performed by the Rrakgadi's (aunts of the biological father of the child). In her response, BB P2 stated that "… *in our family there was a cultural tradition of slaughtering of an animal … welcoming the new baby … name of the baby born was presented by elders … share its meaning … origin within the family …*". Whereas BB P4 mentioned that "*In our times when you were expecting your first child … you were supposed to be with your biological parents … they could take care of you … when the other children were expected your in-laws could look after you …*". Echoing almost the same sentiment, BB P5 said

"... *when the first born was born ... word was sent to my in-laws by my parents ... to communicate the good news ... my in-laws with their delegation came to my place ... they brought gifts ... they named the child ... cultural procedures were performed ...*". In probing this Responded about the specific cultural practices that were performed on the newly born baby. The BB P5 stated "... *in our culture we are not supposed to share the specifics ... it remains within the families ...*". Gen X P4 mentioned that "... *when they received the message that I have a baby boy ... they came to celebrate and give the child a name ... explaining the family history ...gifts were also brought by them ... accepting their presence ... my parents slaughter a goat ... acknowledging that my in-laws also agrees that my child is their grandchild .... also accepting of the name given by my in-laws*". In so far as these cultural procedures that is expected to be followed Gen X P3 cemented it as "... *since I was the first Makoti (bride) to this family ... when they came to visit and confirm the child ... they brought the instructions .... that needs to be followed ... after the child's umbilical cord was broken ... they also share why it is important to name the child after her grandmother ....*".

The results revealed that the common cultural practices that are followed by the Millennials and Generation Z in so far as name giving is the baptism of the child, acknowledgment of the family clan and the ancient tradition of bridal shower whereby the expectant parent is shown with gifts. In his response Mille P1 has this to say "... *there were no cultural procedures that were performed ... these names are reflections from my background ... and upbringing as I grew up ...*". Whereas Mille P2 alluded that "... *the cultural procedures when my parents named our 1st born child after the clan's totem ... is the tradition in our family ... first born child (son) when is having a boy child .... the totem name should be given to that child ...*".

Gen Z P1 indicated that "... *their emphasis was that if you have a child out of wedlock ... that child's name should come from family history ... named her after my aunt ... my boyfriend and I ignored that name ... we gave her our own preferred name ... even now our parents are calling her by their preferred name ... even though is not an official name ...*". In addition, Gen Z P2 indicated that "... *we welcomed them in a western way by throwing baby showers ... to celebrate both of them ... our parents were not part of that occasions ...*". In supporting both Generation Z P1 and P3, Gen Z P3 alluded to this "... *what was done was that my cousins did organize baby shower parties for each one of the children ... were already knowing the sex of our babies before they were born ... we discussed some possible names ... agreed about Setswana name ... no European ones ...*".

This was said by Gen Z P4 "... *no cultural procedures were followed ... we wanted to break the stereotype of having everyone involved in our families when naming children ... we wanted to avoid the influence ...*". Lastly, the same notions are presented by the Generation Z respondent was shared by Gen Z P5 who said "... *parents were not even consulted ... they were told about our child's name immediately ... I was discharged from the hospital ... was no way I could be discharged without giving the child a name ... had to register him ... could apply for a birth certificate on spot ...*".

CONTRIBUTION TO BODY KNOWLEDGE

This chapter will contribute to the body of knowledge in so far as the indigenous knowledge systems are concerned. The research indicates that there have been numerous studies that have been conducted across different continents regarding how names are given to the newly born children. This study will be the first in providing empirical research in how names are given to newly born babies across the different generations within the African continent.

FUTURE RESEARCH DIRECTIONS

In this study, the research focused mainly on investigating on how the Batswana people in Gauteng province give names to their newly born babies across different generations. The findings of the study are insightful and could assist future researchers to build upon them in expanding name giving as an Indigenous Knowledge System across communities.

As a future study, interested researchers can extend their investigation beyond one province (Gauteng) to other provinces such as Northwest and Northern Cape in which the Batswana people are located. Another angle that the future researcher can take, they may go beyond generational mix and investigate names giving across intermarriages or relationships in different races and ethnicities.

CONCLUSION AND RECOMMENDATIONS

It is clear from this discussion that life evolves across generations. This results in some of the traditional and cultural norms being eroded or not compiled too as they used to be. Societies also experience drastic urbanization, and the communities are now becoming inclusive and not catering to one clan. It is understandable to say that in so far as name giving as per generations, there is a need to accommodate each side's point of view instead of holding of the expected traditional and cultural practices in so far as name giving. Bringing this embodiment of the Setswana proverb "*Tshika fa eya baneng e a kgaoga*" will assist in better understanding why the Euro-western names are gradually fading out and why generations like Generation Z are more liberal and not consultative in areas like name giving which earlier generations like the Baby Boomers see this process as binding since it strengthens the family trees.

It is noticeable that in so far as Generation Z are indicating that the names that were given to their children are not inherited. This contrasts with the other generations, especially Generation X and Millennials, though they have their own names which they gave to their children, they were within the family pedigree.

Based on the results and the discussion, the researcher recommended that:

- The traditional norms and values regarding name giving are to be upheld across all generations.
- Name giving be a collaborative and split responsibilities amongst all family members hence this will reduce dominance.
- An in-depth discussion amongst all the interested parties should be held regarding giving inherited names to the newly born babies since this will assist in reducing the negative psychological bearing in the near future (lead to lower self-esteem).

REFERENCES

Alverson, M. (1999). The New Generation Gap. *Woman in Business, 51*(3), 14–17.

Baranov, A., Albitsky, V. & Ivanova, A. (2012). The incidents trends and the state of health of the children population of the Russian Federation. *The Russian paediatric journal*, 6, 4–9.

Broderick, A. (1998). Managing Diversity. *Women in Business, 44*(3), 13–14.

Campbell, M., & Campbell, T. (1996). Behind the name. https://www.behindthename.com/name

Creswell, J. W. (1998). *Qualitative Inquiry and Research Design: Choosing Among Five Traditions* (2nd ed.). Sage Publications Inc.

Cultures, P. (n.d.). Naming children: traditions in 13 different countries. https://pocketcultures.com/2011/04/13/children-naming-traditions/

De Klerk, V. (1999). Beauty or Buhle? On changing one's name. *Nomina Africana, 13*(1&2), 41–44.

Diyal, M. (2021). Rethinking African Indigenous Knowledge and Practices for Environmental Sustainability in Africa. In V. Gumede., M. Muchie & A. Shafi (Eds.), *Indigenous Systems and Africa's Development* (pp. 48-57). Capital Press.

Govitvatana, W.V. (2001). Generation gap in the workplace between Baby Boomers and Generation X. [Unpublished masters' dissertation, University of Wisconsin-Stout, Madison Wisconsin].

Guma, M. (2001). The cultural meaning of names among Basotho of Southern Africa: A historical and linguistic analysis. *Nordic Journal of African Studies, 10*(3), 265–279.

Hedden, J. (2000). Twenty Somethings: Managing the New Breed of Employee.

Hoorens, V. (2015). What's really in a Name-letter Effect? Name-letter preferences as indirect measures of self-esteem. *European Review of Social Psychology, 25*(1), 228–262. Advance online publication. DOI: 10.1080/10463283.2014.980085

Howe, N., & Strauss, W. (1997). *The Fourth Turning: What the cycles of History tell us about Americas' Next Rendezvous with destiny*. Broadway Books.

Kiema, K. (2010). *Tears for my Land: A Social History of the Kva of the Central Kalahari Game Reserve, Tc 'amnquoo*. Mmegi Publishing House.

Koma, H. M. (2012). *An analysis of given and inherited names among the Northern Sotho speaking people in Moletjie and Sekhukhune, Limpopo Province: An onomastic perspective* [Unpublished master's dissertation, University of Limpopo, Limpopo]. http://pocketcultures.com/2011/04/13 children-naming-traditions/

Makoni, B., Makoni, S., & Mashisi, P. (2007). Naming practices and language planning in Zimbabwe. *Current Issues in Language Planning, 8*(3), 437–467. DOI: 10.2167/cilp126.0

Manion, J. D. (2005). The expression of happiness in naming practices. *Journal of Onomastics Research, 12*(1), 45–62.

Mannheim, K. (1952). The problem of generations: Essays of the Sociology of Knowledge. In Kecskemeti, P. (Ed.), *Essays* (pp. 276–322). Oxford University Press.

Ogunniyi, J. (2021). The Development Sustainability of Indigenous Technologies in Africa. In V. Gumede., M. Muchie & A. Shafi (Eds.), *Indigenous Systems and Africa's Development* (p. 37). Capital Pres Okrent, D. (2000). Twilight of the boomers. *Time*, 155(24), 68-72.

Raphelson, S. (2014). From GIs to Gen Z (or is it iGen?): How generations get nicknames. National Public Radio. Available: https://www.npr.org/2014/10/06/349316543/don-t-label-me-origins-of-generational-names-and-why-we-use-them

Republic of South Africa. (1996). *The Constitution: Bill of Rights*. Government Printers.

Saldanha, J. (2009). *The coding manual for qualitative researchers* (2nd ed.). Sage Publication Inc.

Tomanek, P. (2023). Birth order, Gender and Naming Preferences in Contemporary Poland. Names: A Journal of Onomastics, 71(1). DOI: https://doi.org/DOI: 10.5195/names.2023.2387

United Nations. (1989). *Constitution on the Rights of the Child*. UNESCO Publication.

Wikipedia. (n.d.). The Free Encyclopedia. https://en.wikipedia.org/wiki/Proverbs

Yin, R. K. (2003). *Case Study Research: Designed Methods* (3rd ed.). Sage Publication Inc.

Zetlin, M. (1995). Is it worth keeping older salespeople? *Sales and Marketing Management*, *147*(4), 144–147.

ADDITIONAL READING

Bruhn, A., Huschka, D. & Wagner, G.G. (2012) Naming and war in modern GermanDOI: https://doi.org/DOI: 10.1179/0027773812Z.00000000011

Godrington, G. Z. (2008). Detailed Introduction to Generational Theory: Tomorrow Today. Tomorrow Today Ltd. https://www.tomorrowtoday.uk.com

Kanyane, F. M., & Rakgogo, T. (2023). An Analysis of Personal Naming Approaches by Younger and Older Generations among the Bapedi Nation: An Onomastic Comparison. *Gender & Behaviour*, *20*(4).

Mahwasane, M.M. & Tshifaro, T.T. (2019). New-born Baby Naming Practices of the Vhavenda: A

Perspective, S. A. (•••).. . *South African Journal of African Languages*, *39*(2).

Zungu, E. B., & Zulu, S. P. (2018). Elimination by Substitution: Magnifying the Importance of Self through Personal Name Choice. *Indilinga*, *17*(2).

KEY TERMS AND DEFINITIONS

Indigenous practices: Are traditional knowledge, skills and techniques that are passed from generation to generation in communities.

Inherited names: Are hereditary and passed down from generation to generation and they are used for identifying and distinguishing an individual within the family pedigree.

Generations: Is a group of individuals of different ages who have lived during the same period and shaped by the same events, hence their values are common.

Generational theory: Argues that people of similar age or within the same generation share key characteristics due to similar experiences.

Given names: Are names that are given to the newly born babies by their parents, and they are not associated with any inherited names. They are not common, and they are unique, hence they distinguish an individual from other members of the family.

Chapter 14

Integrating Tradition and Innovation:

The Role of Indigenous Knowledge Systems in Modern Agriculture

Sanjeev Kumar
https://orcid.org/0009-0007-6634-020X
Lovely Professional University, India

N. D. Chethan Patil
https://orcid.org/0009-0001-2977-6422
Centre for Aromatic Plants, India

Harpinder Singh Sandhu
https://orcid.org/0000-0003-1104-5274
Lovely Professional University, India

Sapna Jarial
https://orcid.org/0000-0003-2660-9036
Lovely Professional University, India

ABSTRACT

This chapter examines the integration of Indigenous Technical Knowledge (ITK) with modern agricultural practices to promote sustainability, productivity, and resilience in farming systems. Rooted in centuries of adaptation to local environments, ITK encompasses diverse methods such as water harvesting, soil health management, animal health practices, and biodiversity conservation. Highlighting practices like the khadin system, johads, panchagavya, and sacred groves, this chapter showcases ITK's relevance in addressing contemporary agricultural challenges. It explores the potential of combining ITK with advanced technologies like the Internet of Things (IoT), Geographic Information Systems (GIS), and Machine Learning (ML) to optimize resource use, improve soil fertility, and conserve biodiversity. The chapter addresses the challenges of merging traditional wisdom with modern innovations, emphasizing this integration as a vital strategy for agricultural sustainability in the face of climate variability and resource constraints.

DOI: 10.4018/979-8-3373-4124-8.ch014

INTRODUCTION

To thrive and advance, humans seek knowledge. Understanding and navigating our environment teaches us how to survive but finding meaning and purpose is necessary for advancement. Indigenous cultures have preserved natural landscapes for thousands of years by creating knowledge systems for ecological stewardship, survival, values, and purpose. Sustainability and cultural heritage are harmed by the loss of wisdom. We must make use of all the resources and knowledge systems at our disposal to meet the Sustainable Development Goals by 2030. Globally as well in India centuries-long accumulation of knowledge, customs, and inventions is known as its indigenous knowledge systems (IKS). Numerous agricultural and animal agriculture disciplines are covered in this collection. Even though these knowledge systems have the potential to significantly advance modern science, they are currently frequently overlooked (Kumar, 2024).

Understanding and sharing indigenous knowledge(IK) is essential for the society's farming progress, but it is still hard to record and share. It's possible for agriculture to grow faster with IK. Therefore, for IK to have a good effect on agricultural growth, effective steps should be taken to reduce the problems (Ekobi et al.,2023).This includes greater urgency to address gaps in integrating Indigenous Technical Knowledge (ITK) with advanced artificial intelligence(AI) technologies. Despite, emerging artificial technologies revolution such as Internet of things (IoT), Machine learning (ML) and modern technologies such as Geographic Information Systems (GIS), are transforming knowledge systems, ITK remains marginalized. ITK can help achieve global sustainability by understanding and negotiating the intricacies and interdependencies of sustainable social development. Modern techniques can be combined with traditional agricultural practices to encourage sustainable farming practices and protect natural resources. This will undoubtedly bolster the acceptability of technologies for sustainability and guarantee the livelihood security of farmers (Pandey et al, 2025). Therefore, this chapter aims to provide documentation of ITK related to agricultural and animal agricultural practices along with integration with advanced technologies. Therefore, the research question(s) were:

1. What are the ITK related to agricultural and animal agricultural practices?
2. How can the integration of Indigenous Technical Knowledge (ITK) with advanced emerging technologies can enhance agricultural resource management
3. What are the challenges in blending ITK with modern technologies

Using a review of literature methodology, it explores the potential of combining ITK with advanced technologies like the IoT, GIS, and ML to optimize resource use, improve soil fertility, and conserve biodiversity. The chapter addresses the benefits and challenges of merging with modern innovations, emphasizing this integration as a vital strategy for life agricultural sustainability in the face of climate change.

CONCEPTUAL AND THEORETICAL FRAMEWORK

The blending of IKS with modern technologies such as the IoT and ML forms a transformative approach to sustainable development. This theoretical framework bridges traditional wisdom and cutting-edge technology, promoting inclusive, sustainable, and culturally sensitive solutions.

People usually say that ITK is different from western scientific knowledge in terms of what it is and how it knows things. However, many of these differences have been called into question as Western science has been shown to be situational and performative(Goldman & Lovell, 2017). The inquiry that arises is: what is the meaning of "ITK?" ITK is one of term of local knowledge. Local knowledge has various terms or synonyms (see Table 1). ITK is the knowledge possessed by Indigenous people. The hardest part is defining ITK with so many terminology and synonyms. Setting ITK in context solves this puzzle. ITK, or local knowledge, is influenced by its immediate surroundings but not determined by it. In India, skilled and knowledgeable specialists utilize indigenous techniques, but their expertise is not extensively documented and is passed down orally.

Table 1. Synonyms of indigenous technical knowledge

Term	Synonyms/Contexts
Indigenous Knowledge	Endogenous knowledge; Native knowledge, expertise; Local knowledge; Sustainable knowledge; Traditional knowledge; Autochthonous knowledge; People's knowledge; Folk knowledge; folk science; folk competence; Little tradition; Community knowledge; Cultural knowledge; cognition (restricted sense);Ethnic knowledge; Culturally specific knowledge
Ethnoscience	Local knowledge; previously a field of research on cultural knowledge systems
Cultural Knowledge System	Replacing 'ethnoscience'; less scientific connotation
Cultural Belief System	Cultural meaning system
Everyday Knowledge	Practical knowledge, mundane cognition, vernacular, common sense, science of the concrete
Experiential Knowledge	Experimental knowledge
Farmers' Knowledge	Peasant knowledge
Culturally Integrated Knowledge	Knowledge of small, marginal, non-western groups; of internal origin
Knowledge of Nature	Closeness to nature; rooted in religion, culture, or society
Sustainable Knowledge	Sustainable within natural and cultural environments
Oral Tradition	Handed down, implying static, with low levels of change
Political Resistance Knowledge	Broadly disseminated, opposed to elite knowledge
Traditional Rural Knowledge	Tied to rural or industrial societies; oral knowledge
Specific Knowledge	Linked to small social units; culturally integrated and practice-oriented
Ethnic Knowledge	Related to ethnic 'we' groups; implies specificity and singularity
Scientific Knowledge	Ethnobotany, ethno-sociology, ethnomedicine, etc.; systematic rules and structures
Informal Knowledge	Practical, applied knowledge; contrasts with academic, expert, or ritual knowledge

continued on following page

Table 1. Continued

Term	Synonyms/Contexts
Trial-and-Error Knowledge	Knowledge gained through experimentation, contrasted with controlled studies
Farm-Based Knowledge	Relating to farms as economic units
Dependent Knowledge	Implies experiences of dependency, contrasted with elite knowledge

(Source: Antweiler, 1998).

ITK is dynamic, is changing and will continue to change. ITK is tacit, intuitive, experimental, informal uncodified knowledge, which will always be needed and generated, since there will always be an interface between expert knowledge and real-world situations, no matter how much we rely on literate knowledge with authority, technical expert validation, and systematic availability. It will always need to be translated and adapted to local contexts and reconfigured culturally by folks without formal or book expertise.

Figure 1. Conceptual and theoretical frameworki

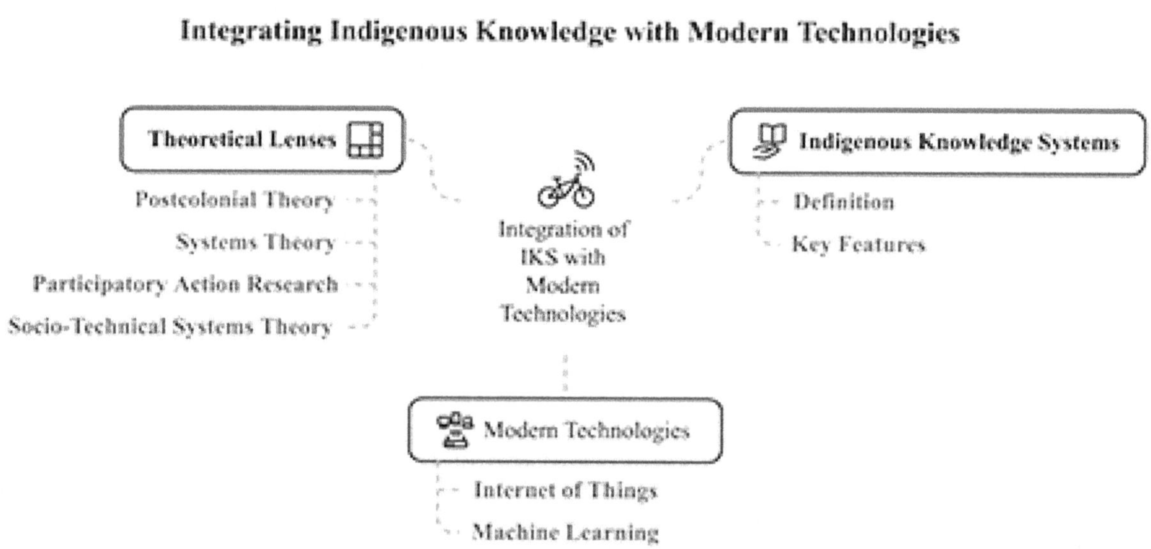

MODERN TECHNOLOGICAL FRAMEWORK

A network of linked devices known as the IoT gathers, sends, and evaluates data to facilitate effective decision-making (Gillis, 2021). Without explicit programming, machine learning (ML) algorithms allow machines to recognize patterns in data, learn from them, and make predictions or decisions (Mitchell,

1997). Another framework for collecting, organizing, and evaluating geographical and spatial data is GIS (DeMers, 2009).

Several theoretical frameworks can be used to effectively understand how IKS is integrated with contemporary technologies. Decolonizing knowledge is promoted by postcolonial theory, which also challenges the predominance of Western-centric scientific paradigms in technological interventions and emphasizes the value of indigenous perspectives (Briggs & Sharp, 2004). According to Segueda and Banerjee (2024), systems theory sees technology and knowledge systems as interrelated parts that support mutual learning and holistic integration. Active community involvement in technological development is emphasized by Participatory Action Research (PAR), which also promotes knowledge co-creation and makes sure that indigenous voices are in the forefront of technological adaptation (Chilisa, 2012). Last but not least, the Socio-Technical Systems Theory acknowledges the interdependence of society and technology and promotes the creation of innovative technologies that are also socially and culturally acceptable (Walker et al., 2008). When combined, these lenses offer a thorough framework for integrating IKS with contemporary technological developments.

METHODOLOGY

A narrative review was conducted on the research questions- 1).What are the challenges in blending ITK with modern technologies; 2).How can the integration of Indigenous Technical Knowledge (ITK) with advanced emerging technologies can enhance agricultural resource management and 3). What are the challenges in blending ITK with modern technologies?

In addition to their potential enhancements through cutting-edge technological applications like IoT, ML, PA, and GIS, the scope encompassed conventional practices like biodiversity conservation, soil health management, animal health practices, and water harvesting. Using the aforementioned research questions, a thorough analysis of scholarly databases including Google Scholar, EBSCO, PubMed, and SCITE.ai was carried out. Book chapters, reports, research, and review papers were among the requirements for inclusion, which included only English-language literature. Grey literature from agricultural organizations like the National Institute of Agricultural Extension Management (MANAGE) and the Indian Council of Agricultural Research (ICAR), Government of India websites were taken into consideration. As a result, 77 scientific publications were included. Only the title was read at first, and then the abstract. Full papers were read after the abstract was linked to keywords. To give a thorough picture of how traditional knowledge and contemporary technology can interact to produce more sustainable agricultural systems, the review ended with implications for real-world applications in agriculture policy recommendations as well as possible directions for further study.

FINDINGS AND DISCUSSION

ITK related to Agriculture and Animal Agriculture Practices

Surface water harvesting systems, soil fertility and health, biodiversity conservation, and animal health management in India are the topics covered in the ITK documentation pertaining to agricultural practices.

In India, check dams, ponds, tanks, and canals are examples of surface water harvesting systems that support sustainable water management and conservation. Tankas, johads, and khadins are designed to store runoff water for irrigation and yield enhancement in arid regions (Hassan et al., 2023). ITK pertaining to water-saving technology practices in India's semi-arid regions is shown in Table 2.

Table 2. Surface water harvesting in indigenous agricultural systems

Category	Description	Historical Origins	Benefits	Challenges	References
Khadin System	Runoff farming method using a large bund to capture water and soil sediments, creating a reservoir for cultivation.	Originated in the 14th century, Rajasthan. Similar to ancient practices in the Middle East and historical systems in Iraq, Negev Desert, and Colorado.	Enhances soil fertility and agricultural yields, reduces salt problems.	Location-specific, water loss due to evaporation.	Rani & Paul, 2023; Francis et al., 2024; Baibout et al., 2022; Verma et al., 2024
Johads	Crescent-shaped earthen check dams built over sloping terrain to collect and store monsoon runoff, improving groundwater and soil moisture.	-	Recharges aquifers, supports biodiversity, and revitalizes local ecology.	Encroachment by urbanization, reduced community involvement.	Yadav, 2023; Das, 2024; Mahato et al., 2020
Tankas	Underground cisterns collecting rainwater from various sources, designed to store water efficiently and sustainably.	Popular in Rajasthan's Thar Desert.	Provides water security, reduces reliance on external water sources.	Dependent on catchment area efficiency and construction materials.	Eckardt et al., 2023; Yadav, 2023

Soil Fertility and Health

The values of ITK and organic farming (OF) are based on the ideas of sustainability, reverence for the natural world, and peaceful coexistence with the environment (Tripura et al., 2024). ITK solutions, such as Tank Silt Application, Jeevamrutha, Panchagavya, and Compost Pit, offer cost-effective and sustainable alternatives to modern chemical fertilizers and irrigation methods. Table 3 below lists the ITK techniques used by various Indian tribes for soil fertility and health.

Table 3. Traditional practices for improving soil fertility and health

ITK	Location and Community	Purpose	Source
Panchagavya	Tamil Nadu, Traditional Farmers	Application of a mixture of cow dung, urine, milk, curd, and ghee to enhance soil health and promote plant growth	Rijal et al., 2023
Jeevamrutha	Karnataka, Rural Communities	Using a fermented blend of cow dung, urine, jaggery, pulse flour, and soil to improve soil fertility	Khan et al., 2022
Compost Pit	Various regions, Multiple Communities	Constructing compost pits to decompose organic waste, resulting in nutrient-rich compost for soil enrichment	Siddique et al., 2024
Tank Silt Application	Andhra Pradesh, Rural Farmers	Spreading tank silt on fields to boost soil fertility and improve the soil's water retention ability	Jain et al., 2024

Biodiversity Conservation

As a social and political process rather than merely a technical or biological endeavor, biodiversity conservation encompasses a variety of values and discussions. It is not only influenced by biological sciences, but also by global economic forces and power dynamics (Martin, 2023). Biodiversity is a fundamental component of sustainable farming methods in India. Ecosystem stability is improved by the blending of various cultures, which results in a diversity of livestock and crop species. In order to effectively manage and conserve natural resources, ITK practices are essential (Lakshmypriya & Varghese, 2024; Barman et al., 2024). Table 4 below lists the different ITK techniques used for biodiversity conservation by various communities throughout India.

Table 4. Traditional practices for biodiversity conservation

ITK	Purpose	Source
Sacred Groves	Protecting forests due to cultural and religious significance, ensuring biodiversity conservation.	Singh et al., 2023
Traditional Agroforestry Systems	Integrating trees, crops, and livestock in a manner that maintains ecological balance.	Kumar et al., 2024
Vrikshayurveda	Maintaining plant health, soil fertility, and pest control using natural, sustainable methods.	Singh et al., 2023
Zabo Farming System	Combining forestry, agriculture, and livestock management to support sustainable farming systems.	Pandey et al., 2022
Community Seed Banks	Preserving indigenous seed varieties to maintain genetic diversity and promote agricultural resilience.	Westengen et al., 2023
Participatory Forest Management	Encouraging community involvement in forest conservation to ensure sustainable management.	Sekar, 2024

Animal Health Management

In India, there are numerous ITK approaches for managing animal health. These traditional methods and treatments are employed by various groups to treat common illnesses and preserve the health of cattle. A thorough summary of the varied ITK procedures used by various tribes throughout India for the management of animal health is given in Table 5 below.

Table 5. ITK for animal health management

Traditional Practice	Region & Community	Application	Reference
Herbal Remedies for Deworming	Various areas, rural populations	Utilizing herbs such as neem, garlic, and pumpkin seeds to eliminate intestinal worms in livestock	Sharifian et al., 2022
Turmeric and Mustard Oil Treatment	Multiple regions, rural populations	Applying a blend of turmeric and mustard oil to wounds to prevent infections and aid in healing	Sharma et al., 2022
Indigenous Veterinary Methods	Various locations	Employing herbs and other natural materials for animal healthcare	Bhat et al., 2023
Local Knowledge of Wildlife	Various areas, diverse communities	Indigenous understanding of animal behaviors and habitat preferences	Aboorva et al., 2024
Cow Dung Cake Fumigation	Himachal Pradesh, rural populations	Burning cow dung cakes to create smoke that repels flies and mosquitoes around livestock	Barman et al., 2024

INTEGRATION OF ITK WITH MODERN TECHNOLOGY

By combining ITK with modern farming practices, there is a significant opportunity to increase agricultural productivity, resilience, and sustainability. Issues and problems related to declining food and nutrition security and agricultural yield failure may be further resolved by the integration. This adaptability is crucial, especially in light of climate change, where traditional approaches may have an impact on modern resilience strategies (Kamakaula, 2024). For example, by enabling people to make decisions based on local climate conditions, indigenous knowledge helps South African communal farmers lower their risk of drought (Muyambo et al., 2017). Farmers have long relied on ITK for weather forecasting, which aids in planting and irrigation decisions (Bheemappa et al., 2022). Such integration may lead to more sustainable farming practices that utilize both traditional knowledge and new scientific findings.

ITK is dynamic, allowing for the incorporation of new ideas and technological advancements to create a hybrid knowledge system that can enhance farming practices. Integrating outside technologies into native knowledge systems can lead to better agricultural outcomes (Barker & Beckford, 2006). By highlighting the importance of hybrid knowledge in assisting the Tharu people of Nepal in implementing climate-resilient farming practices, Chaudhary et al. support this point of view (Chaudhary et al., 2022). By recognizing the value of both scientific and indigenous knowledge, farmers can develop strategies that are better suited to their particular environmental circumstances.

Rain Water Management

One example of how traditional knowledge can be reinforced to address contemporary water issues is India's rainwater management project, which blends Indigenous Technical Knowledge (ITK) with modern technology. Developed over many generations, ITK provides sustainable methods such as contour bunding, stepwells, and tanks and reflects a deep understanding of the local climate and hydrology. Modern technologies like machine learning, Internet of Things sensors, and GIS mapping are now being added to these systems to increase their efficacy and adaptability. Traditional water management systems, like stepwells in Rajasthan and tanks in South India, are being revived with the use of GIS and remote sensing technologies. This makes it possible to precisely identify and repair damaged structures (Liladhar Rane et al., 2024; Pande et al., 2024). GIS-based design has been used to restore tanks in Rajasthan's watershed projects, enhancing their ability to hold and replenish water (Vivek, 2016).

In urban environments, ITK practices are being modernized through the use of IoT sensors and predictive analytics. In cities like Bangalore and Chennai, IoT systems have been integrated into percolation pits and rainwater harvesting units, enabling real-time monitoring of water level and quality (Malik et al., 2007; Vivek, 2016). ML further enhances these systems and increases ITK's resilience to climate change by predicting storage needs based on weather patterns.

Modern hydraulic modeling has improved the design of check dams, a structure inspired by ITK. Computational tools like fluid dynamics models enhance groundwater recharge and water retention, which is especially beneficial for drought-prone regions like Bundelkhand (Vivek, 2016). Drone mapping and soil moisture sensors have significantly increased agricultural yield and water retention in traditional micro-catchments in Andhra Pradesh and Telangana. By integrating ITK with cutting-edge technologies like satellite imagery and the IoT, India is creating adaptable, sustainable water management solutions for the future in addition to revitalizing outdated systems.

Soil Conservation

Modern technology is transforming ITK to offer improved soil conservation solutions that address problems like climatic variability, nutrient depletion, and erosion. ITK methods like contour bunding, terracing, and vegetative barriers have long been used to improve fertility and decrease soil erosion (Siddappa et al., 2024). Modern technological advancements like drones, GIS mapping, and Internet of Things-based systems are elevating these methods by making them more precise and adaptable to changing environmental conditions. Geospatial methods like GIS and remote sensing are crucial for planning for soil conservation and evaluating landscapes for areas vulnerable to erosion. These technologies aid in the construction of precise interventions such as contour bunds and check dams based on slope gradients and flow patterns. In Madhya Pradesh, the use of GIS-enabled watershed projects has improved groundwater recharge and reduced erosion more effectively than traditional terracing methods (Paroda, 2019).

Drone Mapping and Digital Elevation Models (DEMs)

Drones provide high-resolution topographical data that allows traditional soil stabilizing methods to be deployed precisely. In regions like Andhra Pradesh, drones have been used to identify vulnerable areas, facilitating the integration of modern contouring techniques with ITK customs like vegetative barriers (Vishweshwaran & Sujatha, 2024). Drones and remote sensing technologies are also revolutionizing ITK methods by providing high-resolution imagery for crop health monitoring and weed infestation detection. Thanks to these technologies, traditional intercultural practices like mulching and row spacing can now be planned and executed more effectively, ensuring resource efficiency and yield increases.

IoT Technology

This technology makes it possible to track soil moisture and health in real time, which improves the precision of ITK methods. When combined with traditional techniques like field bunding, IoT-based soil sensors maximize irrigation, reduce water waste, and prevent over-saturation—all of which are critical for maintaining soil structure. These linkages have worked well in Telangana, where sensor technology is being used to better control traditional bunds. IoT-based sensors and real-time data systems are enabling smarter multicultural management. Sensors that monitor soil moisture, nutrient levels, and weed growth can help farmers perform manual weeding and intercropping more accurately (Singh et al., 2022). By providing notifications about when it's best to weed and apply nutrients, IoT systems in conjunction with ITK techniques have enhanced cross-cultural operations in Andhra Pradesh and reduced dependency on chemical inputs.

Machine Learning Algorithms

Machine learning (ML), which uses environmental and historical data to make predictions to reduce erosion risks, is another innovation. By ensuring its relevance and effectiveness in tackling today's soil conservation concerns, these observations aid in the modernization of ITK (Chakraborty et al., 2020). ITK offers scalable, long-term solutions for preserving soil health and increasing agricultural productivity when paired with cutting-edge machinery. Machine learning algorithms are creating ITK and ensuring that these techniques remain effective in modern, high-demand farming systems by analyzing historical crop data to recommend the best tillage schedules and intercultural practices. Scalable and sustainable solutions in agriculture are becoming possible thanks to the combination of ITK and technology.

Tillage and Intercultural Management

By combining ITK with modern technology, tillage methods and intercultural management are being transformed, providing more sustainable and efficient agricultural solutions. ITK practices like hand weeding, low tillage, and intercropping have long been used in conventional farming systems to improve soil health and crop yield. These methods are being enhanced by modern innovations such as precision agricultural technologies, data analytics, and Internet of Things-based monitoring, which enable farmers to maintain traditional knowledge while maximizing resources (Beevi et al., 2023).

Precision Agriculture

Precision agricultural technologies, such as GPS-guided equipment and variable-rate technology (VRT), are improving the accuracy and efficiency of tillage operations. These tools minimize soil disturbance by carefully guiding ploughing to specific areas, which lowers erosion and preserves soil structure. Strip tillage, an ITK-based method that increases water retention and reduces energy inputs, is being practiced by farmers in Madhya Pradesh with GPS-enabled tractors.

Biodiversity Conservation

Combining ITK with contemporary technological advancements is significantly improving biodiversity protection in India. ITK incorporates centuries-old practices like holy groves, natural pest control, and traditional farming methods that protect ecosystems and species diversity. Modern innovations that offer precise and scalable solutions, such as GIS mapping, remote sensing, and data analytics, are supporting these traditional techniques.

Sacred woods, which indigenous tribes have historically conserved as biodiversity hotspots, are now mapped and monitored using GIS and remote sensing technologies. By doing this, these crucial ecosystems are successfully managed and encroachments and biodiversity loss are prevented (Gaikwad et al., 2004). ITK-based farming practices like intercropping and agroforestry, which preserve a range of crop and tree species and thereby promote biodiversity, are also being improved by digital technologies. By employing IoT-based sensors to track soil and microclimate variables, farmers can implement practices that enhance ecosystem health while preserving indigenous knowledge (Parkash et al., 2009).

ITK is used to assess the impact of agricultural practices on regional biodiversity in conjunction with modern conservation techniques like deep machine learning and predictive modeling. These tools help communities enhance traditional practices while ensuring they are in line with environmental goals. Without putting non-target species in danger, the use of contemporary biological agents in conjunction with traditional pest control methods has improved pest management (Sharma et al., 2024). Participatory Forest Management (PFM), a community-driven approach to forest restoration, has benefited from GIS mapping and advanced modeling tools. These technologies enhance the efficacy and flexibility of conservation initiatives at the local level by providing useful insights (Barrett et al., 2001). By combining ITK and technology, India is developing workable, culturally sensitive solutions for biodiversity conservation that ensure the sustainability of ecosystems and the life they support.

Animal Health Management

Traditional ethno-veterinary treatments that use plant-based therapies to treat cattle illnesses have been scientifically validated and scaled up by modern pharmacological investigations. Two common plants that are used to treat parasites and mastitis are Cedrus deodara and Ocimum tenuiflorum. The effectiveness and safety of these treatments for widespread usage have been confirmed by biochemical analysis (Chaudhary et al., 2011). This integration promotes sustainable cattle management by significantly reducing the need for synthetic drugs. ITK practices in livestock breeding strongly emphasize selective breeding based on traits like coat color and physical robustness, especially for local breeds like Poda Thurpu cattle. Modern biometric technologies and genetic testing are currently being used to improve these selection criteria, increasing the genetic purity and productivity of these breeds. This approach, which combines traditional knowledge with modern methods, has improved the draught power and disease resistance of cow populations in Telangana (Siripurapu et al., 2022).

Animal husbandry is being revolutionized by real-time health monitoring enabled by IoT technology. In order to support ITK strategies like using natural diets to prevent illness, sensors track the vital signs, feed consumption, and activity levels of livestock. These technologies improve early illness diagnosis and resource allocation, ensuring better animal welfare (Zhang et al., 2024). ITK combined with modern technology not only improves animal husbandry but also boosts climate variability resistance, improves rural livelihoods, and promotes sustainable livestock production in India.

Traditional ethno-veterinary treatments that use plant-based therapies to treat cattle illnesses have been scientifically validated and scaled up by modern pharmacological investigations. Two common plants that are used to treat parasites and mastitis are Cedrus deodara and Ocimum tenuiflorum. The effectiveness and safety of these treatments for widespread usage have been confirmed by biochemical analysis (Chaudhary et al., 2011). This integration promotes sustainable cattle management by significantly reducing the need for synthetic drugs. ITK practices in livestock breeding strongly emphasize selective breeding based on traits like coat color and physical robustness, especially for local breeds like Poda Thurpu cattle. Modern biometric technologies and genetic testing are currently being used to improve these selection criteria, increasing the genetic purity and productivity of these breeds. This approach, which combines traditional knowledge with modern methods, has improved the draught power and disease resistance of cow populations in Telangana (Siripurapu et al., 2022).

Animal husbandry is being revolutionized by real-time health monitoring enabled by IoT technology. In order to support ITK strategies like using natural diets to prevent illness, sensors track the vital signs, feed consumption, and activity levels of livestock. These technologies improve early illness diagnosis and resource allocation, ensuring better animal welfare (Zhang et al., 2024). ITK combined with modern technology not only improves animal husbandry but also boosts climate variability resistance, improves rural livelihoods, and promotes sustainable livestock production in India.

CHALLENGES OF INTEGRATION OF ITK WITH MODERN AGRICULTURE

There are still challenges to successfully combining ITK with contemporary farming methods. Due to their restricted access to resources and contemporary technology, many smallholder farmers continue to rely significantly on traditional practices (Wang, 2018).

Table 6. Challenges and opportunities in integrating ITK with modern agriculture

Key Opportunities	Description	References
Documentation and Sharing of ITK	Supporting regulations for documentation and sharing of ITK can help preserve valuable knowledge and ensure its application in modern agriculture.	Abioye et al., 2014
Recognition and Acknowledgment of ITK	ITK is often undervalued by the scientific community, leading to reluctance among farmers to adopt integrated practices	Mohan, 2024; Abeywardana et al., 2019
Socioeconomic Constraints	Economic challenges push smallholder farmers towards modern practices, risking the loss of traditional methods.	Yamunarani, 2023; Kamakaula, 2024
Diversity of Indigenous Traditions	The diversity in ITK across different communities complicates integration, requiring localized and context-specific approaches.	Michael, 2015
Institutional Constraints	Institutional gaps and lack of engagement with indigenous communities hinder effective integration of ITK.	Kamakaula, 2024; Jaramillo, 2023
Impact of Industrialization and Globalization	Globalization and industrialization have marginalized traditional food systems, affecting cultural heritage and ITK sustainability.	Ellena & Nongkynrih, 2017; Teo, 2024

The impact of globalization and industrialization on traditional agricultural systems cannot be disregarded. This shift affects food security and diminishes the importance of ITK in contemporary farming practices because younger generations may become disconnected from their cultural heritage (Ellena &

Nongkynrih, 2017; Reddy, 2024). There are numerous challenges to overcome even though integrating ITK into modern farming practices holds great promise for enhancing sustainability and resilience. We must collaborate to value and recognize ITK, promote inclusive decision-making, and develop regional strategies that respect the diversity of indigenous customs in order to overcome these challenges. By removing these barriers, a more all-encompassing agricultural system that benefits from both traditional and modern knowledge can be created, which calls for government assistance.

Indian Government agricultural policies and programs

Through its ten regional centers and the National Bureau of Plant Genetic Resources (NBPGR), New Delhi, ICAR collects, conserves, characterizes, and assesses various germplasms. In its National Genebank (NGB), NBPGR currently has 4,69,793 germplasm accessions, including 96,964 traditional/folk varieties and land races (Tyagi et al., 2021). To carry out its mandated activities, it maintains connections with national crop-based institutes in India and international organizations of the Consultative Group on International Agricultural Research (CGIAR). ITK practices are studied and validated by projects like National Innovations on Climate Resilient Agriculture (NICRA) using contemporary technologies like machine learning and the Internet of Things (Zong et al., 2022).

Established during the Union Budget 2021–2025, the Digital Agriculture Mission (2021–2025) uses IoT, ML, and GIS technologies to modernize agriculture while subtly facilitating ITK integration through data-driven farming. promotes the development of digital platforms and a farmer database that can combine ITK-based insights with contemporary technology (PIB, 2024a).

As part of the National Action Plan on Climate Change, the National Mission on Sustainable Agriculture (NMSA) encourages sustainable farming practices that use ITK to adjust to climate variability. supports precision agriculture by combining ITK practices with satellite data, IoT-based irrigation systems, and GIS to map local agro-ecological zones. Traditional water-saving techniques like "johads" enhanced with IoT sensors are mapped by GIS (GOI, 2025).

The ITK-based Paramparagat Krishi Vikas Yojana (PKVY) promotes organic farming through the use of conventional techniques like composting and biopest control. employs IoT for real-time soil health monitoring and GIS to locate organic farming clusters in order to guarantee the successful implementation of ITK techniques. Example: Using IoT sensors to track the traditional use of panchagavya and evaluate its effect on crop yield (PIB, 2024b).

The National Livestock Mission (NLM) acknowledges the importance of ITK in animal husbandry and focuses on sustainable livestock development. maps traditional grazing areas and tracks livestock health by integrating IoT and GIS; machine learning examines genetic data and traditional breeding preferences. For instance, communities have historically used IoT-enabled collars for health monitoring and GIS mapping of grazing areas (GOI, 2014).

Agri-Tech Startups and Public-Private Partnerships: Under programs such as the Agriculture Accelerator Fund, these companies work with private entities and startups to integrate ITK with contemporary tools. They do this by using ML and IoT to provide predictive analytics that integrate ITK data, thereby combining traditional farming knowledge with contemporary precision farming methods (India, 2016).

Under the "10,000 FPO Scheme," digital platforms and Farmer Producer Organizations (FPOs) encourage the digitization of farm collectives, allowing them to effectively integrate ITK and technology. For resource efficiency, FPOs embrace Precision Agriculture and share ITK practices via mobile apps and IoT devices (India, 2016).

FUTURE RESEARCH DIRECTIONS

ITK has proven to be a robust storehouse of sustainable methods based on centuries of local knowledge and adaptation. To make sure that these practices are not lost to modernization, future research should concentrate on methodically documenting them. The focus should be on creating approaches that combine ITK with current scientific studies, utilizing fields like climate modeling, biotechnology, and predictive analytics. Through this integration, the creation of hybrid systems that preserve the ecological harmony inherent in ITK while leveraging the accuracy of contemporary techniques may be further explored. Additionally, a more thorough grasp of how ITK can be improved and expanded to benefit larger communities may be possible through interdisciplinary research involving climatologists, agronomists, ethnobotanists, and technology specialists.

CONCLUSION AND RECOMMENDATIONS

ITK, which skilfully blends ecological stewardship with sustainable methods tailored to local conditions, is the embodiment of generations of traditional wisdom. ITK has proven beneficial in areas like biodiversity, soil health, water conservation, and animal care, but it faces challenges from modernization, globalization, and a lack of legislative support. Combining ITK with modern technologies offers a huge opportunity to increase agricultural sustainability and resilience. But doing so requires addressing socioeconomic limitations, cultural marginalization, and institutional gaps. By valuing the diversity of ITK, empowering individuals, and putting inclusive, regional strategies into practice, we can develop a comprehensive agricultural system that strikes a balance between traditional knowledge and scientific advancement for a sustainable future. ITK provides insightful information about environmentally friendly and culturally sensitive sustainable farming methods. ITK continues to play a crucial role in the livelihoods of rural and indigenous communities in spite of the difficulties presented by globalization and technological advancement. In addition to increasing agricultural sustainability and productivity, integrating ITK with contemporary agricultural technologies guarantees that traditional knowledge will continue to develop and adapt rather than be marginalized. In order to maintain agriculture in the face of climate change and growing human populations, this integration promotes biodiversity, conserves resources, strengthens soil health, and improves water management.

It is advised to create extensive national and international databases for recording ITK practices to facilitate the integration of ITK with contemporary agricultural technologies and guarantee accessibility for upcoming studies and applications. It is recommended that governments and international organizations implement policies that acknowledge and provide financial and subsidizing incentives for ITK. To educate the next generation of farmers and scientists about the importance of ITK, educational programs should include it in agricultural curricula. Additionally, encouraging community involvement through NGOs and local leaders will guarantee culturally aware technology adoption. Innovations ought to concentrate on developing flexible technologies for ITK procedures, like IoT devices specifically designed for conventional farming systems. Encouraging interdisciplinary cooperation will result in solutions that are both culturally appropriate and scientifically sound. Last but not least, coordinating these efforts with the Sustainable Development Goals will help the agricultural sector become more resilient and sustainable while also advancing global goals in biodiversity, climate action, and water conservation.

REFERENCES

Abeywardana, N., Schütt, B., Wagalawatta, T., & Bebermeier, W. (2019). Indigenous agricultural systems in the dry zone of Sri Lanka: Management transformation assessment and sustainability. *Sustainability (Basel), 11*(3), 910. DOI: 10.3390/su11030910

Abhay, R. K. (2024). Evaluating the Role of Indigenous Methods of Rainwater Harvesting for Water Sustainability of Local Ecosystems in India. In *Sacred Landscapes, Indigenous Knowledge, and Ethno-culture in Natural Resource Management: Understanding Multiple Perspectives of Nature Conservation* (pp. 121–134). Springer Nature Singapore. DOI: 10.1007/978-981-97-4206-6_6

Abioye, A., Zaid, Y., & Egberongbe, H. (2014). Documenting and disseminating agricultural indigenous knowledge for sustainable food security: The efforts of agricultural research libraries in Nigeria. *Libri, 64*(1). Advance online publication. DOI: 10.1515/libri-2014-0007

Aboorva, T., & Mani, R. (2024). Contribution of Environmentalists in Preserving Global Environment. Quing. *International Journal of Multidisciplinary Scientific Research and Development, 3*(2), 1–18.

Antweiler, C. (1998). Local knowledge and local knowing: An anthropological analysis of contested cultural products in the context of development. [Paper presented at the seminar on Pastoralism and Environment organized by JET, Arusha.]. *Anthropos: Studie z Oboru Anthropologie, Paleoethnologie, Paleontologie a Kvartérní Geologie, 93*(4-6), 469–494.

Baibout, M., Corcket, E., Kothari, S. L., & Fievet, V. (2022). Ecosystem services provided by Prosopis cineraria (L.) Druce in the drylands of Southern and Western Asia. *Botany Letters, 169*(1), 30–42. DOI: 10.1080/23818107.2021.1981437

Barker, D., & Beckford, C. (2006). Plastic yam and plastic yam sticks – perspectives on indigenous technical knowledge among Jamaican farmers. *Tijdschrift voor Economische en Sociale Geografie, 97*(5), 535–546. DOI: 10.1111/j.1467-9663.2006.00362.x

Barman, B., Ghosh, B., Ranjan, A., & Quader, S. W. (2024). The Potential of Indigenous Technological Knowledge for Sustainable and Climate-Resilient Agriculture. *Int. Journal. Environment Climate Change, 14*(8), 490–501. DOI: 10.9734/ijecc/2024/v14i84369

Barrett, M., Sahay, S., & Walsham, G. (2001). Information technology and social transformation: GIS for forestry management in India. *The Information Society, 17*(1), 5–20. DOI: 10.1080/019722401750067397

Beevi, C. A., Rani, B. R., Shankar, K. R., Nirmala, G., & Nagasree, K. (Eds.). (2023). Innovative Extension Strategies for Sustainable Natural Resource Management in Rainfed Agriculture [E-book] Hyderabad: National Institute of Agricultural Extension Management (MANAGE) and ICAR-Central Research Institute for Dryland Agriculture.

Beniwal, S. P. S., Nene, L. Y., & Pandey, S. T. (2020). Relevance of Vrikshayurveda and Traditional Knowledge for Ecofriendly Sustainable Agriculture to Meet SDGs in India. *Asian Agri-History, 24*(1).

Bhat, N. A., Jeri, L., Karmakar, D., Mipun, P., Bharali, P., Sheikh, N., Nongkynrih, C. J., & Kumar, Y. (2023). Ethnoveterinary practises of medicinal plants used for the treatment of different cattle diseases: A case study in East Khasi Hill district of Meghalaya, North East India. *Heliyon*, *9*(7), e18214. DOI: 10.1016/j.heliyon.2023.e18214 PMID: 37501975

Bheemappa, A., Shruthi, S., Maheshwari, K., & Biradar, N. (2022). Indigenous technical knowledge-based rainfall prediction: a review. *Agricultural Reviews*, (Of). https://doi.org/DOI: 10.18805/ag.R-2347

Briggs, J., & Sharp, J. (2004). Indigenous Knowledges and Development: A Postcolonial Caution. *Third World Quarterly*, *25*(4), 661–676. https://www.jstor.org/stable/3993739. DOI: 10.1080/01436590410001678915

Chakrabortty, R., Pal, S. C., Sahana, M., Mondal, A., Dou, J., Pham, B. T., & Yunus, A. P. (2020). Soil erosion potential hotspot zone identification using machine learning and statistical approaches in eastern India. *Natural Hazards*, *104*(2), 1259–1294. DOI: 10.1007/s11069-020-04213-3

Chaudhary, A. K., Ahmad, S., & Mazumder, A. (2011). Cedrus deodara (Roxb.) Loud.: A review on its ethnobotany, phytochemical and pharmacological profile. *Pharmacognosy Journal*, *3*(23), 12–17. DOI: 10.5530/pj.2011.23.2

Chaudhary, B., Erskine, W., & Acciaioli, G. (2022). Hybrid knowledge and climate-resilient agriculture practices of the Tharu in the western Tarai, Nepal. *Frontiers in Political Science*, *4*, 969835. Advance online publication. DOI: 10.3389/fpos.2022.969835

Chilisa, B. (2012). *Indigenous Research Methodologies*. Sage Publications.

Das, S. (2024). Water management: Community participation. *Journal of the Geological Society of India*, *100*(4), 463–466. DOI: 10.17491/jgsi/2024/173862

DeMers, M. (2009). *Fundamentals of Geographic Information Systems* (4th ed.). John Wiley & Sons, Inc.

Eckardt, N. A., Ainsworth, E. A., Bahuguna, R. N., Broadley, M. R., Busch, W., Carpita, N. C., Castrillo, G., Chory, J., DeHaan, L. R., Duarte, C. M., Henry, A., Jagadish, S. V. K., Langdale, J. A., Leakey, A. D. B., Liao, J. C., Lu, K.-J., McCann, M. C., McKay, J. K., Odeny, D. A., & Zhang, X. (2023). Climate change challenges, plant science solutions. *The Plant Cell*, *35*(1), 24–66. DOI: 10.1093/plcell/koac303 PMID: 36222573

Ekobi, G. A., Mboh, L., & Tanga, P. T. (2023). *Contribution of Indigenous Knowledge to Agricultural Growth in South Africa: A Case of Disaneng Community in the Ratlou Local Municipality*. African Journal of Inter/Multidisciplinary Studies.

Ellena, R., & Nongkynrih, K. (2017). Changing gender roles and relations in food provisioning among matrilineal Khasi and patrilineal chakhesang indigenous rural people of north-east India. *Maternal and Child Nutrition*, *13*(S3), e12560. Advance online publication. DOI: 10.1111/mcn.12560 PMID: 29359434

Francis, D. V., Subhan, A., & Mourad, I., A. H., K. Abdalla, A., & FR Ahmed, Z. (. (2024). Optimizing germination conditions of Ghaf seed using ZnO nanoparticle priming through Taguchi method analysis. *Scientific Reports*, *14*(1), 15946. DOI: 10.1038/s41598-024-67025-6 PMID: 38987397

Gaikwad, S. S., Paralikar, S. N., Chavan, V., & Krishnan, S. (2004). *Digitizing Indian sacred groves–an information model for web interfaced multimedia database*. Focus on Sacred Groves and Ethnobotany: Prisam publications, Mumbai, 123-128.

Gillis, A. (2021). *What is internet of things (IoT)?* IoT Agenda.

GOI. (2014). National Livestock Mission. https://nlm.udyamimitra.in/Home/SchemePage

GOI. (2025). National Mission for Sustainable Agriculture. https://nmsa.dac.gov.in/

Goldman, M. J., & Lovell, E. (2017). Indigenous technical knowledge. In D. Richardson, N. Castree, M. F. Goodchild, A. Kobayashi, W. Liu, & R. A. Marston (Eds.), *International Encyclopedia of Geography: People, the Earth, Environment, and Technology*. DOI: 10.1002/9781118786352.wbieg0253

Hassan, F. E., Alyafei, M. A., Kurup, S., Jaleel, A., Al Busaidi, N., & Ahmed, Z. F. (2023). Effective priming techniques to enhance ghaf (Prosopis cineraria L. Druce) seed germination for mass planting. *Horticulturae*, *9*(5), 542. DOI: 10.3390/horticulturae9050542

India, S. (2016). About Startup India Initiative. https://www.startupindia.gov.in/content/sih/en/about-startup-india-initiative.html

Jain, S., Srivastava, A., Vishwakarma, D. K., Rajput, J., Rane, N. L., Salem, A., & Elbeltagi, A. (2024). Protecting ancient water harvesting technologies in India: Strategies for climate adaptation and sustainable development with global lessons. *Frontiers in Water*, *6*, 1441365. DOI: 10.3389/frwa.2024.1441365

Jaramillo, P. (2023). Agricultural identity of indigenous salasacas in Ecuador. *Alternative*, *19*(4), 882–891. DOI: 10.1177/11771801231197979

Kamakaula, Y. (2024). Ethnoecology and climate change adaptation in agriculture. *Global*, *2*(2), 473–485. DOI: 10.59613/global.v2i2.99

Khan, M. R., Raja, W., Bhat, T. A., Mir, M. S., Naikoo, N. B., Amin, Z., Nazir, A., Mir, S. A., Mohammad, I., Wani, A. A., & Patyal, D. (2022). Zero budget natural farming: A way forward towards sustainable agriculture. *Current Journal of Applied Science and Technology*, *41*(13), 31–43. DOI: 10.9734/cjast/2022/v41i1331712

Kumar, B. M., Kunhamu, T. K., Bhardwaj, A., & Santhoshkumar, A. V. (2024). Subcanopy light availability, crop yields, and managerial implications: A systematic review of the shaded cropping systems in the tropics. *Agroforestry Systems*, *98*(8), 1–26. DOI: 10.1007/s10457-024-00957-0

Kumar, M. J. (2024). Forging Connections: Integrating Indian Knowledge Systems in Higher Education. *IETE Technical Review*, *41*(3), 271–273. DOI: 10.1080/02564602.2024.2342625

Lakshmypriya, K., & Varghese, B. (2024). *Indigenous Tribes and Inclusive Engagement: An Integrated Approach for Sustainable Livelihood into the Future.* In Future Tourism Trends Volume 1: Tourism in the Changing World (pp. 99-114). Emerald Publishing Limited. DOI: 10.1108/978-1-83753-244-520241008

Liladhar Rane, N., Purushottam Choudhary, S., Saha, A., Srivastava, A., Pande, C. B., Alshehri, F., Roy, R., Katipoğlu, O. M., & Abdo, H. G. (2024). Delineation of environmentally sustainable urban settlement using GIS-based MIF and AHP techniques. *Geocarto International*, *39*(1), 2335249. DOI: 10.1080/10106049.2024.2335249

Mahato, S., Mishra, A. K., & Debnath, S. (2020). Impact of rainwater harvesting structures on groundwater recharge potential in India. *International Journal of Current Microbiology and Applied Sciences*, *9*(6), 66–74. DOI: 10.20546/ijcmas.2020.906.008

Malik, A., Singh, V., & Singh, K. (2007). Occurrence and distribution of persistent trace organics in rainwater in an urban region (India). *Bulletin of Environmental Contamination and Toxicology*, *79*(6), 639–645. DOI: 10.1007/s00128-007-9290-8 PMID: 17924039

Martin, D. W. (2023). Biodiversity conservation. In *Dictionary of Ecological Economics* (pp. 31–31). Edward Elgar Publishing. DOI: 10.4337/9781788974912.B.15

Michael, Y.Yohannes Gebre Michael. (2015). Conflicts of indigenous farming practices and modern agriculture in konso community, southern Ethiopia. *Journal of Agricultural Science and Technology B*, *5*(9). Advance online publication. DOI: 10.17265/2161-6264/2015.09.001

Mishra, P., Rai, A., Abdelrahman, K., Chand, S., & Tiwari, A. (2021). Analyzing challenges and strategies in land productivity in Sikkim Himalaya, India. *Sustainability (Basel)*, *13*(19), 11112. DOI: 10.3390/su131911112

Mitchell, T. M. (1997). *Machine learning*. McGraw-Hill. Available at: https://www.cs.cmu.edu/~tom/mlbook.html

Mohan, D. (2024). Wisdom of the ancestors: Harnessing indigenous knowledge for contemporary challenges. *International Journal of Applied Research*, *10*(3), 112–119. DOI: 10.22271/allresearch.2024.v10.i3b.11591

Muyambo, F., Bahta, Y., & Jordaan, A. (2017). The role of indigenous knowledge in drought risk reduction: A case of communal farmers in South Africa. *Jàambá*, *9*(1). Advance online publication. DOI: 10.4102/jamba.v9i1.420 PMID: 29955345

Panday, S. C. Shikha, Mondal, T., Dev, R., Nath, S., Singh, A. K., ... & Paliwal, A. (2025). Blending Traditional Knowledge of Farmers in Agriculture with Modern Scientific Technologies in the Hills of Uttarakhand State. In Blending Indian Farmers' Traditional Knowledge in Agriculture with Modern Scientific Technologies: A Way Forward (pp. 311-330). Singapore: Springer Nature Singapore.

Pande, C. B., Srivastava, A., Moharir, K. N., Radwan, N., Mohd Sidek, L., Alshehri, F., Pal, S. C., Tolche, A. D., & Zhran, M. (2024). Characterizing land use/land cover change dynamics by an enhanced random forest machine learning model: A Google Earth Engine implementation. *Environmental Sciences Europe*, *36*(1), 84. DOI: 10.1186/s12302-024-00901-0

Pandey, D. K., Momin, K. C., Dubey, S. K., & Adhiguru, P. (2022). Biodiversity in agricultural and food systems of jhum landscape in the West Garo Hills, North-eastern India. *Food Security*, *14*(3), 791–804. DOI: 10.1007/s12571-021-01251-y PMID: 35069938

Paroda, R. S. (2019). Report on policies and action plans for a secure and sustainable agriculture. Committee Report submitted to the Principal Scientific Adviser to the Government of India, New Delhi, 198p.

PIB. (2024a). Digital Agriculture Mission: Tech for Transforming Farmers' Lives. https://pib.gov.in/pib.gov.in/Pressreleaseshare.aspx?PRID=2051719

PIB. (2024b). Paramparagat Krishi Vikas Yojana. https://pib.gov.in/pib.gov.in/Pressreleaseshare.aspx?PRID=2082783

PIB. (2025). 10,000 FPOs Achieved under Government's Flagship Scheme. https://pib.gov.in/pib.gov.in/Pressreleaseshare.aspx?PRID=2106913

Prakash, N., Roy, S. S., & Ngachan, S. V. (2009). Role of ITK in conservation agriculture: Blending indigenous and scientific knowledge. *ICAR Research Complex for NEH Region, Umaiam, Meghalaya, 793*, 103.

Rani, J., & Paul, B. (2023). Challenges in arid region reclamation with special reference to Indian Thar Desert—Its conservation and remediation techniques. *International Journal of Environmental Science and Technology*, *20*(11), 12753–12774. DOI: 10.1007/s13762-022-04746-z

Reddy, M. (2024). Merging interest for sustainable agenda: Is there a link between sustainable agriculture practices and farm efficiency? *Sustainable Agriculture Research*, *13*(1), 91. DOI: 10.5539/sar.v13n1p91

Rijal, R., Kumar, A., & Bhetwal, S. (2023). Open-field and pot studies on application of ancient liquid organic manures for the growth and development of tomato (Solanum *lycopersicum*).

Segueda, W. E., & Banerjee, A. (2024). *Indigenous Knowledge Systems beyond biological diversity: Harnessing Indigenous Knowledge Systems for global knowledge cooperation*. The Current Column, German Institute of Development and Sustainability (IDOS), 28 October 2024. Bonn, Germany.

Sekar, T. (2024). Efficacy of Forest Acts and Forest Policies in India. In *Forest Policies, Laws, and Governance in India: Conservation Challenges in the Face of Climate Change* (pp. 95–156). Springer Nature Singapore. DOI: 10.1007/978-981-97-3862-5_3

Senanayake, S. (2006). Indigenous knowledge as a key to sustainable development. *Journal of Agricultural Sciences – Sri Lanka*, 2(1), 87-94. https://doi.org/DOI: 10.4038/jas.v2i1.8117

Sharifian, A., Fernández-Llamazares, Á., Wario, H. T., Molnár, Z., & Cabeza, M. (2022). Dynamics of pastoral traditional ecological knowledge: A global state-of-the-art review. *Ecology and Society*, *27*(1), art14. DOI: 10.5751/ES-12918-270114

Sharma, I., Kumar, N., & Verma, H. (2024). Deep Learning-Based Soil Nutrient Content Prediction for Crop Yield Estimation. In *Futuristic e-Governance Security With Deep Learning Applications* (pp. 85–111). IGI Global. DOI: 10.4018/978-1-6684-9596-4.ch005

Sharma, M., Bithel, N., & Sharma, M. (2022). Ethnoveterinary studies of medicinal plants used to treat livestock in the Haridwar region of Uttarakhand, India. *Current Biotica*, *13*, 53–63. DOI: 10.25081/cb.2022.v13.7234

Shukla, S., Barkman, J., & Patel, K. (2016). Weaving indigenous agricultural knowledge with formal education to enhance community food security: School competition as a pedagogical space in rural anchetty, India. *Pedagogy, Culture & Society*, *25*(1), 87–103. DOI: 10.1080/14681366.2016.1225114

Siddappa, C. (2024). Exploring the relevance and impact of ITKs on soil and water conservation in Bagalkot district. *International Journal of Research Agronomy*, *7*(4S), 276–280. DOI: 10.33545/2618060X.2024.v7.i4Sd.622

Siddique, F., Ali, Z., Arshad, M., Mubeen, K., & Ghazala, A. (2024). Exploration of novel eco-friendly techniques to utilize bioactive compounds from household food waste: Special reference to food applications. *Frontiers in Food Science and Technology*, *4*, 1388461. DOI: 10.3389/frfst.2024.1388461

Singh, A., Mishra, H. S., & Agrawal, A. K. (2023). A review on Vrikshayurveda-Upavanavinoda: Unlocking the potential of sustainable agriculture and environmental health. *Journal of Ayurveda and Integrated Medical Sciences*, *8*(12), 212–217. DOI: 10.21760/jaims.8.12.31

Singh, H. B., Yaipharembi, N., Huidrom, E., & Devi, C. A. (2023). Traditional Knowledge, Beliefs, and Practices Associated with Ethnic People of Manipur, North East India in Conservation of Biodiversity. In *Traditional Ecological Knowledge of Resource Management in Asia* (pp. 61–75). Springer International Publishing.

Singh, R., & Singh, G. (2017). Traditional agriculture: A climate-smart approach for sustainable food production. *Energy, Ecology & Environment*, *2*(5), 296–316. DOI: 10.1007/s40974-017-0074-7

Singh, R. B., Paroda, R. S., & Dadlani, M. (2022). Science, technology and innovation. Indian agriculture towards, 2030(821), 51.

Siripurapu, K. K., Sharma, S., Sainger, A., Das, S., & Kotamraju, K. R. (2022). Validation of the Indig enous Technical Knowledge of Indigenous Livestock Breeding and Management.

Teo, T. (2024). Will sustainable food sovereignty research be sustainable in the future? *Agricultural Sciences*, *15*(01), 165–186. DOI: 10.4236/as.2024.151010

Tripura, P., Devi, C. A., Manzoor, U., & Kapri, A. (2024). *Waste management for sustainable and restored agricultural soil: Indigenous technical knowledge integration in organic farming*. Retrieved from https://www.sciencedirect.com/science/article/abs/pii/B9780443184864000191

Tyagi, V., Brahmi, P., Yadav, S., Ranjan, P., Singh, S., Singh, S., & Singh, K. (2021). Germplasm Access from ICAR-NBPGR and Use within India. *Indian Journal of Plant Genetic Resources*, *34*(2), 216–220. DOI: 10.5958/0976-1926.2021.00020.6

Verma, A., Naorem, A., Keerthika, A., & Kumar, S. (2024). Biomass and carbon partitioning across different components of agroforestry systems in the arid zones of India. *Agroforestry Systems*, *98*(8), 1–20. DOI: 10.1007/s10457-024-01076-6

Vishweshwaran, M., & Sujatha, E. R. (2024). A Review on Applications of Drones in Geotechnical Engineering. *Indian Geotechnical Journal*, 1-15.

Vivek, V. (2016). Rainwater harvesting in Chennai: What made it work? IIM Kozhikode *Society & Management Review*, *5*(1), 91–106. DOI: 10.1177/2277975215617272

Walker, G. H., Stanton, N. A., Salmon, P. M., & Jenkins, D. P. (2008). A review of sociotechnical systems theory: A classic concept for new command and control paradigms. *Theoretical Issues in Ergonomics Science*, *9*(6), 479–499. DOI: 10.1080/14639220701635470

Wang, J. (2018). Integrating indigenous with scientific knowledge for the development of sustainable agriculture: Studies in Shaanxi province. *Asian Journal of Agriculture and Development*, *15*(2), 41–58. DOI: 10.37801/ajad2018.15.2.3

Warrier, K., Warrier, R. R., & Thangavel, V. (2023). Status of Sacred Groves in India: A Review. *International Journal of Environment and Climate Change*, *13*(8), 170–181. DOI: 10.9734/ijecc/2023/v13i81942

Westengen, O. T., Dalle, S. P., & Mulesa, T. H. (2023). Navigating toward resilient and inclusive seed systems. *Proceedings of the National Academy of Sciences of the United States of America*, *120*(14), e2218777120. DOI: 10.1073/pnas.2218777120 PMID: 36972436

Yadav, S. C. (2023). Traditional rain water harvesting systems in Rajasthan: water resources conservation and its sustainable management–a review. *Sustainability, Agri, Food and Environmental Research*, 11.

Yamunarani, K., Sangeetha, S., Dhanushkodi, V., Chitra, K., & Dhanalakshmi, K. (2023). Application of Indigenous Technological Knowledge (ITK) For Plant Disease Management In Tamil Nadu, India. *Asian Journal of Agricultural Extension Economics & Sociology*, *41*(12), 310–315. DOI: 10.9734/ajaees/2023/v41i122334

Zhang, L., Guo, W., Lv, C., Guo, M., Yang, M., Fu, Q., & Liu, X. (2024). Advancements in Artificial Intelligence Technology for Improving Animal Welfare: Current Applications and Research Progress. *Animal Research and One Health*, *2*(1), 93–109. DOI: 10.1002/aro2.44

Zhang, W., Su, K., Wang, Q., Yang, L., Sun, W., Ranjitkar, S., Shen, L., Kindt, R., Ji, Y., Marshall, P., Pisey, P. S., & El-Kassaby, Y. A. (2023). Agroforestry species selection for forest rehabilitation in the asia-pacific region: A meta-analysis on high-level taxonomy. *Forests*, *14*(10), 2045. DOI: 10.3390/f14102045

Zong, X., Liu, X., Chen, G., & Yin, Y. (2022). A deep-understanding framework and assessment indicator system for climate-resilient agriculture. *Ecological Indicators*, *136*, 108597. DOI: 10.1016/j.ecolind.2022.108597

ADDITIONAL READING

Acharya, C. L., Chaudhary, R. S., Gurumurthy, P., & Subba Rao, A. (Eds.). (2025). *Blending Indian farmers' traditional knowledge in agriculture with modern scientific technologies: A way forward.* Springer Singapore., DOI: 10.1007/978-981-96-1020-4

Chhetri, B., & Belbahri, L. (2024). Documentation of Indigenous Technical Knowledge and their application in pest management in Western Mid Hill of Nepal. *International Journal of Agricultural Sciences.* Retrieved from https://www.researchgate.net/publication/385134567

Melash, A. A., Bogale, A. A., Migbaru, A. T., Chakilu, G. G., Percze, A., Ábrahám, É. B., & Mengistu, D. K. (2023). Indigenous agricultural knowledge: A neglected human based resource for sustainable crop protection and production. *Heliyon*, *9*(1), e12978. DOI: 10.1016/j.heliyon.2023.e12978 PMID: 36711305

Misra, S., & Ghosh, A. (2024). Agriculture paradigm shift: A journey from traditional to modern agriculture. In Singh, K., Ribeiro, M. C., & Calicioglu, Ö. (Eds.), *Biodiversity and Bioeconomy* (pp. 113–141). Elsevier., DOI: 10.1016/B978-0-323-95482-2.00006-7

Siakwah, P., Ablo, A. D., Sheburah-Essien, R., Zaami, M., & Yaro, J. A. (2025). Adapting to climate change amidst innovation diffusion and declining indigenous agricultural knowledge and practices in Ghana. *African Journal of Science, Technology, Innovation and Development*, *17*(2), 1–12. DOI: 10.1080/20421338.2025.2459355

KEY TERMS AND DEFINITIONS

Indigenous Technical Knowledge: Is tacit, intuitive, experimental, informal uncodified knowledge.

Artificial Intelligence: Systems that continuously learn by imitating the human mind.

Internet of Things: Is a network of interconnected devices that collect, transmit, and analyze data for efficient decision-making.

Machine Learning: Are algorithms that enable machines to learn from data, identify patterns, and make predictions or decisions without explicit programming.

Geographical Information System: Is a framework for gathering, managing, and analyzing spatial and geographical data.

Chapter 15
Exploring Lepcha Fairs, Festivals, and Events for Enriching Travel Experiences

Akhilesh
https://orcid.org/0009-0009-4406-7868
Assam University, India

Pradip Kumar
https://orcid.org/0000-0002-5138-736X
Assam University, India

Nimemit Lepcha
Independent Researcher, India

Garsemit Lepcha
Independent Researcher, India

ABSTRACT

This chapter talks in depth about the Lepcha people's festivals, events, and traditional background in the Eastern Himalayas. The Lepchas, who were among the first people to settle in the area, are known for their ecological wisdom, spirituality, and cultural identity, which they exhibit via vivid rituals that honor the elements of Mother Earth. Chyu Rum Faat, Pang Lhabsol, and Tendong Lho Rum Faat are examples of festivals that have changed from animistic rituals to important parts of society. Travelers introduced native practices to the region, while foreign practices such as Tibetan Buddhism, British colonialism, and modern state governance created a fascinating combination. Modern festivals support the assertion of Lepcha heritage and provide a one-of-a-kind opportunity for cultural tourists to learn about Lepcha traditions and mythology. Lepcha festivals keep their traditional identity while changing with the times in terms of history and politics. The chapter also investigates the manner in which visitors perceive these events, as well as their engagement and enjoyment.

DOI: 10.4018/979-8-3373-4124-8.ch015

INTRODUCTION

"Earlier the Lepcha used to have such magical powers, now they are weak and **vanishing"---Jenny Bentley (University of Zürich)**

Fairs and festivals are not only occasions for celebration, but they also serve as a means for the Lepcha community to demonstrate its spiritual, culturally aware, and environmentally conscious nature. The Lepchas are the earliest inhabitants of the Eastern Himalayas, and they refer to it as *Rong* (Verma, 2013). Overtime, the Lepcha community developed fairs and festivals as their cultural practice and heritage underpinned by their philosophical orientation of their spirituality and the environment they live in, including, rivers, mountains and forests (Verma, 2013). This culture exhibits a high ecological awareness and a symbiotic relationship with their environment. Festivals speak of resilience and flexibility as they have absorbed outside cultural influences from Tibetan Buddhism, British colonialism, and modern state governance without losing their identity.

Basnet et al., (2017) highlighted that lepcha festivals like *Chyu Rum Faat* (Worship of the Peak) and *Pang Lhabsol* (Worship of Mount Kanchenjunga) originated as propitiatory rituals to nature spirits but later transformed into major forms of appeal to those powers for the well-being, prosperity, and security of the community from Social Calamities. Example are Loss of Traditional Lands and Resources, Cultural Erosion due to Modernization, Natural Disasters, Social Disintegration due to Political Conflicts. The socio-political landscape of the region, particularly with British colonialism and the merger of Sikkim with India, further drew the shape of these festivals. While some of the practices declined under the influence of colonialism and modernity, there has been a resurgence of interest in preserving and promoting indigenous culture over the last few decades. Celebrated in honour of Mount Tendong for its mythical role in saving the Lepchas from a great flood (Vandenhelsken, 2011).

Vandenhelsken (2011) stated that to tourists, these festivals represent opportunities to engage with an ancient culture that remains dynamically relevant in a modern context. The tourists get an opportunity to experience ways in which the Lepcha community depicts and exchange their indigenous knowledge and beliefs through rituals, traditional dances, communal feasts, storytelling sessions, guide nature, walks, and many more (Das, 2024). The chapter focusses on the Lepcha fairs, festivals, and events that have travel experiences attached to them. The chapter explores the Lepcha's fairs and festivals are double-edged-sword. On one hand, are primarily used to live in the world not to conquer it. They give them a sense identity and a sense of belonging. Understanding the cultural and ecological underpinnings of the festivals emphasizes their importance not only for the Lepcha community but also for fostering broader cultural awareness and respect among the visitors (Basnet et al., 2017).

RESEARCH OBJECTIVES

- To provide a detailed description of Lepcha festivals.
- To explore the tourist experience of Lepcha festivals.
- To identify key elements for enhancing festival experiences.

METHODOLOGY

This chapter uses a descriptive research methodology based on secondary data to provide a detailed description of Lepcha festivals, explore the tourist experience of these festivals, and identify key elements for enhancing the festival experience. Siedlecki (2020) defined that descriptive research methodology aims to methodically characterize the features of a population, phenomena, or circumstance. It concentrates on what is occurring rather than why or how to alter it. This approach allows for a comprehensive analysis of the cultural, social, and experiential dimensions of these festivals, ensuring their significance to the local community and tourists.

FINDINGS AND DISCUSSIONS

Historical Background - Origins and Evolution of Lepcha Festivals

The community's animistic beliefs and close connection with nature form the roots of the Lepcha festivals. In their traditional way of life, the Lepchas, also known as the *Rong*, saw nature as a living thing. Rahaman, (2014) stated that every mountain, river, and forest was holy to them and home to spirits. These natural factors were important to their spiritual practice and way of life. In the beginning, the Lepcha fairs were worship ceremonies for the natural gods that brought peace, safety, and wealth to the community. Chyu Rum Faat, which means "worship of the peak," is one of the oldest and most important Lepcha festivals. People believe the event began when the Lepchas relied on the rivers for everything, including water, food, and farming. During Chyu Rum Faat, people pray and do rituals for the spirits of the peaks in the hopes that this will make them happy and bring them health, good crops, and safety from natural disasters. These behaviors demonstrate the community's deep biological awareness and respect for the natural world. These are important parts of their cultural identity (Plaisier, 2005).

Early Buddhist and Christian contacts inspired Lepcha celebrations. To promote unity, Chakdor Namgyal, the third Chogyal monarch of Sikkim, organized Pang Lhabsol, a Lepcha-Bhutia festival in the 17th century. Initially a Lepcha ritual to commemorate Mount Kanchenjunga, the festival later incorporated Buddhist ceremonies and prayers to celebrate the harmonious coexistence of the local religious communities. As per Malhotra, (2024) this syncretism demonstrates how Lepcha festivals have evolved over time by integrating indigenous practices with influences from other cultures and religions. Changes in the area's society and government have influenced the rise of Lepcha celebrations. Sikkim's mid-20th-century merger with India and 19th-century British colonization changed the cultural landscape. During this time, the number of Lepcha events decreased due to the spread of new faiths and the influence of other cultures. People are now more interested in preserving and supporting these native events because they know more about the Lepchas' cultural history and identity (Little, 2010).

Chaudhuri et al. (2020) explained that as time has gone on, Lepcha fairs have grown into important cultural events and tourist attractions. Tendong Lho Rum Faat, which honors Mount Tendong for saving the Lepcha from a devastating flood, has been resurrected and promoted. These festivities reveal the Lepcha mythology, customs, and ecological practices. Today's return and growth of these events show how important they are to the community's cultural identity and how they can attract cultural tourists, which is good for local businesses. Lepcha celebrations show how current dynamics, foreign influences, and indigenous traditions interact. These festivals, deeply rooted in primordial practices and beliefs, have evolved to accommodate changing social, religious, and political contexts, demonstrating the Lepcha community's adaptability and resilience. Lepchas and tourists may now experience a rich legacy and emerging identity as dynamic representations of cultural continuity (Malhotra, 2024).

DESCRIPTION OF KEY FAIRS AND FESTIVALS

Being full of cultural heritage and inextricably linked to nature, the Lepcha celebrate a lot of fairs and festivals all through the year. These have now assumed central importance in their cultural identity. Also, each of these festivals is unique in terms of its traditional practices, beliefs, and lifestyle. Each of the festivals had its significance, rituals, and modes of celebration, reflecting this rich tapestry of Lepcha culture. Some of the major fairs and festivals that are part of the Lepcha community include the following:

Pang Lhabsol

Pang Lhabsol is one of the most important and unique festive celebrations of the Lepchas, unique to Sikkim state. Traditionally, the Pang Lhabsol, a bright festival in August, commemorates the consecration of Mount Kanchenjunga, a guardian deity to the Lepchas kno. The festival can be traced back to an offshoot of a worship ritual to honor the protective spirit of Mount Kanchenjunga, believed to be the guardian deity of Sikkim and her people. Pang Lhabsol is a vibrant and colourfully celebrated festival with various rituals, offerings, and traditional dances like Chyu-Faat, whereby its participants carry the deity's offering in the form of sacred plant leaves, calling on peace, prosperity, and protection of lands under the blessings of the great deity. This festival also represents the bond of brotherhood between the Lepchas and the Bhutia, further established by a historic blood brotherhood treaty signed in the 17th century. The masked dances, lively processions, and prayers led by the Buddhist lamas, which exhibit the syncretic fusion between indigenous Lepcha animistic traditions and influences from Tibetan Buddhism, are a characteristic of this carnival (Molommu, 2018).

Chyu Rum Faat

Chyu Rum Faat, or "worship of the peak," is another important festival among the Lepchas that demonstrates their high level of respect for water bodies and other natural elements. The festival, which lasts throughout the monsoon season, involves many rituals with respective offerings to river spirits. According to (Chettri & Sharma, 2021) They pray for a bountiful harvest, the absence of natural disasters, and the prosperity of the community Lepchas celebrate the festival along the peaks, offering food, flowers, and other sacred items to their rivers' spirits, regarded as the lifelines of their agrarian society. Traditional dances, songs, and storytelling sessions mark the festivities, narrating tales of myths and legends associated with rivers and water bodies in Lepcha folklore. Chyu Rum Faat is the highest point of ecological knowledge among the Lepcha people, as well as their hundreds of years-old customs. It offers a one-of-a-kind chance for visitors to experience living in harmony with nature and the health of the community (Lepcha, 2021).

Tendong Lho Rum Faat

Tendong Lho Rum Faat is a holiday about Mount Tendong, a holy hill that the Lepcha people think saved them from a big flood. The event takes place in August, in honor of and loyalty to the mountain. In Lepcha mythology, Mount Tendong was a symbol of resilience and protection as it ascended to rescue the people during a tremendous deluge (Pradhan, 2016). During Tendong Lho Rum Faat, people pray, do rituals, and make gifts on top of the hills to honour the mountain and thank it for protecting them. Stories, folk songs, and dances depict epic flood stories and Mount Tendong, the hill that saves them. It also provides an opportunity to reconfirm their commitment to environmental protection. This event offers a wide range of traditional activities for tourists to enjoy, demonstrating the Lepchas' deep connection to their land and history (Chettri & Sharma, 2021).

Namsoong Festival

The Namsoong Festival, which commemorates the Lepcha New Year, is a time of tremendous joy and pleasure. People celebrate it with reverence and pray for benefits next year. It often falls around the harvest season in December and January (Chettri & Sharma, 2021). A variety of events occur, such as a communal banquet, dancing performances, and athletic contests. People make traditional Lepcha foods and put flowers and holy leaves around their homes to bring happiness into them. The Zo-Mal-Lok is a traditional dance form of Sikkim and is performed during festivals and celebrations. It is also known as the Yak dance as it involves dancers wearing yak skin costumes and masks. The dance is performed to the beats of traditional Sikkimese musical instruments like the damphu and the lingam. The event reminds the Lepchas of the importance of unity, peace, and community spirit, and it's an open call to see how they live in a joyful and welcoming setting.

Muk Zek Ding Rum Faat

The Lepchas also celebrate Muk Zek Ding Rum Faat, another traditional festival. According to religious rites and customs, the Lepchas celebrate it to thank and ask for help from the spirits of their ancestors. The Lepchas believe that the departed spirits of their progenitors continue to provide protection and guidance to the community's surviving members. During this period, the community gathers for rituals, folk melodies, and dances, fueled by the rhythm of traditional Lepcha music, creating a vibrant communal atmosphere. Muk Zek Ding Rum Faat is founded on the value of family and community ties. It sets a new standard for how visitors can experience the Lepcha people's cultural continuity, as they have kept the practices alive (Chettri & Sharma, 2021).

RITUALS, CEREMONIES, AND TRADITIONAL PRACTICES INVOLVED

The Lepcha community's fairs and festivals, deeply entrenched in their cultural and spiritual life, stand out due to their traditional practices, ceremonies, and intricate rituals. They are not only places to celebrate but also important ways for them to show what they believe, value, and who they are as native people. Each event features a unique set of rites and routines that demonstrate the community's deep connection with nature, the importance of spiritual animism, and the societal organization. This section investigates several of the primary rituals, ceremonies, and traditional practices that are integral to Lepcha fairs and festivals.

Sacred offerings to the gods and nature spirits

According to Rahaman (2014), Lepcha festivals are characterised by the creation of offerings to the deities, ancestors, and nature spirits. For example, during Chyu Rum Faat, the Lepchas present rice, flowers, fruits, and sacred water to the river gods. The elders of the community or the shaman conduct this offering, chanting hymns and prayers, in the spiritual belief that it will appease the spirits, bestow blessings in the form of a good harvest, and safeguard them from natural calamities. Tendong Lho Rum Faat, an offering to Mount Tendong, is believed to symbolise gratitude and reverence for nature, as it is believed to have saved the Lepcha from drowning in a catastrophic flood.

Traditional dances and music performances

Lepcha festivals include music and dance, which are both religious and fun. One of the dances is the Chyu-Faat dance, which is performed during Pang Lhabsol, a time when people worship Mount Kanchenjunga and thank the gods. The dancers wear a variety of colourful traditional clothes and move in ways that show respect for nature. Instruments like the Tungbuk (a three stringed musical instrument played like a bass mandolin)(Chettri & Sharma, 2021) and Tung-darr (drum) play in the background, giving the evening a lively sound and connecting the dancers to their cultural roots.

Feasting and communal meals

Most Lepcha festivals include large meals for the whole community. Traditional foods like Suklyang Sar (fermented soy), Chee (local millet beer), and different wild foods are cooked and shared by the community. It is an offering to the gods and ancestors that brings back the benefits of wealth and plenty. In addition to strengthening social ties and community cooperation, feasting makes people feel more connected to each other and helps everyone in the community (Bruner, 1991).

Storytelling and oral traditions

The Lepcha events include storytelling that involves telling tales, fables, and real stories. During Muk Zikding Rum Faat, which is about passing on the community's oral traditions from their roots, elders tell stories about creation, brave acts, and moral lessons. Telling older people's stories in Lepcha, also known as Rongring, is a way to keep the language alive by teaching traditional ideals to younger people. Lepcha people often tell these stories in the language Rongring, which protects their language heritage and teaches younger generations about traditional values., These traditions provide an essential space for marginalized communities to assert their unique identity amidst the dominance of global cultures. In this way, the Lepcha's storytelling tradition, deeply embedded in Rongring, serves as a safeguard for their cultural heritage, ensuring that their voice continues to resonate across generations. Moreover, storytelling in the Lepcha community can be viewed through the lens of narrative theory, which suggests that stories are instrumental in shaping how individuals and groups make sense of their lives and histories (Bruner, 1991).

The creation of ritual structures and shrines

Festivals of this nature always require the erection of temporary structures or shrines, usually made of bamboo, wood, and other materials. This ensures a secure location for religious activities and the offering of prayers, including sacrifices. In Chyu Faat, for example, people build holy temples near lakes, rivers, and streams and leave gifts for water spirits on them. Communities construct and adorn the structure to symbolise unity and shared spiritual commitment.(Kirkpatrick, 2007).

Purification ceremonies and fire rituals

According to Rahaman (2014), Fire rituals form a common aspect of most Lepcha festivals and are symbolic of purification and renewal. During festivals like Namsoong, community members light bonfires and dance around them, known as the Circle Dance, to burn away past misfortunes and welcome new opportunities. Chanting fire rituals, along with invocational prayers and offerings of grains, fruit, and incense, serves as a mediating force between the human and the divine.

ROLES OF VARIOUS COMMUNITY MEMBERS IN FESTIVALS: SIGNIFICANCE OF TRADITIONAL PRACTICES IN FESTIVALS

The Lepchas deeply integrate festivals into their social and spiritual system, making them more than just an event to celebrate. Everyone, regardless of age or gender, gets a chance to participate and contribute, upholding community values of unity, cooperation, and cultural preservation. Community members are expected to take on various roles during festivals to ensure that traditional practices, rituals, and ceremonies are performed in a way that aligns with cultural and spiritual significance. These roles show how important events are as places for bringing people together, sharing information between generations, and strengthening community identity.

Elders and community leaders

Elders and community leaders, such as the Padem (village chief) and senior members, are instrumental in the organisation and supervision of festival activities. In addition to ensuring the accurate observance of all traditional practices, elders coordinate rituals and conduct communal prayers. As protectors of Lepcha culture, elders narrate mythological stories, conduct rituals, and provide blessings during festivals such as Tendong Lho Rum Faat and Pang Lhabsol. Their participation is critical for the preservation of oral traditions, the transmission of cultural knowledge, and the community's spiritual and moral values (Kirkpatrick, 2007).

Religious leaders: Bongthings and Muns

Rahaman, (2014) stated that the Bongthings (male priests) and Muns (female shamans) are spiritual leaders who participate in significant ceremonies and rituals during Lepcha festivals. They serve as intermediaries between people and the spiritual world, leading prayers, songs, and gifts to nature, spirits, and gods. The purification rituals, healing ceremonies, and divinations they conduct are particularly significant in Chyu Rum Faat (peak worship) and Sukpum Rum Faat (forest worship). To support the Lepchas' animistic views and ecological knowledge, these leaders read nature's signs, guess what will happen in the future, and give spiritual advice to the community (Lepcha, 2020).

Women: preserving tradition and cultural knowledge

In the Lepcha community, women are responsible for a variety of tasks, including the preparation of sacred offerings, such as rice, fruits, flowers, and other symbolic items used in rituals. They are also in charge of organizing communal feasts, brewing millet beer, and cooking traditional foods during festivals such as Namsoong (new year) and Nambun Faat (rice harvest festival). Women also participate in traditional dances and music and create ritual artifacts such as prayer flags and bamboo baskets, which are essential for numerous festival activities (Rahaman, 2014).

Men: performers and designers of rituals

Chaudhuri et al. (2020) highlighted that Rituals and ceremonies are also carried out with significant involvement by men. They often play music and dance from their own culture, like the Chyu-Faat dance during Pang Lhabsol. They are responsible for the tangible preparations, such as the construction of temporary monuments, the establishment of festival grounds, and the coordination of processions. To keep society safe and healthy, men help spiritual leaders perform complicated rites like fire ceremonies and animal sacrifices.

Youth: bearers of cultural continuity

At Lepcha events, young people play a big role. Not only do they dance and sing, but they also learn about and share culture. They are encouraged to learn traditional songs, dances, and rituals, ensuring that cultural knowledge is passed down to future generations (Gergan, 2014). Through traditional games, music, and stories at events like Dung lap Faat, young people build national pride and a sense of who they are. Their participation strengthens social relationships and promotes communal responsibility.

Children: responding to cultural learning

At a young age, Lepcha children begin to participate in events. This gives them a chance to learn about their roots right away. They may contribute to the preparation of tributes, assist in the decoration of festival venues, or participate in performances and processions. Children may see and imitate elders practicing cultural beliefs, tales, and customs, which promotes informal learning. Little (2008) highlighted that the events protect Lepcha practices by giving people of all ages important chances to learn about their culture and share what they've learned.

Artisans and craftsmen: cultural producers

Craftspeople and artists showcase their talents during Lepcha festivities by creating religious artifacts, musical instruments, and decorations for the celebrations to display their abilities. They produce traditional items, including handwoven textiles, wooden masks, and bamboo receptacles, for use in rituals and processions. Namsoong and Chyu Faat festivals showcase these crafts, encouraging economic growth and communal culture. Their work preserves traditional craftsmanship and enhances festival aesthetics (Little, 2010).

Performers: dancers, musicians, and storytellers

Chaudhuri et al. (2020) stated that Lepcha festivals include musicians, dancers, and storytellers. Traditional dance companies, which consist of both men and women, execute significant performances at significant occasions such as Chyu-Faat and Gyen Lak. Indigenous instruments, such as the Tungbuk (A three stringed musical instrument played like a bass mandolin)(Chettri & Sharma, 2021) and Tung-darr (drum), are played by musicians to provide rhythmic accompaniment to rituals and performances. Storytellers, who are usually older people, tell myths and stories, especially during Muk Zikding Rum Faat. They do this to keep oral customs alive and strengthen cultural memory.

HOW TOURISTS TYPICALLY INTERACT WITH LEPCHA FESTIVALS: COMMON ACTIVITIES AND INTERACTIONS FOR TOURISTS

A lot of people come to the Lepcha fairs and festivals to see the rich cultural history, unique customs, and spiritual practices that make them special. During these celebrations, tourists connect with the Lepcha community through immersion activities. These activities provide tourists with unique experiences that help them learn more about the Lepcha way of life. This part talks about some of the most common things tourists do and people they meet at Lepcha fairs and festivals, going into more detail about how these things make travel more fun and help people understand each other's cultures better.

Rituals and ceremonies

People may witness or participate in Lepcha festival customs. For example, during Pang Lhabsol, the Lepchas invite guests to join them in the dances and prayers that honor Mount Kanchenjunga, their guardian god. Tourists may worship rivers in Rum Faat by donating flowers, fruits, and grains to water spirits. For tourists, this is a way to personally experience Lepcha spiritual practices, which helps them learn more about the community's animistic views and love of nature (Kapp, 1986).

Experience traditional dance and music shows

Many people come to the Lepcha fairs to see the lively dance and music performances that are so typical of them. Namsoong and Tendong Lho Rum Faat include Chyu-Faat dances. Indigenous instruments like the Tungbuk (A three stringed musical instrument played like a bass mandolin)(Chettri & Sharma, 2021) and Tung-darr (drum) complement the melody. Tourists who observe these performances and participate in communal routines gain a deeper understanding of the Lepchas' artistic heritage and cultural manifestations. These energetic performances often leave a lasting impression on the audience.

Taste traditional Lepcha cuisine

Lepcha et al. (2024) explained that Lepcha festivals place a significant emphasis on food, often encouraging visitors to sample traditional Lepcha dishes during communal feasts. Festivals like Namsoong and Shyaprum Rum Faat feature Suklyang Sar (fermented (soybean), Chee (local millet beer). Tourists have the opportunity to participate in social contact and cultural exchange by eating with members of the community. This not only gives them the chance to enjoy the special flavors of Lepcha cuisine, but it also benefits the community.

Learning about traditional crafts and artistry

Numerous Lepcha festivals frequently showcase traditional crafts in exhibitions or seminars, such as wood carving, bamboo basket making, mask making, and weaving. Tourists can learn about different crafts and their cultural significance from local artists and try to make their own. Lepcha artists often set up booths at events like Chyu Faat to sell their goods. This lets tourists buy real Lepcha crafts and support the local economy. When people meet the artists, they realize how skilled they are and how important their work is to the Lepcha society (Chakraborty, 2016).

Joining guided nature walks and pilgrimages

Many festivals in the Lepcha community involve guided nature treks or pilgrimages to sacred locations, including rivers, lakes, and mountains, due to the community's profound connection to nature. During Tendong Lho Rum Faat, the Lepchas are believed to have been spared from a great inundation at Mount Tendong, a location of great spiritual significance. For instance, Chakraborty (2016) explained that travellers can participate in pilgrimages to the site with local interpreters. These excursions frequently include explanations of the local flora and fauna, as well as the spiritual beliefs associated with the natural landscape, which provide visitors with a more profound understanding of the ecological knowledge and environmental ethics of the Lepchas (Bentley, 2021).

Storytelling sessions and cultural talks

An important custom is for adults to tell tales, legends, and historical stories at Lepcha gatherings. People are usually welcome to come to these speaking events, which usually happen in public places. Hearing tales about creation, heroes, and ancient traditions at Muk Zikding Rum Faat may teach people about Lepcha oral history (Molommu, 2018). Additionally, leaders and researchers within the community may conduct cultural presentations or lectures to facilitate the acquisition of a deeper understanding of the Lepcha people's history, culture, and contemporary challenges.

Taking part in environmental conservation activities

Molommu (2018) stated that Lepcha festivals promote sustainability and conservation, reflecting the community's respect for nature. The community may encourage travellers to attend sustainable agricultural courses, clean waterways, or plant trees. Holy tree protection and environmental awareness campaigns are common during forest worship festivals, such as Sukpum Rum Faat. At such events, visitors may learn about Lepcha sustainability and contribute to the community.

TOURIST PERCEPTIONS OF THE FESTIVALS: AUTHENTICITY, ENGAGEMENT, ENJOYMENT.

The perception of Lepcha festivals by travellers is influenced by the authenticity of the cultural experience, community participation or engagement, and the enjoyment of the event. Due to the growing growth of cultural tourism, it's important to get visitor input at each place to analyse Lepcha festivals' impact on visitors and their potential for sustainable tourism. This section will examine how visitors see Lepcha festivals' authenticity, participation, and pleasure via responses and recorded experiences.

The perception of authenticity

Most tourists perceive these festivals as highly authentic and genuine cultural events, providing a genuine glimpse into the people's traditions and lifestyle During Pang Lhabsol and Chyu Rum Faat, visitors can witness rituals, dances, music, and traditional crafts that are unique and devoid of commercialization, allowing them to witness centuries-old cultural practices. Tourists consider these experiences very valuable because of their spontaneous and unprocessed nature; they seem almost the opposites of the scripted or commercially available cultural experiences in other travel destinations. For instance, tourists have expressed that the experience of attending shamanic ceremonies, witnessing truly "authentic" traditional dances, and interacting directly with local community members significantly enhances the "aura" of authenticity (Jauhari & Munjal, 2015).

Engagement with the local community

At festivals, tourists tend to form positive perceptions of their level of involvement with the Lepcha community. Unlike several other cultural events, where visitors are mere spectators, the Lepcha festival involves full participation in almost every form of activity: communal dances, rituals, craft workshops, and sharing meals with locals (Munjal & Jauhari, 2015). Many tourists regularly report on this inclusive approach, engaging in and feeling involved in the community instead of merely being passive observers. Such engagements enable tourists to have meaningful relationships with the local people and foster a deep appreciation of Lepcha culture, thus improving their general travel experience.

Enjoyment of the cultural experience

Lepcha et al. (2024) mentioned that Tourists' primary perceptions of the attraction include interest, motivation, and fun derived from participation in Lepcha festivals. There is more often than not a high level of enjoyment because the festivities are vibrant and colorful, with high-spirited dances, music, and community celebrations. Events such as Namsoong and the Lepcha New Year attract tourists due to their merriment and gaiety, as well as the warmth and friendliness of the local hosts. The tours involving dining with indigenous foods or discovering how indigenous crafts are done, apart from nature walks or pilgrimages, provide added value to the tours and thus become one of the most memorable events for many tourists.

Challenges and improvement areas

Many tourists express positive perceptions of Lepcha festivals, but they also highlight challenges and areas for improvement. For example, most visitors may feel that language barriers limit their understanding and realization because most of the rituals and stories are performed in the Lepcha language. Other scholars have referred to a lack of better infrastructure and facilities for tourists, such as signage, restrooms, and seating arrangements in outdoor festivals. It has also been recommended that providing background information about the significance of the festivals through bro Chyures, guided tours, or cultural interpreters would enhance the experience further by adding value to their knowledge on Lepcha traditions and practices (Bradley, 2008).

Concerns about authenticity versus commodification

Some of the tourists did express concerns about the commodification of Lepcha festivals if they indeed gain in popularity in the future. Little (2010) highlighted that there is a thin line between retaining the originality of such cultural phenomena and tweaking it to meet the expectations of a growing visitor population. While many tourists cherish Lepcha festivals for being uninhibited and authentic, they also realise that an influx of tourists could encourage festival organisers to stage festivals differently and cause these festivals to lose their cultural authenticity. Visitors often convey through their positive reports that it is crucial for festivals to remain community-based and culturally sensitive, free from excessive commercialization and tourist dominance.

Positive impacts on cultural awareness and respect

Basnet et al. (2017) mentioned that Participating in Lepcha festivities helps many tourists appreciate indigenous customs and cultural understanding. Many guests found the trip informative, learning about Lepcha spirituality and ecological principles. Many travelers have cited conservation initiatives, narrative sessions, and cultural exchanges as highlights that helped them understand and appreciate Lepcha culture. It argues that Lepcha festivals might promote intercultural understanding, visitor respect, and community cultural heritage (Little, 2008).

TRADITIONAL FOODS SERVED DURING FESTIVALS

Lepcha tradition revolves around food, not only for its practical use but also as an integral part of their culture and community. Chaudhuri et al. (2020) highlighted that these events notably showcase the customary Lepcha cuisines, intricately linked to the region's resources and farming methods. Most of the time, the cooked and given foods reflect the land, faith, and history of a society. Learn about the cultural significance, preparation techniques, and religious functions of traditional foods commonly served at Lepcha Festivals in this region.

- **Zutdop (Rice and Vegetable preparation):** Rice and local vegetables such as bamboo shoots, nettles, fiddlehead ferns, and wild mushrooms combine to make the traditional food known as zutdop. The Lepchas delicately season the curry to preserve the natural flavour of the ingredients, reflecting their reverence for the purity of nature's abundance. People offer Zutdop during festivals such as Namsoong (Lepcha New Year) and *Tendong Lho Rum Faat*, symbolizing fertility, wealth, and the abundance of Mother Earth. For example, people in the community often share it at feasts, which helps individuals get along with each other.
- **Chee (Local millet beer):** Chee, known as Chi or Chaang, is a traditional, local, and fermented beverage made out of millet, rice, or maize. Lepcha festivals centralize Chee, serving it as part of ritual offerings to deities and as a collectively made drink by participants. The process of making cheese entails fermenting the grains with yeast in a bamboo container. As a result, Chee is mildly alcoholic and consumed through a bamboo straw. People highly revere Chee and frequently utilize it in rituals to receive blessings from their ancestors and spirits. During festivals like Chyu Rum Faat, people offer Chee as a tribute to river spirits, praying for prosperity and good health. Chee sharing with others in the community reflects unity, hospitality, and life celebration.(Kirkpatrick, 2007).
- **Suklyang Sar:** a fermented soybean, is a common food item at most Lepcha festivals. People prepare Suklyang Sar by fermenting soybeans, then cooking them with other spices, garlic, and chillies to create a tasty relish for rice and other staple food items. Suklyang Sar, therefore, is more than just taste and nutrition; it has cultural value. Despite being biological, we can interpret the fermentation process as a process of change or transformation, connecting it to themes of spiritual rebirth and communal renewal frequently observed during festivals like Namsoong. In addition, the women of the community typically prepare Suklyang Sar together, demonstrating the Lepcha people's emphasis on cooperation and shared labor.

FUTURE RESEARCH DIRECTIONS

Lepcha festivals provide study possibilities in terms of how culture, spirituality, tourism, and environment interact. Examining the social, environmental, and economic impacts of festivals might help to show how indigenous cultures adapt to modernity while preserving traditions given rich culture and rising Lepcha societal rank. Commodification during Lepcha's festivals, which are important for displaying culture for the society, threatens the culture of his area. As more visitors come, these events' authenticity may suffer as they change to meet their needs. Research may be done to see how an increase in visitors affects the festivals' authenticity and the local culture over time.

Different kinds of alternative tourism, such as eco-tourism or community-based tourism, might provide insights on sustainable ways to promote festivals without sacrificing their culture. Modern-day media, social networks, and worldwide travel have also affected traditional rituals, hence influencing how people celebrate, participate in, and think about such festivals. Examining the evolving function of women in festival preservation may contribute to more general discussions on gender, culture, and power in indigenous societies.

Reflecting adoration of nature, ecology, and sustainability, Lepcha celebrations have great ecological relevance. Future research might investigate whether such rituals maintain biodiversity in local settings, for example, in worship of river spirits in Chyu Rum Faat. Such studies may also investigate how such environmental practice relates to the more general conservation of nature that may be performed all around. Touristic experiences at Lepcha festivals provide a notable platform for study on intersocietal interactions, e.g., visitors' views on the genuineness of events and their influence on general tourist experience. Finding challenges visitors face, such as language barriers and cultural misconceptions, might help to improve the tourist experience without sacrificing festival integrity. A manifestation of communal resilience, the Lepcha festivals offer a venue for the Lepcha people to confirm themselves against outside influences.

By instilling a sense of belonging in both the younger generation and those outside their ancestral lands, festivals help to reinforce ties inside the community. By means of culture, they also serve as tools of resistance, providing the community an avenue to confirm itself in defiance of outside influences. Lepcha festivals' syncretic character and, in particular cases, the incorporation of Tibetan Buddhist beliefs and practices provide a fascinating study. Changes in ritual practice regarding Lepcha festivals and their adaptation to bigger socio-political shifts might provide insight on the flexibility of indigenous culture in reaction to changing religious and cultural settings.

Safeguarding Lepcha history and language depends on oral traditions. Future studies can examine the significance of oral traditions and stories in relation to festivals, highlighting their crucial role in cultural transmission. Future studies should examine the economic significance of Lepcha celebrations, taking into account the financial benefits to local communities, the money reinvested from cultural tourism, and the role of cultural legislation in safeguarding aboriginal festivals. Understanding such processes might help to guide Lepcha culture's long-term conservation.

CONCLUSION

Lepcha fairs and festivals teach people about their beliefs, history, and environment. First to settle in the eastern Himalayas, the Lepchas have long loved their natural surroundings, celebrating rivers, mountains, woodlands, and other elements on festivals and rituals. Lepchas are powerful because of their lengthy cultural past and ability to adapt to Tibetan Buddhism, British control, and contemporary state governance. Now a days, Lepcha festivals such as Pang Lhabsol, Chyu Rum Faat, and Tendong Lho Rum Faat are more than community gatherings. They let people express their culture, participate in politics, and learn about the environment. These activities may teach people about the current culture. Participating in rituals, dances, community feasts, storytelling, and environmental actions may teach tourists about Lepcha culture. These activities also encourage long-term tourism by fostering cultural awareness, respect, and love. By keeping these celebrations authentic, the Lepchas can attract more visitors who desire genuine cultural fiestas. But we must address language barriers, poor infrastructure, and cultural commercialization to protect hosts and visitors.

In conclusion, Lepcha festivals are vital for cultural identity, historical preservation, cultural awareness, and sustainable tourism. These live examples of how new influences may coexist with old practices while respecting their ecological and spiritual integrity are essential. These festivals help the Lepcha preserve their tradition and exhibit their might. They make visitors' trips more pleasurable and foster mutual respect. In a rapidly changing world, this dynamic balance between maintaining culture and drawing visitors honors the Lepchas' unique past.

ACKNOWLEDGMENTS

The authors thanks to Mr. Sanjeev Kumar Mishra, Mr. Rajiv Kumar Mishra, Mr. Rohit Bhandari and Mr. Pankaj Kumar for their assistance their tremendous help. My sincere thanks also go to the esteemed faculty members and my colleagues at the Department of Hospitality and Tourism Management, Assam University, for their encouragement.

REFERENCES

Basnet, D., Dorji, T., Ali, G., Lepcha, R., Bhuchar, S., & Shakya, B. (2017). Promoting innovative livelihoods (ecotourism and value chains) for sustainable landscape management—Regional training of trainers (0 ed.). International Centre for Integrated Mountain Development (ICIMOD). DOI: 10.53055/ICIMOD.688

Bentley, J. (2021). Protectors of the land and water: Citizenship, territory, and vulnerability among the Lepcha in Sikkim and West Bengal. *Asian Ethnicity*, *22*(2), 330–352. DOI: 10.1080/14631369.2020.1793295

Bradley, D. (2008). Languages of the Greater Himalayan Region. Vol. 5, A Grammar of Lepcha. By Heleen Plaisier. Leiden: Brill, 2007. 272 pp. $139.00 (cloth). *The Journal of Asian Studies*, *67*(02). Advance online publication. DOI: 10.1017/S002191180800096X

Bruner, J. (1991). The narrative construction of reality. Critical Inquiry. *Critical Inquiry*, *18*(1), 1–21. DOI: 10.1086/448619

Chakraborty, S. (2016). The nature of Lepcha traditional festival: An overview. *The Journal of the Anthropological Survey of India*, *65*(1), 39–54. DOI: 10.1177/2277436X20160104

Chaudhuri, S. K., Maiti, S., & Lepcha, C. K. (Eds.). (2020). The cultural heritage of Sikkim. Routledge, Taylor & Francis Group. DOI: 10.4324/9781003052487

Chettri, L., & Sharma, B. K. (2021). Some traditional folk songs and dances of Sikkim Himalayas. *International Research Journal*, *8*(2), 30–44.

Das, B. (2024). Community-based tourism approach for sustainable development: A study of the Lepcha community, Kalimpong District, India. *Journal of Architectural/Planning Research and Studies (JARS)*, 21(2), 405–420. https://doi.org/DOI: 10.56261/jars.v21.264845

Kapp, D. B. (1986). A parallel motif in Lepcha and Barela-Bhilala mythology. *Asian Folklore Studies*, *45*(2), 259. DOI: 10.2307/1178620

Kirkpatrick, J. (2007). A Lepcha village in Sikkim. *Visual Anthropology*, *20*(1), 87–89. DOI: 10.1080/08949460601064580

Lepcha, C. K., & Lal, U. (2021). *Communities, institutions and histories of India's northeast* (1st ed.). Routledge., DOI: 10.4324/9781003245865

Lepcha, K. (2020). Culture kept alive through oral traditions: A peep into the world of the Lepchas. In Chaudhuri, S. K., Maiti, S., & Lepcha, C. K. (Eds.), *The cultural heritage of Sikkim* (p. 109). Routledge.

Lepcha, R. (2021). *Dynamics of Material Culture of Lepchas in Sikkim, Klimpong and Darjeeling* (Doctoral dissertation).

Lepcha, R. P., Lepcha, M., & Foning, A. (2024). The confluence narratives: Reflections through water. *Social Sciences & Humanities Open*, *9*, 100841. DOI: 10.1016/j.ssaho.2024.100841

Little, K. (2008). Lepcha narratives of their threatened sacred landscapes. *Transforming Cultures eJournal, 3*(1).

Little, K. (2010). From the villages to the cities: The battlegrounds for Lepcha protests. Transforming Cultures eJournal, 5(1). DOI: 10.5130/tfc.v5i1.1579

Malhotra, N. (2024). Introduction to indigenous tribes. In Malhotra, N. (Ed.), *Sustainable pathways* (pp. 1–39). Emerald Publishing Limited., DOI: 10.1108/978-1-83549-490-520241003

Molommu, S. P. (2018). Traditional Beliefs and Practices of the Lepchas in Eastern India. *Asian Man (The)-. International Journal (Toronto, Ont.)*, *12*(2), 149–160.

Munjal, S., & Jauhari, V. (2015). Leveraging the cultural and economic value of fairs and festivals in India. *Worldwide Hospitality and Tourism Themes*, *7*(4), 417–423. DOI: 10.1108/WHATT-03-2015-0006

Plaisier, H. (2005). A brief introduction to Lepcha orthography and literature. *Bulletin of Tibetology*, *41*(1), 7–24.

Pradhan, P. (2016). Culture and Tourism: A Study of Sikkim (Doctoral dissertation, Sikkim University).

Rahaman, H. (2014). An inside view of Kechopalri Lake in Sikkim: A sociological study. *Asian Journal of Multidisciplinary Studies*, *2*(6), 35–43.

Siedlecki, S. L. (2020). Understanding descriptive research designs and methods. *Clinical Nurse Specialist CNS*, *34*(1), 8–12. DOI: 10.1097/NUR.0000000000000493 PMID: 31789957

Vandenhelsken, M. (2011). The enactment of tribal unity at the periphery of India: The political role of a new form of the Pang Lhabsol Buddhist ritual in Sikkim. *European Bulletin of Himalayan Research*, 84.

Verma, R. (2013). Sikkim: *A Guide and Handbook*. New Delhi Narendra Bhatia & Company.

ADDITIONAL READING

Aden, E. R. (2020). *Ethnicity and Identity: A Study of the Lepchas of Darjeeling Hills* (Doctoral dissertation).

Bahuguna, S. (n.d.). *Water & forest conservation message for Sikkim.*

Bain, W. K. (n.d.). Traditional wisdom and practices involved in bamboo-based crafts of the Lepcha community of North Sikkim—A case study from Dzongu Reserve Area.

Basumatary, P. (2020). *Ambuvachi Festival:-A Socio-Cultural Study* (Doctoral dissertation).

Bhattarai, B. (2020). The contemporary process of economic development, state power and Lepcha. In *Fixed borders, fluid boundaries: Identity, resources and mobility in Northeast India.*

Freshfield, D. W. (1903). *Round Kangchenjunga: a narrative of mountain travel and exploration* (Vol. E). Arnold.

Gergan, M. D. (2014). Precarity and possibility: On being young and indigenous in Sikkim, India. *HIMALAYA, the Journal of the Association for Nepal and Himalayan Studies*, 34(2), 10.

Gurung, A. S. (2014). Two waves of transformation and construction of Lepcha (Mon) identity in Sikkim: An assessment of intra-community divide. *The Asian Conference on Cultural Studies,* Osaka, Japan.

Jauhari, V., & Munjal, S. (2015). Fairs and festivals in India: The cultural and economic potential. *Worldwide Hospitality and Tourism Themes*, *7*(4), 324–330. DOI: 10.1108/WHATT-03-2015-0012

Mu, Y. (2016). *Local Perspectives of sacred landscapes and tourism: Exploring the linkages in Sagarmatha (Mt. Everest) National park, Nepal* (Master's thesis, University of Waterloo).

Nepal, P. (2014). Reclaiming spaces: Politics of place, locality and identity of Lepcha community in the Eastern Himalayas. *Dynamics of Public Administration, 31*(2), 113–124. DOI: 10.5958/0976-0733.2014.00791.3

Rai, B. (2017). *Socio-Economic Impact of Hydro-Power Projects: A Study on Lepcha Community in Dzongu Block in Sikkim, 2001-2015* (Doctoral dissertation).

Smith, L. T. (2019). *Decolonizing methodologies: Research and indigenous peoples*. Zed Books.

KEY TERMS AND DEFINITIONS

Lepcha: The Lepcha, an indigenous group in the Eastern Himalayas, are renowned for their rich cultural heritage, animistic spiritual beliefs, and close connection to nature.

Rituals and Ceremonies: Lepcha festivals involve sacred rituals that honour deities, ancestors, and nature spirits through offerings, prayers, dances, and songs, ensuring spiritual harmony and community well-being.

Sacred Offerings: Lepcha festivals involve offering ritual items like rice, flowers, fruits, and water to deities, ancestors, and nature spirits for blessings, prosperity, and community-nature harmony.

Animism: The Lepcha community practices animistic beliefs, valuing natural elements such as mountains, rivers, and forests as sacred and claiming their spiritual essence or consciousness.

Rong: The Lepcha people's name, referring to themselves and their homeland, the Eastern Himalayas, signifies their profound connection to the land and environment.

Chapter 16
Rekindling the Fire – Vuselela Ilangabi:
Revitalizing Indigenous Storytelling for Literacy Development in South Africa

Kellie Jo-Anne Steinke
https://orcid.org/0009-0007-6704-8673
University of Mpumalanga, South Africa

ABSTRACT

This book chapter investigates the integration of traditional storytelling into the 'Learning to Read: Reading to Learn' approach to literacy teaching for students at tertiary level. Indigenous knowledge systems highlight the concept of relationality, forming and valuing relationships within the community, as well as respecting indigenous knowledge. Therefore, the methodological design consists of community-based projects where the participants take ownership of the research. It makes use of mixed methods, multiple case studies, and participatory action research to explore the potential effects of incorporating traditional storytelling into existing literacy teaching approaches. This design also allows for the collection of quantitative data in the form of literacy assessments. Recommendations are made for future research that can explore the long-term effects of traditional storytelling combined with teacher training and curriculum development to effectively implement storytelling-based interventions

INTRODUCTION AND BACKGROUND

Historically in South African education, traditional storytelling has been largely neglected. However, in pre-colonial times, especially in African communities, storytelling was common and a means of both educating and entertaining younger family members (Nomlomo & Sosibo, 2016). Not only was storytelling central to the preservation and transmission of cultural heritage, but it also played a role in shaping moral values and addressing social dilemmas. However, beyond cultural transmission, storytelling also significantly contributes to language and cognitive development (Boya, 2023).

Presenting a theoretical perspective, this chapter will explore the potential of revitalizing indigenous storytelling practices as a means of promoting literacy development among South African tertiary students by drawing on the 'Learning to Read: Reading to Learn' (RtL) approach to the teaching of reading,

DOI: 10.4018/979-8-3373-4124-8.ch016

which was originally developed for marginalized Aboriginal students in Australia (Rose, 2021). RtL is a practical approach to literacy teaching based upon scaffolded learning, systemic functional grammar (SFG), and socio-cultural theory (Vygotsky, 1978). It is an effective literacy-learning tool that facilitates the continued teaching of reading and writing at any grade level. In addition, it has been successful throughout the past few decades in improving literacy levels in various countries around the world, including Kenya, Botswana, and South Africa (Acevedo, Rose, & Whittaker, 2023). An important principle underlying RtL is the role of storytelling and book reading in the early development of literacy skills in young children (Steinke & Wildsmith-Cromarty, 2019). By including traditional storytelling methods into existing RtL literacy pedagogy, it is possible to create a bridge between the indigenous knowledge students bring into the classroom and the formal education system (Acevedo et al., 2023).

Storytelling does not replace book reading. They are both complementary approaches that can cater to particular learning styles (Griffin, Burns, & Snow, 1998). For example, whereas book reading can provide a structured framework for the understanding of genre structures, and promotes self-directed learning, oral storytelling connects learners to their cultural heritage and creates an engaging and interactive learning experience. Both folktales and books can form unique and essential components of literacy acquisition (Brice-Heath, 1983). In addition, where the lack of access to print is exacerbated by the lack of reading culture within poorer South African communities, storytelling can form an accessible and enjoyable home and community-based support system for formal literacy acquisition at school (Makone, 2021).

The chapter outlines the South African literacy crisis and its impact on higher education, justifying urgent interventions. Following this, it discusses the underlying theories, illustrating storytelling's integration into the six-stage RtL cycle (Rose & Martin, 2012). A literature review defines 'indigenous knowledge' and explores storytelling's role in improving literacy. Case studies examining the benefits of traditional storytelling are included. The chapter then details the methodology, considers future research, and provides conclusions and recommendations. An outline of the current literacy situation in South Africa follows.

CHALLENGES IN SOUTH AFRICAN LITERACY EDUCATION

South Africa faces a significant challenge in developing literacy skills at the foundational levels. The latest PIRLS results reveal that some 80% of South African Grade 6 learners cannot read for meaning (Roux, Van Staden, & Tshele, 2023). Learners who do not learn to read by the end of Grade 3 are not likely to catch up. The reasons for this literacy crisis are varied (Klapwijk, 2015). An underemphasis on reading skills in early grades contributes to high dropout rates in both primary and tertiary institutions, there is insufficient pre-service teacher training (Steinke & Wildsmith-Cromarty, 2019) and a disconnect between indigenous home languages and the language of instruction (Henning, 2012). However, a major contributing factor is the lack of access to print in early childhood by many learners from poorer to highlight economic areas, an aspect that will form the focus of this discussion.

Children who have the advantage of being able to read at home before they start formal education tend to be more successful in school. It is estimated that children in literate, middle-class families reportedly spend up to 1000 hours in parent-child reading before they even begin formal schooling (Rose, 2011). Those early, formative years of story reading within the warm and protected environment of a trusted caregiver foster much of the cognitive, emotional and psychological development of a child. Learners who do not receive this generally enter formal schooling at a severe disadvantage (Unsworth & Christie, 2000).

In traditional African culture, the role of book-reading was occupied by storytelling. These two modalities, book-reading and traditional storytelling, share many similarities in their potential to foster literacy development (Maureen, van der Meij, & de Jong, 2020). Like book-reading, storytelling takes place in a supportive, secure environment. Both scaffold learning, and create excitement, stimulating the child's imagination. Storytelling also develops background knowledge or schemas and models literacy behaviours, such as the sharing of narratives, and the use of language. Storytelling not only has distinct characteristics that make it a unique and powerful tool for literacy learning but also forms an important part of traditional African and South African culture (Lebaka, 2022). Thus, by integrating traditional storytelling with RtL, educators can facilitate the much-needed development of literacy skills, the goal of which is comprehension (Rose, 2017). The research objectives follow.

RESEARCH OBJECTIVES

The objective of this chapter is to investigate the possibility of improving South Africa's critically low literacy skills by revising indigenous traditional storytelling. It aims to do this by integrating traditional storytelling into the RtL approach to literacy teaching, focusing primarily on tertiary level education but with application to other grade levels, including foundation level. The research objectives are formulated below:

- The enhancing of literacy levels at tertiary institutions by introducing traditional storytelling into the existing RtL literacy coursework, including indigenous methodology and relationality.
- To investigate how storytelling could increase learning by building a connection between the more formal education system and the indigenous knowledge of students.
- To utilise decolonisation theory and participate reaction research combined with case studies that will incorporate and value indigenous knowledge systems, to investigate the possible impact of traditional storytelling on the literacy levels of students.
- To make sound recommendations regarding the teaching of reading, teacher training and curricular development based on the findings on the use of traditional storytelling in the teaching of reading.

THEORETICAL FRAMEWORK

The following section provides a theoretical framework for incorporating indigenous storytelling into existing coursework. It examines indigenous knowledge systems, decolonisation theory (Gobena, Hean, Heaslip, & Studsrød, 2024), and integrates social cultural and code theory, as well as RtL (Rose, 2021).

Indigenous Knowledge Systems and Decolonisation Theory

The term 'Indigenous Knowledge Systems' is a relatively modern, broad concept that includes the knowledge of indigenous people worldwide. It is grounded in relationality in which the knowledge systems of indigenous people are both recognised and valued (Wilson, 2020). As it is community-based, it encompasses oral tradition and provides a culturally relevant and engaging environment for students in which to learn (Billawer & Nel, 2024). Furthermore, it is related to decolonisation theory which serves as a counter to the hegemony of Western knowledge and historical power imbalances. Storytelling becomes a form of resistance and a decolonisation process that reasserts indigenous epistemologies (Chilisa, 2012; Smith, 2019). Ultimately, decolonisation theory, RtL and IKS align in their shared goals of viewing language as a social semiotic and learning as an activity that takes place in social contexts (Halliday, 1985).

Code Theory, SFG and Genre

Bernstein's code theory posits that language use varies by social context, influencing language registers. Restricted codes, used in familiar settings and often in poorer communities, feature simple structures and implicit meanings (Bernstein, 1995). Elaborated codes are more complex and are used in formal education. They better prepare middle-class children for formal schooling. However, storytelling can expand both codes: for restricted codes, it introduces vocabulary and concepts, providing safe language practice; for elaborated codes, it promotes complex structures, logical thinking, and enhanced vocabulary and grammar via story-based discussion (Donnelly & Abbas, 2018).

Meanwhile, Halliday's SFG, which underpins RtL, views language as a social semiotic, emphasizing context in learning. Storytelling provides meaningful context, connecting listeners to characters and plot. When students listen to a story, they 'see' what is happening. Language creates meaning through field, tenor, and mode, shaping folktale interpretations: for example, what the story iabout, where it is taking place and what is happening. Storytelling builds relationships and community. In addition, storytelling and writing are connected. Stories also teach how to write and to communicate (Halliday, 1985). From SFG, the genre approach developed, with a focus on the teaching of writing in context. An understanding of how stories are structured, which occurs through storytelling, is important for improving comprehension and developing the ability to create one's own story. It also develops both thinking and communication skills (Rose & Martin, 2012).

Learning to read: Reading to Learn

RtL was originally developed from the Genre approach to teaching writing in context. David Rose added a reading component to the teaching of genres, based upon the principle that students often cannot write effectively because they cannot read for meaning (Rose & Martin, 2012). In RtL, comprehension is taught through a six-stage cycle that uses an interaction-feedback cycle runs parallel to that of a parent who reads to a young child (O'Madagain & Tomasello, 2021). The child is assisted in a non-threatening environment as they read together. The child gives feedback to the adult's initial engagement, and the adult then gauges where the child is and elaborates to extend learning (Christie, 2005). This constitutes meaningful interaction (Steinke, 2023). Storytelling can be substituted for book reading within existing RtL-based literacy course work by adapting and making use of the six-stages of RtL for the teaching of reading (Wildsmith-Cromarty & Steinke, 2014). These are illustrated below in Figure 1:

Figure 1. The RtL scaffolding cycle (Rose and Acevedo, 2006)

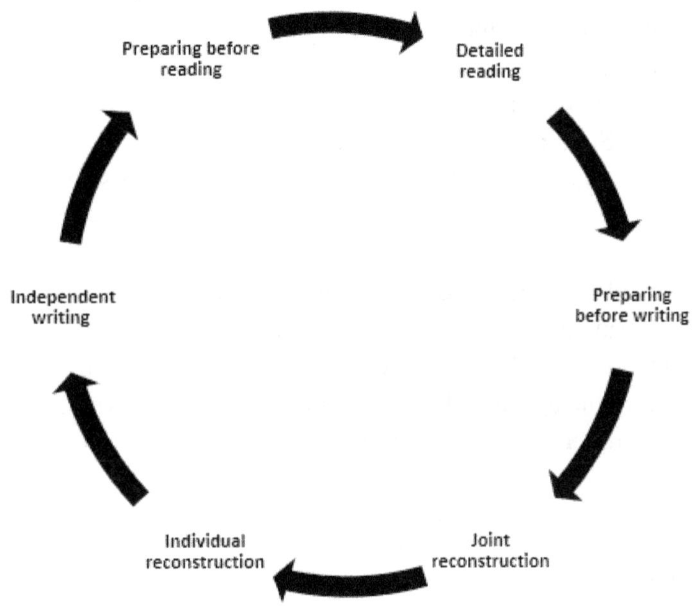

An explanation of the six stages in the above Figure is as follows (Acevedo et al., 2023):

Stage A: Preparing Before Reading

- Activating Prior Knowledge: Students connect the new material to their existing knowledge base.
- Setting Purpose: Students establish clear goals for their reading, such as answering specific questions or finding key information.

Stage B: Detailed Reading

- Strategic Reading: Students use various reading strategies, like summarizing, questioning, and predicting, to enhance comprehension.
- Note-taking: Students record important information and ideas from the text.

Stage C: Preparing Before Writing

- Discussing and Analysing: Students engage in discussions to deepen their understanding and identify key themes.
- Organizing Ideas: Students structure their thoughts and plan for their writing.

Stage D: Joint Reconstruction

- Collaborative Writing: Students work together to create a shared text or summary based on their understanding.
- Peer Review: Students provide feedback on each other's work to improve clarity and coherence.

Stage E: Individual Reconstruction

Independent Writing: Students write their own individual pieces based on their understanding and the collective work.

- Revising and Editing: Students refine their writing by making necessary changes to improve clarity, organization, and style.
  Stage F: Independent Writing
- Finalizing and Sharing: Students complete their writing and prepare it for sharing with others.
- Reflecting on Learning: Students reflect on their learning process and the effectiveness of the reading-to-learn cycle.

Ultimately, the key principles of RtL are engagement, collaborative learning, and the development of meta-cognition (Rose & Martin, 2012; Van Aswegen, Swart, & Oswald, 2019), that integrate to develop strong reading comprehension skills. The RtL cycle that is illustrated in Figure 1 above can be adapted in the classroom, according to the age and skill level of the students, as well as the needs of the teacher, a concept that will be explained in more detail in the methodology section of this chapter. Meanwhile, a literature review follows.

LITERATURE REVIEW

This review will define 'indigenous knowledge' and discuss storytelling, examining literacy development. It will then explore case studies on the effects of storytelling on literacy across various education levels. These studies establish a foundation for this chapter's framework and highlight a research gap: the lack of data on storytelling combined with book-reading in RtL, indicating a need for further research (Damayanti, 2017).

Indigenous Storytelling

In the light of storytelling, it is necessary to begin with a definition of 'indigenous knowledge'. For the sake of this chapter, indigenous knowledge will be seen as relationality, in which knowledge belongs to a community, and is centred within a network of social relationships between people in that community The implications of this are that the focus of the research shifts to relationships within the participating community, and what is referred to as 'relational accountability' (Wilson, 2020), in which the researcher holds a responsibility towards those relationships and must consider the possible impact of her actions on existing relationship networks. In addition, the researcher becomes responsible for how information is collected, utilised, shared and disseminated (Tynan, 2021).

With a similar focus on the importance of relationality and relationships, Kovach (2021) also challenges the hegemony of westernised methodological paradigms, while emphasising the importance of storytelling and story within indigenous research. She sees storytelling as going beyond 'mere' transmission of knowledge, history and cultural values, to being an actual methodology. For example, storytelling can be used to collect data, analyse the findings and disseminate research, while allowing the researcher to engage with the community in a way that builds trust and reciprocity. It can centre indigenous voices and experiences, as opposed to the traditional Western paradigm that prioritises objectivity and detachment, while conveying important information in a manner that is both relatable and memorable (Datta, 2018; Kovach, 2021).

The fact is that indigenous storytelling has served as a source of history, knowledge, and transmission of cultural values for generations. Examples of this may be the ever-popular 'trickster' tales from isiZulu folklore, such as *Chikijana* and *Nogwaja* (Turner, 1994), where characters may serve to address social dilemmas and themes of everyday life, such as the triumph of intelligence over intimidation. Folktales include particular numbers, along with their specific structures (e.g., those of trickster tales, fables, or epic narratives) and teach the listeners important lessons about behavior, relationships and the natural world (Wood, 2010). Relevant stories can hold learner's attention, while building heritage (Canonici, 1994).

Beyond the above advantages, storytelling significantly contributes to language and cognitive development. As storytelling takes an oral form, it serves to enhance oral language skills, including listening, speaking, and vocabulary acquisition (O'Neill, Banoobhai, & Smith, 2016). The use of vivid imagery, rhythmic patterns, and figurative language assists in developing a love for language and develops phonemic awareness. Moreover, storytelling develops memory association along with critical thinking skills. As learners engage with the story or narrative, they learn to draw inferences (Bara, 2021). Storytelling techniques align with concepts in literacy development in specific ways, such as the repetition of phrases, while the sequencing of events in traditional folk tales can foreshadow the use of the structures and patterns used in written language.

Furthermore, because storytelling is interactive, it has call-and-answer phases which follow a similar pattern to the dialogue and interaction that is found in many narrative texts (Boya, 2023). In addition, storytelling allows for the embodiment of the 'knowledgeable other' or mentor, who takes the young listener further along in his or her development than if they were left to themselves (Vygotsky, 1978). To illustrate, in traditional African culture the 'mentor' is often the grandmother. She is a caregiver, who forms the link between past and present, for in oral tradition, what is not passed on is often not remembered. Her animated telling of stories stimulates the memory and recall of her audience, and develops their listening skills, while passing on a sense of belonging and cultural heritage (Canonici, 1994).

Additionally, folk tales, with their recurring patterns and conventions, form distinct genres. Indigenous stories, for instance, typically employ a narrative structure with a clear beginning, middle, and end, often involving a conflict, climax, and resolution. These are elements that align with formal book reading and stories, including the use of literary devices such as repetition, figurative language, and dialogue (Rose, 2017). The telling of stories teaches the importance of register variation, adapting language use to the context and audience, by employing different registers depending on the story's purpose and audience. For instance, a story aimed at children might use simpler language and more vivid imagery compared to a story meant for adults (Martin, 1999). Meanwhile, stories from learners' own cultural backgrounds build on their existing knowledge and experiences, making them more invested in the learning process. When learning is active, it fosters engagement and promotes deeper comprehension compared to passive reading. This aligns with the framework of social constructivism, that asserts that dramatic play enhances the development of children, who should be active participants in their own learning (Pinter, 2017). A definition of literacy follows.

Literacy Development

For the purposes of this chapter, literacy development is seen as a multi-faceted process that includes understanding the written word, expressing ideas through writing and effectively making use of language to both communicate and learn. Importantly, literacy development forms a lifelong pursuit (Zoll et al., 2023). While reading and literacy are not identical concepts, they are closely related. For example, while reading involves decoding words and letters and understanding, literacy includes reading and writing but also listening, speaking, as well as problem solving and critical thinking (Pilgrim & Martinez, 2013).

The ability to read for meaning is generally assumed to have been acquired by the end of the foundation phase of education. However, the reality is that learners are rarely taught to read after grade three. On the contrary, from grade four onwards, they are given a textbook and expected to learn *from* reading. This could provide a rationale as to why many students who enter into tertiary level education in South Africa have not progressed much beyond foundation reading skills (Steinke & Wildsmith-Cromarty, 2019). By way of illustration, it is generally understood that the literacy levels of foundation phase learners act as predictors of later school achievement. Thus, the early and foundation grades of schooling are of the utmost importance for literacy acquisition (Pretorius & Klapwijk, 2016).

Considering this, the employment of skilled storytelling in the early foundation grades can be employed to help develop the reading and language skills that are crucial for academic success. For example, it can develop listening and interactive ability (Spencer & Pierce, 2023). It can also promote communication skills, while incorporating all four crucial language skills, namely: listening, speaking, reading and writing (Wessels, 2011).

This review will now examine four specific case studies conducted in South Africa, as well as a fifth international research project, that highlight the value of incorporating storytelling into the development of literacy and language development. These studies represent foundation, intermediate, secondary and tertiary education level research, and provide evidence that storytelling can assist learners and students across all schooling levels. The case studies selected also involve indigenous South African languages, such as isiXhosa and isiNdebele (Jiyana, 2022; Makaluza, 2018). The South African research will be discussed first, followed by an international study on English foreign language learners (EFL).

Case Study One

This case study investigated the potential of using traditional isiXhosa folktales to enhance early literacy development among Grade 1 learners in the Western Cape. The research drew upon socio-cultural theory and Indigenous Knowledge Systems (IKS) to understand how storytelling can promote literacy skills and reinforce cultural identity (Makaluza, 2018).

Specifically, the study utilised Vygotsky's socio-cultural theory, which emphasizes the crucial role of community and social interaction in children's intellectual development. The concept of the zone of proximal development was central, highlighting how a 'knowledgeable other' or mentor can guide a learner further than they could progress independently. This theory underscores the value of teachers and caregivers in children's education, as well as the significance of preserving cultural heritage through storytelling and folktales. It acknowledges that learning occurs through social interaction with trusted mentors, teachers, or caregivers (Vygotsky, 1978).

The study employed a qualitative research design, using semi-structured interviews and document analysis. Conducted at a primary school in a disadvantaged township area in Philippi, Western Cape, the study aimed to improve isiXhosa literacy among Grade 1 learners. The importance of children acquiring strong literacy and language skills in their home language was recognized, as these skills are foundational for the acquisition of second and third languages (Taylor & von Fintel, 2016). A child who is not literate in their home language faces a significant disadvantage, particularly in South Africa, where learners transition to English as the language of instruction from Grade 4 onwards (Steinke, 2024).

The research findings revealed that the use of folktales effectively promoted oral communication skills, enhanced understanding of story structure, and improved listening comprehension. The study also reported a positive impact on writing skills. Researchers recommended that teachers acknowledge the valuable role of traditional folktales in valuing and incorporating indigenous knowledge within the classroom. They also suggested that traditional folktales be adapted into texts that can be both read and written, which is significant for the incorporation into RtL pedagogy (Makaluza, 2018).

Case Study Two

The next case study examines the role of storytelling in IsiNdebele literacy development with grade three learners (Jiyana, 2022). Also using socio-cultural learning theory, the study used qualitative research with interviews, observations and document analysis to examine the importance of storytelling in providing a supportive environment for the acquisition of literacy skills for foundation learners. The study involved grade three isiNdebele learners, teachers and their parents from two primary schools in Siyabuswa. Interviews were held with the teachers and the parents, while observations were carried out in both the classroom and home environments. Learners' classwork and activity books were examined for data, with thematic analysis used for analysis.

Findings were that storytelling is effective for developing both listening and speaking skills in isiNdebele. It also engaged the learners and provided them with opportunities for language practice. The researchers state that as storytelling is deeply rooted within traditional isiNdebele culture, it can be effectively used as a tool to connect learners to their heritage. Furthermore, they found that parental involvement in the storytelling significantly impacted the development of the children and that, if parents could support their children within the community and home environment via storytelling (Jiyana, 2022), it could be used as a support for the educational environment. In addition, they mention that storytelling aligns with both the current South African DBE curriculum as well as the Curriculum Assessment Policy Statement (CAPS) for grade three, which stresses the importance of both listening and speaking skills (Department of Basic Education, 2011).

Case Study Three

The third case study was conducted at intermediate level and investigated the use of storytelling to promote reading in primary schools within the Lebopo Circuit, Mankweng Cluster of the Capricorn District, Limpopo province. It aimed to understand the practices, benefits, challenges, and potential of storytelling in fostering a reading culture among learners (Mojapelo, 2020).

This study made use of qualitative data to investigate storytelling initiatives in 10 primary schools, with the participants being the intermediate level teachers. Structured interviews were used. These were based upon storytelling activities, with teachers providing information on reading materials, possible challenges, and suggestions for improvement. The researchers used thematic analysis to identify themes and patterns within the interview data.

Findings were that the schools involved in the study were already engaged in storytelling and held regular reading competitions for the learners. However, they found that these schools lacked sufficient reading materials, particularly in in English. The participating teachers believed that storytelling enhances language skills, builds the self-confidence of the learners and improves their vocabulary. The schools involved encouraged the involvement of parents by implementing reading activities and organizing competitions in the schools, which are beneficial to the community as a whole. However, there are limited library facilities and the teachers in the schools found that the multicultural environment can provide challenges for literacy teachers (Paton-Ash & Wilmot, 2015). The study emphasized the importance of storytelling and promoting literacy but also highlighted the urgent need for increased resources and support from the department of basic education in the way of reading materials. It's significance lies particularly in the indications of the initiatives that teachers are taking research projects but the necessity of urgent support from the DBE for these teachers (Mojapelo, 2020).

Case Study Four

In the last example of a South African case study, research has explored the potential of digital storytelling in the teaching of academic writing at a higher education institution in the Western Cape (Masoga, 2022). The researcher wanted to examine the significance of storytelling for education and learning as well as its effectiveness in strengthening social cohesion and promoting cultural understanding, keeping in mind the difficulty many tertiary students in South Africa have with the use of English as the language of learning and teaching, and their tendency to struggle with academic writing as a result. This project was an autoethnography, which involves personal storytelling under social and cultural analysis, allowing the researcher to explore her own personal experiences whilst connecting them to the broader issues in society and autoethnography is effectively a case study involving one participant, who also happens to be the researcher (Chang, 2016).

The aim of the research was to promote storytelling as an important resource in higher education. The study made use of digital storytelling, which is a combination of traditional storytelling and technology, to scaffold academic writing. This also meant that students could, for example, create images and sounds that could potentially increase their engagement in the learning process. The read a- response theory was used as a theoretical underpinning to emphasize the active role of the reader interpreting and understanding text. Furthermore, the study used narrative analysis to focus on literary genre and storytelling techniques that can be used within these. The researcher drew on personal experiences to illustrate the significance and the impact of storytelling. Structural elements of the story were examined as well as plot, character and the types of language used in the various genres (Martin, 1999).

Findings were that integrating traditional storytelling with digital media offered exciting new opportunities for both teaching and learning. Storytelling has the potential to enhance comprehension, critical thinking skills and cultural understanding, while aligning with a social and community-centered approach to education. The growing recognition of the importance of IKS in higher education and the role of storytelling in achieving this is acknowledged (Watson-Verran, Turnbull, & Jasanoff, 2005). Recommendations were made that storytelling be incorporated into higher education, along with a need for further research and the development of effective storytelling programmes that can facilitate a shift towards an inclusive approach. Such an approach would incorporate and value local knowledge and expertise (Masoga, 2022).

Case Study Five

The final case study under discussion involves an international research paper that explored the potential of using storytelling to improve English language acquisition for English foreign language learners (EFL) in Indonesia, using existing RtL pedagogy (Damayanti, 2017). The participants were secondary school learners who ranged in age from 13 to 14 years. Like South Africa, traditional storytelling is quite deeply rooted within Indonesian culture. In addition, language teaching methods tend to be teacher-centered and focus mainly on grammar and rote learning, while learners are not generally motivated to learn English. The intention of this study was to increase student engagement, increase the language input and assist students to learn and understand structure, genre, and language patterns. The study mentions the limited research that currently exists on the use of storytelling in EFL context and that the few stories that do exist tend to show mixed results in terms of effectiveness (Damayanti, 2017).

The research involved two English teachers from a secondary school in Bandung and a total of 42 students in their classes. The methodology consisted of action orientated field work, with the researcher using cycles of plan, observation and reflection to observe eight classroom lessons in total. Writing samples from the learners were assessed, both pre- and post- intervention. The researcher was already actively using the RtL approach in the classroom in order to improve the reading and writing of the learners in the target language, as well as to assist them in their ability to write narratives (Rose, 2017).

The research analyzed the writing of the students using systemic functional grammar (Halliday, 1985), with the intention of examining purpose, stages, phases, and meaning. the control of vocabulary, grammar, and the overall cohesion and coherence of the text. Acting in the role of both a participant observer and as a teacher, the research made use of children's stories which were referred to as 'mentor texts'. Using the RtL cycle, oral storytelling was included in the first stage which consists of' preparing for reading'. This study is relevant because it integrates active storytelling into RtL, providing potential empirical data, though it focuses on EFL, not reading. Despite the benefits of storytelling for literacy, a research gap exists regarding its integration into RtL (Acevedo et al., 2023), raising the question of effective integration, and its potential to support formal education in resource-limited environments (Makone, 2021). The methodology follows.

METHODOLOGY

This section on methodology and design will provide relevant and compatible examples of the various frameworks and paradigms that can be applied in integrating storytelling into the RtL approach. Seeing as the majority of research papers investigated in the literature review involved case studies, it will begin with a discussion of the case study as a choice for research methodology. It will then go on to examine participatory action research as well as various indigenous research methodologies and how they could be applied within an RtL framework. An example of an indigenous folk tale will be provided, along with an English translation and an illustration of how storytelling can be successfully integrated into the Existing RtL teaching and learning cycle.

Case Study Designs and Participatory Action Research

A case study design is a good choice for investigating the relationship between storytelling and literacy. For example, it can provide a solid contextual understanding, whether it be school, community or an individual learner. In addition, it allows for various data collection methods, such as interviews, observations, document analysis, and surveys (Burton & Bartlett, 2009). This triangulation of data provides a more comprehensive picture of storytelling's effects on literacy and allows the research to be sensitive to the needs in context of the participants. It also provides for autonomy and agency and that it can empower the participants to share their own stories and perspectives (Stake, 1995).

The type of case study envisaged would depend on whether or not the researcher wishes to include a group or cohort of students as one particular case study, thus investigating the effect on the class environment as a whole, or whether the researcher wishes to include each participating student as an individual case study, meaning that the effect on learning of an individual participant would be investigated (Yin, 2013). As previously mentioned, the focus of this chapter is to discuss the potential use of storytelling and resulting literacy development at both foundation and tertiary education levels. Thus, the selected sample, participants and unit of analysis may differ depending on the grade level of the learners or students. For example, the author is a lecturer at a tertiary institution and therefore her selection of a particular group or classroom of students would be as a result of her access to a particular teaching module, as well as the design of the relevant materials.

Each module has an average of 200 registered students per class, and each student would consist of one case study, resulting in a multiple case study (Kivunja & Kuyini, 2017). In the case of foundation or intermediate school learners, the research on the effects of traditional storytelling on the literacy levels of learners could be extended to the larger community and thus move into the area of participatory action research. Participatory action research, or PAR, can provide a solid research framework and aligns, with the principles of collaboration, community engagement and social change that are vital for the revitalization of indigenous cultural practices, in this case indigenous storytelling (Kemmis & McTaggart, 2003).

Returning to foundational concepts of indigenous research as relationality, it is important in the methodology to focus on building and maintaining respectful relationships with communities and participants. There is a necessity of ongoing dialogue and collaboration, and it is important that PAR is guided by relational accountability (Wilson, 2020). In addition, the needs of the community must be prioritised and the researcher needs to be able to self-reflect and be aware of any potential power dynamics or biases that she may hold. The community retains control and ownership of the research process (Smith, 2019). The research can focus on specific needs of a community and involve elders, storytellers, parents and communities. It could also integrate teachers, who can incorporate storytelling into the existing curricula. One of the main advantages of PAR is that it can allow the process to extend beyond these specific period of research and become an important community-based initiative (Baum, MacDougall, & Smith, 2006). Figure 2 below illustrates how PAR would work within the context of using storytelling to increase literacy effectiveness:

Figure 2. A framework for research design

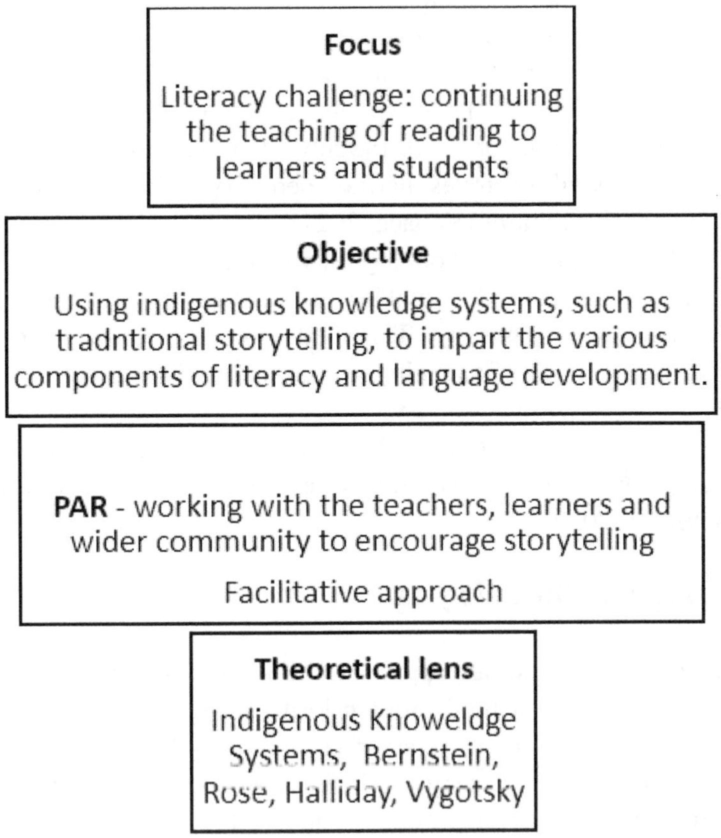

The above illustration begins with the purpose of the research, which is to address the current low literacy levels in the South African education system by incorporating storytelling into existing teaching and learning of literacy approaches. The methodology is envisaged as PAR which will involve teachers, learners as well as the community. The theoretical framework draws from into traditional knowledge systems as well as solid theories of language development, while the approach to the research would be collaborative and facilitative.

An epistemological and an ontological approach are included in the illustration in that the learners would be included in the storytelling process. Their experiences and cultural reality should be at the forefront of the research process. Though much of such a research project would likely fall within a qualitative research paradigm, it would be necessary to assess literacy levels both pre- and post-study and therefore would include some quantitative research as well. Ultimately it provides a holistic and culturally relevant approach to improving literacy levels that seeks to harness the power of both community and indigenous knowledge (Da Silva et al., 2024).

Data Collection

While the data collection consisted mainly of qualitative instrumentation, such as interviews, questionnaires and observations, indigenous epistemologies must be kept in mind. This means that to acknowledge the validity of indigenous epistemology and ontology, this research incorporated storytelling, oral histories and other culturally relevant methods of data collection. The research will not just be inclusive of the learners' experiences but will also view those experiences and as valid forms of knowledge (Windchief & Cummins, 2022). Storytelling will not just be the subject of the study but powerful tool of knowledge creation that will involve storytelling circles, narrative enquiry and other methods on the role of story and storytelling and indigenous cultures (Kovach, 2021).

The focus was on African-centred methodologies, which involves culturally responsive evaluation, meaning that for example if the research is making use of RtL it must be useful and of benefit to the community as well as culturally responsive. This methodology means that storytelling becomes not simply a tool for data but also a way of engaging in community discussions and building relationships (Chilisa et al., 2017).

The principle of relational accountability will be the focus of the methodology and will guide the research to ensure that all interactions with communities and participants are carried out in a respectful and simple manner (Wilson, 2020).

Indigenous Research Methodologies

Indigenous research methodologies are approaches to knowledge production that are deeply rooted in specific cultural contexts. These methods often prioritize community engagement, storytelling, and experiential learning. They can offer unique perspectives and valuable insights into research processes that may not be captured by traditional Western methods. Besides being community centered, these research methodologies are often hands on and involve experiential learning (Kolb, Boyatzis, & Mainemelis, 2014). These methodologies can take the form of, among others, community meetings, or 'imbizo', and 'Ukwabelana', which is an isiZulu term used to describe the practice of sharing information and feedback with the community, oral history or ethnography (Keikelame & Swartz, 2019),

In addition, multilogicality can provide a practical theoretical framework for IKS by providing, for example, purposeful sampling and face-to-face interviews with participants. It would mean that the approach should involve exploring the cultural world from various perspectives and acknowledging the legitimacy of different ways of knowing and being. This would allow the researcher to gain a deeper understanding of the experiences as well as the perspectives of the participants theory (Kincheloe, 2006; Smith, 2021). A practical illustration of how storytelling can be incorporated into the existing RtL cycle of teaching and learning follows.

Incorporating Storytelling into the RtL Cycle

This section will provide a practical explanation of how the existing RtL six-stage cycle could be adapted to incorporate traditional storytelling. I have selected a folktale titled: '*The Story of the Man and the Leopard*', originally from Masoga (2022: 5-6) for the lesson (see Appendix). Masoga provides the text both in its original Sepedi language, along with an English translation.

A diagram of the initial RtL cycle has been provided already in this text (see 4.5), Figure 3 below provides a detailed illustration of how the storytelling can be integrated into the proposed RtL literacy lesson:

Figure 3. Integrated storytelling RtL cycle

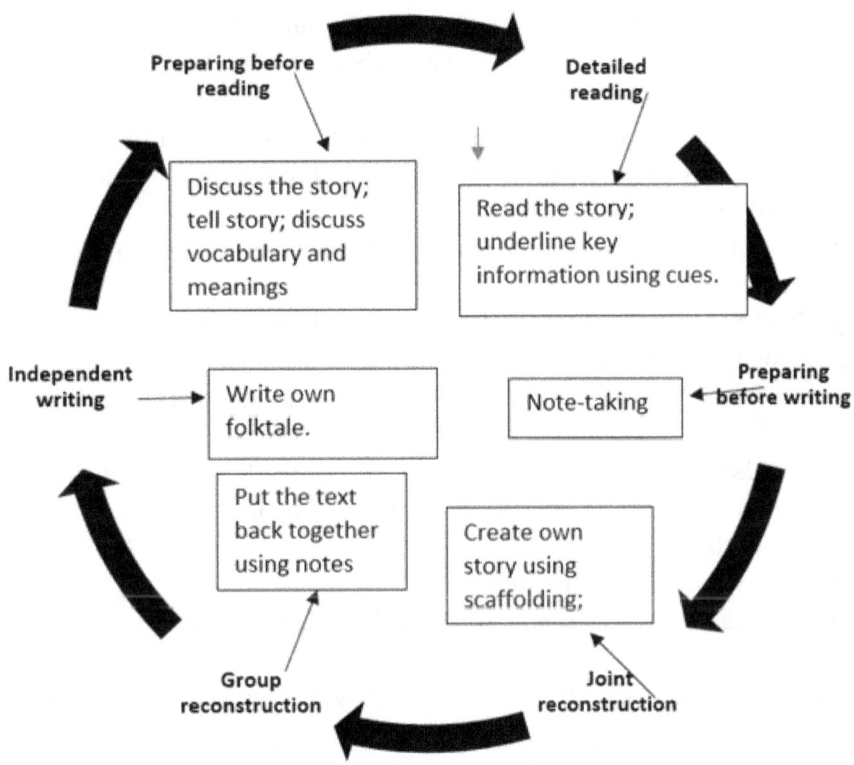

An explanation of the six stages in Figure 3 above is as follows (Acevedo et al., 2023):

Stage A: Preparing Before Reading

- Activating Prior Knowledge: Students connect the new material to their existing knowledge base. In this case they can discuss whether anybody ever told them folk tales when they were younger, and any folk tales that they may remember. They may discuss how those folk tales made them feel and what they learned from them.

- Setting Purpose: Students establish clear goals for their reading, such as answering specific questions or finding key information. The staging and sequencing of the folk tale can be discussed as well as any vocabulary that may be of significance in the passage.
- The teacher tells the story using intonation and animation to make it as interesting as possible.

Stage B: Detailed Reading

Using the aforementioned highlighted portion of the text, the following takes place:

- Strategic Reading: Students use various reading strategies, like summarizing, questioning, and predicting, to enhance comprehension.
- The teacher, along with the class, will read the text. The teacher will select a section from the folktale for detailed reading. In this case selected the goal, problem, temporary solution sections of the passage have been chosen. This tends to run on loop as most narratives would have a problem, a temporary solution and then likely another problem until the final resolution. The teacher will select words and phrases and relevant information beforehand which the students will locate in the text and underline.
- Note-taking: Students record important information and ideas from the text. This will consist of the information that has been underlined while going through the detailed reading.

Stage C: Preparing Before Writing

- Discussing and Analysing: Students engage in discussions to deepen their understanding and identify key themes.
- Organizing Ideas: Students structure their thoughts and plan for their writing.

Stage D: Joint Reconstruction

- Collaborative Writing: Students work together to create a shared text or summary based on their understanding.
- Peer Review: Students provide feedback on each other's work to improve clarity and coherence.
- Using their notes and their learning, the students will create their own story in groups and they will act this out for the class.

Stage E: Individual Reconstruction

- Independent Writing: Students write their own individual pieces based on their understanding and the collective work.
- Revising and Editing: Students refine their writing by making necessary changes to improve clarity, organization, and style.
- The students will use the original notes that they took from the detailed information that was underlined and there will recreate a text that is close to the original version. In this way they reconstruct the text that has been previously deconstructed in order to assist them in meaning making.

Stage F: Independent Writing

- Finalizing and Sharing: Students complete their writing and prepare it for sharing with others.
- Reflecting on Learning: Students reflect on their learning process and the effectiveness of the reading-to-learn cycle.
- As part of their formal assessment task, the students will write their own folktale individually.

Ultimately, the use of the RtL cycle promotes the key principles which are engagement, collaborative learning, and the development of metacognition (Rose & Martin, 2012; Van Aswegen et al., 2019). The main goal is to develop strong reading comprehension skills and teach effective writing in context, and this can be done through the exciting and engaging medium of storytelling whilst revitalizing a lost heritage. There is a further principle that runs throughout this social and community engagement and this is known as 'joint attention' (O'Madagain & Tomasello, 2021).

This joint attention is based on the premise that children develop a cultural mode of learning from an early age that forms the basis of their academic or classroom learning. As Rose (2011:8) states: *"... adults direct children's attention, or follow their attention to things and activities, then name them, evaluate, demonstrate, explain and so on. ... shared emotion is critical as adult and child exchange evaluations of things and actions."* This classroom interaction is known as the 'Initiation-Response-Feedback cycle' (IRF) and provides a solid foundation for the rationale behind the lesson plan.

Joint attention and the IRF Cycle are essential components of effective storytelling in education. Joint attention and the shared focus on a narrative, fosters a cultural learning experience. The IRF cycle, involving teacher initiation, student response, and teacher feedback, promotes active engagement, critical thinking, and language development (Rose, 2021). Together, these elements create a more engaging and effective learning experience that aligns with the "Reading to Learn" framework. By incorporating them, teachers can enhance students' understanding of stories, language skills, and critical thinking abilities (O'Madagain & Tomasello, 2021). Thus, this section demonstrates how traditional storytelling, informed by indigenous knowledge and PAR can enhance the RtL literacy teaching approach by fostering engagement and cultural understanding. The findings and related discussion follow.

FINDINGS AND DISCUSSION

Based on the theoretical framework presented in the paper, it can be concluded that traditional storytelling has the potential to significantly enhance literacy development among South African tertiary students when integrated into the RtL approach. Key findings include that storytelling is important for an holistic approach that can encompass epistemologies. In addition, community-based learning can strengthen the connection between 'home' and the formal education environment. Whilst preserving and strengthening cultural heritage. Finally, storytelling can facilitate literacy development, including the development of vocabulary, comprehension and critical thinking, among others.

The findings suggest that incorporating traditional storytelling into tertiary literacy programmes can create a more inclusive and culturally relevant learning experience for students. It offers a unique opportunity to bridge the gap between indigenous knowledge and formal education, promoting a deeper understanding of language, culture, and identity. By addressing factors such as teacher training, curriculum integration, resources, and community collaboration, tertiary institutions can effectively implement storytelling-based interventions to improve literacy outcomes.

The findings of this theoretical paper have significant relevance for South African education and literacy development. Given the challenges faced by many South African students in acquiring literacy skills, particularly those from marginalized communities, the integration of traditional storytelling into tertiary literacy programmes offers a promising solution. By leveraging the power of storytelling, tertiary institutions can create more engaging, culturally relevant, and effective learning environments that support students' academic success.

FUTURE RESEARCH DIRECTIONS

While this paper provides a strong theoretical foundation, empirical research it is necessary to validate the findings and explore the potential benefits of incorporating traditional oral storytelling into tertiary literacy programmes. Identified areas of future research are comparative studies, where the effectiveness of traditional oral storytelling is compared with more traditional methods of literacy teaching; implementing programmes in teacher training courses that can equip educators to include storytelling in their teaching of reading; and curriculum development, where culturally relevant curriculum materials could be incorporated into existing literacy programmes, particularly those that utilise the RtL approach. By integrating traditional storytelling into the RtL approach, students can benefit, among others, from enhanced comprehension, vocabulary development, a greater sense of belonging and heritage, the development of critical thinking skills, an increased level of motivation and engagement, and stronger writing skills. In other words, storytelling is the potential to enhance the learning experience of students.

CONCLUSION AND RECOMMENDATIONS

This chapter has explored the potential of using traditional storytelling practices within existing RtL coursework to increase literacy levels in South African students, bridging the gap between cultural heritage and a Western-based education system. Grounded in decolonization theory, the research advocates for the use of indigenous knowledge and storytelling, that goes beyond entertainment to build vocabulary, comprehension skills, as well as community relationships. It becomes a powerful vessel for passing on traditional values and heritage. In addition, relationality and incorporating African methodologies and epistemologies, it can secure cultural relevance and ownership of the project by the participating community. In an educational context, combining storytelling with a practical approach to the teaching of reading, such as RtL, can assist in addressing the current literacy crisis in South Africa. As a result, both educators and researchers should be encouraged to incorporate traditional indigenous storytelling into existing literacy coursework to meet the critical needs of South African students.

REFERENCES

Acevedo, C., Rose, D., & Whittaker, R. (2023). *Reading to Learn, reading the world*. Equinox Publishing.

Bara, M. (2021). *Early literacy development in IsiXhosa: Fostering grade 3 learners' imagination and critical thinking through folktales.* (MA). University of the Western Cape, Baum, F., MacDougall, C., & Smith, D. (2006). Participatory action research. *Journal of Epidemiology and Community Health, 60*(10), 854. 10.1136%2Fjech.2004.028662

Bernstein, B. (1995). Code theory and its positioning: A case study in misrecognition. *British Journal of Sociology of Education, 16*(1), 3–19. DOI: 10.1080/0142569950160101

Billawer, W., & Nel, V. B. (2024). *The exclusion of indigenous knowledge systems in planning open spaces in Namibia: The case of Havana, Windhoek.* Paper presented at the Urban Forum. DOI: 10.1007/s12132-024-09510-9

Boya, T. B. (2023). The implications of Setswana hare folktales contemporary for South African children. University of Pretoria, Brice-Heath, S. (1983). *Ways with words: Language, life and work in communities and classrooms*. Cambridge: Cambridge University Press.

Burton, D., & Bartlett, S. (2009). *Key issues for Education Researchers*. Sage. DOI: 10.4135/9781446269480

Canonici, N. (1994). The trickster in Zulu folktales. *Alternation (Durban), 1*(1), 43–56.

Chang, H. (2016). *Autoethnography as method*. Letf Coast Press Inc. DOI: 10.4324/9781315433370

Chilisa, B. (2012). Postcolonial indigenous research paradigms. *Indigenous research methodologies*, 98-127.

Chilisa, B., Major, T., & Khudu-Petersen, K. (2017). Community engagement with a postcolonial, African-based relational paradigm. *Qualitative Research, 17*(3), 326–339. DOI: 10.1177/1468794117696176

Christie, F. (2005). The pedagogic device and the teaching of English. In Christie, F. (Ed.), *Pedagogy and the Shaping of Consciousness* (pp. 156–184). Continnuum.

Da Silva, C., Furtado, K., Soares, S., & De Almeida, J. (2024). Dialogue of knowledges: The importance of the partnership between scientists and indigenous peoples for the preservation of traditional knowledge. *DELOS: Desarrollo Local Sostenible, 17*(53), e1368–e1368. DOI: 10.55905/rdelosv17.n53-030

Damayanti, I. (2017). From storytelling to story writing: The implementation of reading to learn (R2L) pedagogy to teach English as a foreign language in Indonesia. *Indonesian Journal of Applied Linguistics, 6*(2), 232–245. DOI: 10.17509/ijal.v6i2.4870

Datta, R. (2018). Decolonizing both researcher and research and its effectiveness in Indigenous research. *Research Ethics, 14*(2), 1–24. DOI: 10.1177/1747016117733296

Department of Basic Education. South Africa. (2011). *Curriculum and assessment policy statement Grades R-3: English home language*. Pretoria: Government Printing Works

Donnelly, M., & Abbas, A. (2018). Using the sociology of Basil Bernstein in higher education research. In *Theory and Method in Higher Education Research* (Vol. 4, pp. 1–17). Emerald Publishing Limited. DOI: 10.1108/S2056-375220180000004002

Gobena, E., Hean, S., Heaslip, V., & Studsrød, I. (2024). The challenge of western-influenced notions of knowledge and research training: Lessons for decolonizing the research process and researcher education. *Journal of Ethnic & Cultural Diversity in Social Work*, *33*(5), 276–287. DOI: 10.1080/15313204.2023.2197272

Griffin, P., Burns, M., & Snow, C. (1998). *Preventing reading difficulties in young children*. National Academy Press.

Halliday, M. (1985). *An introduction to functional grammar*. Edward Arnold.

Henning, E. (2012). Learning concepts, language, and literacy in hybrid linguistic codes: The multilingual maze of urban Grade 1 classrooms in South Africa. *Perspectives in Education*, *30*(3), 69–77.

Jiyana, M. (2022). *The Role of Traditional Storytelling in the Teaching and Learning of Isindebele in Grade 3*. University of Pretoria (South Africa), Keikelame, M., & Swartz, L. (2019). Decolonising research methodologies: Lessons from a qualitative research project, Cape Town, South Africa. *Global Health Action*, *12*(1), 1–7. DOI: 10.1080/16549716.2018.1561175

Kemmis, & McTaggart. (2003). Participatory action research. In Denzin & Lincoln (Eds.), *Strategies of Qualitative Enquiry* (pp. 336-396). Thousand Oaks, C.A.: Sage.

Kincheloe, J. (2006). Postformalism and the Socio-cultural and Political Challenge: Taking Multilogicality Seriously. In *Reading, Writing, and Thinking* (pp. 141-167): Brill.

Kivunja, C., & Kuyini, A. (2017). Understanding and applying research paradigms in educational contexts. *International Journal of Higher Education*, *6*(5), 26–41. DOI: 10.5430/ijhe.v6n5p26

Klapwijk, N. (2015). EMC^2 = comprehension: A reading strategy instruction framework for all teachers. *South African Journal of Education*, *35*(1), 1–10. DOI: 10.15700/201503062348

Kolb, D., Boyatzis, R., & Mainemelis, C. (2014). Experiential learning theory: Previous research and new directions. In *Perspectives on Thinking, Learning, and Cognitive Styles* (pp. 227-247): Routledge.

Kovach, M. (2021). *Indigenous methodologies: Characteristics, conversations, and contexts*. University of Toronto press.

Lebaka, M. (2022). Adverse impact of the flow of African storytelling between older and younger generations as a result of the movement of people from rural to urban areas: The case of Bapedi story telling tradition. *Dialogo*, *9*(1), 95–104. DOI: 10.51917/dialogo.2022.9.1.7

Makaluza, N. (2018). *Storytelling for early literacy development in isiXhosa: A case study of a grad one class in the Western Cape*. (MA). University of the Western Cape, Makone, I. (2021). A good reading culture: Potential game changer in Africa. *African Books Collective*. Retrieved from https://www.readafricanbooks.com/opinion/a-good-reading-culture-potential-game-changer-in-africa/#:~:text=A%20reading%20culture%20is%20an,has%20a%20poor%20reading%20culture

Martin, J. (1999). Mentoring semogenesis: 'Genre-based' literacy pedagogy. In Christie, F. (Ed.), *Pedagogy and the Shaping of Consciousness: Linguistic and Social Processes* (pp. 123–155). Continuum.

Masoga, M. (2022). Storytelling Research: Implications for the broader Higher Education System in South Africa. *Southern African Journal for Folklore Studies, 32*(2), 14 pages-14 pages.

Maureen, I., van der Meij, H., & de Jong, T. (2020). Enhancing storytelling activities to support early (digital) literacy development in early childhood education. *International Journal of Early Childhood, 52*(1), 55–76. DOI: 10.1007/s13158-020-00263-7

Mojapelo, S. (2020). Storytelling Initiatives and reading in schools in South Africa: A case from Limpopo Province. *Mousaion, 38*(3). Advance online publication. DOI: 10.25159/2663-659X/7873

Nomlomo, V., & Sosibo, Z. (2016). Indigenous knowledge systems and early literacy development: An analysis of IsiXhosa and IsiZulu traditional children's folktales and songs. *Studies of Tribes and Tribals, 14*(2), 110–120. DOI: 10.1080/0972639X.2016.11886738

O'Madagain, C., & Tomasello, M. (2021). Joint attention to mental content and the social origin of reasoning. *Synthese, 198*(5), 4057–4078. DOI: 10.1007/s11229-019-02327-1

O'Neill, B., Banoobhai, M., & Smith, C. (2016). Teaching literacy through dramatic storytelling in foundation phase. *Australasian Journal of Early Childhood, 41*(2), 95–102. DOI: 10.1177/183693911604100213

Paton-Ash, M., & Wilmot, D. (2015). Issues and challenges facing school libraries in selected primary schools in Gauteng Province, South Africa. *South African Journal of Education, 35*(1). doi:http://www.sajournalofeducation.co.za

Pilgrim, J., & Martinez, E. (2013). Defining Literacy in the 21st Century: A guide to Terminology and Skills. *Texas Journal of Literacy Education, 1*(1), 60–69. DOI: 10.5281/zenodo.10537410

Pinter, A. (2017). Teaching young language learners. In *Oxford Handbook for Language Learners*. Oxford University Press.

Pretorius, E., & Klapwijk, N. (2016). Reading comprehension in South African schools: Are teachers getting it, and getting it right? *Per Linguam: a Journal of Language Learning, 32*(1), 1–20. DOI: 10.5785/32-1-627

Rose, D. (2011). Implementation and outcomes of the professional learning programme, 2010: Report for Western NSW Region, NSW Department of education and children's services Retrieved from http://www.readingtolearn.com.au

Rose, D. (2017). Evaluating the task of language learning. In *Writing Development in Struggling Learners* (pp. 161–181). Brill. DOI: 10.1163/9789004346369_010

Rose, D. (2021). Reading metaphor: Symbolising, connoting and abducing meanings. *Linguistics and Education, 64*, 100932. DOI: 10.1016/j.linged.2021.100932

Rose, D., & Martin, J. (2012). *Learning to Write, Reading to Learn: Genre, knowledge and pedagogy in the Sydney School*. Equinox.

Roux, K., Van Staden, S., & Tshele, M. (2023). *Progress in International Reading Literacy Study 2021: South African main report* (1431535338). Retrieved from Pretoria: Smith, L. (2019). *Decolonizing research: Indigenous storywork as methodology*: Bloomsbury Publishing.

Smith, L. (2021). *Decolonizing methodologies: Research and indigenous peoples*. Bloomsbury Publishing. DOI: 10.5040/9781350225282

Spencer, T., & Pierce, C. (2023). Classroom-based oral storytelling: Reading, writing, and social benefits. *The Reading Teacher, 76*(5), 525–534. DOI: 10.1002/trtr.2144

Stake, R. (1995). *The art of case study research*. Thousand Oaks, C.A. *Sage (Atlanta, Ga.)*.

Steinke, K. (2023). The role of dialogue in the teaching of reading in South Africa. *International Journal of Education, 15*(1), 1–22. DOI: 10.5296/ije.v15i1.20784

Steinke, K. (2024). Capturing classroom practice using a mixed methods design. *South African Journal of Education, 44*(1), 1–11. DOI: 10.15700/saje.v44n1a2331

Steinke, K., & Wildsmith-Cromarty, R. (2019). Securing the fort: Capturing reading pedagogy in the Foundation Phase. *Per Linguam: a Journal of Language Learning, 35*(3), 29–58. DOI: 10.5785/35-3-806

Taylor, S., & von Fintel, M. (2016). Estimating the impact of language of instruction in South African primary schools: A fixed effects approach. *Economics of Education Review, 50*, 75–89. DOI: 10.1016/j.econedurev.2016.01.003

Turner, N. (1994). A brief overview of Zulu oral traditions. *Alternation (Durban), 1*(1), 58–67.

Tynan, L. (2021). What is relationality? Indigenous knowledges, practices and responsibilities with kin. *cultural geographies, 28*(4), 597-610. doi:DOI: DOI: 10.1177/14744740211029287

Van Aswegen, S., Swart, E., & Oswald, M. (2019). Developing metacognition among young learners by using stories. *South African Journal of Education, 39*(2), 1–12. DOI: 10.15700/saje.v39n2a1531

Vygotsky, L. (1978). *Mind in Society: The Development of Higher Psychological Processes*. Cambridge, M.A.: Harvard.

Watson-Verran, H., Turnbull, D., & Jasanoff, S. (2005). Science and other indigenous knowledge systems. In Stehr, N., & Grundmann, R. (Eds.), *Knowledge: Critical concepts* (pp. 345–369). Routledge.

Wessels, E. (2011). *Teacher knowledge and implementation of phonological awareness in Grade R*. (Doctor of Philosophy in Education and Teaching). North West University, Unpublished Doctoral Dissertation.

Wildsmith-Cromarty, R., & Steinke, K. (2014). The write approach: Can R2L help at tertiary level? *Per Linguam: a Journal of Language Learning, 30*(1), 38–54. DOI: 10.5785/30-1-570

Wilson, S. (2020). *Research is ceremony: Indigenous research methods*. Fernwood publishing.

Windchief, S., & Cummins, J. (2022). Considering Indigenous research methodologies: Bicultural accountability and the protection of community held knowledge. *Qualitative Inquiry, 28*(2), 151–163. DOI: 10.1177/10778004211021803

Wood, F. (2010). The shape-shifter on the borderlands: A comparative study of the trickster figure in African orality and in oral narratives concerning one South African trickster, Khotso Sethuntsa. *English in Africa*, 71-90. doi:https://www.jstor.org/stable/23074935

Yin, R. (2013). *Case study research: Design and methods*. Thousand Oaks, C.A.: Sage publications.

Zoll, S., Feinberg, N., & Saylor, L. (2023). *Powerful literacy in the Montessori classroom: Aligning reading research and practice*. Teachers College Press.

ADDITIONAL READING

Bangeni, B., & Kapp, R. (2017). *Shifting lenses: Multilingualism in South African schools*. Van Schaik Publishers.

Battiste, M. (2002). Indigenous knowledge and pedagogy in First Nations education. *International Education Journal*, *3*(2), 24–3.

Brockmeier, J. (2010). Narrative and identity: Theoretical and methodological perspectives. *Research in Narrative and Human Development*, *7*(3), 215–232.

Dube, B., & Mhlambi, S. (2020). Using indigenous knowledge to enhance teaching and learning in South African schools. *Indilinga*, *19*(1), 1–12.

Fleisch, B. (2008). *The knowledge economy and the crisis in South African education*. HSRC Press.

Manyike, T. V., Lemmer, E. M., & Mashiangoanc, M. R. (2021). Exploring foundation phase teachers' perceptions of using indigenous games to promote literacy in rural Limpopo province, South Africa. *South African Journal of African Languages*, *41*(1), 89–99.

Mullis, I. V. S., Foy, P., Fishbein, D. D., & Kelly, D. L. (Eds.). (2023). *PIRLS 2021 International Results in Reading. International Association for the Evaluation of Educational Achievement (IEA)*. Retrieved from. [https://pirls2021.org/results/]

Pretorius, E. B. (2002). Reading and the challenge of diversity in South African primary schools. *Perspectives in Education*, *20*(1), 99–113.

KEY TERMS AND DEFINITIONS

Traditional Storytelling: Oral sharing of stories for education, entertainment, and cultural transmission, historically significant in South African communities.

Literacy Development: The process of learning to read, write, and use language effectively for communication and learning.

Learning to Read: Reading to Learn (RtL): A literacy pedagogy emphasizing reading comprehension as foundational for writing, utilizing a six-stage cycle.

Indigenous Knowledge Systems (IKS): The knowledge held by indigenous peoples, valued for its community-based nature and cultural relevance in learning.

Socio-cultural Theory: Learning theory highlighting the role of social interaction and cultural context in cognitive development.

South African Literacy Crisis: The significant challenge of low literacy skills among South African learners at foundational levels.

Lack of Access to Print: A major factor in the South African literacy crisis, referring to the limited availability of reading materials, especially in poorer communities.

Scaffolded Learning: A teaching method providing temporary support that is gradually removed as learners gain proficiency.

Oral Tradition: The passing of knowledge and culture through spoken word, such as storytelling.

Decolonisation Theory: A perspective challenging Western knowledge dominance and advocating for the recognition of indigenous knowledge and learning methods.

Chapter 17
Exploring the Traces of Linguistic Imperialism:
Towards the Decolonisation of South African Sign Language

Petri Du Toit
Durban University of Technology, South Africa

Tlou Maggie Masenya
https://orcid.org/0000-0001-5400-0549
Durban University of Technology, South Africa

ABSTRACT

Indigenous languages in South Africa including South African Sign Language (SASL) are still undermined and neglected, underpinned by a traumatising history of oppression, ethnic divide and colonialism. The study reviewed literature concerning the decolonisation of SASL as a meaningful way of reclaiming Deaf communities' identity and their sign language. An Afrocentric paradigm that places the interests of African people at the centre of the analysis was employed in this study. This chapter further discusses the strategies being established in South Africa and other parts of the world to decolonise SASL for the benefit of Deaf communities. It was recommended that decolonial strategies and language development policy be implemented and be the subject of authentic Deaf-led training and awareness-raising programme for community members.

INTRODUCTION

> *It is clear that we must find an African solution to our problems, and that this can only be found in African unity. Divided we are weak: united, Africa is the greatest force for good in the world.*
> **- Kwame Nkrumah**

Unlike other African countries, South Africa has a mixed ethnical population and it is rich in culture boasting 12 official languages of which 10 of these are considered indigenous languages, namely: Sepedi, Setswana, siSwati, Tshivenda, Xitsonga isiNdebele, isiXhosa, isiZulu, Sesotho and South African Sign Language (SASL). Indigenous languages fulfil the goals and aspirations of an individual, the entire

DOI: 10.4018/979-8-3373-4124-8.ch017

community and society at large, and it is through these languages that the self and identity are known and expressed (Higgins, 2019). However, it took rigorous language planning to enhance the status of these Indigenous languages. Alexander (2004) notes that in South Africa, language planning was inextricably linked with the racist social engineering of the apartheid era, particularly given the damaging impact of Bantu education. Indigenous languages are essential for expressing self-identity and cultural values, external forces such as globalization, Western culture, and cultural assimilation threaten these foundations. South Africa continues to experience traces of colonial oppression (Jeewa & Bhima, 2021), and these ongoing challenges underscore the importance of Indigenous languages in reclaiming self-expression and identity by revalorizing, restoring and reclaiming these languages within a society that is still grappling with the remnants of apartheid and colonialism. Marjorie (2010) also stated that South Africa is a multilingual society with distinct linguistic challenges resulting from its apartheid policy. As noted by Ouedraogo (2000) the complexity of the language question in Africa has been compounded by factors such as the commercial importance of the colonial language, the perception that multilingualism is a problem, the political value of English and French, parental demands, globalization, structural and cultural factors, technological changes, wars and riots, and economic cycles. Ouedraogo (2000) further argues that for the sake of political expediency, many African states chose colonial languages as lingua franca and media of instruction. The Deaf community does not exist in isolation and this marginalized community also forms part of the broader society, and as stated by Aarons and Akach (1998:2) social, political, and cultural factors (equally) shape the evolution of their language and identity.

Historically, due to South Africa's political context, all the languages have acquired socio-political meanings, and English and Afrikaans have become highly dominant languages even among Black people with no economic or educational value (Mekoa, 2020). Historically, language such as Afrikaans has been used as an instrument of oppression or domination, as evidenced during the apartheid and colonial era (Van Heerden, 2017). Language issues in South Africa have since become the centre of politics and a victim of political onslaught (Mekoa, 2020). In post-apartheid South Africa, multilingualism is a core value in the Constitution of the country (Republic of South Africa, 1996). The democratic government has devised a language policy development framework that encompasses various aspects, including the South African Schools Act (1996), which recognizes SASL as a medium of instruction, the Language in Education Policy (1997), and the White Paper on the Rights of Persons with Disabilities (2015). Van Herreweghe and Vermeerbergen (2010: 127) asserted that "SASL has a lower status than English and Afrikaans", reflecting the broader linguistic inequalities rooted in South Africa's history.

In Oyedemi's (2021: 214) view, "the struggle for political decolonisation (freedom) was such a long and painful journey that the wounded still bear the scars," which is also reflected in the ongoing efforts to achieve linguistic equality and elevate the status of SASL. In South Africa today, the colonial pain is continuous, and cultural decolonisation is still an ongoing struggle (Oyedemi, 2021: 214). Oyedemi (2021) further reported 40 years have passed since learners protested against Afrikaans used as a medium of instruction in schools, because of the perception that the language was used as a tool of oppression. However, black tertiary students continue to feel socially and culturally isolated in historically white universities through the language of academic instruction (Oyedemi, 2021: 224). These feelings are echoed in the South African Deaf community, where students at tertiary institutions are often isolated from the mainstream learning and teaching environment due to the limited number of qualified signed language interpreters available (cf. WASLI, 2013).

Stander and McIlroy (2017: 85) noted that the Deaf community consider Sign Language as the most important part of their culture and community. However, as mentioned earlier, Deaf culture has historically been seen as inferior, leading to oppression and discrimination (Woodward, 1982: cited in Stander & McIlroy 2017: 85), which continues to affect the social and academic integration of Deaf individuals. According to Lane (1992), there are significant similarities between the cultural oppression experienced by colonised peoples of Africa and that endured by Deaf communities. Deaf people in South Africa and other African countries have undeniably been affected by colonization and westernization and it is thus imperative to include SASL in the discussions on decolonization to allow for a deeper insight into multilingualism and language development. Although most people conceive colonialism as formed around economic power visited upon cultures less able to defend themselves, there is undeniably a case to be made for the concept of linguistic colonialism in the Deaf community (Ladd, 2003). This chapter thus discusses linguistic oppression and explores the status of SASL through decolonisation lenses. It further explores the historical development of sign language in South Africa, focusing on its status before and post-apartheid. This provides a foundation for understanding our views on linguistic imperialism and the influence of colonialism on SASL. In this chapter we aimed to:

- Assess the impact of linguistic imperialism on SASL
- Examine the effect of colonialism and apartheid on SASL in a Neo-colonial Era
- Establish the strategies for decolonization and of SASL

LITERATURE REVIEW

The following section examined the literature review, specifically sources that explored the historic overview and the development of SASL, linguistic Influences and educational practices, challenges in Deaf Education and language barriers in post-apartheid era and recognition of SASL: Moving towards linguistic justice.

Historical overview and development of SASL

SASL is constructed through the medium of space, using the hands, face, head and upper torso to represent the human experience, and it is neither a universal language nor based on any spoken language (Aarond & Akach, 1998). It arises naturally through the use of a community in a specific context, evolving and developing over time as it is passed down from generation to generation, perhaps more fluid than spoken languages that have written forms (Aarons & Akach, 1998: 24). Every country has its own country-specific signed language i.e. SASL, American Sign Language (ASL), Irish Sign Language (ISL), etc. Ngwinika (2016) notes that historically, SASL has been influenced by British Sign Language (BSL), American Sign Language (ASL) and Irish Sign Language (ISL). Aarons and Akach (1998:4) reported that the South African Schools Act of 1996 declared SASL as the medium of instruction in Schools for the Deaf, however, the consequences of apartheid can be seen in the way that education is carried out in South Africa. Humphries (2013) argues that the dominance of speech has created a sense of supremacy among the hearing community, encouraging an audism attitude, which is a form of discrimination towards the Deaf community in a hearing-dominant society. Such audiological prejudice, or more specifically

Audism, defined as discrimination on the grounds of a person's auditory status, is considered the same as racism, which is discrimination based on race and ethnicity (Bauman, 2004).

Deaf people have been affected by Apartheid's educational and linguistic consequences profoundly and comprehensively (Aarons & Reynolds, 2003). Stander and McIlroy (2017: 86) describe that until 1994 poorly resourced Deaf schools for black children made greater use of sign language than white children in privileged schools for the Deaf. Many Deaf people write English following a South African Sign Language sentence construction. Storbeck (2023) notes that most teachers in schools for the Deaf are hearing, also noting that there is no data available. "They are not required to have any form of specialised training or qualification in Deaf education or in SASL before they are appointed." (Storbeck, 2023). Hearing teachers learn SASL from their learners and often rely on the coded phenomenon of initialized signs, which are formed using the one-hand fingerspelling alphabet handshape corresponding to the initial letter of the English word. It is commonly known anecdotally that some of these signs are generally accepted within the Deaf community, while others are considered to be English-influenced and are rejected. However, in recent years, we have witnessed great strides in the promotion and standardisation of SASL (PanSALB, 2020) and now that the language is declared a 12th official language, we have seen efforts to fast-track its standardisation, which in our opinion should be led by Deaf people themselves. The Centre for Deaf Studies at the University of Witwatersrand reports that there are approximately 4 million Deaf people, with 300,000 to 400,000 using SASL as their first language (Childline, 2025). In a 2013 country report to the World Association of Sign Language Interpreters (WASLI), South Africa noted that only 12 out of 64 schools for the Deaf offered Grade 12, with these schools located in just three of the nine provinces. However, in 2023, Storbeck reported that there are 43 Schools for the Deaf in the country, 38 of which use SASL as the medium of instruction.

Ms Nyeleti Ngwinika has been regarded as the first Deaf person to submit her master's dissertation in SASL and obtained her degree from the University of the Witwatersrand in 2016 (Morgan & Van der Spuy, 2017). Ngwinika (2016) examined the concept of language borrowing in SASL, and her study set out to determine to what extent South African Sign Language comprises lexical items from British, Irish and American Sign Language, the Paget-Gorman Sign System and Initialised signs. Ngwinika (2016) reported that South African Sign Language comprises 25% Irish Sign Language, 28% British Sign Language, 16% American Sign Language, 3% Paget-Gorman Sign and 8% Initialised signs. The findings by Ngwinika (2016) also echo the historic developments of South Africa and Deaf education in South Africa influenced by missionary efforts during the colonial era.

Linguistic Influences and Educational Practices

Branson and Miller (1998: 15) observe that "while education ensured the survival of the Deaf community it also brought with it a range of processes which worked to destroy the linguistic and cultural heart of the community. In the late 19th century, hearing professionals took control of the schools and the language environment began to change, not only through the introduction of oralism but through the transformations of signing towards English grammatical structures and simultaneous speaking and signing (Branson & Miller, 1998). In the 1850s, catholic nuns from the Irish Dominican Order and missionaries from the Dutch Reformed Church were sent to South Africa to assist in formalizing Deaf education (Aarons & Akach, 1998). These missionaries started the first school for the Deaf in South Africa in 1863, known as Dominican Grimley Institute for the Deaf using ISL as the language of instruction (Cleall, 2020). Later German Dominican nuns came to South Africa using German Signs and spoken language, with an emphasis on teaching via oral methods known as "oralism". The oral teaching method later grew in popularity in schools across the country, as it was thought to encourage integration with hearing society (Storbeck, 2010).

According to Regan et al. (2006), Total Communication (TC) or Simultaneous Communication (Simcom), which involves the simultaneous use of speech and signs was adopted as a teaching philosophy in most schools for the Deaf towards the end of apartheid. Regan et al (2006) noted that this was done to establish a single educational system for Deaf people (Reagan, Penn & Ogilvy, 2006). TC philosophy, made up of hand signals and spoken language, has almost no communicative effect at all and violates the grammatical rules of a signed language (Fischer, 2015). The views on TC are contentious as linguists explain the grammatical violation of signed language through the use of Exact Signed English or any other spoken language, while some education experts believe that this teaching philosophy provides equal opportunity if exposed at a young age to learn English, by using figurative speech and understanding idiomatic expression (Stephenson & Zawolkow, 1980; Rendel, et al., 2018). It was designed to bridge the gap between a spoken linear language and a visual, gestural, and three-dimensional language. In addition, TC was also favoured by hearing people who wanted to communicate with Deaf people (Aarons & Akach, 1998). However, the use of this philosophy perpetuates the idea that there is a sign language for every spoken language in South Africa, which is an idea from hearing people who have not learnt the natural sign language of the Deaf community (Aarons & Akach, 1998).

TC philosophy has the potential to not only undermine SASL but also to create a barrier to quality education, if it is not used consistently by well-trained, skilled signers (Magongwa, 2010). However, most teachers in schools for the Deaf are hearing people, and they do not require specialised training such as a qualification in Deaf education or in SASL before they are appointed and their signing skills are not evaluated before they are appointed (Storbeck, 2023). Ngobeni et al. (2020) explain that teachers who do not have the required skills in SASL lack crucial knowledge when they enter the system, leading to teaching and learning being negatively affected. These effects can be seen in the news reports of a 0% matric pass rate of a school for the Deaf in 2023 (TaungDailyNews, 2023). To bridge the audiological and visual gaps, educators of the Deaf often employed manual signs that relied on the initial letter of English words, accompanied by mouth movements that corresponded to the spoken form (Tevenal & Villanueva, 2009).

Figure 1. SASL: Initialised signs (Real South African Sign Language, 2013-2024)

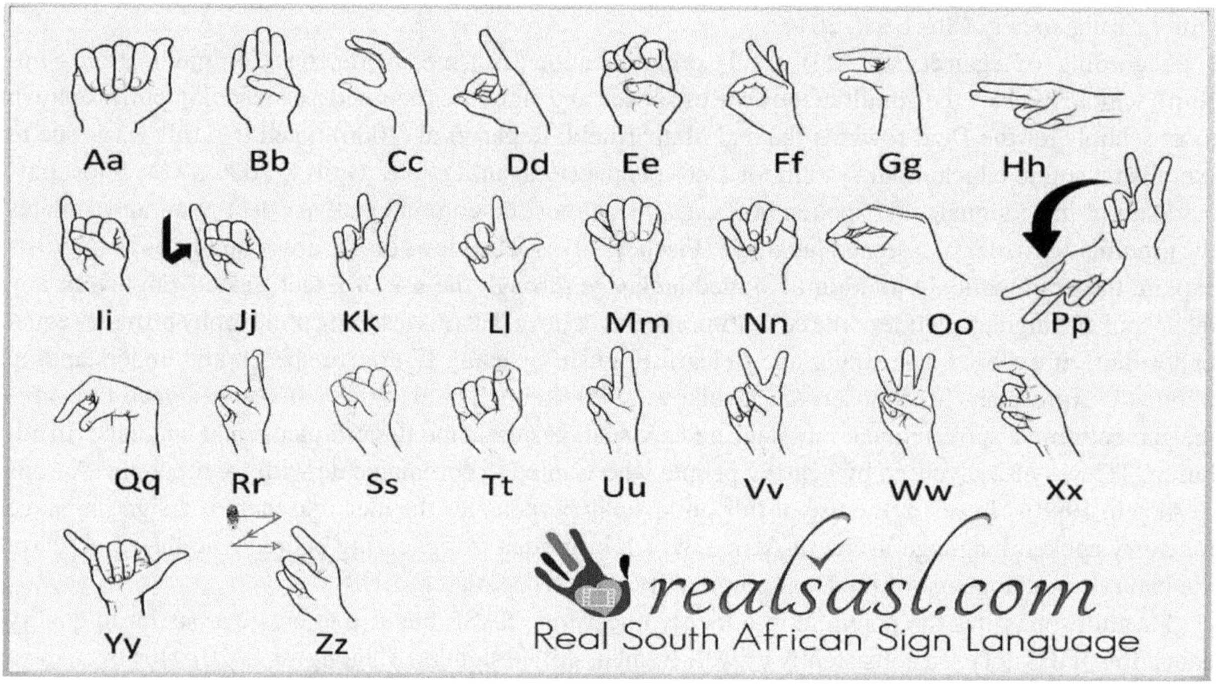

The following examples are presented as illustrative snapshots rather than linguistically annotated data. They serve to provide a general sense of the phenomena under discussion rather than a rigorous linguistic analysis. Figures 2-7 illustrate examples of initialised signs in SASL, showing traces of linguistic imperialism. The examples presented illustrate the initialized sign. The handshape used corresponds to the first letter of the word. The initialized sign is followed by a non-initialized sign (natural sign), which utilizes a handshape that does not resemble the English word. It is important to recognize that some initialized signs, such as the one discussed, have been widely adopted and accepted within the South African Deaf community. However, some initialised signs are rejected by the Deaf community.

Figure 2. Initialized sign for "mother" – The Irish Fingerspelling alphabet "M" (Researchers)

Figure 3. Signs for "mother" - Natural sign of the deaf community (Researchers)

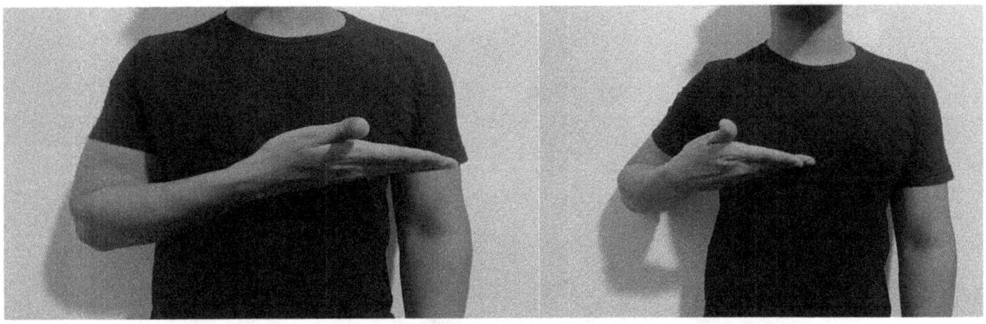

Figure 4. Initialized sign for "toilet" - The one-hand fingerspelling alphabet "T" (Researchers)

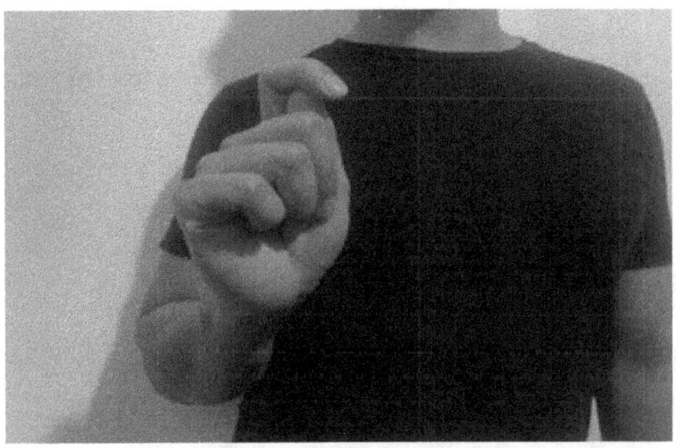

Figure 5. Sign for "toilet" - Natural sign of the deaf community (Researchers)

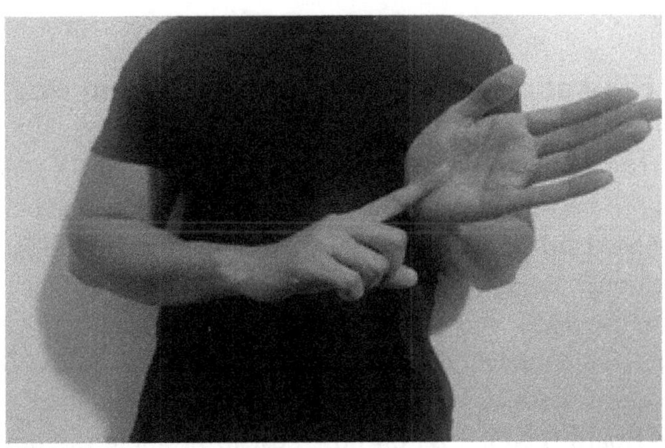

Figure 6. Initialized sign for "work"- W handshape (Researchers)

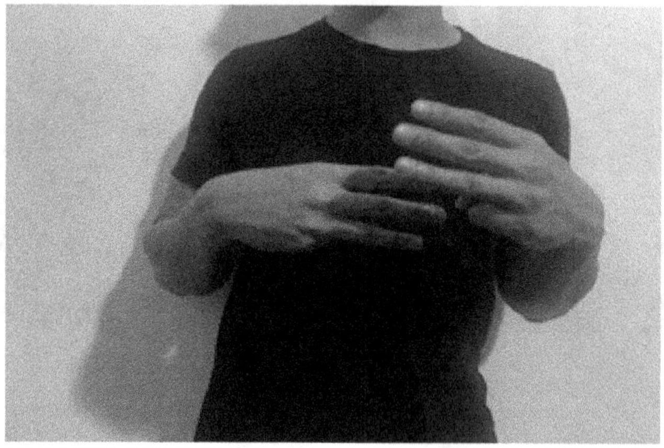

Figure 7. Sign for "work" - Natural sign used by the deaf community (Researchers)

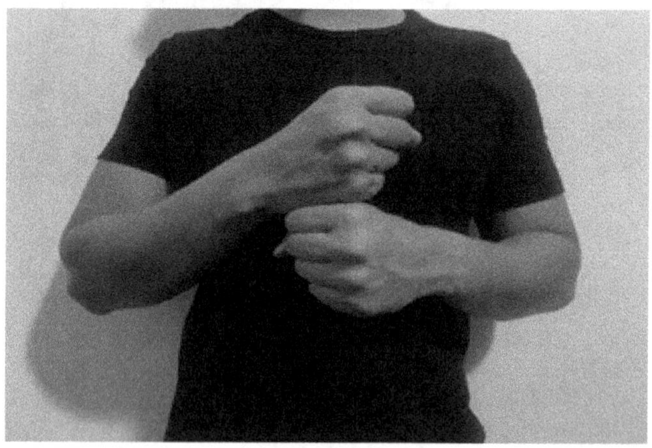

Challenges in Deaf Education and Language Barriers in Post-Apartheid era

Although the South African Schools Act (1996) had granted recognition for education purposes to SASL, the language was still not official and is yet to be offered as a First Language Subject in the curriculum of today. As a result, Deaf and hard-of-hearing people in South Africa found themselves struggling with inaccessibility problems through the lack of official recognition of SASL. During apartheid, White Deaf people were not allowed to use SASL and were forced to speak as South African Deaf Schools adopted oralism1 as a policy. In the new democratic dispensation post 1994, DeafSA, a now Deaf led federation for the Deaf opposed the oralist policies and lobbied with the Department of Arts and Culture and Pan South African Language Board (PanSALB) on how best to promote the development of the SASL. DeafSA also worked with the South African Qualifications Authority (SAQA) in developing standards for SASL registered in 2012 (SAQA, 2012) and assisted the Department of Education in formulating a management plan in South African schools between 2010 and 2014. According to the Department of Basic Education's Curriculum Assessment Policy Statement, the Department undertook the development of Curriculum and Assessment Policy Statements (CAPS) for the South African Sign Language (SASL) Grades R-12 in 2010. It was subsequently approved as policy in 2014. However, in a memorandum towards SASL as the 12th official language, DeafSA (2007) argued that policies addressing Deaf needs were often developed for window dressing purposes. In this case a false impression is being created to the rest of the world that South Africa is a very inclusive country in this era of democracy while the Deaf community finds that the truth is the opposite. In a report to WASLI in 2013, DeafSA states that the average Deaf school-leaver has a reading age of 8 (WASLI, 2013). Ongoing literacy challenges and language barriers in teaching and learning at schools for the Deaf remain a top priority for Deaf rights activists. Organisations around the country, like Sign Language Education and Development (SLED). Wits Centre for Deaf Studies play an imperative role through their interventions that aim to overcome the current challenges that confront Deaf education and language development in South Africa. The then minister of Basic Education, Mrs. Motshekga stated that:

"For a long time, learners who are Deaf were persistently marginalised as they had been deprived of a home language, resulting in their education not being one of quality. This has impacted the extent of their retention within the system and their opportunities of studying further up to and including higher education institutions".

Since the dawn of democracy, DeafSA and Deaf community stakeholders have actively been lobbying government to have South African Sign Language recognised as the twelfth official language of South Africa. In recognizing South African Deaf communities, the President, Mr Cyril Ramaphosa finally enacted Sign Language as 12th Official Language on 19 July 2023. The National Assembly thus approved that Section 6 of the Constitution of the Republic of South Africa, 1996 be amended to include SASL as an official language to promote the rights of Deaf persons and hard-of-hearing people.

Recognition of SASL: Moving Towards Linguistic Justice

As mentioned earlier, Deaf community activists and linguists began campaigning towards the development and officialization of South African Sign Language (South African Schools Act 1996). Marking a significant shift from past marginalization towards broader social acceptance and institutional support, the Pan South African Language Board (PanSALB) published the SASL Charter in September 2020. The SASL Charter contains nine key pledges that should serve as a framework to promote the fair and courteous treatment of Deaf people and be adopted by all spheres of government and civil society. The Charter's (PanSALB 2020: 1- 20) nine pledges include:

- there must be a better understanding, nationally, of SASL as a language in its own right
- there should be self-determination by Deaf people
- there should be a promotion of learning and high-quality teaching of SASL
- multilingual Deaf education for Deaf children should be encouraged and promoted,
- there should be guaranteed accessibility to services and information through SASL
- persons working directly with deaf people should meet the minimum standards of competency in SASL
- local deaf communities should be consulted on a regular basis on all matters affecting their lives

- professional SASL interpreting and translation services should be readily available,
- the South African Police Service (SAPS) and the Director of Public Prosecutions should respect Deaf people

The Deaf community in South Africa saw the first ever interpreted televised public address broadcast on 10 May 1994. Claudine Storbeck, now a professor in Deaf education explains her experience, "The South African Broadcast Corporation (SABC) called me to interpret the inauguration of Nelson Mandela and there was no one else available." Therina Wentzel, a disability activist, also reflects on the milestone years of the Deaf community and added that: "for the first time we had Deaf led leadership in structures." (Wentzel, 2021, personal communication). When South Africa became a democratic country, Deaf self-representation and leadership in Deaf structures began to take shape, with Mr Kobus Kellerman and Mr Nico Beauraine playing a crucial role in mobilising the Deaf community. Mr Kobus Kellerman became the first Deaf Director of DeafSA and Mr Nico Beauraine the first Deaf chairperson. A democratic South Africa was allowed to join the international community for the first time from 1994 onwards. The status of SASL was to be promoted and in 1996 DeafSA appointed Mr. Philemon Akach, a Kenyan national, academic in Deaf education and top international interpreter at the time to go into the Deaf community and promote the status of South African Sign Language through Deaf self-actualisation and language awareness. Claire Penn published a South African Sign Language dictionary in 1992, based on the hypothesis that South Africa has many different signed languages (Aarons & Akach 1998). As stated by Aarons and Akach (1998), Penn's hypothesis is to be doubted because Deaf people interacted with one another, regardless of the ethnicity lines stating the anecdotal fact that all Deaf South Africans seem to be able to communicate easily with one another despite the varieties in sign (Vermeerbergen et al., 2007). Penn's Dictionary (1992) promoted the belief that there are different Sign Languages in South Africa. In response to the dictionary, Aarons and Akach (1998) argue that: "The Dictionary of Southern African Signs is based on a false hypothesis about the effects of apartheid on the signed language used in South Africa, as well as a simplistic understanding of the relationship between words and signs, and a failure to recognise variation within different contexts of use."

In recent years, significant progress has been made in the promotion and standardization of SASL (PanSALB, 2020). This progress was paved by earlier efforts, such as the first Deaf indaba that brought together Deaf community leaders across the country. The indaba was hosted in the Free State in 1998 whereby issues around language development, support and rights were topics of discussion that paved the way for Deaf empowerment. A central tenet of the conference was the development of SASL and negative stereotypes that oppress the Deaf community post-Apartheid (Wentzel, 2021, personal communication). The recognition of Sign language as South Africa's 12th official language, efforts to accelerate its standardization have gained momentum. However, in the pursuit of standardising language to meet the political agenda of country, Branson and Miller (1998) provide a roadmap. When it comes to nationalising signed languages, they argue that such a national approach may be appropriate for first world countries such as America and Europe, but such policies need to be critically examined for non-Western countries. Branson and Miller (1998) further explained that Indonesia, for example, has many geographically dispersed communities where many sign languages are developing. However, in many Western societies' current sign language policies focus exclusively on assumed national sign languages and overlooks the sign languages of indigenous communities (Branson & Miller 1998). As noted by Aarons and Reynolds (2003) the abolishment of Apartheid did little to change the years of profound damage, and such negative heritage will be with us for many years to come.

METHODOLOGY

This chapter was based on the African Ubuntu philosophy and Sankofaism, the African centred Philosophies to decolonise SASL "Ubuntu" is Nguni language, literally translated to English and simply means showing humanity to others. Nobel Prize winner and former president of the Republic of South Africa, Nelson Mandela describes Ubuntu as a philosophy constituting a universal truth, a way of life, which underpins an open society (Mandela, 2006). Ubuntu can also be described as the capacity of an African culture to express compassion, reciprocity, dignity, humanity and mutuality in the interest of building and maintaining communities with justice and mutual caring (Khoza, 2006). The Ubuntu philosophy believes in group solidarity, which is central to the survival of African communities (Mbigi & Maree, 2005). It does not mean that people should not address themselves to a problem, but it does imply that they should look at whether what they are doing will enable or empower the community around them and help them to improve. The Ubuntu philosophy also implies that if people are treated well, they are likely to perform better. African Ubuntu philosophy shares the same characteristics as qualitative research methods, and it is suitable to this study as it resonates with the interpretivist paradigm focusing on both comprehending and apprehending social phenomenon. This philosophy assumes that people use interpretive schemas which must be understood and that the character of the local context must be articulated. Deaf persons, like all other community members in South Africa, are protected by the Constitution (1996) to ensure meaningful realisation of the right to communicate in their own language. The African Ubuntu philosophy was thus adopted in this study to analyse the impact of linguistic imperialism and colonialism on South African Sign Language with a focus on the decolonization of SASL for recognition and restoring the dignity of the Deaf communities in South Africa. In the African Ubuntu philosophy, Africans are thus expected to adapt their realities, history and ultimately their destiny to fit their unique context. However, the African Ubuntu philosophy should be guided by key principles including truth, justicc, harmony, community and commitment. Any inquiry concerning African experiences must thus satisfy the above-mentioned rules or canons to be truly legitimate, and employing all these new African Ubuntu philosophy perspectives will necessarily provide more useful and honest research criteria. African Ubuntu approach or philosophy, centres around humanness, and emphasizes shared values, including deeds of kindness, compassion, caring, solidarity, empathy and generosity towards others (Mnyaka & Motlhabi, 2009). Ntihirageza and Ibrahima (2022) also proposed an Ubuntu-based model of education, composed of the following major processes that integrate multiple languages, namely: revitalization, SASL fresh vitality; revalorization, reassigning merit or value to SASL and decolonization-challenging and eliminating colonial vestiges by acknowledging and integrating SASL and practices in the system of education. Practicing the Ubuntu philosophy thus unlocks the capacity of an African culture in which individuals express compassion, reciprocity, dignity, humanity and mutuality in the interests of building and maintaining communities with justice and communality (Poovan, Du Toit & Engelbrecht, 2006).

This chapter also adopted Sankofaism philosophy or paradigm that stresses reclaiming past knowledge to shape the present and future, and the present would also determine and greatly influence the future (Quan-Baffor, 2012). The philosophy has recently gained traction in African decolonisation studies and has been implemented by scholars such as Kissi (2020) who applied it to disability studies. Kissi, (2020) explains that using Sankofaism as a method will provide new historical awareness and realities that will allow him to bring past knowledge to the present and future generations. The Sankofaism philosophy means "Go back and get it" or "Go back and take it." and it thus informs the chapter's call for the decolonization and revitalization of SASL. Sankofaism is referred to as a revaluation of our indigenous way of life and a new search for our true identity as people and it is founded on the premises and belief that human beings are part of the past that shape the present (Slater, 2019). Slater (2019) further described the Sankofa symbol as a mythical bird flying forward with its head turned backward and the thought behind this is that there is wisdom in learning from the past, both to understand the present and shape the future. The forward and backward gaze of the Sankofa bird is based on the Akan proverb, 'Se wo wen? ti na wosankofa a yenki which translates to 'It is not wrong or shameful to go back for something you have previously forgotten (Quan-Bafour, 2012). The philosophy has also been used as a symbol lessons can be learnt from the experience, the wrong can be rectified whereby a mistake has been made, and also whereby knowledge of the past can be utilised to map out the future (Slater, 2019).

DISCUSSION OF FINDINGS

The chapter explored how our understanding of the Ubuntu and Sankofaism philosophies connects with the ideas and perspectives as put forward by other scholars in this area. The findings were presented in three main themes, namely: the impact of linguistic imperialism on SASL, the effect of colonialism and apartheid on SASL in the Neo-colonial Era and strategies for decolonisation of SASL.

The impact of linguistic imperialism on South African Sign Language

Colonialism influenced the development of SASL, as evidenced by Nkwinika (2016)'s findings that SASL contains a significant percentage of ISL and BSL lexicon. This in turn, impacted the group identity of the Deaf community and their natural language development, which hindered the standardization of SASL (Holness, 2016). Language is regarded as the primary means of group identity formation and a shield to protect this identity, and it is also linked to ideology and power (Makalela, 2005). However, linguistic imperialism or linguicism has been used to criticize and oppress other languages i.e. SASL. Phillipson, (2012) described linguicism as the tendency to attribute favourable characteristics to a particular language whilst devaluing other languages (Phillipson, 2012). It has been ideological and encourages beliefs that the dominant language is more prestigious and superior to others. Linguicism discrimination is thus based on language that unfairly treats certain linguistic communities, or unfairly advantages some languages over others (Galloway & Rose, 2015). In the context of linguicism, ideas are hegemonic, internalized and naturalized as being 'normal'. In linguistically plural African societies, linguistic imperialism has resulted in the marginalization of indigenous languages and English being positioned favourably as a language of prosperity and progress. Skutnabb-Kangas (2016a) further noted that the education of Deaf children, who are a linguistic minority, via other dominant languages constitutes a form of linguicism, especially when education is via submersion (sink-or-swim) practices as previously discussed. In a South African context, the existence of linguistic imperialism has thus been seen as being less disputable, as shown through the historical and current treatment of SASL in comparison to the more dominant languages i.e. English. However, as pointed out by Makalela (2005) the tendency for post-colonial governments to continue implementing their colonial masters and language policies has not resulted in linguistic reform in post-colonial societies. This can be attributed to the following myths surrounding African languages:

- many of the languages are not developed, so they cannot be used in education

- the costs of developing African languages are very high
- the exclusive use of African languages will block the window to the world and result in exclusion from participation in the international community
- some speakers of African languages do not wish to see their languages used in education because they have a total lack of confidence in the languages in these domains
- European languages are neutral and have a potential for creating national cohesiveness among speakers of competing languages

These myths reinforce the historical and ongoing challenges faced by SASL in its development and standardization, which is a direct result of colonial language policies and attitudes towards African languages.

The effect of colonialism and apartheid on SASL in the Neo-colonial Era

Eurocentric ideologies became the epicentres of colonialist thought, thereby excluding African sign languages. Colonial libraries became institutions of knowledge organized in a way that perpetuates colonialism. Western systems perpetuated colonialism and the miseducation, misattribution, and harmful stereotypes about SASL. As Brook, Ellenwood and Lazzaro (2015) note, only the white perspective has been considered and translated into a narrowed perspective within research, literature, and history, while the structure of institutions has led to workflows, policies, and procedures that best serve the White majority. Using decolonial thinking, scholars have critiqued the colonial-modern assumptions and knowledge forms deeply embedded in Western social science (Seth, 2013). Despite the importance of knowledge in determining people's destinies, Western-centered modernity negated alternative ways of knowing Western ways of seeing, imagining, and understanding (Ndlovu, 2018). For example, academics and scientists have disregarded indigenous knowledge systems, so the value of primary knowledge was strategically rejected among Western academics (Mji, 2019). The non-European-centered forms of knowledge, such as those of indigenous peoples from Asia, America, and the Pacific, their classification systems, technologies, and codes of social life, were first preserved by Western science in the 18th Century and were referred to as "discoveries" of Western society (Tabares, 2022).

Consequently, African countries have adopted Western education, knowledge production methodologies and epistemologies, thereby preventing knowledge from spreading beyond national borders. As a result, African libraries relied on books and journals published in Europe and North America in European languages (Blume & Roylance, 2020). The level of Africa's knowledge production and Africa's share of global cultural goods trade, including heritage goods, recorded media, and visual arts, therefore remained very low in neocolonialism. The infamous 1880 Milan Conference on the Education for the Deaf was devoid of Deaf representation (Moore, 2010). It exacerbated the perception that Deaf people need to be fixed to fit into the hearing world as Aarons and Akach (1998: 3) put it: "More common, world over, is a deep and unfounded prejudice against signed languages, and a consequent marginalization of Deaf people and their human rights." By perpetuating the medical model of Deafness, the Milan Conference of 1880 introduced oralism, which exacerbated audist sentiments.". Oralism refers to the method of teaching deaf people to communicate by the use of speech and lipreading and not sign language. Deaf people have been oppressed by the effects of oralism, similar to the effects of indigenous people in colonized countries (Ladd, 2003). Audism is the belief that the ability to hear makes one superior to those with hearing loss.

In response to the Milan 1880 declaration that enforced the use of oralism, more schools for the Deaf in South Africa emerged such as, in the Western Cape in 1881, a combined oral and manual method school and in the Eastern Cape in 1884, a strictly oral method school (Storbeck and Martin 2010). The first School for Black Deaf children was opened in the then Transvaal in 1941 where teachers used signs invented in Britain known as the Paget-Gorman System. Storbeck and Martin (2010: 4) explain that "although the Paget-Gorman System did not correspond to the natural signs of the Deaf community, it did allow for the development of a strong visually based communication code that facilitated rather than repressed a strong Deaf culture." Under the apartheid regime and its segregation policies, more schools were founded along racial lines. Aarons and Akach (1998: 6) explain that little is known about the Deaf community before colonisation, however with the advent of public-provided education, the state authorities took little responsibility in opening schools for the Deaf, which was left entirely to the church. Under Apartheid rule, the oral approach to Deaf education (globally considered the superior approach) was enforced in white schools. Schools for non-white Deaf children, under Bantu education, taught using manual methods and were considered inferior (Akach & Aarons, 1998; Aarons & Reynolds, 2003; Storbeck & Martin, 2010). Apartheid hurt the formal development of SASL where oralism, especially in the white Deaf community, was enforced in schools for the Deaf and learners were hit on their hands if they dared use sign language (Aarons & Akach, 1998).

However, the Deaf community's resoluteness meant that learners used signed language on the playground where hearing educators would not see (anecdotal evidence). Aarons and Akach (1998) reported that Deaf people use natural sign language when communicating with one another. According to them, manual and verbal codes are favoured by hearing people who want to communicate with Deaf people, but such systems "perpetuates the idea that there is a sign language for every spoken language in South Africa, which is an idea from hearing people who have not learnt the natural sign language of the Deaf community." (Aarons & Akach, 1998: 16). In an attempt by educators of the Deaf to bridge the complex auditory and visual gap, manual signs often relied on the initial letter of the word in the English language. The one-hand fingerspelling alphabet handshape of the initial letter of the English word would be used to form the sign (Spence and Woll, 1999). As is evident from anecdotes, these initialised signs often carry with them the remembrance of childhood linguistic trauma as Fischer (2015: 450) explains many initialised signs violate the symmetry and dominance conditions of signed language.

Strategies for decolonization of SASL

The history of colonialism is one of brutal subjugation of indigenous peoples and most of the African continent is still underdeveloped and recovering from colonization. Blakemore (2019) described colonialism as defined as control by one power over a dependent area or people and it occurs when one nation subjugates another, conquering its population and exploiting it, often while forcing its own language and cultural values upon its people. Recovering and maintaining indigenous worldviews, philosophies and ways of knowing and applying those teachings in a contemporary context represents liberation strategies. Literature explains that Deaf education was formalised through missionary efforts that shaped the trajectory of SASL. During the colonial era, Sign language was regarded as inferior, which led to audism, oralism and sign language oppression. Signing codes that negate the natural sign language of Deaf people were used as a strategy to normalise the Deaf community into mainstream society. This study thus argues that the debate around the decolonisation of SASL should extend its scope to include strategies around the eradication of audism that has negatively impacted SASL. The negative impact of audism has been referred to as the oralisation of SASL and the chapter further argues that decolonisation strategies in spoken languages should be extended to signed languages. Indigenist researchers have also advocated for the implementation and adoption of decolonial strategies as an important process in decolonizing indigenous languages. Therefore, engaging in decolonial strategies for the revitalization and restoration of SASL such as the enactment of SASL policies; raising awareness on Deaf communities and education; decolonizing African indigenous languages (including SASL) and school curricula, promoting Deaf identity, self-representation and language ownership through Sankofaism (returning to our roots) is essential for epistemic liberation. These efforts empower African communities to reclaim their status and dignity, serving as both a form of resistance and a pathway to self-determination. By actively participating in this process, academics, indigenous researchers, community leaders, and policymakers will contribute to dismantling colonial structures in their ongoing forms. Decolonial strategies identified in this chapter are further discussed below:

- **Revitalization of SASL**

Language is the foundation and vehicle of a culture and it holds the stories, songs, dances, protocols, family histories and connections. Language proficiency is declining in African communities as they are compelled to embrace Western culture and civilisation and when a language dies, the link to the cultural and historical past also disappears. Many practices throughout history put pressure on ethnocultural communities to abandon the use of their languages. For example, when the federal government set up boarding schools in the 19th century to assimilate Native American children into American culture, one of the objectives was to get them to turn away from the use of their native languages. This also included the forced cultural assimilation that resulted from the Indian Civilization Act of 1819. This act established Indian boarding schools to teach subjects such as math and science while suppressing the use of languages and cultures. The development of boarding schools has also led to cultural assimilation and their effect was devastating for communities and their languages. Some of the communities have lost their languages due to assimilation efforts.

Additionally, the forced relocation of the communities from their places of origin has also caused these communities to become fragmented due to some families remaining behind or being exempt from relocation. These factors also increased the stress on ethnocultural communities to simply survive. Indigenous languages such as SASL continue to be threatened and endangered due to colonization and imperialism. Indigenous peoples have the right to speak in their indigenous languages and in a way that is respectful to their local communities. However, some indigenous languages are under-resourced due to a lack of textual speech data or standardized language forms. Resources to learn indigenous languages such as a lack of trained or proficient teachers are often limited. One common goal for indigenous language revitalization initiatives is to promote intergenerational language transmission and use in multiple social domains such as homes and schools. However, black parents are still resorting to taking their children to English Medium schools and most of the children no longer speak their home languages (indigenous languages) and have switched to English as their primary language. This example shows that some African people still believe that white people are superior hence aspirations for a white man's language. Similarly, audist attitudes contribute to Deaf individuals feeling inferior to their hearing counterparts. Most Higher Education Institutions in South Africa are also not offering indigenous language courses as part of their curriculum.

- ## Decolonization and Africanization of SASL

Colonialism is one of the primary ways that continue to influence people's cultures and identities, in the Global South and African continent, while they can also be marginalized through various forces (Das & Semaan, 2022). Deaf communities were robbed of decades of opportunities such as self or language development through colonialism. Arowolo (2010) described colonialism as an imposition of foreign rule over traditional political settings and foreign dominance and subjugation of African people in all spheres of their social, political, cultural, economic and religious civilisations. Antles-Osortia (2018) described the history of colonization as one defined by the enslavement, rape, and genocide of local peoples. As noted by Blakemore (2019) the impact of colonialism on knowledge systems includes environmental degradation, economic instability, ethnic rivalries and human rights violations issues that can outlast one group's colonial rule. Fanon (2008) also draws attention to colonial racism, where colonists often viewed native cultures as inferior to those of the colonial rulers, thus actively working to replace native cultures with those of the colonial power. Decolonization thus pays explicit attention to removing colonizers, exploring and rejecting their influence, and shifting towards re-establishing independent nations (Das & Semaan, 2022). Decolonization can also be achieved through implementing a language development policy. Although, in post-apartheid South Africa, several policies, aimed at redressing the imbalances of the past, were promulgated after the democratic elections in 1994, however, as stated by Brock-Utne (2001:129) some African countries, such as South Africa, Namibia, and Uganda, are still not implementing their progressive language policies. One of the enacted policies was the 1997 multilingual language in education policy which guaranteed an equal status to all the eleven languages.

Language replacement has also led to language assimilation denying deaf communities, particularly in Southern Africa, opportunities to speak their own language and preventing them from developing an independent sense of identity. Smith (1999) further argues that we still live in a world that continues to portray whiteness as an indication of superiority. Cultural assimilation equates blackness with inferiority and leads colonized people to see themselves as subhuman, and perpetually experience psychological trauma as it also affects their identities while equating whiteness with superiority and pureness (Fanon, 2008). The language of colonisers, which is English is still used as an official language in educational and administrative contexts despite colonial rule ending in 1947. This dominant language is increasingly used by many people as a preferred mode of communication, and in many African countries, English remains an official language, which continues to marginalize regional or indigenous languages (Joseph, 2011). Indigenous and Deaf people thus lose their sense of identity and belonging if there is no crucial connection to their linguistic and cultural history.

Masaka (2018) further argued that the epistemicide instituted by agents of neo-colonialism has undersized the growth and success of the epistemological paradigm of the African indigenous languages. The total exclusion of indigenous languages from the school and university curricula confirms the deliberate and cantankerous initiatives as evidenced by the predominance of Western ways of knowing or finding solutions to problems because the dominant culture is presented as the only authentic producer of knowledge (Masaka, 2018). Quijano and Ennis (2000) highlight the frequent calls for higher education institutions to be less Eurocentric and described "euro-centricity" as a perspective of knowledge whose systematic formation began in Western Europe before the middle of the seventeenth century, although its roots are without doubt much older. It is instead a specific rationality or perspective of knowledge that was made globally hegemonic, colonizing and overcoming other previous or different conceptual formations as much in Europe as the rest of the world (Quijano, 2000). Neocolonial onslaughts will thus severely hamper our capacity to make systematic ventures beyond our current knowledge horizons if we, as Africans, rely exclusively on those aspects of the Western archive that disregard other epistemic traditions (Mbembe, 2015).

- ### Revalorization of SASL

According to Deaf academic and activist, Ladd (2003), Deaf people have been oppressed by the effects of oralism, similar to the effects of indigenous people in colonised countries. Based on the assertion and the specific historical context of South Africa, this study asserts that colonialization and missionary intervention shaped South African Sign Language. This ties in with Nkwinika's (2016) findings that South African Sign Language comprises a high percentage of ISL and BSL lexicon. Based on our observations and anecdotal evidence in the Deaf community, we understand language development has taken centre stage since the abolishment of Apartheid. However, we cannot argue the negative impact of hearing dominance over SASL development. In supporting the process of revalorization of local or indigenous languages, President Cyril Ramaphosa, enacted the South African Sign Language as official indigenous language, and signed into law the South African Sign Language Bill, in 19 July 2023. The recognition of South African Sign Language as the 12th official language is an important step towards revalorization of South African Sign Language.

- ### Reclaiming and restoring SASL through the Sankofaism approach

Some Africans have consequently become contemptuous and ashamed of the indigenous African way of life and have thus come to reject it. The corollary to this attitude is that the white man's way of life is a model to be copied and appropriated by Africans (Dzobo, 1976). Sankofaism, an African philosophical thought, which means 'go back and take it' can be used as a strategy to reclaim SASL. This philosophy is based on the premise and belief that human beings are part of the past and that this past shape the present and it also focuses on promoting African and indigenous ways of knowledge and skill development (Dei, 2012). Indigenous people and their communities need to take steps backwards to reclaim the past, to understand the present, to realise why and how they have come to be where they are and who they are, through Sankofaism approach. As noted by Quan-Baffour (2008) the Sankofa symbol, is like a mythical bird flying forward with its head turned backwards and the thought behind this is that there is wisdom in learning from the past, both to understand the present and shape the future. Sankofa is a revaluation of our indigenous way of life and a new search for our true identity as people. Quan-Bafour (2021). Decolonization of SASL thus needs to be performed in line with Sankofaism, which seeks to provide the foundations that promote methodologies that are consistent with realities in Africa. Bastos (2009) further described Sankofaism as a pivotal theme in Afrocentric education, philosophy and cultural milieus as it presents platforms for confronting mischaracterisations of African creative and pedagogical spaces.

FUTURE RESEARCH DIRECTIONS

This chapter highlighted the absence of decolonization studies intersecting with SASL and emphasized the need for empirical research on the impact of linguistic imperialism on SASL in the post-colonial era by introducing a conceptual framework grounded in an Afrocentric paradigm (Ubuntu Philosophy and Sankofaism). We argue that qualitative studies exploring Deaf-led advocacy in language policy reforms will contribute to a more scientifically grounded, decolonized approach to language planning, development, and standardization.

CONCLUSION AND RECOMMENDATIONS

We encourage further research about the colonial impact of sign language in South Africa and the African continent. We do believe, however, that historic events under colonialism and the missionary approach to Deaf education contributed to the formalisation of Deaf education in our country. However, colonialism and the subsequent effects of hearing dominance over the Deaf community had a tremendous impact on the development of SASL. In this chapter we argued that colonialism, through fostering oralism, negatively impacted natural language development. Historically, oralism engendered hearing dominance over Deaf people and negatively influenced the trajectory of their language. There is a need for further investigation on the impact of colonialism in the Deaf community specifically focusing on what we term 'de-oralisation' of South African Sign Language.

REFERENCES

Aarons, D., & Akach, P. (1998). South African Sign Language - one language or many? A sociolinguistic question . *Stellenbosch papers in Linguistics 31*, 1 - 28.

Aarons, D., & Reynolds, L. (2003). South African Sign Language: changing policies and practice. *Many ways to be deaf: International variation in deaf communities Gallaudet University Press*, 194 - 210.

Africa, S. (2013). *South Africa Country Report.* Retrieved from WASLI: https://wasli.org/wp-content/uploads/2013/06/254_south-africa-country-report.pdf

Alexander, N. (2004). The politics of language planning in post-apartheid South Africa. *Language Problems and Language Planning*, *28*(2), 113–130. DOI: 10.1075/lplp.28.2.02ale

Branson, J., & Miller, D. (1998). Nationalism and the linguistic rights of Deaf communities: Linguistic imperialism and the recognition and development of sign languages. *Journal of Sociolinguistics*, *2*(1), 3–34. DOI: 10.1111/1467-9481.00028

Childline. (2025). *Information for D/deaf children.* Retrieved from Childline: https://www.childlinesa.org.za/about/how-we-help/help-for-the-deaf/

Cleall, E. (2022). Decolonising deaf history: Harlan Lane, Postcolonialism, and Critical Colonial History. . *Zeithistorische Forschungen/Studies in Contemporary History, 19(2)*, 380-387.

Committee, J. C. (2007). *Recognition of South African Sign Language as Official Language: briefing by Deaf Federation of South Africa*. Parliamentary Monitoring Group.

Deaf, S. A. (2007, February 1). *Memorandum towards recognition of SASL as the 12th official language.* Retrieved from Static PMG: https://static.pmg.org.za/docs/2007/070216memorandum.htm?

Deaf, S. A. (2018). *Deaf Education*. DeafSA.

Department of Basic Education. South Africa. (2014). *Curriculum Assessment Policy Statement.* Pretoria: Department of Basic Education.

Fischer, S. D. (2015). Sign languages in their historical context. In *The Routledge handbook of hisotrical linguistics* (pp. 442–465). Routledge.

Higgins, N., & Maguire, G. (2019). Language, indigenous peoples, and the right to self-determination. *New England Journal of Public Policy*, *31*(2), 8.

Holness, W. (2016). The development and use of sign language in South African schools: The denial of inclusive education. *Afr. Disability Rts. YB*, *4*, 141.

Jeewa, T., & Bhima, J. (2021). Discriminatory Language: A Remnant of Colonial Oppression. *Constitutional Court Review*, *11*(1), 1–17. DOI: 10.2989/CCR.2021.0012

Khoza, R. J. (2006). *Let Africa Lead: African Transformational Leadership for 21st Century Business*. Vezubuntu.

Kissi, E. (2020). *Sankofa: Disability and the door of return.*

Ladd, P. (2003). Understanding Deaf Culture. In *Search of Deafhood.* Multilingual Matters. DOI: 10.21832/9781853595479

Magongwa, L. (2010). Deaf Education in South Africa. *American Annals of the Deaf, 155*(4), 493–496. DOI: 10.1353/aad.2010.0042

Marjorie, L. (2010, February 8). *Language Policy and Oppression in South Africa.* Retrieved from Cultural Survival: https://www.culturalsurvival.org/publications/cultural-survival-quarterly/language-policy-and-oppression-south-africa

Mbigi, L., & Maree, J. (2005). *Ubuntu: The spirit of African transformation man agement.* Knowledge Resources.

Mekoa, I. (2020). The Politics and Nuances of Language in South Africa: A Critical Appraisal. *Journal of African Languages and Literary Studies, 1 (1)*, Sabinet Online.

Moores, D. (2010). Partners in Progress: The 21st International Congress on Education of the Deaf and the Repudiation of the 1880 Congress of Milan. *Gallaudet University Press 155(3)*, 309-310.

Morgan, R., & Van der Spuy, A. (2017, March 1). *A degree with a difference: using South African sign language instead of the written word.* Retrieved from The Conversation: https://theconversation.com/a-degree-with-a-difference-using-south-african-sign-language-instead-of-the-written-word-72899

Ngobeni, W., Maimane, J., & Rankhumise, M. (2020). The effect of limited sign language as barrier to teaching and learning among Deaf learners in South Africa. *South African Journal of Education, 40*(2), 1–7. DOI: 10.15700/saje.v40n2a1735

Oyedemi, T. (2021). Postcolonial casualties: 'Born-frees' and decolonisation in South Africa. *Journal of Contemporary African Studies, 39*(2), 214–229. DOI: 10.1080/02589001.2020.1864305

PanSALB. (2020, September 4). *South African Government.* Retrieved from www.gov.za: https://www.gov.za/speeches/pan-south-african-sign-language-board-launches-south-african-sign-language-charter%C2%A0%C2%A0-4-sep

PanSALB. (2020, September 4). *South African Sign Language Charter.* Retrieved from www.gov.za: https://www.gov.za/speeches/pan-south-african-sign-language-board-launches-south-african-sign-language-charter%C2%A0%C2%A0-4-sep

Poovan, N., Du Toit, M. K., & Engelbrecht, A. S. (2006). The effect of the social values of Ubuntu on team effectiveness. *South African Journal of Business Management, 37*(3), 17–27. DOI: 10.4102/sajbm.v37i3.604

Quan-Baffour, K. P. (2012). Sankofa: 'Gazing Back' to Indigenous Knowledge and Skills for Socio-economic Development of Ghana. *Studies of Tribes and Tribals, 10*(1), 1–5. DOI: 10.1080/0972639X.2012.11886637

Reagan, T., Penn, C., & Ogilvy, D. (2006). From policy to practice: Sign language developments in post-apartheid South Africa. *Language Policy, 5*(5), 187–208. DOI: 10.1007/s10993-006-9002-y

Rendel, K., Bargones, J., Blake, B., Luetke, B., & Stryker, D. S. (2018). Signing Exact English: A Simultaneously spoken and signed communication option in Deaf education. *Journal of Early Hearing Detection and Intervention*, *3*(2), 18–29.

Republic of South Africa. (1996). *The Constitution of the Republic of South Africa*. Government Printer.

SAQA. (2012). *Perform everyday communicative tasks using SASL.* Retrieved from South African Qualifications Authority: https://regqs.saqa.org.za/showUnitStandard.php?id=115079&utm

Slater, J. (2019). Sankofa-the need to turn back to move forward: Addressing reconstruction challenges that face Africa and South Africa today. *Studia Historiae Ecclesiasticae*, *45*(1), 1–24. DOI: 10.25159/2412-4265/4167

Stander, M., & McIlroy, G. (2017). Language and culture in the Deaf community: A case study in a South African special school. *Per Linguam: a Journal of Language Learning= Per Linguam. Tydskrif vir Taalaanleer*, *33*(1), 83–99.

Stephenson, P., & Zawolkow, E. (2023, October 7). *Hands and Voices*. Retrieved from Hands and Voices: https://www.handsandvoices.org/comcon/articles/pdfs/see.pdf

Storbeck, C. (2023, June 21). *Sign language is now official in South Africa - how this will help education in Deaf schools.* Retrieved from The Conversation: https://theconversation.com/sign-language-is-now-official-in-south-africa-how-this-will-help-education-in-deaf-schools-207126

Storbeck, C., & Martin, D. (2010). South African Education and the Deaf Community. *American Annals of the Deaf*, *155*(4), 488–490. DOI: 10.1353/aad.2010.0034 PMID: 21305983

Storbeck, P. C. (2019, February 20). History of the Deaf Community in South Africa . (P. d. Toit, Interviewer)

Tevenal, S., & Villanueva, M. (2009). Are you getting the message?: The effects of SimCom on the message received by deaf, hard of hearing, and hearing students. *Sign Language Studies*, *9*(3), 266–286. DOI: 10.1353/sls.0.0015

Van Heerden, M. (2017). Afrikaans: The Language of Black and Coloured Dissent. *South African History Online, 20.*

Van Herreweghe, M., & Vermeerbergen, M. (2010). Deaf perspectives on communicative practices in South Africa: institutional language policies in educational settings. *Text & Talk - An Interdisciplinary Journal of Language Discourse Communication Studies 30(2)*, 125-144.

Wentzel, T. (2021, July 13). History of the Deaf community in South Africa . (P. d. Toit, Interviewer)

KEY TERMS AND DEFINITIONS

Linguistic Imperialism: It is the tendency to attribute favourable characteristics to a particular language whilst devaluing other languages, and it has been ideological and encourages beliefs that the dominant language is more prestigious and superior to others.

Decolonisation: It pays explicit attention to removing colonizers, exploring and rejecting their influence, and shifting towards re-establishing independent nations.

Oralism: It can be defined as the system of teaching Deaf people to communicate using speech and lipreading rather than sign language.

Audism: It is defined as discrimination on the grounds of a person's auditory status, is considered the same as racism, which is discrimination based on race and ethnicity.

Ubuntu philosophy: It be described as the capacity of an African culture to express compassion, reciprocity, dignity, humanity and mutuality in the interest of building and maintaining communities with justice and mutual caring.

Sankofaism: It is the philosophy founded on the premises and belief that human beings are part of the past and that this past shape the present that would also determine and greatly influence the future.

Revalorization: It is the act of redefining the importance, status, or worth of a language, culture, or concept, particularly in response to historical oppression or marginalization.

ENDNOTES

1 Oralism can be defined as the system of teaching deaf people to communicate by the use of speech and lipreading rather than a natural signed language.

Chapter 18
Umtfwalo:
The Traditional Load

Jabulani Sabelo Jarreth Pato
Examinations Council of Eswatini, Eswatini

Remah Joyce Lubambo
https://orcid.org/0000-0003-1990-0984
University of South Africa, South Africa

ABSTRACT

This chapter discusses the significance of 'umtfwalo' (traditional load) in the culture of emaSwati. Umtfwalo is witnessed in diverse cultural events within the society of emaSwati, and it presents different meanings at the different times. In each of the instances, it has its symbol. This study, therefore, seeks to explain the concept of 'umtfwalo' (traditional load) in the society of emaSwati. It is underpinned by Afrocentricity theoretical framework which advocates for traditions and practices of African societies such as emaSwati. The study uses a qualitative research method of data collection and analysis. Interviews and textual analysis are the types of tools used in this study to collect data. The study's findings explain the concepts of umtfwalo. Moreover, the study discusses the instances where umtfwalo is expected in the society of emaSwati. The study concludes by pointing out that emaSwati and their umtfwalo are inseparable.

INTRODUCTION

Members of the emaSwati society have seen, in certain cultural events, person(s) carrying *'umtfwalo'* (cultural load) which is a grass mat, wooden pillow and a blanket (sometimes with a calabash), and really ask themselves why that is done. Even the load-bearers sometimes are ignorant of the importance of the load they carry. Not that they are afraid to ask but it is a known thing that you never ask adults why certain things are done in the culture of the people of Eswatini (Dlamini & Masuku, 2013). Why does a young maiden carry the *'umtfwalo'* (load) and walk in front of the bride when she (the bride) leaves her parental home for her in-laws' home? Women have been reminded several times that if they misbehave in their in-law's home, they shall have their 'load tied up' *kuboshelwa umtfwalo* (load tied up / being chased) and, what is the significance of all that? This is in line with that Gumede cited in Times of Es-

DOI: 10.4018/979-8-3373-4124-8.ch018

watini dated 12/08/2012, who notes that divorce exists in the Swazi Law and Custom. "Every marriage can be divorced, what is why even according to Swazi Law and Custom, a wife can be referred back to her parental homestead if has committed adultery or witchcraft."

EmaSwati (in funerals) again wish to know why someone is carrying an *umtfwalo* (load) when there are burial proceedings in a home? Again, when a male, usually, a young warrior wants to join a Swazi regiment, he comes with *umtfwalo* (load), the question is, why? What does this cultural load have to do with the people of the culture of the people of Eswatini? It is argued that there is a great cultural significance in the people of Eswatini having their '*umtfwalo*' (load) in certain traditional events. It is not merely done but there is a great meaning in having and doing it religiously (though in some quarters certain aspects of it have been forsaken). Dlamini and Masuku (2013) mention that the bride's family should prepared and provide a maiden to carry the bride's load before the dispatching of '*umtsimba*' (bridal party), and this shows the significance of the traditional load.

Umtfwalo is part and parcel of the culture of emaSwati. Culture is, according to Valera (2016), the characteristics and knowledge of a particular group of people, defined by everything from language, religion, cuisine, social habits, music and arts. Pato (2018) says that culture is the similar traits displayed by people in the way they live. Culture

...is shared by a group of people who embrace common aspects of life. It is transmitted from generation to generation. Because of this scenario, culture tends to be followed by a group of people for a long time (Mtumane, 2014).

The foregoing definitions make it easier for one to understand why the Swazi people treasure this cultural aspect. EmaSwati are a particular group that has their religion still intact in some parts of their society, and they embrace their way of living despite the changing times they are experiencing. They have inherited this cultural practice, and they are transmitting it to the next generation.

While the practice of carrying the load in the above manner is noticeable among emaSwati, so far, no study has been conducted to reveal and express the motivation of the practice. This is probably the first attempt at doing that. This study, therefore, seeks to explain the concept of '*umtfwalo*' (traditional load) in the society of emaSwati. It discusses the instances of '*umtfwalo*' in the traditional setting of emaSwati as means to appreciate the culture of the people.

BACKGROUND OF RELEVANT STUDIES FROM OTHER SOCIETIES

The subject of traditional load is prevalent in African societies. The Zulu traditional people practiced Umhlanga long before King Shaka's reign. This ceremony was discontinued, although emaSwati continued to practise it without disruption. The purpose of this event was to urge young ladies to remain pure while also acknowledging and accepting their fragility (Nxumalo, 2014). In these ceremonies, the young maidens carry their own "umtfwalo," or reed. It is mandated that the reeds be handled with care because they are fragile and easily broken.

The Umhlanga festival is thought to be one of the tourist attractions in South Africa, hence the impact of closing it during the COVID-19 pandemic was felt by many (Ndaba,2022). *Umtfwalo* is also realised in AmaZulu marriage ceremonies, where the young girl carries *umtfwalo* in the form of a 'Kist. The Kist is purchased by the bride's family as a symbol of her coffin, indicating that they acknowledge that the young lady is no longer theirs own but belongs to the family she is marrying into. When the young girl arrives at the groom's family, she checks to see if her *umtfwalo* (Kist) is with her. If the girl is having his Kist bridegroom's family is certain that the bride will never go back home but will die and be buried with their ancestors.

umtfwalo can also be recognised in burial traditions, particularly when the deceased died from an unnatural cause such as an accident, being stabbed, or being shot outside the home. The disease's body will not be let inside, and the disease will be buried with his clothes, indicating that he should take his things (*Umtfwalo wakhe*) with him to the grave. His clothes cannot be distributed to other family members since it is believed that they may trigger another comparable death in the family (Ngubane, 2019).

THEORETICAL FRAMEWORK

Afrocentricity

Afrocentricity underpins the study as it is all about the life of African people. Afrocentricity is defined as an exercise in knowledge and a new historical perspective (Chawane, 2016). Chawane (2016) further points out, that the theory of Afrocentricity dates back to the 1970s and was popularised by an African scholar named Asante in the 1980s. The belief is that Afrocentricity is a transformation of attitude, beliefs, values, and behaviour results, suggesting that it is the first and only reality for African people – a simple rediscovery. Essentially Afrocentricity is a way of thinking and acting that prioritises African interests, values, and viewpoints (Gebremikael, 2022). This theory is relevant to the study in the sense that the issue of *'umtfwalo'* is of interest to the lives of emaSwati, an African society, and it speaks to the transformation of the attitudes and values which are areas very key for their preservation as it is the reality of emaSwati.

METHODOLOGY

The study was carried out using qualitative methods whereby two siSwati short stories which have the element of this cultural practice have been purposively selected and used in order to get an in depth understanding and interpretation of the data collected on the research phenomenon (Dewi, 2021). The study also adopted textual analysis since it provides a systematic strategy to organize mainly explicit thematic dimensions of language and describe the content, structure, and functions of messages contained in texts while identifying salient topics within a body of texts (Alejandro & Zhao, 2024; Dawadi, 2020). More to that unstructured interviews with traditionalists familiar with the culture of emaSwati were interviewed to give the insights of the tradition as Alsaawi (2014) as interviews are a basic mode of inquiry with an interest in other individuals' stories because they are worth. The unstructured interviews were key as they allowed the interviewees to elaborate with the topic leading to unpredicted directions and providing more information needed for the research (Elhami & Khoshnevisan, 2022).

DISCUSSION AND FINDINGS

A brief overview of the concept of '*umtfwalo*'

Umtfwalo is a load that is carried by the head most of the time. It could be a load of firewood in the Eswatini context. It could be a bucket of water. It could be a bag with clothes. It is something one carries when one is on a journey moving from point A to point. *Umtfwalo* is defined as:

Umtfwalo ngumgodlagodla lolungisiwe lokungaba timphahla, sishumpa sekudla, inyandza yetinkhuni, tjani noma ngabe yini leboshelwe ndzawonye lengalungela kutsi umuntfu akhone kukubeka enhloko noma akutjatse ahambe aye lapho ayakhona.

A load is a package that is prepared for one. It could be a package of clothes, food, firewood, grass or anything that is tied together for one to be able to lift and carry either by head or shoulders when one embarks on a journey.

SMI-1 (18 October 2018 – personal interview)

SDI-2 (7 November 2018 – personal interview) defines *umtfwalo* as a package one carries when one has to embark on a journey. She states that it contains items that one may need when on the said trip. The *umtfwalo* could be food and beer that one may need when one gets hungry and needs refreshing. She was quick to point out that before globalization and civilization where there were no buses and cars, people would travel miles for days and one had to have *umtfwalo* to sustain one along the journeys.

Most of the time women are the people that are mainly associated with carrying a load. In the culture of emaSwati, a woman would be the one carrying a load and the man would be having his *tindvuku* (sticks) clearing the path for his woman. He has to see to it that nothing harms his woman carrying a load so should be alert to deal with attackers (other people or dangerous animals who could trouble his family).

In the case where a man had no option but to carry a load when embarking on a journey, he would have a rod and would have the rod horizontally across his chest. He would then tie the load at the tip of the rod and allow it (load) to hang loosely at his back. Then, he would hit the path. In such instances, the man would be moving away to a faraway place. Such an act has come up with a proverb in the society of emaSwati. '*Umtfwalo usebhokweni'* (the load is on the rod) which means that one is about to or is already on a journey.

Figuratively, *umtfwalo* could be a burden – something that troubles one's life. One may refer to his / her burden as *umtfwalo* but this time not as a load. South African born songbird, Zahara, sings *'Umthwalo wam' uyangesinda'* (in SiSwati – *umtfwalo wami uyangesindza*) meaning her burden (troubles) is troubling her. In contrast, when a woman is pregnant, the pregnancy is referred to as *umtfwalo* (load) in respect of the unborn baby. In the culture of emaSwati, one never *counts chicks before the eggs hatch* so one cannot refer to the pregnancy as *umntfwana* (baby). This is respect to the mother, baby itself, family and to society. Again, referring to pregnancy as a 'load' shows the value and utmost respect one gives to the phenomenon.

One must note that *umtfwalo* is a load that has to be carried on the head, not on any other part of the body. When using one's hands it is called *'kulengisa'* (hanging loosely), or when holding something close to one's chest it is called *'kugodla'* (holding firmly close to the chest), and when carrying the item using the back of the body it is referred as *'kumema'* (having something on the back). Usually, men use their shoulders to lift and carry items and this is called *'kutjatsa'* (carrying using one's shoulders). The word *'umtfwalo'* is a noun and is derived from the verb is *'tfwala'* (carry using your head).

Simelane (2014) concurs with Gama and Dlamini (undated radio programme) that in Swazi culture a person has two instances where one should have the cultural load. Women have their *umtfwalo* when they leave for marriage and when they leave the present world for the spiritual world to join the family's ancestors. On the other hand, Simelane elaborates that men have their *umtfwalo* which they tie, unlike in all the other instances where the family ties one's load to carry when he joins the regiments. The second instance is when his body is taken to the gravesite with the belief that his spirit will join his family's ancestors.

Umtfwalo in the cultural context cannot, therefore, be said to be a burden since it is not a problem but something to be treasured, and part and parcel of the richness and beauty of the culture of the emaSwati. The *licansi* (grass mat), *ingubo* (blanket), and *sicamelo* (wooded pillow) have not only significance in the culture of the emaSwati but the cultural symbol, pride and moral gestures that are greatly appreciated in the culture of the emaSwati.

Seeing the cultural load – *umtfwalo* triggers different feelings and emotions. '*Umtfwalo*' separates emaSwati from the rest of the other nations. '*Umtfwalo*' in Swazi culture signifies and symbolizes quite a lot. It is brought to different cultural events but, as rightly pointed out earlier, it entails the very same components (grass mat, blankets and wooden pillow).

THE PRACTICE OF THE TRADITIONAL LOAD*(UMTFWALO)*

a) Umtfwalo – kuphuma kwemtsimba

Cultural Load in Traditional Wedding Ceremonies

Different societies have different ways of uniting two people. Each society has its traditional patterns of marriage and the members of the society are always expected to uphold their cultural ways. EmaSwati, too like most if not all cultures, have their peculiar way of conducting marriage. Swazi people have what is called *Umtsimba* (traditional wedding ceremony). *Umtsimba* 'traditional wedding' is:

> *...ngumgidvo wesiSwati lophelelisa bufati. ngesiSwati umtsimba kufanele uphume ngemva kwekutsi intfombi icelwe ngenkhomo nobe-ke nase imekezisiwe.*
>
> ...is a Swazi traditional ceremony that completes womanhood. This shall happen after the groom's family has asked for the bride's hand in marriage or when she has been *teka-ed.* (a ritual that takes place in the kraal when she meets and swears to the groom's ancestors) (Dlamini & Masuku, 2013)

Simelane (2014) agrees with the above definition, but he further states that it could also be defined as:

> ...ludvwendvwe loluhamba nemtsimba lufaka ekhatsi; makoti, gozolo, intfombi yemtfwalo, badlalisi, tintfombi, emajaha, bobabe, bomake, tingcungcu tamakoti, lophetse umtsimba.
>
> ...delegation that includes; bride, intermediator, maid-load bearer, dancers, maidens, young male regiments, men, women, engaged maidens and leader of delegation. (Simelane, 2014)

Umtsimba, the ceremony, is significant in the culture of emaSwati since it is an event where the two families (bride and groom) are joined together and relations are forged. Again, it is a ceremony where the bride is given words of advice possible for the last time by her maternal family. It is also a key event in the sense that to the bride's new family it depicts that she (the bride) has the blessing of her parents. *Umtsimba,* the delegation, is important as they show that Swazi society is one. The delegation shows communal unity as the people (community members) come together to support the two involved families and witness the ceremony. The community members must accompany the bride to indicate that she comes from society.

According to Vilakati and Sibanda (1997), the song *'Sitocela Inkhonto'* (We seek a place to live) that is led in singing by the bride is important as it is an indication to everyone witnesses the event that no one is forcing her to give her hand in marriage.

> *Sitocel' inkhonto Nabo singani* We seek a place to live mother-in-law
> *Sitocel' inkhonto* We seek a place to live
> *Singamandabende* We come from afar
> *Sesite kuwe* We have come to you, mother-in-law
> *Singamandabende* We come from afar. (Vilakati & Sibanda, 1997)

In SiSwati culture a bride does not just leave her home to her groom's home but there are rituals which have to be performed. As stated earlier, the bride is given words of wisdom on how to handle herself, given concoctions, and cleansed with a family *'muti'* (black magic) to enhance safe travel and, a safe and close-to-perfect life with her in-laws. As Simelane (2014) mentions, there are several people who are part of the delegation accompanying the bride on her journey to her new home and amongst them, there is a young maiden who is carrying the significant cultural load, *umtfwalo*.

In this sense, *umtfwalo* is defined as *'lokuboshelwa makoti nakaphuma kabo ayokwendza.'* (A cultural load is tied together for the bride when she leaves her parental home on her new life as a wife with her in-laws) (SMI1- personal interview, October 18, 2018). The load as stated earlier comprises of *licansi* – sleeping mat, *sicamelo* – wooden pillow, and *ingubo* – blanket. All these three are evidence enough that that the bride is leaving her home and would no ways return home. This load depicts that the bride will have no place in her maternal home as the sleeping materials she was using whilst in her home had been taken away. It is a sign that she is leaving her home for good. Even on her visitations or when there are important cultural functions like burials, wedding ceremonies, she would have to return quickly to her in-laws' home as soon as possible. To echo this, there is a traditional wedding song that is sung with regards to the leaving of the bride:

> *Nguye uyedvwa kabo* She is the only one in her home.
> *Nguye* She is her
> *Yemtsimba lokhashane* Wedding delegation that is far
> *Bambiz'izwe lonkhe* Keep the whole country together.
> *Nango-ke asaphuma* There she leaves
> *Siyamsunduzela.* We are shifting her.

The song she sings when she leaves, with the young maiden carrying the load, says it all that she is leaving the home for good and she would not be coming to live in the home she is leaving at the moment. She is expected to 'hold' and keep her new family together with her good morals. *'Nango-ke aphuma'* (there she departs) shows that she is leaving in front of everyone and there are witnesses that she is relocating with her belongings. The word, *'Siyamsunduzela'* (we are shifting on her behalf) says it all that the wedding delegation is shifting or moving her permanently from the parental home to her new home. The *umtfwalo* represents all her belongings that are being shifted and the young maiden carrying the 'load' supports the shifting that the bride is being helped on. The song is in line with the traditional load symbolizing the change of the location of the bride.

Figuratively, the *umtfwalo* means that the bride is ready to leave her home and will be strong enough to shoulder burdens that she may have to bear in her new life. It is a way of telling the family, relatives and neighbours that she is mature enough to stand against all strong oppositions. '*Umtfwalo,*' in this case is no more a 'traditional load' but 'adulthood troubles.' In SiSwati it is said '*Emendvweni kukamkhatsali*' (Marriage is extremely difficult), so the 'burden' signifies that she is ready for that responsibility in life.

Gama and Dlamini (undated radio programme) say that despite all challenges, a woman is not expected to go back to her maternal home. They state that if she contemplates that her family should not allow her as she is the one who chose willingly and, as parents, they permitted her to get married. Gama and Dlamini further allude to the fact carrying the 'load' and the act of 'crying' are significant in the procession of the *umtsimba* ceremony. They list two instances where the *umtfwalo* is carried, the first being in the traditional wedding ceremony. The *umtfwalo* is prepared and 'tied' in the most sacred hut that is *endlini kagogo* (grandmother's hut). This shows the importance of seeking the blessings of the dearly departed family members in the cultural occasion. Gama is quoted giving this view:

> Uyesuka etandleni teyihlo nenyoko uyohlala nebantfu longabati nekutsi bakuphi lebete nengati yakho, uyati yini loko? Kepha ngalokutsi kushito wena sikubophela nankhu umtfwalo sitsi hamba khona.
>
> You leave you parents' home and you shall be living with unknown people whom you do not have same 'blood' with, are you aware that? But that you have said it yourself, we are tying this load, here it is, leave as you desire.

This is said to the bride. She is made to sit with the family members and given words of advice. The bride is made to sit next to the pillar '*insika*' which is very substantial. It instils the idea to the bride that she must being a strong pillar for her family once she gets to her in-laws. The *umtfwalo* is placed next to her as the family bids her farewell. This shows that she is beginning adulthood and she would face challenges on her own with no help from her maternal parents. The bride would be expected to preserve all kinds of obstacles and if she succeeds in that (withstanding challenges) she would be then called a woman. Gama says that *nakatsatsa lomtfwalo kukhala unina* (when she takes the load her mother just cries) as the 'marriage journey' is known to be difficult for a young girl and the mother understands what lies ahead on the journey her daughter is embarking on as it is proverbially said, *emedvweni kukamkhatsali* (marriage is extremely difficult).

The bride does not carry the load herself but someone does. It is always '*litjitji*' (young maiden) as it has been mentioned earlier. She is chosen by the family and must be related to the bride. SMI1(personal interview, October 18, 2018) states that it is significant to the young maiden as it gives her responsibility and acknowledgement that she not only witnessed but also a huge role in culturally sending elder sister to her in-laws.

Again, the elders who have chosen her show that they also wish she, the young maiden, could follow on the footsteps of the present bride and do good so that one day she could be the one having her *umtfwalo* carried for. It is more of motivation and encouraging good moral standards, not only to the young maiden but to the girls (maidens) in society who are witnessing all the happenings.

It should also be noted that this young maiden tasked with the responsibility of carrying the emaSwati cultural 'load' should be one with no child. Ages ago, one tasked with the 'load' would be a virgin since having sex before marriage was and is still discouraged. The young maiden walks in front leading the way showing the pride not only for the bride, but of the family too. The young maiden symbolizes purity and it is a way of encouraging other young maidens to stay pure so they earn community's respect.

Having the young virgin maiden lead is associated with wishing the bride the best of luck as a virgin is associated with spotlessness.

The young maiden carrying the load is given strict instructions. Gama and Dlamini (undated radio programme) warn:

> Umtfwalo wawuphuma endlini lolowutfwele akabuki emuva ngalomhambo. Yini yakho lesele ngoba konkhe kwakho naku enhloko?
>
> When the one carrying the load leaves the hut, she must not look back as the tradition dictates. Why would she because all her belongings are on her head?

This is also another important part. The young maiden is warned not to look back once she has the load on her head. This would mean pronouncing bad luck to the bride who may fail once challenges block her way in marriage. Turning or looking back whilst carrying the load would be bad omen not for the bride but for the whole family too. From the quote again, it is noted that the three items making up the *umtfwalo* represents all her belongings. This means there would be no reason to look back as she is departing with all she needs for her future life. *Umtfwalo* stands for all 'basic' needs the bride would need upon arriving at her in-laws.

The bride's neighbours are expected to come witness the leaving of the bride and the *'umtfwalo'* as it is a sign that 'indeed she left home with all her belongings.' The neighbourhood enjoys *'inhlabisamtsimba'* (a cow slaughtered for the delegation accompanying the bride). These people are witnesses that *umtfwalo* was made and carried for the bride under the blessings of the family's ancestors since other rituals are performed in the kraal and in the sacred hut, *indlu yakagogo*.

The *umtfwalo* also shows that the father and the whole family are affirming the departure of their daughter. *'Umtfwalo'* is not obtained nor gotten from the bush (where there are no moral upright people) but from a home (where there are elders). This vividly shows that the family is fully supporting this (the traditional wedding). *Umtfwalo* is given by parents, which is why after rituals have been performed the bride's father says: *"Hamba ntfombatana uyongona."* (Go and spoil my reputation / name). This is said to encourage the bride to uphold the right teachings that had been instilled in her since birth at *egumeni* (women's hut where young girls and taught all female chores and life expectations). Doing wrong would be spoiling her father's reputation for failing to teach her the right and acceptable standards.

The female elders are the ones who tie the three materials entailed in the Swazi cultural load together. There is also a *'sitja'* - bowl the bride used at her home that is destroyed to show that she no longer has a plate (bowl) in the family since she would no longer have meals at her parental home but with her in-laws. If certain circumstances prevail and force her to come home, her stay shall be temporary. She would not even have a sleeping-mat, a blanket and a wooden pillow, as these would have been kept at her 'own' home (in-laws). Makhanya emphasized that *umtfwalo awutitfwaleli nawungumakoti unless...* (If you're a bride you do not carry your own load unless...)

This *'umtfwalo,'* upon arrival at the new homestead, would be a sign that she has been allowed to come and offer her hand in marriage. If there is no 'load' it would be an indication to the in-laws that she was not given permission back home and she came on her own without blessings from the family. Again, it would also show that she did not come to them wholly, and in the long run, when there is trouble, she would go back to her parental home. She would not gain respect. In times of hardships, she could be harshly treated and insulted, and be sadly reminded time and again that she ran away from her home and forced her way to them. Without *'umtfwalo'* she could be belittled at her in-laws' home. *'Umtfwalo'* shows to society that the bride got the consent of her parents and family (a child belongs to the whole family in Swazi culture). With the ill-treatment that could be given on her, it would not be easy for her to go back to her home that she had deserted.

Upon arrival, the *umtsimba* sings the very song they sang when they left their home. This time around, however, they change one line of the traditional song.

Nguye uyedvwa kabo *She is the only one in her home*
Nguye *She is her*
Yemtsimba lokhashane *Wedding delegation that is far*
Bambiz'izwe lonkhe *Keep the whole country together*
Nango-ke angena *There she comes / arrives*
Siyamsunduzela. *We are shifting her.*

This is the same song sung on departure from parental home. Now it is sung upon arrival at the in-laws' home announcing the coming of the bride and the wedding delegation. *'Nango angena'* (there she enters) is the only line that has been altered that indicates the action that the bride is arriving for good as they have helped her shift from her parental home, *'Siyamsunduzela.'* To the in-laws, the sight of the *'umtfwalo'* is an indication that indeed the bride is coming with all her belongings to live with them. It also gives them a message that she is ready to share whatever troubles *'umtfwalo'* (burden) that may come in her life with the in-laws and she would not be expected to *'chicken out'* as she comes with all her belongings.

The young maiden carrying the load would not bring it down nor offload it once the traditional wedding delegation has arrived at her 'sisters' new home. As she would be leading the delegation the *'umtfwalo'* should remain on her head until a woman from the groom's family comes to take it down. That would be a very important indication that the bride and her delegation are welcomed in the home. The groom's family does not just take down the Swazi cultural load but also offer the young maiden who has been bearing the load, a token of appreciation. This kind gesture is called *'kwetfula umtfwalo'* (offloading the load). The token of appreciation is usually a village fowl. It is believed that it shall also open doors for her own marriage when the time has come. The offloading is performed by a maiden from the groom's side (Dlamini & Masuku, 2013). She also carries the load to their own grandmother's hut where it shall be placed for the duration of the wedding ceremony.

The song is also proof enough to that. Gama and Dlamini (undated radio programme) state that it is believed that those that are not wedded in this way will one way or the other have trouble in their marriages. For no reason would the bride be expected to go back to her home unless she has been requested by her family to come for a certain event or discussion.

Mention was earlier made of *'kuboshelwa umtfwalo'* (packing one's load). Of late, the saying has been used figuratively to mean being sent back to one's maternal home (referred to a woman). In Swazi culture only a woman is sent back to her home. There are certain conditions that warrant that. Even so, there are parties involved when she is sent packing. There are procedures to be followed when *'aboshelwa umtfwalo'* (sent back to her parental home).

It is also pointed out that there should be senior members of *'umphakatsi'* (chiefdom) to witness all the happenings. As *'umphakatsi'* (chiefdom) was alerted when there was going to be *'umtsimba'* (wedding delegation / wedding itself) coming to the home and they probably witnessed the coming of the *umtfwalo* and they should be told that *'umtfwalo'* of the culprit – woman is taken back to her home for good after she was found to have done any of the three offences mentioned above. It must be emphasized that this *umtfwalo* is the same load that the woman, then bride, the young maiden was carrying leading the delegation on that night of arrival. When she arrived, she asked for *'inkhonto'* (a place to stay / hand in marriage). There is, however, a notable material that is tied on this *umtfwalo*. There is *'umtfongwane / umcandvo / imvunulo* (traditional underwear that is used to cover only the tip of the manhood). This symbolizes Swazi cultural divorce. The *umtfongwane* belongs to the former husband and it is a sign that no more shall they have a sexual relationship as husband and wife. SMI1(18 October 2018 – personal interview) states that the *umtfwalo* is sometimes carried without having tied the *umtfongwane* and that means that the husband no more wants to keep the wife but he may still have sexual encounters with her unlike when the *umtfongwane* is tied on the *umtfwalo*.

They are traditionally divorced and the delegation accompanying the woman carrying *'umtfwalo'* are witnesses to the sad event. Culturally, she has to do without objection because she knows her crime. By seeing their daughter carrying the *umtfwalo* with the *umtfongwane,* the family would know that their daughter has brought them shame. The father who had said *'hamba ntfombatana uyongona'* (go girl and spoil my name) would know very well that his daughter would have failed dismally, hence, her return with the *'umtfwalo.'*

Kubopha umtfwalo in the above context is done to shame the evil doing and to discourage bad deeds. This is done in the open to shame the evil doer and other potential doers would be unconsciously discouraged to do something that would warrant them being made to carry the 'load' back to their home.

b) Umtfwalo – kubekwa kwemntfwanenkhosi

Cultural Load in Prince assuming Chieftaincy

Simelane (2013) and Adams et al. (1999) state that there are two types of land tenure in the Kingdom of Eswatini, that is the Title Deed Land (TDL) and the Swazi Nation Land (SNL). The SNL is further divided into two, land held under customary tenure which Adams claim that this kind of land is under the control of chiefs, and cannot be sold, mortgaged or leased. The Constitution of Swaziland (article 233, 135) clearly states that chiefs are the footstool of the king and the king rules through the chiefs. This means that such land is under the king of the land, and the chiefs only monitor the land on his behalf.

Another duty for the chiefs is governing the land and making it accessible through traditional structure called *kukhonta* (seeking residence in SNL or in chiefdom by approaching local authorities). Dlamini *et al* (1986) elaborates that chiefs have the power to evict and re-allocate a household in the community upon evidence of crime and witchcraft.

There are rituals performed when a chief is put into position. The focus of this study is on *umfwalo* and it is part of *sibhimbi* (ceremony held confirming the appointment of a chief). When a prince has been given chieftaincy by the king and royalty, he also has *umtfwalo* prepared and 'tied.' His *umtfwalo* is prepared by the royalty elders to show that he has their blessings, and indeed has been given the task by the king to look after his subjects (people) in that assigned area. The *umtfwalo* is a sign that he is sent away from home and he shall be an 'eye' of the king in that particular chiefdom that he has been assigned to. It is a sign of growth that he shall be leaving home and will be living on his on making decisions that shall benefit him, the king, royalty and the people he shall be leading. The *umtfwalo* in this case signifies responsibility on the shoulders of the prince. His conduct is not an issue but his household is appreciated for whatever cultural role they perform in the royal circles. The load is more of saying no more shall he be seen in the present household but in his new consigned area.

A prince who is made a chief who has his load is chosen considering his birthright. His mother amongst other king's wives has that special position or certain part to play in the royal circles (especially traditional) that is why his son is then given the important position (that of being a chief) in the Swazi setting. SDI2(11 November 2019 – personal interview) mentions that the *umtfwalo* is a way of having the whole household of the newly elected chief leave the royal palace and have their own residence where he shall be leading the king's subjects. She states that the son (the chosen prince) leaves with his mother and other siblings. His new residence will automatically be the king's new home.

The *umtfwalo* is also given to avoid *imisindvo* (trouble in the home) that could be brought by having too many senior people in one place as the royal home becomes bigger and bigger for in siSwati *'tinkunzi letimbili atihlalelani esibayeni'* (one cannot have two bulls in one kraal). Giving the senior princes *umtfwalo* is also a way of preventing and avoiding tension in the royal household so, it helps giving them responsibility elsewhere. It solves unforeseen yet possible outrages before they actually happen.

c) UMTFWALO – UMHLANGA

Cultural Load in Reed Dance Ceremonies

Umhlanga (reed dance), according to Masuku quoted by Sone (2010) is an annual dramatized ceremony whereby maidens or unmarried girls gather at the royal residence to cut reeds to rebuild the '*emaguma*' or royal kraals, to honour and pay homage to the King *inkhosi* and to the Queen Mother *indlovukati*.Simelane (2014) says that the reed dance is the foundation and pillar of the Swazi society. The reed dance is divided into two; for unmarried girls and married women. However, the main reed dance is the one for the unmarried girls who are expected to be pure. All girls from various chiefdoms in the country are expected to come and attend the function.

The reed dance is important in many ways in the lives of emaSwati. First and foremost, this kind of a ceremony encourages the culture of paying homage to the king and to the queen of the land. By joining in shows respect, and the reeds coming in bundles is used in rebuilding *emaguma* (women's huts). It is used as a windbreaker. In addition, *umhlanga* allows the embers of the society, the girls, to express their social experiences through music and dance. Matters that pertain the society are expressed through song. Again, *umhlanga* encourages unity amongst the girls from different chiefdoms. Vilakati and Msibi (2006) state that this cultural aspect is also, above all, a way of keeping purity amongst the unmarried girls. It is meant to preserve their virginity.

Vilakati and Msibi (2006) further mention that, the maidens are also exposed to different lessons on good conduct, perseverance, trustworthy and respect among other key issues in life. Sone (2010) declares that enjoyment of the performance during the reed dance is another vital factor for its realization. Having said that, when the king, who is one of the main characters in this kind of a traditional drama, has commissioned the *Imbali* (flower – a name given to the maidens) to go cut the reed, there should be male regiments carrying the *umtfwalo* for the king's princesses. In this instance, the sign of the *umtfwalo* would be an indication that the king has commissioned the unmarried female regiments to perform the duty before them which is part of paying homage to royalty. And it also shows that they would be sleeping away from the palace for a number of days so they need to be prepared. More to that, it is a sign that they should not come back until they have completed their mission set for them which is to cut and bring the read home. The princesses are the ones leading the maidens, and they should be provided for.

SDI1* explains that the *umtfwalo* at this cultural event is also a sign of unity as the king's daughters (princesses) share its usage. They come from different households and are expected to be together and do many things in harmony. The *umtfwalo* unifies the princesses from all the different households which means it is a symbol of oneness. It is also a symbol of oneness to everyone attending. When they are out in the journey to get the reed, they use it to sleep along the way when the sun sets, and they share the food prepared for them. When the maidens carrying reed return to the royal residence from the different places where they had cut it (reed), the *umtfwalo* returns too. This is a sign that mission has been accomplished.

Even at chiefdoms, the chiefs' princesses have their load carried by regiments. When the king has called the *imbali* regiments, all chiefs should see to it that the maidens from their chiefdoms attend the event without fail. The *umtfwalo* for the chiefs' princesses, carried by the male regiments, is a sign that the chiefs are adhering to the king's call and order. Like the king's princesses' cultural load, the chiefs' princesses' loads return because the maidens are not getting married but are expected to come back once the reed dance ceremony has come to an end.

Unlike the previously discussed instances on the cultural load, in this instance the *umtfwalo* returns. MDI2* (11 November 2019 – personal interview) reveals that this kind of *umtfwalo* should be brought back as it is a sign that the maidens have accomplished the mission they were tasked with. She states that the modernization has changed a lot in the culture of emaSwati. Long time ago where there were no cars, the load would play a huge significance. The load carriers would move out of the compound after being commissioned by the king. On return, the load carriers would enter the compound first before the maidens to indicate that they had done what they were expected to. In the presence age, the male regiments bearing the load move out first and they walk not a very long distance then they board cars. Even on their return, they disembark from their vehicles before they come closer to the royal residence. They then walk towards the royal residence carrying the *umtfwalo*. This shows that despite the changing times, the *umtfwalo* tradition has been kept intact.

d) Umtfwalo – kubutseka

Cultural Load in Giving in to the King

Kubutseka is another cultural practice performed by Swazi males (boys and men). This is giving in or surrendering yourself to the king that you (as a boy / man) shall remain faithful to him at all times. This is joining the regiments to be the king's warrior. Vilakati and Msibi (2006) state that the only way in which a male could join any of the regiments available is through *kubutseka*.Pato (2018) defines *kubutseka* as '*kugana inkhosi*' (which is loosely translated to giving your hand to the king. He sees this as ironic as usually women are the ones who give their hands in marriage yet in this case men give their hand to their king. Simelane (2014) views '*kubutseka*' as a way of paying homage to the king which is a decision taken by a male. He further views *kubutseka* as:

> …kutidzela, kuletsa emandla enkhosi. Kusho kwenta sincumo lesisho kutsi uyabashiya bakini kanye nebangani bakho uyokwendza enkhosini. Umfana nakayobutseka usuke ayaliwe nguyise kanye namkhulu wakhe. Bamyala kuhle kwemuntfu wesifazane nakayalelwa kuyokwendza.
>
> …to give up oneself and give service to the king. It means making a decision of leaving one's family, relatives and friends to 'marry' the king. When a boy surrenders to the king, he is given words of advice by his father and grandfather. They advise him like a bride is advised before she departs home for her marriage.

When a man joins the regiments through *kubutseka* (the only way of joining regiments) he comes with *umtfwalo* too. This *umtfwalo* has the same material / contents as the one carried by women. It must be noted that men, on an ordinary day, never use their heads in carrying objects but only carry using their shoulders, *'kutjatsa'* mostly. When it comes *kubutseka,* the load is carried on the head. Men use their heads, like women carry, when they give their love to their king.

Men, too, have their own *umtfwalo* in the Swazi cultural society. Mabuza, in Msibi (2002), presents the kind of a traditional load in her short story, *'Giyakabi'* (Falling to dance).

'…nakaphuma nalomcicimba wakhe welicansi lelinengubo ekhatsi.'

'…when he left with his load that consists of a sleeping grass mat with a blanket in it.'

The modern short story has been made to depict the processes one undergoes when joining the traditional regiments of the king of Eswatini. The traditional load is short of the wooded pillow that has been replaced with the changing times. It is evidence enough that emaSwati have tried to preserve the traditional load though there are changes that have been experienced.

The regiments' traditional load is strongly linked to the king of Eswatini. Men use the items *'licansi* (sleeping mat), *ingubo* (blanket) and *sicamelo'* (wooden pillow) in *'emalawini'* (men private cultural huts found in royal chiefdoms). A man coming to give-in to the king carries the SiSwati cultural load to show that he is leaving his own comfort at his own home to live his life for the king of Eswatini. Coming with the sleeping mat, wooden pillow and blanket to the *emalawini* is proof enough that the warrior is deserting his home like the bride does when she leaves her home with the *umtsimba* delegation. The *umtfwalo* remains there at the men's private huts for the warrior's usage when there are royal activities taking place or at any time, he feels like going to the royal chiefdom.

Kubutseka is significant in the culture of the emaSwati, and the *umtfwalo* symbolizes a lot as one decides to leave one's home and surrender to the service of the king of the land. The items that make up the *umtfwalo* and the words of advice given before one departs his home depict a lot in the culture of thc Swazi people.

SDI2 (7 November 2018) points out that if a male regiment misbehaves, like the misbehaved wife who has her load tied, he is also sent packing. There are instances whereby a male regiment who has *butseka-ed* (given his love to the king) would warrant that he be told to go back to his home for good. Before he is allowed to be part of the regiments, he is given orders on how to behave, and if he crosses the line, the senior regiments will send him home. Such an act is termed as *'utenyanyile'* (he has hated himself) meaning that he has let himself down by failing to abide by the set standards one is expected to depict.

This *umtfwalo* the *sibutseki* (the male who has come to join the regiments) came with would be the same *umtfwalo* he would be departing back home with. It would be a sign that he no more deserves to be part of the regiments and would be standing for all his belongings he came with. It is also a way of discouraging misbehaving amongst the members of the different regiments under the ruling king. It is a sign of a journey back home one should take. The boys' parents would be shamed the minute they see their boy-child carrying the load entering the family's premises. In fact, the discarded boy shall be accompanied to the chief's home *umphakatsi* by the *emajaha* (male regiments). The king shall fine the chief, and the chief shall fine the boy's family for discrediting his name and the name of the chiefdom. The returning *umtfwalo* would also mean that the boy never heeds the words of advice given by his family before he was allowed to go give his love to the king. This part shows that *kubutseka* is king of an emotional and spiritual marriage of the king and the regiments

e) Umtfwalo – lusekwane

Cultural Load in Sacred Shrub Activity

Luselwane comes in two-fold. Firstly, it is a sacred shrub, a kind of a tree that has thistles, green leaves and it grows in the low-veld type of a vegetation. It is also a cultural ceremony that precedes *incwala ceremony*. The ceremony was named after the sacred tree that plays a pivotal role during the activity. What could be noted about the tree is that it quickly grows after it had been cut. A cut branch takes time to dry. Vilakati and Msibi (2006) claim that emaSwati believe that the tree 'sleeps' like human beings because after it has been cut, the leaves would still be looking beautifully even the day after. They say:

> Nakutsi phasha inyeti lusekwane luyalala, nalitsi phasha lilanga lusekwane luyavuka, nakulo sekusile, emacembe alo bese ayacwatimula.
>
> When the moon appears, the sacred tree 'sleeps,' and when the sun rises, the tree it 'wakes' and to it the sun has risen, the leaves get green.

With that said, it shows how the shrub is valued amongst the Swazi society as it has a very critical role to play during the *Incwala* ceremony. It is a known thing in the Swazi circles that the shrub is cut by pure boys who had not engaged in sexual matters. Vilakati and Msibi (2006) state that:

> Lusekwane lukhiwa bafana nemajaha lacwengekile, lahlobile langekemitsisi, amajaha langazange sekaye ecansini (alale) nemfati noma intfombi lese itele, noma lekhiphe sisu.
>
> The shrub is cut by pure boys who haven't impregnated nor had sex with someone's wife or a woman who has committed abortion.

The cultural practice by the boys encourages purity amongst the boys and it does with *umhlanga* for the maidens. It is believed that if a boy who is not pure cut the sacred shrub, his quickly withers and he is shamed that way. Long time ago, such a boy would be beaten by his peers for shaming them. Again, the shrub is used to cover the place inside the royal kraal where the king performs the expected rituals during the Swazi sacred prayer, *incwala*.

Just like when the king commissions *imbali* during *umhlanga*, he also commissions *ingatja* (unmarried pure males) to go fetch the *lusekwane*. Before the boys leave the royal residence there should be other boys carrying a traditional load. The load, just like in the other instances, is an indication that the boys will be sleeping away from the royal home for a number of days. The shrub is cut at night when the moon rises as it is believed that if one cuts the shrub before the moon comes out the shrub shall not blossom, yet when cut after the moon had risen it blossoms and would serve its purpose perfectly. So, the *umtfwalo* is very much needed for the benefit of the king's sons, princes. The boys are not expected to return without the shrub. The load will come handy as they shall need the sleeping mat and blanket. The dish will be used when eating whatever food prepared for their journey.

On their return, there shall be boys leading the regiments carrying the *umtfwalo* as they enter the premises. Again, this is a sign that they have been successful in bringing home what they were tasked to get. This kind of *umtfwalo* is expected to be brought back as it is convenience's sake. One must note that the boys sleep in the open, unlike the maidens where nowadays facilities are provided for them. With the boys, they show the manhood and bravery that they can stand against all odds.

SUMMARY OF FINDINGS

This chapter argues that traditional load *umtfwalo* are the most imperious legacies of African societies, including emaSwati as a nation. The study highlighted the importance of African people being aware and respectful of *umtfwalo*, as it will continue to shape their African identity, knowledge production, and individuality on a global scale. Most important to this is the assumption that traditional load *umtfwalo* is more commonly practiced by women than *umtfwalo* linked with their male counterpart. However, the traditional load *umtfwalo*, must be protected in order to preserve emaSwati culture, traditions, and indigenous heritage (Nhlabathi, 2024). In addition to the word *umtfwalo* as a concept, the research discovered numerous practices where *umtfwalo* is conducted, demonstrating that most emaSwati, including those residing in South African regions, still fully observe the practice of *umtfwalo* in their daily lives. The research discovered that *umtfwalo* can be multifaced in their use, which is explored further below.

Umtfwalo as a symbol of pride

Young maidens participating in the reeds dance ceremony strive to remain pure, unaware that the reeds they are carrying are their traditional load" umtfwalo". The significance of the reeds lies in the fact that, they are also fragile and can break at any time (Ndaba, 2022). The traditional load they are carrying has more significance than their involvement in the ceremony. Aside from basic participation, this tradition is a source of pride for both the nation and the young maidens' parents. This pride emanates from the fact that the reeds they carry serve an important purpose in the king's household. Another source of pride among the girls is their purity, which is backed by the Afrocentric belief that African values should be preserved and observed even as time passes (Balakrishnan, 2020). The Swazi young girls preserves their culture as well as their virginity.

Swati males also have their own *umtfwalo* as "*lingedla*" to show their pride as Swati traditional men. Even though "lingedla" can be used as a weapon in times of danger, they have a sense of pride that they can overcome any obstacle. The "*lusekwane*" ceremony also displays pride of the pureness of emajaha displaying their traditional load *"lusekwane*" to the king.

Umtfwalo as a symbol of unity and faithfulness

As previously stated, the concept has spread its tentacles to the feminine figure. *Umtfwalo* becomes apparent as well when the king summons unmarried girls (imbali) to go and cut the reeds. As a symbol of unity, both the maidens and the regiment remain away from their homes until the task is completed. In this situation male individuals carry the traditional load even though the *umtfwalo* reeds were cut by the young maidens. Afrocentric philosophy encourages the spirit of "*ubuntu*", which values respect and unity (Dlamini, 2018). As the king has several wives, the princesses come from various households, demonstrating unity. In this situation, *umtfwalo* represents unity because all of the princes from different households gather, work together and travel a long distance together. This unity requires united hearts. It is worth emphasising that the regiments carrying the reeds remain with the young maidens and must demonstrate their loyalty to bring the princesses safely back to the monarch. It is a source of pride to accomplish the work entrusted by the king; therefore, they submit the load to the king first to demonstrate their trustworthiness (Dlamini et al., 2015).

Umtfwalo as a symbol of commitment

It is worth noting that when women get married, they likewise participate in the *umtfwalo* practice given that the young lady who walks in front of the bride carries *umtfwalo*. This traditional load *umtfwalo* has been packed by the bride's parents as a symbol of surrendering their daughter to the groom's family (Dlamini & Masuku 2013). In this instance, the bride is not carrying his load, but someone else is. It is customary mandated that the load the carrier must be a young, unmarried female, preferably a pure girl (Dlamini, 2009). Furthermore, the chapter demonstrates that *umtfwalo* practice can diverge from women as people who commit themselves to men. Young men "emabutfo" in "kubutseka" are required to carry umtfwalo as an indication of their commitment to the king and his people. Their load is likewise carried on their heads, symbolising "umtfwalo," which is less physical than it was for women. The regiments have to carry their load deep inside their hearts by remaining devoted and protecting the king in all times. They demonstrate dedication and commitment by leaving their homes and come to serve the king with their hearts and minds; the regiments commit to be loyal to the king and the nation as a whole.

Umtfwalo as assigned responsibility

The chapter discovered that *umtfwalo* can also be noticed when the prince assumes his responsibility as a king; this load is provided by the family elders to equip him to lead the nation; however, he must also have *umtfwalo* in his heart to lead the nation. The umtfwalo bestowed upon him will support him and grant him the dignity he deserves as a monarch. This load consists of a grass mat (licansi), a spear (lingedla), and a shield (lihawu). Aside from the workload, the prince should demonstrate his willingness to accept responsibility by breaking his father's rod. Taking responsibility is an important aspect of Afrocentric thought (Chawane, 2016). The theory denotes that African values include responsibility to every member of the community and that African believes are observed at all times.

The discussions above demonstrate that siSwati short stories have the capacity to expose culture beneath its entertaining and educational features. Short stories were the primary source of information on the traditional *umtfwalo*. This confirms that the process of fostering indigenous knowledge can help provide knowledge and self-identification as emaSwati, taking into account that siSwati cultures are their identity and cannot be separated from them.

FUTURE RESEARCH DIRECTIONS

The research findings on the traditional load *umtfwalo* as part of Swazi culture may benefit scholars who want to learn more about Swazi culture, particularly in terms of understanding the concept and its meaning. Various ceremonies that reveal the facts of traditional load *umtfwalo* as practiced by emaSwati are explained to provide a clear image of what *umtfwalo* means in a Swazi setting. The description of ceremonial processes such as *umtsimba, kubutseka, lusekwane*, and *umhlanga* will help to preserve these cultures for future generations. The cultures associated with *umtfwalo* will be available online for researchers from other societies to explore.

The chapter incorporates language and culture because the terminology employed in this research is rooted in deep Siswati, which will help to maintain the language and keep it from vanishing. Other scholars will benefit from this research, which can be utilised as a literature review to identify a gap in the domain of culture.

CONCLUSION AND RECOMMENDATIONS

In this chapter, the significance of *umtfwalo* in the Eswatini cultural context has been discussed. *Umtfwalo* one gets immediately one is born. This is so because all people have their load of trouble that is why in siSwati culture there is a saying that goes, *'Akukho geza lingenasiyela'* (no place has got no pond). Mainly, we cannot run from the fact that *umtfwalo* is carried by women. When they join their in-laws, they carry the cultural load. When women are sent packing back to their maternal home for the stipulated and unpardonable offences, they carry the cultural load returning to her original home. Similarly, when men join the regiments, they come with *umtfwalo.* There are other cultural events that demand the presence of *umtfwalo* as it signals that the regiments responsible will be sleeping away from home. It is an indication that *labadzala* (royalty and the country's traditional leaders) expect that particular party to accomplish the desired mission then return home. When one joins the world of the ancestors, one carries the load on behalf of the deceased. This is evidence enough that emaSwati and their cultural loads are inseparable. This shows that traditionally emaSwati are nomadic. It is their belief.

The tradition of *'umtfwalo'* (the traditional load) should be preserved as practising as it has a huge role in the issues of connecting with those living in the ancestral world. Traditional events which *'umtfwalo'* is entailed in them should be done to the core (*umtfwalo* used) as the symbol entailed make the whole cultural event meaningful to all spheres of life to emaSwati. Scholars are encouraged to write on the Indigenous Knowledge Systems as means to preserve such for future generations and for reference purposes.

REFERENCES

Adams, M., Sibanda, S., & Turner, S. D. (1999). *Land Tenure Reform and Rural Livelihoods in Southern Africa*. Overseas Development Institute.

Alejandro, A., & Zhao, L. (2024). Multi-method qualitative text and discourse analysis: A *methodological framework.Qualitative Inquiry*, *30*(6), 461–473. DOI: 10.1177/10778004231184421

Balakrishnan, S. (2020). Afrocentrism revisited: Africa in the Philosophy of Black Nationalism. A Critical Journal of Black Politics. *Cultura e Scuola*, *22*(1), 71–88.

Chawane, M. (2016). The development of Afrocentricity: A historical survey. *Yesterday and Today*, (16), 78–99. DOI: 10.17159/2223-0386/2016/n16a5

Dawadi, S. (2020). Thematic Analysis Aproach: A Syep by Step Guide for ELT Research Practitioners. *Journal of NELTA*, *25*(1-2), 62–71. DOI: 10.3126/nelta.v25i1-2.49731

Dewi, I. G. A. A. O. (2021). Understanding data collection methods in qualitative research: The perspective of interpretive accounting research. *Journal of Tourism Economics and Policy*, *1*(1), 23–34. DOI: 10.38142/jtep.v1i1.105

Dlamini, D. D., & Masuku, M. B. (2011). Land tenure and land productivity: A case of maize production in Swaziland. *Asian Journal of Agricultural Sciences*, *3*(4), 301–307.

Dlamini, M., Masuku, M., & Raufu, M. (2015). Determinants of foreign direct investment inflows in Swaziland's agricultural sector. *Journal of Economics and Sustainable Development*, *6*(15), 150–158.

Dlamini, M., & Masuku, N. (2013). Emagugu EmaSwati. Manzini: Imisebe Publishers.

Dlamini, T. (2018). *Using Afrocentrism to explore African Values in selected short stories prescribed for Grade 10 English First Additional Language Learners in South Africa*. University of Kwazulu-Natal.

Elhami, A., & Khoshnevisan, B. (2022). Conducting an Interview in Qualitative Research: The Modus Operandi. *Mextesol Journal*, *46*(1), n1. DOI: 10.61871/mj.v46n1-3

Gama, J & Dlamini, A. not dated. Khala Mdumbadumbane. Eswatini Broadcasting Information Service.

Gebremikael, L., Sicchia, S., Demi, S., & Rhoden, J. (2022). Afrocentric approaches to disrupting anti-Black racism in health care and promoting Black health in Canada. *Canadian Medical Association Journal*, *194*(42), E1448–E1450. DOI: 10.1503/cmaj.220456 PMID: 36316022

Khokholkova, N. (2016). Worlds of color. *Asia and Africa Today*, *5*, 72–76.

Motsa, Z. (2004). *Khulumani Sive*. Macmillan.

Msibi, T. (2002). Incwadzi Lengenakheli. Manzini: Macmillan.

Mtumane, Z. (2014). The Significance of Ancestors as a Cultural Component in M. Mbambo's *Amakroza* (1992). *Southern Africa Journal for Folklore Studies*. Volume 24(1), pp 22 to 39.

Ndaba, Z. F. (2022). The Covid-19 pandemic and cultural event (s) cancellation in South Africa: views of the royal reed dance in KwaNongoma, KwaZulu Natal (Doctoral dissertation, Cape Peninsula University of Technology).

Ngubane, S. (2019). Death and burial practices in contemporary Zulu culture, South Africa. *Death across cultures: Death and dying in non-western cultures*, 119-131.

Nhlabathi, M. C. (2024). Kulondvolotwa kwemasiko, imihambo, nemagugu ebuve lokutfolakala etindzabeni letimfishane teSiswati letikhetsiwe. (Unpublished Masters dissertation University of South Africa.)

Nxumalo, G. (2014). Young African Women and the Umhlanga Ceremony: A Case Study in Historical Consciousness (Doctoral dissertation, University of KwaZulu-Natal, Durban).

Pato, J. S. J. (2018). Role Played by Swazi Literature in Preserving and Promoting the Culture of Swazi People. Unpublished Dissertation. Stellenbosch University.

Simelane, H. Y. (2013). The Interplay of Urban Land Tenurial Systems and its Effects on the Poor: A Case Study of Manzini in Swaziland. Unpublished PhD Dissertation. University of Sussex.

Simelane, L. (2014). Imvelaphi YemaSwati. Ezulwini: Sebenta National Institute.

Sone, E. (2010, August). Aesthetics from the Swazi Perspective: The Case of the Swazi Umhlanga/Reed Dance. *Journal of Music Research.*, *7*(1), 130–139.

The Constitution of the Kingdom of Swaziland Act, 2005.

Valera, E. (2016). Preserving the Tinguin Cultural Heritage. *Mother Tongue-based Multilingual Education as Tool.*, *2*(1), 79.

Vilakati, T. S., & Msibi, T. (2006). Giya Sigiye. Manzini: Macmillan.

Vilakati, T. S., & Sibanda, E. S. (1997). Chaza Ngive. Manzini: Macmillan.

ADDITIONAL READING

Makaudze, G., & Magongo, V. M. (2023). The dialectics of sport and history: Commemorative naming of selected stadia in the Kingdom of Eswatini. *International Journal of Language and Culture*, *10*(2), 191–208. DOI: 10.1075/ijolc.00050.mak

Mdhluli, N., & Mphasha, L. E. (2024). A Critical Analysis of the Swati Reed Dance. *Gender & Behaviour*, *22*(1), 22201–22208.

Shabangu, T. P. (2021). *A comparative study of Swazi and Zulu traditional plant use at Mkhondo and eDumbe, South Africa*. University of Johannesburg.

Takada, A. (2021). Emerging funeral culture in Swaziland in response to HIV/AIDS.

KEY TERMS AND DEFINITIONS

EmaSwati: People from a Kingdom of Eswatini (and some from South Africa) in the Southern Africa region.

Licansi: A grass mat.

Lusekwane: An event attended by the boy-child to keep him pure / a shrub used for traditional purposes during *incwala* (national prayer) ceremony.

Sicamelo: A wooden pillow.

Umgcwembe: A wooden dish used when dishing meat.

Umhlanga: Reed dance aimed at keeping the girl-child pure / reed tree.

Umngcwabo: Funeral procession.

Umtfwalo: A load that one carries using one's head.

Umtsimba: A traditional wedding / a bridal party.

INTERVEWEE IDENTIFIERS: SMI1 - Interviewee 1, SMI2 _ Interviewee 2, SMI3 – interviewee 3

Chapter 19
Safeguarding Swazi Way of Life:
Swazi Culture in Swazi Oral Literature

Jabulani Sabelo Jarreth Pato
Examinations Council of Eswatini, Eswatini

Remah Joyce Lubambo
https://orcid.org/0000-0003-1990-0984
University of South Africa, South Africa

ABSTRACT

The prime aim of this study is to give evidence that Swazi oral literature (narratives, poetry, drama, proverbs and idioms, and riddles) has been used to preserve and promote the culture of the Swazi people. Nowadays, Swazi oral literature has been written in a drive to safe-guard the culture of the people of the Swazi society. The traditional life of the emaSwati just like their oral literature has been passed through the generations orally. The study focuses mainly on Swazi oral literature that has traces of traditional life of the Swazi people. The genres of oral literature explored in this study are: prose, drama and poetry. Proverbs and idioms, and riddles with bits of Swazi culture are also discussed in this chapter. The chapter concludes by giving recommendations about other possible ways in which could be done to preserve and promote the culture of the Swazi people.

INTRODUCTION

Nie (2021) defines oral literature as the literature that is delivered by the word of mouth. Taking a step back, literature is lamely defined as work of art that is either written or spoken. It is expressed in a very delicate way as one voices out one's thoughts, concepts and ideas. Literature can be subdivided into two, oral literature and modern literature. In this chapter, focus is on the former. Kamera (2001) refers to oral literature as fictional texts which are composed, kept alive and passed on from generation-to-generation by word of mouth. With the changing times, oral literature has been put down to paper

DOI: 10.4018/979-8-3373-4124-8.ch019

for preservation and its promotion. In contemporary setting oral literature covers any volume of written and printed text, which also keeps the kind of literature (Okpewho,1992)

Vilakati and Sibanda (1997) concur with the above assertion that oral literature is also written, unlike before where people relied on transmitting it (oral literature) by the word of mouth. They define oral literature as:

> Lolu luhlobo lwemibhalo lengakabhalwa phansi kepha yendluliselwa etitukulwaneni iye kuletinye iye kuletinye ngemlomo. Kuloluhlobo lwemibhalo kukhulunywa ngetintfo betenteka kadzeni, kungaba yimilandvo nato tonkhe tinkhulumo tebuciko letingakabhalwa phansi. Kodvwa kuletikhatsi tanamuhla temdzabu setiyabhalwa phansi, loko kwentelwa kutsi titogcineka kahle netitukulwane letitako tifundze kutsi kadzeni kwakuphilwa njani nangempilo nje yesintfu.

This is the kind of writings that are not written down but have been passed through generations through the word of mouth. In this kind of writing mentioning is on what was happening in the past such as history and speeches that were expertly said. However, in the present times oral literature has been written, and this is done to preserve it for future generations so they learn how people in the past lived, and about the traditional life too.

This shows that oral literature can also be termed traditional literature in the society of emaSwati since it reflects on the traditional life of emaSwati. This kind of literature contains the history of the society which the people find pride in it. The life of the people was shared verbally through the generations then in the present times some have been put to paper so it (oral literature) is conserved. After all, Finnegan (1998) observes that Africa possesses both written and unwritten traditions. In other spheres, it is also referred to as folklore. Sone (2018) states that this kind of literature is delivered by the word of mouth and is aimed at the ear to move the whole body. It is also noted that oral literature (Africa's) has principles that however differ from one society to the other. This is because of the different cultures of the different societies.

LITERATURE REVIEW

Defining culture

Pato (2018) defines culture as a way in which a certain group lives their lives. This is all about the practises in their way of life. The shared patterns of behaviours are learned by socialisation. Ekwelen et.al (2007) defines culture as an extremely valuable heritage which has been descended down from generation to generation, and they further state that culture appears in all aspects of life which are emotional culture, exchange culture, business culture, oral culture etc. culture can be referred as public rules that we draw upon when we think and act(Patterson, 2014; Cerulo et al, 2021)This depicts that within a society there are many aspects involved, after all culture is shared behaviour of a group of people living together as a society. The treasured aspects of culture attained from a society's fore-fathers are kept within the generations.

The Swazi society is known for its richness in culture. Many societies are losing their way of living because of different aspects. Eswatini, on the contrary, still has aspects of culture that have been preserved. With the changing times, it is evident that the culture of emaSwati has changed. Some elements of the different traditions, customs, norms and beliefs are lost even though the culture itself would still have its traces. A quick example would be that of *umgcwabo* (funeral). In the ancient times, it was conducted in a certain way involving the bereaved family's ancestors. These days, not all families still believe in ancestors, but they still do certain rituals that have to do with the dead, for example, when walking the corpse to the gravesite, those carrying the corpse should have the feet-side leading the way not the head side. There is a belief that it should be that way to prevent bad omen of the person's spirit visiting the home. This shows that emaSwati still respect their fore-fathers way of living even though they have lost a lot in the way of doing things the Swazi way.

In the rural settlements of the Swazi society, the culture is still kept. According to the Swazi administration order, 1998, act of 1998, the royal chiefs are the loyal soldiers in the society seeing to it that cultural ways are not forsaken lest that annoy the ancestors (another cultural belief) who may then punish the living. Such people are the strongholds in the society of emaSwati who make sure that despite colonization, globalization, civilization, ignorance, religion and education taking centre stage (amongst other factors killing Swazi culture) the way of life of emaSwati is still intact and it does not change much from what was inherited from the country's fore-fathers. The richness of Swazi culture should be preserved and promoted, no wonder the traditional ceremonies such as *incwala* (national sacred prayer) and *umhlanga* (maidens' reed dance), to mention only two, draw throngs of people (tourists) from neighbouring countries and worldwide to witness the cultural events keeping the society together.

It is fundamental that the Swazi culture is kept intact. Valera (2016) believes that culture is important since it connects people to their past, and Kuli (2012) concurs with that and further explains that culture gives society members a connection to certain societal values, beliefs, religion and customs. Knowing one's roots is vital in life, and Valera points out that knowing one's origin is vital in life because it makes it easier for one to live according to certain and expected standards. Culture is defined as shared patterns and Spencer-Oatey, (2016) agrees with that as she states that culture offers communal support as the people in the society live together. With the importance of culture highlighted, it cannot be overemphasized that it is imperative that the culture of emaSwati be preserved and promoted so that emaSwati do not lose their identity and the people that they are. EmaSwati live and do many things together and this affirms what Kuli attests about culture being a communal support as members interact on daily basis.

Defining language

Language is the main means of human communication, which includes the use of words in a structured and traditional manner and can be expressed through writing, gesture, or speech. Language is another key element of culture. Language, just like culture, is endangered and should be well-kept. Pato (2018) states that languages are part of a culture of a society. He mentions that similarly to culture, language is inherited and transmitted from one generation to the next. Dlamini (2016) stresses the importance of safeguarding the cultural heritage (language) of the emaSwati and further states that:

Languages are who we are. Our mother tongues in particular are who we are.

She goes on to quote Belarus Minister of Foreign Affairs:

Language, traditions, national values are the foundation without which the existence of any nation and a state would be impossible.

Dlamini (2016) further alludes:

Mother tongue constitutes an essential part of an ethnic community. It is a carrier of values of knowledge very often in the practice and transmission of imperceptible cultural heritage. The spoken word is important in the enactment and transmission of virtually all intangible heritages, especially in oral traditions and expressions, songs and most rituals.

The above quotes are evidence that language is part of culture and should be safe guarded so one's sense of belonging is strengthened. From the assertions it is evident that a culture and language are the foundation and the reasons for the existence of a society. Moreover, language is the main tool that keeps society together. Language is the vital tool that carries the culture through the generations orally (spoken word). The written word is an advancement of keeping culture.

With that said, siSwati is the only indigenous language that is used by emaSwati. It perfectly transmits the culture of the people. The knowledge of the language enhances one to know the values, norms, beliefs and customs of emaSwati, and rituals performed by the members of the Swazi society. Preserving the society's only indigenous language is preserving the culture of emaSwati.

SiSwati should not be cut off in the Swazi society as it is the language that connects the members and the society. Dlamini (2016) likens cutting off a language as cutting off the umbilical cord which shall result in having social misfits and outcasts in the society. Without siSwati language the society would have lost their culture, tradition and sense of belonging. The writing of Swazi oral literature (and modern literature) in siSwati helps in preserving and promoting the culture of emaSwati.

THEORETICAL FRAMEWORK OF THE STUDY

This study employs Afrocentricity and Functionalism theories.

Afrocentricity

Chawane (2016) uses MK Asante definition who defines Afrocentricity as a manner of thought and action in which the centrality of African interest, values, and perspective predominate. He further defines Afrocentricity as an exercise in knowledge and a new historical perspective. He believes that Afrocentricity is a transformation of attitude, beliefs, values, and behaviour results, suggesting that it is the first and only reality for African people a simple rediscovery. Essentially Afrocentricity is a way of thinking and acting that prioritises African interests, values, and viewpoints (Gebremikael, 2022).

Objectives of Afrocentricity

Schiele (2016) says Afrocentricity, amongst three objectives he mentions, seeks to promote a worldview that will facilitate human societal transformation toward spiritual, moral, and humanistic ends and that will persuade people from different cultural and ethnic groups that they share a mutual interest in this regard. Karanga (1993) believes Afrocentricity speaks to specific liberation needs of people of African descent and the spiritual and moral development of the world.

This theory is relevant to the study in the sense that the issue of safeguarding the culture of emaSwati is of interest to the African society which is of great value to the people. The study is all about preserving the identity of the people which aligns with the objectives of Afrocentricity.

Functionalism

Malinowski (1949) argues that cultural practices / norms like tales have a function / role in society. Functionalism speaks to the utilitarian / practical function of cultural norms. Cultural values reflect way of life, attitudes & interests. This approach is applicable to the study in the as there are oral narratives which are part of the study. Those tales used safeguard the life of emaSwati showing their usefulness in the society.

METHODOLOGY

The study uses qualitative research approach since it sought to detail written description and explanation (Creswell, 2009). Qualitative research has its roots in cultural anthropology as it aims to address questions concerned with developing an understanding of the meaning and experiential aspect of human existence (Johnson & Christensen, 2017). Qualitative research is suitable in this study because focus is on details of different literature texts as well as the cultures embedded on them (Choongwa, 2018). Data was collected from a variety of textual sources, including databases, books, journals, reports, and publicly available materials. The researcher purposely sampled different literature books and analysed them in order to disclose the elements of culture, tradition, rituals and other practices found in the carefully chosen textbooks. Textbooks mostly used in schools were also purposively sampled for analysis (Alejandro & Zhao, 2024; Dawadi, 2020). Textual analysis method was used to analyse the selected data. This method of analysis describes the content, structure, and functions of the message contained in texts (Frey et al, 1999). The collected data was carefully organized and integrated into meaningful themes for manageable analysis (Karppinen & Moe, 2012). All findings were recorded down as the final step of the research process.

DISCUSSIONS

Significance of oral literature

The fact that oral literature was solely transmitted by the word of mouth and now it is written, and that is proof enough that it is very significant in the lives of different societies. Oral literature is very important. First and foremost, as it is said, it is transmitted through generations by word of mouth. Finnegan (1998) says that oral literature is by definition reliant on a performer who articulates it on a specific occasion. This is how the concepts, ideas, traditions and culture amongst other key societal values are carried through the generations. The values, concepts etc. are believed to be important that is why they are passed on. Knowledge and traditions are imparted by the oral literature.

In addition to that, Sone (2009) believes oral literature, like written literature, helps individuals to develop socially accepted forms of behaviour. The values passed through the generations are societal expectations. Good morals are part of the values expected from members of society to depict, so oral literature is used to teach that. Also, literature exposes generations to language. Oral literature comes in indigenous languages so natives are exposed to language of their own. With the changing times, culture losing its sting, language lost, oral literature (in native language) helps expose younger generations not only to the way of living of their fore-fathers but also to the language of their fore-fathers.

Just like modern literature, oral literature is a good tool for entertainment. Even though this kind literature is rich with societal values, it is told in light atmosphere usually around the evening fires (in African societies). Oral literature comes with fun. Fun is in oral literature different genres (to be discussed later). Okpewho (2012) states that oral literature presented by African writers advertise the greatness of their indigenous culture. One pens a better piece of what best understands. Again, the lack of knowledge on the African way of life makes the African writers teach other cultures how their lives were lived and ought to be lived for peace's sake. In a way, oral literature markets the rich culture of the African people.

The great area of focus in this chapter is proving that Swazi oral literature indeed preserves the culture of emaSwati. Oral literature, according to Sone (2011) and Sone (2009), is the most appropriate tool for reviving or promoting cultural heritage. The traditions, societal values, norms, and cultures were and are still passed by the word of mouth. Presently, through writing of these (fiction and non-fiction), they are passed, preserved and promoted for the present and future generations. Kamera (2001) concurs with Sone that oral literature passes culture and traditions since it covers all traditional verbal utterances which are either spoken, sung, chanted or recited because they are traditionally (and orally) composed, transmitted and preserved. In short, one cannot divorce culture from oral literature.

Sone (2011) quotes Brunvand (1978) who says:

Folklore is the traditional, unofficial, non-institutional part of culture. It encompasses all knowledge, understanding, values, attitudes, assumptions, feelings and beliefs transmitted in traditional forms by word of mouth or by customary examples. Many of these habits of thought are common to all human beings, but they always interact with and are influenced by the whole cultural context that surrounds them. Folklore manifests itself in many oral and verbal forms (mentifacts), in kinesiciological forms (customary behavior or social facts) and in material forms (artifacts), but folklore itself is the whole traditional complex of thought, content, and process which ultimately can never be fixed or recorded in its entirety; it lives only in performance or communication as people interact with one another.

The above assertion shows the importance of oral literature and the need to use it to preserve and promote culture (that of emaSwati included). Oral literature as stated encompasses knowledge, beliefs and values. It is what society is and this brings the importance of oral literature and culture in society. Oral literature relied on oral performance in the past as people in different societies interacted, but with the changing times it has assumed written forms as to be passed through the generations.

Oral literature also helps contemporary writers have something to base their contemporary writing on. Okpewho (2012) mentions that oral literature is studied together with modern literature because modern African writers borrow techniques and ideas from their oral traditions in constructing works dealing essentially with modern life. In that way, comparison is brought how the lives of the different generations have changed over time. Okpewho (2012) further notes that studying oral literature helps in finding answers about the nature of literature and culture. In the different writings many questions are answered on how the past generations have lived their lives and how they artistically find pleasure in their art. In such a way, cultures of different societies are compared and appreciated in their richness.

Oral literature in Eswatini

The coming up of a nation naturally means that it shall have its own culture. Since Swazi oral literature is manifested on Swazi culture, it means that the Swazi society had its own culture at birth. The genres of literature (poetry, prose and drama) have been there. They are intertwined with the culture of the Swazi society as culture is depicted and displayed in the oral literature of the emaSwati.

To clarify the above statement, oral literature is defined as the literature that is passed through the generations orally. That means the oral proses were shared through the generations verbally. The generations also shared the oral poetry (music and praises) in their gatherings and that way the societal values, history and many more aspects that were passed through the generations. Swazi drama comes in the form of traditional events such as *umhlanga, incwala, umtsimba,* amongst other cultural functions the society is blessed with. Those are not merely new events but were taken and adopted from the lives of the society's fore-fathers. This is adequate proof that oral literature for emaSwati existed when the nation was founded and raised.

Sone (2011) says that literature that existed in pre-colonial Swaziland was oral, and was sustained by the only indigenous language spoken in the country, and that is SiSwati. He further states that the oral literature took form of narratives (folktales, myths, legends) that were on the lives of the past heroes and the culture of emaSwati. Sone goes on to mention that oral poetry came in the form of songs and praises that were sung and recited, respectively, during traditional festivals and ceremonies. Kamera (2001) says oral literature in Swaziland is verbal art which is lived, experienced, produced and observed in every village in the country. Kamera (2001) elaborates that the type of literature (verbal art) is woven into cultural and social fabric of the society. With that said, it is vivid that oral literature in Eswatini has traces of the cultural life of emaSwati people, and the culture was presented well orally as it was not written anywhere. It was just lived in the society.

Sone (2011) views:

Through the form of literature, Swazi history and culture was passed on by word of mouth to future generations through oral tradition. The Swazi personality was therefore cultivated by bringing up men and women with a Swazi spirit who had respect for elders and traditional norms, and a deep knowledge of the love of the land as well as a deep veneration of the king.

This is oral literature that was verbally imparted to generations before the nation was colonized by the colonizers (Whites as history dictates). This tallies with one of the importance of oral literature which is that of the preservation of the country's history. Oral literature of the Swazi society preserves the extremely rich culture of emaSwati.

With the changing times, oral literature has been written. The various forms of oral literature have been stored in different books and this has been made easier for coming generations to come in contact with the traditional literature that is slowly losing its power. Writers who have put the oral literature on paper have greatly helped in preserving and promoting the kind of literature that was only relied the word of mouth to be passed through the generations. The writing of oral literature has in-turn helped with the preservation and promotion of the culture of emaSwati that its know-how has declined because of various factors such as globalization, civilization, Christianity amongst other factors. The written oral literature has helped the presently living generation and those generations to come to know their culture. It is preserving the culture of emaSwati.

Traditional/oral literature has been preserved in a number of textbooks that include Kamera's *Swazi Oral Literature* (2001), Vilakati and Sibanda's *Chaza Ngive* (1996), Vilakati and Msibi's *Giya Sigiye* (2006), Dlamini's *Sinkayinkayi* (1986), Magagula's *Tisho TesiSwati* (undated) and *Live Liyengcayelwa* (1987), Dlamini and Garb's *Bekukhona* (2008) amongst other materials shading light on the oral literature of the Swazi society. The above listed published textbooks have taken turns in the Swazi school syllabus in the different schools' grades in a drive to teach the oral literature to the present generations that lack the knowledge of the richness and abundance of Swazi culture.

Known genres of literature are poetry, prose and drama. Oral literature also has these genres but they come in different fashion. Vilakati and Sibanda (1997) state that oral/ traditional literature includes songs (ceremonial, dance, work) lullabies, children's rhymes/games, praise names and praise poetry. Looking closely at this, one may conclude that these fall under oral poetry since it all comes in verse rather than in dialogue (drama) or continuous unity (prose). Kamera (2001), on the other hand, mentions that oral literature covers all traditional verbal utterances which are either spoken, sung, chanted or recited which includes prose and verse narratives, poems and songs, drama and myths, riddles, proverbs and idioms. Kamera (2001) further states that all these:

...involve and are concerned with speaking, singing, listening and activity, and with speaking on a living culture. Retentive memory and skilful tradition bearer who take pride in transmitting the various forms of this art to their children.

The above assertion brings all the three known genres of literature (prose, poetry and drama). It shows that Swazi oral literature has both the performer and the audience, and its presentation involves the culture, traditions and norms of the generation and that shall be a cycle. This is how the generations get to be in touch with the culture in the oral literature or the oral literature that is part and parcel of the Swazi culture.

Swazi oral literature entails the lives of the people of the Swazi culture. Issues that involve nature and livestock (amongst other topics) are revealed in the oral literature. Topics such as marriages, deaths, love, disappointment and regrets, and praises poetry for kings and heroes are part and parcel of oral literature of emaSwati. Work, dance, ceremonial, and modern songs are amongst other kinds of oral poetry where traces of culture are depicted as they are evident in them.

Again, it should be noted that most of the Swazi oral narratives (myths, legends, and folktales) are about the lives of emaSwati. They depict the way emaSwati live their lives. This shows that oral literature, in general, mirrors the way of the lives of emaSwati, so it (oral literature) should not be discarded but should be made to be part and parcel through all generations of emaSwati as it reveals how the ancient Swazis lived their way and how the present generation should shape their lives not forsaking their indigenous roots.

As it has been said earlier, oral literature has been written for preservation's sake. Its writing promotes the knowledge, and more to that exposes the present and coming generations to the ways of the past. It cannot be overemphasised that the culture of oral literature is part of the culture of emaSwati and that is why the oral literature has traces of the Swazi culture, tradition, customs and beliefs. This is done to preserve and promote the culture of emaSwati.

It is true that Swazi oral literature entertains, gives information and knowledge but one cannot run away from the fact that it preserves the way of living of emaSwati. As it has been mentioned earlier, Swazi oral literature develops one socially. One is exposed on how one is expected to interact and behave in one's society. That is the culture of the people. Swazi oral literature comes in the indigenous language of emaSwati and its (language) is promoted. One must bear in mind that language is another tool that cannot be divorced nor separated from culture. In short, Swazi oral literature and Swazi culture are one and inseparable.

This chapter seeks to give evidence that oral literature has been used, and is one of the right tools to help preserve and promote the culture (way of life) of emaSwati. The writings of such shall help the present and coming generations to know how their forefathers lived. Despite the changing times of civilisation, globalisation and other religions taking centre stage in the lives of emaSwati, oral literature, especially that of displaying and depicting Swazi culture, traditional norms and customs, should be made readily available for generations to know and appreciate. After all, it is what emaSwati are.

The different genres of oral literature will be treated differently in the chapter. Below, the chapter will first look at the oral prose with elements of culture, then oral poetry with aspects culture, and oral drama with cultural practices. Riddles and proverbs rich in the culture of emaSwati shall follow suit. It will then briefly discuss modern literature with traces of oral literature. This will solely do to show that oral literature is still key in the writing of modern literature as it preserves and promotes the way emaSwati lived and should live their lives.

Swazi culture in Swazi oral prose

Prose, according to Abrams and Harphams (2009), is an inclusive term for all discourse, spoken or written, which is not patterned into lines (verse). Oral literature has oral prose that is narrative presented orally, only now that is presented in writing. However, it must be noted that the narrative contains traditions, values, beliefs, customs and rituals of the African Man, emaSwati, in this case. Narrative is a story, whether told in verse or prose, involving events, characters, and what the characters say or do (Abrams & Harphams, 2009). The focus on the oral part of literature, so the discussion is on oral narrative at this moment.

Vilakati and Sibanda (1997) state that oral narratives are:

Lolu luhlobo lwendzaba lendlulisela etitukulwaneni ngemlomo, iyatekwa. Isuka endvulo icocelwe situkulwane naso siyichubela kulesinye ngekuyiteka ngemlomo. Kodvwa-ke lamuhla sewutfola kutsi linyenti letinsimu tibhalwe phansi, loko kwentelwa kutsi titogcineka kahle nalabetako batitfole.

This type of narrative is passed verbally through the generations. It comes from the past and is told through the generations verbally. However, presently one finds a number of oral narratives written for preservation so the coming generations find them.

This shows that the transmission of oral narratives is of the spoken and written word. Part of the spoken word has been written, and this is done to safeguard the oral narratives. If the oral narratives are neglected the coming generations would not know anything about them. It shall be an unknown history. Kamera (2001) observes that Swazi oral narratives are:

...marked by a high degree of hilarity and the typically imaginative staple diet for children who enthusiastically identify with the characters who are either gulls or dupes in the narratives.

The above assertion shows that in oral narratives there is entertainment and amusement for the young ones and it helps develop the imaginative capabilities of the audience (children usually). The audience always identifies themselves with the oral narratives' characters. In this chapter, oral narrative, 'Khabonina' in Dlamini's *Sinkayinkayi* (1988) shall be used to identify an aspect of Swazi culture that has been preserved, and of course its writing also promotes it to present and future generations. The happening of *Umtsimba* (traditional wedding) is presented in the said oral narrative.

Dlamini (1988) narrates a folktale, 'Khabonina' who is a young character whose siblings have been murdered by her mother's co-wives upon birth out of jealousy. After her birth, Khabonina was taken away to be hidden from her father's other wives who were suspected of the killings. Polygamous families define the ancient emaSwati. Even nowadays, there are still traces of polygamous families in the Swazi setting and this shows that it has taken through the generations. Still on that, Khabonina was taken to live with her mother's family for protection's sake. In Swazi society, a child has two homes, one is paternal home and the other is the maternal home. If trouble arises at the paternal home, it is cultural to run away to your maternal home that is why Khabonina was whisked away. *"Wamtsasa wayomfihla ekhaya kubo."* (She took her to her own home to hide her there). This shows that Khabonina's mother resolved to hide her daughter at her maternal home for she knew that she was going to be protected over there. Even the name given to Khabonina shows that emaSwati preserve the naming of children and in a way appreciate the significance, history and meaning surrounding the name of an individual (sometimes relating to the family's forefathers).

"Khabonina" (belongs to your mother's family) and in this case the name given to the child shows that she was raised at her mother's place, and it is believed that a child follows his / her name that connects him / her to his / her family's ancestors, and Jayaraman (2005) asserts that the connotation attached to name plays a substantial role in the meaning of human self, since it is believed that a given name does not only serve as a social identity, but it also influences several aspects of human living.

Again, when the time is right, a Swazi child is expected to *'kugana'* (give love), and it is encouraged that one gets a lifetime partner from afar hence the siSwati idiomatic expression, *"Indvuku lenhle igawulwa etiveni"* (a good stick is fetched from afar). Khabonina gets herself a fiancé from afar, and in good time, she was ready to give her hand in marriage. Dlamini (1988:1) narrates:

Nasakhulile wagana khashane. Bamlungiselela kutsi ayokwendza ekhakhakhe... Bamcola sicholo sakhe saba sihle, naye wabamuhle.

When she grew old, she fell in love with someone from a distance place. They made necessary preparations. Her traditional basket-hat was well prepared. It was beautiful, and she was cute too.

This is evidence enough that the society expects that once one is old enough one should then give one's hand in marriage and the whole society cherishes and embraces this that is why members of the society play different roles in the fruition of the traditional wedding. Preparation prior the commissioning of *umtsimba* (traditional wedding cast) are made. *'Makoti'* (bride) should have *'sicholo'* (traditional basket-hat) as it is in the culture of emaSwati that married women should have their hair covered. Dlamini continues to give more on the Swazi way of life in relation to the customs and rituals surrounding the society's traditional wedding, and to the culture in general:

Uyise waKhabonina wahlaba inkhabi lenkhulu, wababita bonkhe labatakumpheketela Khabonina, nesidzandzane lesitawutfwala umtfwalo wamakoti wabayala bonkhana.

Khabonina's father slaughtered a big ox, he invited all those to accompany Khabonina, and the maiden to carry Khabonina's traditional load. He gave all of them words of advice on how to conduct themselves.

Waphuma-ke umtsimba...

The traditional wedding cast departed...

Before the traditional wedding cast leave the bride's home, the father of the bride should slaughter a beast to feed those that shall be accompanying his daughter. It is also food packs that they shall be feeding on along the way before they get to the groom's home. Amongst all the cast, there is also a maiden carry a traditional load on behalf of the bride which is also a significant indication that the bride is leaving her home have been given blessings by her father. Good conduct is also emphasized by the father or elders commissioning the traditional party. Once all necessary rituals have been performed the *umtsimba* then leaves the bride's homestead for the groom's homestead.

Umtsimba awubuyeli emuva lichilo lelo…

The wedding party never goes back, that's a bad omen…

The above quote is an indication that *umtsimba* once set out, it is not expected that they come back. They need to accomplish their mission and nothing should prevent them from achieving that feat. Even death in both families does not stop *umtsimba.* Shall that happen, then it is believed that it is a bad omen to the families involved no wonder even in the folktale the wedding party continued with their journey despite the trouble that befell the bride.Dlamini (1988) narrates:

Umtsimba wahamba wangenisa kulomunye umuti balala khona.

The wedding party walked a very long distance and spent a night in a certain homestead.

Ekuseni befika, bakhandza umtsimba se usemfuleni uhlabe sahhukulu.

When they arrived in the morning, the wedding party was at the river, having their goat (*sahhukulu*) slaughtered.

…wavunula sidvwaba sakhe…

…she dressed on her cow-hide dress…

…bahlabela wakhuphuka umtsimba…

…the wedding party sang and went up to the groom's home…

…kwagidvwa ekhakha Khabonina, bajabula bonkhe…

…there a dance at Khabonina's in-laws…

…nababuyela ekhaya yabikwa indzaba kuyise waKhabonina. Wayibita lendvodza leyamkhokha eweni, wayipha tinkhomo…

…when they went back to their home, they reported all that happened to Khabonina's s father. He called the man who had rescued Khabonina from the steep, and he gave him a herd of cattle…

The above citations are a true indication of the unfolding of the Swazi traditional wedding that its inscription is a way of preserving it, and definitely promoting it. These show some of the rituals and happenings of the *umtsimba* as a way of safeguarding the way of life of emaSwati. Going to the river for hours by the river is an essential part in the procession of *umtsimba.* Again, *sahhukulu* is the goat or cow that the bride's family (*umtsimba* in this case) is given to slaughter by the river (Dlamini and Masuku, 2013). This is evidence enough that oral prose is used to preserve and promote the life of emaSwati.

Kwachamuka indvodza igawula…

A man approached cutting wood.

The oral narrative also mentions the masculine chores in the Swazi setting. It is part of the men's responsibilities to see to it that there is log used to make a traditional fence around the homestead for protection of some sort. The narrator, by mentioning this, is teaching the coming generations men's chores expected to be performed by males. Oral prose is one way that has been used to preserve the culture of emaSwati. The prose is not only for entertainment, passing time, but it is used to instil societal expectations and is also used as a social control mechanism for the Swazi society. This is purely the way of living of emaSwati.

Swazi culture in Swazi oral poetry

Abrams and Harphams (2009) declare that oral poetry is composed and transmitted by singers or reciters. They go on to state that the recitation in the past days was sometimes accompanied by instruments such as harp, drum and other musical instruments. Kamera (2001) defines oral poetry as all important events whether personal or communal are announced, passed on and preserved, by and in poetic form. Kamera further states that births, marriages and deaths are surrounded by poetry.

Swazi oral poetry is blessed with different kinds of poems. The different kinds of traditional poems are: traditional songs (work songs, ceremonial songs and dance songs), praise names, lullabies, children's rhymes, games, and praise poems. According to Kamera (2001) traditional songs express hopes and aspirations of the Swazi nation. More to that, the songs are sometimes sacred and devotional, adulatory, satirical or condemnatory. Some songs carry the history of emaSwati, for example, *Mswati Uyinkhosi KaHhohho* mentions the elements of history of the Swazi royal kings. There are moral lessons that are acquired from the oral / traditional songs. Cultural aspects are also evident in the different kinds of traditional songs. The chapter will make use of one song. Vilakati and Sibanda (1997) writes:

Siyawufela kaNgwane We will die at the royal place
Solo mane siyahamba We are still walking
Singemasotja aDlamini We are soldiers
Solo mane siyandzindza We shall wander
Sindzindz' enkhosini yetfu. We will wander at our king.

'Kuhlehla' (paying allegiances) to one's leaders is also part of the culture. Swazis are expected to offer their services to royalty and to their chiefs, the oral poem (song) recited by male regiments assuring anyone who cares to listen that they will die loving the king. They say that they are forever king's soldiers and shall wander at their king's place. This shows that they have given up their lives to be the king's servants. This is respect to royalty and that is pure Swazi culture. The oral song written value the way of life and societal expectations of emaSwati.

Lullabies are also part of oral poetry, and the persona (babysitter) through reciting lullabies also reveals the way of life of emaSwati. Pato (2018) recites:

Uyawukhal' ukhal' udzinwe You'll cry and cry till you're tired
Uyawukhal' ukhake' udzinwe, You'll cry and cry till you're tired,
Unyoko ngob' akalimanga, Because your mother never cultivated
Watitjalel' ematsanga, She sowed just pumpkins,
Wahlal' esicelwini, She basked in the sun,
Welukana nemacansi, She was just making grass mates,
Echobana netintfwala, She crushed lice,
Tintfwala esidvwabeni, Lice on her cow-hide dress
Ephotsana nemiyeko, She was twisting her dreads,
Eyigcoba nemafutsa, Rubbing oil on them,
Yabats'iyamvus' inyoni, The bird tried waking her up,
Itsi: etikwemkhono, Saying: Go and work
Isho ihlet' esidvulini, Saying this seated on an anthill,
Watinabela emtfuntini, She just relaxed under the shade,
Watibuka esitfuntini, She looked her reflection in the shade,
Etilungiselela sicholo, She fixed her traditional-basket hat,

Lamuhl' uyatiphonsa nje, Today she is wandering,
Nak' indlal' ibhokile, There is starvation,
Likati lilal' etiko, There is utter no food,
Silala ngemanti, We only have water,
Ungatihluphi ngekukhala, Don't you dare cry,
Utibangela sibibitfwane, You'll have a hiccup,
Utawukhal' ukhal' udzinwe. You'll cry and cry till you're tired.

From the lullaby above, one gets to observe the societal values and traditional life of emaSwati as the persona vents her anger on the baby she is minding as the baby's mother appears irresponsible as she fails to provide food for the citing baby. The line *'Unyok' akalimanga'* (your mother never sowed) reveals the expectations of women as they have to cultivate their fields in support of the families. More to that, the traditional regalia of women in the society is also displayed with the mentioning of *'sidvwaba'* (cow-hide dress) and *'sicholo'* (traditional-basket hat). Again, the making of grass mates *'Welukana nemacansi'* is one of the ways in which woman is expected to do to bring income to her home, and the grass mates are what the family members uses to sit on when they are at the compound. Swazi food is also mentioned in the oral poem, *'ematsanga'* (pumpkins).

The beauty of the language is also preserved as siSwati (society's indigenous language) idiomatic expressions are used in the oral poem. *'Naku indlala ibhokile* (we are troubled by severe starvation), *'Likati lilala etiko'* (cat sleep at the fire-place), and *'Silala ngemanti'* (we only drink water to sleep) are the different lines that emphasize on the hunger and starvation the family is facing which has made the baby minder to vent her anger. The language she uses is deep and full of wisdom, which is the richness of the culture of emaSwati.

Oral praise poetry known as *'Tibongo netibongelelo'* in siSwati also help in the preservation and promotion of the culture of emaSwati. The Swazi society is blessed with oral poets that not only sing praises to the king, queen mother, heroes, but also to inhuman things such as cows, dogs, and to inanimate things such as rivers, trees, cars, buses to mention only a few. This is no wonder why Vilakati and Sibanda (1997) define oral praise poetry as traditional poems that are composed in praise of someone who is important in the Swazi community, and in admiration of inhuman and inanimate things that are held in high esteem for displaying unmatched characteristics. History of the Swazi nation is also mentioned in the praise poetry and is passed through the generations. Vilakati and Msibi (2006) in the second stanza of the king's praise poetry, *Tibongo TeNkhosi Mswati III* recites:

Kwaye kwetfuka bakaGama nebakaMnisi
Nangu naMantjolo umangele Wena waphakathi
Laye lana langakhawuli, Tati tatsi ngumzamcolo.
Wena-ke mbekwa kuphuma lilanga netindzaba talo
Lona latsi liyawushona kwabe kunhhiinhhitela tikhulu Kanye neMalangeni.
Juba lakutsatsa lakuyisa eNgiland'esikolweni
Unele kufika-ke wena Mswati wase uba yiNdlaluselwa
Wase uyayibamba impeniseli
Ngisho logwazana ngayo Mgwaz'egijima.
Bhube leyabhodla ngesheya kwetilwandle
Ingani kwatamatama live laMswati
Timbiba taphuma busakalala
Tingwe tacela empunzini

Lapha eMdzimba
Tatibona iNgwenyama eHlane yehlula umbutfo.
It shocked the Gamas and the Mnisi
Mantjolo is amazed your majesty
It rained without stopping, the said it was cyclone
You who was crowned when the sun rose with its news
It set with rumours amongst chiefs and royal members
The plane took you to a school in England
You held tight to the pencil
You stabbed / pierced using it
When you arrived, you became the king
The lion that roared overseas
And the country Eswatini shook
The rock-rabbits scattered
The leopards ran helter-skelter
At the Mdzimba (mountain)
They were seeing the lion defeating regiments at Hlane.

From the recited stanza of the oral praise poem about the reigning king, one may observe that history of the Swazi nation is put bare for the generations to know and appreciate. The 1982 cyclone that troubled the kingdom of Eswatini is made known, the inauguration of the king shocked the nation. The king was flown in the plane *'Lijubandzendzele'* as he attended school in England, the country's colonizers. Gama and Mnisi are key surnames that play a certain role that no one can do in the royalty of the Swazi society. Significant places in the society of emaSwati are mentioned in the praise poem. Mdzimba is the mountain where kings are put to rest, and Hlane is where the queen mother comes from, and it is a game reserve where another traditional drama, *Butimba* (hunting expedition) takes place where the male regiments display their hunting skills, and the king is said to have outsmarted the male regiments, *Tatibona iNgwenyama eHlane yehlula umbutfo.* Hunting is part of the culture of emaSwati that is practiced by males. The stanza contains information which is history of the society, and culture information that should be known by the country's generations that are still to come so they have less or no unanswered questions on the way of life of the Swazis i.e. why the Gama and Mnisi people play the significant roles in the royalty?

As it has been alluded above, praise poetry also entails composition of inanimate. Pato (2018) and Kamera (2001):

Jamludi Jamludi
Jamludi lobovu Jamludi who is red
Lonjenge ntolwane Who is like a 'ntolwane'
Luphondvo lolunjenge nhlavu yesibhamu Horn like a gun bullet
Usihlahla semdlebe You are a 'mdlebe' tree
Lesingahlalwa yinyoni On which birds never rest on
Yahlala khon'ibhamuka emehlo. It rests on it, its eyes burst.
Nguye Jamludi lobovu It is Jamludi the red one
Losacedze tinkunzi Who has defeated other bulls
Tebantfu netebelungu People's and Whites' bulls
Sigacamabhande You wearing ropes

Inkunzi bayayihlokolota A bull that is troubled

Bafana basePhayandane. Boys from Phayandane.

Only two stanzas of the long praise poem have been used to present the Swazi culture in oral literature (poetry in this case). The lives of emaSwati are displayed in the sense that it is brought to one's attention that cows are very important in the Swazi society as the 'bull' is praised for its unmatched qualities. Pato (2018) notices that emaSwati take pride in their livestock especially cattle as it proves the wealth of a man in praising the 'king' of the kraal. It must be noted again that there is tree that is important in the lives of Swazi boys who are still learning the trick of hunting (as mentioned above). *'Umdlebe'* is a traditional tree that is believe to 'blind' birds, so hunting boys skilfully direct the birds to the 'blinding' spot.

Pato (2018) hints that the lines, *Nguye Jamludi lobovu / Losacedze tinkunzi / tebantfu netebelungu* metaphorically mean that the country got its independence as Eswatini *(Jamuludi – Nguni* cow) was colonized by the Whites (*…tebelungu* - Brahman*)*. This shows that emaSwati freed themselves from the hand of the Europeans hence the praises afforded to the cow. In a way, such history is conserved and upheld for future generations to know and appreciate.

Inanimate objects, as mentioned above, are also praised poetically. An attachment to a material one uses regularly leads to one singing its praises for the utmost qualities, satisfaction and services it dutifully provides. Kamera (2001) pens a praise poem about sour milk (inanimate object):

YeloBhalaza! Sour Milk!

Kholwane kholwane kamanyatsela Kholwane of Manyatsela

Uphetseni ngemlomo? What is your mouth carrying?

Ngiphets'emasi ebantfwana I've children's sour milk

Ngihamba ngiyawahabula I sip all the way

Ngetindlela tabonina Following the mothers' paths.

Ngitsi yeloBhalaza! I say: Hey! Lobhalaza

Wena wentelani Why did you do that?

Kuba mhlophe To be white

Kantsi umnandzi Yet you are nice!

Etandleni! Etandleni! Etandleni! In hands, in hands, in hands

Vuleka Mphimbo Open throat

Kwehl' emasi khehle Let down sour milk, 'khehle!'

Sour milk is part and parcel of the Swazi food. Children rarely ate it. It was mainly used by adults unless the home had plenty of cows to milk. The cited oral poem is evidence that emaSwati identify themselves with inanimate objects that are part of their daily lives. Pato (2018) state that the provision of the milk (Swazi food) is another reason why cows are kept as domestic animals in the Swazi society. The line *'Ngetindlela tabonina'* means that women in the society has to take charge, lead and teach their children how to live their lives meeting societal expectations. The poem is not just about the enjoyment of the Swazi food but about upholding societal values. Furthermore, the line recited by the mother (persona), *"Etandleni! Etandleni! Etandleni!"* vividly displays how 'sour milk' should be eaten. The praise poem has been inscribed to help preserve the way of the life of emaSwati.

The different oral poems are also an indication that their recitations carry the traditional life of emaSwati. The songs, lullabies, praise poetry are part of the culture of emaSwati. More to that, the messages in the different kinds of Swazi oral poetry depicts how Swazis live their lives. Their inscription enhances their preservations which in turn help safeguard the simple life of emaSwati.

Swazi culture in Swazi oral drama

Kamera (2001) defines traditional drama as an ensemble performance for the community. It is not the known story of conflict unfolding in a dialogue between characters performed on stage before an audience – group of people watching. Kamera (2001) further views:

It is rather a purposeful enactment of aspects of life through verbal content, music, dance, song, scenery costume.

Traditional drama is more of an original drama that the known drama was made from. It is a purposeful performance that involves rituals that need to be done for the completion of whatever theme is intended. Swazi culture is blessed with traditional drama in the form of ceremonies such as *Incwala,* (national prayer), *Umhlanga* (reed dance), *Butimba* (hunting expedition), *Lusekwane* (sacred shrub), *Buganu* (drinking spree), *Umtsimba* (wedding), *Umngcwabo* (funeral) and many more. These are ceremonies and events that define emaSwati. Without such, Swazi culture is incomplete. The society of emaSwati comes together during such ceremonies. The setting of these types of traditional drama is either in family compounds, *sibaya* (kraal), *endlini kagogo* (grandmother's hut), *emathuneni* (gravesite, and the characters usually are the people who play a pivotal role in performing the necessary rituals or those that the traditional event is happening because of them.

In between the plot of the unfolding traditional drama, music, song, dance and poetry punctuates the plot. Most, in not all, happenings in the events are sacred and should be performed for proper completion of that particular event. The events should follow a certain order that has been laid bare to be followed, and that has been passed on through the generations by word of mouth. Elders (traditional leaders) are the ones leading and directing what should be done.

It should also be noted that the types of traditional drama have been diluted by other forces such as foreign religion and civilization (amongst other forces) and they are minus certain rituals. In other spheres of the Swazi society the performance of these events is still performed to the core. In the writing of oral literature, traditional drama is also mentioned as part of the narratives and poetry. So far, there is no writing where there is solely unfolding of the plot of the oral drama, like in the oral narrative discussed earlier, *Khabonina.* Traditional drama is also a possible feature in contemporary literature.

Swazi culture in siSwati proverbs / idioms

Swazi oral literature has another genre which is proverbs and idioms. Okpewho (2012) simply defines a proverb as a piece of folk wisdom expressed with terseness and charm. According to Mohammed et al (2009: 57) proverbs and idioms are figures of speech that make an utterance very interesting as there are words of wisdom presenting deep knowledge. Dlamini et al (2008) define proverbs and idioms as the power and technique of the word of mouth that has been passed from one generation to the next by the society's fore-fathers. The sharing of proverbs and idioms happens orally, and one must note that they (proverbs and idioms) contain the way of life of emaSwati. The fore-fathers would coin the words of wisdom observing the way of life of emaSwati, nature, livestock, history over time, and from other things (materials and items) used in the Swazi society.

Proverbs and idioms using foreign material foreign material to Swazi culture were very rare. Finnegan (1998) state that proverbs are a rich source of imagery and succinct expression on which more elaborate forms can draw. So, the use of known material makes it easier for one to understand the proverb though it may be a little bit challenging to give explanation unless engaging high level of comprehension. The proverbs and idioms are part and parcel of oral literature, and their usage safe guard the culture of emaSwati. Pato (2018) mentions that emaSwati are able to relate their state such as hunger using idiomatic expression:

Insika seyiwile – Sengilambile.

The pillar is fallen – I'm hungry.

Insika is the stronghold in the Swazi hut which is the most sacred hut in the Swazi society. Denotatively, it is the log that holds firm the hut. Without it the hut would fall. So, the coinage of the Swazi idiom springs from the lives of emaSwati that once one is hungry there is nothing one could do. That is the connotative part of the idiom. Pato further writes:

Atimane tiphelela – Ingcondvo yakhe ayikasi kahle.

They were not complete – His mind is not stable.

This proverb depicts that in the Swazi culture when a child is born there are certain rituals ought to be performed. Family elders need to find certain plants suitable or used by that family to see to it that nothing harms the new born. Pato (2018) goes on to mention that each family has its plants, trees and ashes, known as *tinyamatane* (wild life game), to use on their new born, so it would cause insanity if one family would miss a certain tree or plant whilst preparing the concoction to be burnt in the presence of the baby to inhale. If traces of insanity are evident, emaSwati came up with the above stated idiom as it reflects on their way of life. Most Swazi proverbs and idioms are culturally derived and their preservation is very important. Present and coming generations may not understand how their forefathers lived their lives, but through studying proverbs and idioms used in the society may play a pivotal role in having the know-how of the culture of emaSwati.

Swazi culture in siSwati riddles

The last genre of Swazi oral literature that is also used in preserving and promoting the culture of emaSwati is riddles. Riddles, according to Okpewho (2012) and Kamera (2001), are word games that intellectually exercise the mind, and they demand verbal skill when expressing them (riddles). Kamera further explains that the riddling session is characterised by questions, statements and puzzles posed by an individual who expects a response which is a solution to the posed riddle. Pato (2018) adds that other than keeping and passing leisure time, riddles also present the culture of emaSwati. In the siSwati riddles, the statements and questions are dominated by life of the members of the Swazi society. The riddling helps in developing the intellectually capabilities of the young ones as they reason deeply in finding the solution to the problem (riddle posed), and it must be noted that many of the riddles are culturally influenced indirectly preserving and promoting the culture of emaSwati. Pato (2018) riddles:

Question: *Ngikuphica ngetinkhomo tami letintsatfu, lebovu ihlaba lemnyama, lemnyama ihlaba lemhlophe, lemhlophe yeca sibaya.*

Question: I quiz you about my three cows. The red one gores the black one, the black gores the white which then escapes the kraal.

Answer: *Ngumlilo* (fire) *libhodo* (three-legged pot) and *lubisi* (boiling milk).

The riddle presents a mental picture that reflects the lives of emaSwati. Swazis and cattle are one and, in this riddle, cattle are used to drive a message across. The picture of a cow goring another cow, and that gored cow goring another cow is a familiar picture. A cow jumping off the kraal is another known sight in the Swazi society. Even the answer also presents the lives of emaSwati. Swazis used the natural fire when preparing food. The three-legged pot is still use in other quarters in the Swazi society especially in rural villages and during big ceremonies.

Another riddle used by Pato is that of the Swazi society preserving their royalty. The Swazi society has a king who rules over all the members of the society. Pato quizzes:

Question: *Ngikuphica ngenkhosi yami letsetsa kudzinwe kukhafule emabutfo.*

Question: I riddle you about my king who reprimands until regiments spit.

Answer: *Litulu nalidvuma bese liyana.* Thunder and rain.

As stated earlier, the society is ruled by a king. There are regiments who are the king's subject and are expected to abide by all the king's orders. When the king shouts, the regiments only shout *'Bayethe!'* (Long live). The king is the thunder, and the regiments is the rain that usually 'spit' after the thundering.

Through the two riddles used one may conclude that riddles are used to depict the culture of emaSwati. Their usage tricks the mind and as one tries to find the solution to the riddles one gets to know how emaSwati live their lives. The mental picture enhances the imagination of how the forefathers lives and how the present and future generations are expected to live life the Swazi way.

FUTURE RESEARCH DIRECTIONS

The research will serve as a hub of information on literature since almost all genres are discussed in these research. It will function as a hub for literature education. Scholars of literature will benefit from the study because it analyses language and culture as portrayed in many literary genres, as well as didactic proses such idioms and proverbs. Linguists and language developers will find it useful because it discloses language and culture for anyone interested in the literature genre. Teachers and students will use the findings and recommendations to expand research where there is currently a gap. Additionally, the language inherent in this literary form will be conserved for future language development.

CONCLUSION AND RECOMMENDATIONS

The chapter discussed the role that is played by oral literature in safeguarding the way of life of emaSwati. Traditional literature had been passed verbally through the generations but, of late the literature of the past has been written to advertise the richness of the Swazi culture in them. Few oral prose, oral poetry and traditional drama used in the preservation and promotion of the culture of emaSwati has been discussed in this chapter to ascertain that indeed the genres of literature play a pivotal role in the drive to safeguard Swazi culture. Proverbs and idioms, and riddles are other genres in the Swazi context that are part of the Swazi culture and most are derived from their context. With the Swazi culture diminishing over time, it is recommended that the modern literature be used to preserve Swazi oral literature and the culture of emaSwati so the coming generations know how their fore-fathers lived their lives. It is also recommended that the Swazi oral literature and the culture of emaSwati be digitalised so that natives find it whenever they want to know more about the lives of the dearly departed members of the Swazi society. This shall also help people from other cultures to know more about the abundance of the Swazi culture.

REFERENCES

Abrams, M. H., & Harpham, G. G. (2009). *A Glossary of Literary Terms* (9th ed.). Wadsworth Cengage Learning.

Alejandro, A., & Zhao, L. (2024). Multi-method qualitative text and discourse analysis: A methodological framework. *Qualitative Inquiry*, *30*(6), 461–473. DOI: 10.1177/10778004231184421

Cerulo, K. A., Leschziner, V., & Shepherd, H. (2021). Rethinking culture and cognition. *Annual Review of Sociology*, *47*(1), 63–85. DOI: 10.1146/annurev-soc-072320-095202

Choongwa, G. H. (2018). *Fundamentals of Applied Research Methodology*. Webster Print.

Dawadi, S. (2020). Thematic Analysis Aproach: A Syep by Step Guide for ELT Research Practitioners. *Journal of NELTA*, *25*(1-2), 62–71. DOI: 10.3126/nelta.v25i1-2.49731

Dlamini, B. S. (2015). *UmSamaliya Lolungile. Manzini*. Macmillan.

Dlamini, M. (2016). *Buciko Bemlomo. Manzini*. Imisebe Publishers.

Dlamini, M., & Masuku, N. (2013). *Emagugu EmaSwati. Manzini*. Imisebe Publishers.

Dlamini, O. S. (1988). *Sinkayinkayi. Manzini*. Longman.

Dlamini, P. A. (2016). NgiliSwati NgesiSwati (I am a Swazi because of siSwati: Preserving/safeguarding the collective identity/intangible cultural heritage of the Swazi community through our mother tongue. *Department of English and Literature*, 1-4.

Finnegan, R. (1998). *Oral Literature in Africa*. Oxford University Piers.

Gebremikael, L., Sicchia, S., Demi, S., & Rhoden, J. (2022). Afrocentric approaches to disrupting anti-Black racism in health care and promoting Black health in Canada. *Canadian Medical Association Journal*, *194*(42), E1448–E1450. DOI: 10.1503/cmaj.220456 PMID: 36316022

Jayaraman, R. (2005). Personal identity in a globalized world: Cultural roots of Hindu personal names and surnames. *Journal of Popular Culture*, *38*(3), 476–490. DOI: 10.1111/j.0022-3840.2005.00124.x

Johnson, R., & Christensen, L. B. (2017). *Educational Research: Quantitative, qualitative and mixed approaches*. Sage.

Kamera, W. D. (2001). *Swazi Oral Literature. Manzini*. Ruswenda Publishing Bureau.

Kuli, H.C. (2012). Literature Preserves Papua New Guinea (PNG) Cultural Heritage. L*iterature Heritage*, 1-7.

Nie, Z. (2021). Ethical literary criticism: Oral literature and the formation mechanism of brain text. *Kritika Kultura*, *1*(37), 20.

Okpewho, I. (1992). *African Oral Literature. Backgrounds Character and Continuity. Bloomington and Indiana Polis*. Indiana University Press. DOI: 10.2979/1034.0

Pato, G. (1985). *Umtsango. Manzini*. Swaziland Academic Services.

Pato, J. S. (2018). Role Played by Swazi Literature in Preserving and Promoting the Culture of Swazi People. Master of Arts Dissertation. Cape Town: University of Stellenbosch.

Patterson, O. (2014). Making sense of culture. *Annual Review of Sociology*, *40*(1), 1–30. DOI: 10.1146/annurev-soc-071913-043123

Sone, E.M. (2009). What Kind of Literature for Ethical Education in Africa? Lwati: A Journal of Contemporary Research, 6(1): 157-165.

Sone, E. M. (2011). Swazi Oral Literature Studies. Yesterday and Today, The Way Forward. *Southern African Journal for Folklore Studies*, *6*(1), 39–53.

Sone, E. M. (2014). Nature in Crisis and Ecological Construction and Conservations of the Environment in Swazi Oral Literature. *South African Journal of African Languages*, *34*(2), 195–205. DOI: 10.1080/02572117.2014.997056

Sone, E. M. (2018). African Oral literature and their humanities: Challenges of Prospects. *Humanities (Washington)*, *7*(30), 1–20. DOI: 10.3390/h7020030

Spencer-Oatey, H. (2012). *What is Culture? A Compilation of Quotations*. Global Pad Core Concepts.

Swazi Administration, (1998). (Laws Africa Legislation Common), Act 6 of 1998

Valera, E. (2016). Preserving the Tinguian Cultural Heritage. *Mother Tongue Based Multi-Lingual Education*, *2*(1), 79.

Vilakati, B., & Sibanda, E. S. (1997). *Chaza Ngive. Manzini*. Macmillan.

Vilakati, F., & Msibi, T. (2016). *Giya Sigiye. Manzini*. Macmillan.

ADDITIONAL READING

Baldwin, J. (2024). *Everybody's Protest Novel: Essays*. Beacon Press.

Djumaboyeva, M. (2022). In folklore-Oral Creativity and Traditionalism. Asia Pacific Journal of Marketing & Management eview Issn: 2319-2836 impact factor: 8.071, 11(12), 306-312.

Elmira, U., Omarov, N., Konyrbayeva, S., Jazdykbayeva, M., Orazbayeva, E., Tautaeva, G., & Eskaraevag, A. (2021). Forming cognitive activity in primary school students with samples of folk oral literature. *Cypriot Journal of Educational Sciences*, *16*(5), 2844–2857. DOI: 10.18844/cjes.v16i5.6377

Hawthorn, J. (2022). *Studying the novel. Bloomsbury Publishing. Thompson, A. R. (2022). The anatomy of drama*. University of California Press.

KEY TERMS AND DEFINITION

Safeguarding: Protect against injury or damage by taking the proper precautions.

EmaSwati: The people from the Swazi society.

Culture: A given people or society's beliefs, traditions, and social conduct are referred to as its culture.

Oral literature: A literary form that is transmitted orally-that is, passed down by spoken or sung traditions as opposed to being written down-is referred to as oral literature, orature, or folk literature.

Chapter 20
Religion as a Catalyst for Conservation of Sacred Spaces and Endangered Species Among the Shona People of Masvingo, Zimbabwe

Menard Musendekwa
https://orcid.org/0000-0002-6644-8727
Reformed Church University, Zimbabwe

ABSTRACT

Among the Shona people of Zimbabwe, religious and cultural beliefs significantly contribute to the conservation of natural resources and species. This paper explores how indigenous belief systems can be harnessed to protect the environment. The Shona worldview reveres certain trees, forests, springs, animals, and shrines as sacred, ensuring their conservation. This sacredness deters tampering and exploitation, promoting environmental preservation. By respecting these sacred places, the Shona people demonstrate a traditional yet effective approach to natural resource conservation.

INTRODUCTION

African traditional religions play a positive role in the conservation of sacred places and certain animal species, as seen in the Shona culture and belief system. These religions attribute psychic powers to the natural environment, viewing it as the abode of gods and goddesses, which in turn promotes their protection and encourages conservation. Despite the erosion of these traditional practices due to Western influence they remain crucial in preserving the environment by protecting sacred areas from exploitation (Eneji et al., 2016).

The indigenous belief systems, as demonstrated in the Shona worlds hold certain places and species sacred and encourage their conservation. These systems, which ascribe supernatural powers to specific parts of the environment, act as explicit protection areas from exploitation. Taboos and forbidden areas associated with worship further promote conservation (Muzondi, 2014). Despite the erosion of these

DOI: 10.4018/979-8-3373-4124-8.ch020

belief systems due to western cultural influence incorporating them into contemporary conservation efforts can foster sustainable use of natural resources.

The Shona people, among others, hold religious beliefs regarding certain trees, forests, springs, animals, and shrines as sacred, which contributes to their conservation. These sacred sites and entities are believed to house deities and ancestral spirits, making them untouchable and deserving of respect (Mahohoma, & Gundani, 2020). Consequently, the traditional taboo systems, including restrictions on access and extractive practices, emerged to protect these sacred spaces from pollution and exploitation. Although these cultural practices are evolving, the fear of divine reprisals continues to deter people from damaging sacred sites, thus ensuring their preservation.

The Shona culture and religious belief system in Africa place great importance on conserving sacred places and certain animal species. This conservation ethic, rooted in the belief that natural environments are the abodes of gods and goddesses is reflected in practices like protecting sacred forests, forbidden areas, and totemic objects (Eneji et al., 2012). However, the influence of Western religions and cultural practices has led to the erosion of these traditional strategies for natural resource conservation (Banda, 2022). African traditional religion, therefore, is not destructive to the environment but to a greater extend aims at conservation it. Among the Shona people of Zimbabwe exists beliefs and taboos that when properly utilized would be of importance in conserving the environment.

PROBLEM QUESTIONS

The study seeks to address several critical questions. Firstly, it aims to explore how sacralizing non-sacred species and spaces can enhance environmental conservation among the Shona people. This question addresses whether extending the sacred status to more species and areas could amplify conservation efforts and ensure broader environmental protection. Secondly, the research investigates the roles that religious and community leaders can play in promoting and enforcing conservation practices. Understanding the influence of these leaders could provide insights into how traditional authority and religious guidance can be leveraged to bolster environmental conservation. Finally, the study examines the effectiveness of the current legal framework in protecting sacred species and spaces, and identifies necessary improvements. This question seeks to assess whether existing laws are adequate and what changes could enhance their efficacy in safeguarding these vital ecological and cultural assets.

OBJECTIVES

To address these questions, the study has several key objectives. The first objective is to investigate the potential benefits of sacralizing non-sacred species and spaces for environmental conservation among the Shona people. This involves examining whether assigning sacred status to additional species and areas could lead to more comprehensive conservation practices and better protection of natural resources. The second objective is to assess the roles of religious and community leaders in promoting and enforcing conservation practices. This objective focuses on understanding how these leaders can influence community behavior and adherence to conservation principles, and how their involvement could strengthen conservation efforts. The third objective is to evaluate the effectiveness of the current legal framework in protecting sacred species and spaces and to recommend necessary improvements. This involves analyzing the current laws related to environmental protection and sacred spaces, identifying any gaps or weaknesses, and suggesting changes to improve their implementation and enforcement.

THEORETICAL FRAMEWORK

• Eco-theology

The relationship between religious convictions and environmental preservation is essential to sustainable activities. Indigenous Knowledge Systems provide efficient management techniques, while eco-theology encourages stewardship. According to the Sacred Ecology Theory, sacred locations serve as both spiritual havens and ecological preserves. When combined, these frameworks demonstrate how spiritual principles can motivate effective conservation initiatives. The study of eco-theology looks at how religious convictions and environmental preservation interact. Religious worldviews, especially those that emphasize stewardship, are said to influence ecological ethics and motivate conservation efforts (Habel & Balabanski, 2002). This viewpoint will aid in the analysis of the ways in which religious narratives impact Masvingo conservation practices.

• Indigenous Knowledge Systems (IKS)

IKS underscores that traditional knowledge is important for sustainable environmental management. According to academics like Berkes (2012), indigenous groups have long-standing ecological knowledge ingrained in their cultural customs. The holy prohibitions, totems (mitupo), and taboos of the Shona people serve as a check on over-exploitation of natural resources (Chirimuuta, 2021). According to this viewpoint, oral traditions and customary rules serve as a bridge between indigenous spirituality and conservation

• Sacred Ecology Theory

According to Fikret Berkes' (2008) Sacred Ecology Theory, sacred locations serve as both ecological reserves and spiritual spaces. Because access and use are restricted, these areas are hotspots for biodiversity. Sites like Great Zimbabwe and sacred forests serve as both conservation areas and cultural heritage sites for the Shona. This idea offers a framework for investigating the ways in which these locations' spiritual value aids in the preservation of threatened species and their environments.

LITERATURE REVIEW

The Shona people of Masvingo, Zimbabwe, rely heavily on religion to preserve sacred areas and threatened animals. Natural habitats and biodiversity are preserved as a result of the close connection between their indigenous religious beliefs and practices and environmental management. Several aspects of how religion serves as a catalyst for conservation in this community are examined in this review of the literature.

a) Indigenous Religious Beliefs and Environmental Ethics

The Shona cosmology, which gives sacred meaning to natural objects including trees, rivers, mountains, animals, and birds, intricately combines environmental ethics with spiritual beliefs. Because some places are thought to be the homes of ancestor spirits, maintaining them is a spiritual obligation, which encourages sustainable environmental management methods (Mabvurira et al., 2021).

The interdependence of all living things and the environment is also emphasized by the Shona philosophy of ukama, which strengthens the community's duty to preserve ecological equilibrium. This holistic ethic encourages cooperation with nature by examining each person's well-being in light of the overall cosmic existence network (Mawere, 2013a).

Furthermore, the Shona place a high value on environmental preservation through customs like taboos. By controlling how people behave toward the environment, these taboos safeguard wildlife, vegetation, and water sources, ensuring that natural resources are used sustainably (Chemhuru & Masaka, 2010).

b) Role of Sacred Groves and Forests

Locally referred to as "rambatemwa," sacred groves are forest sections that have been protected because of their spiritual value. These places are protected by local customs and traditional authorities who enforce laws that forbid hunting and tree-cutting. These methods have successfully preserved the groves' biodiversity, making them home to local plants and animals. These holy forests, which combine cultural legacy with environmental preservation, are essential to Zimbabwe's natural and cultural landscape (Byers et al., 2001).

Sacred groves have significant cultural and historical significance in addition to their ecological significance. For example, the UNESCO World Heritage site Matobo National Park is well known for its many rock art monuments, some of which are over 2,000 years old. These locations, which are frequently found inside sacred groves, highlight the close relationship between cultural practices and environmental preservation by offering insights into the everyday life and spirituality of the early occupants (Matobo National Park, n.d.).

Furthermore, by supplying fruits, medicinal plants, and other non-timber forest products, these groves are essential to the survival of nearby people. Traditional knowledge systems frequently govern the appropriate use of these resources, making sure that extraction doesn't jeopardize the groves' natural integrity. This mutually beneficial relationship demonstrates how well traditional conservation methods support community livelihoods while fostering biodiversity (Mawere, 2013a).

c) Totemic Systems and Species Protection

Traditional religious beliefs have been used to preserve mangrove ecosystems in Benin. Zangbeto and other voodoo deities are called upon to declare specific mangrove regions sacrosanct, which forbids logging, fishing, and hunting. This sacralization has been successful in protecting these important ecosystems since people's fear of divine vengeance frequently overcomes laws (Le Monde, 2024).

Totemism also functions as a conventional approach to environmental preservation. These activities contribute to environmental protection by designating more than 100 plant and animal species as totems among tribes in Uganda, Congo, and the Central African Republic (Wilderness Safaris, 2015). By lessening the strain of hunting on particular species, this method helps to preserve them.

d) Traditional Leadership and Environmental Stewardship

Chiefs and spirit mediums are examples of traditional leaders who are essential in upholding cultural standards pertaining to environmental preservation. In addition to making sure that taboos and rituals that encourage sustainable engagement with the environment are followed, they supervise the safeguarding of sacred locations. In order to preserve ecological equilibrium among their populations, their authority is essential. These leaders play a crucial role in tying cultural customs to environmental preservation initiatives, according to Byers et al. (2001).

Mangrove ecosystems in Benin have been preserved by utilizing indigenous religious beliefs. Certain mangrove areas are designated as sacred by calling upon voodoo deities, like Zangbeto, which forbids logging, fishing, and hunting. Because the dread of heavenly vengeance frequently surpasses legal restrictions, this sacralization has been successful in protecting these important ecosystems (Le Monde, 2024).

e) Integration of Traditional Practices with Modern Conservation Efforts

Integrating indigenous conservation methods with contemporary environmental strategies offers a holistic approach to sustainable environmental management. Recognizing the value of traditional beliefs and practices can enhance biodiversity conservation initiatives, leading to more culturally appropriate and effective outcomes. Byers et al. (2001) suggest that integrating traditional knowledge systems with modern conservation efforts can result in more sustainable environmental management practices.

Recent case studies highlight the effectiveness of such integrations. For instance, in Benin, the preservation of mangrove forests has been significantly enhanced by incorporating voodoo traditions, where deities like Zangbeto protect these ecosystems. This cultural approach has led to the expansion of mangrove areas and improved biodiversity conservation (Le Monde, 2024). Similarly, the Miccosukee Tribe in Florida has combined traditional ecological knowledge with modern environmental strategies to protect the Everglades, demonstrating the potential of such collaborations in addressing climate change impacts (Associated Press, 2025).

Furthermore, scholars like Hernandez (2022) advocate for the decolonization of conservation efforts, emphasizing the importance of integrating Indigenous science and perspectives into environmental management. This approach not only acknowledges the deep-rooted connection indigenous communities have with their lands but also leverages their time-tested practices for ecological sustainability. By embracing both traditional and contemporary methods, conservation initiatives can become more inclusive and effective in preserving biodiversity.

f) Impact of Christianity on Traditional Conservation Practices

Incorporating traditional conservation methods with modern environmental approaches enhances cultural identity while promoting sustainable resource management. Biodiversity has always been protected in Zimbabwe thanks to the Shona people's traditional practices, which include maintaining sacred forests and following totemic systems. But these native customs have been both challenged and enhanced by the arrival of Christianity. Despite the decline of some historic practices, there is an increasing chance for interfaith discussion to work together to address ecological issues and use common moral values to encourage environmental care (Beyers, 2022).

Beyers (2022) examines how Christianity and Shona religion interact, emphasizing both areas of conflict and areas where ecological conservation efforts might work together. In a similar vein, Mugambi (2001) stresses the significance of combining Christian theology with indigenous knowledge in order to promote environmental responsibility. The potential of interfaith activities in advancing environmental sustainability is further illustrated by groups such as GreenFaith (2023), which show how various religious traditions may come together for ecological concerns. Communities may create more culturally relevant and successful conservation plans by incorporating both traditional and modern approaches, guaranteeing the preservation of the environment and cultural heritage.

g) Case Studies of Sacred Sites in Masvingo Province

Certain sacred locations in Masvingo Province, like mountains, rivers, and woods, serve as prime examples of how Shona religious beliefs and environmental preservation connect. Because of their spiritual value, these locations are frequently preserved, which helps to preserve rare species and natural environments. For example, the Shona people consider the Njelele Shrine, which is situated in the Matobo Hills, to be a sacred place for rainmaking. Its spiritual significance has helped to preserve the environment, emphasizing the link between ecological management and religious rituals (Bhebhe, 2019).

However, the breadth of Shona knowledge on sacred sites is not well conveyed by the previous discussion. Natural components are given spiritual meaning in Shona cosmology, which is a rich tapestry of beliefs. Ancestral spirits are thought to reside in mountains like Malindidzimu, which translates to "Hill of the Ancestral Spirits," and are an essential part of the Shona people's spiritual environment. These locations and the biodiversity they sustain are preserved as a result of the deep regard for nature that this belief system cultivates (Mubaya, 2020).

Conservation efforts are further strengthened by the Shona custom of linking clans to certain animal totems. By representing a clan's ancestry with specific animals, these totems encourage a sense of duty to preserve these species. Through common ethical values, this cultural mechanism not only protects biodiversity but also fortifies ties within the community (Chiwara, 2015).

h) Challenges to Traditional Conservation Methods

The effectiveness of ancient religious rituals in preserving sacred areas and species among the Shona people is seriously threatened by modern issues including urbanization, globalization, and the spread of Western ideals. Indigenous beliefs are frequently marginalized as a result of the intrusion of Western norms, which lowers respect for holy natural locations and diminishes traditional conservation efforts. Mabvurira et al. (2021) address how Western hegemony and the penetration of Christianity have altered African traditional beliefs and behaviors, affecting environmental conservation efforts. Similarly, research from Southern Nigeria shows that traditional forest protection methods have been abandoned as a result of exposure to new information through education and civilization, which has contributed to environmental degradation (Okoji, 2001).

Adaptive approaches that respect cultural heritage and embrace innovations that encourage environmental stewardship are required to address these issues. Biodiversity has been successfully preserved by integrating indigenous knowledge into contemporary conservation techniques. As an illustration of the possibilities of fusing traditional beliefs with modern environmental initiatives, the integration of voodoo customs with mangrove protection in Benin has resulted in notable ecological restoration (Le Monde, 2024). Communities may create conservation plans that are both environmentally sound and culturally appropriate by encouraging interfaith communication and respect for one another. This will guarantee the preservation of holy places and the biodiversity they nourish. Raising interfaith understanding and integrating traditional knowledge with contemporary conservation techniques will help the Shona in Masvingo, Zimbabwe, become more environmentally conscious.

METHODOLOGY

The Shona people of Masvingo, Zimbabwe, use religion as a catalyst for the conservation of sacred sites and endangered animals. This study uses a qualitative research approach, including a literature review. The study explores the relationship between traditional ecological knowledge, religious beliefs, and environmental protection in the Shona society using an interpretive qualitative methodology.

The process of gathering data involves a thorough examination of secondary sources, such as academic literature on African indigenous religious beliefs and how they affect conservation efforts (e.g., Mavhunga, 2017; Chanza & de Wit, 2016); historical records describing traditional Shona practices regarding sacred spaces and wildlife conservation (e.g. Beach, 1994; Ranger, 1999); legal and policy documents like Zimbabwe's Environmental Management Act (Government of Zimbabwe, 2013) and international frameworks on indigenous conservation techniques (United Nations, 2019); and environmental reports from institutions like the Zimbabwe Parks and Wildlife Management Authority and UNESCO regarding the condition of sacred groves and endangered species in Masvingo (e.g., Matowanyika, 1991; Muguti & Mapira, 2014).

In order to classify data into categories like religious motivations for conservation, community-based environmental stewardship, and conflicts between traditional and modern conservation measures, thematic analysis is used to find recurrent patterns and themes. Acknowledging indigenous knowledge systems as valid sources of environmental wisdom, avoiding plagiarism, and maintaining the integrity of sources are the main ethical considerations. One significant drawback is the use of secondary sources, which would not account for recent changes in community attitudes toward conservation. To improve the results, ethnographic fieldwork and interviews with local guardians of sacred areas could be incorporated into future studies.

THE FINDINGS

The Shona People of Zimbabwe

The Shona people are an ethnic group in Zimbabwe who constitute about 85% of the Zimbabwean population (Tatira, 2004). Although they are generally found in most parts of the country, they are predominantly located in the provinces of Masvingo, Mashonaland West, Mashonaland East, Mashonaland Central, Manicaland, and parts of Midlands. Before 1890, the term "Shona" was not used to describe these people. Ranger (1985) contends that before this period, no one identified themselves as Shona. The term was coined by Doke (1931) to describe a group of people who spoke mutually intelligible dialects, namely Ndau, Manyika, Korekore, Zezuru, and Karanga. Doke then called the language they speak Shona.

The Shona people's history and culture are rich and complex. According to Bourdillon (1976), the Shona are known for their agricultural practices and intricate social structures. Their traditional religious practices are centered around the veneration of ancestors, which plays a crucial role in their cultural identity and community cohesion (Gelfand, 1973).

The colonization of Zimbabwe significantly impacted the Shona people, leading to the restructuring of their social, economic, and political systems. The introduction of Western education, Christianity, and colonial governance altered the traditional ways of life. As noted by Beach (1994), the colonial period brought about significant changes in land ownership and usage, which had lasting effects on the Shona's agrarian lifestyle.

In contemporary Zimbabwe, the Shona continue to play a vital role in the nation's cultural and political landscape. Their language, Shona, is one of the official languages of Zimbabwe and is widely used in education, media, and daily communication (Chimhundu, 1992). Despite the challenges posed by modernity and globalization, the Shona have managed to preserve many aspects of their traditional culture, adapting them to fit the contemporary context.

Research area

The province of Masvingo and the nation of Zimbabwe are depicted on the maps below. Masvingo, situated in the southeast of the nation (figure 1), is bordered by Matebeleland South Province to the southwest, Midlands Province to the northwest, Manicaland Province to the northeast, and Mozambique to the southeast (Chigumira, 2003). The province's tall savannah grass becomes green throughout much of the summer when it receives heavy rains (Vincent & Thomas, 1960). This seasonal rainfall results in lush vegetation during the wet season, which typically lasts from November to March (Moyo et al., 1993).

However, the region is highly vulnerable during the dry season, which runs from August through late November each year. During this period, water scarcity becomes a significant issue, affecting both agriculture and daily living conditions (Matarira et al., 1992). The contrast between the wet and dry seasons underscores the challenges of managing water resources and sustaining agricultural productivity in Masvingo (Manzungu, 1999).

The climate of Masvingo, like much of Zimbabwe, is characterized by its variability, with periods of drought and erratic rainfall patterns becoming more common due to climate change (Unganai, 1996). This has significant implications for food security and the livelihoods of the local population, who largely depend on rain-fed agriculture (FAO, 2006). Efforts to mitigate these challenges include the development of irrigation infrastructure and the promotion of drought-resistant crop varieties (Nyamapfene, 1991).

Figure 1. Masvingo province situated in Zimbabwe

Zimbabwean Laws on Conservation

Before dedicating time to the contribution of religion to the conservation of sacred spaces and natural resources in Zimbabwe, it is critical to consider the laws that govern the protection of endangered species, heritage sites, and the conservation of natural resources. The Constitution of Zimbabwe Amendment (No 20) Act 2013, Chapter 4 Part 2 Section 73 (1)(b), states that every person has the right "to have the environment protected for the benefit of present and future generations, through reasonable legislative and other measures that i) prevent pollution and ecological degradation; ii) promote conservation; and iii) secure sustainable development and use of natural resources while promoting economic and social development" (Constitution of Zimbabwe, 2013). This legal framework underscores the obligation of every citizen to ensure environmental protection for the benefit of both present and future generations.

Environmental pollution and degradation are explicitly prohibited, reinforcing the importance of sustainable practices. Religious communities, deeply rooted in local communities, can play a vital role in environmental conservation. Chapter 4 Part 1 Section 60.4 of the same Constitution stipulates that "any religious community may establish institutions where religious instruction may be given, even if the institution receives a subsidy or other financial assistance from the State" (Constitution of Zimbabwe, 2013). This provision suggests that the state might have an obligation to subsidize gatherings related to instructions on the conservation of natural resources, given that religion is communal, institutional, and instructive.

Considering the Zimbabwean constitution, it is everyone's obligation to conserve and protect the environment, including the protection of sacred species and spaces. At this juncture, it is pertinent to explore how religious beliefs among the Shona people can promote the conservation of sacred spaces and species. Religious beliefs among the Shona people emphasize the sanctity of nature, which is integral to their cosmology. Traditional Shona religion holds that certain natural features, such as mountains, rivers, and forests, are sacred and inhabited by spirits (Gelfand, 1962). These beliefs contribute to the conservation of these areas, as they are protected and preserved for religious reasons. The concept of "Zvikwambo" (sacred groves) is an example where specific trees or forests are left untouched due to their spiritual significance (Chimuka, 2001).

Moreover, the Shona people's reverence for ancestors (Vadzimu) also fosters environmental stewardship. Ancestors are believed to reside in natural sites and can bless or curse the living based on their treatment of these sacred spaces (Daneel, 2001). This belief system encourages the preservation of the environment as a form of respect and appeasement to the ancestors. Religious institutions in Zimbabwe have also been active in advocating for environmental conservation. For instance, the Zimbabwe Council of Churches (ZCC) has engaged in various environmental campaigns, highlighting the theological basis for caring for creation (Chitando & Togarasei, 2010). These initiatives demonstrate the potential for religious communities to contribute significantly to environmental conservation efforts.

Shona religious beliefs

Shona religious beliefs are deeply intertwined with their agricultural and natural life. Before consuming vegetable leaves at the onset of the rainy season, it is customary to express gratitude to the spirits of the ancestors. This practice reflects the Shona religion's emphasis on maintaining a close connection with nature (Gelfand, 1970). Such a profound connection proposes the reverence of nature, which informs various prohibitions and the sacralization of different sites, species, and the environment (Bourdillon, 1987; Daneel, 1970).

The Shona people believe in Mwari, the supreme being, who is seen as the ultimate creator and overseer of the world. Ancestor worship plays a significant role, with the spirits of the deceased considered intermediaries between the living and the divine (Bourdillon, 1987). These ancestral spirits are believed to protect their descendants and ensure the fertility of the land. Rituals and ceremonies are a vital part of Shona religious practices. For instance, the Bira ceremony is performed to communicate with the ancestors and seek their guidance and blessings. During these ceremonies, offerings such as beer and food are made to the spirits (Lan, 1985). These practices highlight the Shona's belief in the interconnectedness of the spiritual and natural worlds.

The reverence for nature among the Shona is evident in the way certain sites and species are considered sacred. Forests, mountains, and rivers are often seen as dwelling places for spirits and are thus protected and preserved (Chimuka, 2001). This sacralization of the environment promotes conservation and sustainable use of natural resources, reflecting a deep ecological consciousness. The Shona people believe that the spirits of their ancestors, known as "vadzimu," play a crucial role in their everyday lives, guiding agricultural activities and ensuring harmony with the environment (Chavunduka, 1977). This spiritual framework not only dictates agricultural practices but also emphasizes the importance of environmental conservation. For instance, certain trees and animals are considered sacred and are protected as a way of honoring the spirits (Kahari, 1990).

The reverence for nature among the Shona is evident in their rituals and taboos. Specific sites such as mountains, rivers, and forests are often regarded as sacred spaces where spirits dwell. These areas are protected from exploitation and serve as conservation zones, demonstrating a deep ecological awareness rooted in spiritual beliefs (Daneel, 2001). This sacred relationship with the environment highlights the integral role of Shona religious beliefs in promoting sustainability and ecological balance.

Makamure and Vengesai (2015) refer to the Karanga people, a subgroup of the Shona, who hold a deep belief in the sacrality of the land. The Shona people, in general, consider the land to be the backbone of the ancestral spirits, carrying both nature and humanity. This belief is rooted in the understanding that the land belongs to the ancestors, as it bears their remains, especially their graves (Makamure & Vengesai, 2015). The Karanga people's reverence for the land is mirrored across other Shona subgroups. For instance, Bourdillon (1987) explains that the land is seen as a living entity, imbued with the spirits of the ancestors, who are regarded as the guardians of the community's well-being. This connection is not merely symbolic but manifests in the rituals and taboos that govern land use and environmental conservation.

In Shona cosmology, the land's sacrality is also linked to agricultural practices. Gelfand (1970) notes that before any agricultural activities commence, rituals are performed to honor the ancestors and seek their blessings for a fruitful harvest. This practice underscores the belief that the ancestors' spirits influence the fertility of the land and the success of the crops. Furthermore, Lan (1985) highlights that during the Bira ceremony, which is a key ritual among the Shona, offerings are made to the spirits to ensure their continued favor and protection over the land. These ceremonies reinforce the idea that the land is a sacred space, intimately connected with the spiritual and physical well-being of the community.

The belief in the sacrality of the land extends to environmental conservation. Chavunduka (1977) and Chimuka (2001) discuss how certain areas, such as forests, rivers, and mountains, are designated as sacred and protected from exploitation. These areas are often associated with ancestral spirits, and their conservation is seen as a way of maintaining harmony with the natural and spiritual worlds. Thus, the Karanga people, like other Shona subgroups, uphold the sacrality of the land through various cultural and religious practices that emphasize respect for the ancestors and the environment. This deep-seated belief system illustrates the Shona people's holistic view of nature, humanity, and the spiritual realm.

The indigenous belief systems in Zimbabwe significantly contribute to the conservation of natural resources and species. The Shona people of Zimbabwe have various religious and cultural beliefs which, when properly utilized, can lead to the conservation of natural resources and species (Chavunduka, 1977). There are geographical places and animals which are regarded as sacred among the Shona people. Regarding places and species as sacred protects them from being harmed or destroyed by people (Chimuka, 2001).

Religious belief systems are so strong among many Shona people that once something is considered sacred, it is taken as such (Bourdillon, 1987). For example, sacred forests, rivers, and mountains are preserved due to the belief that they are dwelling places of spirits (Daneel, 2001). Similarly, certain animals, like the lion and the python, are often revered and protected because they are believed to have spiritual significance (Gelfand, 1970).

The conservation benefits stemming from these beliefs are evident in the ways the Shona manage their natural environment. Sacred groves and other protected areas serve as conservation zones, preserving biodiversity and maintaining ecological balance (Bourdillon, 1987). The deep reverence for nature ingrained in Shona spirituality fosters a strong community commitment to environmental stewardship.

Rivers and Pools

Among the Shona, where wells are not protected, there are taboos that discourage adventurous children from polluting them through the use of taboos (Tatira, 2000). It is a taboo to urinate in a well. If an individual urinates in a well, there is a belief that such an individual would contract bilharzia. Because everyone needs good health, no one would desecrate the water source, thereby preventing illness (Mawere, 2013b). Additionally, Makamure and Vengesai (2015, 12) state that the perpetrator "will be struck by diarrhea which will only stop when the offender has consulted a traditional healer and payment done to appease the ancestors. The payment can be in the form of a goat which will be ritually killed and dedicated to the ancestors so as to solicit forgiveness."

These taboos and beliefs serve a crucial function in protecting water sources and ensuring the health of the community. Mawere (2013b) notes that the consequence of polluting water sources could "cause health problems to the entire community who use the urinated water for domestic purposes, it means that people were obliged to avoid vicious characters that may result in ill health" (Mawere, 2013, p. 8). This system of traditional beliefs and taboos effectively regulates behavior and promotes the sustainable use of natural resources.

Moreover, the reverence for water bodies among the Shona extends to rivers and pools, which are often considered sacred. These water bodies are believed to be inhabited by spirits, and any form of pollution is seen as a direct offense to these spirits (Chavunduka, 1977). This spiritual belief ensures that rivers and pools are kept clean, supporting both environmental conservation and public health. The role of traditional healers and rituals in addressing pollution offenses highlights the interconnectedness of health, spirituality, and environmental stewardship in Shona culture. By adhering to these taboos, the Shona people not only protect their water sources but also maintain social harmony and respect for ancestral traditions (Bourdillon, 1987; Gelfand, 1970).

Chemhuru and Masaka (2010) reveal that the Shona community heavily depends on open wells, springs, and rivers for domestic, agricultural, and industrial use. Fetching water with a sooty pot is prohibited. It is believed that once a dirty container, like a sooty pot, is used to fetch water, the well will dry up (Chemhuru & Masaka, 2010). This belief ensures that water sources are kept clean and free from pollutants, promoting environmental sustainability (Mawere, 2013b).

Makamure and Vengesai (2015) refer to Shona people including pools, water lands, and springs among sacred sources that should be approached carefully. "Pools like Gonawapotera, Mupwanezi, Rupati, and Chirinda springs and wells like Vadore, Chimusvina, and Negari are thought to be inhabited by Njuzu (mermaids) spirits. They are also thought to be inhabited with other different spirit forces" (Makamure & Vengesai, 2015, p. 12). These beliefs serve to protect these water sources from contamination and overuse, ensuring their conservation.

Chemhuru and Masaka (2010) further explain that Shona environmental awareness takes a spiritual dimension, as ancestral spirits (midzimu) are said to be the custodians of nature. Therefore, they have a conscious interest in the way the living interacts with the environment. It is believed that the environment, including water sources, should be treated with respect since their misuse may provoke ancestral spirits who may in turn punish the human community with droughts and floods (Chemhuru & Masaka, 2010; Chavunduka, 1977).

There are specific beliefs regarding the Shashe River in the Chirumanzi area in the Midlands province of Zimbabwe. There are some pools where it is prohibited for someone to bathe or wash. If people visit such places, they may find clothes laid on the riverbed or close to the water sources. If one goes into the river to bathe, they may be taken by a mermaid. The Shona believe that the mermaid is a water spirit that can take someone to the underworld (Bourdillon, 1987).

If one has been taken by the mermaid, it is believed that they will be trained on the use of medicine. Once someone is taken, people should not mourn; otherwise, the disappeared individual would not return alive. A religious ceremony would be required to be undertaken at the riverbed, pool, or spring by the relatives of the disappeared person. The rituals are meant to appease the mermaid so that it would release the person alive. After a certain period, the disappeared person is believed to return endowed with supernatural powers such as healing or divination (Gelfand, 1970; Chavunduka, 1977).

One would be endowed with the understanding of various medicines. It is believed that most medicine would be provided by the mermaid once a client comes for consultation. Since people who return after their disappearance are generally few, it becomes prohibitive to fetch water or play around such places, thereby protecting them from being destroyed by people (Mawere, 2013b). Some springs are found on hills or just below them, often located in wetlands. These springs or wells are sometimes considered sacred. There are several taboos related to fetching water from these wells and springs. One taboo is that one is prohibited from using dirty pots or pots blackened with soot. It is believed that the water will spill out of the spring with force to signal that it is not allowed to use such a vessel (Mawere, 2013b). Additionally, anyone fetching water is not allowed to make any negative comments at the site (Chemhuru & Masaka, 2010).

In some large rivers, there are pools where people are prohibited from using engines to draw water. Numerous beliefs surround this matter. For instance, on a farm owned by the writer of this paper, located near the mountain shrine of Chirumanzi in the Rusvinga hill area, there is a pool named Vabere (of old hyenas). This pool is believed never to go dry. If it does, it is considered a sign of an impending year of hunger. The pool, about 20 meters wide and 60 meters long, has dark water, and local people are afraid to draw water from it, though cattle can drink from it (Chavunduka, 1977).

During an interview with a local resident, various stories about the pool were recounted. One story tells of a man who came fishing with a net and caught enough fish to fill a twenty-liter bucket. Confident, he returned the next day to catch more. However, when he came back, he found his net heaped outside the water. It was interpreted that the water spirit prohibited him from catching more because he had already been given enough (Gelfand, 1970).

In another incident, in the Chihambakwe area in the Gutu district of Masvingo province, there was a spring on a rock regarded as sacred from time immemorial. People observed all the taboos provided for fetching water. After Zimbabwe's independence in 1980, a project aimed at protecting water sources built a wall at the spring using cement and bricks. However, a few days later, the water is said to have moved and created a new opening, leaving the original site dry (Mawere, 2013a; Chimuka, 2001).

These stories and taboos reflect the Shona people's deep respect for their natural water sources and their belief in the spiritual forces that protect these resources. Such beliefs play a significant role in conserving these vital resources and maintaining the ecological balance in their communities.

Totemism

Kasere (2010) notes that totemism in Zimbabwe was not initially intended as a conservation strategy but rather as a system of social organization. Totem groups represented different clans and were associated with specific animals, which became symbols of these groups. Although the primary purpose of totemism was not environmental protection, it inadvertently contributed to the preservation of wildlife as these animals were regarded with reverence and protected from harm.

The belief system of the indigenous people of Zimbabwe, particularly their adherence to totemism, intertwined their survival with the well-being of certain species. The Shona people, for instance, have a profound cultural connection to wildlife, making them more inclined to protect these animals compared to foreign influences. According to Chemhuru and Masaka (2010), it can be argued that the concept of totem (mutupo) is a wildlife conservation strategy. They cite a Shona saying, "ukadya mutupo wako, unozobva mazino," which means "if you eat your totem animal, you will lose your teeth." This illustrates the deep-seated cultural prohibitions against harming totem animals, effectively serving as a conservation measure.

Taboos regarding totems were institutionalized wildlife conservation measures intended to preserve various animal species from extinction due to unchecked hunting. Mawere (2013a) provides an example of pangolins, a rare species protected by such taboos. Killing pangolins is strictly prohibited, thereby preventing their extinction through cultural means.

Tatira (2000) highlights the use of zviera (taboos) among the Shona people as a means of transmitting crucial social values to the younger generation. Zviera encourages conformity to societal norms and prohibit actions that violate these norms, thereby indirectly fostering wildlife conservation. For instance, taboos forbid the eating of certain animals and the unsustainable use of environmental resources.

Makamure and Vengesai (2015) define taboos as "avoidance rules" that forbid members of the human community from performing certain actions, such as eating specific foods, visiting sacred sites, cruelty to animals, and unsustainable environmental practices. Among the Karanga people, these taboos are specific rules that, when violated, are believed to bring misfortune to the individual and the community, including bad luck, disease, drought, and death.

Gelfand (1979) categorizes taboos into six groups: those concerning living correctly, successful pregnancy, avoidance of danger, good behavior, healthy living, and religious teachings. Mawere (2013b) extends this list to include prohibitions against cutting indigenous fruit trees. Violating this taboo is believed to anger ancestors, causing fruit trees to stop bearing fruit in future seasons. Perpetrators might also be pursued by an ancestral lion, a territorial spirit believed to bring misfortune or death when angered.

Protection of Vegetation and Animal Species

Among the Shona people, it is a taboo to defile or visit sacred sites, which include certain mountains and forests. Activities such as tree cutting and hunting wild animals in these sacred areas are strictly prohibited. The belief is that visiting or defiling these places can result in severe consequences for the perpetrator, such as getting lost in the forest, disappearing forever, or even becoming insane. For instance, there is a myth in Shurugwi in the Midlands Province where members of an apostolic sect purportedly disappeared when they ascended to the summit of Guruguru Mountain to conduct a church service. The Shurugwi community believes this mountain to be sacred (Chemhuru & Masaka, 2010).

This cultural prohibition on disturbing sacred sites serves as an effective conservation measure. By maintaining the sanctity of these areas, the Shona people inadvertently protect the vegetation and animal species within them. The prohibition against tree cutting in sacred groves ensures that these areas remain untouched, and biodiversity is preserved. Sacred groves often serve as refuges for various plant and animal species that might be endangered or vulnerable in other areas due to human activities (Mawere, 2013a).

In addition to preserving flora, the Shona people's taboos also protect fauna. Certain animals are considered sacred or are associated with totems, and hunting these animals is forbidden. This practice helps to maintain healthy populations of these species and ensures their survival. For example, the pangolin, a rare and endangered species, is protected by cultural taboos among the Shona (Mawere, 2013a). Violating these taboos is believed to bring misfortune not only to the individual but also to the community, thus reinforcing adherence to these conservation practices.

The connection between religion, culture, and environmental conservation among the Shona people is further reinforced by their belief in ancestral spirits. These spirits are thought to reside in sacred natural sites, and any disturbance to these areas is seen as a direct affront to the ancestors (Daneel, 2001). This belief fosters a sense of responsibility and stewardship towards the environment, as protecting these sites is seen as a way to honor and appease their ancestors.

Furthermore, the use of taboos and myths to instill fear and respect for the environment is a common strategy among many indigenous communities. It serves as a means of environmental education and ensures that younger generations learn the importance of conservation from an early age (Gelfand, 1979). The transmission of these values through oral traditions, rituals, and community practices helps to preserve the natural heritage and promotes sustainable living.

Protection of Domestic Animals

In Shona culture, the pouring of libation on domesticated animals, known as kupira, is a significant religious ceremony, particularly after the ritual of returning the spirit of the dead, or magadziro. The bull, which is dedicated to the spirit of the deceased family leader, is believed to become the head of the kraal, providing protection to all the animal wealth of the family. This belief system encourages the preservation of cattle, as killing one would be akin to killing a household member, thereby promoting the conservation of domestic animals (Bourdillon, 1976).

When a bull is poured libation, it begins to exhibit unique behaviors. It may graze alone, take care of other cattle, and prevent them from grazing in the fields. Anecdotal evidence suggests that some bulls refrain from entering the fields, grazing only on contour ridges without setting foot in cultivated areas. This behavior underscores the deep respect for these animals and the cultural practices that prevent random killing of domestic animals. Portions of meat from ceremonial killings are often shared among community members, reinforcing the communal and sustainable use of animal resources (Aschwanden, 1982).

When the dedicated bull grows old, it is killed ceremonially. The consumption of its meat by family members is believed to bring the ancestor's spirit into the lives of the successors, allowing it to dwell among them. This ritual underscores the deep connection between religious practices and conservation strategies, demonstrating how spiritual beliefs can foster a culture of preservation (Gelfand, 1973).

Religious Space and Conservation

Musoni (2016) highlights how the Johane Masowe Chishanu yeNyenyedzi Church appropriated African indigenous sacred places such as hills, shades, and dams for religious purposes like all-night prayers and water baptisms. This integration of Shona traditional religion into Christian practices emphasizes the continued religious significance and conservation of these sacred sites.

Chemhuru and Masaka (2010) explain that Shona culture reserves forests and mountains as sacred through the implementation of religious taboos. These taboos guard against unceremonial visitation, contributing to the conservation of natural resources. Sacred places and shrines, revered for their spiritual significance, are often feared and respected by the community. These sites are believed to be the domain of spiritual forces, and visiting them, especially for hunting or resource collection, is thought to invoke ancestral reprisals (Chemhuru & Masaka, 2010).

For example, in the Mwenezi district of Masvingo province, there is a belief that members of the Zion church disappeared after attempting to conduct a church service at the summit of Hariyeumba Mountain, which is considered sacred (Makamure & Vengesai, 2015). In the Chihambakwe area of Chief Makore in Gutu district, a densely forested hill named Chirora is off-limits for walking, climbing, or firewood collection, further demonstrating how religious beliefs protect certain places.

Certain trees are also regarded as sacred. Makamure and Vengesai (2015) describe rituals held under trees such as the muhacha (mobola plum), muonde (fig tree), mumvee (sausage tree), and mushozhowa (Pseudolachnostylis maprouneifolia). These rituals have served as environmental conservation measures, as tall trees are generally considered sacred and are believed to be inhabited by spiritual beings, thus not allowed to be cut down.

ANALYSIS OF THE FINDINGS

Historical, social, spiritual, and legal factors interact intricately in the debate over the Shona people, their cultural identity, and environmental preservation. The following major themes emerge from a comprehensive examination of the data:

a) Colonial Disruptions and Cultural Resilience

The Shona people's historical trajectory demonstrates how colonial interventions affected religious practices, economic activity, and land ownership. Shona customs have shown to be remarkably resilient in the face of these upheavals. A dynamic rather than static cultural identity is indicated by the adaptation of collective land management and ancestral worship to contemporary situations. However, colonial policies' long-term effects-like environmental degradation and land dispossession-continue to influence today's socioeconomic issues.

b) Environmental Challenges and the Need for Sustainable Practices

The environmental hazards that Shona communities face are best illustrated by Masvingo province. The need for sustainable farming methods is highlighted by the dependence on rain-fed agriculture and the unpredictable nature of the environment. Although the implementation of irrigation systems and drought-resistant crops is required, the results indicate that traditional agricultural knowledge, which is based on ecological understanding, may support contemporary scientific methods. The viability of large-scale agriculture using just traditional methods is still up for dispute, though, especially in light of the rapidly accelerating effects of climate change.

c) Legal Frameworks vs. Indigenous Conservation Ethics

The 2013 Zimbabwean Constitution strengthens the state's role in ecological protection by establishing a legal mandate for environmental conservation. The results, however, demonstrate that Shona traditional beliefs provide a different but effective conservation ethic that functions outside of official legal frameworks. An informal regulatory mechanism is the belief in spiritual repercussions for environmental degradation, such as ancestor wrath or misery. In order to establish an inclusive framework that preserves cultural heritage while guaranteeing environmental sustainability, it is difficult to reconcile statutory conservation legislation with indigenous customs.

d) Religion as a Conservation Mechanism

One important discovery is the contribution of Shona spirituality to environmental preservation. Biodiversity is protected because woods, rivers, and wildlife are considered sacred. Despite being largely a societal construct, totemism unintentionally promotes wildlife conservation by deterring the hunting of totemic animals. Additionally, spiritual taboos that govern water use serve as unofficial guidelines to keep water sources clean. In rural areas, these religious and cultural practices offer a system of collective enforcement that may be more successful than outside legal requirements. The viability of these customs is threatened by modernity and the decline of traditional beliefs.

e) The Role of Religious Institutions in Bridging Traditional and Modern Conservation Approaches

The role that Shona spirituality has in protecting the environment is one significant finding. Because forests, rivers, and wildlife are revered, biodiversity is safeguarded. By discouraging the hunting of totemic animals, totemism inadvertently supports wildlife conservation, even though it is mostly a social construct. Furthermore, unofficial rules to maintain the cleanliness of water sources are based on spiritual taboos that regulate water use. These cultural and religious customs provide a method of communal enforcement in rural regions that might be more effective than extralegal regulations. Modernity and the waning of traditional beliefs pose a threat to the survival of cultural traditions.

f) Indigenous Knowledge as a Model for Conservation

According to the research, native conservation methods can provide important information about how to manage resources sustainably. The usage of taboos to safeguard biodiversity and water sources is similar to modern ecological concepts like pollution prevention and ecosystem preservation. However, in favor of Western scientific methods, contemporary conservation initiatives frequently ignore traditional knowledge systems. Integrating traditional conservation techniques into official environmental policy without reducing their cultural importance is a difficult task.

The analysis emphasizes how crucial it is to acknowledge and incorporate indigenous knowledge into Zimbabwe's environmental protection plans. The results show that a strong, community-based approach to sustainability is offered by Shona spiritual and cultural values. But contemporary issues like urbanization, climate change, and the breakdown of traditional belief systems call for a hybrid paradigm that incorporates scientific methods, legal frameworks, and indigenous knowledge. Working together, local communities, religious institutions, and government agencies could preserve cultural heritage and improve environmental resilience.

DISCUSSION

The largest ethnic group in Zimbabwe, the Shona, have a rich cultural and historical legacy that has been influenced by a special fusion of post-colonial adaptations, colonial disruptions, and traditional customs. Despite the fact that the term “Shona” was created in the 20th century as a linguistic classification rather than an indigenous identity, the land-based spirituality, agricultural practices, and a strong sense of social cohesion are all firmly ingrained in the people's cultural heritage. Their identity and existence have always depended heavily on their close relationship with the land. However, their customs of religious rites, economic activities, and land ownership were all upset by colonial meddling. Shona cultural traditions have proven remarkably resilient in the face of these difficulties, adjusting to the contemporary sociopolitical environment while upholding fundamental principles.

The Shona people make up the majority of the population of Masvingo, an area that confronts severe environmental problems, especially droughts and irregular rainfall patterns. The sustainability of agriculture is threatened by these issues, which are made worse by the wider consequences of climate change. The area has become more dependent on adaptable tactics to deal with these issues, like installing irrigation systems and growing crops that can withstand drought. Nonetheless, the province's reliance on rain-fed farming emphasizes how urgently sustainable water management techniques are needed. In order to protect food security and increase agricultural systems' resilience to climate uncertainty, such actions are crucial.

Environmental conservation is highly valued in Zimbabwe's legal system, especially the 2013 Constitution, which acknowledges ecological protection as a fundamental obligation and right. Particularly in regions where traditional ecological knowledge is still essential to day-to-day existence, this legal obligation presents a critical chance for cooperation between the state and indigenous populations. The country's environmental sustainability initiatives may be greatly strengthened if the Shona people's traditional conservation methods were incorporated into official policy, given their long-standing respect for the natural world. Furthermore, with the backing of constitutional provisions, faith-based groups are in a good position to actively promote environmental stewardship and awareness, bridging the gap between conventional and contemporary conservation methods.

The foundation of the Shona people's environmental protection initiatives is their religious beliefs. Spiritual taboos safeguard sacred areas like woods, rivers, and mountains, guaranteeing their conservation and avoiding environmental exploitation. These hallowed locations are thought to be the homes of ancestor spirits, whose guardianship acts as a disincentive to destructive human behavior. This mentality of stewardship is strengthened by ancestor reverence, which creates a deeply ingrained cultural norm of honoring and protecting the land. Rituals such as the Bira ceremony, which asks ancestors for advice, highlight the value of natural resources and strengthen the notion that the land has been committed to the people by God for prudent stewardship.

Beyond sacred sites, biodiversity is unintentionally protected by the Shona people's totemism system. Totemism, which is essentially a social structuring system, forbids the hunting or eating of some animals, especially those associated with a clan's totem. By preventing overexploitation of fragile species, this cultural ban aids in the protection of biodiversity as a whole. By encouraging the sustainable use of natural resources, taboos concerning the preservation of vegetation-such as the ban on felling sacred trees—also support contemporary conservation techniques.

In instance, Shona spiritual beliefs are strongly linked to water conservation. Taboos that forbid contaminating essential water sources, such rivers, pools, and wells, have ecological and spiritual benefits. For example, the idea that peeing in a well causes diarrhea or bilharzia serves as an unofficial but efficient environmental control, keeping water sources pure and preventing contamination. Since people revere these holy creatures and refrain from abusing or contaminating their habitats, the idea that spiritual beings, like njuzu (mermaids), live in water bodies also deters pollution.

These results demonstrate the complex interrelationships among environmental preservation, spirituality, and Shona cultural identity. An indigenous model for sustainable resource management that is in line with contemporary environmental conservation ideas is provided by the Shona people's longstanding spiritual love for nature. An all-encompassing strategy for environmental stewardship in Zimbabwe is provided by combining these customs with official conservation regulations. Zimbabwe can develop a more inclusive and culturally relevant model of sustainability-one that honors both environmental and cultural heritage and ensures a sustainable future for future generations-by fusing legislative frameworks with traditional ecological principles.

FUTURE RESEARCH DIRECTION

Future comparative research should examine the ways in which different indigenous cultures incorporate spirituality into conservation initiatives, highlighting both shared and distinctive customs. It will be instructive to compare the success of religious and secular endeavours in comparable situations. To evaluate how religious participation in conservation changes over time and its long-term effects on biodiversity and community customs-especially when socioeconomic circumstances shift-long-term research is required.

Multidisciplinary methods are essential; in order to develop comprehensive conservation plans, research should combine ecological science and cultural anthropology. It's also critical to look at how sacred place conservation affects community identity and wellbeing psychologically. In the twenty-first century, adopting technology is essential. Research should look into how mobile apps and social media may spread religious teachings about conservation and provide resources for using GIS to document sacred locations. Another important aspect is policy formation, which focuses on the efficacy of current laws pertaining to endangered species and sacred areas as well as the influence of religious leaders on environmental policy.

Last but not least, examining opposition and difficulties can reveal obstacles to combining spirituality with conservation, and a global viewpoint can encourage collaborations to address the effects of climate change on holy places. Enhancing knowledge of the complex connections of religion, culture, and conservation is the goal of these directions.

CONCLUSION AND RECCOMMENDATIONS

By using a religious and cultural perspective, the Shona people have been able to preserve large areas of their natural environment through the practice of sacralizing particular species and locations. However, it is crucial to apply the same safeguards to non-sacred species and areas in order to stop more ecological deterioration. Involving religious and community leaders is essential to this effort because communities can be inspired to embrace more ecologically friendly behaviors by incorporating conservation themes into religion teachings. Furthermore, increasing local participation and accountability in ecosystem protection will come from giving chiefs, village headmen, and other traditional authorities the authority to implement environmental conservation regulations.

To safeguard sacred species and areas, the state should create a clear legal framework with consequences for infractions. Traditional authorities should be instrumental in enforcing this framework. Additionally, educational initiatives that highlight the ecological and cultural significance of both sacred and non-sacred regions are required to increase public awareness of the need to conserve them. It's also critical to include species and regions that are very vulnerable to exploitation to the list of sacred species and habitats. This could entail taking pragmatic steps like fencing off specific areas and designating them as sacred.

Lastly, encouraging cooperation among government agencies, religious organizations, and community leaders will result in a cohesive strategy for environmental preservation. The Shona people can continue to preserve their environment for future generations while respecting their cultural and religious traditions by forming partnerships that can guarantee that conservation efforts are appreciated and supported across many societal sectors.

REFERENCES

Aschwanden, H. (1982). *Symbols of Life: An Analysis of the Consciousness of the Karanga*. Mambo Press.

Banda, E. (2022). *Dust to Dust: Ritualization of an Ongoing Relationship with the Land in Shona Christian Burial Practice*. Scholar Commons.

Beach, D. (1994). *The Shona and Zimbabwe: 900-1850*. Heinemann.

Beach, D. N. (1994). *A Zimbabwean Past: Shona Dynastic Histories and Oral Traditions*. Heinemann.

Berkes, F. (2008). *Sacred ecology* (3rd ed.). Routledge. DOI: 10.4324/9780203928950

Berkes, F. (2012). *Sacred ecology: Traditional ecological knowledge and resource management* (4th ed.). Routledge. DOI: 10.4324/9780203123843

Beyers, J. (2022). Christianity and Shona religion and ecology: An ethical-practical perspective. *Hervormde Teologiese Studies*, *78*(4), 1–10.

Bhebhe, S. (2019). Understanding the traditional and contemporary purpose of the Njelele Rainmaking Shrine through the oral testimonies of local people in Matobo. *Oral History Journal of South Africa*, *7*(2), 45–60. DOI: 10.25159/2663-6670/4015

Bourdillon, M. F. C. (1976). *The Shona Peoples: An Ethnography of the Contemporary Shona, with Special Reference to Their Religion*. Mambo Press.

Bourdillon, M. F. C. (1987). *The Shona peoples: An ethnography of the contemporary Shona, with special reference to their religion*. Mambo Press.

Braun, V., & Clarke, V. (2006). Using thematic analysis in psychology. *Qualitative Research in Psychology*, *3*(2), 77–101. DOI: 10.1191/1478088706qp063oa

Byers, B. A., Cunliffe, R. N., & Hudak, A. T. (2001). Linking the conservation of culture and nature: A case study of sacred forests in Zimbabwe. *Human Ecology: an Interdisciplinary Journal*, *29*(2), 187–218. DOI: 10.1023/A:1011012014240

Chanza, C., & de Wit, A. (2016). Enhancing climate governance through indigenous knowledge: Case in sustainability science. *South African Journal of Science*, *112*(3-4), 1–7. DOI: 10.17159/sajs.2016/20140286

Chavunduka, G. L. (1977). *Traditional healers and the Shona patient*. Mambo Press.

Chemhuru, M., & Masaka, D. (2010). Taboos as sources of Shona people's environmental ethics. *Journal of Sustainable Development in Africa*, *12*(7), 121–133.

Chigumira, E. (2003). Masvingo Province. *The Geographical Journal*, *169*(2), 155–162.

Chimhundu, H. (1992). *The Vernacularization of African Languages after Independence. The Center for Advanced Studies of African Society*. CASAS.

Chimuka, T. A. (2001). *Traditional Environmental Knowledge and African Cultural Values: A Case Study of the Shona in Zimbabwe*. University of Zimbabwe.

Chirimuuta, C. (2021). Totemic beliefs and conservation among the Shona people of Zimbabwe: An indigenous knowledge perspective. *International Journal of African Renaissance Studies, 16*(1), 74–89. https://doi.org/xxxx

Chitando, E., & Togarasei, L. (2010). Ethical Implications of Climate Change for Intra- and Intergenerational Justice: The Case of Zimbabwe. *Journal of Theology for Southern Africa, 136*, 6–18.

Chiwara, M. (2015). Totemic linkages and biodiversity conservation among the Shona people. *Journal of Environmental Management, 16*(3), 123–130.

Constitution of Zimbabwe Amendment (No 20) Act 2013. (2013).

Creswell, J. W. (2018). *Research design: Qualitative, quantitative, and mixed methods approaches* (5th ed.). SAGE Publications.

Daneel, M. L. (1970). *The god of the Matopo hills: An essay on the Mwari cult in Rhodesia*. Mouton.

Daneel, M. L. (2001). *African Earthkeepers: Environmental Missions in Africa*. Orbis Books.

Doke, C. M. (1931). Report on the Unification of the Shona Dialects. Government of Southern Rhodesia. *Educational Research Journal, 12*(2), 144–159.

Eneji, C. V. O., Ntamu, G. U., Unwanade, C. C., Godwin, A. B., Bassey, J. E., & Willaims, J. J. (2016). *Joseph Ignatius*. Traditional African Religion in Natural Resources Conservation and Management in Cross River State.

Eneji, C. V. O., Ntamu, G. U., Unwanade, C. C., Godwin, A. B., Bassey, J. E., Willaims, J. J., & Ignatius, J. (2012). Traditional African religion in natural resources conservation and management in Cross River State, Nigeria. *Environment and Natural Resources Research*, 2(4), 45–53.

FAO. (2006). *Climate Change and Food Security: A Framework Document*. Food and Agriculture Organization of the United Nations.

Fournier, A. (2011). Consequences of wooded shrine rituals on vegetation conservation in West Africa: A case study from the Bwaba cultural area (West Burkina Faso). *Biodiversity and Conservation, 20*(9), 1895–1910. DOI: 10.1007/s10531-011-0065-5

Gelfand, M. (1962). *Shona Religion: With Special Reference to the Makorekore*. Juta.

Gelfand, M. (1970). *The spiritual beliefs of the Shona. C.* Struik Publishers.

Gelfand, M. (1973). *The Genuine Shona: Survival Values of an African Culture*. Mambo Press.

Gelfand, M. (1979). *Growing Up in Shona Society*. Mambo Press.

Government of Zimbabwe. (2013). Environmental Management Act (Chapter 20:27). Harare: Government Printer.

Habel, N. C., & Balabanski, V. (Eds.). (2002). *The earth story in theology and ethics*. Sheffield Academic Press.

Hernandez, J. (2022). *Fresh banana leaves: Healing Indigenous landscapes through Indigenous science*. North Atlantic Books.

Kahari, G. P. (1990). *The moral vision of Patrick Chakaipa: A study in Shona traditional religion and Western Christianity*. Mambo Press.

Kasere, M. (2010). Totemism in Zimbabwe: Social and Environmental Implications. In Zimbabwe. *Journal of Environmental Studies (Northborough, Mass.)*, *15*(2), 56–68.

Lan, D. (1985). *Guns & rain: Guerrillas & spirit mediums in Zimbabwe*. University of California Press.

Mabvurira, V., Matlakala, F. K., & Mashumba, A. (2021). Indigenous knowledge systems and environmental conservation: Exploring the role of Shona culture in Zimbabwe. *African Journal of Social Work*, *11*(2), 45–53.

Mabvurira, V., Muchinako, G. A., & Smit, E. I. (2021). Shona traditional religion and sustainable environmental management: An Afrocentric perspective. *African Journal of Social Work*, *11*(3), 111–118.

Mahohoma, T. & Gundani, Paul H. (2020). Experiencing the sacred. UNISA Press. DOI: 10.25159/2412-4265/3363

Makamure, K., & Vengesai, E. (2015). Taboos as Avoidance Rules: The Karanga People's Perspective. In Zimbabwean Cultural Heritage Journal, 8(1), 7-15.

Manzungu, E. (1999). *Strategies of Smallholder Irrigation Management in Zimbabwe*. University of Zimbabwe Publications. DOI: 10.18174/192721

Matarira, C. H., Makadho, J. M., & Mukaratirwa, S. (1992). *Drought and Sustainable Development in Zimbabwe*. University of Zimbabwe Publications.

Matobo National Park. (n.d.). In Wikipedia. Retrieved March 13, 2025, from https://en.wikipedia.org/wiki/Matobo_National_Park

Matowanyika, J. Z. Z. (1991). Indigenous resource management and sustainability in Zimbabwe: Towards a political ecology of environmental management. *Environment and Urbanization*, *3*(2), 32–48. DOI: 10.1177/095624789100300204

Mavhunga, C. C. (2017). *Transient workspaces: Technologies of everyday innovation in Zimbabwe*. MIT Press.

Mawere, M. (2013a). *Culture, indigenous knowledge and development in Africa: Reviving interconnections for sustainable development*. Langaa RPCIG.

Mawere, M. (2013b). Traditional Ecological Knowledge and Its Use in Modern Conservation: The Case of the Shona of Zimbabwe. *International Journal of Modern Anthropology*, *1*(7), 8–15.

Monde, L. (2024, September 2). Au Bénin, ONG et divinités vaudoues au chevet des précieuses mangroves.

Monde, L. (2024, September 3). Benin's voodoo deities take care of precious mangroves. Le Monde.

Moyo, S., O'Keefe, P., & Sill, M. (1993). *The Southern African Environment: Profiles of the SADC Countries*. Earthscan Publications Ltd.

Mubaya, C. P. (2020). Malindidzimu Hill: A sacred site and its role in Shona cosmology. *Zimbabwean Journal of Cultural Studies*, *5*(1), 15–22.

Muguti, T., & Mapira, J. (2014). Indigenous knowledge systems and their role in environmental conservation in Zimbabwe. *Journal of Environmental Science and Water Resources*, *3*(8), 207–211.

Musoni, P. (2016). Contestation of "the holy places in the Zimbabwean Religious Landscape": A study of the Johane Masowe Chishanu yeNyenyedzi Church's sacred places. *Hervormde Teologiese Studies*, *72*(1), a3269. DOI: 10.4102/hts.v72i1.3269

Muzondi, N. (2014). The contribution of traditional belief systems in biodiversity conservation among the Ndau people of Nyagadza community in Chipinge (Doctoral dissertation).

Nyamapfene, K. (1991). *Soils of Zimbabwe*. Nehanda Publishers.

Okoji, M. A. (2001). Depletion of forest resources in southeastern Nigeria: Who loses? *The Environmentalist*, *21*(3), 213–218. DOI: 10.1023/A:1017979303340

Ranger, T. (1999). *Voices from the rocks: Nature, culture, and history in the Matopos Hills of Zimbabwe*. James Currey. DOI: 10.2979/3921.0

Ranger, T. O. (1985). *Peasant Consciousness and Guerrilla War in Zimbabwe: A Comparative Study*. University of California Press.

Wilderness Safaris. (2015, June 22). African totems & kinship. Wilderness Safaris.

Smith, L. T. (2012). *Decolonizing methodologies: Research and indigenous peoples* (2nd ed.). Zed Books.

Tatira, L. (2000). The role of taboos in conservation of natural resources. *Shona Culture Journal*, *5*(1), 28–33.

Unganai, L. S. (1996). Climate change impacts on agriculture in Zimbabwe: A regional assessment. *Climate Research*, *6*, 137–145. DOI: 10.3354/cr006137

United Nations. (2019). *Global biodiversity outlook 5*. UN Environment Programme.

Vincent, V., & Thomas, R. G. (1960). *An Agricultural Survey of Southern Rhodesia: Part I: Agro-ecological Survey*. Government Printer.

ADDITIONAL READING

Chibango, S. (2019). Environmental Preservation in Zimbabwe: A religio-cultural nexus. American Journal of Humanities and Social Sciences Research (AJHSSR), 3(3), 01-07.

Mafukidze, J. (2017). Perspectives on land and water politics at Mushandike Irrigation Scheme, Masvingo, Zimbabwe (Doctoral dissertation, RHODES UNIVERSITY).

Mahohoma, T., & Gundani, P. H. (2020). Experiencing the sacred. *Studia Historiae Ecclesiasticae*, *46*(1), 1–17. DOI: 10.25159/2412-4265/3363

Manyonganise, M., Chitando, E., & Chirongoma, S. (Eds.). (2023). Women, Religion and Leadership in Zimbabwe, Volume 1: An Ecofeminist Perspective. Springer Nature.

Mapara, J. (2016). The environmenT as signifiCanT oTher: The green naTure of shona indigenous religion. Natures of Africa: Ecocriticism and animal studies in contemporary cultural forms, 77-96.

Musasa, G. (2013). Challenges for Rural Tourism Development in Zimbabwe: A Case of the Great Zimbabwe Masvingo Area (Doctoral dissertation, University of Fort Hare). Gunda, M. R. (2018). At the crossroads: A call to Christians to act in faith for an alternative Zimbabwe (Vol. 21). University of Bamberg Press.

KEY TERMS AND DEFINITIONS

Shona: The Shona are the Bantu-speaking ethnic group indigenous to Zimbabwe renowned for their rich cultural history, which includes traditional beliefs, dancing, music, and art.

Conservation: The term conservation describes the responsible management of the environment and natural resources to stop their depletion or degradation and maintain the sustainability and health of ecosystems for coming generations.

Sacred Spaces: Sacred spaces are areas that Areas that have particular meaning for religious or spiritual activities which can include man-made buildings like temples or shrines as well as natural locations like mountains, forests and water bodies.

Indigenous Beliefs: Indigenous beliefs include the spiritual and cultural customs, values, and worldviews of a region's original inhabitants. A strong bond with the land and the natural world is frequently emphasized in these beliefs.

Natural Resources: Natural resources include things like water, minerals, forests, and wildlife that are found in the environment naturally and can be used for food or profit.

Cultural Preservation: Cultural preservation is the practice that safeguard and preserve a community's customs, languages, artwork, and historical locations in order to ensure that they are passed down to future generations.

Zimbabwe: Zimbabwe is a landlocked country in Southern Africa, renowned for its varied wildlife, scenery, and cultural legacy. In 1980, it became independent of British colonial control.

Traditional Practices: Traditional practices are the customs, rituals, and methods that are handed down through a culture's generations are known as traditional practices. Agricultural methods, crafts, religious rituals, and social conventions are a few examples.

Chapter 21
Tailoring Indigenous Economic Systems for New Generation Agripreneurship Through Indigenous Modernism

Sthokozile Mamba
https://orcid.org/0009-0005-3789-4030
University of KwaZulu-Natal, South Africa

Andrisha Beharry Ramraj
University of KwaZulu-Natal, South Africa

ABSTRACT

The agricultural landscape in South Africa presents a unique opportunity to leverage the wisdom of the past with the innovations of the present. This chapter aimed to explore modern techniques that can be used to tailor indigenous economic systems to best suit the new generation of agripreneurs and promote market access. Furthermore, exploring the challenges that could accompany this tailoring. The study used a desktop method to collect secondary data using keywords related to the study to obtain relevant publications. This study found that farming stokvels can be modernised through modern digital tools and precision farming technologies. However, the digital divide, loss of traditional communal values, increased complexity and costs associated with technology are some of the major shortfalls that could undermine the success of modernising these indigenous economic systems. It was recommended that stakeholders, actively promote the integration of modern technologies with traditional practices to enhance the effectiveness of tailoring indigenous economic systems.

INTRODUCTION

Indigenous economic systems, such as farming stokvels, livestock mafisa, Community Supported Agriculture (CSA), and resource pooling, have been foundational to rural livelihoods in South Africa for generations. These systems are not just economic mechanisms but are deeply rooted in community cohesion, reciprocal support, and an intimate understanding of local ecological conditions (Grey &

DOI: 10.4018/979-8-3373-4124-8.ch021

Patel, 2015). They enable communities to share risks, pool resources, and collectively tackle the challenges of rural life. However, while these systems are rich in cultural and social capital, they often lack the financial management, technological integration, and market access necessary to thrive in today's globalized economy (Sansom, 2024).

To address these gaps, indigenous modernism offers a promising pathway. By strategically adapting these traditional systems with modern techniques, such as financial literacy training, digital platforms, and value addition strategies, indigenous modernism seeks to enhance their effectiveness and sustainability (Trendov, Varas & Zeng, 2019). This approach not only preserves the cultural integrity and communal strengths of these systems but also equips young agripreneurs with the essential tools to navigate the increasingly complex and competitive agricultural landscape.

By embracing Indigenous modernism, these age-old practices can be transformed into dynamic, resilient, and market-savvy systems that continue to support rural communities while opening new opportunities for growth and prosperity in the modern economy. This is not just a preservation of heritage, it can also be an evolution that aligns traditional wisdom with the innovations of today, empowering a new generation of youth agripreneurs to thrive in the agriculture of tomorrow.

BACKGROUND OF THE STUDY

South Africa possesses a significant amount of arable land, predominantly located in rural areas, which presents a critical opportunity for agricultural development (Fox, Senbet & Simbanegavi, 2016). Despite this potential, there is a noticeable trend of urban migration among young people, leading to a declining interest in rural agricultural careers (Geza, Ngidi, Slotow & Mabhaudhi, 2022). The South African National Development Plan identifies agriculture as a key economic driver in rural areas, with the potential to create substantial employment opportunities and stimulate economic growth (Cousins, 2015). Yet, despite these prospects, there remains a strikingly low percentage of young people engaging in the agricultural sector (Magagula & Tsvakirai, 2020).

The current participants in this sector are predominantly those who hold extensive indigenous knowledge, gained from generations of lived experience and traditional practices (Grey & Patel, 2015). However, as the world rapidly transforms, young people are increasingly drawn to technology-driven, instant, and more efficient methods (Chipfupa & Tagwi, 2021). This divergence between the traditional and the modern poses a significant challenge which is the risk of losing valuable indigenous knowledge as the older generation exits the sector.

To address this gap, there is a need to blend indigenous agricultural practices with modern technological advancements, creating a hybrid approach that resonates with the aspirations of young people. By incorporating digital tools, precision farming, and value-added strategies into indigenous methods, it is possible to make agriculture more attractive to the youth. This approach not only holds the promise of revitalizing the agricultural sector but also ensures the preservation and enhancement of indigenous knowledge systems, making them relevant and accessible in a rapidly evolving world. The integration of these elements could be the key to drawing young people back into agriculture, ensuring the sustainability of rural economies and the continued legacy of indigenous agricultural wisdom.

PROBLEM STATEMENT

Rural youth, who are essential for the future sustainability of agriculture, are increasingly drawn away from indigenous systems due to their perceived lack of modern relevance. The challenge lies in adapting these indigenous economic systems to align with the needs and aspirations of young agripreneurs who, according to Lohento and Ajilore (2015), prefer technology-based, efficient, and market-responsive approaches. Indigenous modernism offers a framework for this adaptation by integrating modern tools and practices with traditional systems. However, the specific pathways and strategies for effectively tailoring these systems to meet the expectations of a new generation have not been sufficiently explored or implemented. This gap not only limits the appeal of agriculture as a viable career for rural youth but also jeopardizes the continuity and evolution of indigenous economic practices.

PURPOSE OF THE STUDY

The purpose of this chapter is to provide young agripreneurs, particularly those in the early stages of their entrepreneurial journey, with insights into alternative agricultural methods rooted in indigenous practices. By highlighting these traditional systems, the chapter aims to demonstrate how indigenous approaches, when integrated with contemporary technologies and strategies, can offer viable pathways for navigating the agricultural landscape. This exploration not only equips young agripreneurs with diverse tools and knowledge but also seeks to influence policymakers to incorporate these blended methods into agricultural policies that target youth engagement. By advocating for the inclusion of indigenous-modern hybrid models, the chapter aims to enhance policy frameworks that support sustainable and culturally relevant agricultural development. Furthermore, this chapter addresses a critical research gap in the literature on rural development and agriculture by examining how indigenous economic systems can be modernized to remain relevant and effective in today's evolving agricultural context. In doing so, it seeks to contribute valuable insights into the role of indigenous modernism in fostering a vibrant, inclusive, and innovative agricultural sector that resonates with the aspirations of the new generation.

OBJECTIVE OF THE STUDY

The main objective of this chapter is to explore modern techniques that can be used to tailor indigenous economic systems to be best suited for the new generation of agripreneurs and the challenges associated with it.

LITERATURE REVIEW

The agricultural landscape in South Africa presents a unique opportunity to leverage the wisdom of the past with the innovations of the present. This can be achieved by tailoring indigenous economic systems, like farming stokvels, livestock mafisa, CSA, and resource pooling, for the new generation of agripreneurs through strategic integration of modern techniques. This section of the chapter will delve into the different modern techniques that can be integrated into each of the aforementioned indigenous economic systems.

Definition of the indigenous economic systems

In the South African context, indigenous economic systems like farming stokvels, livestock mafisa, CSA, and resource pooling have historically underpinned the livelihoods of rural communities (Cameron, 2015; Mokomane, 2018; Koenane, 2019). Warren (1996), as cited by Raheem and Bako (2015), stated that these systems are deeply intertwined with cultural practices and communal values, emphasizing shared resources, mutual support, and sustainable practices, which are critical in addressing the socio-economic challenges faced by rural agripreneurs today.

Farming stokvels function as cooperative savings and credit societies where members contribute regular, fixed amounts into a communal pool. This pooled capital is then loaned out to members for various needs (Verhoef, 2020), including agricultural investments. In South Africa, where access to formal financial services is often limited in rural areas, stokvels provide a vital source of capital, fostering financial inclusion and reducing dependency on predatory lending practices (Zondo, 2017). According to Koenane (2019), the collaborative nature of farming stokvels promotes risk-sharing and community resilience, allowing members to collectively navigate economic challenges such as droughts, fluctuating market prices, or unexpected crop failures.

Livestock mafisa is a traditional livestock sharing system wherein a senior community member loans livestock to a younger or less resourced individual. In return, the borrower shares a portion of the livestock's offspring or produce (Mokomane, 2018). Lutoombi (2015) asserts that this system not only facilitates wealth distribution and economic participation among youth but also ensures the continuity of livestock resources and genetic diversity within the community. In the South African rural context, livestock mafisa could support young agripreneurs by providing them with a low-risk entry into livestock farming, fostering mentorship, and preserving traditional knowledge related to animal husbandry and land management.

Community Supported Agriculture (CSA) directly connects consumers with farmers, offering a model where consumers pay in advance for a share of the farm's produce. In South Africa, where smallholder farmers often struggle with market access and price volatility, CSA provides a reliable income stream and reduces the risks associated with traditional market dependencies (Cheteni & Umejesi, 2023). According to Chen, Gao, Chen and Zhang (2019), this model encourages sustainable farming practices as farmers are directly accountable to their consumers, often leading to reduced pesticide use and a greater focus on organic and eco-friendly methods, aligning with the growing demand for healthy, locally produced food.

Resource pooling, involving the collective management of shared resources such as land, water, or agricultural equipment (Jayne, Chamberlin & Headey, 2014), is another indigenous practice that can be modernized to meet the needs of South African agripreneurs. By pooling resources, communities can achieve economies of scale, reduce individual costs, and invest in larger, more efficient agricultural projects like communal irrigation systems or shared processing facilities (Musavengane & Simatele, 2016). Collier and Dercon (2014) stated that this approach promotes equitable access to essential resources, enhances productivity, and strengthens community ties, all of which are crucial for sustainable rural development.

These indigenous economic systems provide valuable lessons for contemporary agripreneurship in South Africa. Drawing from this, integrating the principles of collaboration, sustainability, and community orientation into modern agricultural business models, young agripreneurs can address critical challenges such as food security, rural unemployment, and climate resilience. For instance, adapting the stokvel model to cooperative farming ventures, implementing livestock sharing initiatives inspired by mafisa, developing CSA models to link directly with local consumers, and engaging in resource pooling to optimize land and water use can all contribute to building a robust and inclusive agricultural sector. Tailoring these indigenous economic systems to the modern era offers a unique opportunity to create innovative and sustainable agribusinesses that not only drive economic growth but also preserve cultural heritage and strengthen community resilience (Thierfelder, Rusinamhodzi, Ngwira, Mupangwa, Nyagumbo, Kassie & Cairns, 2015). By blending traditional wisdom with contemporary approaches, South African agripreneurs can develop business models that are both economically viable and socially responsible, ensuring that agriculture remains a vibrant and integral part of rural life.

Indigenous modernism in indigenous economic systems

Farming Stokvels

Indigenous modernism is the concept of modernizing traditional practices without losing the cultural and communal essence that defines them (Campbell, 2020). Drawing from this, in the context of farming stokvels, this means leveraging modern digital tools while preserving the collaborative and community-driven spirit that has always underpinned these systems. One way to modernize farming stokvels is through the implementation of digital platforms designed specifically for cooperative financial management (Masoka, 2022). Kariuki and Ofusori (2017) state that secure online systems can streamline contributions, record-keeping, and financial oversight, making the management of funds more transparent and efficient. This transparency is vital for maintaining trust among members, especially as the younger generation becomes more involved. Furthermore, according to Menze and Tsibolane (2019), these platforms can facilitate the organization of virtual meetings, ensuring that even members in remote areas can participate in decision-making processes. Beyond financial management, digital forums and chat functionalities offer a space for continuous learning and knowledge exchange (Masoka, 2022). These platforms can connect stokvel members with agricultural experts, enabling them to stay informed about the latest farming techniques, market trends, and available resources. This is particularly important for youth agripreneurs who may lack the experience of older farmers but are more tech-savvy and open to innovative farming practices.

The integration of precision farming technologies, such as soil sensors and weather monitoring systems, is another aspect of modernizing farming stokvels. These technologies can provide real-time data on soil health, water usage, and weather patterns, enabling stokvel members to make informed decisions (Ncube, Mupangwa & French, 2018). For example, real-time data can help optimize irrigation practices by ensuring that water is used efficiently, reducing waste and conserving a vital resource (Griffin, Shockley & Mark, 2018). Similarly, targeted nutrient application based on soil health data can enhance crop yields, reduce costs, and minimize environmental impact (Magocha, Soundy, Muchie & Magocha, 2019). For youth agripreneurs, access to these technologies can be a game-changer. According to Geza et al. (2022), many young people in rural areas face barriers to entry into the agricultural sector due to a lack of resources and knowledge. By participating in a modernized farming stokvel, they can benefit from shared access to precision farming tools and the collective wisdom of the group, helping to level the playing field and making agriculture a more attractive and viable career option.

Community Supported Agriculture (CSA)

Indigenous modernism seeks to modernize traditional systems like CSA by incorporating contemporary tools and methods that enhance efficiency, accessibility, and resilience, without losing the core values of the system (Buthelezi & Hughes, 2014). In the context of CSA, this can involve the use of user-friendly online subscription models that simplify the consumer experience. Trendov et al. (2019) asserts that digital platforms can allow customers to choose their preferred farm, select subscription options, manage accounts, and receive updates directly from farmers. This not only streamlines enrolment but also fosters customer engagement by providing a transparent view of the food supply chain and allowing consumers to feel more connected to the source of their food.

For youth agripreneurs in rural South Africa, these platforms offer a new way to market their produce, reach a wider audience, and secure a more predictable income stream (Rosegrant, Koo, Cenacchi, Ringler, Robertson, Fisher & Sabbagh, 2014). By guaranteeing a market for their products, CSA can mitigate the risks associated with traditional farming, such as fluctuating market prices and unpredictable demand. This financial stability is crucial for young farmers who often face barriers to accessing credit and other financial services (Fletschner & Kenney, 2014). Moreover, Hvitsand (2016) states that digital platforms can enhance the operational efficiency of CSAs by optimizing delivery systems. According to Campbell, Boucquey, Stoll, Coppola and Smith (2014), freshness and quality are paramount in CSA, and effective logistics are essential to meet consumer expectations. Exploring partnerships with local delivery companies, investing in refrigerated transport options, or establishing a network of pick-up points closer to consumers can address logistical challenges. For youth agripreneurs, these modernized delivery systems provide an opportunity to differentiate their offerings by ensuring that produce reaches consumers in peak condition, thus building a reputation for quality.

Livestock Mafisa

One of the primary ways to modernize livestock mafisa is through the adoption of digital tracking technologies such as Global Positioning System (GPS) collars and mobile apps. Ilyas and Ahmad (2020) and Aquilani, Confessore, Bozzi, Sirtori and Pugliese (2022) stated that these tools can revolutionize the monitoring and management of livestock, offering real-time location data that allows for improved oversight and faster responses to emergencies such as theft, disease outbreaks, or livestock straying into restricted areas. For youth agripreneurs, who may be managing livestock for the first time, digital tracking provides an invaluable layer of security and control over their herds. According to Turner and Schlecht (2019), it reduces the risks associated with traditional, manual oversight methods, which are often labour-intensive and less precise.

Furthermore, data on animal movement can be used to map optimal grazing routes, helping to prevent overgrazing and ensure the sustainable use of pasture resources. Overgrazing is a significant challenge in many rural areas, leading to soil degradation and reduced forage availability (Aquilani et al., 2022). By using digital tracking data to inform grazing decisions, mafisa participants can promote healthier pastures and maintain herd productivity over the long term. This is particularly beneficial for young farmers, who are more likely to adopt innovative, data-driven approaches to herd management.

Data-driven decision-making extends beyond tracking animal movement. By collecting and analyzing data on animal health, weight gain, breeding cycles, and market trends, mafisa participants can make informed decisions about their livestock operations (Gumbo, 2018). For example, insights from data analysis can guide breeding strategies to improve herd genetics (Neethirajan & Kemp, 2021), optimize feeding regimens to enhance growth rates (Aquilani et al., 2022), and identify the best times to sell livestock to maximize profits. For youth agripreneurs, who may have limited experience in livestock management, access to this data can significantly improve their decision-making and help them achieve better outcomes.

Resource Pooling

Community-owned financial institutions, such as credit unions or cooperative banks, are pivotal in modernizing resource pooling. These institutions can provide secure financial services like savings accounts, loans, and microfinancing options tailored to the needs of indigenous economic systems (Fletschner & Kenney, 2014), such as stokvels, mafisa groups, and CSA initiatives. By offering these services, community financial institutions can enhance financial inclusion, enabling youth agripreneurs to access the capital needed to start or expand their agricultural ventures. According to Hilson (2016), this financial support is crucial in rural areas, where conventional banking services are often inaccessible or prohibitively expensive. Furthermore, these institutions can facilitate collective investments in critical infrastructure or shared equipment, such as irrigation systems, renewable energy sources, or processing facilities. For youth agripreneurs, access to such shared resources can significantly reduce the cost barriers associated with starting a business, allowing them to scale operations more quickly and sustainably (Chinaka, 2016). This collective approach to investment not only amplifies the economic impact of individual contributions but also strengthens the overall resilience of the community against economic shocks or climate-related challenges.

Online platforms can further modernize resource pooling by enabling the efficient sharing of agricultural equipment and facilities within and across indigenous economic systems. Digital platforms can provide a centralized space where community members can list and book equipment like tractors, harvesters, or cold storage facilities, optimizing resource utilization and reducing the need for every individual or group to own expensive machinery (Larson, Muraoka & Otsuka, 2016). Jayne et al. (2014) concur that this shared access model can dramatically lower operational costs, making modern agricultural technologies more accessible to youth agripreneurs who may lack the capital to purchase their own equipment. Moreover, digital platforms can facilitate better coordination and collaboration among different indigenous economic groups. For example, drawing from this, stokvels, mafisa groups, and CSAs can use these platforms to synchronize activities, share market information, and jointly plan for collective actions such as bulk purchasing of inputs or joint marketing efforts. Hilson (2016) affirms that this interconnectedness strengthens the bargaining power of small-scale producers, allowing them to negotiate better prices for their products and inputs, thereby increasing their profitability and market presence.

Modernizing indigenous farming systems with technology in Africa and Asia: case studies

Traditional community-based economic systems in agriculture, from South African stokvel savings clubs to Asian community-supported farms and livestock sharing like mafisa, are being updated with modern tools. Across Africa and Asia, communities are integrating digital platforms, financial services, and precision farming into these indigenous models. The following case studies highlight how these innovations were implemented, the challenges encountered, and the outcomes achieved, often supported by research, government programs, or non-profit initiatives.

Digital Farming Stokvels in South Africa

In South Africa, stokvels have long helped communities pool money for various goals. Now, tech startups are bringing stokvels into the digital age. For example, the AgriCrowdy platform enables a farming stokvel approach where it lets traditional stokvel groups collectively invest in real farm assets online as virtual farmers, while experienced farmers do the actual work (AgriCrowdy, n.d.). The platform handles contributions, tracks investments, and disburses returns through a unified digital system. This modern twist provides stokvel members an opportunity to fund agricultural projects for profit, moving their pooled savings into productive farming ventures.

Groups register on a mobile/web app, contribute funds electronically, and purchase shares in crops or livestock ventures. They can then monitor progress and yields via the app, receiving periodic updates. Another example is the Circles360 toolkit, which digitizes stokvel administration by managing payments, recording member shares, and automating payouts. It drastically reduces the manual admin work and errors that traditional stokvels faced by moving everything to an online ledger (SME South Africa, 2024). These tools also offer greater transparency, where every member can see the fund's status, and security, where funds are held safely, not as cash under a mattress.

Despite the benefits, getting communities to trust digital systems has been a hurdle. Many stokvels still prefer cash and face-to-face operations due to mistrust of formal institutions and habit (Ipsos, 2024). Members worry about bank fees or fraud, and some are unfamiliar with apps. To overcome this, platforms like StokFella and Circles360 focused on user education and trust-building, emphasizing secure transactions, simplifying the interface, and even incorporating familiar tools like WhatsApp for communication (Ipsos, 2024). Ensuring affordability is also key, keeping fees low so as not to erode the collective savings. Gradually, younger and more tech-savvy members have driven adoption, and technology is transforming how groups operate, with many now handling record-keeping and even meetings online (Ipsos, 2024).

Early results show improved efficiency and larger pooled funds. Digital stokvel platforms report that groups can scale up easily, managing hundreds of members and bigger contributions, which would be impractical with pen-and-paper bookkeeping (SME South Africa, 2024). Automation has reduced conflicts over calculations and increased transparency, strengthening trust among members. By linking informal groups to formal financial services, some clubs have gained access to bank accounts and even insurance products tailored for stokvels. In farming stokvel cases, the pooled investments have funded livestock purchases and crop planting that individual members alone couldn't afford, leading to profits shared by the group. For instance, AgriCrowdy advertises returns in the range of 8–20% for ventures like pig or cattle virtual farming, demonstrating the potential financial uplift (AgriCrowdy, n.d.). These successes have attracted interest from government and NGOs as a means to boost rural entrepreneurship and food security by leveraging communal finance. Academic research also points out that formalizing and scaling such group structures, potentially through technology, could amplify their impact on food systems (Lukwa, Wayas, Lambert, "Savings for Health" IDRC collaborators & Alaba, 2024).

Community Supported Agriculture (CSA) Goes Online in Asia

CSA has roots in traditional cooperation. In Asia, this concept is being modernized through e-commerce and digital engagement. A notable case comes from northern Thailand, where researchers piloted an e-CSA system in Chiang Rai Province to connect organic farmers with consumers via an online platform (Morepje, Sithole, Msweli & Agholor, 2024). This electronic CSA allowed urban consumers to order produce from local farms through a website/app, cutting out middlemen and providing farmers a guaranteed market.

The project, by Wicha et al., (2017), built a web-based platform for organic farmers' cooperatives. Farmers would list their seasonal produce and consumers subscribed or placed orders online for regular deliveries. The system integrated digital payments and information sharing, consumers could pay through e-wallets, and farmers received orders and feedback through SMS and app notifications. The Thai government's broader "Thailand 4.0" digital economy policy supported such efforts with grants and training in e-commerce (Lilavanichakul, 2020). The long-term strategy was to shift these farmers from purely traditional markets to a hybrid of on-farm and online sales (Lilavanichakul, 2020).

Initially, many small farmers were reluctant to use the new platform. Surveys found that digital literacy and internet access were major barriers in rural areas (Morepje et al., 2024). Older farmers preferred familiar face-to-face selling at local markets, and there was a *"comfort zone"* effect slowing adoption (Morepje et al., 2024). To tackle this, the project involved extensive training and hands-on support, demonstrating how the e-CSA could increase their income. Younger, tech-savvy farmers and those with niche products, such as organic and pesticide-free, were the first to participate, showing others the benefits (Morepje et al., 2024). Another challenge was infrastructure, some remote villages had poor connectivity, so the system had to allow offline operation with periodic syncing, such as using SMS when internet was down).

The e-CSA pilot showed promising results in Chiang Rai. Farmers who participated managed to access new customers in nearby cities and command better prices, since they bypassed intermediaries (Morepje et al., 2024). Consumers benefited by conveniently getting fresh, organic produce and felt more connected to local growers. However, adoption was modest at first, the program highlighted that a long-term transition plan is needed to truly bring traditional farmers online (Lilavanichakul, 2020). This includes improving rural broadband and gradually building farmers' e-commerce skills. The project did succeed as a proof of concept, it demonstrated that even in a rural Thai community, a digital platform can facilitate CSA-style partnerships. Its findings have informed government e-agriculture strategies in Thailand, emphasizing the need to close the digital divide and provide incentives for farmers to join such platforms (Morepje et al., 2024). Similar digital CSA initiatives are now spreading, for example, other Southeast Asian provinces and parts of China have developed apps that connect consumers with local farm co-ops, inspired by this model (Morepje et al., 2024).

Modernizing Livestock Mafisa and Pastoral Systems in Africa

Communal livestock sharing practices, such as mafisa in Southern Africa, are also evolving with modern interventions. *Mafisa* traditionally refers to a system where wealthy cattle owners loan animals to poorer families, who care for them and share in the offspring or milk. This indigenous form of resource pooling provided a social safety net and a way to grow herds collectively. Recent efforts seek to preserve the spirit of mafisa while improving it with formal agreements, better inputs, and digital tracking.

In Botswana, the government revived the mafisa concept as a formal program to boost agro-tourism and farming livelihoods. In 2022, the Ministry of Environment launched a scheme loaning out livestock (including wildlife like impala and antelope) to aspiring farmers, echoing the mafisa idea of "lend a starter herd" (Sunday Standard Reporter, 2022). Recipients breed the animals and then pay forward an equivalent number of offspring to the next farmer, under supervision of wildlife officials (Sunday Standard Reporter, 2022). All transactions require permits and are monitored in writing, effectively modernizing mafisa through government regulation to ensure fairness and prevent abuse. This program integrates traditional knowledge, community mentoring in animal husbandry, with formal oversight and aims to expand farmers' access to high-value livestock for tourism and meat production.

Non-profit initiatives have also emerged. In Zambia, a social enterprise named Mafisa works with traditional cattle-herding communities on the fringes of the Kalahari in the Western Province, in the southwestern part of Zambia. Their goal is to make these pastoral livelihoods resilient and profitable in today's climate. The Mafisa project introduces climate-smart management practices, like controlled grazing and improved veterinary care, and links herders to better markets and finance (African Food Changemakers, n.d.). By doing so, it hopes to restore the viability of cattle culture after years of drought and disease. The program, expanding to 120 communities over seven years, provides training in improved breeding, rangeland management, and business skills for youth and women in the community (African Food Changemakers, n.d.). Early support from donors and government extensions has helped communities adopt new techniques, such as supplementary feeding in dry season and vaccination campaigns, resulting in healthier herds and higher offtake rates. Mafisa in Zambia also encourages supplemental livelihoods, like selling dairy products or crafts, to reduce over-reliance on cattle, blending tradition with diversification.

Technology plays a growing role in livestock management. A standout example is Botswana's Livestock Identification and Trace-back System (LITS), a nationwide program to electronically tag and track cattle. Starting in the 2000s to meet European Union export standards, Botswana equipped each cow with an RFID microchip linked to a central database (Monyatsiwa, Ferguson & Phumaphi, 2013). This was a massive modernization leap for a country where cattle often roam vast communal lands. Veterinary officers were given mobile handheld computers that could scan tags in the field and sync data to the database, even in connectivity-poor areas (Monyatsiwa et al., 2013).

Infrastructure was the biggest hurdle, many remote cattle posts had no network coverage or electricity. The government tackled this by investing in solar-powered equipment and offline data collection that could be uploaded later (Monyatsiwa et al., 2013). There was also an adjustment period for farmers to accept the new system, some ranchers were wary of having their cattle "electronically watched." Through outreach and demonstrating how LITS helps quickly trace diseases or recover stolen animals, participation grew.

Botswana's cattle traceability has been among the most advanced in Africa, allowing the country to protect its EU beef export market (Mooketsi, 2013). An unexpected benefit is that the system's data could be repurposed for farmers' own use, for example, helping manage herds or prove ownership, though it was noted that this potential is not fully realized yet (Mooketsi, 2013). Nonetheless, LITS shows how a traditional pastoral economy can be connected to global value chains through technology. Likewise, digital tools for small-scale herders are emerging elsewhere. In South Africa and Zimbabwe, a Cattle Grading mobile app now assists small farmers in evaluating their cows' body condition and market value. Using image recognition, the app grades a cow by analysing photos of its teeth and body, then recommends feeding or veterinary actions to improve its grade (Solidaridad Network, 2019). This helps farmers treat livestock as market assets and not just a status symbol, commanding better sale prices. Notably, the app works offline and syncs when internet is available, a crucial feature for rural users (Solidaridad Network, 2019). By 2019, early adopters reported more informed herd management and fewer instances of underpriced cattle sales, breaking the cycle of poverty in some communities (Solidaridad Network, 2019). These examples illustrate how everything from remote sensing to mobile apps are enhancing the age-old cattle economies in Africa.

Pooling Resources via Shared Mechanization and Finance

Another area of innovation is resource pooling for inputs and equipment, where communities jointly access what an individual could never afford alone. Traditional examples include neighbours collectively buying an ox or sharing labour to plough each other's fields. Today, digital platforms are augmenting this concept, especially for farm mechanization:

- Hello Tractor in Nigeria: Often called the "Uber for tractors," Hello Tractor is an agritech service that connects small farmers with tractor owners via a mobile app and SMS system. In Nigeria, where there is roughly one tractor per 5000 farmers, the high cost of machinery has long limited productivity. Hello Tractor's solution was to create a network where a farmer can text a request for tractor services, and the platform matches them with a nearby tractor operator for hire (Harvard Ministerial Leadership Program, 2024). The tractor owners use GPS-enabled devices to track usage, so the service is transparent and billed per hectare. This innovation drew support from Nigeria's Federal Ministry of Agriculture and partners like John Deere. In 2021, a public-private initiative deployed 10,000 smart tractors equipped with Hello Tractor telematics to serve smallholders across the country (Harvard Ministerial Leadership Program, 2024). The program also introduced a pay-as-you-go financing model for youth to become tractor entrepreneurs, since the platform data proved machine usage, banks felt confident giving loans to buy tractors (Harvard Ministerial Leadership Program, 2024).

Building two-sided trust was critical, farmers had to trust that a tractor would actually show up and do quality work, and tractor owners had to trust they would be paid. Hello Tractor addressed this with a rating system and secure payments, as well as training for both operators and farmers (Harvard Ministerial Leadership Program, 2024). Connectivity in rural areas was another hurdle, solved by making the system SMS-compatible, no smartphone needed, and providing agents who aggregate requests in very remote villages. As a result, Hello Tractor has dramatically increased land cultivated in some regions. By efficiently sharing tractors, farmers can prepare land on time and plant more, boosting yields. The program projected an additional 37 million metric tons of food could be produced and over 2 million jobs created once the 10 000 tractors are fully operational (Harvard Ministerial Leadership Program, 2024). While those are forward-looking estimates, field reports already show higher cropping intensity and new custom-hire businesses emerging. This model of shared asset utilization is now spreading to Kenya, Bangladesh, and other countries.

- Custom Hiring Centres App in India: In India, the government launched a similar initiative to modernize its existing network of Custom Hiring Centres (CHC), local cooperatives or entrepreneurs who rent out farm equipment. In 2019 the Ministry of Agriculture unveiled the CHC Farm Machinery app, an Uber-like platform for renting farm implements (Future Farming, n.d.). Through this app, farmers anywhere in India can find and order machinery within a 50 km radius, from tractors to advanced equipment like laser land-levellers (Future Farming, n.d.). By 2020, over 38 000 hiring centres were linked on the app, offering more than 100 000 pieces of equipment for hire (Future Farming, n.d.). The app allows farmers to compare rates and book what they need at affordable prices set by the centres and government subsidies (Future Farming, n.d.).

Initial challenges included onboarding thousands of small rental providers onto a digital system, where many required training to use the app and smartphones. The government conducted nationwide workshops and leveraged its extensive extension network to enrol CHCs. Another challenge was awareness where millions of farmers had to learn that such a service exists. This was tackled with radio campaigns and integration of the app into the popular local-language agriculture apps. The impact has been significant. Farmers who previously had to do without equipment or rely on outdated, inefficient tools can now rent modern machines easily. For example, a laser-guided land leveller, which is far too expensive for an individual smallholder, can be rented via the app, enabling the farmer to level their field in hours instead of days, improving yield by 10–15% credits to better water management (Future Farming, n.d.). The initiative thus increases technology adoption by lowering cost barriers. As of its launch, the app was quickly populated with a huge number of rentals, and it facilitated over 250 000 equipment hiring transactions per year by linking supply and demand more transparently (Future Farming, n.d.). India's success with the CHC app has become a template for other countries looking to implement digital marketplaces for shared farm tools.

Beyond equipment, modern finance is being woven into resource pooling. In Mali and Tanzania, for instance, NGOs have helped village savings groups use mobile money and apps for their pooled funds, allowing members to deposit savings via phone and receive digital loans. Studies in Tanzania found that groups using a digital record-keeping app had greater transparency and fewer disputes over the fund's money (Global Communities, 2020). In Kenya, the MasterCard Foundation's Community Pass program is linking small farmers, often organized in cooperatives or savings circles, to formal banks using a single digital ID and wallet (African Development Bank Group, 2024). This integration means farmers can leverage their pooled savings history to get credit or insurance offers previously out of reach. For example, some dairy co-ops now use digital platforms to track members' milk deliveries and pay them via mobile banking, which creates an electronic record that banks use to extend micro-loans for buying cows or feed.

By pooling resources and connecting to formal services, communities reduce risk and scale up their operations. Shared tractors and tools lead to more timely planting and harvesting, directly raising productivity and incomes. Digital savings and credit give farmers a cushion and capital to invest in better seeds or livestock. Perhaps most importantly, these innovations preserve the collective ethos of indigenous systems, neighbours helping each other, while overcoming their old limitations, such as the lack of capital, information, or market access. The result is more resilient local food systems. In Indonesia's digital farmer ecosystem project, for example, once farmers had real-time price info and could team up through online groups, they negotiated better prices and planted crops aligned with market demand, significantly boosting their incomes despite the COVID-19 pandemic (Asian Development Bank, 2023). Women's participation also rose, as seen when nearly half the users were women in that project, they gained a stronger bargaining position and a voice in community resource decisions (Asian Development Bank, 2023).

These case studies from Africa and Asia underscore a common theme which is blending traditional cooperation with modern innovation can yield impressive gains. Whether it's a Maasai herder using a smartphone app to manage cattle or a farmers' club in Java pooling e-wallet funds to buy a tractor session, technology is reinforcing, not replacing, the social structures that underlie indigenous economic systems. By addressing challenges like digital literacy, trust, and infrastructure, many of these initiatives have achieved greater food production, financial inclusion, and community empowerment. Farmers are increasingly leveraging precision farming tools, mobile finance, and online marketplaces on their own terms, adapting them to fit local customs. The continued support of governments (through policy and investment), researchers (through studies and training), and non-profits (through on-the-ground facilitation) is vital to scale up these successes. As these examples show, empowering communities with the right tools can transform traditional practices into engines of modern prosperity (Lukwa et al., 2024).

Market access through modernizing indigenous economic systems

Farming stokvels

Online platforms have the potential to revolutionize market linkages for farming stokvels. Traditionally, small-scale farmers have had to rely on intermediaries to sell their produce, often resulting in lower profits due to the cuts taken by these middlemen (Iwara & Netshandama, 2021). Digital platforms can connect stokvels directly with retailers and consumers, ensuring fairer prices for their produce. Rotz, Gravely, Mosby, Duncan, Finnis, Horgan and Fraser (2019) concur that this direct connection can also improve market access, allowing farmers to reach a wider audience and potentially tap into new revenue streams such as pre-harvest sales or subscription boxes, which are becoming increasingly popular in CSA models. For youth agripreneurs, who are often more attuned to digital trends, these online marketplaces offer an opportunity to innovate and expand their businesses beyond traditional models. They can experiment with different marketing strategies, engage directly with consumers, and create value-added products that cater to niche markets, thereby increasing their profitability and sustainability (Khapayi & Celliers, 2016).

Community Supported Agriculture

Building strong connections between farmers and consumers is at the core of CSA's success. Digital channels such as social media, email newsletters, and virtual farm tours can play a significant role in fostering these connections (Rosegrant et al., 2014). According to Hvitsand (2016), these tools allow farmers to showcase their practices, educate consumers about the benefits of seasonal consumption, and keep customers informed about what is happening on the farm. For youth agripreneurs, who are often more comfortable with digital communication, these channels offer a way to personalize the customer experience, build a loyal customer base, and differentiate themselves in a competitive market (Fletschner & Kenney, 2014; Fabregas, Kremer & Schilbach, 2019). By leveraging digital tools, young farmers can share their unique stories, highlighting their commitment to sustainable practices, organic farming, or indigenous crop varieties. This transparency builds trust and encourages consumers to support local producers, which can lead to a more resilient and sustainable agricultural ecosystem. Additionally, digital engagement allows farmers to gather feedback, respond to consumer needs, and adapt their offerings accordingly, creating a more dynamic and responsive CSA model (Aker & Fafchamps, 2015).

Livestock mafisa

Similar to the modernization of stokvels, online platforms can play a transformative role in livestock mafisa by connecting participants directly with buyers such as abattoirs, tanneries, local consumers, or online meat marketplaces. Khapayi and Celliers (2016) states that this direct access to markets broadens the buyer base, allowing mafisa participants to secure more competitive prices and reducing their dependence on traditional middlemen, who often take a significant portion of the profits. For young farmers, this direct market access opens doors to value-added activities like selling meat, hides, or other livestock products directly to consumers, thereby increasing their revenue potential (Mbatha, 2021). According to Larson et al. (2016), online platforms can also facilitate the pooling of resources and the sharing of best practices among mafisa participants. Forums, chat groups, and digital workshops can connect young farmers with experienced mentors, livestock experts, and peers facing similar challenges. This networked approach can enhance knowledge transfer, provide support, and foster a sense of community, even in the absence of physical proximity.

Resource pooling

Collective marketing initiatives are a powerful tool for modernizing resource pooling. By pooling resources, members of indigenous economic systems can create a unified online brand, participate in farmers' markets collectively, or negotiate bulk sales with larger buyers (Aker & Fafchamps, 2015; Fabregas et al., 2019). Hilson (2016) asserts that this collective approach can significantly amplify market visibility and bargaining power, enabling small-scale producers to access markets that would otherwise be dominated by larger competitors. For youth agripreneurs, collective marketing provides a platform to showcase their products alongside established producers, enhancing their market credibility and enabling them to benefit from shared promotional activities (Khapayi & Celliers, 2016). Drawing from this, digital marketing tools, such as social media, e-commerce platforms, and online marketplaces, can be particularly effective in this regard. These tools allow indigenous economic groups to reach a broader audience, including urban consumers who are increasingly interested in supporting local and sustainable food systems. By leveraging digital marketing, youth agripreneurs can build direct relationships with consumers, educate them about the unique value of their products, and create a loyal customer base that values the social and environmental benefits of indigenous agriculture (Jayne et al., 2014).

Shortfalls in the modernization of indigenous economic systems

Farming Stokvels

While the modernization of farming stokvels holds great promise, there are several challenges and potential shortfalls that must be considered. One significant challenge is the digital divide in rural areas. Limited access to the internet and digital devices can hinder the full adoption of online platforms (Neethirajan & Kemp, 2021), leaving some members of the stokvel excluded from the benefits. This could exacerbate existing inequalities within the group, particularly between older members who may be less familiar with technology and younger members who are more digitally literate. Another potential shortfall is the risk of losing the communal and participatory essence of traditional stokvels. Rotz et al. (2019) states that the shift to online platforms could lead to reduced face-to-face interactions, weakening the social bonds that are fundamental to the success of these groups. It is important to find a balance between embracing digital tools and maintaining the strong sense of community that defines a stokvel.

Moreover, the adoption of precision farming technologies may require significant upfront investment (Turner & Schlecht, 2019), which could be a barrier for some stokvels, particularly those in resource-constrained areas. Barrett, Christiaensen, Sheahan and Shimeles (2017) put forward that without proper training and support, there is also a risk that these technologies could be underutilized or misused, leading to suboptimal outcomes. Neethirajan and Kemp (2021) suggests that it is essential to provide ongoing education and technical assistance to ensure that all members can fully benefit from these innovations. Lastly, while bypassing intermediaries through direct-to-consumer platforms can result in higher profits, it also requires farmers to take on additional responsibilities, such as marketing, packaging, and distribution (Larson et al., 2016). These tasks can be time-consuming and may require skills that farmers do not currently possess, potentially leading to burnout or disengagement.

Community Supported Agriculture

Despite the modernization of CSA through indigenous modernism offering numerous benefits, there are also significant challenges and potential drawbacks that must be critically considered. One of the main issues is the limited access to reliable internet and digital devices in rural areas (Aker & Fafchamps, 2015). This can hinder the full adoption of online CSA platforms, potentially excluding some farmers and consumers from participating. This digital exclusion could perpetuate existing inequalities, particularly affecting older farmers or those in more remote areas who may struggle to adapt to new technologies. Additionally, Hvitsand (2016) states that the shift to digital platforms can alter the community-centric nature of CSA. Traditionally, CSA relies on personal relationships and direct interactions between farmers and consumers, fostering a strong sense of community and mutual support (Jayne et al., 2014). Moving these interactions online can risk diminishing the personal connection that is central to CSA's appeal, leading to a more transactional relationship that undermines the communal values of the system.

The adoption of modern delivery systems, while crucial for maintaining product quality, also presents challenges. Investing in refrigerated transport or establishing new logistics networks can be costly (Campbell et al., 2014), and these expenses may be prohibitive for small-scale or start-up CSA operations. For youth agripreneurs, who often face financial constraints, these costs could pose a significant barrier to scaling their operations (Chinaka, 2016). Moreover, Campbell et al. (2014) states that the increased reliance on third-party delivery services introduces another layer of complexity, as farmers must now manage relationships with delivery partners and ensure that these partners uphold the standards of freshness and quality that consumers expect.

Furthermore, according to Rotz et al. (2019), the use of digital marketing and engagement tools requires a set of skills that not all farmers possess. While younger agripreneurs may be more comfortable with social media and online communication, there is still a need for training and support to effectively leverage these tools for business growth. Mismanagement of digital platforms or inadequate customer engagement strategies can lead to missed opportunities and a failure to build the strong consumer relationships (Barrett et al., 2017) that are essential for CSA success. While the modernization of CSA can provide a more reliable income stream, it can also create new pressures for farmers to meet the demands of a subscription-based model. This can include the need to consistently deliver high-quality produce, manage customer expectations, and deal with the complexities of customer retention in a competitive market (Aker & Fafchamps, 2015). For youth agripreneurs, who may still be learning the nuances of agricultural production and business management, these pressures can be daunting.

Livestock Mafisa

There are potential challenges presented by the modernization of livestock mafisa through indigenous modernism offers that must be addressed. The limited access to reliable internet, electricity, and digital devices in rural areas can hinder the adoption of GPS tracking and online platforms (Ilyas & Ahmad, 2020). For many rural youth agripreneurs, the upfront costs of purchasing GPS collars, smartphones, and data subscriptions can be prohibitive. Additionally, there may be a lack of technical skills required to effectively use these tools, necessitating ongoing training and support (Turner & Schlecht, 2019). Without addressing these barriers, the benefits of modernization may be unevenly distributed, potentially excluding some participants from the opportunities that digital tools provide.

Modernizing mafisa through technology can risk disconnecting the practice from its traditional roots, which are deeply embedded in communal values, trust, and reciprocal relationships. The introduction of digital platforms and data-driven decision-making could shift the focus from communal benefit to individual profit maximization, undermining the collective spirit that is central to mafisa (Larson et al., 2016; Gumbo, 2018). For example, if younger participants prioritize maximizing market prices through online sales, they may be less inclined to share resources or provide assistance to less fortunate community members, thereby eroding the social fabric that sustains the mafisa system. Ilyas and Ahmad (2020) assert that the integration of data-driven decision-making and digital tracking adds a layer of complexity to livestock management that may be overwhelming for some participants, particularly those new to farming. The need to interpret data accurately and make decisions based on that data requires a certain level of expertise. Incorrect use or interpretation of data can lead to suboptimal decisions, such as overgrazing or poor breeding choices, which could negatively impact herd health and profitability (Gumbo, 2018). Youth agripreneurs, in particular, may require substantial training and ongoing support to navigate these complexities effectively.

There is also the risk of becoming overly reliant on technology, which can be vulnerable to failures such as GPS signal loss, app malfunctions, or data breaches (Aquilani et al., 2022). In rural areas, where technical support may be limited, such failures could have significant consequences, including loss of livestock or financial losses due to mismanagement. Moreover, the reliance on technology could create a disconnect from the hands-on, observational skills that are traditionally valued in livestock management. For instance, while GPS can track location, it cannot replace the nuanced understanding of animal behaviour and health that comes from close, personal observation (Gumbo, 2018). While direct access to markets via online platforms can provide better prices, it also exposes mafisa participants to market fluctuations and economic volatility. Neethirajan and Kemp (2021) states that young farmers who are heavily reliant on online sales may find themselves vulnerable to shifts in market demand, price drops, or competition from larger, more established producers. This increased exposure to market risks could undermine the financial stability that mafisa aims to provide, particularly for new entrants who lack the resources to absorb economic shocks.

Resource Pooling

Modernizing resource pooling through indigenous modernism presents potential setbacks even though it offers numerous advantages. One of the most significant challenges in modernizing resource pooling is the inadequate access to reliable internet, digital devices, and digital literacy skills (Fabregas et al., 2019). This can hinder the effective adoption of online platforms for resource sharing and collective marketing. For many rural youth, the upfront costs of smartphones, data plans, and training can be prohibitive, creating barriers to participation (Barrett et al., 2017). Without targeted efforts to bridge this digital divide, the benefits of modernized resource pooling may not reach the most marginalized members of the community, potentially exacerbating existing inequalities.

Modernization through digital tools and financial institutions risks eroding the traditional practices and values that underpin resource pooling. Indigenous systems of resource pooling are built on trust, reciprocity, and communal relationships, which may be difficult to replicate in a digital environment (Rotz et al., 2019). For instance, while online platforms can facilitate resource sharing, they may not capture the nuanced social dynamics that govern traditional resource pooling, such as the prioritization of those in greatest need or the reciprocal obligations that come with resource use. Mbatha (2021) reported that there is a risk that the digitalization of resource pooling could lead to a more transactional and less community-oriented approach, undermining the social cohesion that is central to indigenous economic systems. The integration of digital platforms and community-owned financial institutions can introduce new complexities into the management and decision-making processes of resource pooling. Hilson (2016) adds that coordinating resource sharing, managing collective investments, and navigating the financial services landscape require a level of organization and governance that may be challenging for communities that have traditionally operated on more informal bases. For youth agripreneurs, who may be less experienced in collective management, this added complexity can be overwhelming, potentially leading to conflicts or inefficiencies in resource use.

While digital platforms and collective marketing can expand market access, they also expose indigenous economic systems to the volatility of external markets and the risks associated with technological dependence (Fabregas et al., 2019). For instance, fluctuations in market demand, changes in consumer preferences, or disruptions in digital services could negatively impact the income stability of resource pooling participants. According to Jayne et al. (2014), the over-reliance on these external factors could undermine the self-sufficiency that is a characteristic of indigenous economic systems, making them more vulnerable to external shocks. Furthermore, Hilson (2016) states that access to loans and microfinancing through community-owned financial institutions can provide critical support for resource pooling, but it also introduces financial risks. The pressure to repay loans or meet financial obligations can be particularly challenging for youth agripreneurs, who may face unpredictable farming conditions or market uncertainties. Without careful financial management and support, there is a risk that some participants could fall into debt, undermining the economic stability that resource pooling aims to provide.

The Role of Women and Gender Dynamics in Indigenous Economic Systems

Farming Stokvels

Women have historically played pivotal roles within farming stokvels in South Africa, primarily due to their responsibilities in household financial management and subsistence agriculture. According to Verhoef (2020), women are central to the operation of stokvels, often driving community savings initiatives that provide critical capital for agricultural investments. This system has traditionally offered women financial autonomy and empowerment in otherwise male-dominated rural economies. However, Zondo (2017) argues that despite their leadership in stokvel management, women's contributions remain undervalued, reflecting broader gender inequalities. The modernization of stokvels through digital platforms could empower women by improving their access to formal financial services, yet there is a risk that without deliberate efforts to bridge the digital divide, rural women may be marginalized due to lower digital literacy and restricted access to technology (Menze & Tsibolane, 2019).

Community Supported Agriculture

The CSA model inherently relies on active participation and leadership from women, who often manage the direct interactions between farms and consumers. Hvitsand (2016) asserts that CSA practices empower women agripreneurs by enabling them to control production decisions and directly engage with markets, reducing gender disparities in market access and economic power. However, as highlighted by Fletschner and Kenney (2014), the shift toward digital platforms may inadvertently reinforce gender inequities, as women in rural areas often face significant barriers in accessing technology. Therefore, while CSA modernization can empower female agripreneurs by expanding market opportunities, it must be coupled with targeted initiatives aimed at enhancing women's digital literacy and access to resources to avoid deepening existing gender inequalities.

Livestock Mafisa

In livestock mafisa, women traditionally engaged in livestock management, contributing significantly to decision-making and animal care. Although, men were responsible for negotiating and trading. Mokomane (2018) asserts that women's involvement in mafisa systems enhances household resilience and promotes knowledge transfer across generations. However, according to Larson et al. (2016), the introduction of digital technologies such as GPS tracking and mobile applications, though beneficial, may inadvertently marginalize women if they are not provided equitable access to these technologies or necessary training. Furthermore, Neethirajan and Kemp (2021) explain that technology adoption may shift emphasis from communal resource-sharing toward individualized profit-making, potentially disrupting traditional gender dynamics and cooperative relationships crucial to the system's success. Thus, inclusive modernization strategies must explicitly prioritize gender-sensitive interventions to avoid undermining women's traditional roles within mafisa.

Resource Pooling

Resource pooling has historically been beneficial to women in rural South African communities by allowing them collective access to agricultural resources, thereby reducing individual financial burdens and enhancing economic participation (Hilson, 2016). Nevertheless, Jayne et al. (2014) assert that as resource pooling moves towards modernization, challenges emerge around equitable governance and decision-making, potentially marginalizing women whose voices are already underrepresented in formal institutional contexts. Fabregas et al. (2019) further explains that digital platforms intended for resource management can disproportionately disadvantage women if they fail to account for gender-specific barriers, including lower rates of technological adoption among rural women. Consequently, successful modernization must include women-specific support mechanisms such as targeted training and digital empowerment programs, ensuring their active and equitable participation in modernized resource pooling initiatives.

RESEARCH METHODOLOGY

This study examined how tailoring indigenous economic systems for new generation agripreneurship through indigenous modernism can serve as a pathway to revitalising rural agriculture and fostering sustainable development. By integrating traditional practices with modern techniques, the study aligns with broader goals of enhancing agricultural efficiency, inclusivity, and economic resilience among young agripreneurs in South Africa. A secondary data approach was employed, utilizing existing literature and research findings to explore the potential impacts of modernizing indigenous economic systems, rather than conducting primary research in specific communities (Healey, 2017; Roulston & Choi, 2018). As noted by Rassel, Leland, Mohr and O'Sullivan (2020), this method enables the comprehensive synthesis of prior studies to build a deeper understanding of how indigenous modernism can bridge the gap between traditional knowledge and contemporary agricultural practices.

The secondary data collection focused on reviewing past research to investigate how integrating modern elements into indigenous systems can support new generation agripreneurs, enhance market access, and promote sustainable agricultural practices. This review included examining the impact of blending indigenous methods with digital tools, value addition, and collaborative strategies within frameworks such as farming stokvels, livestock mafisa, resource pooling and CSA. To gather relevant academic publications, electronic databases like Google Scholar, Science Direct, ResearchGate, and the University of KwaZulu-Natal online library were systematically searched using terms such as "indigenous economic systems," "agripreneurship," "farming stokvel," "livestock mafisa," "Community Supported Agriculture," "indigenous modernism," and "youth in agriculture."

This approach provided valuable insights into the ways indigenous modernism can transform traditional agricultural systems to better align with the needs and aspirations of young agripreneurs. The findings highlight the critical role of modernizing indigenous economic systems in supporting youth engagement in agriculture, preserving cultural heritage, and contributing to the sustainable development of rural economies in South Africa.

FINDINGS AND DISCUSSION

The evolving agricultural landscape in South Africa presents a critical need to integrate traditional indigenous economic systems with modern techniques to better serve the next generation of agripreneurs. These indigenous systems, farming stokvels, livestock mafisa, CSA, and resource pooling, are deeply rooted in the socio-cultural structure of rural communities, fostering a sense of community, mutual support, and sustainability. However, to make these systems relevant and effective for young agripreneurs, who are increasingly inclined towards technology-driven and efficiency-oriented approaches, it is essential to tailor these systems through the lens of indigenous modernism. The strategic integration of modern digital tools, precision farming technologies, and community financial institutions into these systems can significantly enhance their effectiveness, accessibility, and appeal, addressing the key challenges faced by the agricultural sector in South Africa.

Modernizing Farming Stokvels

As mentioned in the literature, farming stokvels, as a form of cooperative savings and credit societies, provide a critical financial lifeline in rural areas where formal banking services are limited or inaccessible. The modernization of farming stokvels through digital platforms is essential for meeting the needs of contemporary agripreneurs. Implementing secure, user-friendly online systems can streamline financial contributions, record-keeping, and management, enhancing transparency and accountability among members (Kariuki & Ofusori, 2017). This modernization is not merely about digitizing existing processes but also about reinforcing trust within the group, which is paramount as younger, more tech-savvy generations take an active role in these cooperatives. Furthermore, digital platforms can facilitate remote participation, enabling members from different geographic locations to engage in decision-making processes, thus broadening the reach and inclusivity of stokvels.

The integration of precision farming technologies, such as soil sensors and weather monitoring systems, as noted by Ncube et al. (2018), can further modernize farming stokvels by providing real-time data on key agricultural variables. These technologies empower agripreneurs with actionable insights that can optimize resource use, improve crop yields, and reduce operational costs. For instance, data-driven irrigation scheduling based on real-time soil moisture levels can conserve water, a critical resource in South Africa's often drought-prone regions. Similarly, predictive analytics based on weather data can help farmers plan planting and harvesting schedules more effectively, reducing the risks associated with weather variability. For young agripreneurs, who may lack the practical experience of their older counterparts, access to such technologies through a stokvel can level the playing field, making agriculture a more viable and attractive career path.

Revitalizing Livestock Mafisa

The literature of this study mentioned that livestock mafisa, a traditional system of livestock sharing, offers an effective means of wealth distribution and knowledge transfer within rural communities. To adapt mafisa for modern agripreneurs, integrating digital tracking technologies like GPS collars and mobile apps is crucial. These technologies offer enhanced oversight, allowing agripreneurs to monitor livestock health, movement, and grazing patterns with precision. Therefore, this could be particularly beneficial in South Africa, where issues such as livestock theft, disease outbreaks, and overgrazing pose significant challenges. By providing real-time data on livestock location and health, digital tracking reduces the labour-intensive and less reliable nature of traditional monitoring methods, offering a more secure and efficient management approach, especially for young agripreneurs.

Additionally, data analytics can inform strategic decisions related to breeding, feeding, and market timing. As aforementioned by Gumbo (2018) and Aquilani et al. (2022), tracking weight gain and health indicators can help optimize feeding regimens, while market data can guide the timing of livestock sales to maximize profitability. This data-driven approach not only enhances the productivity and profitability of livestock farming but also promotes sustainable practices by preventing the overuse of grazing lands. For youth agripreneurs, who may be new to livestock management, these technologies provide a critical support system that combines the traditional benefits of mafisa with the precision and efficiency of modern farming practices.

Enhancing Community Supported Agriculture

As noted by Chen et al. (2019), CSA offers a direct link between farmers and consumers, providing agripreneurs with a stable and predictable income stream. To modernize CSA, the implementation of digital platforms for subscription management and consumer engagement is essential. These platforms can simplify the process for consumers, allowing them to easily select farm shares, manage their subscriptions, and receive regular updates from the farmers. This not only improves customer satisfaction but also builds a stronger relationship between the consumer and the producer, aligning with the values of transparency and accountability that are central to CSA. Additionally, this can promote word-of-mouth marketing among the customers.

For young agripreneurs, digital platforms can open new markets and reduce the reliance on traditional, often volatile, market channels. The predictability of income through CSA subscriptions could help mitigate financial risks, enabling better planning and investment in sustainable farming practices. Furthermore, modern logistics solutions, such as optimized delivery routes and partnerships with local distribution services, can address the challenges of getting fresh produce to consumers efficiently. By ensuring high-quality produce reaches the market in peak condition, young agripreneurs can establish a strong reputation and build a loyal customer base, thus driving the growth and sustainability of their CSA operations.

Advancing Resource Pooling

Hilson (2016) mentioned that resource pooling, a communal approach to managing shared resources like land, water, and equipment, is a fundamental practice in indigenous economic systems. Modernizing resource pooling through the establishment of community-owned financial institutions, such as cooperative banks or credit unions, could significantly enhance the financial resilience of rural agripreneurs. These institutions can provide tailored financial services that cater specifically to the needs of indigenous economic systems, offering savings accounts, loans, and microfinance options that are accessible and affordable. This could be similar to the tailoring of farming financial services offered by the South

African Land Bank. For young agripreneurs, access to such financial services could be transformative, providing the capital needed to invest in agricultural ventures, purchase equipment, or scale operations.

Digital platforms can further revolutionize resource pooling by facilitating the efficient sharing of agricultural equipment and infrastructure. A centralized online system can enable community members to list, book, and share resources such as tractors, harvesters, or cold storage facilities. This not only maximizes the utilization of these assets but also reduces the cost burden on individual agripreneurs, making advanced agricultural technologies more accessible. By pooling resources, communities can achieve economies of scale, invest in larger projects, and strengthen their collective bargaining power, all of which contribute to a more resilient and inclusive agricultural sector.

The integration of modern techniques into indigenous economic systems is pivotal for enhancing market access for agripreneurs in South Africa. Traditional systems have historically played a critical role in supporting rural economies through shared resources and communal decision-making. However, these systems often face challenges such as limited market access, dependency on intermediaries, and insufficient visibility in competitive markets. To address these issues, modernizing indigenous economic systems through digital platforms, direct marketing, and online engagement can offer transformative opportunities for the new generation of agripreneurs. By leveraging these tools, young farmers can enhance their market presence, secure better prices for their products, and build resilient and sustainable agricultural businesses.

Modernizing Market Access for Farming Stokvels

Farming stokvels, traditionally focused on pooling resources and sharing financial risks among small-scale farmers, can greatly benefit from digital transformation, particularly in expanding market access. Historically, these farmers have been reliant on intermediaries to reach markets, which often results in reduced profitability due to the margins taken by middlemen. By adopting online platforms, stokvels could directly connect with retailers and consumers, thereby eliminating intermediaries and securing fairer prices for their produce. Digital marketplaces provide an opportunity for agripreneurs to tap into new revenue streams, such as pre-harvest sales or subscription-based models similar to those seen in CSAs. This approach not only broadens the market reach of farming stokvels but also introduces greater flexibility and innovation in how these farmers conduct business.

For young agripreneurs, who are often more comfortable with digital technologies, online platforms represent a gateway to explore diverse marketing strategies and engage directly with consumers. They can experiment with digital marketing, social media outreach, and direct-to-consumer sales models, which allow them to personalize the customer experience and cater to niche markets. This direct engagement could also facilitate feedback loops, where consumer preferences and demands are quickly relayed back to the farmers, enabling them to adjust their production strategies accordingly. By integrating digital tools into farming stokvels, young agripreneurs can build more dynamic, consumer-responsive, and profitable agricultural enterprises.

Enhancing Community Supported Agriculture with Digital Tools

CSA models inherently rely on strong, transparent connections between farmers and consumers. Digital tools such as social media, newsletters, and virtual farm tours can significantly enhance these connections, providing platforms for agripreneurs to educate and engage with their customers. These tools allow farmers to showcase their farming practices, emphasize the benefits of supporting local agriculture, and foster a deeper understanding of seasonal and sustainable food production among consumers. For young farmers, who are often adept at navigating digital communication channels, these tools offer a way to differentiate themselves in a competitive market by sharing their unique stories, values, and commitments to sustainable practices.

Furthermore, digital engagement allows for the creation of personalized customer experiences, which can build loyalty and encourage continuous business. Young agripreneurs could use these channels to highlight their commitment to organic farming, indigenous crop varieties, or innovative farming techniques that align with consumer values. This transparency and authenticity build trust and could significantly enhance the appeal of CSA models. Additionally, the feedback mechanisms enabled by digital engagement allows farmers to quickly adapt their offerings to meet consumer needs, making CSA more dynamic and resilient. By leveraging these digital tools, CSA participants can create a more interactive, responsive, and consumer-centric agricultural ecosystem that not only boosts market access but also strengthens the overall sustainability of local food systems.

Digital Market Linkages in Livestock Mafisa

Livestock mafisa, a traditional livestock sharing system, offers an effective means of wealth distribution and risk-sharing within rural communities, as mentioned in the literature. Modernizing mafisa through online platforms could significantly broaden market access for participants by connecting them directly with buyers, as noted by Khapayi and Celliers (2016), such as abattoirs, tanneries, and consumers. This direct access reduces dependence on traditional middlemen, who often take a considerable share of the profits, and allows farmers to secure more competitive prices for their livestock products. For young agripreneurs, this digital transformation presents opportunities to engage in value-added activities, such as direct-to-consumer sales of meat, hides, or other livestock products, which can significantly enhance their revenue potential.

In addition to market access, digital platforms can facilitate knowledge sharing and resource pooling among mafisa participants. Online forums, chat groups, and digital workshops could create virtual spaces for young farmers to connect with experienced mentors, livestock experts, and peers. This networked approach not only enhances the sharing of best practices and innovative techniques but also fosters a sense of community and support, which is particularly valuable for young agripreneurs who may lack extensive experience in livestock management. By integrating digital tools into livestock mafisa, participants can build stronger, more connected communities that are better equipped to navigate the challenges of modern livestock farming.

Collective Marketing Through Resource Pooling

Resource pooling, which involves the communal management of shared resources like land, equipment, and financial capital, can be significantly enhanced through collective marketing initiatives. By pooling resources, members of indigenous economic systems can create unified online brands, participate collectively in farmers' markets, and negotiate bulk sales with larger buyers. This collective approach amplifies market visibility and bargaining power, enabling small-scale producers to compete more effectively with larger agricultural enterprises. For youth agripreneurs, collective marketing offers a platform to showcase their products alongside more established producers, enhancing their credibility and market presence.

Digital marketing tools, such as social media, e-commerce platforms, and online marketplaces, could be particularly effective in expanding the reach of collective marketing initiatives. These tools enable indigenous economic groups to connect directly with urban consumers who are increasingly interested in supporting local and sustainable food systems. By leveraging these digital channels, agripreneurs can educate consumers about the unique value of their products, including the cultural significance, sustainability, and environmental benefits of indigenous agriculture. This direct engagement not only builds a loyal customer base but also fosters a deeper appreciation of the social and environmental contributions of indigenous economic systems. Through collective digital marketing, young agripreneurs can build resilient, consumer-centric businesses that are well-positioned to thrive in a competitive market landscape.

The modernization of indigenous economic systems presents an opportunity to enhance the economic participation of youth agripreneurs in South Africa. However, this transition is not without its challenges and potential pitfalls. While modern techniques offer the promise of increased efficiency, market access, and profitability, they also pose significant risks that should be carefully navigated to ensure that these systems remain inclusive, equitable, and true to their communal roots. The digital divide, loss of traditional communal values, and increased complexity and costs associated with technology are some of the major shortfalls that could undermine the success of modernizing these indigenous economic systems. A nuanced approach that balances innovation with the preservation of traditional values is essential to tailor these systems to the needs of the new generation of agripreneurs.

Challenges in Modernising Farming Stokvels

The transition of farming stokvels into the digital age highlights several critical challenges that could impede their success. A significant issue is the digital divide that persists in many rural areas, where limited access to reliable internet and digital devices poses a barrier to the adoption of online platforms. This digital exclusion risks exacerbating existing inequalities within stokvels, as older members or those in remote locations may struggle to participate fully in digital transformations. This divide could not only limit the reach and effectiveness of modernized stokvels but could also risk alienating members who are less familiar with the technology. Consequently, it is crucial to implement strategies that address these access gaps, such as providing affordable internet access, digital literacy training, and ongoing technical support to ensure that all members can participate equally in the benefits of modernization.

Another critical concern is the potential erosion of the communal and participatory essence that defines traditional stokvels. The shift to digital platforms could reduce face-to-face interactions, weakening the social bonds that are fundamental to the success of these groups. Farming stokvels are not just economic cooperatives, they are deeply rooted in community values and collective decision-making. If the shift to digital platforms leads to a transactional, impersonal mode of operation, the very structure that sustains these groups could be undermined. It is essential to strike a balance between embracing digital tools and maintaining the strong sense of community that defines a stokvel. Hybrid models that combine digital transactions with regular in-person meetings could help preserve the communal spirit while still reaping the benefits of modernization.

Additionally, adopting advanced farming technologies, such as precision agriculture, introduces significant financial and technical barriers. The upfront investment required for such technologies could be prohibitive for many stokvels, especially those in resource-constrained areas. Without adequate financial support, training, and education, these technologies risk being underutilized or misused, leading to suboptimal outcomes. Furthermore, bypassing intermediaries through direct-to-consumer platforms, while potentially increasing profits, places additional burdens on farmers who must now handle marketing, packaging, and distribution. These tasks require skills and time that many farmers do not currently possess, potentially leading to burnout or disengagement. It is vital to provide comprehensive support, including skills training and resource pooling, to ensure that farmers can manage these additional responsibilities without compromising their well-being or the quality of their produce.

Complexities in Modernising Community Supported Agriculture

The modernization of CSA through indigenous modernism also faces significant challenges that need to be critically addressed to ensure the model remains sustainable and inclusive. One of the primary issues is the digital divide, which limits access to online CSA platforms for both farmers and consumers in rural South Africa. This digital exclusion can perpetuate existing inequalities, particularly affecting farmers who may struggle with new technologies. The shift to digital platforms also risks altering the community-centric nature of CSA. Traditionally, according to Jayne et al. (2014), CSA thrives on personal relationships and direct interactions between farmers and consumers, fostering a strong sense of community and mutual support. Moving these interactions online could diminish these personal connections, transforming what was once a deeply communal activity into a more transactional relationship. This shift could undermine the values of trust and mutual support that are central to CSA, making it crucial to find ways to maintain these connections in the digital realm.

Modern delivery systems, essential for maintaining product quality and reliability, present another layer of complexity. Investing in refrigerated transport or establishing new logistics networks can be financially burdensome, particularly for small-scale or start-up CSA operations. For youth agripreneurs who often face financial constraints, these costs can be a significant barrier to scaling their operations. Furthermore, the reliance on third-party delivery services adds another layer of risk, as farmers must manage these partnerships and ensure that the standards of freshness and quality expected by consumers are consistently met. Any lapses in delivery quality can harm the reputation of the CSA, making it essential for farmers to have robust systems in place to manage these logistics effectively.

Furthermore, the use of digital marketing and engagement tools requires a skill set that not all farmers possess. While younger agripreneurs may be more comfortable with social media and digital communication, Barrett et al. (2017) and Rotz et al. (2019) stated that effective utilization of these tools for business growth requires training and ongoing support. Mismanagement of digital platforms or inadequate engagement strategies can lead to missed opportunities and a failure to build strong consumer relationships. Additionally, the pressures of a subscription-based model, which requires consistent delivery of high-quality produce and effective customer retention strategies, can be daunting. For youth agripreneurs still learning the nuances of agricultural production and business management, these pressures can lead to fatigue. To mitigate these challenges, it is essential to provide tailored support that addresses both the technical and emotional aspects of managing a modern CSA.

Barriers in the Modernisation of Livestock Mafisa

The modernization of livestock mafisa through digital tools offers substantial benefits but also introduces several challenges. The digital divide remains a significant barrier, as limited access to reliable internet, electricity, and digital devices in rural areas can hinder the adoption of technologies such as GPS tracking and online platforms. The upfront costs associated with these technologies, including purchasing GPS collars, smartphones, and data subscriptions, can be unaffordable for many rural youth agripreneurs. Additionally, the technical skills required to effectively utilize these tools necessitate ongoing training and support. By addressing these barriers, the benefits of modernisation may be evenly distributed, potentially excluding some participants and exacerbating inequalities within the community.

Moreover, the modernisation of mafisa risks disconnecting the practice from its traditional roots, which are deeply embedded in communal values, trust, and reciprocal relationships. The shift towards digital platforms and data-driven decision-making could shift the focus from communal benefit to individual profit maximization, weakening the collective spirit that is central to mafisa. For instance, younger participants who prioritize maximizing market prices through online sales may be less inclined to share resources or assist less fortunate community members, thereby eroding the social fabric that sustains the mafisa system. The complexity of integrating data-driven decision-making and digital tracking also adds a layer of complexity to livestock management that may be overwhelming for some participants, particularly those new to farming. The need to interpret data accurately and make informed decisions based on that data requires a level of expertise that many young farmers may not yet possess. Incorrect use or interpretation of data can lead to suboptimal decisions, such as overgrazing or poor breeding choices, which could negatively impact herd health and profitability. Substantial training and ongoing support are essential to help youth agripreneurs navigate these complexities effectively.

There is also the risk of becoming overly reliant on technology, which can be vulnerable to failures such as GPS signal loss, app malfunctions, or data breaches. In rural areas, where technical support may be limited, such failures could have significant consequences, including the loss of livestock or financial setbacks due to mismanagement. Moreover, an overreliance on technology could create a disconnect from the hands-on, observational skills that are traditionally valued in livestock management. While GPS can track location, it cannot replace the nuanced understanding of animal behaviour and health that comes from close, personal observation. This emphasises the need for a balanced approach that incorporates technology while maintaining sight of traditional skills and knowledge.

Challenges in Modernising Resource Pooling

Modernising resource pooling through indigenous modernism presents a unique set of challenges, primarily driven by the digital divide and the risk of eroding traditional practices. Limited access to reliable internet, digital devices, and digital literacy skills can hinder the effective adoption of online platforms for resource sharing and collective marketing. For many rural youth, the upfront costs of smartphones, data plans, and training can be expensive, creating barriers to participation. Without targeted efforts to bridge this digital divide, the benefits of modernized resource pooling may not reach the most marginalized members of the community, potentially exacerbating existing inequalities.

Modernisation through digital tools and financial institutions risks undermining the traditional practices and values that underpin resource pooling. Indigenous systems of resource pooling are built on communal relationships, which may be difficult to replicate in a digital environment. Online platforms may facilitate resource sharing but may fail to capture the nuanced social dynamics that govern traditional resource pooling, such as prioritizing those in greatest need or upholding reciprocal obligations. This could lead to a more transactional and less community-oriented approach, threatening the social cohesion that is central to indigenous economic systems. Additionally, the integration of digital platforms and community-owned financial institutions introduces new complexities into the management and decision-making processes of resource pooling. Coordinating resource sharing, managing collective investments, and navigating the financial services landscape require a level of organization and governance that may be challenging for communities that have traditionally operated on more informal bases. For youth agripreneurs, who may be less experienced in collective management, this added complexity can be overwhelming, potentially leading to conflicts or inefficiencies in resource use.

While digital platforms and collective marketing can expand market access, they also expose indigenous economic systems to the volatility of external markets and the risks associated with technological dependence. Fluctuations in market demand, changes in consumer preferences, or disruptions in digital services could negatively impact the income stability of resource-pooling participants. Over-reliance on these external factors could undermine the self-sufficiency that is a characteristic of indigenous economic systems, making them more vulnerable to external shocks. Access to loans and microfinancing through community-owned financial institutions can provide critical support for resource pooling but also introduce financial risks. The pressure to repay loans or meet financial obligations can be particularly challenging for youth agripreneurs, who may face unpredictable farming conditions or market uncertainties. Without careful financial management and support, there is a risk that some participants could fall into debt, challenging the economic stability that resource pooling aims to provide. To mitigate these risks, it is crucial to provide robust financial education and support mechanisms that help youth agripreneurs navigate the complexities of modernized resource pooling without losing sight of the communal values that underpin these systems.

RECOMMENDATIONS

To enhance the effectiveness of tailoring indigenous economic systems for new generation agripreneurship through indigenous modernism, it is recommended that stakeholders, including policymakers, community leaders, and agricultural organizations, actively promote the integration of modern technologies with traditional practices. Providing targeted financial and technical support, such as grants, training programs, and access to digital tools, can empower young agripreneurs to adopt these hybrid systems. Additionally, developing policy frameworks that recognize and support indigenous-modern approaches could help institutionalise these practices and encourage broader participation. It is also recommended to establish mentorship programs that connect experienced practitioners of indigenous methods with young agripreneurs to facilitate knowledge transfer and foster a collaborative environment. Finally, ongoing research and data collection should be prioritized to monitor the impacts of these initiatives, ensuring they are adaptive and responsive to the evolving needs of the youth in agriculture.

FUTURE RESEARCH DIRECTIONS

Future research could focus on assessing the impact of integrating indigenous modernism on youth engagement in agriculture. This could involve longitudinal studies that track the participation rates, success stories, and retention of young agripreneurs in agricultural sectors where indigenous and modern practices have been blended. Further research could also explore the policy implications of incorporating indigenous modernism into agricultural development strategies. This could include developing policy frameworks that support the integration of indigenous practices with modern methods, ensuring that such frameworks are inclusive, culturally sensitive, and designed to empower young agripreneurs. Examining the long-term sustainability and resilience of indigenous economic systems that have been modernized could provide valuable insights into their effectiveness. Future research could explore how these systems adapt to external pressures such as climate change, market fluctuations, and socio-economic changes, and the role of indigenous modernism in enhancing their resilience. Conducting an economic impact analysis to quantify the benefits of modernized indigenous economic systems on rural communities, youth employment, and overall economic development could provide robust evidence of the value of this approach. This research could help to build a strong case for further investment and support from both the public and private sectors.

CONCLUSION

The integration of modern techniques into indigenous economic systems holds significant promise for the new generation of agripreneurs in South Africa. By incorporating digital platforms, precision technologies, and community financial institutions, these systems can be revitalised to meet contemporary challenges while maintaining their cultural and communal foundations. This approach not only enhances the efficiency, profitability, and sustainability of agripreneurship but also ensures the preservation and continuity of indigenous knowledge and practices. For young agripreneurs who navigate the intersection of tradition and innovation, this tailored strategy offers a viable pathway to establishing resilient, inclusive, and sustainable agricultural enterprises that align with the evolving dynamics of rural South Africa.

However, the modernisation of indigenous economic systems also presents significant challenges that require careful management. Issues such as the digital divide, potential erosion of communal values, and the need for adequate support and training must be addressed to ensure that these modernized systems are inclusive, equitable, and resilient. A balanced approach that thoughtfully integrates modern technology with traditional wisdom can transform systems like farming stokvels, CSA, livestock mafisa, and resource pooling, allowing them to thrive in today's market while preserving their essential communal and cultural values.

Ultimately, the fusion of traditional practices with modern innovations provides a powerful framework for young agripreneurs to build innovative, resilient, and inclusive agricultural enterprises. This approach can drive sustainable rural development, offering agripreneurs the tools and support they need to thrive in a rapidly changing agricultural landscape. By navigating these complexities with care, indigenous economic systems can be effectively tailored to meet the needs of the next generation, ensuring their long-term success and contribution to South Africa's rural economy.

REFERENCES

African Development Bank Group. (2024). World Food Day: How the new MADE Alliance will use digital technologies to help farmers in Africa feed the continent. https://www.afdb.org/en/news-and-events/world-food-day-how-new-made-alliance-will-use-digital-technologies-help-farmers-africa-feed-continent-74856#:~:text=,economy%20via%20Mastercard%20Community%20Pass

African Food Changemakers. (n.d.). https://afchub.org/entrepreneurs/charlotte-scott-02344#:~:text=Mafisa%20helps%20traditional%20cattle,management%2C%20and%20better%20market%20access

AgriCrowdy. (n.d.). https://agricrowdy.com/agri-crowdy-savings-club/#:~:text=How%20does%20Agri%20Crowdy%E2%84%A2%C2%A0%C2%A0Savings%20Club,work

Aker, J. C., & Fafchamps, M. (2015). Mobile phone coverage and producer markets: Evidence from West Africa. *The World Bank Economic Review*, *29*(2), 262–292. DOI: 10.1093/wber/lhu006

Aquilani, C., Confessore, A., Bozzi, R., Sirtori, F., & Pugliese, C. (2022). Precision Livestock Farming technologies in pasture-based livestock systems. *Animal*, *16*(1), 100429. DOI: 10.1016/j.animal.2021.100429 PMID: 34953277

Asian Development Bank. (January 2023). How Farmer-Led Digital Ecosystems Can Boost Yields and Incomes in Indonesia. https://development.asia/case-study/how-farmer-led-digital-ecosystems-can-boost-yields-and-incomes-indonesia#:~:text=Despite%20the%20challenges%20created%20by,crops%20in%20line%20with%20market

Barrett, C. B., Christiaensen, L., Sheahan, M., & Shimeles, A. (2017). On the structural transformation of rural Africa. *Journal of African Economies*, *26*(suppl_1), i11–i35. DOI: 10.1093/jae/ejx009

Buthelezi, N. N., & Hughes, J. C. (2014). Indigenous knowledge systems and agricultural rural development in South Africa: Past and present perspectives. *Indilinga*, *13*(2), 231–250.

Cameron, J. (2015). Enterprise innovation and economic diversity in community supported agriculture: Sustaining the agricultural commons. *Making other worlds possible: Performing diverse economies*, 53-71.

Campbell, H. (2020). *Farming inside invisible worlds: Modernist agriculture and its consequences*. Bloomsbury Academic. DOI: 10.5040/9781350120570

Campbell, L. M., Boucquey, N., Stoll, J., Coppola, H., & Smith, M. D. (2014). From vegetable box to seafood cooler: Applying the community-supported agriculture model to fisheries. *Society & Natural Resources*, *27*(1), 88–106. DOI: 10.1080/08941920.2013.842276

Chen, J., Gao, Z., Chen, X., & Zhang, L. (2019). Factors affecting the dynamics of community supported agriculture (CSA) membership. *Sustainability (Basel)*, *11*(15), 4170. DOI: 10.3390/su11154170

Cheteni, P., & Umejesi, I. (2023). Application of indigenous knowledge and sustainable agriculture by small-scale farmer households and entrepreneurs in South Africa. *International Journal of Development and Sustainability*, *12*(1), 1–18.

Chinaka, M. (2016). *Blockchain technology--applications in improving financial inclusion in developing economies: case study for small scale agriculture in Africa* (Doctoral dissertation, Massachusetts Institute of Technology).

Chipfupa, U., & Tagwi, A. (2021). Youth's participation in agriculture: A fallacy or achievable possibility? Evidence from rural South Africa. *Suid-Afrikaanse Tydskrif vir Ekonomiese en Bestuurswetenskappe, 24*(1), 1–12. DOI: 10.4102/sajems.v24i1.4004

Collier, P., & Dercon, S. (2014). African agriculture in 50 years: Smallholders in a rapidly changing world? *World Development, 63*, 92–101. DOI: 10.1016/j.worlddev.2013.10.001

Cousins, B. (2015). Through a glass darkly': Towards agrarian reform in South Africa. *Land divided, land restored. Land reform in South Africa for the 21st century*, 250-269.

Fabregas, R., Kremer, M., & Schilbach, F. (2019). Realizing the potential of digital development: The case of agricultural advice. *Science, 366*(6471), eaay3038. DOI: 10.1126/science.aay3038 PMID: 31831641

Fletschner, D., & Kenney, L. (2014). *Rural women's access to financial services: credit, savings, and insurance*. Springer Netherlands.

Fox, L., Senbet, L. W., & Simbanegavi, W. (2016). Youth employment in Sub-Saharan Africa: Challenges, constraints and opportunities. *Journal of African Economies, 25*(suppl_1), i3–i15. DOI: 10.1093/jae/ejv027

Future Farming. (n.d.). India launches Uber-like app for farm equipment. https://www.futurefarming.com/smart-farming/tools-data/india-launches-uber-like-app-for-farm-equipment/#:~:text=The%20Indian%20government%20has%20launched,app%20for%20renting%20farm%20equipment

Geza, W., Ngidi, M. S. C., Slotow, R., & Mabhaudhi, T. (2022). The dynamics of youth employment and empowerment in agriculture and rural development in South Africa: A scoping review. *Sustainability, 14*(9), 5041. DOI: 10.3390/su14095041 PMID: 37691717

Global Communities. (March 2020). New Study Explores Impact of Digital Savings Groups in Tanzania. https://globalcommunities.org/blog/new-study-explores-impact-of-digital-savings-groups-in-tanzania/#:~:text=New%20Study%20Explores%20Impact%20of,an%20increase%20in%20financial

Grey, S., & Patel, R. (2015). Food sovereignty as decolonization: Some contributions from Indigenous movements to food system and development politics. *Agriculture and Human Values, 32*(3), 431–444. DOI: 10.1007/s10460-014-9548-9

Griffin, T. W., Shockley, J. M., & Mark, T. B. (2018). Economics of precision farming. *Precision agriculture basics*, 221-230.

Gumbo, B. G. (2018). The dynamics of cattle farming, bovine disease and state intervention in North West District of Botswana, 1966-2000. *Botswana Notes and Records, 50*, 72–84.

Harvard Ministerial Leadership Program. (2024). Empowering Farmers: Hello Tractor's Digital Revolution in Nigeria. https://ministerialleadership.harvard.edu/leadership-resources/tech-database/hello-tractor-case-study/#:~:text=Uber,rent%20tractors%20on%20demand

Healey, N. M. (2017). *Reflections on the value of insider research as a qualitative research methodology.* SAGE Publications Ltd. DOI: 10.4135/9781526401489

Hilson, G. (2016). Farming, small-scale mining and rural livelihoods in Sub-Saharan Africa: A critical overview. *The Extractive Industries and Society*, *3*(2), 547–563. DOI: 10.1016/j.exis.2016.02.003

Hvitsand, C. (2016). Community supported agriculture (CSA) as a transformational act—Distinct values and multiple motivations among farmers and consumers. *Agroecology and Sustainable Food Systems*, *40*(4), 333–351. DOI: 10.1080/21683565.2015.1136720

Ilyas, Q. M., & Ahmad, M. (2020). Smart farming: An enhanced pursuit of sustainable remote livestock tracking and geofencing using IoT and GPRS. *Wireless Communications and Mobile Computing*, *2020*(1), 6660733. DOI: 10.1155/2020/6660733

Ipsos. (March 2024). Stokvels remain the untapped 'human banks' of South Africa (https://www.ipsos .com/en-za/stokvels-remain-untapped-human-banks-south-africa#:~:text=,App%20transforming%20 how%20groups%20operate

Iwara, I. O., & Netshandama, V. O. (2021). Small-enterprise capital mobilization and marketing in rural areas: A review synthesis of Stokvel model. *Academy of Entrepreneurship Journal*, *27*(2), 1–11.

Jayne, T. S., Chamberlin, J., & Headey, D. D. (2014). Land pressures, the evolution of farming systems, and development strategies in Africa: A synthesis. *Food Policy*, *48*, 1–17. DOI: 10.1016/j.foodpol.2014.05.014

Kariuki, P., & Ofusori, L. O. (2017, March). WhatsApp-operated stokvels promoting youth entrepreneurship in Durban, South Africa: experiences of young entrepreneurs. In *Proceedings of the 10th International Conference on Theory and Practice of Electronic Governance* (pp. 253-259). DOI: 10.1145/3047273.3047397

Khapayi, M., & Celliers, P. R. (2016). Factors limiting and preventing emerging farmers to progress to commercial agricultural farming in the King William's Town area of the Eastern Cape Province, South Africa. *South African Journal of Agricultural Extension*, *44*(1), 25–41. DOI: 10.17159/2413-3221/2016/ v44n1a374

Koenane, M. L. (2019). Economic development in Africa through the stokvel system: 'our' indigenous way or 'theirs'. *Filosofia Theoretica: Journal of African Philosophy. Culture and Religion*, *8*(1), 109–124.

Larson, D. F., Muraoka, R., & Otsuka, K. (2016). Why African rural development strategies must depend on small farms. *Global Food Security*, *10*, 39–51. DOI: 10.1016/j.gfs.2016.07.006

Lilavanichakul, A. (2020). Development of agricultural e-commerce in Thailand. *The FFTC Journal of Agricultural Policy*, *1*, 7–17.

Lohento, K., & Ajilore, O. (2015). ICT and Youth in Agriculture. *Africa agriculture status report*, 118-42.

Lukwa, A. T., Wayas, F. A., Lambert, E. V., & Alaba, O. (2024). Stakeholder mapping to explore social and economic capital of Rotating Savings and Credit Associations (ROSCAs) to increase demand for and access to healthy food. *Frontiers in Sustainable Food Systems*, *8*, 1301578. DOI: 10.3389/ fsufs.2024.1301578

Lutoombi, M. (2015). *Cattle Husbandry and trade in Bulozi, a historical perspective C. 1880-1973* (Doctoral dissertation, University of Zambia).

Magagula, B., & Tsvakirai, C. Z. (2020). Youth perceptions of agriculture: Influence of cognitive processes on participation in agripreneurship. *Development in Practice*, *30*(2), 234–243. DOI: 10.1080/09614524.2019.1670138

Magocha, M., Soundy, P., Muchie, M., & Magocha, B. (2019). Reviewing the applications of indigenous knowledge systems (IKS) in innovative crop production. *Indilinga*, *18*(2), 229–244.

Masoka, X. (2022). *Social Media used by selected stokvels to engage with stakeholders in South Africa* (Doctoral dissertation, University of Johannesburg).

Mbatha, C. N. (2021). Livestock production and marketing for small emerging farmers in South Africa and Kenya: Comparative lessons. *South African Journal of Agricultural Extension*, *49*(1), 141–161. DOI: 10.17159/2413-3221/2021/v49n1a10783

Menze, A., & Tsibolane, P. (2019). Online Stokvels: the use of social media by the marginalized.

Mokomane, Z. (2018). Indigenous social security systems in Botswana. *Indigenous Social Security Systems in Southern and West Africa, 13.*

Monyatsiwa, C., Ferguson, D., & Phumaphi, N. (2013, May). *Bridging the IT divide—Case study on the Botswana livestock industry. In 2013 IST-Africa Conference & Exhibition*. IEEE.

Mooketsi, B. E. (2013). Optimisation of livestock identification and trace-back system LITS database to meet local needs: Case study of Botswana. *The Journal of Community Informatics*, *9*(4). Advance online publication. DOI: 10.15353/joci.v9i4.3143

Morepje, M. T., Sithole, M. Z., Msweli, N. S., & Agholor, A. I. (2024). The Influence of E-Commerce Platforms on Sustainable Agriculture Practices among Smallholder Farmers in Sub-Saharan Africa. *Sustainability (Basel)*, *16*(15), 6496. DOI: 10.3390/su16156496

Musavengane, R., & Simatele, D. M. (2016). Community-based natural resource management: The role of social capital in collaborative environmental management of tribal resources in KwaZulu-Natal, South Africa. *Development Southern Africa*, *33*(6), 806–821. DOI: 10.1080/0376835X.2016.1231054

Ncube, B., Mupangwa, W., & French, A. (2018). Precision agriculture and food security in Africa. *Systems analysis approach for complex global challenges*, 159-178.

Neethirajan, S., & Kemp, B. (2021). Digital livestock farming. *Sensing and Bio-Sensing Research*, *32*, 100408. DOI: 10.1016/j.sbsr.2021.100408

Raheem, W. M., & Bako, A. I. (2015). Indigenous Practices for Sustainable Development Planning.

Rassel, G., Leland, S., Mohr, Z., & O'Sullivan, E. (2020). *Research methods for public administrators*. Routledge. DOI: 10.4324/9780429319860

Rosegrant, M. W., Koo, J., Cenacchi, N., Ringler, C., Robertson, R. D., Fisher, M., & Sabbagh, P. (2014). *Food security in a world of natural resource scarcity: The role of agricultural technologies*. Intl Food Policy Res Inst.

Rotz, S., Gravely, E., Mosby, I., Duncan, E., Finnis, E., Horgan, M., LeBlanc, J., Martin, R., Neufeld, H. T., Nixon, A., Pant, L., Shalla, V., & Fraser, E. (2019). Automated pastures and the digital divide: How agricultural technologies are shaping labour and rural communities. *Journal of Rural Studies, 68*, 112–122. DOI: 10.1016/j.jrurstud.2019.01.023

Roulston, K., & Choi, M. (2018). Qualitative interviews. *The SAGE handbook of qualitative data collection*, 233-249.

Sansom, B. (2024). Traditional economic systems. *In The Bantu-speaking peoples of southern* [Routledge.]. *Africa*, •••, 135–176.

Solidaridad Network. (April 2019). Cattle in Africa get graded with mobile monitoring. https://www.solidaridadnetwork.org/news/cattle-in-africa-get-graded-with-mobile-monitoring/#:~:text=The%20application%20makes%20use%20of,loin%2C%20rib%20cage%20and%20rump

South Africa, S. M. E. (August 2024). Circles 360 Is Launching In South Africa as Digital Stokvel Solution. https://smesouthafrica.co.za/circles-360-is-launching-in-south-africa-as-digital-stokvel-solution/#:~:text=Circle%20360%20is%20Software%20to,Benefit%20South%20African%20SMEs

Sunday Standard Reporter. (February 2022). Botswana revives Mafisa system to prop up agro-tourism. https://www.sundaystandard.info/botswana-revives-mafisa-system-to-prop-up-agro-tourism/#:~:text=Botswana%20government%20has%20embarked%20on,boost%20the%20country%E2%80%99s%20pastoral%20agriculture

Thierfelder, C., Rusinamhodzi, L., Ngwira, A. R., Mupangwa, W., Nyagumbo, I., Kassie, G. T., & Cairns, J. E. (2015). Conservation agriculture in Southern Africa: Advances in knowledge. *Renewable Agriculture and Food Systems, 30*(4), 328–348. DOI: 10.1017/S1742170513000550

Trendov, N. M., Varas, S., & Zeng, M. (2019). Digital technologies in agriculture and rural areas.

Turner, M. D., & Schlecht, E. (2019). Livestock mobility in sub-Saharan Africa: A critical review. *Pastoralism, 9*(1), 1–15. DOI: 10.1186/s13570-019-0150-z

Verhoef, G. (2020). Stokvels and economic empowerment: the case of African women in South Africa, c. 1930-1998. In *Women and credit* (pp. 91–114). Routledge. DOI: 10.4324/9781003103233-8

Warren, D. M. (1996). Indigenous knowledge systems for sustainable agriculture in Africa. *Sustainable development in third world countries*, 15-24.

Wicha, S., Photiphun, T., Janjaroenpan, P., Taweesak, C., Chainilwan, I., & Rodmanee, P. (2017). Proposed of e-Community Supported Agriculture (e-CSA) system to promote local organic products: The empirical study of Chiang Rai province. *2017 International Conference on Digital Arts, Media and Technology (ICDAMT),* Chiang Mai, 2017, 273-281. DOI: 10.1109/ICDAMT.2017.7904976

Zondo, N. M. (2017). *Self-help economic empowerment amongst women in Blaauwbosch, KwaZulu-Natal* (Doctoral dissertation).

ADDITIONAL READING

Adedeji, A. (Ed.). (2023). *Indigenization of African economies*. Taylor & Francis.

Campbell, H. (2020). *Farming inside invisible worlds: Modernist agriculture and its consequences*. Bloomsbury Academic. DOI: 10.5040/9781350120570

Cheteni, P., & Umejesi, I. (2023). Application of indigenous knowledge and sustainable agriculture by small-scale farmer households and entrepreneurs in South Africa. *International Journal of Development and Sustainability*, *12*(1), 1–18.

Diko, M. (2023). Advancing agriculture through Indigenous Knowledge Systems (IKS) in South African indigenous or black communities. *International Journal of Research in Business & Social Science*, *12*(2), 267–277. DOI: 10.20525/ijrbs.v12i2.2333

Magaya, S. (2021). *Indigenous Knowledge Systems cues essential in agricultural management and tracking seasonality in KwaZulu-Natal and Eastern Cape, South Africa*. University of the Witwatersrand.

Mamba, S. G., & Beharry-Ramraj, A. (2025). Psycho-Economic Dynamics of Black Women within the South African Rural Agricultural Landscape. *Zhongguo Kuangye Daxue Xuebao*, *30*(1), 124–130.

Mawere, M. (2024). *Culture, Indigenous Knowledge and Development in Africa*. African Books Collective.

Richards, P. (2023). *Indigenous agricultural revolution: ecology and food production in West Africa*. Routledge. DOI: 10.4324/9781003383734

KEY TERMS AND DEFINITIONS

Agripreneurship: Agripreneurship refers to entrepreneurial activity within the agricultural sector, where individuals innovate, manage, and assume risks associated with agricultural enterprises. It plays a critical role in revitalizing rural economies by generating employment and promoting sustainable food systems (Magagula & Tsvakirai, 2020).

Community Supported Agriculture (CSA): CSA is a direct farm-to-consumer model where consumers pay in advance to receive a share of the farm's produce. This arrangement ensures a stable income for farmers and fosters sustainable agricultural practices (Chen et al., 2019).

Digital Divide: The digital divide signifies the gap between individuals with access to digital technologies and those without. In South African rural contexts, it impedes the adoption of online tools necessary for modernizing indigenous economic systems (Neethirajan & Kemp, 2021).

Farming Stokvels: Farming stokvels are cooperative financial groups where members regularly contribute to a communal pool used to fund agricultural activities. These systems foster financial inclusion, reduce dependence on formal banking, and promote community-based risk-sharing (Verhoef, 2020; Zondo, 2017).

Indigenous Economic Systems: Indigenous economic systems are traditional, community-oriented economic arrangements involving shared resources and collective decision-making (Warren, 1996; Raheem & Bako, 2015). In this chapter, these include systems like stokvels, CSA, livestock mafisa, and resource pooling, rooted in sustainable practices and social solidarity.

Indigenous Modernism: Indigenous modernism is the strategic integration of modern technologies into traditional systems without eroding their cultural and communal values. It aims to improve the efficiency and relevance of indigenous practices while preserving their historical and social significance (Campbell, 2020).

Livestock Mafisa: Livestock mafisa is a customary system where livestock is loaned to another community member in exchange for shared benefits from the animal's productivity. This promotes mentorship, resource sharing, and intergenerational wealth transfer (Lutoombi, 2015; Mokomane, 2018).

Precision Farming: Precision farming is a modern agricultural approach using GPS, soil sensors, and data analytics to manage farming inputs efficiently. It enables farmers to optimize water, fertilizer, and pesticide use, leading to better yields and reduced environmental impact (Ncube, Mupangwa & French, 2018).

Resource Pooling: Resource pooling involves the shared use of essential, in this chapter, agricultural assets such as land, machinery, and water among community members. This collective model increases efficiency, reduces individual costs, and enables investment in infrastructure that would otherwise be unaffordable (Jayne, Chamberlin & Headey, 2014; Musavengane & Simatele, 2016).

Chapter 22
Incorporating Vernacular Tribal Literature Into the English Language Classroom:
A Study of the Gond Tribe of District Mandala in Madhya Pradesh

Shibani Basu Dubey
The Bhopal School of Social Sciences, India

Tanuja Khan
https://orcid.org/0000-0002-0013-4717
The Bhopal School of Social Sciences, India

Vineeta Saluja
Mangalayatan University, India

ABSTRACT

The Holistic Integration and Appreciation Model (HIAM) was developed to enhance educational outcomes for students from the Gond tribes by integrating their cultural heritage into the curriculum. This study aims to evaluate the effectiveness of the HIAM model in improving language proficiency, student engagement, cultural understanding, and community impact.An experimental research method was employed using a group post-test design. A purposive sample of one section of Class 9 students from Brain Child Academy in Mandla, Jabalpur, Madhya Pradesh, was selected as the intervention group. The conceptual framework of the tool was tested using Smart-PLS 4 software.The post-intervention data indicated significant improvements in language proficiency and student engagement. Students demonstrated a greater appreciation and understanding of their cultural heritage, reflected in increased self-esteem and sense of identity. The community also reported positive impacts, including heightened educational aspirations and greater involvement in school activities.

DOI: 10.4018/979-8-3373-4124-8.ch022

INTRODUCTION

Cultural diversity and inclusivity have been given more prominence in educational institutions in the past few years. The renewed focus on the significance of Indigenous heritage and Indigenous peoples' knowledge has spurred researchers, educators, and other stakeholders to search for new ways to incorporate Indigenous perspectives into school programs or mainstream curricula. The use of vernacular tribal literature in English utilizes English language classrooms to expose students to this cultural heritage (International Experiences in Language Testing and Assessment, 2015). The present study is based on the use of Gond Tribal literature among the Gond Tribe of Madhya Pradesh, district Mandla, India. The Gond Tribe, with its rich cultural heritage and distinct literary traditions, offers a valuable reservoir of knowledge and storytelling that can enrich the educational experiences of students. By incorporating Gond tribal literature into the English language curriculum, educators have the opportunity to foster cultural understanding, linguistic proficiency, and appreciation for indigenous perspectives among learners (Deogaonkar, 2008).

The Gond Tribe population in the district of Mandla in Madhya Pradesh is quite high and there are educational institutions that are trying to emphasize inclusive and culturally sensitive education for the first-grade children (Stauss et al., 2021). By conducting a critical analysis of the integration process and its effects on the student's learning, this study aims to enhance the existing literature on Multicultural Education and Curriculum Integration (Koreti, 2016). This study also seeks to explore the difficulties, advantages, and strategies involved in using the vernacular tribal literature in English language classes with the hope of helping shape educational policies and practices for cultural diversity and equality in the education system. Finally, the conclusions made in this research may contribute to the establishment of multicultural learning environments that focus on indigenous cultures such as the Gond Tribe (Thamanna & Subramani, 2023).

It will be important to understand how to merge the vernacular tribal literature used in the English language classroom besides taking some particular consideration of the indigenous Gond Tribe of Mandla district in Madhya Pradesh, India. Cultural context therefore, encompasses the social, historical, economic, and political framework that defines the culture of a certain group of people group (Gupta et al., 2010).

THEORETICAL FRAMEWORK AND DESIGNING INSTRUCTIONAL STRATEGIES

Designing Instructional Strategies using the Holistic Integration and Appreciation Model (HIAM) involves a purposeful approach to incorporating Vernacular Tribal Literature (VTL) into educational contexts. HIAM is a concise yet potent framework tailored to seamlessly integrate VTL, with a strong emphasis on cultural integration, cultural engagement, and the implementation of effective teaching strategies supported by comprehensive teachers' training.

Cultural Integration

HIAM prioritizes the seamless infusion of Vernacular Tribal Literature into educational settings, ensuring that it becomes an integral part of the curriculum. This involves carefully selecting VTL texts that authentically represent the cultural identity, values, and traditions of the indigenous community.

Educators leverage VTL to enhance students' understanding of diverse cultures, fostering inclusivity and respect for cultural diversity within the classroom (Koreti, 2016).

Cultural Engagement

Central to HIAM is the active engagement of students with Vernacular Tribal Literature, encouraging them to explore, analyze, and appreciate the cultural nuances embedded within the texts. Through interactive activities, discussions, and experiential learning opportunities, students are immersed in the rich tapestry of indigenous cultures, enabling them to develop a deeper connection and appreciation for VTL. Cultural engagement fosters empathy, critical thinking, and cross-cultural communication skills among students (Idrus et al., 2023).

Effective Teaching Strategies and Teachers' Training

HIAM underscores the importance of equipping educators with the knowledge, skills, and resources needed to effectively integrate Vernacular Tribal Literature into their teaching practices. Comprehensive teachers' training programs provide educators with guidance on selecting appropriate VTL texts, designing culturally relevant lesson plans, and implementing innovative teaching strategies tailored to meet the diverse needs of students. By empowering educators with the tools and support necessary for VTL integration, HIAM ensures that students receive a culturally responsive education that honors indigenous knowledge systems (Bottiani et al., 2020).

Professional Development for Teachers

Professional development for teachers plays a pivotal role in equipping educators with the necessary tools and skills to effectively integrate Vernacular Tribal Literature (VTL) into educational settings using the Holistic Integration and Appreciation Model (HIAM). This comprehensive professional development initiative is designed to enhance educators' understanding of VTL, foster cultural sensitivity, and empower them with effective teaching strategies (Osei-Owusu, 2022). Emphasis is placed on promoting cultural humility and competence, encouraging educators to recognize and address implicit biases that may impact their interactions with VTL and Indigenous communities (Blyznyuk, 2019).

Educators are introduced to a range of teaching strategies tailored to facilitate VTL integration, including culturally relevant pedagogy and differentiated instruction (Rizk & Hillier, 2022). Professional development for teachers empowers educators to embrace the challenges and opportunities of VTL integration, creating inclusive learning environments where students develop a deep appreciation for Vernacular Tribal Literature and the diverse cultures it represents (Bottiani et al., 2020).

Community Outreach and Celebration

Community outreach and celebration form integral components of the holistic approach to integrating Vernacular Tribal Literature (VTL) into educational settings using the Holistic Integration and Appreciation Model (HIAM). These initiatives are aimed at fostering collaboration, inclusivity, and cultural appreciation both within the school community and with external stakeholders, particularly indigenous communities like the Gond Tribe of Mandla district in Madhya Pradesh, India (Yang et al., 2021).

Community Impact

VTL serves as a bridge between educational institutions and indigenous communities, fostering mutual understanding, respect, and collaboration. By integrating VTL into the curriculum and engaging with local Indigenous communities, schools can positively impact community relationships and contribute to the preservation and promotion of Indigenous cultures and languages. Community members feel valued and empowered when their cultural heritage is acknowledged and celebrated within educational settings, leading to stronger community bonds and a sense of pride in cultural identity.

Language Proficiency

The incorporation of VTL into language instruction provides students with opportunities to develop proficiency in both their native language and the language of instruction, such as English. Through exposure to VTL texts, students enhance their vocabulary, grammar, and comprehension skills in both languages. Learning VTL promotes linguistic diversity and encourages students to appreciate the linguistic richness of indigenous languages, thereby preserving linguistic heritage and fostering bilingualism or multilingualism among students (Osei-Owusu, 2022).

Student Engagement

VTL engages students by offering culturally relevant and meaningful content that resonates with their lived experiences and cultural backgrounds. When students see themselves reflected in the literature they study, they become more motivated and engaged in learning (Tanuja Khan et al., 2021). VTL texts spark curiosity, critical thinking, and empathy among students as they explore themes, characters, and narratives that reflect indigenous cultures and traditions. Furthermore, interactive activities, such as storytelling sessions, role-playing exercises, and creative projects inspired by VTL, foster active participation and deeper engagement in the learning process.

Cultural Understanding

VTL serves as a window into the cultural heritage, values, and traditions of indigenous communities, promoting cultural understanding and appreciation among students. Through the study of VTL, students gain insights into diverse worldviews, belief systems, and ways of life, fostering empathy, respect, and cross-cultural communication skills.

OBJECTIVES OF THE STUDY

a. To investigate the impact of integrating Gond tribal literature into the curriculum on various educational and cultural outcomes among students.
b. To examine the effects of implementing various teaching strategies (TS) into the curriculum on diverse educational and cultural outcomes among students.
c. To examine the effects of implementing various teaching training programs (TT) into the curriculum on diverse educational and cultural outcomes among students.

d. To examine the effects of implementing various Community Engagement (CE) into the curriculum on diverse educational and cultural outcomes among students.

HYPOTHESIS OF THE STUDY

a. Ho: There is no significant effect of the degree of integration of Gond tribal literature (CI: Cultural Integration) into the curriculum on students' academic and cultural outcomes.
b. Ho: There is no significant effect of implementing various teaching strategies (TS) into the curriculum on students' academic and cultural outcomes.
c. Ho: There is no significant effect of implementing various training programs (TT) into the curriculum on students' academic and cultural outcomes.
d. Ho: There is no significant effect of implementing various Community Engagement (CE) into the curriculum on students' academic and cultural outcomes.

SIGNIFICANCE OF THE STUDY AND RESEARCH GAP

The significance of the study lies in its exploration of the effect of the interventions on the student's academic and cultural outcomes. This study contributed deeper understanding of the complex dynamics at play within educational settings. Existing literature has highlighted various socio-economic barriers that hinder the English language. There is a notable gap in research that specifically addresses the effectiveness of Vernacular Tribal Literature in English Language curricula as a means to improve language proficiency and cultural understanding the effectiveness of integrating Vernacular Tribal Literature (VTL) into English language curricula as a means to improve language proficiency and cultural understanding among students from tribal communities. The research gap underscores the need for targeted studies that explore innovative educational strategies, such as the Holistic Integration and Appreciation Model (HIAM), to bridge the gap in English language acquisition among tribal students.

RESEARCH METHODOLOGY

This study utilized an experimental research method, which is a systematic and scientific approach to research where the researcher manipulates one or more variables, and controls and measures any change in other variables. The experimental method is particularly well-suited for this study as it aims to establish a cause-and-effect relationship between the intervention (implementation of the HIAM model) and the educational outcomes of the students. By employing this method, the study seeks to demonstrate the direct impact of the HIAM model on various educational metrics such as language proficiency, student engagement, cultural understanding, and community impact. The research design chosen for this study was the only one-group post-test Design. In this design, only one group (the experimental group) is selected to receive the intervention. Data is collected solely after the intervention to assess its impact.

This design is straightforward and practical, especially when it is difficult to establish a control group due to logistical or ethical reasons.

The sample size for this study comprised one section of Class 9 students from Brain Child Academy in Mandla, Jabalpur, Madhya Pradesh. The specific number of students in this section represented the intervention group. Purposive sampling was used to select the sample. This non-probability sampling method involves selecting participants based on specific characteristics that align with the research objectives.

Intervention model

The Holistic Integration and Appreciation Model (HIAM) is an educational intervention specifically designed to address the unique needs of Gond tribe students. Recognizing the importance of cultural relevance in education, the HIAM model integrates four key components: cultural integration, community engagement, innovative teaching strategies, and specialized teacher training. Cultural Integration involves embedding the rich cultural heritage of the Gond tribes into the curriculum, ensuring that their history, traditions, languages, and values are respected and represented in the educational process.

Figure 1. HIAM model

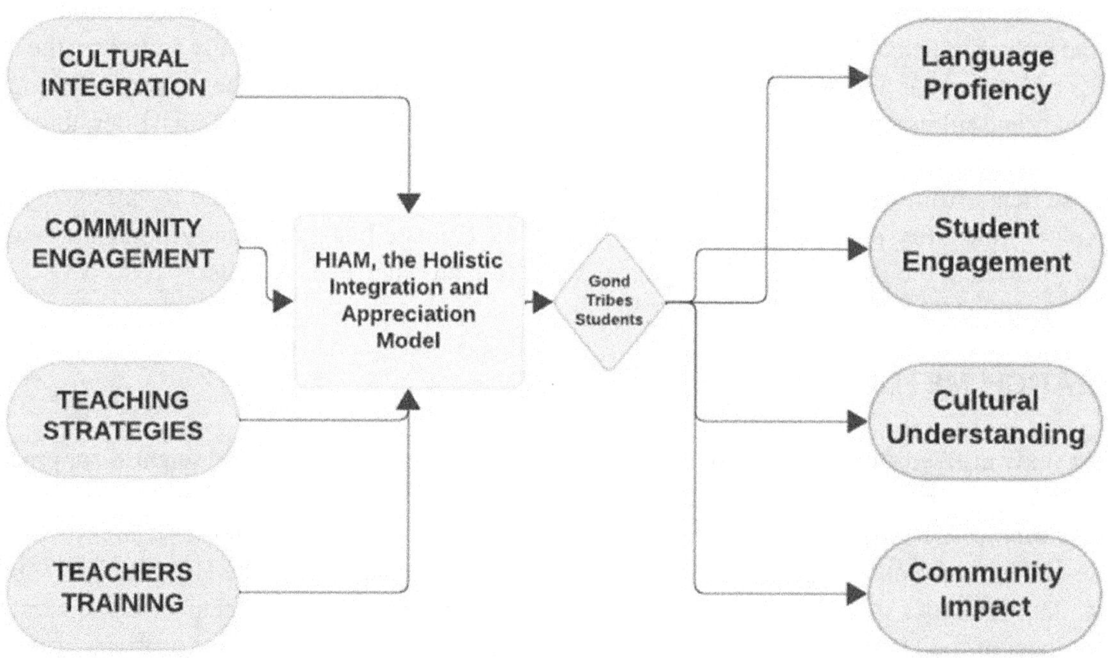

Conceptual Framework

Figure 2. Conceptual framework

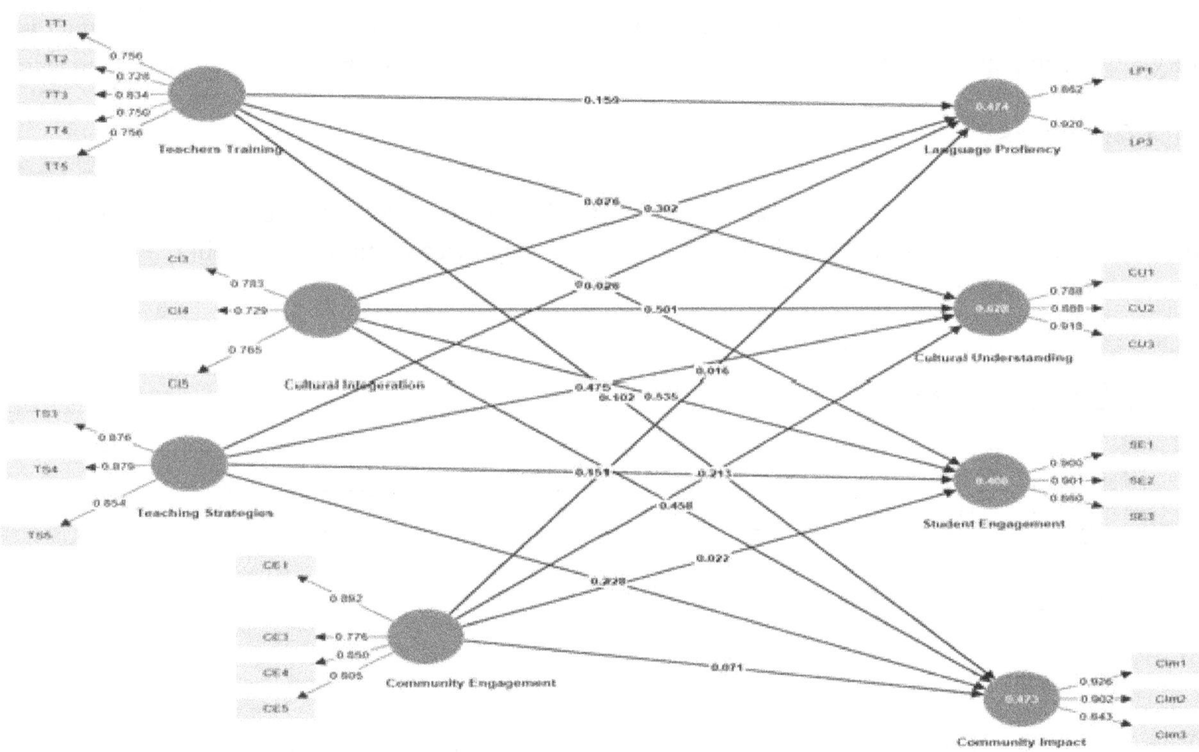

Table 1. Construct reliability and validity

	Cronbach's alpha	Composite reliability (rho_a)	Composite reliability (rho_c)	Average variance extracted (AVE)
Community Engagement	0.851	0.904	0.892	0.624
Community Impact	0.871	0.893	0.920	0.794
Cultural Integeration	0.714	0.737	0.803	0.453
Cultural Understanding	0.832	0.845	0.900	0.750
Language Profiency	0.718	0.803	0.842	0.643
Student Engagement	0.865	0.870	0.917	0.787
Teachers Training	0.837	0.899	0.874	0.581
Teaching Strategies	0.786	0.825	0.855	0.553

Interpretation: Cronbach's alpha coefficients range from 0.714 to 0.871, indicating moderate to high levels of internal consistency. The constructs of Community Impact and Student Engagement demonstrate particularly strong internal reliability, with Cronbach's alpha coefficients of 0.871 and 0.865, respectively. Composite reliability coefficients (rho_a and rho_c) further corroborate the reliability of the constructs, with values ranging from 0.737 to 0.904. Regarding construct validity, the average variance extracted (AVE) values range from 0.453 to 0.794. AVE assesses the amount of variance captured by the construct relative to measurement error, thereby indicating convergent validity. Constructs with higher AVE values, such as Community Impact and Student Engagement, demonstrate stronger convergent validity, suggesting that they effectively measure the intended constructs with minimal measurement error.

Data Analyses

The collected data was then analyzed using statistical methods to determine the effectiveness of the HIAM model. The analysis was conducted to identify any significant improvements in the educational outcomes of the students post-intervention.

Table 2. Paired samples T-test

			Statistic	df	p	Mean difference	SE difference
CI	**LP**	**Student's t**	**-3.16**	**29.0**	**0.004**	**-0.373**	**0.118**
	CU	Student's t	-4.82	29.0	< .001	-0.573	0.119
	SE	Student's t	-1.68	29.0	0.104	-0.218	0.130
	CII	Student's t	-2.84	29.0	0.008	-0.373	0.131

Interpretation:

a. The paired samples t-test statistic is -3.16 with a degree of freedom (df) of 29 and a p-value of 0.004. Since the p-value is less than the conventional significance level of 0.05, we reject the null hypothesis (Ho). This means that there is evidence to suggest that the degree of integration of Gond tribal literature into the curriculum has a significant effect on the improvement of students' English language proficiency.

b. The paired samples t-test statistic is -4.82 with 29 degrees of freedom and a p-value of less than 0.001. Given that the p-value is less than 0.001, which is smaller than the conventional significance level of 0.05, we reject the null hypothesis. This indicates that there is significant evidence to suggest that the degree of integration of Gond tribal literature into the curriculum has a significant effect on the improvement of cultural understanding among students.

c. The paired samples t-test statistic is -1.68 with 29 degrees of freedom and a p-value of 0.104. Since the p-value (0.104) is greater than the conventional significance level of 0.05, we fail to reject the null hypothesis. Therefore, based on this analysis, there is not enough evidence to suggest that the degree of integration of Gond tribal literature into the curriculum has a significant effect on student engagement.Top of Form

d. The paired samples t-test statistic is -2.84 with 29 degrees of freedom and a p-value of 0.008. Since the p-value (0.008) is less than the conventional significance level of 0.05, we reject the null hypothesis. This indicates that there is sufficient evidence to suggest that the degree of integration of Gond tribal literature into the curriculum has a significant effect on community impact.

Table 3. Paired samples T-test

			Statistic	Df	p	Mean difference	SE difference
TS	**LP**	**Student's t**	**2.446**	**29.0**	**0.021**	**0.2400**	**0.0981**
	CU	Student's t	0.417	29.0	0.680	0.0400	0.0960
	SE	Student's t	2.863	29.0	0.008	0.3956	0.1382
	CII	Student's t	1.791	29.0	0.084	0.2400	0.1340

Interpretation:

a. Interpretation: The paired samples t-test statistic for LP is 2.446 with 29 degrees of freedom, and the p-value is 0.021. Since the p-value (0.021) is less than the conventional significance level of 0.05, we reject the null hypothesis. This suggests that there is a significant effect of implementing various teaching strategies on English language proficiency. The positive mean difference (0.2400) indicates an improvement in English language proficiency following the implementation of teaching strategies.

b. Interpretation: The paired samples t-test statistic for CU is 0.417 with 29 degrees of freedom, and the p-value is 0.680. With a p-value greater than 0.05, we fail to reject the null hypothesis. Therefore, there is insufficient evidence to suggest a significant effect of implementing various teaching strategies on cultural understanding. The non-significant mean difference (0.0400) further supports this conclusion.

c. Interpretation: The paired samples t-test statistic for SE is 2.863 with 29 degrees of freedom, and the p-value is 0.008. Similar to the English language proficiency test, the p-value (0.008) is less than 0.05, leading to the rejection of the null hypothesis. This indicates a significant effect of implementing various teaching strategies on student engagement. The positive mean difference (0.3956) suggests an increase in student engagement with the implementation of teaching strategies.

d. Interpretation: The paired samples t-test statistic for CII is 1.791 with 29 degrees of freedom, and the p-value is 0.084. Since the p-value (0.084) is greater than 0.05, we fail to reject the null hypothesis. This implies that there is not enough evidence to suggest a significant effect of implementing various teaching strategies on community impact. The non-significant mean difference (0.2400) supports this conclusion.

Table 4. Paired samples T-test

			statistic	df	p	Mean difference	SE difference
TT	**LP**	**Student's t**	**-2.66**	**29.0**	**0.013**	**-0.420**	**0.158**
	CU	Student's t	-3.72	29.0	<.001	-0.620	0.167
	SE	Student's t	-1.54	29.0	0.136	-0.264	0.172
	CII	Student's t	-2.55	29.0	0.016	-0.420	0.165

a. Interpretation: The paired samples t-test statistic for LP is -2.66 with 29 degrees of freedom, and the p-value is 0.013. Since the p-value (0.013) is less than the conventional significance level of 0.05, we reject the null hypothesis. This suggests that there is a significant effect of implementing various training programs on English language proficiency. The negative mean difference (-0.420) indicates a decrease in English language proficiency following the implementation of training programs.
b. Interpretation: The paired samples t-test statistic for CU is -3.72 with 29 degrees of freedom, and the p-value is less than 0.001. With a p-value less than 0.001, we reject the null hypothesis. Therefore, there is significant evidence to suggest a significant effect of implementing various training programs on cultural understanding. The negative mean difference (-0.620) indicates a decrease in cultural understanding following the implementation of training programs.
c. Interpretation: The paired samples t-test statistic for SE is -1.54 with 29 degrees of freedom, and the p-value is 0.136. Since the p-value (0.136) is greater than 0.05, we fail to reject the null hypothesis. This suggests that there is insufficient evidence to support a significant effect of implementing various training programs on student engagement. The negative mean difference (-0.264) indicates a decrease in student engagement following the implementation of training programs, but this change is not statistically significant.
d. Interpretation: The paired samples t-test statistic for CII is -2.55 with 29 degrees of freedom, and the p-value is 0.016. Since the p-value (0.016) is less than 0.05, we reject the null hypothesis. This suggests that there is a significant effect of implementing various training programs on community impact. The negative mean difference (-0.420) indicates a decrease in community impact following the implementation of training programs.

Table 5. Paired samples T-test

			Statistic	df	P	Mean difference	SE difference
CE	**LP**	**Student's t**	**-2.14**	**29.0**	**0.041**	**-0.360**	**0.168**
	CU	Student's t	-2.94	29.0	0.006	-0.560	0.190
	SE	Student's t	-1.05	29.0	0.301	-0.204	0.194
	CII	Student's t	-2.06	29.0	0.048	-0.360	0.174

Interpretation:

a. The paired samples t-test statistic for LP is -2.14 with 29 degrees of freedom, and the p-value is 0.041. Since the p-value (0.041) is less than the conventional significance level of 0.05, we reject the null hypothesis. This suggests that there is a significant effect of implementing various Community Engagement initiatives on English language proficiency. The negative mean difference (-0.360) indicates a decrease in English language proficiency following the implementation of Community Engagement activities.
b. The paired samples t-test statistic for CU is -2.94 with 29 degrees of freedom, and the p-value is 0.006. With a p-value less than 0.05, we reject the null hypothesis. Therefore, there is significant evidence to suggest a significant effect of implementing various Community Engagement initiatives

on cultural understanding. The negative mean difference (-0.560) indicates a decrease in cultural understanding following the implementation of Community Engagement activities.

c. The paired samples t-test statistic for SE is -1.05 with 29 degrees of freedom, and the p-value is 0.301. Since the p-value (0.301) is greater than 0.05, we fail to reject the null hypothesis. This suggests that there is insufficient evidence to support a significant effect of implementing various Community Engagement initiatives on student engagement. The negative mean difference (-0.204) indicates a decrease in student engagement following the implementation of Community Engagement activities, but this change is not statistically significant.
d. The paired samples t-test statistic for CII is -2.06 with 29 degrees of freedom, and the p-value is 0.048. Since the p-value (0.048) is less than 0.05, we reject the null hypothesis. This suggests that there is a significant effect of implementing various Community Engagement initiatives on community impact. The negative mean difference (-0.360) indicates a decrease in community impact following the implementation of Community Engagement activities.

DISCUSSION

Present research underscores the transformative potential of integrating indigenous knowledge systems into mainstream education. Scholars such as Sanshez (2019) argue that the recognition of indigenous epistemologies enriches the academic environment, fostering inclusivity and cultural responsiveness. The observed improvements in cultural understanding and language proficiency following the integration of Gond tribal literature corroborate these arguments(Sanchez-Ruiz & El Khoury, 2019). As Rose notes, when students encounter curriculum content that reflects their cultural heritage, it not only validates their identity but also enhances their academic motivation and comprehension.

These outcomes align with the broader goals of multicultural education, which aim to create equitable learning spaces by acknowledging and incorporating the lived experiences of marginalized communities (Ross et al., 2022). Research by Marley (2022) emphasizes that the curricular inclusion of community narratives and indigenous texts encourages deeper cognitive engagement and empathy among learners (Marley et al., 2022). By embedding Gond literature into the curriculum, the model likely enhanced students' awareness of community dynamics, thereby nurturing a sense of belonging and agency—a critical component of holistic development.

The mixed outcomes across different intervention categories also reveal the complex nature of educational transformation. HIAM model led to marked gains in language proficiency and cultural understanding; the impact on student engagement was comparatively limited. In tribal or rural settings, external factors such as infrastructural inadequacies, socio-economic constraints, or unfamiliar pedagogical formats can temper the impact of even well-designed interventions. Student engagement might require sustained, multi-pronged efforts that extend beyond content delivery and encompass mentoring, peer collaboration, and emotional support structures (Padilla Rodriguez et al., 2020). The findings from the teaching strategy and training interventions offer further nuance.

The results suggest that while instructional approaches can foster student engagement and improve linguistic outcomes, their efficacy in enhancing cultural understanding and community engagement is less definitive (Crider, 2021). Interestingly, the outcomes associated with community engagement activities indicated some unintended consequences, including declines in certain areas such as cultural understanding and community impact. This paradox aligns with the literature by Lewis (2011), who cautions that community engagement initiatives, if not contextually grounded and participatory, may reinforce hierarchical relationships rather than empower learners. It is possible that the design or implementation of such initiatives lacked a co-constructive dimension, thereby limiting their effectiveness (Lewis et al., 2011). For community engagement to yield meaningful educational outcomes, it must be dialogical, sustained, and genuinely inclusive of students' voices and lived experiences.

FUTURE RESEARCH DIRECTIONS

Future research in the area of incorporating vernacular tribal literature into English language classrooms, particularly within the context of the Gond tribe in the Mandla district, Madhya Pradesh, should aim to expand the scope and scale of the study. The present research focused on a single section of Class 9 students from Brain Child Academy using purposive sampling; future studies should consider including a larger and more diverse sample across multiple schools and districts to enhance the generalizability of findings. Comparative studies between tribal and non-tribal student groups could provide deeper insights into the pedagogical impact of culturally responsive teaching methods. Longitudinal studies would help assess the long-term effects of integrating vernacular tribal literature on students' language acquisition, cultural appreciation, and critical thinking abilities. Exploration is also needed into the development of teaching aids, curriculum frameworks, and teacher training programs that effectively incorporate tribal oral traditions and literary forms into the mainstream English syllabus. Collaborative efforts involving linguists, tribal scholars, educators, and curriculum developers can significantly enrich this research domain, ensuring both the preservation of indigenous knowledge and the promotion of inclusive education practices.

CONCLUSION

Across all sixteen hypotheses tested, the analyses have provided valuable insights into the effects of various interventions on students' academic and cultural outcomes. Regarding the integration of Gond tribal literature into the curriculum, significant positive effects were observed in multiple areas. Specifically, the analyses demonstrated significant improvements in English language proficiency (LP), cultural understanding (CU), and community impact (CII) following the integration of Gond tribal literature. These findings underscore the potential of integrating diverse cultural perspectives into educational settings to enhance students' linguistic skills, cultural awareness, and community engagement.

The analyses provide evidence of the multifaceted impacts of educational interventions on students' academic and cultural development. By leveraging diverse approaches, such as integrating cultural literature, employing teaching strategies, implementing training programs, and engaging with communities, educators can foster comprehensive learning experiences that promote linguistic proficiency, cultural understanding, and community involvement. However, further research is needed to deepen our understanding of the mechanisms underlying these effects and to inform the development of tailored interventions that address the diverse needs of students and communities.

Recommendations

- Integrate Vernacular Tribal Literature into Curriculum: Embed tribal stories, poems, and folklore into English language education to enhance cultural understanding and student engagement among indigenous learners.
- Adopt the HIAM Model in Teaching Practices: Use the Holistic Integration and Appreciation Model (HIAM) to design culturally responsive teaching strategies that honor indigenous knowledge systems.
- Strengthen Teacher Training and Capacity Building: Provide targeted professional development and workshops for teachers to effectively implement contextual and culturally grounded pedagogy.
- Promote Community Participation in Education: Involve local tribal leaders, storytellers, and cultural experts in educational planning and classroom activities to ensure cultural relevance and authenticity.
- Support Multilingual and Culturally Inclusive Policies: Advocate for educational policies that promote multilingual education and the inclusion of indigenous content across subjects to preserve cultural heritage and improve learning outcomes.

REFERENCES

Thamanna, A.& Subramani, R. (2023).Indigenous Communication Of Everyday Life And Philosophy: An Analysis Of Gond Paintings In Madhya Pradesh. *ShodhKosh: Journal of Visual and Performing Arts*, *4*(2). Advance online publication. DOI: 10.29121/shodhkosh.v4.i2.2023.645

Blyznyuk, T. (2019). Formation of Teachers' Digital Competence: Domestic Challenges and Foreign Experience. *Journal of Vasyl Stefanyk Precarpathian National University*, *5*(1), 40–46. Advance online publication. DOI: 10.15330/jpnu.5.1.40-46

Bottiani, J. H., McDaniel, H. L., Henderson, L., Castillo, J. E., & Bradshaw, C. P. (2020). Buffering Effects of Racial Discrimination on School Engagement: The Role of Culturally Responsive Teachers and Caring School Police. *The Journal of School Health*, *90*(12), 1019–1029. Advance online publication. DOI: 10.1111/josh.12967 PMID: 33184890

Crider, J. S. (2021). Teachers Experiences with Positive Emotion, Engagement, Relationships, Meaning, and Achievement (PERMA): A Qualitative Interview Study. In *Dissertation*.

Deogaonkar, S. G. (2008). Traditional handicrafts of the Gond tribes of Vidarbha, Maharashtra. *Indian Journal of Traditional Knowledge*, *7*(1).

Gupta, R., Vairale, M. G., Deshmukh, R. R., Chaudhary, P. R., & Wate, S. R. (2010). Ethnomedicinal uses of some plants used by Gond tribe of Bhandara district, Maharashtra. *Indian Journal of Traditional Knowledge*, *9*(4).

Idrus, F., Hussin, M. E., & Gulca, M. (2023). THE INTEGRATION OF CULTURAL ELEMENTS IN THE ENGLISH LANGUAGE CLASSROOMS: A CASE STUDY OF SELECTED ORANG ASLI SCHOOLS IN PAHANG, MALAYSIA. [JONUS]. *Journal of Nusantara Studies*, *8*(1), 26–45. Advance online publication. DOI: 10.24200/jonus.vol8iss1pp26-45

International Experiences in Language Testing and Assessment. (2015). *International Experiences in Language Testing and Assessment.*, DOI: 10.3726/978-3-653-02511-8

Koreti, S. (2016). Socio-Cultural History of the Gond Tribes of Middle India. *International Journal of Social Science and Humanity*, *6*(4), 288–292. Advance online publication. DOI: 10.7763/IJSSH.2016.V6.659

Lewis, A. D., Huebner, E. S., Malone, P. S., & Valois, R. F. (2011). Life Satisfaction and Student Engagement in Adolescents. *Journal of Youth and Adolescence*, *40*(3), 249–262. Advance online publication. DOI: 10.1007/s10964-010-9517-6 PMID: 20204687

Marley, S. A., Siani, A., & Sims, S. (2022). Real-life research projects improve student engagement and provide reliable data for academics. *Ecology and Evolution*, *12*(12), e9593. Advance online publication. DOI: 10.1002/ece3.9593 PMID: 36514554

Osei-Owusu, B. (2022). Impact of Professional Development Programmes on Teachers' Knowledge and Academic Performance of Senior High School Students in Ghana. *European Journal of Education and Pedagogy*, *3*(2), 60–69. Advance online publication. DOI: 10.24018/ejedu.2022.3.2.276

Padilla Rodriguez, B. C., Armellini, A., & Rodriguez Nieto, M. C. (2020). Learner engagement, retention and success: Why size matters in massive open online courses (MOOCs). *Open Learning*, *35*(1), 46–62. Advance online publication. DOI: 10.1080/02680513.2019.1665503

Rizk, J., & Hillier, C. (2022). Digital technology and increasing engagement among students with disabilities: Interaction rituals and digital capital. *Computers and Education Open*, *3*, 100099. Advance online publication. DOI: 10.1016/j.caeo.2022.100099

Ross, P. M., Scanes, E., Poronnik, P., Coates, H., & Locke, W. (2022). Understanding STEM academics' responses and resilience to educational reform of academic roles in higher education. *International Journal of STEM Education*, *9*(1), 11. Advance online publication. DOI: 10.1186/s40594-022-00327-1 PMID: 35127335

Sanchez-Ruiz, M. J., & El Khoury, J. (2019). A model of academic, personality, and emotion-related predictors of university academic performance. *Frontiers in Psychology*, *10*(OCT), 2435. Advance online publication. DOI: 10.3389/fpsyg.2019.02435 PMID: 31749740

Stauss, K., Koh, E., Johnson-Carter, C., & Gonzales-Worthen, D. (2021). OneCommunity Reads: A Model for Latino Parent-Community Engagement and Its Effect on Grade-Level Reading Proficiency. *Education and Urban Society*, *53*(4), 402–424. Advance online publication. DOI: 10.1177/0013124520928612

Tanuja Khan, M., Joshi, B., & Thomas, S. (2021). Impact of Digital Learning on the Academic Achievement of Students of Government and Private Schools. *Educational Resurgence Journal, 3*(6).

Yang, Y., Qin, L., & Ning, L. (2021). School Violence and Teacher Professional Engagement: A Cross-National Study. *Frontiers in Psychology*, *12*, 628809. Advance online publication. DOI: 10.3389/fpsyg.2021.628809 PMID: 33935880

ADDITIONAL READING

Choudhary, S. (2016). A Literary Approach to Teaching English Language in a Multicultural Classroom. *Higher Learning Research Communications*, *6*(4). Advance online publication. DOI: 10.18870/hlrc.v6i4.352

Jayakar, T. J. R., & Tejaswani, K. (2020). *Challenges to Vernacular Learners (L1) in Acquisition of English Language (L2) at UG Level in Hyderabad Region. Journal of Education and Practice.*

Watterson, K. (2022). Incorporating Vernacular into the Foreign Language Classroom: A Step in Curriculum Development. *Open Journal of Modern Linguistics*, *12*(4), 414–421. DOI: 10.4236/ojml.2022.124029

KEY TERMS AND DEFINITIONS

Vernacular Tribal Literature: Refers to the body of oral and written literary works produced by indigenous tribal communities in their native languages, reflecting their customs, beliefs, values, myths, and local knowledge systems.

English Language Classroom: A learning environment where English is taught as a subject or used as a medium of instruction, aimed at developing language proficiency, literacy skills, and cross-cultural communication.

HIAM Model (Holistic Integration and Appreciation Model): An educational framework that integrates students' cultural heritage, indigenous knowledge, and local context into classroom teaching to enhance engagement, understanding, and inclusivity.

Cultural Integration: The process of incorporating diverse cultural elements, practices, and perspectives into the mainstream education system to create a more inclusive and respectful learning environment.

Cultural Engagement: Active participation of learners with cultural content, traditions, and values that promote a deeper understanding of their identity and heritage within the educational setting.

Teaching Strategies: The methods and techniques educators use to facilitate learning, engage students, and meet diverse learning needs effectively.

Teachers' Training: Structured programs designed to enhance the pedagogical, content, and cultural competencies of teachers to improve classroom instruction and student outcomes.

Professional Development Community: A collaborative group of educators and stakeholders engaged in continuous learning, sharing best practices, and fostering growth through workshops, mentoring, and reflection.

Indigenous Knowledge Systems: Localized knowledge developed by indigenous communities through generations, encompassing ecological wisdom, cultural practices, spirituality, and survival strategies.

Cultural Understanding: The ability to appreciate, interpret, and respond to cultural differences with empathy and respect, fostering mutual respect and inclusivity.

Language Proficiency: The ability to use a language effectively and appropriately for communication in both academic and social contexts, involving speaking, listening, reading, and writing skills.

Student Engagement: The degree of attention, curiosity, interest, and passion that students exhibit in the learning process, leading to higher levels of motivation and achievement.

Community Impact: The broader effects that educational practices and initiatives have on the social, cultural, and economic well-being of the local community.

Curriculum Development: The systematic planning, organization, and revision of educational content, learning outcomes, assessment methods, and instructional materials to enhance learning experiences.

Experiential Learning: A learning process where students acquire knowledge and skills through hands-on activities, reflection, and real-world experiences.

Contextual Pedagogy: An approach to teaching that considers the learners' cultural, social, and environmental background, making education more meaningful and relevant.

Sociocultural Learning: A theory of learning that emphasizes the influence of social interaction, cultural tools, and community practices on a learner's cognitive development.

Chapter 23
South African Universities and Their Responsibility Towards Introducing the Indigenous Knowledge Systems Accredited Programmes:
Where to From Here?

Monicca Thulisile Bhuda
https://orcid.org/0000-0002-8506-9562
University of Mpumalanga, South Africa

Tlou Maggie Masenya
Durban University of Technology, South Africa

ABSTRACT

The call to have Indigenous Knowledge System (IKS) accredited programmes by universities is necessary to challenge existing Western Systems which have posed a threat to IKS in education. The aim of the study was to investigate the implementation of the IKS programmes by universities which were part of the 1996 indigenous technologies audit. The study used qualitative content analysis to extract information from a variety of sources, such as government documents, books, theses, and scholarly journals that are available on websites like Google Scholar, ScienceDirect, ResearchGate and university repositories. The findings of the study revealed that with assistance from government, there are universities who have successfully implemented IKS accredited programmes while others are still lacking behind since the 1996 audit. Therefore, the study recommends that universities which are yet to start this process need to benchmark and collaborate with those that are offering the IKS programmes. IKS postgraduates can also be sourced to be part of this process because they are equipped with expertise to implement these programmes.

DOI: 10.4018/979-8-3373-4124-8.ch023

INTRODUCTION

Since 1994, many scholars in South Africa have placed a high value on the inclusion of the IKS in the curriculum. Subsequent conferences on Africanization and decolonization of education or integration of the IKS into the educational system were prompted by a call for the African Renaissance in the mid-1990s (Dei, 2014). The integration of IKS in universities is part of an educational project that focuses on decolonization of the curriculum in many formerly colonized states in Africa. As Dei (2014) discussed, the inclusion of IKS gives institutions the chance to create a variety of knowledge forms rather than relying just on a superior Eurocentric form of knowledge as a standard and benchmark. This method allows academics to value different types of information rather than just one that is determined by a predetermined study program. Scholars such as Odora Hoppers (2021) and Wane and Wane (2019) highlighted that instead of embracing Eurocentric knowledge, which is so ineffectual and detrimental to Africa's economic drive, they urge for South Africa to build on its historical indigenous capital.

However, as per Kaya and Seleti's (2013) study, there are a number of obstacles to integrating IKS into higher education, including a lack of acceptance and validation in academic institutions, challenges with standardizing and documenting IKS, issues with cultural sensitivity, language barriers, access to indigenous communities, intellectual property rights, and a potential disconnect between traditional knowledge and contemporary academic frameworks. Maditsi and Bhuda (2023) underscored that IKS as a body of knowledge has not yet been granted complementary status to western science, despite its evident benefits. Even after achieving independence, South Africa has maintained its Westernized universities.

The impact of colonization and post-colonial aftereffects are seen by Kaya and Seleti (2013) as significant obstacles to IKS inclusion at South African universities. Colonial powers employed strategies and practices to defeat and subjugate Africans during that time (Bhuda, 2024b). The goal of those programs was to eradicate preexisting knowledge systems and replace them with Western ideas and knowledge systems by continuously oppressing and disparaging indigenous civilizations. Many times, these difficulties call for significant curriculum-structure modification (Chabaya & Chabaya, 2023). It is imperative to prioritize cultural sensitivity, challenge Western-dominated educational structures, embrace interculturalism, and create inclusive learning settings (da Silva et al., 2024). Openness, collaboration, and a readiness to question accepted wisdom are necessary for this journey to maintain the integrity of IKS in education (Yip & Chakma, 2024).

Chinenye-Nwokocha and Legg-Jack (2024) indicated that there are two main objectives of the IKS program at South African universities. One is a pragmatic desire to enable the locals to use their IKS to enhance their own standard of living. The other goal, which is more philosophical but no less significant, is to merge Western scientific knowledge with African Indigenous knowledge (AIK) and authenticate it (Bhuda, 2024a). Such a feat is said to assist South Africans overcome an innate sense of powerlessness or inferiority and embrace a sense of identity in higher education. A sense of identity and purpose in life will be established through the symbolic reappropriation of identity and pride in one's own cultural history, expertise, and abilities.

SCOPE OF RESEARCH

With an aim of investigating the universities that participated in the 1996 audit on indigenous technologies and their further involvement in implementing the IKS accredited programs, this chapter offers a thorough examination of IKS accredited programs in universities. The study examines the steps universities have made to guarantee the implementation and sustainability of the IKS accredited programs as part of teaching, learning, research and community engagement. The achievement of government to use IKS to decolonize South African universities depends on the present advancements in the IKS accredited programs. The study contends that the integration of IKS in universities is a suitable answer to the decolonial agenda and indigenization of higher education, notwithstanding appeals and efforts to decolonize it.

RESEARCH QUESTIONS

The study used specific research questions which assisted in setting the boundaries for data collection and analysis to answer the questions effectively. The study's main research question was: Which South African universities have successfully implemented and sustained the IKS accredited programs? This main research question enabled the researchers to formulate the research questions below that the study intended to address.

a) When were the IKS accredited programs developed in South African Universities?
b) What is the nature of the IKS accredited programs?
c) Which Universities implemented the IKS accredited programs?
d) What are the challenges faced with establishing IKS accredited programs?
e) How can IKS accredited programs be successfully integrated in South African universities?

THE CURRENT STATE OF SOUTH AFRICAN UNIVERSITIES

South African universities have been accused of adopting Western forms of academic organization from the beginning, which generally ignored the wisdom of colonized people (Heleta, 2016; Higgs, 2016; Le Grange, 2016). Prior to 1994, there were significant attempts to alter extremely problematic institutional cultures, as well as their colonial structures and practices, as part of the reform agendas of a number of previous and current higher education programs. South African universities have been accused of adopting Western forms of academic organization from the beginning, which generally ignored the wisdom of colonized people (Heleta, 2016; Higgs, 2016; Le Grange, 2016). Since the end of apartheid, the majority of the changes in higher education have focused on demographics and numbers rather than institutional cultures, politics of knowledge, epistemic decolonization, and the promotion of plurality of knowledge (Badat, 2020; Motala et al. 2021).

Hlatshwayo (2024) argued that people who were previously marginalized under apartheid are now participating in decolonized universities, which choose to embrace and acknowledge their own cultures, narrate their own histories, read books written by African authors, and operate institutions based on African cultural values rather than Eurocentric ones. Since educational institutions like universities are microcosms of society, their decolonization will mark the start of a much larger transition that will ultimately impact South Africa as a whole (Le Grange, 2016). Proposals to decolonize higher education are predicated on a rejection of contemporary colonial education, whose guiding ideology was to transform the colonized into colonial subjects while depriving them of their humanity and potential. According to the decolonial academics da Silva et al. (2024), the knowledge of indigenous peoples, non-Europeans, and colonized communities was either suppressed or deemed inferior. Indigenous knowledge (IK), or the knowledge of indigenous people, receives little to no attention from (colonial) universities (Ajani & Gamede, 2021; Tegegne et al., 2025).

According to this study, the IKS accredited program can offer the essential parameters for developing a comprehensive and integrated epistemological framework. IKS can be used in higher education to decolonize the curriculum and promote the inclusion of local knowledge (Marsandi et al., 2025). Scholars like Bhuda (2024b), Chabaya and Chabaya (2023), Gumbo et al. (2024), Masenya (2022a), and Msila (2025) amongst others have called for the decolonization of education, which includes incorporating IKS into the curriculum. To put it another way, current realities in terms of lived experiences, cultural values, and belief systems. IKS, which are products of the local environment, ought to serve as the cornerstone upon which a society's formal education system is built (Burgess et al., 2022).

THEORETICAL FRAMEWORK

Decolonizing theory

The goal of decolonizing theory is to question how colonialism has shaped the existing power and knowledge structures. Masenya (2022a) elaborated that by challenging colonialism's social structures, it aims to establish more just systems. In the process of decolonization, universities play a crucial role. Decolonizing theory is a critical viewpoint that looks at how colonialism still affects society today (Bhambra, 2021). It makes the case that coloniality and the legacy of colonialism continues to influence our social, political, economic, and knowledge systems. According to this study, a decolonial curriculum based on IKS is an effective way to dismantle colonial influences on education and knowledge. Wa Thiong'o's (1986) and Prah's (2002) writings have been in the forefront of the decolonization model in Africa. In order to restore African dignity and decolonize African thought, African scholars must recover African history.

This study contends that universities must answer to the decolonial call, which places the IKS curriculum at the forefront to dismantle the colonial structures that still exist in South African universities. It is inspired by the work of wa Thiong'o's (1986) and Prah (2002), among other scholars who paved a way for decolonization that provided a platform for IKS. According to this study, universities should react to the decolonization agenda and the restructuring of higher education by implementing the IKS accredited program, as requested by government. IKS, Indigenous knowledge holders, and communities that have historically been underrepresented and misrepresented in academic settings should be championed by these universities. Universities must thus decolonize and embrace the incorporation of IKS as an area of study that will revolutionize higher education as well as a necessary curriculum change.

Postcolonial theory

Postcolonial theory is a critical examination of the oppression of colonized people such as indigenous people and their knowledge. To comprehend how colonial experience has affected the knowledge systems of colonized people, postcolonial theory can be applied to IKS studies. IKS emphasizes that postcolonial theory encompasses more than only the former colonized people's reaction to the colonial heritage, which may involve relocating the center as suggested by wa Thiong'o's (1986) or writing back to the center. As discussed by Mapara (2009), they also focus on the once colonized writing back to the center, showcasing their prior accomplishments and advancements that were occasionally interrupted by the arrival of colonialism. Furthermore, Mapara (2009) clarifies that the theory primarily focuses on how colonizers' literature misrepresents the socioeconomic and governance realities and experiences of the colonized. Mapara (2009) further noted that the literature of the colonizers frequently promotes the superiority of the colonizer while portraying the colonized as inferior. It is evident from the research on postcolonial discourse that the theory's current focus is on how the realities and experiences of the colonized people have been disregarded.

This study's argument, which is informed by postcolonial theory, is that the integration of IKS in higher education addresses the myth of western superiority and corrects the ideology that there is only one knowledge system. By incorporating IKS into universities, the former colonized can regain their humanity and respect that colonization had taken away from them. Universities in South Africa are responsible for establishing a space for Indigenous Knowledge Systems and embracing them within the academic community. In addition to incorporating IKS into universities, academics should endeavor to create formal, accredited qualifications that students can pursue with an aim to become indigenous scholars.

METHODOLOGY

As the main methodology for discussing the institutions that were required to have credentials certified by IKS and the advancements that have been accomplished so far, a qualitative research methodology was employed under the content analysis. According to Hammarberg et al. (2016), qualitative research entails gathering and examining non-numerical data in order to comp4rehend ideas, viewpoints, or personal experiences. By employing a qualitative research approach, the researchers can gain a deeper understanding of how people understand concepts (Hammarberg et al., 2016). In this study, the researchers wanted to understand how IKS has been implemented in diverse universities and how such a shift has been received as part of institutional teaching, learning, research and community engagement. The impact of the IKS in universities will be visible through the literature that will be discussed.

Study Design

The introduction and development of recognized programs in IKS by South African universities were examined using a systematic, descriptive review style (Kim et al., 2017). Books, book chapters, dissertations, and peer-reviewed journal articles that were pertinent to this study were among the information sources used to perform this literature review. Information was sources from university repositories, Google Scholar, ResearchGate, ScienceDirect.

Literature selection process

In order to find material for this study, a systematic search was conducted utilizing particular keywords. "IKS in education," "IKS in higher universities," "IKS curriculum," "decolonization of higher education," and "Indigenous voices in education" were among the search terms that were used. Through the use of a sophisticated search feature, 130 unique articles were found for this study's investigation. After examining the abstracts and titles, 20 papers were deemed irrelevant to the study. As shown in Figure 1, 77 of the 110 remaining publications were judged pertinent and added to the research.

Figure 1. Flowchart of literature selection process (Authors)

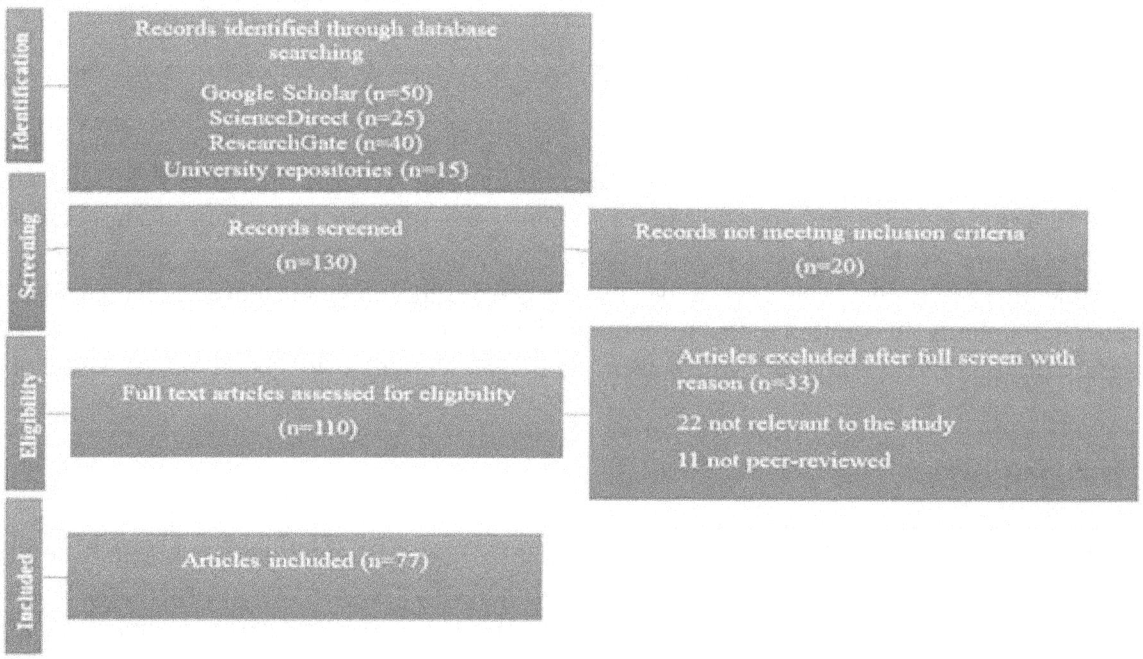

Triangulation Approach

To guarantee the reliability and authenticity of the results, a triangulation technique was applied. In order to find common themes and patterns, data from many sources and viewpoints have to be cross-referenced (Noble & Heale, 2019). The study sought to give a thorough and accurate portrayal of IKS in universities by employing a triangulation approach. This approach was necessary to understand the future of IKS accredited programs in South African universities.

Ethical implications

The researchers made proactive measures to lessen biases in reporting, data interpretation, and source selection. This necessitated determining any restrictions or gaps in the study, evaluating the sources rigorously for possible conflicts of interest or biases, and clearly communicating the results (Arifin, 2018). Additionally, measures were taken to adhere to ethical principles and academic integrity standards throughout the research process, ensuring the validity and integrity of the study.

FINDINGS OF THE RESEARCH ON IKS ACCREDITED PROGRAMMES IN UNIVERSITIES

This section of the study uncovers the inception of the IKS accredited programs in South African Universities. These results shed light on the government's directives about universities' acknowledgment of the value of IKS in teaching, learning, research, and community development. The findings of the initial steps taken in the creation of the IKS accredited programs will shed light on the number of universities that have embraced this initiative, as well as the difficulties and opportunities facing the programs' growth. Furthermore, the results of this study are essential to understand the government's support for IKS at universities, which also seeks to empower knowledge custodians who must ensure that this information is safeguarded, preserved, and disseminated. This study calls for all 26 public universities to recognize the value of IKS in education and adopt frameworks that will enable the implementation of an IKS accredited programs in undergraduate and postgraduate studies.

In order to accomplish the purpose of this article, the following themes were created in accordance with the previously established research questions. These are namely, a) The developments towards recognizing IKS in South Africa universities; b) Universities that participated in the 1996 indigenous technologies audit; c) The nature of the IKS accredited programs to be introduced in South African universities; d) The adoption of the IKS accredited programs by South Africa universities; e) Challenges and opportunities of establishing the IKS accredited programs. Existing literature was used to address each of these themes.

THE DEVELOPMENTS TOWARDS RECOGNISING IKS IN SOUTH AFRICA UNIVERSITIES

IKS inquiry has pushed itself onto the agenda of serious reflection in both academic and practitioner contexts since 1996. With the help of the Portfolio Committee of Arts, Culture, Science, and Technology, the Council for Scientific and Industrial Research (CSIR) commissioned nine historically underrepresented universities to audit indigenous technologies in September 1996. As stated by Mosimege (2004), each university held provincial workshops on IKS as a result of the audits which formed a basis for the 1998 workshop that was hosted at the University of the North West. Moreover, the Department of Arts, Culture, Science, and Technology was directed by the Portfolio Committee of Parliament to begin drafting a policy and bill on the recognition, promotion, and protection of Indigenous Knowledge Systems in South Africa following the National Workshop (Moahi, 2012; Ngulube et al., 2015).

Higges and Van Niekerk (2002) indicated that the goal of this committee was to identify the kind of technologies that South Africans, particularly those from underprivileged and marginalized communities, have long relied on to survive. The goal was to identify such technologies so that they could be documented, developed, and used to the greater advantage of knowledge holders. In Mosimege's (2004) work, it was indicated that in 1998, the first National workshop was organized under the auspices of the Portfolio Committee on Arts, Culture, Language, Science, and Technology at the University of North West (now known as the North-West University). This led to the establishment of a task team to draft IKS legislation and policy by Department of Arts, Culture, Science and Technology. After the 1996 and 1998 initiatives, the former Department of Arts, Culture, Science, and Technology began to take an active interest in IKS issues (Mosimege, 2004).

The task team members drew heavily from their travels to nations that have been heavily involved in IK issues, like China and India (Higge & Van Niekerk, 2002). Following the submission of the draft materials to the Department of Arts, Culture, Science and Technology in 2002, the Department engaged in additional stakeholder discussions to further develop and improve the legislation and policy. The following were set to be implemented by the task team:

- Creating the IKS Policy and IKS legislation to safeguard indigenous knowledge.
- Using countries with extensive histories of working on IKS as examples.
- Work together with various stakeholders, especially those with knowledge, to advance the law and policy.
- To support the establishment of an IKS unit within the Department of Science and Technology as part of their dedication to IKS recognition, affirmation, development, promotion, and protection.
- Support the creation of an IKS program in South Africa, particularly in light of the current discussion about education in higher education.

This study wants to stress that on the initiative to establish IKS programs in South African education by the task team was supported by National Research Foundation (NRF) which also played its role and created a designation of IKS as a field for special funding (Tisani, 2004). Moreover, Masoga (2005, 16) explained that IKS has been recognized by the National Research Foundation (NRF) as a focus area that has drawn researchers to this significant subject. Three research areas

- The production, transmission and use of IK (which covers the nature of IK, IKS and indigenous technology);
- The role of IK in nation-building (which includes issues such as traditional medicine and health, indigenous food systems, socio-cultural systems, as well as arts, crafts and materials), and
- IK at the interface with other systems of knowledge (which explores the intersections of knowledge systems and the kinds of synergies and new knowledge they can generate). The IKS specialty area has given rise to research initiatives that have attracted both domestic and foreign players. (Masoga, 2005:16).

The NRF Act of 1998 served as the basis for allocating the ring-fenced funds, and the organization has reaffirmed its agency function and its role in influencing and executing policy within the National System of Innovation (NSI) in the NRF Strategy 2020- 2025 (Masoga, 2005). NRF Act of 1998 states that the NRF should "support and promote research through funding, human resource development and provision of the necessary infrastructure in order to facilitate the creation of knowledge, innovation and development in all fields of science and technology, including indigenous knowledge, and thereby contributing to the improvement of the quality of life for all the people of the Republic."

Tisani (2004) discussed that the then Department of Science and Technology's which was initially part of the Department of Arts, Culture, Science, and Technology (and currently known as Department of Science and Innovation) focused efforts to promote coordination and research in this field, the creation of a of the IKS of South Africa Trust (iIKSSA) guardian organization for practitioners of indigenous knowledge, the rise in conferences and workshops on IKS, and, last but not least, the launch of the Indilinga journal of Indigenous Knowledge Systems, which focuses exclusively on IKS research, all marked significant advancements in the recognition of IKS in education and other government agencies (Masoga, 2005).

UNIVERSITIES THAT PARTCIPATED IN THE 1996 INDIGENOUS TECHNOLOGIES AUDIT

The 1996 selected universities who participated in the auditing indigenous technologies were not formally mandated to start the IKS accredited programs during that time. However, after the IKS Policy was established in 2004 by the Department of Science and Technology the initial participated universities were looked at to further advancing their interest in IKS. The IKS Policy of 2004 was established with a goal of recognizing, affirming, developing, promoting, and protecting IKS in South Africa. It places strong emphasis on integrating IKS into the curriculum and education system to promote a more comprehensive and culturally appropriate learning environment. While some institutions have lagged behind in their development of IKS accredited programs, others have in fact demonstrated increased commitment.

This section therefore focuses on the nine universities that were part of the 1996 audit and the work that they have done in creating IKS accredited programs. Some of these universities have changed as a result of mergers with other universities and Technikons as follows:

- University of Venda,
- University of the North (now known as the University of Limpopo)
- University of the North West (now known as the North-West University)
- Vista University (now known as the University of Pretoria)
- University of South Africa,
- University of the North – Qwaqwa campus (now part of the University of the Free State)
- University of Zululand,
- University of Transkei (now known as Walter Sisulu University)
- University of Fort Hare

Each university held provincial workshops on IKS as a result of the audits. In detail, the study provides discussions on how the IKS program was implemented in the above universities.

THE NATURE OF THE IKS ACCREDITED PROGRAMME TO BE INTRODUCED IN SOUTH AFRICAN UNIVERSITIES

Kaya and Seleti (2013) emphasized that the crucial role that African universities can play in reversing the recolonization of the African mind by globalizing forces resulting from the overly monocultural dominance of Western Knowledge Systems. In this context, African scholars are urged by Shahjahan et al. (2022) to critically examine the knowledge taught in curriculum and should further understand how the integration of IKS in higher education should recognize that African problems should be located, identified, analyzed and solved by Africans.

On the other hand, Kaya and Seleti (2013) argued that the IKS accredited programs aim to enable marginalized indigenous groups to use their African cultural, technological, artistic, linguistic, and traditional healing legacy to establish and improve their own livelihood. Because indigenous technologies are the foundation of rural, self-sustaining businesses, an IKS approach promises to use them to greatly contribute to the production of wealth in local communities (Higgs & Van Niekerk, 2002). This calls for a fundamental reorientation of current initiatives, particularly at the local government level, and demands that civil society be actively involved in the definition and identification of IKS in South Africa, thus making a tangible contribution to the African Renaissance.

As per Kaya (2013), IKS in higher education focuses on serving as a bridgehead for the universities' objective of community participation. In order to participate in the IKS accredited programs, South African universities must be dedicated to training and improving the South African whole population, not just a select few. The overall goal of the IKS accredited programs is to preserve, develop, and promote local knowledge systems and integrate them into different sectors in order to use IK for sustainable development and economic prosperity. Universities that establish the IKS accredited programs must have the following partners:

As per Kaya and Seleti (2013), the IKS accredited program involves various stakeholders, including:

- **Government:** The Department of Science and Innovation (DSI), the Department of Sport, Arts and Culture (DSAC) and other relevant departments.
- **Research Institutions:** Collaboration with other universities and research councils such as the National Research Foundation (NRF) and the Council for Scientific and Industrial Research (CSIR) among others.
- **Community Organizations:** Traditional leaders, traditional health practitioners, and traditional organizations.
- **Industry:** Private businesses eager to develop products using indigenous knowledge.

The IKS accredited programs are of undergraduate and postgraduate level which according to this study universities must create frameworks with an understanding of goals that they set to achieve on both levels. Below, the study provides the foundation of the IKS accredited programs in undergrad and postgraduate level. The following principles are detailed in the study and serve as the foundation for the IKS accredited program:

The nature of the IKS undergraduate program

Preservation and Promotion

Balogun and Kalusopa's (2022) study argued that the IKS program aims to record, document, preserve, and protect IKS for the benefit of communities, recognizing and affirming the value of this knowledge for social and economic development.

Economic Development

Bhuda and Kgwadi (2023) discussed that the IKS accredited program aims to give students the skills they need to become professionals who will support South Africa's economic growth through entrepreneurship and other projects that use local resources to spur innovation and boost the economy through the creation of services offered by IK practitioners and holders (Githui et al., 2015).

Interfacing the IKS accredited program with other Knowledge Systems

The program promotes the integration of IKS with other knowledge Systems to increase innovation and tackle difficult problems. Integrating IKS accredited program with other knowledge systems, such as scientific knowledge, calls for mutual respect, candid communication, and an appreciation that both systems are beneficial and can enhance rather than replace one another (Moremoholo, 2023).

Community Empowerment

The IKS accredited program acknowledges the value of empowering indigenous communities through the establishment of self-sustaining businesses that leverage their cultural, technological, artistic, linguistic, and traditional healing legacy (Balogun, 2023). Furthermore, Kohsaka and Rogel (2021) in their study mentioned that it is essential to ensure the durability and relevance of program that cater to the needs of indigenous people by involving them in their conception and implementation. Promoting social justice and equity requires acknowledging and appreciating the expertise of IKS custodians and practitioners, who have frequently been neglected. Moreover, universities that implement the IKS program must also make sure to build alliances among various knowledge holders and encourage collaboration and knowledge exchange through teaching and learning (Parthiban et al., 2024).

Human Capacity Development

The program's main goal is to increase the capacity of students/ graduates to successfully manage and apply indigenous knowledge. By acknowledging and appreciating local knowledge, skills, and traditions, IKS program plays a critical role in enhancing human capacity. Additionally, the IKS program equips students with knowledge and skills they need to manage and use IKS in areas like agriculture, health, water management, biodiversity and natural resource protection, cultural practices, rituals, and traditional medicine (Odora Hoppers, 2023).

Holistic and Community-Based Approach

As per Bhuda and Marumo (2022), to establish a connection between the academy and IK holders and practitioners in local communities, IKS accredited programs are intended to be comprehensive and community based. IKS accredited programs are based on the unique circumstances and requirements of local communities and are intended to be all-encompassing, including many facets of IK and customs. IKS's multidisciplinary character makes it distinctive and pertinent to several subjects, such as cultural diversity, decolonization, transformation, and the African Renaissance. IKS accredited programs' community-based approach helps graduates understand the importance of culture in education and to be aware of the difficulties facing local communities' development (Ma et al., 2023).

The nature of the IKS postgraduate program

The postgraduate IKS program is based on the IKS Policy, which was enacted in November 2004 and aims to identify, confirm development, and safeguard South Africa's IKS. The key goals of the IKS program at a postgraduate level are as follows:

Recording and Documentation

The IKS program at a Postgraduate level focuses on collecting and documenting IKS in various formats, including local languages. Documentation of IKS as stated by Balogun, (2023) is crucial for preserving and sharing this valuable knowledge, which is often transmitted orally and at risk of disappearing. It involves translating IKS into tangible forms like reports, maps, and digital formats to ensure accessibility and safeguard against exploitation (Masenya, 2022b).

The following are the benefits of IKS documentation: 1) Preservation: IKS is susceptible to extinction if it is not recorded, as it is frequently transmitted orally across generations. 2) Accessibility: For development and research reasons, documentation makes IKS available to a larger audience, including specialists and policymakers. 3) Safeguard: IKS is shielded from exploitation by individuals outside the community of origin thanks to documentation. 4) Advocating: By laying the groundwork for lobbying, IKS documentation can help indigenous peoples stand up for their rights and have an impact on laws that affect their resources, territory, and cultural legacy. 5) Sustainable Resource Management: Maintaining biodiversity and promoting sustainable utilization of resources and their management (Odora Hoppers, 2021).

Preservation and Protection

The IKS program's goal is to prevent the exploitation of IK while guaranteeing that communities are fairly acknowledged and benefit from its application. As promoted by the IKS Policy of 2004 and the IK Act of 2019, postgraduate students in the IKS should collaborate with communities and the government to safeguard IKS against exploitation. The identification, development, promotion, and protection of IKS are the main objectives of these policy frameworks (Masenya, 2022b).

Community Benefit

The aim is to produce postgraduate projects, source funding and other research related engagements to support sustainable lives and community empowerment through socioeconomic development (Bhuda, 2024b; Verhoef, 2023). The community engagement from researchers is needed to educate knowledge holders about intellectual property rights and the Access and Benefit Sharing (ABS) in case of IKS commercialization. Moreover, Jonas (2010) indicated that The Access and Benefit Sharing (ABS) is an agreement between indigenous communities and research institutions, like the one between the Khoi and San communities and the Council for Scientific and Industrial Research (CSIR) regarding the patents derived from the Hoodia plant. Additionally, initiatives that support sustainable practices, cultural preservation, better health and well-being, and the development of a feeling of pride and identity may benefit indigenous communities even more (Bonde & Pinjari, 2023).

Research and development

Through research, the IKS program in South African universities seeks to develop, cultivate, and safeguard IKS by allowing communities to harness their knowledge and utilizing local resources for innovation, especially in fields like nutrition, health, and beauty. Keane and Khupe (2017) explained that currently in South Africa, most universities' higher research programs are attempting to conducting research, mainstreaming, and integrating courses to include IKS. Government through DST, NRF, CSIR and DSAC has funded the IKS research programs to address both decolonial and development initiatives pertinent to the revival of African nations, and governments are beginning to take the initiative (Balogun & Kalusopa, 2022). IKS that cover marine studies, agriculture, food security, cultural astronomy (Govender 2011), education etc. must be included in order to boost research and development in South Africa.

Research by Balogun (2023) shows that the DSI, the NRF, and CSIR in South Africa play a crucial role in fostering research and innovation, including IKS. They support IKS programs and research in universities through funding, capacity building, and policy development, aiming to integrate IKS into the national science system and promote its use for socio-economic development. In order to improve the pursuit of research excellence and capacity development, the DSI and NRF established Centers of Excellence (CoEs), which are physical and virtual centers of research that focus existing research excellence, capacity, and resources. This allows researchers from different disciplines and institutions to work together on long-term projects that are both locally relevant and globally competitive. Research/Knowledge Production, Education and Training, Information Brokerage, Networking, and Service Rendering are the five CoEs' primary performance areas. IKS practitioners are being trained and integrated into the research workforce as part of the DSI, NRF, and CSIR's efforts to increase IKS's research capability (National Research Foundation, 2024).

The DSI also houses the National IK Registration System (NIKRS) formerly known as the National Recordal System and the National IK Management System (NIKMAS) which serve as an Information and Communication Technology (ICT) platform for managing IKS, as governed by the Protection, Promotion, Development and Management of Indigenous Knowledge Act No 6 of 2019. DSI has further developed the IKS Documentation Centers (IKSDCs) which are available in 9 provinces, of which 6 are housed in public universities such as The North-West University (NWU), The University of Venda (UNIVEN), Walter Sisulu University (WSU), The University of KwaZulu-Natal (UKZN) and the University of Zululand (UNIZULU), Cape Peninsula University of Technology (CPUT) and University of Free State (UFS) as stated by (Van Rooyen, 2011).

Pinning of Indigenous Knowledge Research

Bhuda (2024a) highlighted that the postgraduate IKS program seeks to train aspiring scholars how to carry out Indigenous Knowledge Research (IKR). Seminars, webinars, writing retreats, and workshops are used to do this. Similar studies were done by Datta (2018) which explained that the fact that IKR is research by, for, and for Indigenous peoples, communities, and knowledge systems must be understood by aspiring IKS researchers. They should also stress the significance of Indigenous perspectives, values, and knowledge in influencing research questions, methods, and results.

The main priorities of IKR according to Bhuda (2024b) are as follows: 1) Indigenous Knowledge: It acknowledges that Indigenous ways of knowing are valuable and ought to be respected, placing a high priority on the distinctive knowledge, viewpoints, and values of Indigenous groups. 2) Community Based research: IKR is frequently carried out in partnership with Indigenous people, guaranteeing that the research is pertinent to their interests and needs. 3) Decolonizing Research: IKR seeks to decolonize methods of research, strengthening Indigenous peoples' agency and self-determination while challenging the dominance of Western research methodologies. 4) Data Sovereignty: IKR acknowledges the significance of Indigenous data sovereignty, guaranteeing that Indigenous groups have authority over the use of their data. 5) Principles: IKR frequently makes use of principles like relational accountability, establishing a community-based research foundation, reciprocity, respect, accountability, substantive discussions, and informed consent. 6) Ethical Considerations: IKR emphasizes ethical principles such as responsibility, reciprocity, respect, and informed consent.

THE ADOPTION OF THE IKS ACCREDITED PROGRAMMES BY SOUTH AFRICA UNIVERSITIES

As mentioned above in this study by Mosimege (2004), the former Department of Science and Technology (now known as Department of Science and Innovation), in collaboration with the South African Qualification Authority (SAQA), initially sought to expand the IKS program to nine universities; however, this effort was not as successful (Kaya & Seleti, 2013). There are just two universities which are NWU and UNIVEN that have effectively created IKS programs, and just one of them offers undergraduate and graduate-level courses as per the agreement. The remaining universities are either connected to IKS through IKSDCs or research studies from other diverse fields that incorporate IKS, while at least two universities offer IKS-related modules in other programs/courses.

Therefore, this section of the study provides depth on how IKS accredited programs have been implemented by the universities that were part of the indigenous technologies audit of 1996.

The North-West University

In September 1998, (NWU) conducted a National Workshop on Indigenous Knowledge Systems in South Africa to set the stage for the country's first Indigenous Knowledge Systems (IKS) accredited program. In light of its possible contribution to the African Renaissance, the program's main goal for

IKS was created as follows (National Workshop, 1998:1): "... to unearth, promote and protect the African heritage to shine a light on that which has in the past been dismissed and denied. Our country needs to develop an indigenous knowledge system which supports local economic development and enhances its rich cultural, technological, artistic, linguistic and traditional healing heritage."

As noted by Kaya and Seleti (2013), NWU started developing and integrating the IKS into its undergraduate and graduate research, teaching, learning, and community engagement curricula in 2001. In the same year, the Bachelor of Arts in Indigenous Knowledge Systems (BA.IKS) was introduced and it was sustained until 2012. Since 2001 until 2012, the NWU became the first university in South Africa and the African continent to have accredited IKS accredited programs. In 2013, a revised program was introduced by the NWU which is Bachelor of Indigenous Knowledge Systems (B.IKS) with for streams as follows:

a) Health care stream
b) Arts and culture stream
c) Science and Technology stream
d) Agriculture stream

This program is offered by the IKS Centre which is also responsible for the IKSDC in the North West Province (Kaya & Seleti, 2013). Moreover, The IKS accredited programs extend to postgraduate level, and they are offered as follows: Master of Indigenous Knowledge Systems (M.IKS) and Doctor of Philosophy in Social Sciences with Indigenous Knowledge Systems. North-West University's IKS Centre is in partnership with the following institutions:

- National Research Foundation (NRF)
- Department of Science and Innovation (DSI)
- Department of Sport, Arts and Culture (DSAC)
- Department of Arts, Culture, Sports and Recreation (ACSR)
- Council for Scientific and Industrial Research (CSIR)
- North-West House of Traditional Leadership (NWHTL)
- North-West Dingaka Association (NWDA)
- Mahikeng Museum, North West Province
- North-West Provincial Government (NWPG)
- New Partnership for Africa's Development (NEPAD)
- Freedom Park
- Commission for the Promotion and Protection of Cultural, Religious and Linguistic Communities (CRL)
- Agricultural Research Council (ARC)
- South African Network for Biosciences (SANBio)
- The Centre in Indigenous Knowledge Systems (CIKS) from UKZN now known as DSI-NRF CIKS

Since 2013, the IKS students at an undergraduate and postgraduate level have received consistent funding opportunities from the DSI, CSIR, NRF and DSAC. Job opportunities on IKS also exist from these mentioned institutions. These institutions have indeed lived up to the 1996 auditing project, 1998 workshop on IKS, the IKS policy of 2004 and the IK Act of 2019 in supporting IKS accredited programs in universities (Kaya, 2013). Therefore, the goal of NWU's IKS accredited programs is not to replace western frameworks of knowledge but to challenge their superiority and restore IKS to their proper position as equally legitimate forms of knowledge within the world's diverse array of knowledge systems, thereby improving the ability to solve both local and global issues.

University of Venda

Tasani' (2004) study indicated that following a September 2001 conference, the UNIVEN released the Uneven Declaration on IKS, a statement of commitment involving relevant parties and organizations. This Declaration is essentially an endorsement and acknowledgement of IKS African Perspectives. Among other things, the conference attendees pledged to support the creation and adoption of a core IKS curriculum in all educational establishments. Additionally, they decided to create a committee to use a journal and website to encourage networking efforts (University of Venda, 2025).

Natshandama and Nevhudoli (2021) conducted a study which showed that in 2014, the UNIVEN answered the call to include IKS accredited programs within universities. The B.IKS program was created by UNIVEN and was intended to be comparable to the one that NWU launched in 2013. The two programs are different in that the NWU's B.IKS program offers four streams, but UNIVEN 's B.IKS program only offers the health stream because it has not yet added any more streams in accordance with its frameworks (Nevhudoli & Netshandama, 2023). Furthermore, students pursuing postgraduate studies presently enroll in African studies at the Masters and Postgraduate levels, while UNIVEN has yet to launch the IKS program at the postgraduate level. Since NWU is the only university in South Africa and the entire African continent to provide these IKS degrees at an undergraduate and postgraduate level, some alumni from UNIVEN 's B.IKS program have enrolled for NWU's master's and doctoral programs in IKS.

The students enrolled for the IKS accredited program at UNIVEN receive funding from the DSI, and NRF. As it has been mentioned before, UNIVEN hosts an IKSDC and the Department of Indigenous Knowledge Systems and Heritage Studies which is a home for the B.IKS is actively involved in the management of the IKSDC (Nevhudoli & Netshandama, 2023). In addition, Nevhudoli and Netshandama, (2023) stated that relationships have been made between organizations, stakeholders, businesses and the Department of Indigenous Knowledge Systems and Heritage Studies to strengthen the IKS program. Students participating in work-integrated learning (WIL), which focuses on IKS especially, are hosted by companies and other stakeholders (Martinez-Vargas et al., 2025). WIL is an organized teaching approach that blends classroom instruction with real-world job experience with the goal of improving students' employability and giving them information and skills for the workplace.

Other universities

Besides the NWU and UNIVEN, no other university that was part of the 1996 audit has taken steps to introduce the IKS accredited programs. However, evidence of modules in IKS from other courses are visible from the University of Limpopo (UL) and the University of South Africa (UNISA). For example, the Indigenous Knowledge Systems and African Languages program (also available in indigenous languages) is provided by (University of South Africa, 2025). The UL's Cultural and Political Studies program and the Bachelor of Arts degree both include the African Religions Indigenous Knowledge Systems (HANA031) module. The fourth year (Honors) Anthropology module provided by UL is called

Theories and Indigenous Knowledge Systems (HANB081) (University of Limpopo, 2025a; University of Limpopo, 2025b). These efforts show that there is a positive shift in recognizing IKS as part of the curriculum. However, at the time this report was produced, neither UL nor UNISA had any affiliation with any IKSDCs. Nevertheless, both universities have conducted research studies from a variety of disciplines that have integrated and funded by CSIR, DSI, and NRF.

The University of Free State (UFS), University of Fort Hare (UFH), University of Zululand UNIZULU, University of KwaZulu-Natal (UKZN) and Walter Sisulu University (WSU) do not have IKS accredited programs. All the mentioned universities besides the UFH are hosts of the IKDCs by DSI and are working with indigenous communities (Balogun, 2023). Additionally, UKZN is home to the DSI-NRF CIKS, which provides funding for postgraduate students (honors, master's, and doctoral) at public universities in South Africa. The program by DSI-NRF CIKS focuses on research related to national priority areas and the Science, Technology, and Innovation (STI) Decadal Plan of 2022, which includes IKS. Additionally, IKS-based research is offered at these universities, just like at UL, UNISA, and UNIVEN, and researchers there are supported by CSIR, NRF, DSI, and other funders. This study should highlight that, notwithstanding NWU's IKS accredited programs, researchers from a variety of disciplines continue to use IKS in their work, particularly at the postgraduate level, and that these studies are archived in the NWU library collection (Kaya & Seleti, 2013).

CHALLENGES AND OPPORTUNITIES FOR ESTABLISHING THE IKS ACCREDITED PROGRAMMES

This section of the study focuses on the challenges that have been encountered over the years to establish the IKS accredited programs in universities at both undergraduate and postgraduate levels.

Challenges of establishing IKS in Universities

Part of the reason for the sluggish introduction of IKS accredited programs is the contextual rhetoric and oversimplified perspectives on the use of indigenous knowledge. The discussion up to this point has mostly focused on what may be referred to as IKS discourse, or the claims about its worth and potential cognitive uses in solving real life challenges (Murwira & Higgs, 2023). Furthermore, Cocks et al. (2012) noted that there are issues with IKS in the context of intellectual colonialism, which supports the widespread belief that IKS is inferior. Nkondo's (2012) presentation on IKS on polyepistemic world argues that even if all knowledge systems have a local basis, Western countries have universally imposed their cultures, knowledge systems, and languages on South Africa because of unequal power relations brought about by colonization and other kinds of imperialism. Therefore, IKS represents the resurgence of a long-suppressed indigenous identity in the academy. However, there is still hesitation in academia to take steps in implementing IKS accredited programs.

Kaya and Seleti (2013) argued that African Indigenous ways of knowing and producing knowledge are viewed as unscientific, archaic, and incompatible with formal education, particularly higher education, because of this Eurocentric viewpoint. They are unable to generate human capital that can handle the demands of contemporary globalization-related research and technology. Universities continue to follow the Western legacy of cognitive culture and theoretical conceptualizations. This could have something to do with the lack of resources, but it could also have something to do with ignorance and unwillingness (Nkondo, 2012).

Despite calls for IKS to be used to decolonize education, most South African universities respond by doing research linked to IKS rather than implementing IKS accredited programs at the undergraduate and graduate levels (Mekoa, 2015). Universities have staff members who possess undergraduate and postgraduate degrees from a variety of fields and may not have passion to fight for the existence of IKS in the academy. This may result in them putting their fields first, leaving the IKS agenda behind and ignored. Since DSI, NRF, and CSIR provide money for IKS research, most academics choose the simpler route while ignoring IKS accredited programs. Since academics have their own areas to develop and safeguard, little to no effort has been made to benchmark with universities that already have IKS programs (Martinez-Vargas et., 2025).

Some of the academics who speared the fight to integrate IKS in South African institutions were unable to further develop the program in more universities until they retired, while others have since passed. With no "successors" who shared their passion to carry on this endeavor, it is quite likely that the programs' implementation was abandoned when they left universities (Balogun, 2023). Nkondo (2012) advised that in order to establish IKS accredited programs, academics who wish to advance through benchmarking may encounter difficulties if they fail to properly confer with IKS academics who have completed the programs. Therefore, in order to steer them in the correct direction and prevent failure, they would want guidance and frameworks from scholars who have studied the IKS programs. The successful execution of IKS programs, which were historically disparaged by colonialism and other types of imperialism, will result from this (Nkondo, 2023).

The success of integrating the IKS programmed in South African universities

IKS integration at all South African educational levels will help students by making instruction more relevant and effective and by providing them with a curriculum that aligns with their own cultural backgrounds and worldviews. Shahjahan et al. (2023) provided a perspective that teachers can better prepare students for life in the real world by creating curricula, instructional materials, and textbooks that integrate indigenous knowledge with Western knowledge systems. The NWU and UNIVEN IKS programs have given both students and academics the chance to reconsider the inherent hierarchy of knowledge systems. Furthermore, as per Nkondo (2023) they were able to recognize and admit the existence of different kinds of knowledge rather than a single, standard, benchmark system based on western ideals and ways of knowing with the introduction of the IKS accredited programs.

NRF and DSI currently fund IKS research and promote postgraduate studies and high-caliber research with a focus on alignment with national priority research projects and the STI Decadal Plan (Tshishonga, 2019). The DST-NRF center in IKS (CIKS), located in UKZN, was created by DSI and NRF to support research which shows to be successful. It is an IKS research hub in Africa and works/ partnered with academic institutions such as NWU, UNISA, UL, and UNIVEN to document IKS through research from a variety of disciplines. DSI-NRF CIKS is managed by the Research Centers and Centers for Excellence (RCCE) directorate (Bhuda & Gumbo, 2024).

The DSI is essential to the creation and execution of IKS-related legislation and regulations, such as the IK Act of 2019 and the IKS Policy of 2004 which promote the integration of IKS in education (Mosimege, 2005). The IKS Act of 2019 and the IKS Policy of 2004 both support the advancement of IKS in higher education as a crucial element of human capital and the transformational aspect of higher education to address the nation's developmental concerns. Kaya and Chinsamy (2021). This is because policymakers, researchers, academics, and development organizations are becoming more aware that a nation's capacity to mobilize and expand its citizens' existing knowledge systems is just as crucial to socioeconomic development as the availability of material and monetary resources (Mekoa, 2015).

Initiatives to incorporate IKS into research, teaching, learning, and community involvement programs have been developed by a number of South African higher education institutions. Among them is NWU, which has generated many postgraduate graduates with master's and doctoral degrees. For the purpose of developing the IKS programs in the universities that are currently falling behind in its implementation, NWU alumni can be recruited to work on an as needed or permanent basis. Because they completed the undergraduate program that has four streams and postgraduate degrees, the NWU alumni possess the knowledge and abilities of the IKS programs. They are also active researchers who are contributing to bodies of knowledge in research (see Maditsi, 2018; Mashego et al., 2021; Ndhlovu et al., 2023; Bhuda, 2024a). They can carry out the 1996 mission by implementing, developing, and maintaining the IKS programs.

In order to help the IKS accredited programs become institutionalized in all 26 universities, Prof. Hassan Kaya, the Director of the DSI-NRF CIKS, has formed an IKS institutional committee composed of postgraduate degree holders from NWU and other researchers who have been funded by the DSI-NRF CIKS. This committee was established in 2024, and the universities that were selected in 1996 to participate in the indigenous technologies audit are the ones it plans to visit first. Providing universities with the IKS frameworks and implantation approaches is the committee's goal. Following this process, the universities will have the responsibility of deciding whether to accept or reject the committee's recommendations and proposed frameworks for the successful implementation of IKS accredited programs that will be aligned with each other in all universities.

WHERE TO FROM HERE?

According to this study, the best way to respond to the question "where to from here?" is to first recognize the progress made in recognizing IKS in education and the attempts made to implement IKS accredited programs while also eschewing Western Knowledge Systems. Initiatives must be recognized, and plans for the future must be developed, despite the fact that much work remains. Shortly after two years after South Africa gained independence and democracy, discussions regarding the Indigenous technologies at South African universities began and that was 29 years ago. Even if certain universities started a long time ago with the IKS project and were left behind along the way, there are still chances to work towards building the future of IKS in academia.

A few government agencies that are still willing to assist with the creation of the IKS accredited programs are the DSI, NRF, and DSAC. Although this will not be an overnight success, it is feasible to launch and maintain the programs in the same way that UNIVEN and the NWU have done. Universities must contribute to the IKS implementation process as well. They need to review the IKS frameworks, update their policies, and align themselves to other universities that have already begun this process (Nevhudoli & Netshandama, 2023). Universities that are willing to implement IKS accredited programs must keep in mind that there are IKS graduates in South Africa who possess the expertise and skills needed to support or oversee this process as part of their response to the 1996 Indigenous technologies project.

For example, the NWU began its journey in 2001 and has produced graduates from undergraduate to graduate levels over the years (Mosimege, 2004; Odora Hoppers, 2001). Scholars who have completed both undergraduate and graduate coursework are needed to construct the IKS program because they possess the knowledge and skills necessary to make the programs a reality. Although UNIVEN's IKS curriculum only offers one stream, there are graduates who are experts in it, and others have pursued IKS-focused further education in African studies. Additionally, they can help build IKS programs in other universities (Nevhudoli & Netshandama, 2023). Since there are still universities which are hesitant to create the IKS programs, consulting/ hiring IKS graduates who have gone through the curriculum recommended.

Therefore, the IKS institutional committee had to be established in order to provide universities with practical solutions to their current issues of IKS accredited programs implementation and to oversee this process when required to. It will be up to those universities to decide whether to accept or reject the committee's proposal as a way forward from 1996. NWU and UNIVEN have generated results that demonstrate the programs' effectiveness which is enough for other universities to use as point of reference and proceed. Now is the ideal moment for IKS scholars who have actually done the IKS studies (not just research) to demonstrate their contributions to existing knowledge bases and carry out the 1996 vision, as these two universities mentioned trained them to do.

IMPLICATION FOR FUTURE RESEARCH

This study significantly advances the fields of higher education studies, decolonizing studies, and IKS. It hopes to share light on how the IKS programs are structured in South Africa universities. Moreover, there were few studies with a focus on IKS programs in South African higher education. Therefore, this study provides recent research and progress towards implementing IKS accredited programs in universities. This study fills in some of the gaps in literature that can be addressed by future research, but it also adds to the body of knowledge regarding IKS in education.

CONCLUSION AND RECOMMENDATIONS

The aim of the study was to investigate the implementation of the IKS accredited programs by universities since the 1996 audit of indigenous technologies. In order to identify universities that were pioneers of IKS by starting with auditing Indigenous technologies in 1996, this study first examined historical data regarding the beginning of the engagements of IKS. According to the study's findings, the government has fully committed itself to establishing IKS by working with universities who offer IKS accredited programs and those who conduct IKS integrated research. There is financing available for the IKS accredited programs and research in IKS which are already supported by the DSI, NRF, and CSIR. Moreover, there are universities which are connected to IKDCs that are managed by DSI. The study's findings also provided insights into universities that have gone beyond 1996 in creating the IKS accredited programs. At the time this study was written which was 2025, only NWU and UNVEN offered IKS accredited programs. The study further acknowledged universities that integrate IKS in other disciplines. This demonstrates that while there is a push among universities to acknowledge IKS, frameworks and roadmaps must be put in place to guarantee that universities integrate the IKS accredited programs. These initiatives can be implemented by scholars who have done the IKS accredited programs for they will understand the structure/ curriculum of these programs. Universities must also benchmark and work with UNIVEN and NWU to receive further support during this process. Based on the reports of this study, the following is recommended:

- Universities interested in offering IKS accredited programs ought to benchmark with those that currently offer them to ensure that the structures are aligned.
- Workshops with the other public universities that do not offer IKS accredited programs must be held in order to introduce the relevance of IKS in higher education. Universities that embarked on the IKS journey in 1996 can serve as a starting point of this initiative. DSI-NRF CIKS may facilitate the workshops.
- Use social media and other sites to share the success and activities of IKS students, alumni and academics from NWU and UNIVEN in effort to encourage other universities to consider implementing the programs.
- Through conferences, webinars, seminars, and research, academics and students from UNIVEN and NWU must work together with researchers from other universities. This will open up possibilities for additional IKS engagements, which may eventually capture the interest of other academics who wish to get involved and take the lead in implementing IKS within their universities.
- The Department of Higher Education and Training (DHET) should support the notion of IKS accredited programs by assigning delegates to attend the workshops that will be held in universities.
- To guarantee the success of IKS accredited programs, universities must consider IKS graduates (with master's and doctoral degrees) as participants in the implementation process.

REFERENCES

Ajani, O. A., & Gamede, B. T. (2021). Decolonising Teacher Education Curriculum in South African Higher Education. *International Journal of Higher Education, 10*(5), 121–131. DOI: 10.5430/ijhe.v10n5p121

Arifin, S. R. M. (2018). Ethical considerations in qualitative study. *International journal of care scholars, 1*(2), 30-33.

Badat, S. (2020). Reproduction, transformation and public South African higher education during and beyond Covid-19. *Transformation (Durban), 104*(1), 24–42. DOI: 10.1353/trn.2020.0030

Balogun, T. (2023). Digitization of Indigenous knowledge systems in Africa: The case of South Africa's National Recorded System (NRS). *Records Management Journal, 33*(1), 88–102. DOI: 10.1108/RMJ-10-2022-0029

Balogun, T., & Kalusopa, T. (2022). Web archiving of indigenous knowledge systems in South Africa. *Information Development, 38*(4), 658–671. DOI: 10.1177/02666669211005522

Bhambra, G. K. (2021). Decolonizing critical theory? Epistemological justice, progress, reparations. *Critical Times, 4*(1), 73–89. DOI: 10.1215/26410478-8855227

Bhuda, M. T. (2024a). Integrating Ethnomathematical Activities in the Foundation Phase as Means of Decolonisation: Insights from Indigenous Knowledge Systems. *African Journal of Development Studies, 14*(4).

Bhuda, M. T. (2024b). Decolonising and indigenising African Indigenous research: Why Indigenous ways of knowing, being and living matter in research. In *Higher Education Transformation in Africa* (pp. 50–69). Routledge. DOI: 10.4324/9781003528319-4

Bhuda, M. T., & Gumbo, M. T. (2024). Developments Towards Integrating African Indigenous Knowledge in South African Universities. *UnisaRxiv*. DOI: 10.25159/UnisaRxiv/000082.v1

Bhuda, M. T., & Kgwadi, B. (2023). The economic measures established to rebuild the South African cultural heritage tourism after the COVID-19 impact. *Gender & Behaviour, 21*(1), 21398–21412.

Bhuda, M. T., & Marumo, P. (2022). Ubuntu Philosophy and African Indigenous Knowledge Systems: Insights from decolonization and indigenization of research. *Gender & Behaviour, 20*(1), 19133–19151.

Bonde, A. R., & Pinjari, R. M. (2023). Hoodia Gordonii a herbal plant: A Review. *International Journal of Rural Development Environment and Health Research, 7*(5), 98–106. DOI: 10.22161/ijreh.7.5.8

Burgess, C., Bishop, M., & Lowe, K. (2022). Decolonising Indigenous education: The case for cultural mentoring in supporting Indigenous knowledge reproduction. *Discourse (Abingdon), 43*(1), 1–14. DOI: 10.1080/01596306.2020.1774513

Chabaya, O., & Chabaya, R. A. (2023). Integration of African Indigenous Knowledge systems into higher education curriculum in Zimbabwe universities: Exploring experiences of lecturers. *African Perspectives of Research in Teaching and Learning, 7*(3), 625–638.

Chinenye Nwokocha, G., & Legg-Jack, D.Chinenye -Nwokocha. (2024). Reimagining STEM education in south Africa: Leveraging Indigenous knowledge systems through the m-know model for curriculum enhancement. *International Journal of Social Science Research and Review*, *7*(2), 173–189. DOI: 10.47814/ijssrr.v7i2.1951

Cocks, M. L., Alexander, J., & Dold, T. (2012). Inkcubeko Nendalo: A bio-cultural diversity schools education project in South Africa and its implications for inclusive indigenous knowledge systems (IKS) sustainability. *Journal of Education for Sustainable Development*, *6*(2), 241–252. DOI: 10.1177/0973408212475232

da Silva, C., Pereira, F., & Amorim, J. P. (2024). The integration of indigenous knowledge in school: A systematic review. *Compare: A Journal of Comparative Education*, *54*(7), 1210–1228. DOI: 10.1080/03057925.2023.2184200

Datta, R. (2018). Decolonizing both researcher and research and its effectiveness in Indigenous research. *Research Ethics*, *14*(2), 1–24. DOI: 10.1177/1747016117733296

Dei, G. (2014). Global education from an 'indigenist' anti-colonial perspective. *Journal of Contemporary Issues in Education*, *9*(2). Advance online publication. DOI: 10.20355/C53G6B

Githui, W., David, N., & Maurice, S. (2015). The role of indigenous knowledge in socio-economic development. *International Journal of Scientific Research*, *4*(4), 32–37.

Govender, N. (2012). Educational implications of applying the complexity approach to Indigenous Knowledge Systems (IKS). *Alternation (Durban)*, *19*(2), 112–137.

Gumbo, M. T., Knaus, C. B., & Gasa, V. G. (2024). Decolonising the African doctorate: Transforming the foundations of knowledge. *Higher Education*, *88*(4), 1611–1627. DOI: 10.1007/s10734-024-01185-2

Hammarberg, K., Kirkman, M., & De Lacey, S. (2016). Qualitative research methods: When to use them and how to judge them. *Human Reproduction (Oxford, England)*, *31*(3), 498–501. DOI: 10.1093/humrep/dev334 PMID: 26759142

Heleta, S. (2016). Decolonisation of higher education: Dismantling epistemic violence and Eurocentrism in South Africa. *Transformation in higher education, 1*(1), 1-8.

Higgs, P. (2016). The African renaissance and the transformation of the higher education curriculum in South Africa. *Africa Education Review*, *13*(1), 87–101. DOI: 10.1080/18146627.2016.1186370

Higgs, P., & Van Niekerk, M. P. (2002). The programme for indigenous knowledge systems (IKS) and higher educational discourse in South Africa: A critical reflection: Perspectives on higher education. *South African Journal of Higher Education*, *16*(3), 38–49. DOI: 10.4314/sajhe.v16i3.25215

Higgs, P., & Van Niekerk, M. P. (2002). The programme for indigenous knowledge systems (IKS) and higher educational discourse in South Africa: A critical reflection: Perspectives on higher education. *South African Journal of Higher Education*, *16*(3), 38–49. DOI: 10.4314/sajhe.v16i3.25215

Jonas, H., Bavikatte, K., & Shrumm, H. (2010). Community protocols and access and benefit sharing. *Asian Biotechnology and Development Review*, *12*(3), 49–76.

Kaya, H. O. (2013). Integration of African indigenous knowledge systems into higher education in South Africa: Prospects and challenges. *Alternation (Durban), 20*(1), 135–153.

Kaya, H. O., & Chinsamy, M. (2021). Combating world pandemics through complementarity of indigenous knowledge with other knowledge systems: A paradigm shift in healthcare research and education. *Alternation (Durban), 38*, 254–272. DOI: 10.29086/2519-5476/2021/sp38a11

Kaya, H. O., & Seleti, Y. N. (2013). African indigenous knowledge systems and relevance of higher education in South Africa. *International Education Journal: Comparative Perspectives, 12*(1).

Kim, H., Sefcik, J. S., & Bradway, C. (2017). Characteristics of qualitative descriptive studies: A systematic review. *Research in Nursing & Health, 40*(1), 23–42. DOI: 10.1002/nur.21768 PMID: 27686751

Kohsaka, R., & Rogel, M. (2021). Traditional and local knowledge for sustainable development: Empowering the indigenous and local communities of the world. In *Partnerships for the Goals* (pp. 1261–1273). Springer International Publishing. DOI: 10.1007/978-3-319-95963-4_17

Le Grange, L. (2016). Think piece: Sustainability education and (curriculum) improvisation. *Southern African Journal of Environmental Education, 32*, 26–36.

Ma, C., Qirui, C., & Lv, Y. (2023). "One community at a time": Promoting community resilience in the face of natural hazards and public health challenges. *BMC Public Health, 23*(1), 2510. DOI: 10.1186/s12889-023-17458-x PMID: 38097956

Maditsi, M. E. (2018). *The role of traditional leaders in postapartheid local governance in South Africa: a case study of the Bapong 2 community, North West Province, South Africa* (Doctoral dissertation, North-West University (South Africa)).

Maditsi, M. E., & Bhuda, M. T. (2023). Decolonising and Humanising Pedagogies in South African Postgraduate Education: Lessons From Indigenous Knowledge Systems. In *Digital Preservation and Documentation of Global Indigenous Knowledge Systems* (pp. 282-302). IGI Global.

Mapara, J. (2009). 'Indigenous Knowledge Systems in Zimbabwe: Juxtaposing Postcolonial Theory'. *The Journal of Pan African Studies, 3*(1), 139–155.

Marsandi, F., Sutadji, E., Kuntadi, I., Rizal, F., Rahma, A. B. N., & Fajri, H. (2025). Integrating Ethnobotany and Indigenous Knowledge into Higher Education Curricula: Insights from a Global Biobliometric Analysis. *Ethnobotany Research and Applications, 30*, 1–12. DOI: 10.32859/era.30.17.1-12

Martinez-Vargas, C., Ndou, U. E., Makhanikhe, J., & Madima, T. (2025). Decoloniality, identity and the third space: The case of students' experiences of well-being from an indigenous knowledge system university programme in South Africa. *International Journal of Educational Research, 130*, 102547. DOI: 10.1016/j.ijer.2025.102547

Masenya, T. M. (2022a). Decolonization of indigenous knowledge systems in South Africa: Impact of policy and protocols. [IJKM]. *International Journal of Knowledge Management, 18*(1), 1–22. DOI: 10.4018/IJKM.310005

Masenya, T. M. (2022b). Digital preservation of indigenous knowledge in South African rural communities. In *Handbook of research on protecting and managing global indigenous knowledge systems* (pp. 317–340). IGI Global. DOI: 10.4018/978-1-7998-7492-8.ch017

Mashego, S. L., Maditsi, M. E., & Bhuda, M. T. (2021). A response to Covid-19: Perspectives from African indigenous knowledge systems. *Gender & Behaviour*, *19*(1), 17242–17254.

Masoga, M. (2005). South African research in indigenous knowledge systems and challenges of change. *Indilinga*, *4*(1), 15–30.

Mekoa, I. (2015). Challenges and opportunities for IKS in higher education institutions in South Africa: Politics, ideological, institutional cultures and structural dimensions. *Indilinga*, *14*(1), 12–24.

Moahi, K. (2012). Promoting African indigenous knowledge in the knowledge economy: Exploring the role of higher education and libraries. *Aslib Proceedings*, *64*(5), 540–554. DOI: 10.1108/00012531211263157

Moremoholo, T. P. (2023). The role of culture in shaping the curriculum of higher education in South Africa. *Journal of Curriculum Studies Research*, *5*(2), 37–55. DOI: 10.46303/jcsr.2023.17

Mosimege, M. (2004). Indigenous knowledge systems in South Africa: perspectives from the Department of Science and Technology: news and views. *Indilinga*, *3*(1), 78–84.

Mosimege, M. (2005). National priorities in Indigenous Knowledge Systems: Implications for research and curriculum development. *Indilinga*, *4*(1), 31–37.

Motala, S., & Menon, K. (2023). In search of the holy grail: Social justice and funding in higher education in South Africa. *Southern African Review of Education with Education with Production*, *28*(1), 32–51.

Msila, V. (2025). Transforming university through Africanised research: In search of the indigenous. *African Identities*, •••, 1–14. DOI: 10.1080/14725843.2024.2439440

Murwira, S., & Higgs, P. (2023). Afrocentric epistemic systems and higher education curriculum. The case of a university in Zimbabwe. *South African Journal of Higher Education, 37*(2), 283-297. / https://orcid.org/0000-0003-2563-4336

National Research Foundation. (2024). Centres of Excellence. https://www.nrf.ac.za/core-mandate-business-divisions/risa-directorates/research-chairs-and-centres-of-excellence-rcce/centres-of excellence/#:~:text=The%20Department%20of%20Science%20and,and%20institutions%20on%20long%2Dterm

Natshandama, O. V., & Nevhudoli, N. D. (2021). Creating intergenerational learning spaces: A collaboration between UNIVEN Community Engagement Programme and Dzomo la Mupo. [CriSTaL]. *Critical Studies in Teaching and Learning*, *9*(2), 39–63. DOI: 10.14426/cristal.v9i2.462

Ndhlovu, P. T., Asong, J. A., Omotayo, A. O., Otang-Mbeng, W., & Aremu, A. O. (2023). Ethnobotanical survey of medicinal plants used by indigenous knowledge holders to manage healthcare needs in children. *PLoS One*, *18*(3), e0282113. DOI: 10.1371/journal.pone.0282113 PMID: 36972257

Nevhudoli, N. D., & Netshandama, V. O. (2023). What do Bachelor of Indigenous Knowledge Systems graduates say about their curriculum? A qualitative tracer study at the University of Venda. *Journal of Curriculum Studies Research*, *5*(1), 141–158. DOI: 10.46303/jcsr.2023.11

Ngulube, P., Dube, L., & Mhlongo, M. (2015). Towards a cartography of indigenous knowledge systems in library and information science training and education in Anglophone eastern and southern Africa. *Indilinga, 14*(2), 145–168.

Nkondo, M. (2012). *Indigenous African knowledge systems in a polyepistemic world: The capabilities approach and the translatability of knowledge systems.* Paper presented at the Southern African Regional Colloquium on Indigenous African knowledge systems: methodologies and epistemologies for research, teaching, learning and community engagement in higher education. Howard College Campus: University Of KwaZulu-Natal.

Noble, H., & Heale, R. (2019). Triangulation in research, with examples. *Evidence-Based Nursing, 22*(3), 67–68. DOI: 10.1136/ebnurs-2019-103145 PMID: 31201209

Odora Hoppers, C. (2021). Research on Indigenous knowledge systems: The search for cognitive justice. *International Journal of Lifelong Education, 40*(4), 310–327. DOI: 10.1080/02601370.2021.1966109

Odora Hoppers, C. A. O. (2001). Indigenous knowledge systems and academic institutions in South Africa. *Perspectives in Education, 19*(1), 73–86.

Odora Hoppers, C. A. O. (2023). Knowledge Production, Access and Governance: A Song from the South. In *Decolonizing Education for Sustainable Futures* (pp. 67–82). Bristol University Press.

Parthiban, R., Sun, R., Qureshi, I., & Bandyopadhyay, S. (2024). Empowering rural micro-entrepreneurs through technoficing: A process model for mobilizing and developing indigenous knowledge. *The Journal of Strategic Information Systems, 33*(2), 101836. DOI: 10.1016/j.jsis.2024.101836

Prah, K. K. (2002). The burden of English in Africa: From colonialism to neo-colonialism. *TESL-EJ (Teaching English as a Second or Foreign Language–Electronic Journal), 6*(1), 1-14.

Shahjahan, R. A., Estera, A. L., Surla, K. L., & Edwards, K. T. (2022). "Decolonizing" curriculum and pedagogy: A comparative review across disciplines and global higher education contexts. *Review of Educational Research, 92*(1), 73–113. DOI: 10.3102/00346543211042423

Tegegne, H. R., Tesfay, T., Hailesilassie, M. B., & Gebreyohaness, S. T. (2025). Indigenous Knowledge and Practices of the Tigray People: How Do Teachers and Students Perceive Its Integration in Science and Mathematics Education? *Science & Education*, •••, 1–28. DOI: 10.1007/s11191-024-00608-x

Tisani, N. (2004). African indegenous knowledge systems (AIKSs): another challenge for curriculum development in higher education?: perspectives on higher education. *South African Journal of Higher Education, 18*(3), 174–184.

Tshishonga, N. S. (2019). Incorporating Indigenous Knowledge in the Higher Education Sector for the Advancement of African Scholarship. In *The Formation of Intellectual Capital and Its Ability to Transform Higher Education Institutions and the Knowledge Society* (pp. 212–231). IGI Global. DOI: 10.4018/978-1-5225-8461-2.ch011

University of Limpopo. (2025a). School of Social Sciences: Sociology and Anthropology. https://www.ul.ac.za/faculty-of-humanities/school-of-social-sciences/sociology-and-anthropology/

University of Limpopo. (2025b). School of Social Sciences: Cultural and Political studies. https://www.ul.ac.za/faculty-of-humanities/school-of-social-sciences/cultural-and-political-studies/

University of South Africa. (2025). Register to study through UNISA. https://www.unisa.ac.za/sites/corporate/default/Register-to-study-through-Unisa/Subjects-&-modules/All-modules/Indigenous-Knowledge-Systems-and-African-Languages-(Sesotho)-%E2%80%93-SSO3702

University of Venda. (2025). Department of Indigenous Knowledge Systems and Heritage Studies. https://www.univen.ac.za/faculties/humanities-social-science-and-education/department-of-indigenous-knowledge-systems-and-heritage-studies/

Van Rooyen, B. (2011). Safeguarding the future of indigenous knowledge through ICT: ICT and society. *CSIR Science Scope*, *5*(2), 24–26.

Verhoef, G. (2023). *The public domain and the legal protection of Indigenous Knowledge Systems* (Doctoral dissertation, North-West University (South Africa).). wa Thiong'o, N (1986). Decolonising the mind: *The politics of language in African literature.* London, England: James Currey

Wane, N. N., & Wane, N. N. (2019). African indigenous education. *Gender, democracy and institutional development in Africa*, 99-121.

Yip, S. Y., & Chakma, U. (2024). The teaching of Indigenous knowledge and perspectives in initial teacher education: A scoping review of empirical studies. *Journal of Further and Higher Education*, *48*(3), 287–300. DOI: 10.1080/0309877X.2024.2327029

ADDITIONAL READING

Jima, A. O. (2022). Significance and restraint of indigenous knowledge inclusion in Ethiopian higher education curriculum: In focus Gadaa system. *Cogent Education*, *9*(1), 2046241. DOI: 10.1080/2331186X.2022.2046241

Khan, S., & Sharma, M. (2024). An Overview on Indian Knowledge System. *Integrated Journal for Research in Arts and Humanities*, *4*(4), 42–46. DOI: 10.55544/ijrah.4.4.7

Mahesh, K. M., Aithal, P. S., & Sharma, K. R. S. (2023). Literature review on Indian ancient university in imparting holistic and multidisciplinary: To create Indian knowledge system (IKS). [IJPL]. *International Journal of Philosophy and Languages*, *2*(1), 1–17.

Thapliyal, P. (2023). Indian Knowledge Systems in the Curriculum of Higher Education: A Proposed Model of a PG Course in IKS. *International Journal of Research Publication and Reviews. 2582-7421: 4 (7)*, 3296-3301.

KEY TERMS AND DEFINITIONS

Indigenous Knowledge Systems: Indigenous Knowledge Systems are integrated epistemological frameworks that promote an awareness of the interconnectedness of the world and our roles in it. They are taught using Indigenous pedagogies.

Curriculum: A curriculum includes all from lessons and tests to the general objectives and standards of a course or programme. It is the planned learning experiences and information that students encounter in an educational context.

Decolonization: Documentation is the systematic means of locating, recording, and classifying knowledge inside a company or setting in order to guarantee its availability and preservation.

Integration: The term "integration" in education often refers to bringing different topics, ideas, or learning experiences together to form a coherent learning environment, frequently in order to give students a more comprehensive and pertinent educational experience.

Universities: Universities are educational establishments renowned for their emphasis on scholarships and research that provide a broad range of academic programmes, including undergraduate and graduate degrees.

Compilation of References

Aarons, D., & Akach, P. (1998). South African Sign Language - one language or many? A sociolinguistic question . *Stellenbosch papers in Linguistics 31*, 1 - 28.

Aarons, D., & Reynolds, L. (2003). South African Sign Language: changing policies and practice. *Many ways to be deaf: International variation in deaf communities Gallaudet University Press*, 194 - 210.

Abah, J., Mashebe, P., & Denuga, D. D. (2015). Prospect of integrating African indigenous knowledge systems into the teaching of sciences in Africa. *American Journal of Educational Research*, *3*(6), 668–673.

Abdul, M., Ingabire, A., Lam, C. Y. N., Bennett, B., Menzel, K., MacKenzie-Shalders, K., & Van Herwerden, L. (2024). Indigenous food sovereignty assessment—A systematic literature review. *Nutrition & Dietetics: the Journal of the Dietitians Association of Australia*, *81*(1), 12–27. DOI: 10.1111/1747-0080.12813 PMID: 37154011

Abeywardana, N., Schütt, B., Wagalawatta, T., & Bebermeier, W. (2019). Indigenous agricultural systems in the dry zone of Sri Lanka: Management transformation assessment and sustainability. *Sustainability (Basel)*, *11*(3), 910. DOI: 10.3390/su11030910

Abhay, R. K. (2024). Evaluating the Role of Indigenous Methods of Rainwater Harvesting for Water Sustainability of Local Ecosystems in India. In *Sacred Landscapes, Indigenous Knowledge, and Ethno-culture in Natural Resource Management: Understanding Multiple Perspectives of Nature Conservation* (pp. 121–134). Springer Nature Singapore. DOI: 10.1007/978-981-97-4206-6_6

Abioye, A., Zaid, Y., & Egberongbe, H. (2014). Documenting and disseminating agricultural indigenous knowledge for sustainable food security: The efforts of agricultural research libraries in Nigeria. *Libri*, *64*(1). Advance online publication. DOI: 10.1515/libri-2014-0007

Aboorva, T., & Mani, R. (2024). Contribution of Environmentalists in Preserving Global Environment. Quing. *International Journal of Multidisciplinary Scientific Research and Development*, *3*(2), 1–18.

Abrams, M. H., & Harpham, G. G. (2009). *A Glossary of Literary Terms* (9th ed.). Wadsworth Cengage Learning.

Acevedo, C., Rose, D., & Whittaker, R. (2023). *Reading to Learn, reading the world*. Equinox Publishing.

Adade Williams, P., Sikutshwa, L., & Shackleton, S. (2020). Acknowledging indigenous and local knowledge to facilitate collaboration in landscape approaches—Lessons from a systematic review. *Land (Basel)*, *9*(9), 331. DOI: 10.3390/land9090331

Adams, M., Sibanda, S., & Turner, S. D. (1999). *Land Tenure Reform and Rural Livelihoods in Southern Africa*. Overseas Development Institute.

Adeleye, O. A., Femi-Oyewo, M. N., Bamiro, O. A., Bakre, L. G., Alabi, A., Ashidi, J. S., & Fakoya, G. (2021). Ethnomedicinal herbs in African traditional medicine with potential activity for the prevention, treatment, and management of coronavirus disease 2019. *Future Journal of Pharmaceutical Sciences*, *7*, 1–14.

Adeola, M. O. (1992). Importance of wild animals and their parts in the culture, religious festivals, and traditional medicine of Nigeria. *Environmental Conservation*, *19*(2), 125–134. DOI: 10.1017/S0376892900030605

Afolabi, E. A. A. (2005). Cultural Practices of African Women. In African Women: APolitical Economy (pp. 97-115). Zed Books.

Africa, S. (2013). *South Africa Country Report.* Retrieved from WASLI: https://wasli.org/wp-content/uploads/2013/06/254_south-africa-country-report.pdf

African Development Bank Group. (2024). World Food Day: How the new MADE Alliance will use digital technologies to help farmers in Africa feed the continent. https://www.afdb.org/en/news-and-events/world-food-day-how-new-made-alliance-will-use-digital-technologies-help-farmers-africa-feed-continent-74856#:~:text=,economy%20via%20Mastercard%20Community%20Pass

African Food Changemakers. (n.d.). https://afchub.org/entrepreneurs/charlotte-scott-02344#:~:text=Mafisa%20helps%20traditional%20cattle,management%2C%20and%20better%20market%20access

AgriCrowdy. (n.d.). https://agricrowdy.com/agri-crowdy-savings-club/#:~:text=How%20does%20Agri%20Crowdy%E2%84%A2%C2%A0%C2%A0Savings%20Club,work

Agyei-Mensah, S., & Oteng-Ababio, M. (2012). Perceptions of health and environmental impacts of e-waste management in Ghana. *International Journal of Environmental Health Research*, *22*(6), 500–517. DOI: 10.1080/09603123.2012.667795 PMID: 22428915

Aiswarya Thamanna, . (2023). Indigenous Communication Of Everyday Life And Philosophy: An Analysis Of Gond Paintings In Madhya Pradesh. *ShodhKosh: Journal of Visual and Performing Arts*, *4*(2). Advance online publication. DOI: 10.29121/shodhkosh.v4.i2.2023.645

Ajani, O. A., & Gamede, B. T. (2021). Decolonising Teacher Education Curriculum in South African Higher Education. *International Journal of Higher Education*, *10*(5), 121–131. DOI: 10.5430/ijhe.v10n5p121

Aker, J. C., & Fafchamps, M. (2015). Mobile phone coverage and producer markets: Evidence from West Africa. *The World Bank Economic Review*, *29*(2), 262–292. DOI: 10.1093/wber/lhu006

Akhtar, S., & Farooq, N. (2023). Folk Dance and Music of Rajasthan: A Study of their Role in Fairs and Festivals. *Social Science Journal for Advanced Research*, *3*(4), 59–64. DOI: 10.54741/ssjar.3.4.9

Alejandro, A., & Zhao, L. (2024). Multi-method qualitative text and discourse analysis: A *methodological framework.Qualitative Inquiry*, *30*(6), 461–473. DOI: 10.1177/10778004231184421

Alexander, N. (2004). The politics of language planning in post-apartheid South Africa. *Language Problems and Language Planning*, *28*(2), 113–130. DOI: 10.1075/lplp.28.2.02ale

Alhojailan, M. I., & Ibrahim, M. (2012). Thematic analysis: A critical review of its process and evaluation. *West east journal of social sciences, 1*(1), 39-47.

Al-Khusaibi, M., Al-Habsi, N., & Rahman, M. (2019). Traditional Foods History, Preparation, Processing and Safety: History, Preparation, Processing and Safety. https://doiDOI: 10.1007/978-3-030-24620-4

Alverson, M. (1999). The New Generation Gap. *Woman in Business, 51*(3), 14–17.

Amechi, E. P. (2018). Traditional knowledge on the medicinal uses of plants, biopiracy and national patent measures in Africa: Exploratory reflections and comparative experiences. *Journal of Comparative Law in Africa, 5*(1), 73–109.

Amusan, O. O., Sukati, N. A., Dlamini, P. S., & Sibandze, F. G. (2007). Some Swazi phytomedicines and their constituents. *African Journal of Biotechnology, 6*(3), 267–272.

Anand, S., Kumar, H., Gagan, A., & Mayala, V. K. (2025). Indigenous Knowledge and Water Conservation Practices in South Africa: A Systematic Literature Review. *Journal of Environment and Earth Science.*

Anderson, R. (2004). Intuitive inquiry: An epistemology of the heart for scientific inquiry. *The Humanistic Psychologist, 32*(4), 307–341. DOI: 10.1080/08873267.2004.9961758

Andrew, M., & Ainslie, A. C. & Shackleton. (2003). Land use and livelihoods. Evaluating Land and Agrarian Reform in South Africa, No. 8. Bellville: University of the Western Cape.

Anene, C. P., & Njoku, C. U. (2020). *African indigenous knowledge systems: Problems and perspectives. Communal oral historians as reservoirs and curators of indigenous historical scholarship.* Retrieved May 10, 2023, from https://acjol.org/index.php/jassd/article/view/1497/1479

Antweiler, C. (1998). Local knowledge and local knowing: An anthropological analysis of contested cultural products in the context of development. [Paper presented at the seminar on Pastoralism and Environment organized by JET, Arusha.]. *Anthropos: Studie z Oboru Anthropologie, Paleoethnologie, Paleontologie a Kvartérní Geologie, 93*(4-6), 469–494.

Apple, M. W. (2004). *Ideology and curriculum* (3rd ed.). Routeledge Taylor & Francis. DOI: 10.4324/9780203487563

Aquilani, C., Confessore, A., Bozzi, R., Sirtori, F., & Pugliese, C. (2022). Precision Livestock Farming technologies in pasture-based livestock systems. *Animal, 16*(1), 100429. DOI: 10.1016/j.animal.2021.100429 PMID: 34953277

Arafat, A. A. D. (2024). Introduction: Theoretical Framework. In *Human Security in the Middle East and North Africa* (pp. 1–46). Springer Nature Switzerland. DOI: 10.1007/978-3-031-70196-2_1

Araia, M. G. (2019). Ecological and Socio-Cultural Potential of Human Modified Forest Landscapes in Conservation of Tree Species Diversity: The Case of Vhembe Biosphere Reserve, South Africa (Doctoral dissertation, University of Pretoria (South Africa)).

Ardill, A. (2013). Australian sovereignty, indigenous standpoint theory and feminist standpoint theory: First peoples' sovereignties matter. *Griffith Law Review, 22*(2), 315–343. DOI: 10.1080/10383441.2013.10854778

Arifin, S. R. M. (2018). Ethical considerations in qualitative study. *International journal of care scholars, 1*(2), 30-33.

Armitage, D. (2008). Governance and the commons in a multi-level world. *The International Journal of the Commons*, *2*(1), 7–32. DOI: 10.18352/ijc.28

Asante, M. K. (1991). The Afrocentric Idea in Education, Spring. *The Journal of Negro Education*. 60(2), 170-180. Retrieved June, 11, 2024, from https://www.jstor.org/stable/pdf/2295608.pdf

Asante, M. K. (2006). *The Afrocentric Paradigm*. Africa World Press.

Aschwanden, H. (1982). *Symbols of Life: An Analysis of the Consciousness of the Karanga*. Mambo Press.

Ashforth, A. (2002). An epidemic of witchcraft? The implications of AIDS for the post-apartheid state. *African Studies*, *61*(1), 121–143. DOI: 10.1080/00020180220140109

Asian Development Bank. (January 2023). How Farmer-Led Digital Ecosystems Can Boost Yields and Incomes in Indonesia. https://development.asia/case-study/how-farmer-led-digital-ecosystems-can-boost-yields-and-incomes-indonesia#:~:text=Despite%20the%20challenges%20created%20by,crops%20in%20line%20with%20market

Asomani-Boateng, R. (2016). Local networks: Commodity queens and the management of organic solid waste in indigenous open-air markets in Accra, Ghana. *Journal of Planning Education and Research*, *36*(2), 182–194. DOI: 10.1177/0739456X15604445

Assembly, U. G. (2007). United Nations declaration on the rights of indigenous peoples. *UN Wash*, *12*, 1–18.

Assuah, A. (2023). "A lot of people ignore our culture when it comes to waste management": Examining the impacts of culture on solid waste management in two Canadian First Nations. *Alternative*, *19*(2), 466–474. DOI: 10.1177/11771801231163635

Asuquo, M. E., Okenjom, G. P., Okpa, O. E., & Eyiene, A. E. (2023). Opportunities and challenges in digitization of indigenous knowledge and implication for educational management in the Nigerian Context. *Digital preservation and documentation of global indigenous knowledge systems*, 87-104.

Ávila-Bello, C. H., Hernández-Romero, Á. H., Vázquez-Luna, D., Lara-Rodríguez, D. A., Martínez-Jerónimo, A., Meneses-García, B. N., & Sánchez-Sandoval, X. M. (2024). Design of complex agroecosystems: Traditional and formal knowledge to conserve agrobiodiversity in the Santa Marta Mountains, Veracruz, México. *Environment, Development and Sustainability*, *26*(3), 7129–7161. DOI: 10.1007/s10668-023-03002-9

Aworh, O.C. (2023). African traditional foods and sustainable food security. *Journal Food Control*, 145.

Awour, P. (2007). *Indigenous knowledge and sustainable crop production: A study of the millennium village in Nyanza province Kenya*. LAP Lambert Academic publishing.

Azubuike, F. C., & Aji, T. C. (2021). Role of information and communication technologies (ICTs) in repackaging indigenous knowledge: A 21st century perspective. *Library Philosophy and Practice*, 1-14.

Badat, S. (2020). Reproduction, transformation and public South African higher education during and beyond Covid-19. *Transformation (Durban), 104*(1), 24–42. DOI: 10.1353/trn.2020.0030

Baggio, J. A. (2021). Knowledge generation via social-knowledge network co-evolution: 30 years (1990–2019) of adaptation, mitigation and transformation related to climate change. *Climatic Change, 167*(1), 13. DOI: 10.1007/s10584-021-03146-5

Bagley, M. A. (2019). The fallacy of defensive protection for traditional knowledge. *Washburn Law Journal, 58*, 323–363. DOI: 10.13140/RG.2.2.15115.34088

Baibout, M., Corcket, E., Kothari, S. L., & Fievet, V. (2022). Ecosystem services provided by Prosopis cineraria (L.) Druce in the drylands of Southern and Western Asia. *Botany Letters, 169*(1), 30–42. DOI: 10.1080/23818107.2021.1981437

Bairwa, M., & Patawari, S. (2021). Festivals for Sustainable Tourism Development: A Case Study of Marwar Region of Rajasthan, India. *Local Food and Community Empowerment through. Tourism (Zagreb)*, •••, 82–96.

Balakrishnan, S. (2020). Afrocentrism revisited: Africa in the Philosophy of Black Nationalism. A Critical Journal of Black Politics. *Cultura e Scuola, 22*(1), 71–88.

Balogun, T. (2023). Digitization of Indigenous knowledge systems in Africa: The case of South Africa's National Recorded System (NRS). *Records Management Journal, 33*(1), 88–102. DOI: 10.1108/RMJ-10-2022-0029

Balogun, T., & Kalusopa, T. (2021). A framework for digital preservation of Indigenous knowledge system (IKS) in repositories in South Africa. *Records Management Journal, 31*(2), 176–196. DOI: 10.1108/RMJ-12-2020-0042

Balogun, T., & Kalusopa, T. (2022). Web archiving of indigenous knowledge systems in South Africa. *Information Development, 38*(4), 658–671. DOI: 10.1177/02666669211005522

Baloyi, J. K. (2010). An analysis of constraints facing smallholder farmers in the Agribusiness value chain: A case study of farmers in the Limpopo Province (Doctoral dissertation, University of Pretoria).

Banda, E. (2022). *Dust to Dust: Ritualization of an Ongoing Relationship with the Land in Shona Christian Burial Practice*. Scholar Commons.

Bangura, A. K. (2005). Ubuntugogy: An African educational paradigm that transcends pedagogy, andragogy, ergonagy and heutagogy. *Journal of Third World Studies, 22*(2), 13–53. https://www.jstor.org/stable/45198556

Bara, M. (2021). *Early literacy development in IsiXhosa: Fostering grade 3 learners' imagination and critical thinking through folktales.* (MA). University of the Western Cape, Baum, F., MacDougall, C., & Smith, D. (2006). Participatory action research. *Journal of Epidemiology and Community Health, 60*(10), 854. 10.1136%2Fjech.2004.028662

Baranov, A., Albitsky, V. & Ivanova, A. (2012). The incidents trends and the state of health of the children population of the Russian Federation. *The Russian paediatric journal*, 6, 4–9.

Barker, D., & Beckford, C. (2006). Plastic yam and plastic yam sticks – perspectives on indigenous technical knowledge among Jamaican farmers. *Tijdschrift voor Economische en Sociale Geografie, 97*(5), 535–546. DOI: 10.1111/j.1467-9663.2006.00362.x

Barman, B., Ghosh, B., Ranjan, A., & Quader, S. W. (2024). The Potential of Indigenous Technological Knowledge for Sustainable and Climate-Resilient Agriculture. *Int. Journal. Environment Climate Change, 14*(8), 490–501. DOI: 10.9734/ijecc/2024/v14i84369

Barnhardt, R., & Oscar Kawagley, A. (2005). Indigenous knowledge systems and Alaska Native ways of knowing. *Anthropology & Education Quarterly, 36*(1), 8–23. DOI: 10.1525/aeq.2005.36.1.008

Barrett, C. B., Christiaensen, L., Sheahan, M., & Shimeles, A. (2017). On the structural transformation of rural Africa. *Journal of African Economies, 26*(suppl_1), i11–i35. DOI: 10.1093/jae/ejx009

Barrett, M., Sahay, S., & Walsham, G. (2001). Information technology and social transformation: GIS for forestry management in India. *The Information Society, 17*(1), 5–20. DOI: 10.1080/019722401750067397

Bartlett, C., Marshall, M., & Marshall, A. (2012). Two-eyed seeing and other lessons learned within a co-learning journey of bringing together indigenous and mainstream knowledges and ways of knowing. *Journal of Environmental Studies and Sciences, 2*(4), 331–340. DOI: 10.1007/s13412-012-0086-8

Basnet, D., Dorji, T., Ali, G., Lepcha, R., Bhuchar, S., & Shakya, B. (2017). Promoting innovative livelihoods (ecotourism and value chains) for sustainable landscape management—Regional training of trainers (0 ed.). International Centre for Integrated Mountain Development (ICIMOD). DOI: 10.53055/ICIMOD.688

Battiste, M. (2018). Reconciling Indigenous knowledge in education: Promises, possibilities, and imperatives. *Dissident knowledge in higher education*, 123-148.

Beach, D. (1994). *The Shona and Zimbabwe: 900-1850*. Heinemann.

Beach, D. N. (1994). *A Zimbabwean Past: Shona Dynastic Histories and Oral Traditions*. Heinemann.

Beevi, C. A., Rani, B. R., Shankar, K. R., Nirmala, G., & Nagasree, K. (Eds.). (2023). Innovative Extension Strategies for Sustainable Natural Resource Management in Rainfed Agriculture [E-book] Hyderabad: National Institute of Agricultural Extension Management (MANAGE) and ICAR-Central Research Institute for Dryland Agriculture.

Belani, A., Oladimeji, K. E., Hyera, F. L., Matubatuba, C., & Juzang, I. J. (2022). Can the use of Impepho leaves extend beyond traditional to modern medicine? An opinion. *Africa Journal of Nursing and Midwifery, 24*(2), 1–4. DOI: 10.25159/2520-5293/11872

Belcher, B., & Claus, R. (2020). Theory of Change. td-net toolbox profiles. Swiss Academies of Arts and Sciences: Td-Net Toolbox for Co-Producing Knowledge. https://doi.org/doi.org/10.5281/zenodo.3717451

Beniwal, S. P. S., Nene, L. Y., & Pandey, S. T. (2020). Relevance of Vrikshayurveda and Traditional Knowledge for Ecofriendly Sustainable Agriculture to Meet SDGs in India. *Asian Agri-History, 24*(1).

Bennett, M., & James, P. (2016). *Environmental Management: Principles and Practice*. Routledge.

Bentley, J. (2021). Protectors of the land and water: Citizenship, territory, and vulnerability among the Lepcha in Sikkim and West Bengal. *Asian Ethnicity, 22*(2), 330–352. DOI: 10.1080/14631369.2020.1793295

Berglund, A. I. (1990). *Zulu thought patterns and symbolism.* David Philiph.

Berkes, F., & Davidson-Hunt, I. (2008). The cultural basis for an ecosystem approach. *The ecosystem approach: Complexity, uncertainty, and managing for sustainability*, 109.

Berkes, F. (2008). *Sacred ecology* (3rd ed.). Routledge. DOI: 10.4324/9780203928950

Berkes, F. (2009). Community conserved areas: Policy issues in historic and contemporary context. *Conservation Letters, 2*(1), 20–25. DOI: 10.1111/j.1755-263X.2008.00040.x

Berkes, F. (2012). *Sacred ecology: Traditional ecological knowledge and resource management* (4th ed.). Routledge. DOI: 10.4324/9780203123843

Bernard, R. H. (2002). *Research methods in anthropology: Qualitative methods and quantitative methods.* Altamira Press.

Bernstein, B. (1990). *The structuring of pedagogic discourse: class, codes and control.* Routledge.

Bernstein, B. (1995). Code theory and its positioning: A case study in misrecognition. *British Journal of Sociology of Education, 16*(1), 3–19. DOI: 10.1080/0142569950160101

Bernstein, B. (1996). *Pedagogy, symbolic control and identity: theory, research, critique.* Taylor & Francis.

Bernstein, B. (2000). *Pedagogy, symbolic control and identity: Theory, research, critique.* Rowman & Littlefield. DOI: 10.5771/9781461636205

Beyers, J. (2022). Christianity and Shona religion and ecology: An ethical-practical perspective. *Hervormde Teologiese Studies, 78*(4), 1–10.

Bhambra, G. K. (2021). Decolonizing critical theory? Epistemological justice, progress, reparations. *Critical Times, 4*(1), 73–89. DOI: 10.1215/26410478-8855227

Bhat, N. A., Jeri, L., Karmakar, D., Mipun, P., Bharali, P., Sheikh, N., Nongkynrih, C. J., & Kumar, Y. (2023). Ethnoveterinary practises of medicinal plants used for the treatment of different cattle diseases: A case study in East Khasi Hill district of Meghalaya, North East India. *Heliyon, 9*(7), e18214. DOI: 10.1016/j.heliyon.2023.e18214 PMID: 37501975

Bhebhe, S. (2019). Understanding the traditional and contemporary purpose of the Njelele Rainmaking Shrine through the oral testimonies of local people in Matobo. *Oral History Journal of South Africa, 7*(2), 45–60. DOI: 10.25159/2663-6670/4015

Bheemappa, A., Shruthi, S., Maheshwari, K., & Biradar, N. (2022). Indigenous technical knowledge-based rainfall prediction: a review. *Agricultural Reviews*, (Of). https://doi.org/DOI: 10.18805/ag.R-2347

Bhuda, M. T. & Marumo, P. (2020). African traditional medicine and healing in South Africa: Challenges and prospects before and during COVID 19. *Ife Centre for Psychological Studies/Services, 18*, 16710-16723.

Bhuda, M. T. (2021). Making a case for indigenous education systems in South Africa. *African Journal of Development Studies (AJDS)*, 67-82, https://doi.org/DOI: 10.31920/2634-3649/2021/SIv1a4

Bhuda, M. T. (2021). Making a Case for Indigenous Education Systems in South Africa. *African Journal of Development Studies,* 67-82.

Bhuda, M. T. (2024). Integrating Ethnomathematical Activities in the Foundation Phase as Means of Decolonisation: Insights from Indigenous Knowledge Systems. *African Journal of Development Studies (formerly AFFRIKA Journal of Politics, Economics and Society), 14*(4), 401-424.

Bhuda, M. T., & Marumo, P. (2021). The Ndebele indigenous games pertinent to primary school mathematics learning: why indigenous games are a vital tool for mathematics teaching and learning. *Proceeding on the conference on the Conference on the Implications of Covid-19*, South Africa, 114-125

Bhuda, M. T., & Pudi, T. I. (2020). Indigenising mathematics in schools: Why ethnomathematics matters. *Developing teaching and learning in Africa: Decolonising perspectives*, 21-42.

Bhuda, M. T. (2024). Decolonising and indigenising African Indigenous research: Why Indigenous ways of knowing, being and living matter in research. In *Higher Education Transformation in Africa* (pp. 50–69). Routledge. DOI: 10.4324/9781003528319-4

Bhuda, M. T. (2024). Integrating ethnomathematical activities in the foundation phase as a means of decolonisation: Insights from indigenous knowledge systems. [AJDS]. *African Journal of Development Studies, 14*(4), 401–424. DOI: 10.31920/2634-3649/2024/v14n4a18

Bhuda, M. T. (2024a). Integrating Ethnomathematical Activities in the Foundation Phase as Means of Decolonisation: Insights from Indigenous Knowledge Systems. *African Journal of Development Studies, 14*(4).

Bhuda, M. T., & Gumbo, M. T. (2024). Developments Towards Integrating African Indigenous Knowledge in South African Universities. *UnisaRxiv*. DOI: 10.25159/UnisaRxiv/000082.v1

Bhuda, M. T., & Kgwadi, B. (2023). The economic measures established to rebuild the South African cultural heritage tourism after the COVID-19 impact. *Gender & Behaviour*, *21*(1), 21398–21412.

Bhuda, M. T., & Koitsiwe, M. (2022). The Covid-19 Pandemic and The Challenges Encountered by The Southern Ndebele Crafts Entrepreneurs Due to Closure of The Tourism Sector. *Gender & Behaviour, 20*(1), 19188–19202.

Bhuda, M. T., & Marumo, P. (2020). African traditional medicine and healing in South Africa: Challenges and prospects before and during COVID 19. Ife Centre for Psychological Studies/Services. PO Box 1548. *Ile-Ife Osun State, Nigeria*, *18*(4), 16710.

Bhuda, M. T., & Marumo, P. (2022). Ubuntu Philosophy and African Indigenous Knowledge Systems: Insights from decolonization and indigenization of research. *Gender & Behaviour*, *20*(1), 19133–19151.

Bhuda, M. T., Marumo, P., & Motswaledi, T. (2022). Behavioral Analysis of The Impact of African Traditional Approaches on Curing and Restoring of Health During Covid-19. *Gender & Behaviour, 20*(1), 19152–19166.

Bhuda, M., & Khazamula, E. (2022). Towards finding a cure for COVID-19: The role of African Traditional Medicine as an indicator of development. *African Journal of Development Studies*, *2022*(2, si1), 137–153. DOI: 10.31920/2634-3649/2022/SIv2a8

Bhuda, M., & Koitsiwe, M. (2022). The Importance of Underpinning Indigenous Research Using African Indigenous Philosophies: Perspectives From Indigenous Scholars. In *Handbook of Research on Protecting and Managing Global Indigenous Knowledge Systems* (pp. 223–248). IGI Global. DOI: 10.4018/978-1-7998-7492-8.ch013

Bhuda, T. M., & Mdhluli, T. D. (2024). *African traditional medicine as a COVID-19 treatment and the South African government response*. Pharos Journal of Theology. DOI: 10.46222/pharosjot.105.42

Bhunda, M. T., & Pudi, T. I. (2020). Indigenising mathematics in schools: Why ethnomathematics matters. In Msila, V. (Ed.), *Developing Teaching and Learning in Africa: Decolonising Perspectives* (pp. 21–42). SUN Press., DOI: 10.18820/9781928480716/02

Bienabe, E., Coronel, C., Le Coq, J., & Liagre, L. (2004). *Linking smallholder farmers to markets: Lessons learned from literature review and analytical review of selected projects. Study Report, Final draft, March 2004. CIRAD & IRAM*. World Bank.

Biermann, S., & Townsend-Cross, M. (2008). Indigenous pedagogy as a force for change. *Australian Journal of Indigenous Education*, *37*(S1), 146–154. DOI: 10.1375/S132601110000048X

Billawer, W., & Nel, V. B. (2024). *The exclusion of indigenous knowledge systems in planning open spaces in Namibia: The case of Havana, Windhoek*. Paper presented at the Urban Forum. DOI: 10.1007/s12132-024-09510-9

Blench, R. M. (2001). *Globalisation and policies towards cultural diversity. Natural Resource Paper, 70*. Overseas Development Institute., http://www.oneworld.org/odi/nrp/70.html

Blose, P., & Gumbo, M. T. (2024). Developing a framework for integrating IKS in technology education with sustainable development principles. *African Journal of Research in Mathematics. Science and Technology Education*, *28*(3), 330–342.

Blyznyuk, T. (2019). Formation of Teachers' Digital Competence: Domestic Challenges and Foreign Experience. *Journal of Vasyl Stefanyk Precarpathian National University*, *5*(1), 40–46. Advance online publication. DOI: 10.15330/jpnu.5.1.40-46

Bohensky, E. L., & Maru, Y. (2011). Indigenous knowledge, science, and resilience: What have we learned from a decade of international literature on "integration"? *Ecology and Society*, *16*(4), art6. DOI: 10.5751/ES-04342-160406

Boluk, K. A., Cavaliere, C. T., & Higgins-Desbiolles, F. (2019). A critical framework for interrogating the United Nations Sustainable Development Goals 2030 Agenda in Tourism. *Journal of Sustainable Tourism*, *27*(7), 847–864. DOI: 10.1080/09669582.2019.1619748

Bonde, A. R., & Pinjari, R. M. (2023). Hoodia Gordonii a herbal plant: A Review. *International Journal of Rural Development Environment and Health Research*, *7*(5), 98–106. DOI: 10.22161/ijreh.7.5.8

Bottiani, J. H., McDaniel, H. L., Henderson, L., Castillo, J. E., & Bradshaw, C. P. (2020). Buffering Effects of Racial Discrimination on School Engagement: The Role of Culturally Responsive Teachers and Caring School Police. *The Journal of School Health, 90*(12), 1019–1029. Advance online publication. DOI: 10.1111/josh.12967 PMID: 33184890

Bourdillon, M. F. C. (1976). *The Shona Peoples: An Ethnography of the Contemporary Shona, with Special Reference to Their Religion*. Mambo Press.

Bourdillon, M. F. C. (1987). *The Shona peoples: An ethnography of the contemporary Shona, with special reference to their religion*. Mambo Press.

Bovill, C. (2021). *Curriculum conceptualisation and the University of Edinburg Institute for Academic Development*, University of Edinburg, docs.hss.ed.ac.uk

Boya, T. B. (2023). The implications of Setswana hare folktales contemporary for South African children. University of Pretoria, Brice-Heath, S. (1983). *Ways with words: Language, life and work in communities and classrooms*. Cambridge: Cambridge University Press.

Brabyn, L., & Mark, D. M. (2011). Classifying landscape character. In Stea, D. (Ed.), *Landscape in Language: Transdisciplinary Perspectives; Mark, D.M., Turk, A.G., Burenhult, N* (pp. 395–409). John Benjamins Publishing Co. DOI: 10.1075/clu.4.21bra

Bradley, D. (2008). Languages of the Greater Himalayan Region. Vol. 5, A Grammar of Lepcha. By Heleen Plaisier. Leiden: Brill, 2007. 272 pp. $139.00 (cloth). *The Journal of Asian Studies, 67*(02). Advance online publication. DOI: 10.1017/S002191180800096X

Branson, J., & Miller, D. (1998). Nationalism and the linguistic rights of Deaf communities: Linguistic imperialism and the recognition and development of sign languages. *Journal of Sociolinguistics, 2*(1), 3–34. DOI: 10.1111/1467-9481.00028

Braun, V., & Clarke, V. (2006). Using thematic analysis in psychology. *Qualitative Research in Psychology, 3*(2), 77–101. DOI: 10.1191/1478088706qp063oa

Braun, V., & Clarke, V. (2014). What can "thematic analysis" offer health and wellbeing researchers? *International Journal of Qualitative Studies on Health and Well-being, 9*(1), 26152. DOI: 10.3402/qhw.v9.26152 PMID: 25326092

Braun, V., & Clarke, V. (2019). Reflecting on reflexive thematic analysis. *Qualitative Research in Psychology, 16*(2). Advance online publication. DOI: 10.1080/14780887.2019.1628806

Brayboy, B. M. J., & Castagno, A. E. (2008). Indigenous knowledges and native science as partners: A rejoinder. *Cultural Studies of Science Education, 3*(3), 787–791. DOI: 10.1007/s11422-008-9142-9

Briggs, J., & Sharp, J. (2004). Indigenous Knowledges and Development: A Postcolonial Caution. *Third World Quarterly, 25*(4), 661–676. https://www.jstor.org/stable/3993739. DOI: 10.1080/01436590410001678915

Broderick, A. (1998). Managing Diversity. *Women in Business, 44*(3), 13–14.

Browne, D. (2011), Is the poor economy to blame for the rise in diaper rashes? Retrieved from http://www.cnn.com/2011/09/12/health/pooreconomy-diaper-rash-

Bruce, D. M., Jackson, M. J., & Cantallops, A. S. (2001). PREPARe: A model to aid the development of policies for less unsustainable tourism in historic towns. *Tourism and Hospitality Research*, *3*(1), 21–36. DOI: 10.1177/146735840100300103

Bruchac, M. M. (2014). Indigenous Knowledge and Traditional Knowledge. In Smith, C. (Ed.), *Encyclopedia of Global Archaeology* (pp. 3814–3824). Springer. DOI: 10.1007/978-1-4419-0465-2_10

Bruner, J. (1991). The narrative construction of reality. Critical Inquiry. *Critical Inquiry*, *18*(1), 1–21. DOI: 10.1086/448619

Bryant, A. T. (1966). *The Zulu Medicine and Medicine Men*. Centaur Publishers.

Bryant, A. T. (1970). *The Zulu Medicine and Medicine-Men, C*. Struik.

Bryman, A. (2016). *Social Research Methods* (5th ed.). Oxford University Press.

Burayu, E. T., & Degefa, B. D. (2024). Exploring harmful traditional practices and its associated factors among puerperal mothers who delivered in public health facilities in southwestern Ethiopia. *Scientific Reports*, *14*(1), 27065. https://pmc.ncbi.nlm.nih.gov/articles/PMC11544142/. DOI: 10.1038/s41598-024-75344-x PMID: 39511244

Burgess, C., Bishop, M., & Lowe, K. (2022). Decolonising Indigenous education: The case for cultural mentoring in supporting Indigenous knowledge reproduction. *Discourse (Abingdon)*, *43*(1), 1–14. DOI: 10.1080/01596306.2020.1774513

Bürgi, M., Hersperger, A. M., & Schneeberger, N. (2005). Driving forces of landscape change-current and new directions. *Landscape Ecology*, *19*, 857–868. DOI: 10.1007/s10980-005-0245-3

Burton, D., & Bartlett, S. (2009). *Key issues for Education Researchers*. Sage. DOI: 10.4135/9781446269480

Buthelezi, N. N., & Hughes, J. C. (2014). Indigenous knowledge systems and agricultural rural development in South Africa: Past and present perspectives. *Indilinga*, *13*(2), 231–250.

Buthelezi, S. C., Ocholla, D., & Dlamini, P. (2024). Strategies for documenting and disseminating indigenous knowledge at a South African university. *South African Journal of Information Management*, *26*(1), 1648. DOI: 10.4102/sajim.v26i1.1648

Buxner, S., Davis, L. F., Anides Morales, A., Sandhaus, S. A., Abrell, L., Kilungo, A., . . . Morales, F. (2018, December). Project Harvest: Engaging Diverse Citizen Scientists in Rainwater Quality Monitoring. In *AGU Fall Meeting Abstracts* (Vol. 2018, pp. NH43B-1036).

Bwalya, F. (2022). The Role of Aunts in Family Dynamics: Bridging Generational Gaps in African Contexts. *Journal of Intergenerational Relationships*, *20*(3), 245–260.

Byers, B. A., Cunliffe, R. N., & Hudak, A. T. (2001). Linking the conservation of culture and nature: A case study of sacred forests in Zimbabwe. *Human Ecology: an Interdisciplinary Journal*, *29*(2), 187–218. DOI: 10.1023/A:1011012014240

Cajete, G. A. (2020). Indigenous science, climate change, and indigenous community building: A framework of foundational perspectives for indigenous community resilience and revitalization. *Sustainability (Basel), 12*(22), 9569. DOI: 10.3390/su12229569

Cameron, J. (2015). Enterprise innovation and economic diversity in community supported agriculture: Sustaining the agricultural commons. *Making other worlds possible: Performing diverse economies*, 53-71.

Campbell, H. (2020). *Farming inside invisible worlds: Modernist agriculture and its consequences.* Bloomsbury Academic. DOI: 10.5040/9781350120570

Campbell, L. M., Boucquey, N., Stoll, J., Coppola, H., & Smith, M. D. (2014). From vegetable box to seafood cooler: Applying the community-supported agriculture model to fisheries. *Society & Natural Resources, 27*(1), 88–106. DOI: 10.1080/08941920.2013.842276

Campbell, M., & Campbell, T. (1996). Behind the name. https://www.behindthename.com/name

Canonici, N. (1994). The trickster in Zulu folktales. *Alternation (Durban), 1*(1), 43–56.

Carlile, R., Kessler, M., & Garnett, T. (2021). *What is food sovereignty?* TABLE Explainer Ser. DOI: 10.56661/f07b52cc

Castleberry, A., & Nolen, A. (2018). Thematic analysis of qualitative research data: Is it as easy as it sounds? *Currents in Pharmacy Teaching & Learning, 10*(6), 807–815. DOI: 10.1016/j.cptl.2018.03.019 PMID: 30025784

Cavendish, W. (2000). Empirical regularities in the poverty-environment relationship of rural households: Evidence from Zimbabwe. *World Development, 28*(11), 1979–2003. DOI: 10.1016/S0305-750X(00)00066-8

Cerulo, K. A., Leschziner, V., & Shepherd, H. (2021). Rethinking culture and cognition. *Annual Review of Sociology, 47*(1), 63–85. DOI: 10.1146/annurev-soc-072320-095202

Chabaya, O., & Chabaya, R. A. (2023). Integration of African Indigenous Knowledge systems into higher education curriculum in Zimbabwe universities: Exploring experiences of lecturers. *African Perspectives of Research in Teaching and Learning, 7*(3), 625–638.

Chabaya, O., & Chabaya, R. A. (2023). Integration of African Indigenous Knowledge Systems into Higher Education Curriculum in Zimbabwe Universities: Exploring Experiences of Lecturers. *African Perspectives of Research in Teaching and Learning, 7*(3), 625–638.

Chakrabortty, R., Pal, S. C., Sahana, M., Mondal, A., Dou, J., Pham, B. T., & Yunus, A. P. (2020). Soil erosion potential hotspot zone identification using machine learning and statistical approaches in eastern India. *Natural Hazards, 104*(2), 1259–1294. DOI: 10.1007/s11069-020-04213-3

Chakraborty, S. (2016). The nature of Lepcha traditional festival: An overview. *The Journal of the Anthropological Survey of India, 65*(1), 39–54. DOI: 10.1177/2277436X20160104

Chang, H. (2016). *Autoethnography as method.* Letf Coast Press Inc. DOI: 10.4324/9781315433370

Chanza, C., & de Wit, A. (2016). Enhancing climate governance through indigenous knowledge: Case in sustainability science. *South African Journal of Science, 112*(3-4), 1–7. DOI: 10.17159/sajs.2016/20140286

Chanza, N., Musakwa, W., & Kelso, C. (2024). Overlaps of indigenous knowledge and climate change mitigation: Evidence from a systematic review. *Frontiers in Climate*, *6*, 1344931. DOI: 10.3389/fclim.2024.1344931

Chapman, J. M., & Schott, S. (2020). Knowledge coevolution: Generating new understanding through bridging and strengthening distinct knowledge systems and empowering local knowledge holders. *Sustainability Science*, *15*(3), 931–943. DOI: 10.1007/s11625-020-00781-2

Chatterjee, S. (2017). Indigenous knowledge and cultural tourism in India: Challenges and prospects. *Journal of Heritage Management*, *2*(1), 45–58. DOI: 10.1177/2455929617706439

Chaudhary, A. K., Ahmad, S., & Mazumder, A. (2011). Cedrus deodara (Roxb.) Loud.: A review on its ethnobotany, phytochemical and pharmacological profile. *Pharmacognosy Journal*, *3*(23), 12–17. DOI: 10.5530/pj.2011.23.2

Chaudhary, B., Erskine, W., & Acciaioli, G. (2022). Hybrid knowledge and climate-resilient agriculture practices of the Tharu in the western Tarai, Nepal. *Frontiers in Political Science*, *4*, 969835. Advance online publication. DOI: 10.3389/fpos.2022.969835

Chaudhuri, S. K., Maiti, S., & Lepcha, C. K. (Eds.). (2020). The cultural heritage of Sikkim. Routledge, Taylor & Francis Group. DOI: 10.4324/9781003052487

Chavunduka, G. L. (1977). *Traditional healers and the Shona patient*. Mambo Press.

Chawane, M. (2016). The development of Afrocentricity: A historical survey. *Yesterday and Today*, (16), 78–99. DOI: 10.17159/2223-0386/2016/n16a5

Chemhuru, M., & Masaka, D. (2010). Taboos as sources of Shona people's environmental ethics. *Journal of Sustainable Development in Africa*, *12*(7), 121–133.

Chen, J., Gao, Z., Chen, X., & Zhang, L. (2019). Factors affecting the dynamics of community supported agriculture (CSA) membership. *Sustainability (Basel)*, *11*(15), 4170. DOI: 10.3390/su11154170

Cheteni, P., & Umejesi, I. (2023). Application of indigenous knowledge and sustainable agriculture by small-scale farmer households and entrepreneurs in South Africa. *International Journal of Development and Sustainability*, *12*(1), 1–18.

Chettri, L., & Sharma, B. K. (2021). Some traditional folk songs and dances of Sikkim Himalayas. *International Research Journal*, *8*(2), 30–44.

Chhabra, D. (2010). Authenticity and tourism: Productive debates, constructive dialogues and a way forward. *Tourism Recreation Research*, *35*(3), 255–258. DOI: 10.1080/02508281.2010.11081639

Chigumira, E. (2003). Masvingo Province. *The Geographical Journal*, *169*(2), 155–162.

Chigwada, J., & Ngulube, P. (2024). Librarians' role in the preservation and dissemination of indigenous knowledge. *IFLA Journal*, *50*(2), 242–256. DOI: 10.1177/03400352231217270

Chigwada, J., & Ngulube, P. (2025). Integrating CARE and FAIR Principles in the Digital Preservation of Indigenous Knowledge to Achieve the Sustainable Development Goals. In *Harnessing Indigenous Epistemologies for Sustainable Progress* (pp. 197–230). IGI Global Scientific Publishing. DOI: 10.4018/979-8-3693-8447-3.ch010

Chigwada, J., Ngulube, P., & Dewah, P. (2024). Preserving Indigenous Knowledge in Culture Houses in the Digital Era in Masvingo Province, Zimbabwe. *African Journal of Library Archives and Information Science*, *34*(2). Advance online publication. DOI: 10.4314/ajlais.v34i2.4

Chikaire, J., Osuagwu, C. O., Ihenacho, R. A., Oguegbuchulam, M. N., Ejiogu-Okereke, N., & Obi, K. U. (2012). Indigenous knowledge system: The need for reform and the way forward. *Global Advanced Research Journal of Agricultural Science*, *1*(8), 201–209.

Chikanda, A. (2017). The Role of Aunts in Zimbabwean Families: A Cultural Perspective. *Journal of Southern African Studies*, *43*(2), 345–360.

Chikazunga, D., & Paradza, G. (2012). Can smallholder farmers find a home in South Africa's food-system? Lessons from Limpopo Province. The Institute for Poverty, Land and Agriculture Studies, (PLAAS) Blog. https://www.plaas.org.za/blog/can-smallholderfarmers-find-home-south-africa%E2%80%99s-food-system-lessons-limpopo-province. Accessed: 25 March 2023

Chikazunga, D., & Paradza, G. (2012). Smallholder farming: A panacea for employment creation and enterprise development in South Africa? Lessons from the Pro-Poor Value Chain Governance Project in Limpopo Province. *Strategies to overcome poverty and inequality: Towards Carnegie III.*

Chikozho, C., Managa, R., & Dabata, T. (2020). Ensuring access to water for food production by emerging farmers in South Africa: What are the missing ingredients? *Water S.A.*, *46*(2), 225–233.

Childline. (2025). *Information for D/deaf children.* Retrieved from Childline: https://www.childlinesa.org.za/about/how-we-help/help-for-the-deaf/

Chilisa, B. (2012). Postcolonial indigenous research paradigms. *Indigenous research methodologies*, 98-127.

Chilisa, B. (2012). *Indigenous Research Methodologies*. Sage Publications.

Chilisa, B., Major, T., & Khudu-Petersen, K. (2017). Community engagement with a postcolonial, African-based relational paradigm. *Qualitative Research*, *17*(3), 326–339. DOI: 10.1177/1468794117696176

Chimhundu, H. (1992). *The Vernacularization of African Languages after Independence. The Center for Advanced Studies of African Society*. CASAS.

Chimuka, T. A. (2001). *Traditional Environmental Knowledge and African Cultural Values: A Case Study of the Shona in Zimbabwe*. University of Zimbabwe.

Chinaka, M. (2016). *Blockchain technology--applications in improving financial inclusion in developing economies: case study for small scale agriculture in Africa* (Doctoral dissertation, Massachusetts Institute of Technology).

Chinenye Nwokocha, G., & Legg-Jack, D.Chinenye -Nwokocha. (2024). Reimagining STEM education in south Africa: Leveraging Indigenous knowledge systems through the m-know model for curriculum enhancement. *International Journal of Social Science Research and Review*, *7*(2), 173–189. DOI: 10.47814/ijssrr.v7i2.1951

Chipfupa, U., & Tagwi, A. (2021). Youth's participation in agriculture: A fallacy or achievable possibility? Evidence from rural South Africa. *Suid-Afrikaanse Tydskrif vir Ekonomiese en Bestuurswetenskappe*, *24*(1), 1–12. DOI: 10.4102/sajems.v24i1.4004

Chirimuuta, C. (2021). Totemic beliefs and conservation among the Shona people of Zimbabwe: An indigenous knowledge perspective. *International Journal of African Renaissance Studies*, *16*(1), 74–89. https://doi.org/xxxx

Chitando, E., & Togarasei, L. (2010). Ethical Implications of Climate Change for Intra- and Intergenerational Justice: The Case of Zimbabwe. *Journal of Theology for Southern Africa*, *136*, 6–18.

Chivasa, S. (2022). Aunts as Cultural Custodians: The Role of Tete in Shona Society. *Journal of African Cultural Studies*, *34*(1), 56–70.

Chiwara, M. (2015). Totemic linkages and biodiversity conservation among the Shona people. *Journal of Environmental Management*, *16*(3), 123–130.

Choongwa, G. H. (2018). *Fundamentals of Applied Research Methodology*. Webster Print.

Choudhury, G., & Kalita, D. (2025). Collaborative stewardship: Sustainable approaches to indigenous knowledge in academic libraries. *Journal of Academic Librarianship*, *51*(2), 103007. DOI: 10.1016/j.acalib.2025.103007

Chowdhary, N., & Mishra, S. (2020). Tourism development and environmental sustainability in Pushkar, Rajasthan. *Indian Journal of Tourism and Hospitality Management*, *11*(2), 89–102.

Christie, F. (2005). The pedagogic device and the teaching of English. In Christie, F. (Ed.), *Pedagogy and the Shaping of Consciousness* (pp. 156–184). Continnuum.

Chuvunduku, G. (2001). The African Religion in Zimbabwe Today, Occasional Paper no.1, Crossover Communication dr.merten@web.de Harare/Munster.

Cleall, E. (2022). Decolonising deaf history: Harlan Lane, Postcolonialism, and Critical Colonial History. . *Zeithistorische Forschungen/Studies in Contemporary History, 19(2)*, 380-387.

Clyne, J., Mininni, P., Norton, A., & Rast, M. (2007). Interactive desktop analysis of high resolution simulations: Application to turbulent plume dynamics and current sheet formation. *New Journal of Physics*, *9*(8), 301. DOI: 10.1088/1367-2630/9/8/301

Cocks, M. L., Alexander, J., & Dold, T. (2012). Inkcubeko Nendalo: A bio-cultural diversity schools education project in South Africa and its implications for inclusive indigenous knowledge systems (IKS) sustainability. *Journal of Education for Sustainable Development*, *6*(2), 241–252. DOI: 10.1177/0973408212475232

Cohen, E. (1988). Authenticity and commoditization in tourism. *Annals of Tourism Research, 15*(3), 371–386. DOI: 10.1016/0160-7383(88)90028-X

Cohen, L., Manion, L., & Morrison, K. (2007). *Research methods in education* (6th ed.). Routledge. DOI: 10.4324/9780203029053

Collier, P., & Dercon, S. (2014). African agriculture in 50 years: Smallholders in a rapidly changing world? *World Development, 63*, 92–101. DOI: 10.1016/j.worlddev.2013.10.001

Committee, J. C. (2007). *Recognition of South African Sign Language as Official Language: briefing by Deaf Federation of South Africa*. Parliamentary Monitoring Group.

Constitution of Zimbabwe Amendment (No 20) Act 2013. (2013).

Cooper, C., Booth, A., Varley-Campbell, J., Britten, N., & Garside, R. (2018). Defining the process to literature searching in systematic reviews: A literature review of guidance and supporting studies. *BMC Medical Research Methodology, 18*(1), 1–14. DOI: 10.1186/s12874-018-0545-3 PMID: 30107788

Cordeiro-Rodrigues, L., & Metz, T. (2021). Afro-communitarianism and the role of traditional African healers in the COVID-19 pandemic. *Public Health Ethics, 14*(1), 59–71.

Cousins, B. (2015). Through a glass darkly': Towards agrarian reform in South Africa. *Land divided, land restored. Land reform in South Africa for the 21st century*, 250-269.

Cousins, B. (2013). Land redistribution, populism, and elite capture: New land reform policy proposals under the microscope. *The Journal of the Helen Suzman Foundation, 70*, 11–19.

Creswell, J. W. (1998). *Qualitative Inquiry and Research Design: Choosing Among Five Traditions* (2nd ed.). Sage Publications Inc.

Creswell, J. W. (2018). *Research design: Qualitative, quantitative, and mixed methods approaches* (5th ed.). SAGE Publications.

Creswell, J. W., & Creswell, J. D. (2017). *Research Design: Qualitative, Quantitative,and Mixed Methods Approaches* (5th ed.). SAGE Publications.

Creswell, J. W., & Poth, C. N. (2018). *Qualitative Inquiry and Research Design: Choosing Among Five Approaches* (4th ed.). SAGE Publications.

Crider, J. S. (2021). Teachers Experiences with Positive Emotion, Engagement, Relationships, Meaning, and Achievement (PERMA): A Qualitative Interview Study. In *Dissertation*.

Crumley, C. L. (2017). Historical ecology and the study of landscape. *Landscape Research, 42*(sup1), S65-S73.

CSVR. (2016). *Mapping local gender-based violence prevention and response strategies in South Africa*. Retrieved May 10, 2023, from http://www.csvr.org.za

Cultures, P. (n.d.). Naming children: traditions in 13 different countries. https://pocketcultures.com/2011/04/13/children-naming-traditions/

DA KwaZulu-Natal. (2024). *Debate on the KZN Premier's Report: To combat GBV, we need men to be part of the solution – not the problem*. Retrieved June 20, 2025, from https://kzn.da.org.za/2024/11/debate-on-the-kzn-premiers-report-to-combat-gbv-we-need-men-to-be-part-of-the-solution-not-the-problem

Da Silva, C., Furtado, K., Soares, S., & De Almeida, J. (2024). Dialogue of knowledges: The importance of the partnership between scientists and indigenous peoples for the preservation of traditional knowledge. *DELOS: Desarrollo Local Sostenible*, *17*(53), e1368–e1368. DOI: 10.55905/rdelosv17.n53-030

da Silva, C., Pereira, F., & Amorim, J. P. (2024). The integration of indigenous knowledge in school: A systematic review. *Compare: A Journal of Comparative Education*, *54*(7), 1210–1228. DOI: 10.1080/03057925.2023.2184200

Daley, P. O., & Murrey, A. (2022). Defiant scholarship: Dismantling coloniality in contemporary African geographies. *Singapore Journal of Tropical Geography*, *43*(2), 159–176.

Damayanti, I. (2017). From storytelling to story writing: The implementation of reading to learn (R2L) pedagogy to teach English as a foreign language in Indonesia. *Indonesian Journal of Applied Linguistics*, *6*(2), 232–245. DOI: 10.17509/ijal.v6i2.4870

Daneel, M. L. (1970). *The god of the Matopo hills: An essay on the Mwari cult in Rhodesia*. Mouton.

Daneel, M. L. (2001). *African Earthkeepers: Environmental Missions in Africa*. Orbis Books.

Das, B. (2024). Community-based tourism approach for sustainable development: A study of the Lepcha community, Kalimpong District, India. *Journal of Architectural/Planning Research and Studies (JARS)*, 21(2), 405–420. https://doi.org/DOI: 10.56261/jars.v21.264845

Das, S. (2024). Water management: Community participation. *Journal of the Geological Society of India*, *100*(4), 463–466. DOI: 10.17491/jgsi/2024/173862

Datta, R. (2018). Decolonizing both researcher and research and its effectiveness in Indigenous research. *Research Ethics*, *14*(2), 1–24. DOI: 10.1177/1747016117733296

Davis, M. (2019). *Indigenous Knowledge and Biodiversity*. Cultural Ecology Journal.

Dawadi, S. (2020). Thematic Analysis Aproach: A Syep by Step Guide for ELT Research Practitioners. *Journal of NELTA*, *25*(1-2), 62–71. DOI: 10.3126/nelta.v25i1-2.49731

De Beer, J. J., & Van Wyk, B. E. (2011). An ethnobotanical survey of the Agter–Hantam, Northern Cape Province, South Africa. *South African Journal of Botany*, *77*(3), 741–754. DOI: 10.1016/j.sajb.2011.03.013

De Cock, N., D'Haese, M., Vink, N., Van Rooyen, C. J., Staelens, L., Schönfeldt, H. C., & D'Haese, L. (2013). Food security in rural areas of Limpopo province, South Africa. *Food Security*, *5*(2), 269–282. DOI: 10.1007/s12571-013-0247-y

De Klerk, V. (1999). Beauty or Buhle? On changing one's name. *Nomina Africana*, *13*(1&2), 41–44.

DEA. (2008). National Environmental Management: Waste Act, 2008 (Act No. 59 Of 2008), Draft Health Care Risk Waste Management Regulations. [Online]. Available:https://www.gov.za/sites/default/files/gcisdocument/201409/35405gen452.pdf Accessed 10/02 2022.

Deaf, S. A. (2007, February 1). *Memorandum towards recognition of SASL as the 12th official language.* Retrieved from Static PMG: https://static.pmg.org.za/docs/2007/070216memorandum.htm?

Deaf, S. A. (2018). *Deaf Education.* DeafSA.

DEAT. (2007), Assessment of the Status of waste service delivery and capacity at local government level, Pretoria, Department of Environmental Affairs and Tourism. Accessed at https://www.deat.gov.za

Dei, G. (2014). Global education from an 'indigenist' anti-colonial perspective. *Journal of Contemporary Issues in Education, 9*(2). Advance online publication. DOI: 10.20355/C53G6B

Dei, G. J. S. (2000). Rethinking the role of indigenous knowledges in the academy. *International Journal of Inclusive Education, 4*(2), 111–132. DOI: 10.1080/136031100284849

Dei, G. J. S. (2008). Indigenous knowledge studies and the next generation: Pedagogical possibilites for anti-colonial education. *Australian Journal of Indigenous Education, 37*(S1), 5–13. DOI: 10.1375/S1326011100000326

DeMers, M. (2009). *Fundamentals of Geographic Information Systems* (4th ed.). John Wiley & Sons, Inc.

Deogaonkar, S. G. (2008). Traditional handicrafts of the Gond tribes of Vidarbha, Maharashtra. *Indian Journal of Traditional Knowledge, 7*(1).

Department of agriculture forestry and fisheries. (2010). The Strategic Plan for South African Agriculture. Department of Agriculture, Directorate of Agricultural Information Services, Pretoria

Department Of Agriculture, Forestry and Fisheries (2014). The national policy on food and nutrition security for the republic of South Africa. Government gazette: No. 37915.

Department of Basic Education. South Africa. (2011). *Curriculum and assessment policy statement Grades R-3: English home language.* Pretoria: Government Printing Works

Department of Basic Education. South Africa. (2014). *Curriculum Assessment Policy Statement.* Pretoria: Department of Basic Education.

Department of Education (2002). *Revised national curriculum statement for Grades R–9 (schools) natural sciences.* Pretoria: DoE.

Department of Education (2003). *National curriculum statement: Grades 10 –12 (general) physical sciences.* Pretoria: DoE.

Der Wagen, L. (2001) Event Management: For tourism, cultural, business and sporting events. Sydeny: Hospitality Press Melbourne

Desmarchelier, R. (2016). *Whose knowledge?: Science education, Indigenous knowledges and teacher praxis* (Doctoral dissertation, University of Southern Queensland).

Desmarchelier, R. (2020). *Indigenous Knowledges and Science Education: Complexities.* Considerations and Praxis.

Dewi, I. G. A. A. O. (2021). Understanding data collection methods in qualitative research: The perspective of interpretive accounting research. *Journal of Tourism Economics and Policy*, *1*(1), 23–34. DOI: 10.38142/jtep.v1i1.105

Dinesh, D., Hegger, D. L. T., Klerkx, L., Vervoort, J., Campbell, B. M., & Driessen, P. P. J. (2021). Enacting Theories of Change for Food Systems Transformation under Climate Change. *Global Food Security*, *32*, 100583. DOI: 10.1016/j.gfs.2021.100583

Diver, S., Vaughan, M., Baker-Médard, M., & Lukacs, H. (2019). Recognizing "reciprocal relations" to restore community access to land and water. *The International Journal of the Commons*, *13*(1), 400. DOI: 10.18352/ijc.881

Diyal, M. (2021). Rethinking African Indigenous Knowledge and Practices for Environmental Sustainability in Africa. In V. Gumede., M. Muchie & A. Shafi (Eds.), *Indigenous Systems and Africa's Development* (pp. 48-57). Capital Press.

Dlamini, G. M., & Lupupa, T. (1995). Swaziland: Country report to the FAO international technical conference on plant genetic resources. Food and Agricultural Organization.

Dlamini, M., & Masuku, N. (2013). Emagugu EmaSwati. Manzini: Imisebe Publishers.

Dlamini, N. Z. (2022). Unmasking Christian women survivor voices against gender-based violence: a pursuit for a feminist liberative pastoral care praxis for married women in the Anglican Church of Southern Africa. University of KwaZulu-Natal (Doctoral Dissertation).

Dlamini, P. A. (2016). NgiliSwati NgesiSwati (I am a Swazi because of siSwati: Preserving/safeguarding the collective identity/intangible cultural heritage of the Swazi community through our mother tongue. *Department of English and Literature*, 1-4.

Dlamini, P. M. (2001). Gender, biodiversity and local knowledge systems: Highlights of the seminar discussions. In: Dlamini, P.M. & Vilane, B.R.T. (Eds), *Proceedings of the National Seminar on Institutions Working on Gender, Biodiversity and Local Knowledge Systems in Swaziland*. April, pp. 10-20.

Dlamini, R. M. (2007). *Investigation of sustainable indigenous agricultural practices: A systems approach* [Unpublished master's dissertation]. Stellenbosch University, Stellenbosch, South Africa.

Dlamini, B. S. (2015). *UmSamaliya Lolungile. Manzini*. Macmillan.

Dlamini, D. D., & Masuku, M. B. (2011). Land tenure and land productivity: A case of maize production in Swaziland. *Asian Journal of Agricultural Sciences*, *3*(4), 301–307.

Dlamini, M. (2016). *Buciko Bemlomo. Manzini*. Imisebe Publishers.

Dlamini, M., Masuku, M., & Raufu, M. (2015). Determinants of foreign direct investment inflows in Swaziland's agricultural sector. *Journal of Economics and Sustainable Development*, *6*(15), 150–158.

Dlamini, M., & Masuku, N. (2013). *Emagugu EmaSwati. Manzini*. Imisebe Publishers.

Dlamini, O. S. (1988). *Sinkayinkayi. Manzini*. Longman.

Dlamini, P. N. (2024). Acquiring and transferring of indigenous knowledge among its owners: A literature review. In Ocholla, D., Onyancha, O. B., & Adesina, A. O. (Eds.), *Information, Knowledge, and Technology for Teaching and Research in Africa. Synthesis Lectures on Information Concepts, Retrieval, and Services*. Springer., DOI: 10.1007/978-3-031-60267-2_8

Dlamini, P., & Ocholla, D. N. (2018). Information and communication technology tools for managing indigenous knowledge in KwaZulu-Natal Province, South Africa. *African Journal of Library Archives and Information Science*, *28*(2), 137–153.

Dlamini, T. (2018). *Using Afrocentrism to explore African Values in selected short stories prescribed for Grade 10 English First Additional Language Learners in South Africa*. University of Kwazulu-Natal.

Doke, C. M. (1931). Report on the Unification of the Shona Dialects. Government of Southern Rhodesia. *Educational Research Journal*, *12*(2), 144–159.

Dold, A. P., & Cocks, M. L. (2002). The trade in medicinal plants in the Eastern Cape Province, South Africa. *South African Journal of Science*, *98*(11), 589–597.

Donald, D. R., Lazarus, S., & Lolwana, P. (1997). *Educational Psychology in Social Context: Challenges of Development, Social Issues & Special Need in Southern Africa: a Teacher's Resource*. Oxford University Press.

Donda, G. N. (1997). *A Study of Zulu Concepts, Terms and expression Associated with Umuthi, Master of Arts*. University of Vista.

Donnelly, M., & Abbas, A. (2018). Using the sociology of Basil Bernstein in higher education research. In *Theory and Method in Higher Education Research* (Vol. 4, pp. 1–17). Emerald Publishing Limited. DOI: 10.1108/S2056-375220180000004002

Dube, E. (2021, July 23). King ask Sobhuza 11 to save him from Swazis, present over 15 cows at his cave. *Swati Newsweek*. https://sdnewsweek.wordpress.com/2021/07/23/king-ask-sobhuza-ii-to-save-him-from-swazis-present-over-15-cows-at-his-cave/

Dube, L. (2011). *AFRICAN INITIATIVES in healing ministry*. Unisa Press.

Dube, L., & Guveya, E. (2014). Technical Efficiency of Smallholder Out-grower Tea. (Camellia Sinensis) Farming in Chipinge District of Zimbabwe. *Greener Journal of Agricultural Sciences*, *4*(8), 368–377. DOI: 10.15580/GJAS.2014.8.091114354

Duncan, J., Carolan, M., & Wiskerke, J. S. C. (2021). *Routledge Handbook of Sustainable and regenerative food systems*. Routledge.

Dziva, D., Mpofu, V., & Kusure, L. (2011). Teachers' conception of indigenous knowledge in science curriculum in the context of Mberengwa District, Zimbabwe.

Ebewo, P. (2011). Swazi Incwala: The performative and radical poetics in a ritual practice. *South African Theatre Journal*, *25*(2), 89–100. DOI: 10.1080/10137548.2011.636976

Eckardt, N. A., Ainsworth, E. A., Bahuguna, R. N., Broadley, M. R., Busch, W., Carpita, N. C., Castrillo, G., Chory, J., DeHaan, L. R., Duarte, C. M., Henry, A., Jagadish, S. V. K., Langdale, J. A., Leakey, A. D. B., Liao, J. C., Lu, K.-J., McCann, M. C., McKay, J. K., Odeny, D. A., & Zhang, X. (2023). Climate change challenges, plant science solutions. *The Plant Cell*, *35*(1), 24–66. DOI: 10.1093/plcell/koac303 PMID: 36222573

Edana (2007). Disposable Diapers: A Technical Review. Edana, Brussels, Belgium

Edgar, B. A., Matshidze, P. E., & Lee, K. S. (2022). Exploring Vhavenda indigenous methods of water conservation and management in Tshidzivhe village, Limpopo province, South Africa. *South African Journal of Cultural History*, *36*(2), 1–22. DOI: 10.54272/sach.2022.v36n2a1

Edwards, F., Sonnino, R., & Cifuentes, M. L. (2024). Connecting the dots: Integrating Food Policies Towards Food Systems Transformation. *Environmental Science & Policy*, *158*, 103810. DOI: 10.1016/j.envsci.2024.103810

Ekobi, G. A., Mboh, L., & Tanga, P. T. (2023). *Contribution of Indigenous Knowledge to Agricultural Growth in South Africa: A Case of Disaneng Community in the Ratlou Local Municipality*. African Journal of Inter/Multidisciplinary Studies.

Ekwamu, A. (2021). Building Inclusive National Food Systems: Considerations for the United Nations Food Systems Summit. https://www.ruforum.org. Accessed: 15 Feb 2025

El Yazidi, R., & Rijal, K. (2024). Science learning in the context of'indigenous knowledge'for sustainable development. *International Journal of Ethnoscience and Technology in Education*, *1*(1), 28–41. DOI: 10.33394/ijete.v1i1.10880

Elands, B. H., Vierikko, K., Andersson, E., Fischer, L. K., Goncalves, P., Haase, D., Kowarik, I., Luz, A. C., Niemelä, J., Santos-Reis, M., & Wiersum, K. F. (2019). Biocultural diversity: A novel concept to assess human-nature interrelations, nature conservation and stewardship in cities. *Urban Forestry & Urban Greening*, *40*, 29–34. DOI: 10.1016/j.ufug.2018.04.006

Elhami, A., & Khoshnevisan, B. (2022). Conducting an Interview in Qualitative Research: The Modus Operandi. *Mextesol Journal*, *46*(1), n1. DOI: 10.61871/mj.v46n1-3

Ellena, R., & Nongkynrih, K. (2017). Changing gender roles and relations in food provisioning among matrilineal Khasi and patrilineal chakhesang indigenous rural people of north-east India. *Maternal and Child Nutrition*, *13*(S3), e12560. Advance online publication. DOI: 10.1111/mcn.12560 PMID: 29359434

Elliott, B., Jayatilaka, D., Brown, C., Varley, L., & Corbett, K. K. (2012). "Are We not Being Heard": Aboriginal Perspectives on Traditional Foods Access and Security. *Journal of Environmental and Public Health*, *6*, 130945. https://onlinelibrary.wiley.com/doi.10.1155/2012130945. DOI: 10.1155/2012/130945 PMID: 23346118

Emeagwali, G. (2014). Intersections between Africa's indigenous knowledge systems and history. In *African indigenous knowledge and the disciplines* (pp. 1–17). Brill. DOI: 10.1007/978-94-6209-770-4_1

Emeagwali, G. (2020). African Indigenous Knowledges and the Decolonization of Education in Africa. In *The Palgrave Handbook of African Education and Indigenous Knowledge* (pp. 335–349). Palgrave Macmillan. DOI: 10.1007/978-3-030-38277-3_17

Eneji, C. V. O., Ntamu, G. U., Unwanade, C. C., Godwin, A. B., Bassey, J. E., & Willaims, J. J. (2016). *Joseph Ignatius*. Traditional African Religion in Natural Resources Conservation and Management in Cross River State.

Eneji, C. V. O., Ntamu, G. U., Unwanade, C. C., Godwin, A. B., Bassey, J. E., Willaims, J. J., & Ignatius, J. (2012). Traditional African religion in natural resources conservation and management in Cross River State, Nigeria. *Environment and Natural Resources Research*, *2*(4), 45–53.

Eswatini National Trust Commission. (2024). *Umhlanga (The Reed dance).*https://entc.org.sz/umhlanga-the-reed-dance/

Etikan, I., Musa, S. A., & Alkassim, R. S. (2016). Comparison of convenience sampling and purposive sampling. *American Journal of Theoretical and Applied Statistics*, *5*(1), 1–4. DOI: 10.11648/j.ajtas.20160501.11

Evans, J., & Thomas, C. (2023). *Environmental governance*. Routledge. DOI: 10.4324/9781003334699

Excelsior News., (2017), The ugly truth behind disposable diapers. (Accessed on 06 June 2024)

Eyong, C. T. (2007). Indigenous knowledge and sustainable development in Africa: Case study on Central Africa. *Tribes and Tribals*, *1*(1), 121–139.

Ezeanya, S. N. (1980). The Contribution of African Traditional Religion to Nation Building. In Amucheazi, E. C. (Ed.), *Reading in Social Sciences*. Fourth Dimension.

Fabregas, R., Kremer, M., & Schilbach, F. (2019). Realizing the potential of digital development: The case of agricultural advice. *Science*, *366*(6471), eaay3038. DOI: 10.1126/science.aay3038 PMID: 31831641

Fakudze, M. (2023, May 6). Eating ligusha, chuchuza could kill you – health expert. *Times of Swaziland*. http://www.times.co.sz/sports/140028-eating-ligusha%2C-chuchuza-could-kill-you-%E2%80%93-health-expert.html

Fakudze, C. (2021). The influence of local worldview presuppositions on learners' conceptions of Selected Mechanics topics. *South African Journal of Education*, *41*(2), 1885. DOI: 10.15700/saje.v41n2a1885

Fakudze, C. G. (2004). Learning of science concepts within a traditional socio-cultural environment. *South African Journal of Education*, *24*(4), 270–277.

Falassi, A. (1987). *Time Out of Time: Essays on the Festival*. University of New Mexico Press.

FAO. (2006). *Climate Change and Food Security: A Framework Document*. Food and Agriculture Organization of the United Nations.

FAO. (2013). *Climate-Smart Agriculture Sourcebook. The Food and Agricultural Organisation of the United Nations*. FAO.

Favretto, N., Afionis, S., Stringer, L. C., Dougill, A. J., Quinn, C. H., & Ranarijaona, H. L. T. (2020). Delivering climate-development co-benefits through multi-stakeholder forestry projects in Madagascar: Opportunities and challenges. *Land (Basel)*, *9*(5), 157. DOI: 10.3390/land9050157

Favretto, N., Shackleton, S., Sallu, S. M., & Hoffman, T. (2021). Editorial for special issue:"collaboration and multi-stakeholder engagement in landscape governance and management in Africa: lessons from practice". *Land (Basel)*, *10*(3), 285. DOI: 10.3390/land10030285

Feenstra, G., & Campbell, D. C. (2014). Local and Regional Food Systems. In Thompson, B. P., & Kaplan, D. M. (Eds.), *Encyclopedia of Food and Agricultural Ethic* (pp. 1345–1352). Springer.

Fenwick, L. J., & Lyne, M. C. (1999). The relative importance of liquidity and other constraints inhibiting the growth of small-scale farming in KwaZulu-Natal. *Development Southern Africa*, *16*(1), 141–155. DOI: 10.1080/03768359908440066

Ferronato, N., & Torretta, V. (2019). Waste mismanagement in developing countries: A review of global issues. *International Journal of Environmental Research and Public Health*, *16*(6), 1060. DOI: 10.3390/ijerph16061060 PMID: 30909625

Finnegan, R. (1998). *Oral Literature in Africa*. Oxford University Piers.

Fischer, S. D. (2015). Sign languages in their historical context. In *The Routledge handbook of hisotrical linguistics* (pp. 442–465). Routledge.

Fletschner, D., & Kenney, L. (2014). *Rural women's access to financial services: credit, savings, and insurance*. Springer Netherlands.

Flickr. (2006). Swazi huts. (https://www.flickr.com/photos/50194168@N00/115207528)

Foley, D. (2003). Indigenous epistemology and Indigenous standpoint theory. *Social Alternatives*, *22*(1), 44–52.

Food and Agricultural Organization of the United Nations (FAO). (2014). Sustainable Food Systems Concept and Framework. https://openknowledge.fao.org/server/api/core/bitstreams/b620989c-407b-4caf-a152-f790f55fec71/content. Accessed: 15 Feb 2025

Food and Agricultural Organization of the United Nations (FAO). (2023). Alarm bells ignored as Africa continues to face deepening food crisis. https://www.fao.org/newsroom/detail/alarm-bells-ignored-as-africa-continues-to-face-deepening-food-crisis/en. Accessed: 15 Feb 2025

Food and Agricultural Organization. (2007). FAO Forestry Paper: Vol. 151. *Fire Management-Global Assessment 2006: A thematic study prepared in the framework of the Global Forest Resources Assessment 2005*. Food and Agricultural Organization.

Food and Agriculture Organization (FAO), European Union (EU), CIRAD & DSI-NRF Centre of Excellence in Food Security (CoE-FS). (2022). Food Systems Profile – South Africa. Catalysing the sustainable and inclusive transformation of food systems. Rome, Brussels, Montpellier, France and Bellville, South Africa. DOI: 10.4060/cc0071en

Food and Agriculture Organization of the United Nation (FAO). (2014). Food Sovereignty Systems. https://openknowledge.fao.org/server/api/core/bitstreams/b620989c-407b-4caf-a152-f790f55fec71/content#:~:text=Food%20systems%20(FS)%20encompass%20the,the%20broader%20economic%2C%20societal%2. Accessed: 15 Feb 2025

Food and Agriculture Organization of the United Nations (FAO). (2024). Indigenous Peoples' Systems and Well-Being Interventions and policies for healthy communities. FAO Cine. https://www.fao.org/docrep/018/i3144e/i31144e. Accessed: 15 Feb 2025

Food and Agriculture Organization of the United Nations (FAO). (2024). Rediscovering African Foods: Promoting Native Foods for Improved Nutrition, Health and Food Security. Thirty-third Session Morrocco FAO Regional Conference for Africa. https://openknowledge.fao.org/server/api/core/bitstreams/2b445818-7a24-450f-9b4b-c9e392bf909a/content. Accessed: 15 Feb 2025

Food and Agriculture Organization of the United Nations. (2021). Traditional Food Systems to support healthy and culture-base diet. http://www.fao.orggiahs/news/detail-events

Formica, S. (1998). The development of festivals and special events studies. *Festival Management & Event Tourism*, *5*(3), 131–137.

Fournier, A. (2011). Consequences of wooded shrine rituals on vegetation conservation in West Africa: A case study from the Bwaba cultural area (West Burkina Faso). *Biodiversity and Conservation*, *20*(9), 1895–1910. DOI: 10.1007/s10531-011-0065-5

Fox Gotham, K. (2005a). Theorizing urban spectacles: Festivals, tourism and the transformation of urban space. *City (London, England)*, *9*(2), 225–245. DOI: 10.1080/13604810500197020

Fox, L., Senbet, L. W., & Simbanegavi, W. (2016). Youth employment in Sub-Saharan Africa: Challenges, constraints and opportunities. *Journal of African Economies*, *25*(suppl_1), i3–i15. DOI: 10.1093/jae/ejv027

Fragkias, M., & Boone, C. G. (2016). Modern political economy, global environmental change and urban sustainability transitions. *Current Opinion in Environmental Sustainability*, *22*, 63–68. DOI: 10.1016/j.cosust.2017.04.007

Francia, S., & Stobart, A. (Eds.). (2014). *Critical approaches to the history of Western herbal medicine: From classical antiquity to the early modern period*. Bloomsbury Academic. DOI: 10.5040/9781474210577

Francis, D. V., Subhan, A., & Mourad, I., A. H., K. Abdalla, A., & FR Ahmed, Z. (. (2024). Optimizing germination conditions of Ghaf seed using ZnO nanoparticle priming through Taguchi method analysis. *Scientific Reports*, *14*(1), 15946. DOI: 10.1038/s41598-024-67025-6 PMID: 38987397

Fraser, S. P., & Bosanquet, M. B. (2006). The curriculum? That is just a unit outline, isn't it? *Studies in Higher Education*, *31*(3), 269–284. DOI: 10.1080/03075070600680521

Frequin, S., Anseeuw, W., & D'haese, M. (2012). *Demystifying contract farming: evidence from rural South Africa. Department of Agricultural Economics*. Ghent University.

Friant, S., Bonwitt, J., Ayambem, W. A., Ifebueme, N. M., Alobi, A. O., Otukpa, O. M., Bennett, A. J., Shea, C., Rothman, J. M., Goldberg, T. L., & Jacka, J. K. (2022). Zootherapy as a potential pathway for zoonotic spillover: A mixed-methods study of the use of animal products in medicinal and cultural practices in Nigeria. *One Health Outlook, 4*(5), 1–21. DOI: 10.1186/s42522-022-00060-3 PMID: 35216623

Friedman, C. (2022). *African indigenous knowledge systems in the CAO sector in SA*. Retrieved May 20, 2024, from https://idl-bnc-idrc.dspacedirect.org/server/api/core/bitstreams/5fee0a83-36f0-4237-be0d-4483a169cbeb/content

Frosch, C. A., Schoppe-Sullivan, S. J., & O'Banion, D. D. (2019). Parenting and Child Development: A relational health perspective. *American Journal of Lifestyle Medicine, 15*(1), 45–59. DOI: 10.1177/1559827619849028 PMID: 33447170

Fruitema, M. L. (2015). *A Political Ecology of Solid Waste Management in Niadub, Panama* (Doctoral dissertation, University of Miami).

Future Farming. (n.d.). India launches Uber-like app for farm equipment. https://www.futurefarming.com/smart-farming/tools-data/india-launches-uber-like-app-for-farm-equipment/#:~:text=The%20Indian%20government%20has%20launched,app%20for%20renting%20farm%20equipment

Gaikwad, S. S., Paralikar, S. N., Chavan, V., & Krishnan, S. (2004). *Digitizing Indian sacred groves–an information model for web interfaced multimedia database*. Focus on Sacred Groves and Ethnobotany: Prisam publications, Mumbai, 123-128.

Galang, E. I. N. E., & Vaughter, P. (2020). Generational local ecological knowledge on the benefits of an agroforestry landscape in Mindanao, Philippines. *Asian Journal of Agriculture and Development, 17*(1), 90–108. DOI: 10.37801/ajad2020.17.1.6

Gama, J & Dlamini, A. not dated. Khala Mdumbadumbane. Eswatini Broadcasting Information Service.

Gandure, S., Walker, S., & Botha, J. J. (2013). Farmers' perceptions of adaptation to climate change and water stress in a South African rural community. *Environmental Development, 5*, 39–53. DOI: 10.1016/j.envdev.2012.11.004

Gautam, S. K. (2019). The role of indigenous knowledge in biodiversity conservation: Integrating traditional practices with modern environmental approaches. *Environment Reporter, 1*(2), 1–3. DOI: 10.51470/ER.2019.1.2.01

Gazzola, P., Pavione, E., Grechi, D., & Ossola, P. (2018). Cycle tourism as a driver for the sustainable development of little-known or remote territories: The experience of the Apennine regions of Northern Italy. *Sustainability (Basel), 10*(6), 1863. DOI: 10.3390/su10061863

Gebremikael, L., Sicchia, S., Demi, S., & Rhoden, J. (2022). Afrocentric approaches to disrupting anti-Black racism in health care and promoting Black health in Canada. *Canadian Medical Association Journal, 194*(42), E1448–E1450. DOI: 10.1503/cmaj.220456 PMID: 36316022

Gelfand, M. (1962). *Shona Religion: With Special Reference to the Makorekore*. Juta.

Gelfand, M. (1970). *The spiritual beliefs of the Shona. C.* Struik Publishers.

Gelfand, M. (1973). *The Genuine Shona: Survival Values of an African Culture*. Mambo Press.

Gelfand, M. (1979). *Growing Up in Shona Society*. Mambo Press.

George, E. W., & Reid, D. G. (2005). The power of community tourism in destination development. *Journal of Travel Research*, *43*(5), 523–532. DOI: 10.1177/0047287505278296

Gershman, B. (2022). Witchcraft beliefs around the world: An exploratory analysis. *PLoS One*, *17*(11), e0276872. Retrieved October 12, 2024, from https://pmc.ncbi.nlm.nih.gov/articles/PMC9683553/. DOI: 10.1371/journal.pone.0276872 PMID: 36417350

Getz, D. (1997). *Event Management and Event Tourism*. Cognizant Communications Corp.

Getz, D. (2008). Event Tourism: Definition, Evolution and Research. *Tourism Management*, *29*(1), 403–428. DOI: 10.1016/j.tourman.2007.07.017

Geza, W., Ngidi, M. S. C., Slotow, R., & Mabhaudhi, T. (2022). The dynamics of youth employment and empowerment in agriculture and rural development in South Africa: A scoping review. *Sustainability*, *14*(9), 5041. DOI: 10.3390/su14095041 PMID: 37691717

Ghosh, S., Meyer-Rochow, V. B., & Jung, C. (2023). Embracing Tradition: The Vital Role of Traditional Foods in Achieving Nutrition Security. *Foods*, *12*(23), 4220. DOI: 10.3390/foods12234220 PMID: 38231593

Gibbons, M. (2000). Universities and the New Production of Knowledge: Some Policy Implications for Government. In Kraak, A. (Ed.), *Changing Modes*. Human Sciences Research Council Publishing.

Gillis, A. (2021). *What is internet of things (IoT)?* IoT Agenda.

Githui, W., David, N., & Maurice, S. (2015). The role of indigenous knowledge in socio-economic development. *International Journal of Scientific Research*, *4*(4), 32–37.

Global Communities. (March 2020). New Study Explores Impact of Digital Savings Groups in Tanzania. https://globalcommunities.org/blog/new-study-explores-impact-of-digital-savings-groups-in-tanzania/#:~:text=New%20Study%20Explores%20Impact%20of,an%20increase%20in%20financial

Gobena, E., Hean, S., Heaslip, V., & Studsrød, I. (2024). The challenge of western-influenced notions of knowledge and research training: Lessons for decolonizing the research process and researcher education. *Journal of Ethnic & Cultural Diversity in Social Work*, *33*(5), 276–287. DOI: 10.1080/15313204.2023.2197272

GOI. (2014). National Livestock Mission. https://nlm.udyamimitra.in/Home/SchemePage

GOI. (2025). National Mission for Sustainable Agriculture. https://nmsa.dac.gov.in/

Goldblatt, J. (1997). *Special Events: Best Practice in Modern Event Management*. Van Nostrand.

Goldman, M. J., & Lovell, E. (2017). Indigenous technical knowledge. In D. Richardson, N. Castree, M. F. Goodchild, A. Kobayashi, W. Liu, & R. A. Marston (Eds.), *International Encyclopedia of Geography: People, the Earth, Environment, and Technology*. DOI: 10.1002/9781118786352.wbieg0253

Gómez-Baggethun, E., Corbera, E., & Reyes-García, V. (2013). Traditional ecological knowledge and global environmental change: research findings and policy implications. *Ecology and society: a journal of integrative science for resilience and sustainability, 18*(4), 72.

Govender, N. (2012). Educational implications of applying the complexity approach to Indigenous Knowledge Systems (IKS). *Alternation (Durban), 19*(2), 112–137.

Government of Zimbabwe. (2013). Environmental Management Act (Chapter 20:27). Harare: Government Printer.

Govitvatana, W.V. (2001). Generation gap in the workplace between Baby Boomers and Generation X. [Unpublished masters' dissertation, University of Wisconsin-Stout, Madison Wisconsin].

Graaff, K. (2021). The implications of a narrow understanding of gender-based violence. *Feminist Encounters, 5*(1), 1–12. DOI: 10.20897/femenc/9749

Grange, L. L. (2007). Integrating western and indigenous knowledge systems: The basis for effective science education in South Africa. *International Review of Education, 53*(5-6), 577–591. DOI: 10.1007/s11159-007-9056-x

Grant, C., & Osanloo, A. (2014). Understanding, selecting, and integrating a theoretical framework in dissertation research: Creating the blueprint for your "house". *Administrative issues journal, 4*(2), 4.

Green, E. C., & Makhubu, L. (1984). Traditional healers in Swaziland: Toward improved cooperation between the traditional and modern health sectors. *Social Science & Medicine, 18*(12), 1071–1079. DOI: 10.1016/0277-9536(84)90167-9 PMID: 6463675

Green, L. J. (2008). 'Indigenous knowledge' and 'science': Reframing the debate on knowledge diversity. *Archaeologies, 4*(1), 144–163. DOI: 10.1007/s11759-008-9057-9

Grey, S. (2014). Indigenous knowledge. *Acta Paediatrica (Oslo, Norway), 91*(1), 1–8.

Grey, S., & Patel, R. (2015). Food sovereignty as decolonization: Some contributions from Indigenous movements to food system and development politics. *Agriculture and Human Values, 32*(3), 431–444. DOI: 10.1007/s10460-014-9548-9

Griffin, T. W., Shockley, J. M., & Mark, T. B. (2018). Economics of precision farming. *Precision agriculture basics*, 221-230.

Griffin, P., Burns, M., & Snow, C. (1998). *Preventing reading difficulties in young children.* National Academy Press.

Grydehøj, A. (2019). Cultural sustainability: Indigenous knowledge and island tourism development. *Island Studies Journal, 14*(1), 165–182. DOI: 10.24043/isj.74

Guma, M. (2001). The cultural meaning of names among Basotho of Southern Africa: A historical and linguistic analysis. *Nordic Journal of African Studies, 10*(3), 265–279.

Gumbo, B. G. (2018). The dynamics of cattle farming, bovine disease and state intervention in North West District of Botswana, 1966-2000. *Botswana Notes and Records, 50*, 72–84.

Gumbo, M. T., Gasa, V., & Knaus, C. (2022). Centring African knowledges to decolonise higher education. In Knaus, C. B., Mino, T., & Seroto, J. (Eds.), *Decolonising African higher education* (pp. 21–36). Routledge. DOI: 10.4324/9781003158271-2

Gumbo, M. T., Knaus, C. B., & Gasa, V. G. (2024). Decolonising the African doctorate: Transforming the foundations of knowledge. *Higher Education, 88*(4), 1611–1627. DOI: 10.1007/s10734-024-01185-2

Gupta, S. K., Tiwari, S., & Voda, M. (2021). Stakeholders'satisfaction Towards Sustainable Tourism Development In Pushkar Region Of Rajasthan. *Geographia Technica, 16*(1).

Gupta, A. D. (2012). Way to study indigenous knowledge and indigenous knowledge system. *Research Journal of Recent Sciences ISSN, 2277*, 2502.

Gupta, R., Vairale, M. G., Deshmukh, R. R., Chaudhary, P. R., & Wate, S. R. (2010). Ethnomedicinal uses of some plants used by Gond tribe of Bhandara district, Maharashtra. *Indian Journal of Traditional Knowledge, 9*(4).

Gutierrez, C., von Glascoe, C. S., Bertran, M., Calderon, N., Villada-Canela, M., & Leyva, J. C. (2022). The Kumiai food system: Reconnecting nature, food, and health through ancestral knowledge. *Frontiers in Sustainable Food Systems, 6*, 824264. DOI: 10.3389/fsufs.2022.824264

Habel, N. C., & Balabanski, V. (Eds.). (2002). *The earth story in theology and ethics*. Sheffield Academic Press.

Habermas, J. (1968). *Knowledge and human interests*. Beacon Press.

Hall, R., & Williams, G. (2003). Land reform in South Africa: problems and prospects. *From Cape to Congo: Southern Africa's evolving security architecture*, 97-129.

Hall, C. M. (2019). Constructing sustainable tourism development The 2030 agenda and the managerial ecology of sustainable tourism. *Journal of Sustainable Tourism, 27*(7), 1044–1060. DOI: 10.1080/09669582.2018.1560456

Hall, C. M., & Macionis, N. (1998). Wine tourism in Australia and New Zealand. In Hall, C. M., & Jenkins, J. (Eds.), *Tourism and Recreation in Rural Areas*. Wiley.

Halliday, M. (1985). *An introduction to functional grammar*. Edward Arnold.

Hall, J. (2009). *Sangoma: My odyssey into the spirit world of Africa*. Sterling Publishing Company, Inc.

Hall, R. (2004). A political economy of land reform in South Africa. *Review of African Political Economy, 31*(100), 213–227. DOI: 10.1080/0305624042000262257

Hall, R., Jacobs, P., & Lahiff, E. (2003). *'Final report,' Evaluating Land and Agrarian Reform in South Africa 10*. Programme for Land and Agrarian Studies, University of the WesternCape.

Hammarberg, K., Kirkman, M., & De Lacey, S. (2016). Qualitative research methods: When to use them and how to judge them. *Human Reproduction (Oxford, England), 31*(3), 498–501. DOI: 10.1093/humrep/dev334 PMID: 26759142

Hammersley, M. (2015). Sampling and thematic analysis: A response to Fugard and Potts. *International Journal of Social Research Methodology, 18*(6), 687–688. DOI: 10.1080/13645579.2015.1005456

Handayani, R. D., Wilujeng, I., & Prasetyo, Z. K. (2018). Elaborating Indigenous Knowledge in the Science Curriculum for the Cultural Sustainability. *Journal of Teacher Education for Sustainability, 20*(2), 74–88. DOI: 10.2478/jtes-2018-0016

Hart, A. M. (2010). Indigenous worldviews, knowledge, and research: The development of an indigenous research paradigm. *Journal of Indigenous Voices in Social Work, 1*(1), 1–16. http://hdl.handle.net/10125/12527

Harvard Ministerial Leadership Program. (2024). Empowering Farmers: Hello Tractor's Digital Revolution in Nigeria. https://ministerialleadership.harvard.edu/leadership-resources/tech-database/hello-tractor-case-study/#:~:text=Uber,rent%20tractors%20on%20demand

Hassan, F. E., Alyafei, M. A., Kurup, S., Jaleel, A., Al Busaidi, N., & Ahmed, Z. F. (2023). Effective priming techniques to enhance ghaf (Prosopis cineraria L. Druce) seed germination for mass planting. *Horticulturae, 9*(5), 542. DOI: 10.3390/horticulturae9050542

Haverkort, B., & Reijntjes, C. (2010). Diversities of knowledge communities, their worldviews and sciences: On the challenges of their co-evolution. *Traditional knowledge in policy and practice: Approaches to development and human well-being*, 12-30.

Hawkes, C. (2022). Centre for Food Policy at City, University of London, and Results for Development (R4D), Taking Food System Approach to Policymaking: Resource for Policymaker. https://r4d.org/wp-content/uploads/R4D. Accessed: 15 Feb 2025

Haywood, L. K., Kapwata, T., Oelofse, S., Breetzke, G., & Wright, C. Y. (2021). Waste disposal practices in low-income settlements of South Africa. *International Journal of Environmental Research and Public Health, 18*(15), 8176. DOI: 10.3390/ijerph18158176 PMID: 34360468

Healey, N. M. (2017). *Reflections on the value of insider research as a qualitative research methodology*. SAGE Publications Ltd. DOI: 10.4135/9781526401489

Heaney, D., Padilla-Zakour, O. I., & Chen, C. (2024). Processing and Preservation Technologies to Enhance Indigenous Food Sovereignty, Nutrition Security and Health Equity in North America. *Frontiers In Nutrition*, 19,11, 1395962. https://www.frontiersin.org/journals/nutrition/articles/10.3389/fnut.2024.1395962/full. Accessed: 15 Feb 2025

Hedden, J. (2000). Twenty Somethings: Managing the New Breed of Employee.

Heimall, L. M., Storey, B., Stellar, J. J., & Davis, K. F. (2012). Beginning at the bottom: Evidence-based care of diaper dermatitis. *MCN. The American Journal of Maternal Child Nursing, 37*(1), 10–16. DOI: 10.1097/NMC.0b013e31823850ea PMID: 22072018

Heimer, H. (2005). Topophilia and quality of life: defining the ultimate restorative environment.

Heleta, S. (2016). Decolonisation of higher education: Dismantling epistemic violence and Eurocentrism in South Africa. *Transformation in higher education, 1*(1), 1-8.

Henderson, L. (1993). Interactive multimedia computer courseware and culturally appropriate ways of learning. In Latchem, C., Williamson, J., & Henderson-Lancett, L. (Eds.), *Interactive multimedia: Practice and promise* (pp. 189–203). Kogan Page.

Hendriks, S. L. (2003). The potential for nutritional benefits from increased agricultural production in rural KwaZulu-Natal. *South African Journal of Agricultural Extension*, *32*, 28–44.

Henning, E. (2012). Learning concepts, language, and literacy in hybrid linguistic codes: The multilingual maze of urban Grade 1 classrooms in South Africa. *Perspectives in Education*, *30*(3), 69–77.

Henriques, A. C. (2013). *Environment, Worldview and healing among the Zulu Christians (19th and 20th centuries)*. UKZN.

Hernandez, J. (2022). *Fresh banana leaves: Healing Indigenous landscapes through Indigenous science*. North Atlantic Books.

Hewett, M., Butler-Kruger, L., Foflonker, K., Edge, J., Botha, M. H., & Seedat, S. (2023). Desktop analysis and qualitative investigation of programmes and initiatives for the prevention of and response to gender based violence at South African universities with medical campuses. *South African Journal of Higher Education*, *37*(5), 142–157. DOI: 10.20853/37-5-5181

Hewson, M. G., Javu, T. T., & Holtman, L. B. (2009). The indigenous knowledge of African traditional health practitioners and the South African Science curriculum. *African Journal of Research in Mathematics. Science and Technology Education*, *13*(1), 5–18.

Higgins, N., & Maguire, G. (2019). Language, indigenous peoples, and the right to self-determination. *New England Journal of Public Policy*, *31*(2), 8.

Higgs, P. (2016). The African renaissance and the transformation of the higher education curriculum in South Africa. *Africa Education Review*, *13*(1), 87–101. DOI: 10.1080/18146627.2016.1186370

Higgs, P., & Van Niekerk, M. P. (2002). The programme for indigenous knowledge systems (IKS) and higher educational discourse in South Africa: A critical reflection: Perspectives on higher education. *South African Journal of Higher Education*, *16*(3), 38–49. DOI: 10.4314/sajhe.v16i3.25215

Hill, C. E., & Mathews, H. (2021). Traditional health beliefs and practices among southern rural Blacks: A complement to biomedicine. In *Perspectives on the American South* (pp. 307–322). Routledge.

Hilson, G. (2016). Farming, small-scale mining and rural livelihoods in Sub-Saharan Africa: A critical overview. *The Extractive Industries and Society*, *3*(2), 547–563. DOI: 10.1016/j.exis.2016.02.003

Hlalele, D. J. (2019). Indigenous knowledge systems and sustainable learning in rural South Africa. *Australian and International Journal of Rural Education*, *29*(1), 88–100. DOI: 10.47381/aijre.v29i1.187

Holness, W. (2016). The development and use of sign language in South African schools: The denial of inclusive education. *Afr. Disability Rts. YB*, *4*, 141.

Hoorens, V. (2015). What's really in a Name-letter Effect? Name-letter preferences as indirect measures of self-esteem. *European Review of Social Psychology*, *25*(1), 228–262. Advance online publication. DOI: 10.1080/10463283.2014.980085

Hoppers, C. A. O. (2005). *Culture, indigenous knowledge and development: The role of the university.* Centre for Education Policy Development.

Houghton, A. (2020). Aunts in the Family: The Role of Extended Family in Child Development and Socialization. In Family Relations and Social Capital: New Perspectives (pp. 110-125). Routledge.

Howe, N., & Strauss, W. (1997). *The Fourth Turning: What the cycles of History tell us about Americas' Next Rendezvous with destiny*. Broadway Books.

Hsiao, C. H., Lee, C. Y., & Yang, C. C. (2020). Exploring the effectiveness of WhatsApp for data collection: A systematic review. *International Journal of Social Research Methodology*, *23*(5), 475–486.

Huang, Y., Liu, C., & Chen, Y. (2018). Environmental impact of disposable diapers: A life cycle assessment. *Waste Management (New York, N.Y.)*, *75*, 123–132.

Human Rights Watch. (2024). Eswatini: Address violence against women. Retrieved March 20, 2025, from https://www.hrw.org/news/2024/01/29/eswatini-address-violence-against-women

Hutchings, A. (1996). *Zulu medical plants*. University of Natal Press.

Hvitsand, C. (2016). Community supported agriculture (CSA) as a transformational act—Distinct values and multiple motivations among farmers and consumers. *Agroecology and Sustainable Food Systems*, *40*(4), 333–351. DOI: 10.1080/21683565.2015.1136720

Idrus, F., Hussin, M. E., & Gulca, M. (2023). THE INTEGRATION OF CULTURAL ELEMENTS IN THE ENGLISH LANGUAGE CLASSROOMS: A CASE STUDY OF SELECTED ORANG ASLI SCHOOLS IN PAHANG, MALAYSIA. [JONUS]. *Journal of Nusantara Studies*, *8*(1), 26–45. Advance online publication. DOI: 10.24200/jonus.vol8iss1pp26-45

Igbal Nst, M., & Batubara, N. K. (2024). The importance of ethnomathematics in education: A systematic literature review. *Proceedings of the International Conference on Education Teacher Training & Education Faculty*. Universitas Serambi Mekkah.

Ilyas, Q. M., & Ahmad, M. (2020). Smart farming: An enhanced pursuit of sustainable remote livestock tracking and geofencing using IoT and GPRS. *Wireless Communications and Mobile Computing*, *2020*(1), 6660733. DOI: 10.1155/2020/6660733

Independent News Eswatini. (2020, December 20). King commissions Ingatja, Incwala holiday on Saturday, January 2. https://independentnews.co.sz/local-news/5584/king-commissions-ingatja-incwala-holiday-on-saturday-january-2/

India, S. (2016). About Startup India Initiative. https://www.startupindia.gov.in/content/sih/en/about-startup-india-initiative.html

Indigenous and Northern Affairs Committee. (2017). Report on Indigenous Waste Management.

Indigenous Climate Hub. (2024). Traditional Food Systems and Climate Change: Adapting Indigenous Agriculture. https://indigenousclimatehub.ca. Accessed: 15 Feb 2025

International Commission of Jurists (ICJ). (2015). *Sexual and gender-based violence, fair trial rights and the rights of victims: Challenges in using law and justice systems faced by women human rights defenders*. Retrieved November 5, 2023, https://www.refworld.org/docid/57ee7f724.html

International Development Law Organization. (2023). Rule of Law for Food Systems Transformation: Policy Briefing. https://www.idlo.int/sites/default/files/2023/other/documents/for_web_-_rule_of_law_for_food_systems_transformation. Accessed: 15 Feb 2025

International Experiences in Language Testing and Assessment. (2015). *International Experiences in Language Testing and Assessment*., DOI: 10.3726/978-3-653-02511-8

International Fund for Agricultural Development (IFAD). (2022) Sustainable and resilient Indigenous Peoples' Food Systems for Improved Nutrition. https://www.ifad.org/en/w/publications/sustainable-and-resilient-indigenous-peoples-food-systems-for-improved-nutrition. Accessed: 15 Feb 2025

Ipsos. (March 2024). Stokvels remain the untapped 'human banks' of South Africa (https://www.ipsos.com/en-za/stokvels-remain-untapped-human-banks-south-africa#:~:text=,App%20transforming%20how%20groups%20operate

Isabirye, J. (2022). Reclaiming indigenous epistemes: *Entenga* drums revival at Kyambongo University. In Knaus, C. B., Mino, T., & Seroto, J. (Eds.), *Decolonising African higher education: Practitioner perspectives from across the continent*. Taylor & Francis Group. DOI: 10.4324/9781003158271-6

Ivanova, M. INNOVATION AS A PREREQUISITE FOR SUSTAINABLE BUSINESS DEVELOPMENT. *ИНОВАЦИИТЕ КАТО ПРЕДПОСТАВКА ЗА УСТОЙЧИВО РАЗВИТИЕ НА БИЗНЕСА*, 18.

Iwara, I. O., & Netshandama, V. O. (2021). Small-enterprise capital mobilization and marketing in rural areas: A review synthesis of Stokvel model. *Academy of Entrepreneurship Journal*, *27*(2), 1–11.

Jacobs, T., Lahiff, E., & Hall, R. (2003). *Evaluating land and agrarian reform in South Africa: Land redistribution*). PLAAS, University of the Western Cape.

Jacobs, K. R. (2015). *The classroom implementation of indigenous knowledge in the science curriculum by science teachers in the Western Cape province*.

Jain, T., & Kour, G. (2020) Destination events and campaign planning for Repositioning: Case of Rajasthan, India.

Jain, S., Srivastava, A., Vishwakarma, D. K., Rajput, J., Rane, N. L., Salem, A., & Elbeltagi, A. (2024). Protecting ancient water harvesting technologies in India: Strategies for climate adaptation and sustainable development with global lessons. *Frontiers in Water*, *6*, 1441365. DOI: 10.3389/frwa.2024.1441365

James, P. B., Wardle, J., Steel, A., & Adams, J. (2018). Traditional, complementary and alternative medicine use in Sub-Saharan Africa: A systematic review. *BMJ Global Health*, *3*(5), e000895. Advance online publication. DOI: 10.1136/bmjgh-2018-000895 PMID: 30483405

Jaramillo, P. (2023). Agricultural identity of indigenous salasacas in Ecuador. *Alternative*, *19*(4), 882–891. DOI: 10.1177/11771801231197979

Jayaraman, R. (2005). Personal identity in a globalized world: Cultural roots of Hindu personal names and surnames. *Journal of Popular Culture, 38*(3), 476–490. DOI: 10.1111/j.0022-3840.2005.00124.x

Jayne, T. S., Chamberlin, J., & Headey, D. D. (2014). Land pressures, the evolution of farming systems, and development strategies in Africa: A synthesis. *Food Policy, 48*, 1–17. DOI: 10.1016/j.foodpol.2014.05.014

Jeewa, T., & Bhima, J. (2021). Discriminatory Language: A Remnant of Colonial Oppression. *Constitutional Court Review, 11*(1), 1–17. DOI: 10.2989/CCR.2021.0012

Jeong, J., Franchett, E. E., Ramos de Oliveira, C. V., Rehmani, K., & Yousafzai, A. K. (2021). Parenting interventions to promote early child development in the first three years of life: A global systematic review and meta-analysis. *PLoS Medicine, 18*(5), e1003602.

Jernigan, B. B. V., Maudrie, T. L., Nikolaus, C. J., Benally, T., Johnson, S., Teague, T., Mayes, M., Jacob, T., & Taniguchi, T. (2021). Food Sovereignty Indicators for Indigenous Community Capacity Building and Health. *Frontiers in Sustainable Food Systems*, 5, 704750. https://doiDOI: <ALIGNMENT.qj></ALIGNMENT>10.3389/fsufs

Jessen, T. D., Ban, N. C., Claxton, N. X., & Darimont, C. T. (2022). Contributions of indigenous knowledge to ecological and evolutionary understanding. *Frontiers in Ecology and the Environment, 20*(2), 93–101. DOI: 10.1002/fee.2435

Jiyana, M. (2022). *The Role of Traditional Storytelling in the Teaching and Learning of Isindebele in Grade 3*. University of Pretoria (South Africa), Keikelame, M., & Swartz, L. (2019). Decolonising research methodologies: Lessons from a qualitative research project, Cape Town, South Africa. *Global Health Action, 12*(1), 1–7. DOI: 10.1080/16549716.2018.1561175

Johnson, R. B., & Christensen, L. (2014). *Educational research: Quantitative, qualitative, and mixed approaches (5th Ed.)*. Los Angels: Sage

Johnson, R., & Christensen, L. B. (2017). *Educational Research: Quantitative, qualitative and mixed approaches*. Sage.

Johns, T., Powell, B., Maundu, P., & Eyzaguirre, P. B. (2013). Agricultural biodiversity as a link between traditional food systems and contemporary development, social integrity and ecological health. *Journal of the Science of Food and Agriculture, 93*(14), 3433–3442. DOI: 10.1002/jsfa.6351 PMID: 23963831

Jonas, H., Bavikatte, K., & Shrumm, H. (2010). Community protocols and access and benefit sharing. *Asian Biotechnology and Development Review, 12*(3), 49–76.

Jonathan, S. D., Maracle, K., & Wild, S. (2022). Exploring indigenous foods & food sovereignty. https://ecampusontario.pressbooks.pub/ifs1/

Jones, R. (2019). *The Environmental Impact of Disposable Diapers*. Waste Management Studies.

Justo (2023). Indigenous Views on Landscape Design and Maintenance for Sustainability. *Native Tribe Info*.

Jwara, S. M. (1998). *Healing Ministry in The Zulu Context: The Potential Role of Zulu Traditional Medicine in the Catholic Church's Healing Ministry, with Special Reference to the Marianhill Mission. Unpublished MTh in Theology*. University of Natal.

Kadykalo, A. N., Cooke, S. J., & Young, N. (2021). The role of western-based scientific, indigenous and local knowledge in wildlife management and conservation. *People and Nature*, *3*(3), 610–626. DOI: 10.1002/pan3.10194

Kahari, G. P. (1990). *The moral vision of Patrick Chakaipa: A study in Shona traditional religion and Western Christianity*. Mambo Press.

Kalu, O. U. (2000). *Ancestral Spirituality and Society in Africa*. Crossroad Publishing Company.

Kamakaula, Y. (2024). Ethnoecology and climate change adaptation in agriculture. *Global*, *2*(2), 473–485. DOI: 10.59613/global.v2i2.99

Kamera, W. D. (2001). *Swazi Oral Literature. Manzini*. Ruswenda Publishing Bureau.

Kanter, R., Kennedy, G., & Boza, S. (2023). Local, traditional and indigenous food systems in the 21st century to combat obesity, under-nutrition and climate change. *Frontiers in Sustainable Food Systems*, *7*, 119574. DOI: 10.3389/fsufs.2023.1195741

Kanus, C. B., Mino, T., & Seroto, J. (2022). Decolonising higher education: Definitions, conceptualisations, epistemologies. In Knaus, C. B., Mino, T., & Seroto, J. (Eds.), *Decolonising African Higher Education: Practitioner perspectives from across the continent*. Taylor and Francis Group.

Kapp, D. B. (1986). A parallel motif in Lepcha and Barela-Bhilala mythology. *Asian Folklore Studies*, *45*(2), 259. DOI: 10.2307/1178620

Kariuki, P., & Ofusori, L. O. (2017, March). WhatsApp-operated stokvels promoting youth entrepreneurship in Durban, South Africa: experiences of young entrepreneurs. In *Proceedings of the 10th International Conference on Theory and Practice of Electronic Governance* (pp. 253-259). DOI: 10.1145/3047273.3047397

Karppinen, K., & Moe, H. (2012). What we talk about when we talk about document analysis. *Trends in communication policy research: New theories, methods and subjects*, 177-193.

Kasenene, P. (1993). *Swazi traditional religion and society*. Websters.

Kasere, M. (2010). Totemism in Zimbabwe: Social and Environmental Implications. In Zimbabwe. *Journal of Environmental Studies (Northborough, Mass.)*, *15*(2), 56–68.

Kaya, H. O. (2014). Revitalizing African indigenous ways of knowing and knowledge production. *Restoring indigenous self-determination, 105*.

Kaya, H. O., & Seleti, Y. N. (2013). African indigenous knowledge systems and relevance of higher education in South Africa. *International Education Journal: Comparative Perspectives, 12*(1).

Kaya, H. O. (2013). Integration of African indigenous knowledge systems into higher education in South Africa: Prospects and challenges. *Alternation (Durban)*, *20*(1), 135–153.

Kaya, H. O., & Chinsamy, M. (2021). Combating world pandemics through complementarity of indigenous knowledge with other knowledge systems: A paradigm shift in healthcare research and education. *Alternation (Durban)*, *38*, 254–272. DOI: 10.29086/2519-5476/2021/sp38a11

Kaya, H. O., & Seleti, Y. N. (2013). African indigenous knowledge systems and relevance of higher education in South Africa. *The International Education Journal: Comparative Perspectives*, *12*(1), 30–44.

Kazi, M.R., Chowdhury, N., Chowdhury, M., Turin, T.C. (2021). Conducting comprehensive scoping review to systemically capture the landscape of a subject matter. *Popul. Med*, 3(36). htpps://DOI: 10.18332/popmed/143831

Keane, M. (2015). Why indigenous knowledge has a place in the school science curriculum. Retrieved from https://www.researchgate.net/publication/280216361 Accessed: 10 February 2025

Kemmis, & McTaggart. (2003). Participatory action research. In Denzin & Lincoln (Eds.), *Strategies of Qualitative Enquiry* (pp. 336-396). Thousand Oaks, C.A.: Sage.

Kennedy, G., Kanter, S. C., Covic, N., Delormier, T., Longvah, T., Maundu, P., Omidvar, N., Vish, P., & Kuhnlein, H. (2021). Traditional and indigenous fruits and vegetables for food system transformation. *Current Developments in Nutrition*, *5*(8), nzab092. DOI: 10.1093/cdn/nzab092 PMID: 34423230

Kesa, H., Kamgain, A. D. T., & Zuma, M. K. (2023). Knowledge, perceptions and consumption of indigenous foods in Gauteng Region, South Africa. *International Journal of Environmental Research and Public Health*, *20*(20), 6961. DOI: 10.3390/ijerph20206961 PMID: 37887699

Keske, C., Mills, M., Tanguay, L., & Dicker, J. (2018). Waste management in Labrador and northern communities: Opportunities and challenges. *The Northern Review*, *47*(47), 79–112. DOI: 10.22584/nr47.2018.005

Khalala, G. (2021). Protecting and benefiting from indigenous knowledge: The National Indigenous Knowledge Management System. *CSIR Science Scope*, *17*(1), 56–57.

Khan, M. R., Raja, W., Bhat, T. A., Mir, M. S., Naikoo, N. B., Amin, Z., Nazir, A., Mir, S. A., Mohammad, I., Wani, A. A., & Patyal, D. (2022). Zero budget natural farming: A way forward towards sustainable agriculture. *Current Journal of Applied Science and Technology*, *41*(13), 31–43. DOI: 10.9734/cjast/2022/v41i1331712

Khanyile, N. C., Dlamini, P., & Masenya, T. M. (2023). Techniques in preserving traditional medicinal knowledge in uMhlathuze in KwaZulu Natal, South Africa. *The Serials Librarian*, *84*(5–8), 148–164. DOI: 10.1080/0361526X.2023.2286234

Khapayi, M., & Celliers, P. R. (2015). Issues and constraints for emerging farmers in the Eastern Cape Province, South Africa. *African Journal of Agricultural Research*, *10*(41), 3860–3869. DOI: 10.5897/AJAR2015.9956

Khapayi, M., & Celliers, P. R. (2016). Factors limiting and preventing emerging farmers to progress to commercial agricultural farming in the King William's Town area of the Eastern Cape Province, South Africa. *South African Journal of Agricultural Extension*, *44*(1), 25–41. DOI: 10.17159/2413-3221/2016/v44n1a374

Khokholkova, N. (2016). Worlds of color. *Asia and Africa Today*, *5*, 72–76.

Khoza, R. J. (2006). *Let Africa Lead: African Transformational Leadership for 21st Century Business*. Vezubuntu.

Khupe, C. (2014). *Indigenous knowledge and school science: Possibilities for integration* (Unpublished doctoral dissertation). Johannesburg, South Africa: University of the Witwatersrand

Kiely, T. J. (2025). *What Is Desk Research? Meaning, Methodology, Examples*. Melwater.

Kiema, K. (2010). *Tears for my Land: A Social History of the Kva of the Central Kalahari Game Reserve, Tc 'amnquoo*. Mmegi Publishing House.

Kimani, E. W., Muchiri, J., & Makindi, S. (2015). Soiled diapers disposal practices among caregivers in poor and middle income urban settings.

Kim, H., Sefcik, J. S., & Bradway, C. (2017). Characteristics of qualitative descriptive studies: A systematic review. *Research in Nursing & Health*, *40*(1), 23–42. DOI: 10.1002/nur.21768 PMID: 27686751

Kincheloe, J. (2006). Postformalism and the Socio-cultural and Political Challenge: Taking Multilogicality Seriously. In *Reading, Writing, and Thinking* (pp. 141-167): Brill.

Kiran, A., & Knights, J. (2010). Traditional Indigenous games promoting physical activity and cultural connectedness in primary schools? Cluster Randomised Control Trial. *Health Promotion Journal of Australia*, *21*(2), 149–151. DOI: 10.1071/HE10149 PMID: 20701567

Kirkpatrick, J. (2007). A Lepcha village in Sikkim. *Visual Anthropology*, *20*(1), 87–89. DOI: 10.1080/08949460601064580

Kissi, E. (2020). *Sankofa: Disability and the door of return.*

Kivunja, C., & Kuyini, A. (2017). Understanding and applying research paradigms in educational contexts. *International Journal of Higher Education*, *6*(5), 26–41. DOI: 10.5430/ijhe.v6n5p26

Klapwijk, N. (2015). EMC^2 = comprehension: A reading strategy instruction framework for all teachers. *South African Journal of Education*, *35*(1), 1–10. DOI: 10.15700/201503062348

Knorr, D., & Augustin, M. A. (2023). Preserving the food preservation legacy. *Critical Reviews in Food Science and Nutrition*, *63*(28), 9519–9538. DOI: 10.1080/10408398.2022.2065459 PMID: 35442825

Koenane, M. L. (2019). Economic development in Africa through the stokvel system:'our'indigenous way or 'theirs'. *Filosofia Theoretica: Journal of African Philosophy. Culture and Religion*, *8*(1), 109–124.

Kohsaka, R., & Rogel, M. (2021). Traditional and local knowledge for sustainable development: Empowering the indigenous and local communities of the world. In *Partnerships for the Goals* (pp. 1261–1273). Springer International Publishing. DOI: 10.1007/978-3-319-95963-4_17

Kolb, D., Boyatzis, R., & Mainemelis, C. (2014). Experiential learning theory: Previous research and new directions. In *Perspectives on Thinking, Learning, and Cognitive Styles* (pp. 227-247): Routledge.

Koliyabandara, P. A., Preethika, D. D. P., Cooray, A. T., Liyanage, S. S., Siriwardana, C., & Vithanage, M. (2024). The environmental pressure by open dumpsites and way forward. In *Technical Landfills and Waste Management: Volume 1: Landfill Impacts, Characterization and Valorisation* (pp. 171-204). Cham: Springer Nature Switzerland. DOI: 10.1007/978-3-031-52633-6_7

Koma, H. M. (2012). *An analysis of given and inherited names among the Northern Sotho speaking people in Moletjie and Sekhukhune, Limpopo Province: An onomastic perspective* [Unpublished master's dissertation, University of Limpopo, Limpopo]. http://pocketcultures.com/2011/04/13 children-naming-traditions/

Kom, Z., & Nethengwe, N. S. (2024). Indigenous and Local Knowledge: Instruments Towards Achieving SGD2: A Review in an African Context. *Sustainability (Basel)*, *16*(20), 9137. DOI: 10.3390/su16209137

Kom, Z., Nicolau, M. D., & Nenwiini, S. C. (2024). The Use of Indigenous Knowledge Systems Practices to Enhance Food Security in Vhembe District, South Africa. *Agricultural Research*, *13*(3), 599–612. DOI: 10.1007/s40003-024-00716-8

Kordecki, H., Antrobus-Wuth, R., Uys, M. T., Van Wyk, I., Root, E. D., & Berrian, A. M. (2022). Disposable diaper waste accumulation at the human-livestock-wildlife interface.

Kordecki, H., Antrobus-Wuth, R., Uys, M. T., van Wyk, I., Root, E. D., & Berrian, A. M. (2022). Disposable diaper waste accumulation at the human-livestock-wildlife interface: A one health approach. *Environmental Challenges*, *8*, 100589. DOI: 10.1016/j.envc.2022.100589

Koreti, S. (2016). Socio-Cultural History of the Gond Tribes of Middle India. *International Journal of Social Science and Humanity*, *6*(4), 288–292. Advance online publication. DOI: 10.7763/IJSSH.2016.V6.659

Kosoe, E. A., Diawuo, F., & Osumanu, I. K. (2019). Looking into the past: Rethinking traditional ways of solid waste management in the Jaman South Municipality, Ghana. *Ghana Journal of Geography*, *11*(1), 228–244.

Kovach, M. (2021). *Indigenous methodologies: Characteristics, conversations, and contexts*. University of Toronto press.

Kuhnlein, H. V., & Chotiboriboon, S. (2022). Why and how to strengthen indigenous peoples' food systems with examples from two unique indigenous communities. *Frontiers in Sustainable Food Systems*, *6*, 808670. DOI: 10.3389/fsufs.2022.808670

Kuhnlein, H. V., & Receveur, O. (1996). Dietary change and traditional food systems of indigenous peoples. *Annual Review of Nutrition*, *16*(1), 417–442. DOI: 10.1146/annurev.nu.16.070196.002221 PMID: 8839933

Kuli, H.C. (2012). Literature Preserves Papua New Guinea (PNG) Cultural Heritage. L*iterature Heritage,* 1-7.

Kumalo, S. (2013). *Religion and Politics in Swaziland*. University of Johannesburg Press. DOI: 10.18820/9781920382230

Kumar, R., & Keshar, D. (2017). Effects of Chemical Fertilizers on Human Health and Environment: A review. International Advanced Research Journal Science, Engineering and Technology, 4(6), 204. https://iarjset.com/upload/2017/june-17/IARJSET%2036.pdf

Kumar, A. (2019). Food Preservation: Traditional and Modern Techniques. *Acta Scientific Nutritional Health*, *3*(12), 45–49. DOI: 10.31080/ASNH.2019.03.0529

Kumar, B. M., Kunhamu, T. K., Bhardwaj, A., & Santhoshkumar, A. V. (2024). Subcanopy light availability, crop yields, and managerial implications: A systematic review of the shaded cropping systems in the tropics. *Agroforestry Systems*, *98*(8), 1–26. DOI: 10.1007/s10457-024-00957-0

Kumar, M. J. (2024). Forging Connections: Integrating Indian Knowledge Systems in Higher Education. *IETE Technical Review*, *41*(3), 271–273. DOI: 10.1080/02564602.2024.2342625

Kumar, R. (2014). *Research methodology: A step-by-step guide for beginners*. Sage.

Kuo, C. Y., & Tsai, Y. F. (2020). The role of extended family in emotional support: A study of aunts and uncles. *Journal of Family Studies*, *26*(2), 123–138.

Kuper, H. (1963). *The Swazi: A South African Kingdom*. Holt, Rinehart and Winston.

Ladd, P. (2003). Understanding Deaf Culture. In *Search of Deafhood*. Multilingual Matters. DOI: 10.21832/9781853595479

Laidler, G. (2006). Inuit and Scientific Perspectives on the Relationship Between Sea Ice and Climate Change: The Ideal Compliment? *Climatic Change*, *78*(2–4), 407–444. DOI: 10.1007/s10584-006-9064-z

Lakshmypriya, K., & Varghese, B. (2024). *Indigenous Tribes and Inclusive Engagement: An Integrated Approach for Sustainable Livelihood into the Future*. In Future Tourism Trends Volume 1: Tourism in the Changing World (pp. 99-114). Emerald Publishing Limited. DOI: 10.1108/978-1-83753-244-520241008

Lalonde, A. (1991). African indigenous knowledge and its relevance to environment and development activities. *Canadian International Development Agency*, 1-12.

Lan, D. (1985). *Guns & rain: Guerrillas & spirit mediums in Zimbabwe*. University of California Press.

Larson, D. F., Muraoka, R., & Otsuka, K. (2016). Why African rural development strategies must depend on small farms. *Global Food Security*, *10*, 39–51. DOI: 10.1016/j.gfs.2016.07.006

Lather, P. (1993). Fertile obsession: Validity after poststructuralism. *The Sociological Quarterly*, *34*(4), 33–34. DOI: 10.1111/j.1533-8525.1993.tb00112.x

Le Grange, L. (2018). Decolonising, Africanising, indigenising, and internationalising curriculum studies: Opportunities to (re) imagine the field. *Journal of Education (University of KwaZulu-Natal)*, (74), 4-18.

Le Grange, L. (2016). Decolonizing the university curriculum. *South African Journal of Higher Education*, *30*(2), 1–12. DOI: 10.20853/30-2-709

Le Grange, L. (2016). Think piece: Sustainability education and (curriculum) improvisation. *Southern African Journal of Environmental Education*, *32*, 26–36.

Le Roux, C. (2018). African indigenous knowledge systems and sustainable development. *Journal of Sustainable Development*, *11*(2), 1–12.

Lebaka, M. (2022). Adverse impact of the flow of African storytelling between older and younger generations as a result of the movement of people from rural to urban areas: The case of Bapedi story telling tradition. *Dialogo*, *9*(1), 95–104. DOI: 10.51917/dialogo.2022.9.1.7

Lederman, N. G., & Lederman, J. S. (2015). What is a theoretical framework? A practical answer. *Journal of Science Teacher Education*, *26*(7), 593–597. DOI: 10.1007/s10972-015-9443-2

Leedy, P. D., & Ormrod, J. E. (2005). *Practical research-planning and design*. McMillan Publishing.

Lemus, K., Barrasa, S., & McCall, M. K. (2024). Wildlife and cultural landscape management from a local knowledge approach: A review. *Human Dimensions of Wildlife*, •••, 1–13. DOI: 10.1080/10871209.2024.2384747

Lepcha, R. (2021). *Dynamics of Material Culture of Lepchas in Sikkim, Klimpong and Darjeeling* (Doctoral dissertation).

Lepcha, C. K., & Lal, U. (2021). *Communities, institutions and histories of India's northeast* (1st ed.). Routledge., DOI: 10.4324/9781003245865

Lepcha, K. (2020). Culture kept alive through oral traditions: A peep into the world of the Lepchas. In Chaudhuri, S. K., Maiti, S., & Lepcha, C. K. (Eds.), *The cultural heritage of Sikkim* (p. 109). Routledge.

Lepcha, R. P., Lepcha, M., & Foning, A. (2024). The confluence narratives: Reflections through water. *Social Sciences & Humanities Open*, *9*, 100841. DOI: 10.1016/j.ssaho.2024.100841

Letseka, M. (2000). African philosophy and educational discourse. In Higgs, P., Vakalisa, N. C. G., Mda, T. V., & Assie-Lumumba, N. T. (Eds.), *African voices in education*. Juta.

Lev, E. (2003). Traditional healing with animals (zootherapy): Medieval to present-day Levantine practice. *Journal of Ethnopharmacology*, *85*(1), 107–118. DOI: 10.1016/S0378-8741(02)00377-X PMID: 12576209

Lewis, A. D., Huebner, E. S., Malone, P. S., & Valois, R. F. (2011). Life Satisfaction and Student Engagement in Adolescents. *Journal of Youth and Adolescence*, *40*(3), 249–262. Advance online publication. DOI: 10.1007/s10964-010-9517-6 PMID: 20204687

Liang, Y., Dai, K., & Mathews, K. E. (2024). Followers, customers, or partners in Australian, Mainland Chinese, and Hong Kong universities. *Higher Education*, *88*, 2251–2267. DOI: 10.1007/s10734-024-01215-z

Liladhar Rane, N., Purushottam Choudhary, S., Saha, A., Srivastava, A., Pande, C. B., Alshehri, F., Roy, R., Katipoğlu, O. M., & Abdo, H. G. (2024). Delineation of environmentally sustainable urban settlement using GIS-based MIF and AHP techniques. *Geocarto International*, *39*(1), 2335249. DOI: 10.1080/10106049.2024.2335249

Lilavanichakul, A. (2020). Development of agricultural e-commerce in Thailand. *The FFTC Journal of Agricultural Policy*, *1*, 7–17.

Limpopo Department of Agriculture (LDA). 2008. Basic principles: Policy document. Available online at: www.lda.gov.za

Little, K. (2008). Lepcha narratives of their threatened sacred landscapes. *Transforming Cultures eJournal, 3*(1).

Little, K. (2010). From the villages to the cities: The battlegrounds for Lepcha protests. Transforming Cultures eJournal, 5(1). DOI: 10.5130/tfc.v5i1.1579

Loch, T. K., & Riechers, M. (2021). Integrating indigenous and local knowledge in management and research on coastal ecosystems in the Global South: A literature review. *Ocean and Coastal Management*, *212*, 105821. DOI: 10.1016/j.ocecoaman.2021.105821

Loffler, L., & Loffler, P. (2005). *Swaziland tree atlas - including selected shrubs and climbers.* (Southern African Botanical Diversity Network Report No. 38). SABONET.

Lohento, K., & Ajilore, O. (2015). ICT and Youth in Agriculture. *Africa agriculture status report*, 118-42.

Long, C. (2005). *Eswatini flora - Siswati names and uses*. http://eswatinibiodiversity.com/flora/clusagelist.asp?uid=2&pg=21

Lopez, L. (2013) La Imagen de Santiago de Compostela y del Camino en Italia. Una aproximación desde la geografía cultural. Universidad de Santiago de Compostela. Available at: http://dspace.usc.es/handle/10347/7111 (accessed 20 April 2013).

Luckett, K. (2001). A proposal for an epistemically diverse curriculum for South African higher education in the 21st Century. *South African Journal of Education*, *15*(2), 49–61.

Lukwa, A. T., Wayas, F. A., Lambert, E. V., & Alaba, O. (2024). Stakeholder mapping to explore social and economic capital of Rotating Savings and Credit Associations (ROSCAs) to increase demand for and access to healthy food. *Frontiers in Sustainable Food Systems*, *8*, 1301578. DOI: 10.3389/fsufs.2024.1301578

Lut, M., Linkova, M., & Strid, S. (2023). Theorising gender-based violence policies: A 7P Framework. *Social Sciences (Basel, Switzerland)*, *12*(7), 385. DOI: 10.3390/socsci12070385

Lutoombi, M. (2015). *Cattle Husbandry and trade in Bulozi, a historical perspective C. 1880-1973* (Doctoral dissertation, University of Zambia).

Luvuno, L. B., Kotze, D. C., & Kirkman, K. P. (2016). Long-term landscape changes in vegetation structure: Fire management in the wetlands of KwaMbonambi, South Africa. *African Journal of Aquatic Science*, *41*(3), 279–288. DOI: 10.2989/16085914.2016.1177482

Maasz, D., Winschiers-Theophilus, H., Stanley, C., Rodil, K., & Mbinge, U. (2018). A digital indigenous knowledge preservation framework: the 7C model—repositioning IK holders in the digitization of IK. *Digitisation of culture: Namibian and international perspectives*, 29-47.

Mabhaudhi, T., Chibarabada, T. P., Chimonyo, V. G. P., Murugani, V. G., Pereira, L. M., Sobratee, N., Govender, L., Slotow, R., & Modi, A. T. (2019). Mainstreaming Underutilized Indigenous and Traditional Crops into Food Systems: A South African Perspective. *Sustainability*, *11*(1), 172. DOI: 10.3390/su11010172 PMID: 37681213

Mabvurira, V., Matlakala, F. K., & Mashumba, A. (2021). Indigenous knowledge systems and environmental conservation: Exploring the role of Shona culture in Zimbabwe. *African Journal of Social Work*, *11*(2), 45–53.

Mabvurira, V., Muchinako, G. A., & Smit, E. I. (2021). Shona traditional religion and sustainable environmental management: An Afrocentric perspective. *African Journal of Social Work, 11*(3), 111–118.

Ma, C., Qirui, C., & Lv, Y. (2023). "One community at a time": Promoting community resilience in the face of natural hazards and public health challenges. *BMC Public Health, 23*(1), 2510. DOI: 10.1186/s12889-023-17458-x PMID: 38097956

Macedo, D., Semali, L., & Kincheloe, J. (1999). Decolonizing indigenous knowledge. *What is indigenous knowledge: Voices from the academy*, 134-176.

Machethe, C. L., Mollel, N. M., Ayisi, K., Mashatola, M. B., Anim, F. D. K., & Vanasche, F. (2004). *Smallholder irrigation and agricultural development in the Olifants River Basin of Limpopo Province: Management transfer, productivity, profitability, and food security issues. Report to the Water Research Commission on the Project "Sustainable Local Management of Smaller Irrigation in the Olifants River Basin of Limpopo Province."*. Water Research Commission.

Maditsi, M. E. (2018). *The role of traditional leaders in postapartheid local governance in South Africa: a case study of the Bapong 2 community, North West Province, South Africa* (Doctoral dissertation, North-West University (South Africa)).

Maditsi, M. E., & Bhuda, M. T. (2023). Decolonising and Humanising Pedagogies in South African Postgraduate Education: Lessons From Indigenous Knowledge Systems. In *Digital Preservation and Documentation of Global Indigenous Knowledge Systems* (pp. 282-302). IGI Global.

Maditsi, M. E., & Bhuda, M. T. (2023). Decolonising and Humanising Pedagogies in South African Postgraduate Education: Lessons from Indigenous Knowledge Systems. In Masenya, T. M. (Ed.), *Digital Preservation and Documentation of Global Indigenous Knowledge Systems* (pp. 282–302). IGI Global. DOI: 10.4018/978-1-6684-7024-4.ch015

Madonsela, B. S., Semenya, K., & Shale, K. (2024). A review of indigenous knowledge systems and their application in sustainable solid waste management. *WORLD (Oakland, Calif.)*, *5*(2), 219–239. DOI: 10.3390/world5020012

Madzivhandila, T. S. (2024). Integrating Indigenous Knowledge Systems (IKS) to Climate Change Adaptation in South Africa: Lessons from the "Past". *South African Journal of Higher Education, 38*(5), 159–168. DOI: 10.20853/38-5-6417

Maffi, L., & Dilts, O. (2014). *Biocultural diversity toolkit: An introduction to biocultural diversity*. Terralingua.

Mafukata, A. M. (2015). "Mythical leopards (Panthera Pardus) of the Vhavenda." International. *Journal of Information Research and Review, 2*(7), 836–852.

Magagula, B., & Tsvakirai, C. Z. (2020). Youth perceptions of agriculture: Influence of cognitive processes on participation in agripreneurship. *Development in Practice, 30*(2), 234–243. DOI: 10.1080/09614524.2019.1670138

Magesa, L. (1997). African Religion, USA: Orbis Books, Maryknoll.

Magocha, M., Soundy, P., Muchie, M., & Magocha, B. (2019). Reviewing the applications of indigenous knowledge systems (IKS) in innovative crop production. *Indilinga, 18*(2), 229–244.

Magongwa, L. (2010). Deaf Education in South Africa. *American Annals of the Deaf, 155*(4), 493–496. DOI: 10.1353/aad.2010.0042

Mahato, S., Mishra, A. K., & Debnath, S. (2020). Impact of rainwater harvesting structures on groundwater recharge potential in India. *International Journal of Current Microbiology and Applied Sciences, 9*(6), 66–74. DOI: 10.20546/ijcmas.2020.906.008

Mahlalela, L. S. (2014). *Economic valuation and natural resource rent as tools for wetland conservation in Swaziland: The case of Lawuba wetland* [Unpublished Master's dissertation]. University of Pretoria, Pretoria, South Africa.

Mahohoma, T. & Gundani, Paul H. (2020). Experiencing the sacred. UNISA Press. DOI: 10.25159/2412-4265/3363

Makaluza, N. (2018). *Storytelling for early literacy development in isiXhosa: A case study of a grad one class in the Western Cape.* (MA). University of the Western Cape, Makone, I. (2021). A good reading culture: Potential game changer in Africa. *African Books Collective.* Retrieved from https://www.readafricanbooks.com/opinion/a-good-reading-culture-potential-game-changer-in-africa/#:~:text=A%20reading%20culture%20is%20an,has%20a%20poor%20reading%20culture

Makamure, K., & Vengesai, E. (2015). Taboos as Avoidance Rules: The Karanga People's Perspective. In Zimbabwean Cultural Heritage Journal, 8(1), 7-15.

Makhura, T. M. (2001). Overcoming transaction costs barriers to market participation of smallholder farmers in the Northern Province of South Africa. Unpublished PhD thesis, Pretoria, University of Pretoria.

Makhura, M. N., & Wasike, W. S. (2003). Patterns of access to rural service infrastructure: The case of farming households in Limpopo Province. *Agrekon, 42*(2), 129–143. DOI: 10.1080/03031853.2003.9523616

Makokotlela, M. V., & Gumbo, M. T. (2025). Decolonisation of the curriculum through the integration of indigenous knowledge in higher education. *South African Journal of Higher Education, 39*(1), 237–254. DOI: 10.20853/39-1-6004

Makoni, B., Makoni, S., & Mashisi, P. (2007). Naming practices and language planning in Zimbabwe. *Current Issues in Language Planning, 8*(3), 437–467. DOI: 10.2167/cilp126.0

Malapane, O. L., Chanza, N., & Musakwa, W. (2024). Transmission of indigenous knowledge systems under changing landscapes within the vhavenda community, South Africa. *Environmental Science & Policy, 161*, 103861. DOI: 10.1016/j.envsci.2024.103861

Malhotra, N. (2024). Introduction to indigenous tribes. In Malhotra, N. (Ed.), *Sustainable pathways* (pp. 1–39). Emerald Publishing Limited., DOI: 10.1108/978-1-83549-490-520241003

Malik, A., Singh, V., & Singh, K. (2007). Occurrence and distribution of persistent trace organics in rainwater in an urban region (India). *Bulletin of Environmental Contamination and Toxicology, 79*(6), 639–645. DOI: 10.1007/s00128-007-9290-8 PMID: 17924039

Malli, A., Monteith, H., Hiscock, E. C., Smith, E. V., Fairman, K., Galloway, T., & Mashford-Pringle, A. (2023). Impacts of Colonization on Indigenous Food Systems in Canada and the United States: A scoping review. *BMC Public Health*, *23*(1), 2105. DOI: 10.1186/s12889-023-16997-7 PMID: 37885000

Maluni, J. K. (2020). Disposal of disposable child diapers by caregivers and their environmental health implications in Kenya: A review. *J Res Innovation Implications Educ*, *4*(1), 22–28.

Mandillah, K. L., & Ekosse, G. I. (2018). African totems: Cultural heritage for sustainable environmental conservation. *Conservation Science in Cultural Heritage*, *18*, 201–218.

Manion, J. D. (2005). The expression of happiness in naming practices. *Journal of Onomastics Research*, *12*(1), 45–62.

Mannheim, K. (1952). The problem of generations: Essays of the Sociology of Knowledge. In Kecskemeti, P. (Ed.), *Essays* (pp. 276–322). Oxford University Press.

Manthwa, A. (2024). The transformative nature of indigenous knowledge systems to address gender inequality. *Journal of Law. Social Development*. Advance online publication. DOI: 10.25159/2520-9515/14759

Manyatsi, A. M. (1999). Development of a computerised rangelands resource information system for Swaziland. *University of Swaziland Journal of Agriculture*, *8*(1), 12–22. DOI: 10.4314/uniswa.v8i1.4612

Manyatsi, A. M. (2011). Application of Indigenous Knowledge Systems in Hydrological Disaster Management in Swaziland. *Current Research Journal of Social Sciences*, *3*(4), 353–357.

Manyatsi, A. M., & Hlophe, E. T. (2010). Contribution of sale of firewood towards rural livelihood in Swaziland, and its environmental sustainability. *Current Research Journal of Social Sciences*, *2*(4), 226–232.

Manyele, S. V., Kivela, M., & Maziba, W. (2018). Environmental health risks associated with waste disposal in rural Tanzania. *Journal of Environmental Science and Health.*

Manzungu, E. (1999). *Strategies of Smallholder Irrigation Management in Zimbabwe*. University of Zimbabwe Publications. DOI: 10.18174/192721

Mapara, J. (2009). 'Indigenous Knowledge Systems in Zimbabwe: Juxtaposing Postcolonial Theory'. *The Journal of Pan African Studies*, *3*(1), 139–155.

Maponya, P. (2021). Opportunities and constraints faced by smallholder farmers in the Vhembe District, Limpopo Province in South Africa. *Circular Economy and Sustainability*, *1*(4), 1387–1400. DOI: 10.1007/s43615-021-00028-x

Marcus, T., Eales, K., & Wildschut, A. (1996). *Down to earth: land demand in the new South Africa. Land and Agriculture Policy Centre*. Indicator Press.

Marerro, A., & Mattei, J. (2022). Reclaiming Traditional, Plant-based, Climate-resilient Food System in Small Islands. *The Lancet. Planetary Health*, *6*(2), 2. Retrieved February 15, 2025, from https://www.thelancet.com/journals/lanplh/article/PIIS2542-5196(21)00322-3/fulltext. DOI: 10.1016/S2542-5196(21)00322-3

Marjorie, L. (2010, February 8). *Language Policy and Oppression in South Africa.* Retrieved from Cultural Survival: https://www.culturalsurvival.org/publications/cultural-survival-quarterly/language-policy-and-oppression-south-africa

Marker, M. (2019). Indigenous knowledges, universities, and alluvial zones of paradigm change. *Discourse (Abingdon)*, *40*(4), 500–513. DOI: 10.1080/01596306.2017.1393398

Marley, S. A., Siani, A., & Sims, S. (2022). Real-life research projects improve student engagement and provide reliable data for academics. *Ecology and Evolution*, *12*(12), e9593. Advance online publication. DOI: 10.1002/ece3.9593 PMID: 36514554

Marques, B., Freeman, C., & Carter, L. (2021). Adapting traditional healing values and beliefs into therapeutic cultural environments for health and well-being. *International Journal of Environmental Research and Public Health*, *19*(1), 426.

Marsandi, F., Sutadji, E., Kuntadi, I., Rizal, F., Rahma, A. B. N., & Fajri, H. (2025). Integrating Ethnobotany and Indigenous Knowledge into Higher Education Curricula: Insights from a Global Biobliometric Analysis. *Ethnobotany Research and Applications*, *30*, 1–12. DOI: 10.32859/era.30.17.1-12

Martin, D. W. (2023). Biodiversity conservation. In *Dictionary of Ecological Economics* (pp. 31–31). Edward Elgar Publishing. DOI: 10.4337/9781788974912.B.15

Martinez-Vargas, C. (2020). Decolonizing higher education research: From a university to a pluri-versity of approaches. *South African Journal of Higher Education*, *34*(2), 112–128. DOI: 10.20853/34-2-3530

Martinez-Vargas, C., Ndou, U. E., Makhanikhe, J., & Madima, T. (2025). Decoloniality, identity and the third space: The case of students' experiences of well-being from an indigenous knowledge system university programme in South Africa. *International Journal of Educational Research*, *130*, 102547. DOI: 10.1016/j.ijer.2025.102547

Martin, J. (1999). Mentoring semogenesis: 'Genre-based' literacy pedagogy. In Christie, F. (Ed.), *Pedagogy and the Shaping of Consciousness: Linguistic and Social Processes* (pp. 123–155). Continuum.

Maru, Y., Gebrekirstos, A., & Haile, G. (2023). Indigenous sacred forests as a tool for climate change mitigation: Lessons from Gedeo community. *Southern Ethiopia. J. Sust.Journal of Sustainable Forestry*, *42*(3), 260–287. DOI: 10.1080/10549811.2021.2007490

Masango, L. P. (2008). Reading the Swazi reed dance (Umhlanga) as a literary traditional performance art [Unpublished Doctoral dissertation]. University of the Witwatersrand, Johannesburg, South Africa.

Masango, C. A. (2010). Indigenous traditional knowledge protection: Prospects in South Africa's intellectual property framework? *South African Journal of Library and Information Science*, *76*(1), 74–80. DOI: 10.7553/76-1-88

Masekoameng, M. R., & Makwena, M. C. (2019). The Role of Indigenous Foods and Indigenous Knowledge Systems for Rural Households' Food Security in Sekhukhune District, Limpopo Province, South Africa. *Journal of Consumer Sciences, Special Edition Food and nutrition challenges in Southern Africa*, 4. https://www.ajol.info/index.php/jfecs/article/view/191563. Accessed: 15 Feb 2025

Masenya, T. M. (2021). Trustworthiness of institutional repositories in academic libraries in South Africa. *Library Philosophy and Practice*, 1-36.

Masenya, T. M. (2024). Reterritorialization and Digital Preservation of Indigenous Knowledge Systems: Returning South African Indigenous Communities to Their Cultural Roots. In *Inclusivity and Indigeneity in Education for Sustainable Development* (pp. 64-77). IGI Global.

Masenya, T. M. (2022). Decolonization of indigenous knowledge systems in South Africa: Impact of policy and protocols. [IJKM]. *International Journal of Knowledge Management*, *18*(1), 1–22. DOI: 10.4018/IJKM.310005

Masenya, T. M. (2022b). Digital preservation of indigenous knowledge in South African rural communities. In *Handbook of research on protecting and managing global indigenous knowledge systems* (pp. 317–340). IGI Global. DOI: 10.4018/978-1-7998-7492-8.ch017

Masenya, T. M. (2023). Revitalization and Digital Preservation of Indigenous Knowledge Systems for Sustainable Development of Indigenous Communities in South Africa. *The Serials Librarian*, *84*(5-8), 86–102. DOI: 10.1080/0361526X.2023.2277962

Masenya, T. M., & Ngulube, P. (2019). Digital preservation practices in academic libraries in South Africa in the wake of the digital revolution. *South African Journal of Information Management*, *21*(1), 1–9. DOI: 10.4102/sajim.v21i1.1011

Mashego, S. L., Maditsi, M. E., & Bhuda, M. T. (2021). A response to Covid-19: Perspectives from African indigenous knowledge systems. *Gender & Behaviour*, *19*(1), 17242–17254.

Mashego, S. L., Maditsi, M. E., & Bhuda, M. T. (2021). A Response to Covid-19: Perspectives From African Indigenous Knowledge Systems. *Gender & Behaviour*, *19*(1), 17242–17254.

Mashele, N. M., Thompson, L. J., & Downs, C. T. (2021). Traditional health practitioners' and other community members' perceptions of vultures in the Kruger to Canyons biosphere region, South Africa. *The Journal of Raptor Research*, *55*(3), 340–358.

Masiphephe Network. (2024). *Indigenous knowledge systems and gender-based violence in South Africa*. Retrieved November 10, 2024, from https://www.masiphephe.org.za/wp-content/uploads/2024/04/Perspective-2-Indeginous-Knowledge-systems-.pdf

Masoga, M. (2022). Storytelling Research: Implications for the broader Higher Education System in South Africa. *Southern African Journal for Folklore Studies, 32*(2), 14 pages-14 pages.

Masoga, M. (2005). South African research in indigenous knowledge systems and challenges of change. *Indilinga*, *4*(1), 15–30.

Masoka, X. (2022). *Social Media used by selected stokvels to engage with stakeholders in South Africa* (Doctoral dissertation, University of Johannesburg).

Matarira, C. H., Makadho, J. M., & Mukaratirwa, S. (1992). *Drought and Sustainable Development in Zimbabwe*. University of Zimbabwe Publications.

Materechera, F., & Scholes, M. C. (2021). Characterization of farming systems using land as a driver of production and sustainability in the Vhembe District, Limpopo, South Africa. *Agricultural Sciences, 12*(11), 1352–1373. DOI: 10.4236/as.2021.1211087

Materechera, F., & Scholes, M. C. (2022). Understanding the Drivers of Production in South African Farming Systems: A Case Study of the Vhembe District, Limpopo South Africa. *Frontiers in Sustainable Food Systems, 6*, 722344. DOI: 10.3389/fsufs.2022.722344

Matfunjwa, M. N. (2022). *A pragmatic exploration of Siswati naming practices* [Unpublished Doctoral dissertation]. University of South Africa, Pretoria, South Africa.

Mathinya, V. N., Franke, A. C., Van De Ven, G. W. J., & Giller, K. E. (2022). Productivity and constraints of small-scale crop farming in the summer rainfall region of South Africa. *Outlook on Agriculture, 51*(2), 139–154. DOI: 10.1177/00307270221091839

Mathur, P., & Chowdhary, N. R. (1999). ENVIRONMENTAL DEGRADATION AT PUSHKAR VALLEY (INDIA): RESTRAIN TOURISM OR PROMOTE TOURISM? *Biologica, 37*, 117–123.

Matike, E. (2012). Knowledge and perceptions of educators and learners on the incorporation of indigenous knowledge into the school curriculum. *African indigenous knowledge systems and sustainable development*, 169-184.

Matlala, E. (2019). Long-term preservation of digital records at the University of KwaZulu-Natal archives. *Journal of the South African Society of Archivists, 52*, 95–109.

Matobo National Park. (n.d.). In Wikipedia. Retrieved March 13, 2025, from https://en.wikipedia.org/wiki/Matobo_National_Park

Matowanyika, J. Z. Z. (1991). Indigenous resource management and sustainability in Zimbabwe: Towards a political ecology of environmental management. *Environment and Urbanization, 3*(2), 32–48. DOI: 10.1177/095624789100300204

Matsekoleng, T. K., Mapotse, T. A., & Gumbo, M. T. (2024). The role of indigenous games in education: A Technology and Environmental Education perspective. *Diaspora, Indigenous, and Minority Education, 16*(4), 68–82. DOI: 10.1080/15595692.2022.2160436

Maureen, I., van der Meij, H., & de Jong, T. (2020). Enhancing storytelling activities to support early (digital) literacy development in early childhood education. *International Journal of Early Childhood, 52*(1), 55–76. DOI: 10.1007/s13158-020-00263-7

Mavhunga, C. (2018). Family Dynamics and the Role of Women in Zimbabwean Society. *African Journal of Gender and Development, 6*(1), 45–60.

Mavhunga, C. C. (2017). *Transient workspaces: Technologies of everyday innovation in Zimbabwe.* MIT Press.

Mawere, M. (2013a). *Culture, indigenous knowledge and development in Africa: Reviving interconnections for sustainable development.* Langaa RPCIG.

Mawere, M. (2013b). Traditional Ecological Knowledge and Its Use in Modern Conservation: The Case of the Shona of Zimbabwe. *International Journal of Modern Anthropology*, *1*(7), 8–15.

Mawere, M. (2015). African environmental ethics, sustainable development and the future of humanity. *Journal of Environmental Science and Health. Part B, Pesticides, Food Contaminants, and Agricultural Wastes*, *50*, 1–10.

Mawere, M. (2015). *Humans, other beings and the environment: Harurwa (edible stinkbugs) and environmental conservation in southeastern Zimbabwe*. Cambridge Scholars Publishing.

Mbatha, C. N. (2021). Livestock production and marketing for small emerging farmers in South Africa and Kenya: Comparative lessons. *South African Journal of Agricultural Extension*, *49*(1), 141–161. DOI: 10.17159/2413-3221/2021/v49n1a10783

Mbhenyane, X. G., & Tambe, A. B. (2024). The Influence of Household and Community Food Environments on Food Insecurity in Limpopo Province, South Africa. *International journal of environmental research and public health,* 21(2), 125. https://pubmed.ncbi.nlm.nih.gov/38397616/ Accessed: 15 Feb 2025

Mbigi, L., & Maree, J. (2005). *Ubuntu: The spirit of African transformation man agement*. Knowledge Resources.

Mbiti, J. S. (1969). *African Religion and Philosophy, London.Ibadan*. Heinemann.

McCormick, R. (2004). Issues of learning and knowledge in Technology education. *International Journal of Technology and Design Education*, *14*(1), 21–44. DOI: 10.1023/B:ITDE.0000007359.81781.7c

McCune, L. M., Nuvayestewa, V., & Kuhnlein, H. V. (2019). Why and how to document Traditional Food Systems in your community: Report from breakout discussions at the 2017 Native American Nutrition Conference. *Current Developments in Nutrition*, *3*(2), 19–24. Retrieved February 15, 2025, from https://www.sciencedirect.com/science/article/pii/S2475299122136954. DOI: 10.1093/cdn/nzy089

McGregor, D. (2014). Lessons for Collaboration Involving Traditional Knowledge and Environmental Governance in Ontario, Canada. *Alternative*, *10*(4), 340–353. DOI: 10.1177/117718011401000403

McMichael, P. M. S., & Schneider, M. (2011). Food security politics and the Millennium development goals. *Third World Quarterly*, *32*(1), 119–139. DOI: 10.1080/01436597.2011.543818 PMID: 21591303

McMillan, J. H., & Schumacher, S. (2010). *Research in education: Evidence-based enquiry* (8th ed.). Pearson Education.

Mdhluli, T. D., Mokgoatšana, S., Kugara, S. L., & Vuma, L. (2021). 'Knowledge management: Preserving, managing and sharing indigenous knowledge through digital library'. *Hervormde Teologiese Studies*, *77*(2), a6795. Retrieved February 15, 2025, from https://hts.org.za/index.php/hts/article/view/6795/21040. DOI: 10.4102/hts.v77i2.6795

Meeran, S., Kodisang, S. M., Moila, M. M., Davids, M. N., & Makokotlela, M. V. (2024). *Ethnomathematics in intermediate phaser: Reflections on the Morabaraba game as indigenous mathematical knowledge*. African Journal of Research in Mathematics, Science and Technology Education., DOI: 10.1080/18117295.2024.2340095

Mehta, K., Alter, T. R., Semali, L. M., & Maretzki, A. (2022). AcademIK connections: Bringing indigenous knowledge and perspectives into the classroom. *Journal of Community Engagement and Scholarship*, *6*(2). Advance online publication. DOI: 10.54656/XLUX1060

Mekoa, I. (2020). The Politics and Nuances of Language in South Africa: A Critical Appraisal. *Journal of African Languages and Literary Studies, 1 (1)*, Sabinet Online.

Mekoa, I. (2015). Challenges and opportunities for IKS in higher education institutions in South Africa: Politics, ideological, institutional cultures and structural dimensions. *Indilinga*, *14*(1), 12–24.

Mendoza, B. (2020). Decolonial theories in comparison. *Journal of World Philosophies*, *5*(1), 43–60.

Menze, A., & Tsibolane, P. (2019). Online Stokvels: the use of social media by the marginalized.

Metz, T., & Molefe, M. (2021). Traditional African religion as a neglected form of monotheism. *The Monist*, *104*(3), 393–409.

Michael, Y.Yohannes Gebre Michael. (2015). Conflicts of indigenous farming practices and modern agriculture in konso community, southern Ethiopia. *Journal of Agricultural Science and Technology B*, *5*(9). Advance online publication. DOI: 10.17265/2161-6264/2015.09.001

Mics, F. (2024). The sacral significance of forests and trees in the religions, philosophies, and spirituality. *Opuscula Theologica et Scientifica*, *2*(2), 23–65. DOI: 10.59531/ots.2024.2.2.23-65

Mignolo, W. D. (2011). I am where I think: Remapping the order of knowing. In Lionnet, F., & Shih, S. (Eds.), *The creolization of theory* (pp. 159–192). Duke University Press.

Miles, M. B., & Huberman, A. M. (1994). *Qualitative data analysis*. Sage.

Miller, T. R., Baird, T. D., Littlefield, C. M., Kofinas, G., Chapin, F. S.III, & Redman, C. L. (2008). Epistemological pluralism: Reorganizing interdisciplinary research. *Ecology and Society*, *13*(2), art46. DOI: 10.5751/ES-02671-130246

Ministry of Tourism, Government of India. (2020). *Incredible India: Cultural festivals and heritage tourism*. https://tourism.gov.in

Ministry of Tourism, Government of India. (2020). Sustainable tourism guidelines for cultural destinations. https://tourism.gov.in

Mishra, P., Rai, A., Abdelrahman, K., Chand, S., & Tiwari, A. (2021). Analyzing challenges and strategies in land productivity in Sikkim Himalaya, India. *Sustainability (Basel)*, *13*(19), 11112. DOI: 10.3390/su131911112

Mishra, S., & Kumar, R. (2022). Post-COVID tourism recovery and destination resilience: A case of Pushkar, Rajasthan. *Journal of Tourism and Sustainability*, *4*(1), 44–56.

Mitchell, T. M. (1997). *Machine learning*. McGraw-Hill. Available at: https://www.cs.cmu.edu/~tom/mlbook.html

Mkhize, N. (2019). The importance of extended family in African societies: A focus on aunts and uncles. *African Journal of Social Work*, *9*(1), 45–58.

Mkhize, N., & Ndimande-Hlongwa, N. (2014). African languages, indigenous knowledge systems (IKS), and the transformation of the humanities and social sciences in higher education. *Alternation (Durban), 21*(2), 10–37.

Mkhize, N., & Reddy, P. (2020). Cultural Transmission and Mediation in African Families: The Role of Aunts. In African. *Journal of Family Studies*, *5*(1), 45–58.

Mkhwebane, L. N. (2024). Life sciences teachers' integration of indigenous knowledge: A vision for making science classrooms culturally responsive. *EURASIA Journal of Mathematics, Science and Technology Education, 20*(8), em2483.\ https://doi.org/DOI: 10.29333/ejmste/14859

Mkosi, N. N., Mavuso, M. P., & Olawumi, K. B. (2023). Using Ubuntu values in integrating African indigenous knowledge into teaching and learning: A review of literature. *International Journal of Learning. Teaching and Educational Research*, *22*(5), 140–159. DOI: 10.26803/ijlter.22.5.7

Mlotshwa, N. P., & Tsakeni, M. (2024). Exploring Indigenous Knowledge Systems Integrated in Natural Sciences Practical Work in Selected South African Schools. *International Journal of Learning. Teaching and Educational Research*, *23*(9), 383–400.

Mlotshwa, N., & Tsakeni, M. (2024). Challenges Experienced by Teachers when Integrating Indigenous Knowledge Systems in Natural Sciences Practical Work. *African Journal of Research in Mathematics. Science and Technology Education*, *28*(2), 300–313.

Mmbengwa, V., Nyhodo, B., Myeki, L., Ngethu, X., & van Schalkwyk, H. (2015). Communal livestock farming in South Africa: Does this farming system create jobs for poverty stricken rural areas. *Sylwan*, *159*(10), 176–192.

Mngomezulu, T., Mkhonza, L., & Nzima, D. (2019). Causes of men abuse in the Province of Kwazulu-Natal, South Africa. *Gender & Behaviour*, *17*(4), 14327–14339.

Moahi, K. (2012). Promoting African indigenous knowledge in the knowledge economy: Exploring the role of higher education and libraries. *Aslib Proceedings*, *64*(5), 540–554. DOI: 10.1108/00012531211263157

Moalosi, R. (2007). The impact of socio-cultural factors upon human-centred design in Botswana. Unpublished doctoral thesis. Queensland University of Technology.

Moalosi, R., Popovic, V., & Hickling-Hudson, A. (2007). Product analysis based on Botswana's post-colonial socio-cultural perspective. *International Journal of Design*, *1*(2), 35–43.

Mohan, D. (2024). Wisdom of the ancestors: Harnessing indigenous knowledge for contemporary challenges. *International Journal of Applied Research*, *10*(3), 112–119. DOI: 10.22271/allresearch.2024.v10.i3b.11591

Mohanty, P., & Choudhury, R. (2022). Events tourism in the eye of the COVID-19 storm: Impacts and implications. In *Event Tourism in Asian Countries* (pp. 97–114). Apple Academic Press. DOI: 10.1201/9781003161134-6

Moichela, K. Z. (2017). Integration of indigenous knowledge systems in the curriculum for basic education: possible experiences of Canada. Unpublished doctoral dissertation, University of South Africa, Pretoria

Mojapelo, S. (2020). Storytelling Initiatives and reading in schools in South Africa: A case from Limpopo Province. *Mousaion*, *38*(3). Advance online publication. DOI: 10.25159/2663-659X/7873

Mokhesi, T., & Modjadji, P. (2022). *Usage of traditional, complementary and alternative medicine and related factors among patients receiving healthcare in Lesotho*. Retrieved June 14, 2024, from https://purerims.smu.ac.za/en/publications/usage-of-traditional-complementary-and-alternative-medicine-and-r

Mokhutso, R. J. (2021). The impact of African indigenous knowledge system on healthcare system in South Africa: The Covid-19 perspective. *Gender & Behaviour*, *19*(1), 17713–17728.

Mokomane, Z. (2018). Indigenous social security systems in Botswana. *Indigenous Social Security Systems in Southern and West Africa, 13*.

Molise, H. V. (2025). Teachers' pedagogical knowledge of integrating indigenous knowledge systems in economics education curriculum. *Interdisciplinary Journal of Education Research*, *7*(1), a04–a04. DOI: 10.38140/ijer-2025.vol7.1.04

Moller, H., Berkes, F., Lyver, P. O. B., & Kislalioglu, M. (2004). Combining Science and Traditional Ecological Knowledge: Monitoring Populations for Co-management. *Ecology and Society*, *9*(3), 2. DOI: 10.5751/ES-00675-090302

Moloi, M. J. (2010). *A comparison of socioeconomic characteristics that determine the farm income of emerging lifestock and horticultural farmers in South Africa* (Doctoral dissertation).

Molommu, S. P. (2018). Traditional Beliefs and Practices of the Lepchas in Eastern India. *Asian Man (The)-. International Journal (Toronto, Ont.)*, *12*(2), 149–160.

Monde, L. (2024, September 2). Au Bénin, ONG et divinités vaudoues au chevet des précieuses mangroves.

Monde, L. (2024, September 3). Benin's voodoo deities take care of precious mangroves. Le Monde.

Monyatsiwa, C., Ferguson, D., & Phumaphi, N. (2013, May). *Bridging the IT divide—Case study on the Botswana livestock industry. In 2013 IST-Africa Conference & Exhibition*. IEEE.

Mooketsi, B. E. (2013). Optimisation of livestock identification and trace-back system LITS database to meet local needs: Case study of Botswana. *The Journal of Community Informatics*, *9*(4). Advance online publication. DOI: 10.15353/joci.v9i4.3143

Moores, D. (2010). Partners in Progress: The 21st International Congress on Education of the Deaf and the Repudiation of the 1880 Congress of Milan. *Gallaudet University Press 155(3)*, 309-310.

Morcom, L., & Freeman, K. (2018). Niinwi-Kiinwa-Kiinwi: Building non-Indigenous allies in education through Indigenous pedagogy. *Canadian Journal of Education/Revue canadienne de l'éducation, 41*(3), 808-833.

Moremoholo, T. P. (2023). The role of culture in shaping the curriculum of higher education in South Africa. *Journal of Curriculum Studies Research*, *5*(2), 37–55. DOI: 10.46303/jcsr.2023.17

Morepje, M. T., Sithole, M. Z., Msweli, N. S., & Agholor, A. I. (2024). The Influence of E-Commerce Platforms on Sustainable Agriculture Practices among Smallholder Farmers in Sub-Saharan Africa. *Sustainability (Basel)*, *16*(15), 6496. DOI: 10.3390/su16156496

Morgan, R., & Van der Spuy, A. (2017, March 1). *A degree with a difference: using South African sign language instead of the written word.* Retrieved from The Conversation: https://theconversation.com/a-degree-with-a-difference-using-south-african-sign-language-instead-of-the-written-word-72899

Morrison, D. (2008). BC Food System Network Working Group on Indigenous Food Sovereignty: Final Activity Report (Prepared for: Provincial Health Services Authority-Community Food Action Initiative, Interior Health-Community Food Action Initiative &the B.C. Food Systems Network-Working Group on Indigenous Food Sovereignty.

Moscardo, G. (2007). Analysing the role of festivals and events in regional development. *Event Management, 11*(1-2), 23–32. DOI: 10.3727/152599508783943255

Mosimege, M., & Nkopodi, N. (2017). The use of indigenous game of Morabaraba in mathematics education. In M. T. Gumbo & V. Msila (Eds.). African voices on indigenization of the curriculum: Insights from practice, pp. 155-180. Westville: Reach Publishers (Self-Published).

Mosimege, M. (2004). Indigenous knowledge systems in South Africa: perspectives from the Department of Science and Technology: news and views. *Indilinga, 3*(1), 78–84.

Mosimege, M. (2005). National priorities in indigenous knowledge systems: Implications for research and curriculum development. *Indilinga, 4*(1), 31–37.

Mosimege, M. (2005). National priorities in Indigenous Knowledge Systems: Implications for research and curriculum development. *Indilinga, 4*(1), 31–37.

Mosimege, M. (2020). The use of indigenous games in the teaching and learning of Mathematics. *Revemop, 2*, e202009–e202009. DOI: 10.33532/revemop.e202009

Motala, S., & Menon, K. (2023). In search of the holy grail: Social justice and funding in higher education in South Africa. *Southern African Review of Education with Education with Production, 28*(1), 32–51.

Motsa, Z. (2004). *Khulumani Sive*. Macmillan.

Motsoaledi, P. A. (2003). Supplement to Nufarmer & African Entrepreneur (March 8, 2003, Vol. 2). Press release.

Moyo, T. (2012). Emotional Support and Family Dynamics in African Contexts. In African Family Dynamics: Perspectives and Challenges (pp. 45-60). African Books Collective.

Moyo, S., O'Keefe, P., & Sill, M. (1993). *The Southern African Environment: Profiles of the SADC Countries.* Earthscan Publications Ltd.

Moyo, T. (2019). Cultural Practices and Family Structures in Zimbabwe: The Case of the Shona People. *Zimbabwean Journal of Sociology, 12*(3), 78–92.

Mpandeli, S., Simalenga, T., Siambi, M., Ramugondo, R., Mailula, N., & Liphadzi, K. (2008). Constraints and challenges to agricultural development in Limpopo Province, South Africa. *CGIAR Challenge Program on Water and Food*, 138.

Mpandeli, S., Nesamvuni, E., & Maponya, P. (2015). Adapting to the impacts of drought by smallholder farmers in Sekhukhune District in Limpopo Province, South Africa. *Journal of Agricultural Science (Toronto)*, *7*(2), 115. DOI: 10.5539/jas.v7n2p115

Mpani, P., & Nsibande, N. (2015). *Gender-based violence*. April 21, 2025. https://www.soulcity.org.za/projects/advocacy/gbv/resources/understanding-gender-policy-and-gender-based-violence-in-south-africa-a-literature-review

Msibi, T. (2002). Incwadzi Lengenakheli. Manzini: Macmillan.

Msila, V. (2025). Transforming university through Africanised research: In search of the indigenous. *African Identities*, •••, 1–14. DOI: 10.1080/14725843.2024.2439440

Mtumane, Z. (2014). The Significance of Ancestors as a Cultural Component in M. Mbambo's *Amakroza* (1992). *Southern Africa Journal for Folklore Studies.* Volume 24(1), pp 22 to 39.

Mubangizi, J., & Kaya, H. (2015). African indigenous knowledge systems and human rights: Implications for higher education, based on the South African experience. *International journal of African renaissance studies-multi-, inter-and transdisciplinarity, 10*(2), 125-142.

Mubaya, C. P. (2020). Malindidzimu Hill: A sacred site and its role in Shona cosmology. *Zimbabwean Journal of Cultural Studies*, *5*(1), 15–22.

Mudaly, R., & Ismail, R. (2013). Teacher learning through tapping into indigenous knowledge systems in the science classroom. *Alternation (Durban)*, *20*(1), 178–202.

Mudau, L. S., Thumbathi, A. M. D., & Nkosi, D. V. (2023). Poor disposal of soiled diapers in Sub-Saharan Africa: A gap in environmental health literacy in rural communities. In *Environmental Health Literacy Update-New Evidence, Methodologies and Perspectives*. IntechOpen.

Mudau, T. J., Matshidze, P., & Nthane, T. (2024). Investigating African indigenous ways to mitigate gender-based violence in Bushbuckridge of Mpumalanga Province in South Africa. [ERATS]. *E-Journal of Religious and Theological Studies*, *10*(6), 130–142. DOI: 10.38159/erats.202410613

Mudzengi, B. K., Gandiwa, E., Muboko, N., Mutanga, C. N., & Chiutsi, S. (2021). Ecotourism resilience: The case of Mahenye community project, Chipinge. District, Zimbabwe. *African Journal of Hospitality, Tourism and Leisure*, *10*(2), 459–471. DOI: 10.46222/ajhtl.19770720.111

Muguti, T., & Mapira, J. (2014). Indigenous knowledge systems and their role in environmental conservation in Zimbabwe. *Journal of Environmental Science and Water Resources*, *3*(8), 207–211.

Mukuka, S. (2019). *A quest for embracing indigenous knowledge systems in the united church of Zambia: Pastoral care and imbusa, submitted in fulfilment of the degree requirement for the degree of PhD, in practical theology*. University of KwaZulu-Natal, Pietermaritzburg.

Mukuka, S. (2022). Affirming the traditional teaching of imbusa as a tool against gender-based violence. *Journal of International Women's Studies,* 24(4), (journal). Retrieved November 20, 2024, from https://vc.bridgew.edu/jiws/vol24/iss4/4

Muller, J. (2009). Forms of knowledge and curriculum coherence. *Journal of Education and Work, 22*(3), 205–226. DOI: 10.1080/13639080902957905

Muller, J., & Gamble, J. (2010). Curriculum and structuralist sociology: The theory of codes and knowledge structures. In Petersen, P., Baker, E., & McGaw, B. (Eds.), *International Encyclopedia of Education* (Vol. 1, pp. 505–509). Elsevier. DOI: 10.1016/B978-0-08-044894-7.00090-7

Mulovhedzi, S. A., & Luhalima, T. R. (2023). The role and place of african indigenous knowledge systems in the analyses of inclusive education. In *Using African epistemologies in shaping inclusive education knowledge* (pp. 29–48). Springer. DOI: 10.1007/978-3-031-31115-4_3

Munjal, S., & Jauhari, V. (2015). Leveraging the cultural and economic value of fairs and festivals in India. *Worldwide Hospitality and Tourism Themes, 7*(4), 417–423. DOI: 10.1108/WHATT-03-2015-0006

Munn, Z., Peters, M. D., Stern, C., Tufanaru, C., McArthur, A., & Aromataris, E. (2018). Systematic review or scoping review? Guidance for authors when choosing between a systematic or scoping review approach. *BMC Medical Research Methodology, 18*(1), 1–7. DOI: 10.1186/s12874-018-0611-x PMID: 30453902

Muranko, Ż., Tassell, C., Zeeuw van der Laan, A., & Aurisicchio, M. (2021). Characterisation and environmental value proposition of reuse models for fast-moving consumer goods: Reusable packaging and products. *Sustainability (Basel), 13*(5), 2609. DOI: 10.3390/su13052609

Murovhi, A., Matshidze, P., Netshandama, V., & Klu, E. (2018). Traditional child rearing practices in Vhavenda families South Africa. Journal of Gender [JGIDA]. *Information and Development in Africa, 7*(1), 1–14.

Murwira, S., & Higgs, P. (2023). Afrocentric epistemic systems and higher education curriculum. The case of a university in Zimbabwe. *South African Journal of Higher Education, 37*(2), 283-297. / https://orcid.org/0000-0003-2563-4336

Musakwa, W., & Niekerk, A. V. (2013). Implications of land use change for the sustainability of urban areas: A case study of Stellenbosch, South Africa. *Cities (London, England), 32*, 143–156. DOI: 10.1016/j.cities.2013.01.004

Musakwa, W., & Wang, S. (2018). Landscape change and its drivers: A Southern African perspective. *Current Opinion in Environmental Sustainability, 33*, 80–86. DOI: 10.1016/j.cosust.2018.05.001

Musavengane, R., & Simatele, D. M. (2016). Community-based natural resource management: The role of social capital in collaborative environmental management of tribal resources in KwaZulu-Natal, South Africa. *Development Southern Africa, 33*(6), 806–821. DOI: 10.1080/0376835X.2016.1231054

Musoni, P. (2016). Contestation of "the holy places in the Zimbabwean Religious Landscape": A study of the Johane Masowe Chishanu yeNyenyedzi Church's sacred places. *Hervormde Teologiese Studies, 72*(1), a3269. DOI: 10.4102/hts.v72i1.3269

Muswazi, P. (2001). Indigenous knowledge management in Swaziland: Perspectives. *Information Development, 17*(4), 250–256. DOI: 10.1177/0266666014241079

Musyoki, A. (2012). The emerging policy for green economy and social development in Limpopo, South Africa (No. 8). UNRISD Occasional Paper: Social Dimensions of Green Economy and Sustainable Development.

Muyambo, F., Bahta, Y., & Jordaan, A. (2017). The role of indigenous knowledge in drought risk reduction: A case of communal farmers in South Africa. *Jàambá*, *9*(1). Advance online publication. DOI: 10.4102/jamba.v9i1.420 PMID: 29955345

Muzondi, N. (2014). The contribution of traditional belief systems in biodiversity conservation among the Ndau people of Nyagadza community in Chipinge (Doctoral dissertation).

Mwikamba, C. M. (1991). Search for an African identity. *Wajibu*, *6*(4), 6–10.

Myeza, M. C. (2013). *Ucwaningo lezincazelo zamagama emithi ngesiZulu, umqulu ongashicilelwe weqhuzu leMaster of Arts*. University of South Africa.

Myeza, M. C. (2017). *Imimoya yamathongo. Pinetown*. Pearl Publisher.

Mzimela, J. H., & Moyo, I. (2024). On the Efficacy of Indigenous Knowledge Systems in Responding to the COVID-19 Pandemic: Unsettling Coloniality. *International Journal of Environmental Research and Public Health*, *21*(6), 1–12. DOI: 10.3390/ijerph21060731 PMID: 38928977

Naicker, S. M. (1999). *Curriculum 2005: A space for all: An introduction to inclusive education*. Renaissance.

Naidoo, J. (2021). Integrating indigenous knowledge and culturally based activities in South African mathematics classrooms. *African Journal of Teacher Education*, *10*(2), 17–36. DOI: 10.21083/ajote.v10i2.6686

Naidu, M., &Muchono, W. (2019). Male Perceptions Toward the Cultural Practice of Chinamwali/Khomba Among the Mahenye. Oriental Anthropologist: A Bi-annual International Journal of the Science of Man, 19(1), 25–40. DOI: 10.1177/0972558X19835381

Nakata, M. (2007). *Disciplining the Savages: Savaging the Disciplines*. Aboriginal Studies Press.

Nann, S. (2021). How beliefs in traditional healers' impact on the use of allopathic medicine: In the case of indigenous snakebite in Eswatini. *PLoS Neglected Tropical Diseases*, *15*(9), e0009731. DOI: 10.1371/journal.pntd.0009731 PMID: 34499648

National Research Foundation. (2024). Centres of Excellence. https://www.nrf.ac.za/core-mandate-business-divisions/risa-directorates/research-chairs-and-centres-of-excellence-rcce/centres-of excellence/#:~:text=The%20Department%20of%20Science%20and,and%20institutions%20on%20long%2Dterm

Natshandama, O. V., & Nevhudoli, N. D. (2021). Creating intergenerational learning spaces: A collaboration between UNIVEN Community Engagement Programme and Dzomo la Mupo. [CriSTaL]. *Critical Studies in Teaching and Learning*, *9*(2), 39–63. DOI: 10.14426/cristal.v9i2.462

Ncube, B., Mupangwa, W., & French, A. (2018). Precision agriculture and food security in Africa. *Systems analysis approach for complex global challenges*, 159-178.

Ncube, B. (2018). Insights into indigenous knowledge strategies for coping and adapting to drought in agriculture: A Karoo scenario, South Africa. *Indilinga*, *17*(1), 92–108.

Ndaba, Z. F. (2022). The Covid-19 pandemic and cultural event (s) cancellation in South Africa: views of the royal reed dance in KwaNongoma, KwaZulu Natal (Doctoral dissertation, Cape Peninsula University of Technology).

Ndagala, D. K. (1990). Pastoralists and the state in Tanzania. *Nomadic Peoples*, •••, 51–64.

Ndhlovu, P. T., Asong, J. A., Omotayo, A. O., Otang-Mbeng, W., & Aremu, A. O. (2023). Ethnobotanical survey of medicinal plants used by indigenous knowledge holders to manage healthcare needs in children. *PLoS One*, *18*(3), e0282113. DOI: 10.1371/journal.pone.0282113 PMID: 36972257

Ndlovu, H. L. (2019). Joint worship ceremonies of Africanists and Christians in the kingdom of Swaziland. In C. Lewis & D. Cohn-Sherbok, (Eds.), *Interfaith worship and prayer: We must pray together* (pp 58-67). Jessica Kingsley Publishers.

Ndlovu-Gatsheni, S. J. (2018). *Epistemic freedom in Africa: Deprovincialization and decolonization.* Routledge. DOI: 10.4324/9780429492204

Ndlovu-Gatsheni, S. J. (2021). The cognitive empire, politics of knowledge, and African intellectual productions: Reflections on struggles for epistemic freedom and resurgence of decolonisation in the twenty-first century. *Third World Quarterly*, *42*(5), 882–901. DOI: 10.1080/01436597.2020.1775487

Ndlovu, W., Mwale, M., & Zuwarimwe, J. (2021). Challenges and critical success factors for rural agrarian reforms in limpopo province, south africa. AGRARIS. *Journal of Agribusiness and Rural Development Research*, *7*(2), 191–206. DOI: 10.18196/agraris.v7i2.11272

Ndove, T. S., Whitbread, A. M., Clark, R. A., & Pengelly, B. C. (2004). Identifying the factors that contribute to the successful adoption of improved farming practices in the smallholder sector of Limpopo Province, South Africa. *Tropical legumes for sustainable farming systems in Southern Africa and Australia*, (115), 146.

Ndwandwe, J. (2019, December 15). Incwala third phase indigenous sacredness. *Eswatini Observer.* http://new.observer.org.sz/details.php?id=11125

Neethirajan, S., & Kemp, B. (2021). Digital livestock farming. *Sensing and Bio-Sensing Research*, *32*, 100408. DOI: 10.1016/j.sbsr.2021.100408

Nel, P. (2006). Indigenous knowledge systems, local community and community in the making. *Indilinga*, *5*(2), 99–107.

Nemogá, G. R., Appasamy, A., & Romanow, C. A. (2022). Protecting indigenous and local knowledge through a biocultural diversity framework. *Journal of Environment & Development*, *31*(3), 223–252. DOI: 10.1177/10704965221104781

Nepstad, D., Schwartzman, S., Bamberger, B., Santilli, M., Ray, D., Schlesinger, P., Lefebvre, P., Alencar, A., Prinz, E., Fiske, G., & Rolla, A. (2006). Inhibition of Amazon deforestation and fire by parks and indigenous lands. *Conservation Biology*, *20*(1), 65–73. DOI: 10.1111/j.1523-1739.2006.00351.x PMID: 16909660

Nesamvuni, A. E., Oni, S. A., Odhiambo, J. J. O., & Nthakheni, N. D. (2003). *Agriculture is the cornerstone of the economy of the Limpopo Province. A study commissioned by the Economic cluster of the Limpopo Provincial Government under the Leadership of the Department of Agriculture. University of Venda for Science and Technology*. Thohoyandou.

Neufeld, H. T., Richmond, C. A. M., & Health Access Centre, S. O. A. (2017). Impacts of place and social spaces on traditional food systems in Southwestern Ontario. *International Journal of Indigenous Health*, *12*(1), 93–103. DOI: 10.18357/ijih112201716903

Neuman, W. L. (2014). *Social research methods: qualitative, quantitative approaches* (7th ed.). Pearson Education Limited.

Nevhudoli, N. D., & Netshandama, V. O. (2023). What do Bachelor of Indigenous Knowledge Systems graduates say about their curriculum? A qualitative tracer study at the University of Venda. *Journal of Curriculum Studies Research*, *5*(1), 141–158. DOI: 10.46303/jcsr.2023.11

Ngobeni, W., Maimane, J., & Rankhumise, M. (2020). The effect of limited sign language as barrier to teaching and learning among Deaf learners in South Africa. *South African Journal of Education*, *40*(2), 1–7. DOI: 10.15700/saje.v40n2a1735

Ngubane, S. (2019). Death and burial practices in contemporary Zulu culture, South Africa. *Death across cultures: Death and dying in non-western cultures*, 119-131.

Ngubane, H. (1977). *Body and Mind in Zulu Medicine, An Ethnography of health and disease in Nyuswa-Zulu thought and practice*. University of Edinburg.

Ngulube, P., Dube, L., & Mhlongo, M. (2015). Towards a cartography of indigenous knowledge systems in library and information science training and education in Anglophone eastern and southern Africa. *Indilinga*, *14*(2), 145–168.

Nhalevilo, E. Z. F. A. (2013). Rethinking the history of inclusion of IKS in school curricula: Endeavoring to legitimate the subject. *International Journal of Science and Mathematics Education*, *11*(1), 23–42. DOI: 10.1007/s10763-012-9382-8

Nhlabathi, M. C. (2024). Kulondvolotwa kwemasiko, imihambo, nemagugu ebuve lokutfolakala etindzabeni letimfishane teSiswati letikhetsiwe. (Unpublished Masters dissertation University of South Africa.)

Nie, Z. (2021). Ethical literary criticism: Oral literature and the formation mechanism of brain text. *Kritika Kultura*, *1*(37), 20.

Nkomazana, F. (2017). The impact of modernization on traditional African cultural practices. *Journal of Cultural Studies*, *19*(1), 1–12.

Nkondo, M. (2012). *Indigenous African knowledge systems in a polyepistemic world: The capabilities approach and the translatability of knowledge systems.* Paper presented at the Southern African Regional Colloquium on Indigenous African knowledge systems: methodologies and epistemologies for research, teaching, learning and community engagement in higher education. Howard College Campus: University Of KwaZulu-Natal.

Nkonge, D. K. (2014). The Relationship between Religion and Science in Mission: Reflection from an African Christian Perspective. Research on Humanities and Social Sciences, International Institute for Science, Technology and Education Journal, Vol.4, No.2

Nkopodi, N., & Mosimege, M. (2009). Incorporating the indigenous game of morabaraba in the learning of mathematics. *South African Journal of Education*, *29*(3), 377–392. DOI: 10.15700/saje.v29n3a273

Noble, H., & Heale, R. (2019). Triangulation in research, with examples. *Evidence-Based Nursing*, *22*(3), 67–68. DOI: 10.1136/ebnurs-2019-103145 PMID: 31201209

Nomlomo, V., & Sosibo, Z. (2016). Indigenous knowledge systems and early literacy development: An analysis of IsiXhosa and IsiZulu traditional children's folktales and songs. *Studies of Tribes and Tribals*, *14*(2), 110–120. DOI: 10.1080/0972639X.2016.11886738

Nonaka, I. (1994). A dynamic theory of organizational knowledge creation. *Organization Science*, *5*(1), 14–37. DOI: 10.1287/orsc.5.1.14

Ntloko, N. P., & Masenya, T. M. (2024). Enhancing access to electronic information through digital transformation in KwaZulu-Natal Department of health libraries in South Africa. In *Multidisciplinary Approach to Information Technology in Library and Information Science* (pp. 114–131). IGI Global Scientific Publishing.

Ntshangase, N. L., Muroyiwa, B., & Sibanda, M. (2018). Farmers' perceptions and factors influencing the adoption of no-till conservation agriculture by small-scale farmers in Zashuke, KwaZulu-Natal Province. *Sustainability (Basel)*, *10*(2), 555. DOI: 10.3390/su10020555

Ntshoe, I. M. (2020). Ontological social justice in decolonized and post-apartheid settings. *South African Journal of Higher Education*, *34*(3), 263–280. DOI: 10.20853/34-3-3283

Nwoye, A. (2017). An African Theory of Human Personhood: Psychology in Society, 42-66.

Nxumalo, G. (2014). Young African Women and the Umhlanga Ceremony: A Case Study in Historical Consciousness (Doctoral dissertation, University of KwaZulu-Natal, Durban).

Nxumalo, S. A., & Mncube, D. W. (2018). Using indigenous games and knowledge to decolonise the school curriculum: Ubuntu perspectives. *Perspectives in Education*, *36*(2), 103–118. DOI: 10.18820/2519593X/pie.v36i2.9

Nyahunzvi, D. K. (2015). Negotiating livelihoods among Chivi curio traders in a depressed Zimbabwe tourism trading environment. *Anatolia*, *26*(3), 397–407. DOI: 10.1080/13032917.2014.974065

Nyamapfene, K. (1991). *Soils of Zimbabwe*. Nehanda Publishers.

Nyikai, R. A. (2003). Commercial and subsistence Farming: What is the future for smallholder Kenyan agriculture? *African Crop Science Conference Proceedings* 6:591-596.

Nyoni, T. (2020). The Influence of Extended Family on Child Development in Zimbabwe. *Journal of Child and Family Studies*, *29*(4), 1023–1035.

Nyuyen, H. (2018). Sustainable food systems: Concept and framework. Food and Agriculture Organization of the United Nations. https://openknolwedge.fao.org/handle/20.500.14283/ca2079en. Accessed: 15 Feb 2025

O'Donoghue, R. (2004). Marrying indigenous and western science through environmental education. In School science in Africa: Learning to teach, teaching to learn (pp. 180-190). Lansdowne: Juta Gariep (Pty) Ltd

O'Madagain, C., & Tomasello, M. (2021). Joint attention to mental content and the social origin of reasoning. *Synthese*, *198*(5), 4057–4078. DOI: 10.1007/s11229-019-02327-1

O'Neill, B., Banoobhai, M., & Smith, C. (2016). Teaching literacy through dramatic storytelling in foundation phase. *Australasian Journal of Early Childhood*, *41*(2), 95–102. DOI: 10.1177/183693911604100213

Odhiambo, T., Kamp, J. R., & Kamp, R. (1990). You cannot fix indigenous knowledge. *Informationcentre for Low External Input Agriculture Newsletter*, •••, 11.

Odora Hoppers, C. (2021). Research on Indigenous knowledge systems: The search for cognitive justice. *International Journal of Lifelong Education*, *40*(4), 310–327. DOI: 10.1080/02601370.2021.1966109

Odora Hoppers, C. A. (2002). Indigenous knowledge and the integration of knowledge systems. In Odora Hoppers, C. A. (Ed.), *Indigenous knowledge and the integration of knowledge systems: Towards a philosophy of articulation* (pp. 2–23). New Africa Books.

Odora Hoppers, C. A. (2015). Think Piece: Cognitive Justice and Integration without Duress. The Future of Development Education—Perspectives from the South. *International Journal of Development Education and Global Learning*, *7*(2), 89–106. DOI: 10.18546/IJDEGL.07.2.08

Odora Hoppers, C. A. O. (2001). Indigenous knowledge systems and academic institutions in South Africa. *Perspectives in Education*, *19*(1), 73–86.

Odora Hoppers, C. A. O. (2023). Knowledge Production, Access and Governance: A Song from the South. In *Decolonizing Education for Sustainable Futures* (pp. 67–82). Bristol University Press.

Ogegbo, A. A., & Ramnarain, U. (2024). A systematic review of pedagogical practices for integrating indigenous knowledge systems in science teaching. *African Journal of Research in Mathematics. Science and Technology Education*, *28*(3), 343–361.

Ogundele, O. M., Rapheal, O. M., & Abiodun, A. M. (2018). Effects of municipal waste disposal methods on community health in Ibadan-Nigeria. *Polytechnica*, *1*(1), 61–72. DOI: 10.1007/s41050-018-0008-y

Ogunniyi, J. (2021). The Development Sustainability of Indigenous Technologies in Africa. In V. Gumede., M. Muchie & A. Shafi (Eds.), *Indigenous Systems and Africa's Development* (p. 37). Capital Pres Okrent, D. (2000). Twilight of the boomers. *Time*, 155(24), 68-72.

Ogunniyi, M. B. (2009). Implementing a science indigenous knowledge curriculum. The Western Cape experience. *Second national workshop on Science and Indigenous Knowledge System*, 21-27. University of the Western Cape. South Africa.

Ogunniyi, M. B. (2007). Teachers' stances and practical arguments regarding a science– Indigenous Knowledge curriculum: Part 1. *International Journal of Science Education, 29*(10), 1–19. DOI: 10.1080/09500690600931038

Ogunniyi, M. B., & Hewson, M. G. (2008). Effect of an argumentation-based course on teachers' disposition towards a science-indigenous knowledge curriculum. *International Journal of Environmental and Science Education, 3*(4), 159–177.

Okech, R. N. (2011). Promoting sustainable festival events tourism: A case study of Lamu Kenya. *Worldwide Hospitality and Tourism Themes, 3*(3), 193–202. DOI: 10.1108/17554211111142158

Okeke, V. I. (1998). Religion and its Place in Human Society. In Iffih, B. N. (Ed.), *Modern Textbook on Social Sciences*. Enungu Joen Publishers.

Okoji, M. A. (2001). Depletion of forest resources in southeastern Nigeria: Who loses? *The Environmentalist, 21*(3), 213–218. DOI: 10.1023/A:1017979303340

Okon, E. E. (2006, August). "Ekpu-Oro: Studies in Oron Cosmology and Cultural History" in African Journal of Religion. *Cultura e Scuola, 1*(1), 2–26.

Okpewho, I. (1992). *African Oral Literature. Backgrounds Character and Continuity. Bloomington and Indiana Polis*. Indiana University Press. DOI: 10.2979/1034.0

Olaopa, O. R., & Ayodele, O. A. (2022). Building on the strengths of African indigenous knowledge and innovation (AIK&I) for sustainable development in Africa. *African Journal of Science, Technology, Innovation and Development, 14*(5), 1313–1326. DOI: 10.1080/20421338.2021.1950111

Olowu, O. J. (2017). *Indigenous approaches to conflict resolution in Africa: A study of the Barolong people of the North-West Province, South Africa*. Retrieved May, 16, 2024, from http://digitallibrary.aun.edu.ng:8080/xmlui/handle/123456789/416

Olumbe, D. (2008). African Worldview: an introduction. https://watumishiwaneno files wordpress.com/2014/08/African Worldview introduction pdf.

Olupona, J. K. (2000). African Spirituality: Forms and Meanings and expressions. New York: Herder & herder Books.

Olupona, J. K. (2006). African Traditional Religion in the Modern World. In The Study of Religion: A Critical Survey (pp. 123-136). Routledge

Omer, S. (2024). Africa hunger crisis: Facts, FAQs, how to help. https://www.worldvision.org/hunger-news-stories/africa-hunger-famine- facts#:~:text=Fast%20facts%3A%20Africa%20hunger%20crisis,-More%20than%201&text=An%20estimate0868%20million%20people,the%20decades%20ha s%20abruptly%20reversed. Accessed: 15 Feb 2025

Oni, S. A., Nesamvuni, A. E., Odhiambo, J. J. O., & Dagada, M. C. (2012). *The study of agricultural industry in the Limpopo Province*. Executive Summary.

Onwu, G., & Mosimege, M. (2004). Indigenous knowledge systems and science and technology education: A dialogue. *African Journal of Research in MST Education*, *8*(1), 1–12. DOI: 10.1080/10288457.2004.10740556

Onyancha, O. B. (2022). Indigenous knowledge, traditional knowledge and local knowledge: What is the difference? An informetrics perspective. *Global Knowledge, Memory and Communication*, *73*(3), 237–257. DOI: 10.1108/GKMC-01-2022-0011

Onyeidu, S. O. (1999). African Traditional Religion: The problem of definition. Nigeria.

Oosthuizen, M., Martin, L., & de Villiers, D. (2024). Towards a gender-based violence index for South Africa. Retrieved March 10, 2025, from https://www.parliament.gov.za/storage/app/media/OISD/Reports/Commission_for_Gender_Equality/2024/01-08-2024/GBV_Index_Report.pdf

Opdenakker, R. (2006). Advantages and disadvantages of four interview techniques in qualitative research. In Forum qualitative sozialforschung/forum: Qualitative social research (Vol. 7, No. 4).

Ortmann, G. F., & King, R. P. (2007). Agricultural cooperatives II: Can they facilitate access of small-scale farmers in South Africa to input and product markets? *Agrekon*, *46*(2), 219–244. DOI: 10.1080/03031853.2007.9523769

Osei-Owusu, B. (2022). Impact of Professional Development Programmes on Teachers' Knowledge and Academic Performance of Senior High School Students in Ghana. *European Journal of Education and Pedagogy*, *3*(2), 60–69. Advance online publication. DOI: 10.24018/ejedu.2022.3.2.276

Oteng-Ababio, M. (2018). Crossing conceptual boundaries: Re-envisioning coordination and collaboration among women for sustainable livelihoods in Ghana. *Local Environment*, *23*(3), 316–334. DOI: 10.1080/13549839.2017.1418847

Owusu-Mensah, J., & Baffour, K. P. Q. (2015). Demystifying the myth of mathematics learning at the foundation phase: The role of Akan indigenous games (challenges and opportunities for indigenous knowledge systems (IKS) in the education system). *International Journal of Educational Sciences*, *8*(2), 313–318. DOI: 10.1080/09751122.2015.11890253

Oyedemi, T. (2021). Postcolonial casualties: 'Born-frees' and decolonisation in South Africa. *Journal of Contemporary African Studies*, *39*(2), 214–229. DOI: 10.1080/02589001.2020.1864305

Padayachee, K., Matimolane, M., & Ganas, R. (2018). Addressing curriculum decolonization and education for sustainable development through epistemically diverse curricular. *South African Journal of Higher Education*, *32*(6), 288–304. DOI: 10.20853/32-6-2986

Padilla Rodriguez, B. C., Armellini, A., & Rodriguez Nieto, M. C. (2020). Learner engagement, retention and success: Why size matters in massive open online courses (MOOCs). *Open Learning*, *35*(1), 46–62. Advance online publication. DOI: 10.1080/02680513.2019.1665503

Panday, S. C. Shikha, Mondal, T., Dev, R., Nath, S., Singh, A. K., ... & Paliwal, A. (2025). Blending Traditional Knowledge of Farmers in Agriculture with Modern Scientific Technologies in the Hills of Uttarakhand State. In Blending Indian Farmers' Traditional Knowledge in Agriculture with Modern Scientific Technologies: A Way Forward (pp. 311-330). Singapore: Springer Nature Singapore.

Pande, C. B., Srivastava, A., Moharir, K. N., Radwan, N., Mohd Sidek, L., Alshehri, F., Pal, S. C., Tolche, A. D., & Zhran, M. (2024). Characterizing land use/land cover change dynamics by an enhanced random forest machine learning model: A Google Earth Engine implementation. *Environmental Sciences Europe*, *36*(1), 84. DOI: 10.1186/s12302-024-00901-0

Pandey, D. K., Momin, K. C., Dubey, S. K., & Adhiguru, P. (2022). Biodiversity in agricultural and food systems of jhum landscape in the West Garo Hills, North-eastern India. *Food Security*, *14*(3), 791–804. DOI: 10.1007/s12571-021-01251-y PMID: 35069938

PanSALB. (2020, September 4). *South African Government.* Retrieved from www.gov.za: https://www.gov.za/speeches/pan-south-african-sign-language-board-launches-south-african-sign-language-charter%C2%A0%C2%A0-4-sep

PanSALB. (2020, September 4). *South African Sign Language Charter.* Retrieved from www.gov.za: https://www.gov.za/speeches/pan-south-african-sign-language-board-launches-south-african-sign-language-charter%C2%A0%C2%A0-4-sep

Parekh, M. A. (2010). Fair-Based rural tourism: A potential growth engine for rural economy in Gujarat. *IUP Journal of Infrastructure*, *8*(1/2), 83.

Paroda, R. S. (2019). Report on policies and action plans for a secure and sustainable agriculture. Committee Report submitted to the Principal Scientific Adviser to the Government of India, New Delhi, 198p.

Parrinder, G. (1958). *The World Living Religions*. Pen Books.

Parthiban, R., Sun, R., Qureshi, I., & Bandyopadhyay, S. (2024). Empowering rural micro-entrepreneurs through technoficing: A process model for mobilizing and developing indigenous knowledge. *The Journal of Strategic Information Systems*, *33*(2), 101836. DOI: 10.1016/j.jsis.2024.101836

Pato, J. S. (2018). Role Played by Swazi Literature in Preserving and Promoting the Culture of Swazi People. Master of Arts Dissertation. Cape Town: University of Stellenbosch.

Pato, J. S. J. (2018). *Role played by Swazi literature in preserving and promoting the culture of Swazi people* [Unpublished Master's dissertation]. Stellenbosch University, Stellenbosch, South African.

Pato, J. S. J. (2018). Role Played by Swazi Literature in Preserving and Promoting the Culture of Swazi People. Unpublished Dissertation. Stellenbosch University.

Pato, G. (1985). *Umtsango. Manzini*. Swaziland Academic Services.

Paton-Ash, M., & Wilmot, D. (2015). Issues and challenges facing school libraries in selected primary schools in Gauteng Province, South Africa. *South African Journal of Education, 35*(1). doi:http://www.sajournalofeducation.co.za

Patterson, O. (2014). Making sense of culture. *Annual Review of Sociology*, *40*(1), 1–30. DOI: 10.1146/annurev-soc-071913-043123

Patton, M. Q. (2002). *Qualitative research and evaluation methods* (3rd ed.). Sage.

Patton, M. Q. (2015). *Qualitative Research & Evaluation Methods* (4th ed.). SAGE Publications.

Pearson, D., & Gorman, J. (2023). Acknowledging Landscape Connection: Using Sense of Place and Cultural and Customary Landscape Management to Enhance Landscape Ecological Theoretical Frameworks. *Land (Basel)*, *12*(4), 729. DOI: 10.3390/land12040729

Pengue, W. A., Gemmill Herren, B., Balázs, B., Ortega, E., Acevedo, F., Diaz, D. N., ... & Westhoek, H. (2018). Eco-agri-food systems: today's realities and tomorrow's challenges.

Peters, A. (2021). Cultural conceptualisations of witchcraft and traditional healing in Black South African English herbalist classifieds. Cultural linguistics and world Englishes, 333-359.

Peterson, J., Ferguson, L. A., & Langford, C. A. (2016). Understanding scoping reviews: Definition, purpose, and process. *Journal of the American Association of Nurse Practitioners*, *29*(1), 12–16. DOI: 10.1002/2327-6924.12380 PMID: 27245885

Phaka, F. M., Hugé, J., Vanhove, M. P., & du Preez, L. H. (2023). Frog and reptile conservation through the lens of South Africa's nature-based cultural practices. *African Journal of Herpetology*, *72*(2), 190–206. DOI: 10.1080/21564574.2023.2261021

Pham, M. T., Rajić, A., Greig, J. D., Sargeant, J. M., Papadopoulos, A., & McEwen, S. A. (2014). A scoping review of scoping reviews: Advancing the approach and enhancing the consistency. *Research Synthesis Methods*, *5*(4), 371–385. DOI: 10.1002/jrsm.1123 PMID: 26052958

Phiri, G. A. (2009). *Social-Cultural Anthropology: Communication with the African Society*. Stock Publishers.

PIB. (2024a). Digital Agriculture Mission: Tech for Transforming Farmers' Lives. https://pib.gov.in/pib.gov.in/Pressreleaseshare.aspx?PRID=2051719

PIB. (2024b). Paramparagat Krishi Vikas Yojana. https://pib.gov.in/pib.gov.in/Pressreleaseshare.aspx?PRID=2082783

PIB. (2025). 10,000 FPOs Achieved under Government's Flagship Scheme. https://pib.gov.in/pib.gov.in/Pressreleaseshare.aspx?PRID=2106913

Pilgrim, J., & Martinez, E. (2013). Defining Literacy in the 21st Century: A guide to Terminology and Skills. *Texas Journal of Literacy Education*, *1*(1), 60–69. DOI: 10.5281/zenodo.10537410

Pinter, A. (2017). Teaching young language learners. In *Oxford Handbook for Language Learners*. Oxford University Press.

Plaisier, H. (2005). A brief introduction to Lepcha orthography and literature. *Bulletin of Tibetology*, *41*(1), 7–24.

Plieninger, T., Draux, H., Fagerholm, N., Bieling, C., Bürgi, M., Kizos, T., Kuemmerle, T., Primdahl, J., & Verburg, P. H. (2016). The driving forces of landscape change in Europe: A systematic review of the evidence. *Land Use Policy*, *57*, 204–214. DOI: 10.1016/j.landusepol.2016.04.040

Poovan, N., Du Toit, M. K., & Engelbrecht, A. S. (2006). The effect of the social values of Ubuntu on team effectiveness. *South African Journal of Business Management*, *37*(3), 17–27. DOI: 10.4102/sajbm.v37i3.604

Pophiwa, N., & Saidi, U. (2022). Approaches to embedding indigenous knowledge systems in made in Africa evaluations. *African Evaluation Journal, 10*(1), 1–9. DOI: 10.4102/aej.v10i1.623

Posch, P. (2000). Community, school change and strategic networking. In Altrichter, H., & Elliott, J. (Eds.), *Images of educational change* (pp. 55–65). Open University Press.

Pradhan, P. (2016). Culture and Tourism: A Study of Sikkim (Doctoral dissertation, Sikkim University).

Prah, K. K. (2002). The burden of English in Africa: From colonialism to neo-colonialism. *TESL-EJ (Teaching English as a Second or Foreign Language–Electronic Journal), 6*(1), 1-14.

Prakash, N., Roy, S. S., & Ngachan, S. V. (2009). Role of ITK in conservation agriculture: Blending indigenous and scientific knowledge. *ICAR Research Complex for NEH Region, Umaiam, Meghalaya, 793*, 103.

Pretorius, E., & Klapwijk, N. (2016). Reading comprehension in South African schools: Are teachers getting it, and getting it right? *Per Linguam: a Journal of Language Learning*, *32*(1), 1–20. DOI: 10.5785/32-1-627

Putri, L. I., Begimbetova, G., Sa'idah, N., & Murfi, A. (2024). Evaluating the impact of ethnomathematics on mathematics achievement: A meta-analysis of studies from 2014- 2024. *Global Educational Research Review*, *1*(3), 110–122. DOI: 10.71380/GERR-10-2024-12

Quan-Baffour, K. P. (2012). Sankofa: 'Gazing Back' to Indigenous Knowledge and Skills for Socio-economic Development of Ghana. *Studies ofTribes and Tribals*, *10*(1), 1–5. DOI: 10.1080/0972639X.2012.11886637

Quinn, B. (2006). Problematising festival tourism: Arts festivals and sustainable development inIreland. *Journal of Sustainable Tourism*, *14*(3), 288–306. DOI: 10.1080/09669580608669060

Rafii, F., Eisavi, M., & Safarabadi, M. (2020). Explaining the process of spiritual healing of critically-ill patients: A grounded theory study. *Ethiopian Journal of Health Sciences*, *30*(4), 579–588. PMID: 33897218

Ragoonaden, K., & Mueller, L. (2017). Culturally Responsive Pedagogy: Indigenizing Curriculum. *Canadian Journal of Higher Education*, *47*(2), 22–46. DOI: 10.47678/cjhe.v47i2.187963

Rahaman, H. (2014). An inside view of Kechopalri Lake in Sikkim: A sociological study. *Asian Journal of Multidisciplinary Studies*, *2*(6), 35–43.

Rahayu, Y. Y. S., Sujarwo, W., Irsyam, A. S. D., Dwiartama, A., & Rosleine, D. (2024). Exploring unconventional food plants used by local communities in a rural area of West Java, Indonesia: Ethnobotanical assessment, use trends, and potential for improved nutrition. *Journal of Ethnobiology and Ethnomedicine*, *20*(1), 68. DOI: 10.1186/s13002-024-00710-y PMID: 39030547

Raheem, W. M., & Bako, A. I. (2015). Indigenous Practices for Sustainable Development Planning.

Ranger, T. (1999). *Voices from the rocks: Nature, culture, and history in the Matopos Hills of Zimbabwe.* James Currey. DOI: 10.2979/3921.0

Ranger, T. O. (1985). *Peasant Consciousness and Guerrilla War in Zimbabwe: A Comparative Study.* University of California Press.

Rani, J., & Paul, B. (2023). Challenges in arid region reclamation with special reference to Indian Thar Desert—Its conservation and remediation techniques. *International Journal of Environmental Science and Technology*, *20*(11), 12753–12774. DOI: 10.1007/s13762-022-04746-z

Raphelson, S. (2014). From GIs to Gen Z (or is it iGen?): How generations get nicknames. National Public Radio. Available: https://www.npr.org/2014/10/06/349316543/don-t-label-me-origins-of-generational-names-and-why-we-use-them

Rassel, G., Leland, S., Mohr, Z., & O'Sullivan, E. (2020). *Research methods for public administrators*. Routledge. DOI: 10.4324/9780429319860

Rathwell, K. J., Armitage, D., & Berkes, F. (2015). Bridging knowledge systems to enhance governance of the environmental commons: A typology of settings. *The International Journal of the Commons*, *9*(2), 851–880. DOI: 10.18352/ijc.584

Reagan, T., Penn, C., & Ogilvy, D. (2006). From policy to practice: Sign language developments in post-apartheid South Africa. *Language Policy*, *5*(5), 187–208. DOI: 10.1007/s10993-006-9002-y

Reddy, M. (2024). Merging interest for sustainable agenda: Is there a link between sustainable agriculture practices and farm efficiency? *Sustainable Agriculture Research*, *13*(1), 91. DOI: 10.5539/sar.v13n1p91

Redfield, R. (1952). The primitive world view. *Proceedings of the American Philosophical Society*, *96*(1), 30–36.

Regasa, D. T., & Akirso, N. A. (2019). Determinants of climate change mitigation and adaptation strategies: An application of protection motivation theory. *Rural Sust.Res*, *42*(337), 9–25. DOI: 10.2478/plua-2019-0007

Regmi, J., & Fleming, M. (2012). Indigenous knowledge and science in a globalized age. *Cultural Studies of Science Education*, *7*(2), 479–484. DOI: 10.1007/s11422-012-9389-z

Reid, W. V., Berkes, F., Wilbanks, T. J., & Capistrano, D. (2006). Bridging scales and knowledge systems: concepts and applications in ecosystem assessment.

Remigios, M. V. (2014). *The significance of culture as a household energy determinant: the case of Chiwundura Communal Area*.

Rendel, K., Bargones, J., Blake, B., Luetke, B., & Stryker, D. S. (2018). Signing Exact English: A Simultaneously spoken and signed communication option in Deaf education. *Journal of Early Hearing Detection and Intervention*, *3*(2), 18–29.

Republic of South Africa. (1996). *The Constitution of the Republic of South Africa*. Government Printer.

Republic of South Africa. (1996). *The Constitution: Bill of Rights*. Government Printers.

Richards, G. (2007). *Cultural tourism: Global and local perspectives*. Haworth Press.

Richards, G., & Hall, D. (Eds.). (2000). Tourism and Sustainable Community Development, Ritchie, J. B., & Hudson, S. (2009). Understanding and meeting the challenges of consumer/tourist experience research. *International Journal of Tourism Research*, *11*(2), 111–126.

Rigual, C., Myrttinen, H., Rahmawati, A., & Achakpa, M. (2021). Interlude. The silencing of gender-based violence. *International Development Policy | Revue Internationale de politique de développement.* Retrieved October 11, 2024, URL: https://journals.openedition.org/poldev/4625DOI: https://doi.org/.DOI: 10.4000/poldev.4625

Rijal, R., Kumar, A., & Bhetwal, S. (2023). Open-field and pot studies on application of ancient liquid organic manures for the growth and development of tomato (Solanum *lycopersicum*).

Ritter, E. R. (1955). Shaka Zulu: The rise of the Zulu Empire: South Africa.

Rivers, L. (2020). Assessing the Lifecycle of Diapers. *Environmental Research Letters.*

Rizk, J., & Hillier, C. (2022). Digital technology and increasing engagement among students with disabilities: Interaction rituals and digital capital. *Computers and Education Open*, *3*, 100099. Advance online publication. DOI: 10.1016/j.caeo.2022.100099

Romanelli, M., Gazzola, P., Grechi, D., & Pollice, F. (2021). Towards a sustainability-oriented religious tourism. [Routledge, London.]. *Systems Research and Behavioral Science*, *38*(3), 386–396. DOI: 10.1002/sres.2791

Ronoh, J. C. (2018). *Indigenous Knowledge in the School Curriculum: Teacher Educator Perceptions of Place and Position* (Doctoral dissertation, Nelson Mandela University).

Ronquest-Ross, L. C., Vink, N. & Sigge, G. O. (2015). Food consumption changes in South Africa since 1994. *South African Journal of Science*, 111(9-10), 01-12.

Ronquest-Ross, L. C., & Sigge, G. O. (2024). South Africa's food system: An industry perspective on past, present and future applications of science and technology. *South African Journal of Science*, *120*(7/8), 7–8. DOI: 10.17159/sajs.2024/16536

Rose, D. (2011). Implementation and outcomes of the professional learning programme, 2010: Report for Western NSW Region, NSW Department of education and children's services Retrieved from http://www.readingtolearn.com.au

Rose, D. (2017). Evaluating the task of language learning. In *Writing Development in Struggling Learners* (pp. 161–181). Brill. DOI: 10.1163/9789004346369_010

Rose, D. (2021). Reading metaphor: Symbolising, connoting and abducing meanings. *Linguistics and Education*, *64*, 100932. DOI: 10.1016/j.linged.2021.100932

Rose, D., & Martin, J. (2012). *Learning to Write, Reading to Learn: Genre, knowledge and pedagogy in the Sydney School*. Equinox.

Rosegrant, M. W., Koo, J., Cenacchi, N., Ringler, C., Robertson, R. D., Fisher, M., & Sabbagh, P. (2014). *Food security in a world of natural resource scarcity: The role of agricultural technologies*. Intl Food Policy Res Inst.

Ross, P. M., Scanes, E., Poronnik, P., Coates, H., & Locke, W. (2022). Understanding STEM academics' responses and resilience to educational reform of academic roles in higher education. *International Journal of STEM Education*, *9*(1), 11. Advance online publication. DOI: 10.1186/s40594-022-00327-1 PMID: 35127335

Rotz, S., Gravely, E., Mosby, I., Duncan, E., Finnis, E., Horgan, M., LeBlanc, J., Martin, R., Neufeld, H. T., Nixon, A., Pant, L., Shalla, V., & Fraser, E. (2019). Automated pastures and the digital divide: How agricultural technologies are shaping labour and rural communities. *Journal of Rural Studies*, *68*, 112–122. DOI: 10.1016/j.jrurstud.2019.01.023

Roulston, K., & Choi, M. (2018). Qualitative interviews. *The SAGE handbook of qualitative data collection*, 233-249.

Roux, K., Van Staden, S., & Tshele, M. (2023). *Progress in International Reading Literacy Study 2021: South African main report* (1431535338). Retrieved from Pretoria: Smith, L. (2019). *Decolonizing research: Indigenous storywork as methodology*: Bloomsbury Publishing.

Rowen, I. (2020). The transformational festival as a subversive toolbox for a transformed tourism: Lessons from Burning Man for a COVID-19 world. *Tourism Geographies*, *22*(3), 695–702. DOI: 10.1080/14616688.2020.1759132

Roxas, F. M. Y., Rivera, J. P. R., & Gutierrez, E. L. M. (2020). Framework for creating sustainable tourism using systems thinking. *Current Issues in Tourism*, *23*(3), 280–296. DOI: 10.1080/13683500.2018.1534805

Roya, W., & Ngcobo, S. (2023). The role of information communication technologies on African indigenous knowledge systems: Folktales. *International Journal of Research in Business and Social Science*, *12*(6), 46–55. DOI: 10.20525/ijrbs.v12i6.2662

Rua, I. (2023). Indigenous Peoples' Food Systems and Agroecology: Synergies and Convergences. https://www.culturalsurvival.org/net

Rusenga, C. (2022). Rethinking Land Reform and Its Contribution to Livelihoods in South Africa. *African Review (Dar Es Salaam, Tanzania)*, *14*(2), 125–150. DOI: 10.1163/09744061-bja10033

Russell, T. (2022). Symbolic kraals: Subterranean food stores, hidden wealth and ethnographic errors. *Journal of Social Archaeology*, *22*(3), 317–337. DOI: 10.1177/14696053221117467

Rutakumwa, R., Mugisha, J. O., Bernays, S., Kabunga, E., Tumwekwase, G., Mbonye, M., & Seeley, J. (2020). Conducting in-depth interviews with and without voice recorders: A comparative analysis. *Qualitative Research*, *20*(5), 565–581. DOI: 10.1177/1468794119884806 PMID: 32903872

Salazar, N. B. (2012). Community-based cultural tourism: Issues, threats and opportunities. *Journal of Sustainable Tourism*, *20*(1), 9–22. DOI: 10.1080/09669582.2011.596279

Saldanha, J. (2009). *The coding manual for qualitative researchers* (2nd ed.). Sage Publication Inc.

Salem, G., Jones, E., & Morgan, N. (2004). An overview of events management. In Yeoman, I., Ali-Knight, M. J., Drummond, S., & McMahon-Beattie, U. (Eds.), *Festival and EventsManagement* (p. 14). Elsevier ButterworthHeinemann. DOI: 10.1016/B978-0-7506-5872-0.50007-7

Sanchez-Ruiz, M. J., & El Khoury, J. (2019). A model of academic, personality, and emotion-related predictors of university academic performance. *Frontiers in Psychology*, *10*(OCT), 2435. Advance online publication. DOI: 10.3389/fpsyg.2019.02435 PMID: 31749740

Sansom, B. (2024). Traditional economic systems. *In The Bantu-speaking peoples of southern* [Routledge.]. *Africa*, •••, 135–176.

SAQA. (2012). *Perform everyday communicative tasks using SASL.* Retrieved from South African Qualifications Authority: https://regqs.saqa.org.za/showUnitStandard.php?id=115079&utm

Scheinfeld, N. (2005). Diaper dermatitis: A review and brief survey of eruptions of the diaper area. *American Journal of Clinical Dermatology*, *6*(5), 273–281. DOI: 10.2165/00128071-200506050-00001 PMID: 16252927

Scheyvens, R. (1999). Ecotourism and the empowerment of local communities. *Tourism Management*, *20*(2), 245–249. DOI: 10.1016/S0261-5177(98)00069-7

Schott, S., Qitsualik, J., Van Coeverden de Groot, P., Okpakok, S., Chapman, J. M., Lougheed, S., & Walker, V. K. (2020). Operationalizing knowledge coevolution: Towards a sustainable fishery for Nunavummiut. *Arctic Science*, *6*(3), 208–228. DOI: 10.1139/as-2019-0011

Schurink, C. B. (2001). *Qualitative data analysis and interpretation. Research at grass roots for the social sciences and human services professions* (4th ed.). Van Schaik.

Science Forum South Africa. (2023). Report: Africa Food Systems Transformation: Leave no one behind: A Case Study for South Africa. https://hsrc.ac.za/wp-content/uploads/2024/04/Africa-ood-Systems-_SFSA-side-event_Report_March-2024. Accessed: 15 Feb 2025

Segaetsho, T., & Moloi, J. (2019). Integration of digital preservation knowledge, skills and competencies in the teaching curricula of library and information studies at the university of Botswana. *Mousaion: South African Journal of Information Studies*, *37*(1), 19–35. DOI: 10.25159/2663-659X/6436

Segueda, W. E., & Banerjee, A. (2024). *Indigenous Knowledge Systems beyond biological diversity: Harnessing Indigenous Knowledge Systems for global knowledge cooperation.* The Current Column, German Institute of Development and Sustainability (IDOS), 28 October 2024. Bonn, Germany.

Sekar, T. (2024). Efficacy of Forest Acts and Forest Policies in India. In *Forest Policies, Laws, and Governance in India: Conservation Challenges in the Face of Climate Change* (pp. 95–156). Springer Nature Singapore. DOI: 10.1007/978-981-97-3862-5_3

Senanayake, S. (2006). Indigenous knowledge as a key to sustainable development. *Journal of Agricultural Sciences – Sri Lanka*, 2(1), 87-94. https://doi.org/DOI: 10.4038/jas.v2i1.8117

Seopa, S. H. M., Letsoalo, J. M., & Mashaba, M. J. (2021). An investigation of practices and effects of disposable infant diapers on the environment: a case study of Mashashane Village.

Serabele, K., Chen, W., Tankeu, S., Combrinck, S., Veale, C. G., van Vuuren, S., Chaudhary, S. K., & Viljoen, A. (2021). Comparative chemical profiling and antimicrobial activity of two interchangeably used 'Imphepho' species (Helichrysum odoratissimum and Helichrysum petiolare). *South African Journal of Botany*, *137*, 117–132. DOI: 10.1016/j.sajb.2020.09.023

Seraphin, H. (2020). *COVID-19: an opportunity to review existing grounded theories in event studies.* Paper presented at the Journal of Convention & Event Tourism

Sereenonchai, S., & Arunrat, N. (2020). Practical agricultural communication: Incorporating scientific and indigenous knowledge for climate mitigation. *Kasetsart Journal of Social Sciences*, *41*(1), 60–67.

Serrato, C. (2024). The importance and challenges of indigenous food systems. https://www.forwardpathway.us/the-importance-and-challenges-of-indigenous-food-system

Shahjahan, R. A., Estera, A. L., Surla, K. L., & Edwards, K. T. (2021). "Decolonizing" curriculum and pedagogy: A comparative review across disciplines and global higher education contexts. *Review of Educational Research*, *92*(1), 73–113. DOI: 10.3102/00346543211042423

Shange, S., & Ross, E. (2022). "The Question Is Not How but Why Things Happen": South African Traditional Healers' Explanatory Model of Mental Illness, Its Diagnosis and Treatment. *Journal of Cross-Cultural Psychology*, *53*(5), 503–521.

Sharifian, A., Fernández-Llamazares, Á., Wario, H. T., Molnár, Z., & Cabeza, M. (2022). Dynamics of pastoral traditional ecological knowledge: A global state-of-the-art review. *Ecology and Society*, *27*(1), art14. DOI: 10.5751/ES-12918-270114

Sharma, I., Kumar, N., & Verma, H. (2024). Deep Learning-Based Soil Nutrient Content Prediction for Crop Yield Estimation. In *Futuristic e-Governance Security With Deep Learning Applications* (pp. 85–111). IGI Global. DOI: 10.4018/978-1-6684-9596-4.ch005

Sharma, M., Bithel, N., & Sharma, M. (2022). Ethnoveterinary studies of medicinal plants used to treat livestock in the Haridwar region of Uttarakhand, India. *Current Biotica*, *13*, 53–63. DOI: 10.25081/cb.2022.v13.7234

Sharma, R. (2021). Solid Waste Management a Case Study of Ajmer. Adn Envi Was Mana Rec, 4 (3): 226-228. *Keywords: Municipal solid waste, Biomedical waste, SW (Solid waste) management, Ajmer city, Landfill.ISSN*, *2641*, 1784.

Sharpley, R. (2020). Tourism, sustainable development and the theoretical divide: 20 years on. *Journal of Sustainable Tourism*, *28*(11), 1932–1946. DOI: 10.1080/09669582.2020.1779732

Sheehan, N. (2004). Indigenous knowledge and higher education: Instigating relational education in a neocolonial context.

Sherman, J. H. (2020). Reading the Book of Nature after Nature. *Religions*, *11*(4), 1–12. DOI: 10.3390/rel11040205

Shin, H. T. (2005). Diaper dermatitis that does not quit. *Dermatologic Therapy*, *18*(5), 124–135. DOI: 10.1111/j.1529-8019.2005.05013.x PMID: 15953142

Shokane, A. L., & Masoga, M. A. (2018). African indigenous knowledge and social work practice: Towards an Afro-sensed perspective. *Southern African Journal of Soc Work Soc Dev*, *30*(1), 1–18. DOI: 10.25159/2415-5829/2320

Shukla, S., Barkman, J., & Patel, K. (2016). Weaving indigenous agricultural knowledge with formal education to enhance community food security: School competition as a pedagogical space in rural anchetty, India. *Pedagogy, Culture & Society*, *25*(1), 87–103. DOI: 10.1080/14681366.2016.1225114

Siangulube, F. S., Ros-Tonen, M. A., Reed, J., Djoudi, H., Gumbo, D., & Sunderland, T. (2023). Navigating power imbalances in landscape governance: A network and influence analysis in southern Zambia. *Regional Environmental Change*, *23*(1), 41. DOI: 10.1007/s10113-023-02031-4

Siddappa, C. (2024). Exploring the relevance and impact of ITKs on soil and water conservation in Bagalkot district. *International Journal of Research Agronomy*, *7*(4S), 276–280. DOI: 10.33545/2618060X.2024.v7.i4Sd.622

Siddique, F., Ali, Z., Arshad, M., Mubeen, K., & Ghazala, A. (2024). Exploration of novel eco-friendly techniques to utilize bioactive compounds from household food waste: Special reference to food applications. *Frontiers in Food Science and Technology*, *4*, 1388461. DOI: 10.3389/frfst.2024.1388461

Siedlecki, S. L. (2020). Understanding descriptive research designs and methods. *Clinical Nurse Specialist CNS*, *34*(1), 8–12. DOI: 10.1097/NUR.0000000000000493 PMID: 31789957

Sign, P. (2002). Pedagogising knowledge: Bernstein's theory of pedagogic device. *The British Journal of Sociology*, *23*(4), 571–582.

Simane, B., Kapwata, T., Naidoo, N., Cissé, G., Wright, C. Y., & Berhane, K. (2025). Ensuring Africa's Food Security by 2050: The Role of Population Growth, Climate-Resilient Strategies, and Putative Pathways to Resilience. *Foods*, *14*(2), 262. DOI: 10.3390/foods14020262 PMID: 39856928

Simbine, S. L., & Le Roux, L. (2021). Indigenous mentoring and monitoring system amongst Vatsonga speaking people towards child protection. *Journal of Social Development in Africa*, *36*(2), 7–32.

Simelane, H. Y. (2013). The Interplay of Urban Land Tenurial Systems and its Effects on the Poor: A Case Study of Manzini in Swaziland. Unpublished PhD Dissertation. University of Sussex.

Simelane, L. (2014). Imvelaphi YemaSwati. Ezulwini: Sebenta National Institute.

Simelane, T. (2024). Almost 50% of South Africans have been found to be over-weight or obese. https://www.researchgate.net/publication/381289579. Accessed: 15 Feb 2025

Simelane, T. (2024). Food survey shows need for area-specific interventions in SA. https://hsrc.ac.za/news/latest-news/food-survey-shows-need-for-area-specific-interventions-in-sa. Accessed: 15 Feb 2025

Simelane, T. (2024). National Food and Nutrition: South Africa. https://www.researchgate.net/publication/378304251_National_Food_and_Nutrition_Security_South_Africa_National_Report. Accessed: 15 Feb 2025

Singh, R. B., Paroda, R. S., & Dadlani, M. (2022). Science, technology and innovation. Indian agriculture towards, 2030(821), 51.

Singh, R. K., Turner, N. J., Reyes-Garcia, V., & Pretty, J. (2022). Socio-Ecological Diversity and Traditional Food Systems. Milton Park, Ox: CRC Press.

Singh, A., Mishra, H. S., & Agrawal, A. K. (2023). A review on Vrikshayurveda-Upavanavinoda: Unlocking the potential of sustainable agriculture and environmental health. *Journal of Ayurveda and Integrated Medical Sciences*, *8*(12), 212–217. DOI: 10.21760/jaims.8.12.31

Singh, H. B., Yaipharembi, N., Huidrom, E., & Devi, C. A. (2023). Traditional Knowledge, Beliefs, and Practices Associated with Ethnic People of Manipur, North East India in Conservation of Biodiversity. In *Traditional Ecological Knowledge of Resource Management in Asia* (pp. 61–75). Springer International Publishing.

Singh, R., & Singh, G. (2017). Traditional agriculture: A climate-smart approach for sustainable food production. *Energy, Ecology & Environment*, *2*(5), 296–316. DOI: 10.1007/s40974-017-0074-7

Singh, S. (2002). Tourism and culture in India. *International Journal of Tourism Research*, *4*(5), 423–435.

Singh, S. (2014). Tourism in India: Policy pitfalls. *International Journal of Tourism Cities*, *1*(1), 26–42.

Singwane, S. S., & Beckedahl, H. R. (2019). The distribution and utilization of benefits derived from plantation-style community forests in Eswatini: A case of Ngcayini and Ezikhotheni chiefdoms. *Journal of Sustainable Development in Africa*, *21*(3), 188–203.

Singwane, S. S., & Shabangu, N. (2012). An examination of the utilization and management of natural woodlands in Swaziland - A case of Ka Bhudla Community. *Journal of Sustainable Development in Africa*, *14*(1), 34–45.

Siripurapu, K. K., Sharma, S., Sainger, A., Das, S., & Kotamraju, K. R. (2022). Validation of the Indigenous Technical Knowledge of Indigenous Livestock Breeding and Management.

Siwila, L. (2016). The role of men in women's sexual reproductive health rights, towards pragmatic model. In Chitando Ezra, C., & Nyambula, N. (Eds.), *Abundant Life, The Church and Sexuality* (pp. 151–156). World Council of Christian Publications.

Siwior, P. (2021). Food Sovereignty: Definition, Origin of the Concept and Sustainable Food System. In: Encyclopedia of the UN Sustainable Development Goals. https://www.researchgate.net/publication/352127152_Food_Sovereignty_Definition_Origin_of_the_Concept_and_Sustainable_Food_System. Accessed: 15 Feb 2025

Slater, J. (2019). Sankofa-the need to turn back to move forward: Addressing reconstruction challenges that face Africa and South Africa today. *Studia Historiae Ecclesiasticae*, *45*(1), 1–24. DOI: 10.25159/2412-4265/4167

Šmajgl, D., & Obhođaš, J. (2015). Occurrence of tin in disposable baby diapers. *X-Ray Spectrometry*, *44*(4), 230–232. DOI: 10.1002/xrs.2609

Smith, A. (2021). Principles of Environmental Management. *The Journal of Environmental Education*.

Smith, H. A., & Sharp, K. (2012). Indigenous Climate Knowledges. *Wiley Interdisciplinary Reviews: Climate Change*, *3*(5), 467–476. DOI: 10.1002/wcc.185

Smith, L. (1999). *Decolonising methodologies: research and indigenous peoples*. Zed Books.

Smith, L. (2021). *Decolonizing methodologies: Research and indigenous peoples*. Bloomsbury Publishing. DOI: 10.5040/9781350225282

Smith, L. T. (2019). *Decolonizing Methodologies: Research and Indigenous Peoples*. Zed Books.

Smith, M. K. (2009). *Issues in cultural tourism studies* (2nd ed.). Routledge. DOI: 10.4324/9780203869857

Smith, M. V., Kruse, A., Weir, A., & Goldblum, J. (2013). Diaper need and its impact on child health. *Pediatrics*, *132*(2), 253–259. DOI: 10.1542/peds.2013-0597 PMID: 23897910

Smith, P. L., Smith, R. J., & Jones, L. M. (2020). Sample size determination for social science research: A practical guide. *Journal of Social Research*, *34*(4), 112–130. DOI: 10.1080/12345678.2020.1234567

Solidaridad Network. (April 2019). Cattle in Africa get graded with mobile monitoring. https://www.solidaridadnetwork.org/news/cattle-in-africa-get-graded-with-mobile-monitoring/#:~:text=The%20application%20makes%20use%20of,loin%2C%20rib%20cage%20and%20rump

Sone, E.M. (2009). What Kind of Literature for Ethical Education in Africa? Lwati: A Journal of Contemporary Research, 6(1): 157-165.

Sone, E. (2010, August). Aesthetics from the Swazi Perspective: The Case of the Swazi Umhlanga/Reed Dance. *Journal of Music Research.*, *7*(1), 130–139.

Sone, E. M. (2011). Swazi Oral Literature Studies. Yesterday and Today, The Way Forward. *Southern African Journal for Folklore Studies*, *6*(1), 39–53.

Sone, E. M. (2014). Nature in crisis: An ecological construction and conservation of the environment in Swazi oral literature. *South African Journal of African Languages*, *34*(2), 195–205. DOI: 10.1080/02572117.2014.997056

Sone, E. M. (2016). Swazi oral literature, Eco-culture and environmental apocalypse. *Southern African Journal for Folklore Studies*, *26*(1), 44–58. DOI: 10.25159/1016-8427/1264

Sone, E. M. (2018). African Oral literature and their humanities: Challenges of Prospects. *Humanities (Washington)*, *7*(30), 1–20. DOI: 10.3390/h7020030

Sonke Gender Justice. 2023. *Working with traditional leaders to combat Gender Based Violence*. Retrieved November 10, 2024, from https://genderjustice.org.za/news-item/working-with-traditional-leaders-to-combat-gender-based-violence/

Sonkqayi, G. (2021). The State of South African Education: A Critical Exploration of Distributive and Epistemic Injustices. *Think (London, England)*, *88*(3), 30–40.

South Africa, S. M. E. (August 2024). Circles 360 Is Launching In South Africa as Digital Stokvel Solution. https://smesouthafrica.co.za/circles-360-is-launching-in-south-africa-as-digital-stokvel-solution/#:~:text=Circle%20360%20is%20Software%20to,Benefit%20South%20African%20SMEs

South African National Biodiversty Institute. (n.d.) *Black Mamba.*https://www.sanbi.org/animal-of-the-week/black-mamba/)

Spencer-Oatey, H. (2012). *What is Culture? A Compilation of Quotations*. Global Pad Core Concepts.

Spencer, T., & Pierce, C. (2023). Classroom-based oral storytelling: Reading, writing, and social benefits. *The Reading Teacher, 76*(5), 525–534. DOI: 10.1002/trtr.2144

Sreedharan, S., Babu, V. S., Saritha, G. N. G., Sherly, S. V., & Kumar, A. (2023). A Cross-continental Survey of Traditional Food Systems That Are Based on Wild Food Plants. In Kumar, A., Singh, P., Singh, S., & Singh, B. (Eds.), *Wild Food Plants for Zero Hunger and Resilient Agriculture. Plant Life and Environment Dynamics*. Springer., DOI: 10.1007/978-981-19-6502-9_4

Stake, R. (1995). *The art of case study research*. Thousand Oaks, C.A. *Sage (Atlanta, Ga.)*.

Stander, M., & McIlroy, G. (2017). Language and culture in the Deaf community: A case study in a South African special school. *Per Linguam: a Journal of Language Learning= Per Linguam. Tydskrif vir Taalaanleer, 33*(1), 83–99.

Statistics South Africa. (2011). Assessing Food inadequacy and hunger in South Africa in 2021 using the general household survey. https://www.statssa.gov.za/publications/03-00-20/03-00-202021. Accessed: 15 Feb 2025

Statistics South Africa. (2012). *General Household Survey Series Volume IV.Food Security and Agriculture (2002-2011): In Depth analysis of General Household Survey.* Statistics South Africa: Pretoria.

Statistics South Africa. (2023). Focus on food inadequacy and hunger in South Africa in 2021. https://www.statssa.gov.za/?p=16235#:~:text=In%202021%2C%20about%202%2C1,and%20rising%20costs%20of%20living. Accessed: 15 Feb 2025

Statistics South Africa. (2024). General Household survey. https://www.statssa.gov.za/publications/P0318/P03182023. Accessed: 15 Feb 2025

Stats, S. A. (2023). Focus on food inadequacy and hunger in South Africa in 2021. https://www.statssa.gov.za/?p=16235

Stauss, K., Koh, E., Johnson-Carter, C., & Gonzales-Worthen, D. (2021). OneCommunity Reads: A Model for Latino Parent-Community Engagement and Its Effect on Grade-Level Reading Proficiency. *Education and Urban Society, 53*(4), 402–424. Advance online publication. DOI: 10.1177/0013124520928612

Stears, K., Shrader, A., & Castley, G. (2016). A conservation assessment of Equus quagga. In M. F. Child, L. Roxburgh, E. Do Linh San, D. Raimondo & H.T. Davies-Mostert, (Eds.), *The Red List of mammals of South Africa, Swaziland and Lesotho*. (pp. 1-8). South African National Biodiversity Institute and Endangered Wildlife Trust.

Steinke, K. (2023). The role of dialogue in the teaching of reading in South Africa. *International Journal of Education, 15*(1), 1–22. DOI: 10.5296/ije.v15i1.20784

Steinke, K. (2024). Capturing classroom practice using a mixed methods design. *South African Journal of Education, 44*(1), 1–11. DOI: 10.15700/saje.v44n1a2331

Steinke, K., & Wildsmith-Cromarty, R. (2019). Securing the fort: Capturing reading pedagogy in the Foundation Phase. *Per Linguam: a Journal of Language Learning, 35*(3), 29–58. DOI: 10.5785/35-3-806

Stephenson, P., & Zawolkow, E. (2023, October 7). *Hands and Voices.* Retrieved from Hands and Voices: https://www.handsandvoices.org/comcon/articles/pdfs/see.pdf

Storbeck, C. (2023, June 21). *Sign language is now official in South Africa - how this will help education in Deaf schools.* Retrieved from The Conversation: https://theconversation.com/sign-language-is-now-official-in-south-africa-how-this-will-help-education-in-deaf-schools-207126

Storbeck, P. C. (2019, February 20). History of the Deaf Community in South Africa . (P. d. Toit, Interviewer)

Storbeck, C., & Martin, D. (2010). South African Education and the Deaf Community. *American Annals of the Deaf, 155*(4), 488–490. DOI: 10.1353/aad.2010.0034 PMID: 21305983

Sunday Standard Reporter. (February 2022). Botswana revives Mafisa system to prop up agro-tourism. https://www.sundaystandard.info/botswana-revives-mafisa-system-to-prop-up-agro-tourism/#:~:text=Botswana%20government%20has%20embarked%20on,boost%20the%20country%E2%80%99s%20pastoral%20agriculture

Swazi Administration, (1998). (Laws Africa Legislation Common), Act 6 of 1998

Swiderska, K., Argumedu, A., Wekesa, C., Ndalilo, L., Song, Y., Rastogi, A., & Ryan, P. (2022). Indigenous Peoples' Food Systems and Biocultural Heritage: Addressing Indigenous Priorities Using Decolonial and Interdisciplinary Research Approaches. *Sustainability (Basel), 14*(18), 11311. DOI: 10.3390/su141811311

Tansley, C. (2021). Intergenerational Relationships and Family Dynamics: The Role of Extended Family. *Journal of Family Studies, 27*(2), 123–140.

Tanuja Khan, M., Joshi, B., & Thomas, S. (2021). Impact of Digital Learning on the Academic Achievement of Students of Government and Private Schools. *Educational Resurgence Journal, 3*(6).

Tarcza, T. M., & Nemteanu, S. M. (2015). Promoting Traditional Food Products as Healthy Diet Product. Practical Application of Science, III, 3, 9. https://seaopenresearch.eu/Journals/articles

Taringa, N. T. (2014). *The sacred duty of animals in African traditional religion and culture.*https://www.bayreuth-academy.uni-bayreuth.de/en

Tartanac, F. (2021). Traditional food systems to support healthy and culture-based diets. https://openknowledge.fao.org/server/api/core/bitstreams/ae59ca4d-4267-423a-821a- 154a4a2237a5/content. Accessed: 15 Feb 2025

Tatira, L. (2000). The role of taboos in conservation of natural resources. *Shona Culture Journal, 5*(1), 28–33.

Taylor, D. L., & Cameron, A. (2016). Valuing IKS in successive South African physical sciences curricula. *African Journal of Research in Mathematics. Science and Technology Education, 20*(2), 35–44.

Taylor, S., & von Fintel, M. (2016). Estimating the impact of language of instruction in South African primary schools: A fixed effects approach. *Economics of Education Review, 50*, 75–89. DOI: 10.1016/j.econedurev.2016.01.003

Tefft, J., Jonasova, M., Adjao, R., & Morgan, A. (2017). Food Systems for an Urbanizing World: Knowledge product. https://www.semanticscholar.org/paper/Food-systems-for-an-urbanizing-world-%253A-knowledge-Tefft-Jonasova/ad6aa03d304c7f9e7fb0c24cf. Accessed: 15 Feb 2025

Tegegne, H. R., Tesfay, T., Hailesilassie, M. B., & Gebreyohaness, S. T. (2025). Indigenous Knowledge and Practices of the Tigray People: How Do Teachers and Students Perceive Its Integration in Science and Mathematics Education? *Science & Education*, •••, 1–28. DOI: 10.1007/s11191-024-00608-x

Tembo, E., & Chazireni, E. (2016). The negative environmental impact of disposable diapers: The case of Mberengwa District, Zimbabwe. *International Journal of Health Sciences*, *4*(2), 2158–2161.

Tengö, M., Brondizio, E. S., Elmqvist, T., Malmer, P., & Spierenburg, M. (2014). Connecting diverse knowledge systems for enhanced ecosystem governance: The multiple evidence base approach. *Ambio*, *43*(5), 579–591. DOI: 10.1007/s13280-014-0501-3 PMID: 24659474

Teo, T. (2024). Will sustainable food sovereignty research be sustainable in the future? *Agricultural Sciences*, *15*(01), 165–186. DOI: 10.4236/as.2024.151010

Tevenal, S., & Villanueva, M. (2009). Are you getting the message?: The effects of SimCom on the message received by deaf, hard of hearing, and hearing students. *Sign Language Studies*, *9*(3), 266–286. DOI: 10.1353/sls.0.0015

Thabede, D. (2008). African worldview as the basis of practice in the helping professions. Social Work/ Maatskaplike Werk Journal, 44 (3), 233-245.

Thamaga-Chitja, J. M., & Morojele, P. (2014). The context of smallholder farming in South Africa: Towards a livelihood asset building framework. *Journal of Human Ecology (Delhi, India)*, *45*(2), 147–155. DOI: 10.1080/09709274.2014.11906688

The Constitution of the Kingdom of Swaziland Act, 2005.

Theodorson, G. A. (1969). *Modern Dictionary of Sociology*. Methuen.

Thierfelder, C., Rusinamhodzi, L., Ngwira, A. R., Mupangwa, W., Nyagumbo, I., Kassie, G. T., & Cairns, J. E. (2015). Conservation agriculture in Southern Africa: Advances in knowledge. *Renewable Agriculture and Food Systems*, *30*(4), 328–348. DOI: 10.1017/S1742170513000550

Thinh, D., Mu, K., & Mul, X. (2007). The role of indigenous knowledge In sustainable development: A case study of the Vietnam mountain regions. In *Finds and Results from the Swedish Cyprus Expedition: A Gender Perspective at the Medelhavsmuseet* (p. 262). Kamla-Raj Enterprises.

Thomas, A. (2023). Rural Event Tourism: Developing Sustainable Rural Event Tourism for the Development of the Community. In *Event Tourism and Sustainable Community Development* (pp. 279-300). Apple Academic Press.

Thompson, G. (2022). Sustainable Environmental Practices. *Journal of Sustainability Studies*.

Tilman, D., Fargione, J., Wolff, B., D'antonio, C., Dobson, A., Howarth, R., ... & Swackhamer, D. (2001). Forecasting agriculturally driven global environmental change. *science, 292*(5515), 281-284.

Timothy, D. J., & Boyd, S. W. (2003). *Heritage tourism*. Pearson Education.

Tisani, N. (2004). African indegenous knowledge systems (AIKSs): another challenge for curriculum development in higher education?: perspectives on higher education. *South African Journal of Higher Education, 18*(3), 174–184.

Tolkach, D., & King, B. (2015). Strengthening Community-Based Tourism in a new resource-based island nation: Why and how? *Tourism Management, 48*, 386–398. DOI: 10.1016/j.tourman.2014.12.013

Tomanek, P. (2023). Birth order, Gender and Naming Preferences in Contemporary Poland. Names: A Journal of Onomastics, 71(1). DOI: https://doi.org/DOI: 10.5195/names.2023.2387

Toneva, P. I. (2017). Studying the impacts of event tourism on the host communities in Bulgaria.

Toussaint, S., Sullivan, P., & Yu, S. (2005, March). Water ways in Aboriginal Australia: An interconnected analysis. [). Taylor & Francis Group.]. *Anthropological Forum, 15*(1), 61–74. DOI: 10.1080/0066467042000336715

Trendov, N. M., Varas, S., & Zeng, M. (2019). Digital technologies in agriculture and rural areas.

Tricco, A. C., Lillie, E., Zarin, W., O'brien, K., Colquhoun, H., Kastner, M., Levac, D., Ng, C., Sharpe, J. P., Wilson, K., Kenny, M., Warren, R., Wilson, C., Stelfox, H. T., & Straus, S. E. (2016). A scoping review on the conduct and reporting of scoping reviews. *BMC Medical Research Methodology*, 16 (1). https://doi DOI: 10.1186/s12874-0160116-4

Tripura, P., Devi, C. A., Manzoor, U., & Kapri, A. (2024). *Waste management for sustainable and restored agricultural soil: Indigenous technical knowledge integration in organic farming*. Retrieved from https://www.sciencedirect.com/science/article/abs/pii/B9780443184864000191

Trisos, C. H., Auerbach, J., & Madhusudan, K. (2021). Decoloniality and anti-oppressive practices for a more ethical ecology. *Nature Ecology & Evolution, 5*(9), 1205–1212. DOI: 10.1038/s41559-021-01460-w PMID: 34031567

Tshishonga, N. S. (2019). Incorporating Indigenous Knowledge in the Higher Education Sector for the Advancement of African Scholarship. In *The Formation of Intellectual Capital and Its Ability to Transform Higher Education Institutions and the Knowledge Society* (pp. 212–231). IGI Global. DOI: 10.4018/978-1-5225-8461-2.ch011

Turner, B. L.II, Lambin, E. F., & Reenberg, A. (2007). The emergence of land change science for global environmental change and sustainability. *Proceedings of the National Academy of Sciences of the United States of America, 104*(52), 20666–20671. DOI: 10.1073/pnas.0704119104 PMID: 18093934

Turner, M. D., & Schlecht, E. (2019). Livestock mobility in sub-Saharan Africa: A critical review. *Pastoralism, 9*(1), 1–15. DOI: 10.1186/s13570-019-0150-z

Turner, N. (1994). A brief overview of Zulu oral traditions. *Alternation (Durban), 1*(1), 58–67.

Twala, J. M. (2005). *Tangoma nemisebenti yato ekutfutfukiseni lulwimi lweSiswati.* [Unpublished master's dissertation]. University of Zululand, Richards Bay, South African.

Tyagi, V., Brahmi, P., Yadav, S., Ranjan, P., Singh, S., Singh, S., & Singh, K. (2021). Germplasm Access from ICAR-NBPGR and Use within India. *Indian Journal of Plant Genetic Resources, 34*(2), 216–220. DOI: 10.5958/0976-1926.2021.00020.6

Tyler, R. W. (1981). Specific approaches to curriculum development. In Giroux, H. A., Penna, A. N., & Pinar, W. F. (Eds.), *Curriculum & Instruction*. McCutchan Publishing Corporation.

Tynan, L. (2021). What is relationality? Indigenous knowledges, practices and responsibilities with kin. *cultural geographies, 28*(4), 597-610. doi:DOI: DOI: 10.1177/14744740211029287

UNESCO. (2017). Intangible cultural heritage and sustainable tourism. United Nations Educational, Scientific and Cultural Organization. https://ich.unesco.org

Unganai, L. S. (1996). Climate change impacts on agriculture in Zimbabwe: A regional assessment. *Climate Research, 6*, 137–145. DOI: 10.3354/cr006137

UNICEF. (2019). Diarrhea and wastewater.

UNISA. (2025). Register to study through UNISA. https://www.unisa.ac.za/sites/corporate/default/Register-to-study-through-Unisa/Subjects-&-modules/All-modules/Indigenous-Knowledge-Systems-and-African-Languages-(Sesotho)-%E2%80%93-SSO3702

United Nations Children Fund (UNICEF). (2019). *Gender-based violence in emergencies operational guide*. Retrieved May 20, 2024 from https://www.unicef.org/sites/default/files/2020-05/Gender-Based-Violence-in-Emergencies-Operational-Guide-May-2019.pdf

United Nations Development Group. (2022). Theory of Change: UNDAF Companion Guideline. https://unsdg.un.org/resources/theory-change-undaf-companion-guidance

United Nations Educational, Scientific and Cultural Organization. (2016). Goal 2 End Hunger, Achieve Food Security, and Improved Nutrition and Promote Sustainable Agriculture. https://unstats.un.org/sdgs/report/2016/goal-02/ Accessed: 15 Feb 2025

United Nations. (1989). *Constitution on the Rights of the Child*. UNESCO Publication.

United Nations. (2019). *Global biodiversity outlook 5*. UN Environment Programme.

United Nations. (2019). Indigenous Peoples and the 2030 Agenda.

United Nations. (2024). Sustainable Development Goal Report 2024. https://unstats.un.org/sdgs/report/2024/. Accessed: 15 Feb 2025

Univen. (2025). Department of Indigenous Knowledge Systems and Heritage Studies. https://www.univen.ac.za/faculties/humanities-social-science-and-education/department-of-indigenous-knowledge-systems-and-heritage-studies/

University of KwaZulu-Natal. (2021). *African indigenous knowledge systems, gender dimensions and youth in conflict resolution and peace building highlighted in high-level colloquium*. Retrieved July 10, 2024, from https://ndabaonline.ukzn.ac.za/UkzndabaStory/Vol9-Issue56/African%20Indigenous%20Knowledge%20Systems%20Gender%20Dimensions%20and%20Youth%20in%20Conflict%20Resolution%20and%20Peace%20Building%20Highlighted%20in%20High%20Level%20Colloquium/

University of Limpopo. (2025a). School of Social Sciences: Sociology and Anthropology. https://www.ul.ac.za/faculty-of-humanities/school-of-social-sciences/sociology-and-anthropology/

University of Limpopo. (2025b). School of Social Sciences: Cultural and Political studies. https://www.ul.ac.za/faculty-of-humanities/school-of-social-sciences/cultural-and-political-studies/

University of Mpumalanga. (2025). https://www.ump.ac.za/Study-with-us/Faculties-and-Schools/Faculty-of-Economics,-Development-and-Business-Sci/School-of-Social-Sciences/Bachelor-of-Arts-Honours-in-Culture-and-Heritage-S.aspx

University of South Africa. (2025). Register to study through UNISA. https://www.unisa.ac.za/sites/corporate/default/Register-to-study-through-Unisa/Subjects-&-modules/All-modules/Indigenous-Knowledge-Systems-and-African-Languages-(Sesotho)-%E2%80%93-SSO3702

University of Venda. (2025). Department of Indigenous Knowledge Systems and Heritage Studies. https://www.univen.ac.za/faculties/humanities-social-science-and-education/department-of-indigenous-knowledge-systems-and-heritage-studies/

UNWTO. (2018). Tourism and culture: Investing in future generations. United Nations World Tourism Organization. https://www.unwto.org

Valera, E. (2016). Preserving the Tinguian Cultural Heritage. *Mother Tongue Based Multi-Lingual Education*, *2*(1), 79.

Valera, E. (2016). Preserving the Tinguin Cultural Heritage. *Mother Tongue-based Multilingual Education as Tool.*, *2*(1), 79.

Van Aswegen, S., Swart, E., & Oswald, M. (2019). Developing metacognition among young learners by using stories. *South African Journal of Education*, *39*(2), 1–12. DOI: 10.15700/saje.v39n2a1531

Van Averbeke, W. M., Marete, C. K., Igodan, C. O., & Belete, A. (1998). An investigation into Food Plot Production at Irrigation into Food Plot Production at Irrigation Schemes in Central Eastern Cape. Water Research Commision Report, No. 719/1/98.

Van der Wagen, L. (2007). *Human resource management for events: Managing the event workforce*. Routledge.

Van der Zeijst, M., Veling, W., Makhathini, E. M., Susser, E., Burns, J. K., Hoek, H. W., & Susser, I. (2021). Ancestral calling, traditional health practitioner training and mental illness: An ethnographic study from rural KwaZulu-Natal, South Africa. *Transcultural Psychiatry*, *58*(4), 471–485.

Van Dinter, R., Tekinerdogan, B., & Catal, C. (2021). Automation of systematic literature reviews: A systematic literature review. *Information and Software Technology*, *136*, 106589. DOI: 10.1016/j.infsof.2021.106589

Van Heerden, M. (2017). Afrikaans: The Language of Black and Coloured Dissent. *South African History Online, 20.*

Van Herreweghe, M., & Vermeerbergen, M. (2010). Deaf perspectives on communicative practices in South Africa: institutional language policies in educational settings. *Text & Talk - An Interdisciplinary Journal of Language Discourse Communication Studies 30(2)*, 125-144.

Van Rooyen, B. (2011). Safeguarding the future of indigenous knowledge through ICT: ICT and society. *CSIR Science Scope*, *5*(2), 24–26.

Van Wyk, B.-E., & Gericke, N. (2000). *People's plants*. Briza Publications.

Van Wyk, B., & Higgs, P. (2011). The curriculum in an African context. *Indilinga*, *10*(2), 171–181.

Vandenhelsken, M. (2011). The enactment of tribal unity at the periphery of India: The political role of a new form of the Pang Lhabsol Buddhist ritual in Sikkim. *European Bulletin of Himalayan Research*, 84.

Verhoef, G. (2023). *The public domain and the legal protection of Indigenous Knowledge Systems* (Doctoral dissertation, North-West University (South Africa).). wa Thiong'o, N (1986). Decolonising the mind: *The politics of language in African literature*. London, England: James Currey

Verhoef, G. (2020). Stokvels and economic empowerment: the case of African women in South Africa, c. 1930-1998. In *Women and credit* (pp. 91–114). Routledge. DOI: 10.4324/9781003103233-8

Verma, R. (2013). Sikkim: *A Guide and Handbook*. New Delhi Narendra Bhatia & Company.

Verma, A., Naorem, A., Keerthika, A., & Kumar, S. (2024). Biomass and carbon partitioning across different components of agroforestry systems in the arid zones of India. *Agroforestry Systems*, *98*(8), 1–20. DOI: 10.1007/s10457-024-01076-6

Verma, S., & Summarwar, S. (2012). Analysis of water of Pushkar Lake before, during and after the Pushkar Fair. *International Journal of Life Sciences Biotechnology and Pharma Research*, *1*(4), 35–41.

Vilakati, J. D. (2001). Biodiversity management in Swaziland. In P. M. Dlamini, & B.T. Vilane, (Eds.), *Proceedings of the National Seminar on Institutions Working on Gender, Biodiversity and Local Knowledge Systems in Swaziland* (pp. xxxv-xlii). Food and Agricultural Organization.

Vilakati, T. S., & Msibi, T. (2006). Giya Sigiye. Manzini: Macmillan.

Vilakati, T. S., & Sibanda, E. S. (1997). Chaza Ngive. Manzini: Macmillan.

Vilakati, B., & Sibanda, E. S. (1997). *Chaza Ngive. Manzini*. Macmillan.

Vilakati, F., & Msibi, T. (2016). *Giya Sigiye. Manzini*. Macmillan.

Vilakati, T., & Msibi, T. (2006). *Giya sigiye*. Macmillan Boleswa. Manzini.

Vincent, V., & Thomas, R. G. (1960). *An Agricultural Survey of Southern Rhodesia: Part I: Agro-ecological Survey*. Government Printer.

Vishweshwaran, M., & Sujatha, E. R. (2024). A Review on Applications of Drones in Geotechnical Engineering. *Indian Geotechnical Journal*, 1-15.

Vivek, V. (2016). Rainwater harvesting in Chennai: What made it work? IIM Kozhikode *Society &. Management Review*, *5*(1), 91–106. DOI: 10.1177/2277975215617272

Von Braun, J., Afsana, K., Fresco, L. O., & Hassan, M. H. A. (2023). Food systems: seven priorities to end hunger and protect the planet. In *Science and innovations for food systems transformation* (pp. 3–9). Springer International Publishing. DOI: 10.1007/978-3-031-15703-5_1

Vorster, J. A., & Quinn, L. (2012). Privileging knowledge, creating knowers: an analysis of a formal programme for university lecturers. In Quinn, L. (Ed.), *Re-imagining academic staff development: spaces for disruption*. Sun Press.

Vygotsky, L. (1978). *Mind in Society: The Development of Higher Psychological Processes*. Cambridge, M.A.: Harvard.

Wahab, M. A., Yousuf, M., & Hossain, M. E. (2004). *Some indigenous medicinal knowledge for treating Jaundice in Chittagong Hill Tracts*.

Walentowski, H. (2024). Integrating Indigenous Knowledge and Landscape Ecology for Sustainable Land Management. *Journal Biodiversity Bioprospecting Development*, *10*, 114.

Walker, G. H., Stanton, N. A., Salmon, P. M., & Jenkins, D. P. (2008). A review of sociotechnical systems theory: A classic concept for new command and control paradigms. *Theoretical Issues in Ergonomics Science*, *9*(6), 479–499. DOI: 10.1080/14639220701635470

Wane, N. N., & Wane, N. N. (2019). African indigenous education. *Gender, democracy and institutional development in Africa*, 99-121.

Wang, J. (2018). Integrating indigenous with scientific knowledge for the development of sustainable agriculture: Studies in Shaanxi province. *Asian Journal of Agriculture and Development*, *15*(2), 41–58. DOI: 10.37801/ajad2018.15.2.3

Warren, D. M. (1996). Indigenous knowledge systems for sustainable agriculture in Africa. *Sustainable development in third world countries*, 15-24.

Warrier, K., Warrier, R. R., & Thangavel, V. (2023). Status of Sacred Groves in India: A Review. *International Journal of Environment and Climate Change*, *13*(8), 170–181. DOI: 10.9734/ijecc/2023/v13i81942

Watson-Verran, H., Turnbull, D., & Jasanoff, S. (2005). Science and other indigenous knowledge systems. In Stehr, N., & Grundmann, R. (Eds.), *Knowledge: Critical concepts* (pp. 345–369). Routledge.

Weiss, K., Hamann, M., & Marsh, H. (2013). Bridging Knowledge: Understanding and Applying Indigenous and Western Scientific Knowledge for Marine Wildlife Management. *Society & Natural Resources*, *26*(3), 285–302. DOI: 10.1080/08941920.2012.690065

Welford, R. (2016). *Corporate environmental management 1: Systems and strategies*. Routledge. DOI: 10.4324/9781315825120

Welman, C. F., & Michell, B. (2005). *Research methodology*. Oxford University Press.

Wentzel, T. (2021, July 13). History of the Deaf community in South Africa . (P. d. Toit, Interviewer)

Wessels, E. (2011). *Teacher knowledge and implementation of phonological awareness in Grade R*. (Doctor of Philosophy in Education and Teaching). North West University, Unpublished Doctoral Dissertation.

Westengen, O. T., Dalle, S. P., & Mulesa, T. H. (2023). Navigating toward resilient and inclusive seed systems. *Proceedings of the National Academy of Sciences of the United States of America, 120*(14), e2218777120. DOI: 10.1073/pnas.2218777120 PMID: 36972436

White, J. (2021). Indigenous Practices and Biodiversity. *Global Ecology and Conservation.*

WHO. (2018). Water, sanitation and hygiene.

Wicha, S., Photiphun, T., Janjaroenpan, P., Taweesak, C., Chainilwan, I., & Rodmanee, P. (2017). Proposed of e-Community Supported Agriculture (e-CSA) system to promote local organic products: The empirical study of Chiang Rai province. *2017 International Conference on Digital Arts, Media and Technology (ICDAMT),* Chiang Mai, 2017, 273-281. DOI: 10.1109/ICDAMT.2017.7904976

Wikimedia commons. (n.d.). Nilecrocodile https://commons.wikimedia.org/w/index.php?curid=66978

Wikipedia. (n.d.). The Free Encyclopedia. https://en.wikipedia.org/wiki/Proverbs

Wilder, B. T., O'meara, C., Monti, L., & Nabhan, G. P. (2016). The importance of indigenous knowledge in curbing the loss of language and biodiversity. *Bioscience, 66*(6), 499–509. DOI: 10.1093/biosci/biw026

Wilderness Safaris. (2015, June 22). African totems & kinship. Wilderness Safaris.

Wildsmith-Cromarty, R., & Steinke, K. (2014). The write approach: Can R2L help at tertiary level? *Per Linguam: a Journal of Language Learning, 30*(1), 38–54. DOI: 10.5785/30-1-570

Wilson, D. C., Rodic, L., Modak, P., Soos, R., Carpintero, A., Velis, K., & Simonett, O. (2015). *Global waste management outlook.* Unep.

Wilson, S. (2008). *Research Is Ceremony: Indigenous Research Methods.* Fernwood Publishing.

Wilson, S. (2020). *Research is ceremony: Indigenous research methods.* Fernwood publishing.

Windchief, S., & Cummins, J. (2022). Considering Indigenous research methodologies: Bicultural accountability and the protection of community held knowledge. *Qualitative Inquiry, 28*(2), 151–163. DOI: 10.1177/10778004211021803

Wolfgramm, R. M. (2007). *Continuity and vitality of worldview (s) in organisational culture: Towards a Maori perspective* (Doctoral dissertation, ResearchSpace@ Auckland).

Wood, F. (2010). The shape-shifter on the borderlands: A comparative study of the trickster figure in African orality and in oral narratives concerning one South African trickster, Khotso Sethuntsa. *English in Africa*, 71-90. doi:https://www.jstor.org/stable/23074935

World Health Organisation. (2010). *A handbook of medical waste management guideline.*

World Health Organization. (2018). *Water.* Sanitation, Hygiene, and Waste Management.

World Health Organization., (2016), Population Growth and Diapers. Accessed on 02/06/2024.

World Intellectual Property Organization. (2017). *Documentation of traditional knowledge and traditional cultural expressions.* http:// www.wipo.int/edocs/pubdocs/en/wipo_pub_tk_9.pdf

Wulandari, D. U., Marianna, N., Moh, W., & Amien, S. (2024). Integration of ethnomathematics teaching materials in mathematics learning in elementary school. *IJORER: International Journal of Recent Educational Research, 5*(1), 204-218. https://journal.ia-education.com/index.php/ijorer

Yadav, S. C. (2023). Traditional rain water harvesting systems in Rajasthan: water resources conservation and its sustainable management–a review. *Sustainability, Agri, Food and Environmental Research*, 11.

Yamunarani, K., Sangeetha, S., Dhanushkodi, V., Chitra, K., & Dhanalakshmi, K. (2023). Application of Indigenous Technological Knowledge (ITK) For Plant Disease Management In Tamil Nadu, India. *Asian Journal of Agricultural Extension Economics & Sociology, 41*(12), 310–315. DOI: 10.9734/ajaees/2023/v41i122334

Yang, Y., Qin, L., & Ning, L. (2021). School Violence and Teacher Professional Engagement: A Cross-National Study. *Frontiers in Psychology, 12*, 628809. Advance online publication. DOI: 10.3389/fpsyg.2021.628809 PMID: 33935880

Yin, R. (2013). *Case study research: Design and methods*. Thousand Oaks, C.A.: Sage publications.

Yin, R. K. (2003). *Case study research design and methods* (3rd ed.). Sage.

Yin, R. K. (2003). *Case Study Research: Designed Methods* (3rd ed.). Sage Publication Inc.

Yip, S. Y., & Chakma, U. (2024). The teaching of Indigenous knowledge and perspectives in initial teacher education: A scoping review of empirical studies. *Journal of Further and Higher Education, 48*(3), 287–300. DOI: 10.1080/0309877X.2024.2327029

Young, A. M., Moji, M., Duby, Z., Tenza, S., Atujuna, M., Palanee-Phillips, T., Minnis, A. M., & Montgomery, E. T. (2024). The ideal man: How gender norms and expectations shape South African men's masculinity, sexual identities, and well-being. *Journal of Men's Studies, 0*(0). Advance online publication. Retrieved December 12, 2024, from. DOI: 10.1177/10608265241303734

Yung, K., & Neathway, C. (2019). Community champions for safe, sustainable, and traditional food system. *Current Developments in Nutrition, 4*, 119. PMID: 32258999

Yusuf, M. W., Saidi, I., & Halliru, A. (2010). Ethnomathematics (A Mathematical game in the Hausa culture). *International Journal of Mathematical Science Education, 3*(1), 36–42.

Yu, Y., & Turco, D. (2000). Issues in tourism event economic impact studies: The case of the Albuquerque International Balloon Fiesta. *Current Issues in Tourism, 3*(2), 138–149. DOI: 10.1080/13683500008667870

Zazu, C. (2007). Exploring Opportunities and Challenges for achieving the integration of Indigenous Knowledge Systems into Environmental Education processes: A case study of the Sebakwe Environmental Education programme (SEEP) in Zimbabwe. Unpublished Master thesis. Rhodes University.

Zetlin, M. (1995). Is it worth keeping older salespeople? *Sales and Marketing Management, 147*(4), 144–147.

Zhang, L., Guo, W., Lv, C., Guo, M., Yang, M., Fu, Q., & Liu, X. (2024). Advancements in Artificial Intelligence Technology for Improving Animal Welfare: Current Applications and Research Progress. *Animal Research and One Health, 2*(1), 93–109. DOI: 10.1002/aro2.44

Zhang, W., Su, K., Wang, Q., Yang, L., Sun, W., Ranjitkar, S., Shen, L., Kindt, R., Ji, Y., Marshall, P., Pisey, P. S., & El-Kassaby, Y. A. (2023). Agroforestry species selection for forest rehabilitation in the asia-pacific region: A meta-analysis on high-level taxonomy. *Forests*, *14*(10), 2045. DOI: 10.3390/f14102045

Zimunya, T. (2021). Women's Roles in Zimbabwean Society: A Historical Perspective. *African Studies Review*, *64*(2), 123–140.

Zoll, S., Feinberg, N., & Saylor, L. (2023). *Powerful literacy in the Montessori classroom: Aligning reading research and practice*. Teachers College Press.

Zondo, N. M. (2017). *Self-help economic empowerment amongst women in Blaauwbosch, KwaZulu-Natal* (Doctoral dissertation).

Zong, X., Liu, X., Chen, G., & Yin, Y. (2022). A deep-understanding framework and assessment indicator system for climate-resilient agriculture. *Ecological Indicators*, *136*, 108597. DOI: 10.1016/j.ecolind.2022.108597

Zwane, P. E., Masarirambi, M. T., Seyoum, T., & Nkosi, B. S. (2011). Natural fibre plant resources of economic value found in wetlands of Swaziland: A review. *African Journal of Agricultural Research*, *6*(4), 773–779.

About The Contributors

Tlou Maggie Masenya holds PhD in Information Science from University of South Africa (UNISA) and Masters in Information Technology from University of Pretoria (UP). She has 10 years of working experience as a senior lecturer in the field of Information Science, Information Systems and Technology. She also worked as a Senior Lecturer at University of South Africa and University of Zululand. She is currently a Head of Department in Information Technology at Durban University of Technology. She is a visionary leader and distinguished researcher with extensive and an impressive national and international research standing, and has published 17 book chapters, 19 articles in peer reviewed journals and also presented papers nationally and internationally within the broad niche area of Digital preservation and Data Curation, Records and Knowledge Management,Technopreneurship, Digital Entrepreneurship, Cloud Computing, Innovative Pedagogies, Disruptive Technologies and Indigenous Knowledge Systems, which had a positive impact on her scholarly growth. She recently edited two books, and the first book is entitled "Innovative technologies for enhancing knowledge access in academic libraries" published by IGI- Global International Academic Publisher in July 2022. The second book is entitled "Digital preservation and Documentation of Global Indigenous Knowledge Systems" also published by IGI-Global International Academic Publisher in August 2023.

Monicca Thulisile Bhuda holds PhD holder in Social Sciences with Indigenous Knowledge Systems (IKS) from the North-West University (NWU), with 6 years' experience of teaching within the IKS, culture and heritage fields. I am a coordinator of a programme that focuses on African identities, Heritage and History at the University of Mpumalanga.Furthermore, I am also a culture and heritage lecturer within the same University. Currently as an academic, Ihave 24 academic publications on IKS and presented 31 abstracts in diverse conferences. I am co supervisingstudents who are registered for master's degrees in IKS with NWU. I have also been declared a national expert ofIKS by the Minister of Sport, Arts and Culture in the republic of South Africa and I serve in a national panel ofexperts as the youngest member. In 2023, I have been awarded by the South African National Heritage Council (NHC) under the category of Voice of heritage. The North-West University in 2023 also awarded me with an alumni excellence award within the education category. I have been featured numerous times on TV, radio and newsletters as an IKS and culture expert.

***

Andani Budeli, am an emerging academic with experience in teaching and learning, as well as community engagement in higher education. I am employed as a Lecturer in the Department of Indigenous Knowledge and Heritage Studies (University of Venda). I hold Bachelor of Indigenous Knowledge Systems and Master of Arts in African Studies degrees, both from the University of Venda. Currently, I am enrolled for PhD in Social Sciences with Indigenous Knowledge Systems at the North-West University (Mafikeng). My research interests center around indigenous water conservation, and food security." I have published two articles in accredited journals on indigenous water conservation and management and a book chapter on Vhavenda indigenous food security practices. My academic citizenship is coupled with presenting papers in both local and international conferences and supervising postgraduate students. I am a proud champion of the Vhembe Biosphere Reserve Youth Network where I serve as an advisor. I am also a Secretary for the Library Information Services Committee of the Faculty of Humanities, Social Sciences and Education (University of Venda).

Petros N. Dlamini, (PhD) is a Senior Lecturer in the Department of Information Studies in the Faculty of Humanities and Social Sciences at the University of Zululand, South Africa. He holds a PhD and MLIS in Library and Information Science from University of Zululand in 2016 and 2004. He has published several papers in accredited journals and book chapters. He has supervised several masters and PhD students to completion in the domain.

Mishack T. Gumbo is an NRF C1-Rated Researcher and a Research Professor of Indigenous Technology Knowledge Systems Education in the Department of Science and Technology Education at the University of South Africa. He is Editor-in-Chief of Indilinga: African Journal of Indigenous Knowledge Systems. He is a well-published scholar. He leads research projects and mentors developing academics. He serves on numerous academic committees and boards. He achieved the following between 2019 and 2022: External Recognition Award; NRF C1 Rating; NRF C1 Rating Award; Recognized Among the Top 10 000 Influential Scientists in Africa in the AD Scientific Index 2021; Competitive Programme for Rated Researchers; and Appointed as Research Professor.

Sapna Jarial is an Associate Professor in the Department of Agricultural Economics & Extension, School of Agriculture at Lovely Professional University, Phagwara, Punjab, India.

Aruditya Jasrotia is currently working as an Assistant Professor in Amity Institute of Travel and Tourism at Amity University, Noida, India. He completed his Ph.D. from the Department of Tourism and Travel Management,Central University of Jammu, Jammu and Kashmir, India. His areas of expertise are Smart Tourism, Smart Tourism Destinations, Smart Cities, Tourism Technologies, Urban Planning, Community Based Tourism and Sustainable Tourism.

Tanuja Khan worked as Assistant Professor in Department of Education, Bhopal School of Social Sciences (BSSS), affiliated to Barkatullah University since July, 2015 to June, 2018. They currently work in the Department of Psychology.

Xholiso Ennocent Khazamula is a Heritage Officer at Department of Sport, Arts and Culture with expertise in Cultural Relations and Indigenous knowledge systems. He is an aspiring Lecturer in Intangible Cultural Heritage/Indigenous knowledge Systems.

Mothusiotsile Edwin Maditsi is a PhD candidate at the North-West University (Mahikeng campus). He holds a BA (Hons) in Indigenous Knowledge Systems (IKS) degree from the North-West University (NWU), a Master of IKS degree also from the NWU and currently a PhD candidate at the NWU. During his Master's degree journey he served as a full member of PRIMCO in which he was appointed as a researcher for the project leader. He has participated and facilitated some community engagement activities for and on behalf of the IKS Centre at National and local level. Some of the community engagement activities entails; IK Act of 2019 and BCP public awareness workshops across the country; IK Act regulations public consultations across the country and a host of other IKS related activities.

Oluwasegun. J. Aroba, a Durban University of Technology lecturer, Department of Information Systems. Honorable Research Associate in the Department of Operations and Quality Management. He has a PhD. in Information Technology. A graduate of Information Technology University from the prestigious Coventry University United Kingdom, BSc. Computer Science and Technology (Upper-Class Division) Crawford University whose research inclination focus area are into Wireless Sensor Networks, Hyper-Heuristic, Hybrid-Heuristic, Meta-Analysis, Project Management, Data Science, Machine Learning, SAP Specialist with over decade years of experience in the higher education sector, healthcare industries, government parastatals, a consultant across the globe, a graduate member of IEEE, IITPSA South Africa, Member of IET UK. He has Chaired and Co-Chaired International and National Conferences. He is a global mentor to SMEs and European African, Sisonke NdabaX, and a seasoned guest speaker. He has a vast collaboration network across Spain, France, Morocco, the USA, the UK, South Africa, and Sweden. oluwaseguna@dut.ac.za, aroseglinks@yahoo.co.uk

Muzi Matfunjwa is a senior lecturer with a focus on Siswati at SADiLaR. His research interest and experience lies in pragmatics and sociolinguistics. He focuses on the use of language in society and the socio-cultural traits that manifest through language usage. As a Siswati researcher, his interest is not limited to pragmatics and sociolinguistics, but he is also concerned with other linguistic fields such as semantics, lexicography, morphology and phonology. Muzi would like to play a major role in preserving the Siswati language and culture by conducting research and developing the language. He is involved with the use of language technologies to promote the utilization of digital resources in order to stimulate teaching and learning as well as research in the Siswati language.

Respect Mlambo is a Xitsonga language researcher at the South African Centre for Digital Resources (SADiLaR). His research interests include translation, terminology development, and lexicography, as well as sociolinguistics, morphology, semantics, and syntax. He is also engaged in research and initiatives aimed at utilising language technologies and resources to promote indigenous languages.

Diane Thelma Molokwa is currently serving as a Student Assistant at the Indigenous Knowledge Systems Centre, Mahikeng Campus. She possesses a Master's degree in Indigenous Knowledge Systems from North-West University. Her research interests focus on the realms of indigenous food security, climate change, and preservation of wild genetic resources. Notably, Molokwa has received recognition from esteemed International GoldenKey Organization as a high-achieving student. She has actively participated in scholarly gatherings, including a symposium themed "Taking Indigenous Knowledge Seriously" hosted by the University of Johannesburg and the Indigenous Knowledge Conference hosted by Nelson Mandela University in South Africa, thereby demonstrating her commitment to advancing the field of indigenous knowledge systems.

Menard Musendekwa is a PhD Candidate with the University of South Africa. He is currently the Chair of Theology at the Reformed Church University and also the Vice -Principal of Murray Theological College. He has published papers in theology and interdisciplines.

Taimi Nghikembua is a Language Teacher Educator at the University of Namibia, currently heading the Department: Early Childhood Education and Care in the School of Education. Her background is in Early Language Teaching in Mother Tongue and English. She has 13 years experience in Primary Teacher Education. At the moment, her academic role is primarily in Early Childhood Education and Care with a strong focus on play and outdoor pedagogies as part of the Transformed Curriculum. Dr Nghikembua is involved in various international projects in the field of Early Childhood Education as well as Early Literacy acquisition and learning.

John Nyambe is the Associate Dean of the School of Education at the University of Namibia, Windhoek. Before then, he served in several other higher education leadership positions, which included Vice-Rector of a College of Education, Rector of a College of Education, Deputy Director of a National Institute for Curriculum Development, and two 5-year term Director positions of a learning and teaching centre at the University of Namibia. His research interests are teacher education and development, curriculum development, assessment, learning and teaching, higher education studies and academic development.

Andrisha Beharry Ramraj holds a PhD in Management studies majoring in management, leadership and governance. Further to this she holds a Diploma in Higher Education (UKZN). In 2004 and 2005 she received NRF scholarships to complete her Honors and Masters degrees respectively. She has presented at local and international conference in the areas of Management, Leadership and Entrepreneurship. She has published in the fields of Entrepreneurship and Business and has successfully written two book chapters. She received best presentation awards two years in a row (2016 and 2017 respectively). She serves on the Teaching and Learning committee within the University and is currently the BBA Program Manager (SFTP). Having realized her desire for higher education, Dr Beharry serves as an expert advisor for the Student Society of Entrepreneurship (CAVE) based at UKZN. Further to this she now shares her expertise as a research expert for the Center of Entrepreneurship. She is keen on undergraduate curriculum transformation within the discipline of Entrepreneurship. Redesigning curricular to suite the dynamics of the modern world enabling sustainability is fast becoming her new passion.

Vineeta Saluja is the Pro Vice-Chancellor at Mangalayatan University, Jabalpur. A doctorate in English Literature, she is a research supervisor for both M. Phil, as well as, Ph. D scholars. Presently, she is researching cross-cultural communication and actively training students in effective communication and soft skills. She completed a UGC funded major project on poems and short stories in tribal literature of Madhya Pradesh. Her persevering efforts have been recognized and well rewarded by national awards for academics as well as community service.

Mamokete Venolia Setshego is a dedicated researcher and academic specialising in indigenous knowledge systems. Currently she is pursuing her Doctoral degree in Indigenous Knowledge Systems in North West University (South Africa). She earned her Master's degree in Indigenous Knowledge Systems at North-West University (South Africa) and has contributed significantly to understanding the role of natural resources in traditional cosmeceuticals. Her research highlights the use of plant-based remedies for skin conditions and their cultural significance within rural communities, particularly in South Africa's Vhembe district. Setshego's work emphasises the preservation of ethnobotanical knowledge, exploring sustainable and biodegradable alternatives to synthetic skincare. By integrating both qualitative and quantitative methods, her research bridges scientific inquiry and traditional practices to address global challenges in food security, nutrition, and natural product development.

Kuldeep Singh has a PhD (Tourism Management) from M.D. University, Rohtak. He is also a UGC NET (JRF -2014) in Tourism Mgt. PGDM: IITTM, Gwalior (Topper). His research areas are Service Quality, Tourist Satisfaction, Revisit Intention, Rural Tourism, Eco Tourism, Sustainable Tourism & Pro Poor Tourism

Pindai Sithole is a Director and Senior International Development Consultant at CeDRE International Africa based in Harare, Harare Previously, Pindai was a President at Zimbabwe Evaluation Society. He has Professional Experience in: Research and Writing; Results-Based Strategy Development; Policy Development; Monitoring and Evaluation; Teaching at university level; Gender and Economic Empowerment; Child Rights and Social Protection; Sustainable Development; Community/Stakeholder Engagement; Climate Change and Community Resilience; and Indigenous Knowledge Systems for Development.

Kellie Steinke, an Applied Linguist and have a PhD in Education, specializing in Reading, and currently work as a lecturer in Academic Literacy.

Index

E

F

G

H

I

K

T

U

V

Z

www.ingramcontent.com/pod-product-compliance
Lightning Source LLC
LaVergne TN
LVHW080403011026
849161LV00009B/2338